Bio-inspired Polymers

RSC Polymer Chemistry Series

Editor-in-Chief:
Professor Ben Zhong Tang, *The Hong Kong University of Science and Technology, Hong Kong, China*

Series Editors:
Professor Alaa S. Abd-El-Aziz, *University of Prince Edward Island, Canada*
Professor Stephen Craig, *Duke University, USA*
Professor Jianhua Dong, *National Natural Science Foundation of China, China*
Professor Toshio Masuda, *Shanghai University, China*
Professor Christoph Weder, *University of Fribourg, Switzerland*

Titles in the Series:
1: Renewable Resources for Functional Polymers and Biomaterials
2: Molecular Design and Applications of Photofunctional Polymers and Materials
3: Functional Polymers for Nanomedicine
4: Fundamentals of Controlled/Living Radical Polymerization
5: Healable Polymer Systems
6: Thiol-X Chemistries in Polymer and Materials Science
7: Natural Rubber Materials: Volume 1: Blends and IPNs
8: Natural Rubber Materials: Volume 2: Composites and Nanocomposites
9: Conjugated Polymers: A Practical Guide to Synthesis
10: Polymeric Materials with Antimicrobial Activity: From Synthesis to Applications
11: Phosphorus-Based Polymers: From Synthesis to Applications
12: Poly(lactic acid) Science and Technology: Processing, Properties, Additives and Applications
13: Cationic Polymers in Regenerative Medicine
14: Electrospinning: Principles, Practice and Possibilities
15: Glycopolymer Code: Synthesis of Glycopolymers and their Applications
16: Hyperbranched Polymers: Macromolecules in-between Deterministic Linear Chains and Dendrimer Structures
17: Polymer Photovoltaics: Materials, Physics, and Device Engineering
18: Electrical Memory Materials and Devices

How to obtain future titles on publication:
A standing order plan is available for this series. A standing order will bring delivery of each new volume immediately on publication.

For further information please contact:
Book Sales Department, Royal Society of Chemistry, Thomas Graham House, Science Park, Milton Road, Cambridge, CB4 0WF, UK
Telephone: +44 (0)1223 420066, Fax: +44 (0)1223 420247
Email: booksales@rsc.org
Visit our website at www.rsc.org/books

Bio-inspired Polymers

Edited by

Nico Bruns
University of Fribourg, Fribourg, Switzerland
Email: nico.bruns@unifr.ch

and

Andreas F. M. Kilbinger
University of Fribourg, Fribourg, Switzerland
Email: andreas.kilbinger@unifr.ch

ROYAL SOCIETY OF CHEMISTRY

THE QUEEN'S AWARDS FOR ENTERPRISE: INTERNATIONAL TRADE 2013

RSC Polymer Chemistry Series No. 22

Print ISBN: 978-1-78262-413-4
PDF eISBN: 978-1-78262-666-4
EPUB eISBN: 978-1-78262-923-8
ISSN: 2044-0790

A catalogue record for this book is available from the British Library

Published by The Royal Society of Chemistry,
Thomas Graham House, Science Park, Milton Road,
Cambridge CB4 0WF, UK

Registered Charity Number 207890

For further information see our web site at www.rsc.org

Printed in the United Kingdom by CPI Group (UK) Ltd, Croydon, CR0 4YY, UK

Preface

Life on earth has been evolving for the last 3–4 billion years, in which time it has developed a level of complexity that is most challenging to understand. With major breakthroughs in physics and chemistry in the last century, it is the advances in biological sciences that are gaining momentum in this century. The insights gained into the inner workings of the living world have always inspired scientists from bordering disciplines to copy or mimic nature in abiotic systems. Engineers have long been inspired by the macroscopic living world and have developed bio-inspired technologies ranging from Velcro to wing design for aeroplanes to advanced robotics. Bio-inspiration and biomimetics often go hand-in-hand. While the latter term describes the direct replication of biological working principles in man-made systems, bio-inspiration usually refers to a more general translation of molecular, microscopic, and macroscopic structures and functionalities from living systems into synthetic ones. On the molecular, and especially on the macromolecular level, research into bio-inspired structures has been growing exponentially over the last twenty years and has led to some major scientific and technological breakthroughs, such as self-cleaning surfaces, adhesives that provide strong bonding in wet conditions, adaptive and self-reporting materials and synthetic tools for the preparation of sequence-controlled polymers. Today, research into bio-inspired polymers is being carried out at many research institutes worldwide. In fact, several of the co-authors of this book are collaborating in a *National Center of Competence in Research* (NCCR), a research instrument of the Swiss National Science Foundation (SNSF), working on Bio-Inspired Materials. The number of textbooks addressing this advanced topic, however, are few, and either cover bio-inspired chemistry in a broad sense, or bio-inspired materials with a strong focus on inorganic materials and hybrid materials. Nevertheless, a vast body of literature deals with soft bio-inspired materials based on

---

RSC Polymer Chemistry Series No. 22
Bio-inspired Polymers
Edited by Nico Bruns and Andreas F. M. Kilbinger

Published by the Royal Society of Chemistry, www.rsc.org

synthetic and natural polymers. The different aspects of bio-inspiration in polymer science are fairly varied even within the constraints of a book focusing "only" on bio-inspired polymers. We therefore believed that compiling a book co-authored by some of the leading authorities in biologically inspired polymers would be a worthwhile endeavour. This book aims to provide an essential overview of the field.

We have thematically grouped the 17 chapters of our book into three sections. The first section deals with *polymers based on biological building blocks.* The first chapter in this section, written by Börner, ten Brummelhuis, and co-authors, gives an overview of the techniques available in the "post-click area" to synthesize peptide and protein polymer conjugates. Hartmann's Chapter 2 on glycopolymers similarly addresses pathways to obtain sugar carrying homo- and block copolymers, as well as their applications in biotechnology and biomedicine.

The second section covers *bio-inspired and biomimetic polymeric structures.* The chapter by Lutz and co-authors focuses on the fundamental problem of controlling the primary structure of synthetic polymers inspired by nature's sequence-controlled macromolecules. The team of Berda describe the folding of individual polymer chains into single-chain nanoparticles, in analogy to nature's folding of protein chains. Kilbinger and co-authors review several different ways of obtaining polymeric tubular structures, inspired by tubes in nature that exist on many different length scales. Polymeric membranes are addressed in the chapter by Meier, Palivan, and co-authors. Such membranes take their inspiration from natural cell membranes, but offer increased mechanical stability while allowing the incorporation of functional biomolecules such as proteins. As a result, they allow the preparation of robust nanoreactors and nanocontainers, as well as advanced water purification membranes. Polymeric ionic liquids with micelle-like topologies are presented by Mülhaupt and co-authors. Here, micellar assemblies of amphiphiles are used as an inspiration to go beyond what is observed in nature, *i.e.* polymeric unimolecular micelles. The chapter by Studart, Burgert, and co-authors represents the last in this section on bio-inspired polymeric structures. Here, plants serve as the inspiration for the creation of new composite materials with gradual variations of their mechanical properties. Improved composites with superior tolerance of peak stresses and better load distribution can be obtained by mimicking nature.

The third and last section of our book deals with *polymeric materials with bio-inspired functionality.* The opening chapter by Waite gives a biological perspective of how mussels adhere to interfaces and emphasizes the importance of first understanding the biology before new, truly bio-inspired materials can be achieved. The second chapter in this section, written by Lee, describes the synthetic achievements of mussel glue-inspired polymers, and depicts the polymer chemist's approach to mimic the mussel adhesion described by Waite in the previous chapter. Bruns and co-authors present a chapter on self-reporting polymeric materials with mechanochromic properties. Such materials mimic nature's ability to report, repair, and improve materials exposed to mechanical stress. Inspired by the stiffness change of the sea cucumber dermis, Weder and Moatsou review the general design principles to achieve materials that can mimic this naturally observed behavior. Approaches to mimic natural skeleton muscles using synthetic polymers are reviewed by Naficy, Spinks, and Baughman in the chapter on bio-inspired polymer artificial muscles. Cameron's chapter on materials for tissue engineering and 3D cell culture reviews common methods used to prepare porous three dimensional scaffold materials and how to improve attachment and proliferation of various cell types. Colonialization of surfaces by cells is crucial for cell culture materials. However, if the cells are harmful bacteria, this effect is not wanted. Thus, living organisms have developed ways to stop bacterial colonialization on their surfaces. These antimicrobial properties can be mimicked with synthetic polymers, as highlighted by Tiller and coworkers. The chapter on superwettability of polymer surfaces reported by Jiang and co-authors reviews approaches to mimic structured surfaces common in nature such as those of the Lotus leaf, as found on the leg of the water strider or on the surface of certain water collecting cacti. The final chapter of this section (and last chapter in our book) by Vignolini, Steiner, and co-authors deals with the mimicking of structural colour using biopolymers. Some of the most brilliant colours observed in nature, for example on the wings of certain butterflies, the surface of some plants, or the armour of certain beetles, are caused by periodically organised biopolymers and their interference with visible light. This last chapter describes this phenomenon and reviews current approaches to mimic this behavior.

It is to be expected that in a book on bio-inspired polymers certain chapters will appear "more bio-inspired" than others to some readers at a first glance. The reason for this is, in our opinion, simply the degree to which certain ideas from nature have been incorporated into our daily lives or research activities already. As we increase our understanding of the underlying biology of nature's "box of tricks", our view of bio-inspiration will continuously change as well. It is therefore difficult to determine a "degree of bio-inspiration" and we therefore prefer to see this book as a current snapshot of research activities on bio-inspired polymers. We hope that the book itself serves as inspiration for our readers to develop novel materials with advanced properties that address the numerous challenges and opportunities that polymer chemistry and material sciences are going to face in the future.

We would like to thank all of our authors for contributing such high quality chapters on the topics of their expertise. We would also like to thank the editorial staff at the Royal Society of Chemistry for their support in realizing this book project. Finally, we would like to greatly acknowledge the financial support of the Swiss National Science Foundation through its NCCR Bio-Inspired Materials and the many chapter contributions of our NCCR colleagues.

Nico Bruns and Andreas Kilbinger
Fribourg

Contents

RSC Polymer Chemistry Series No. 22
Bio-inspired Polymers
Edited by Nico Bruns and Andreas F. M. Kilbinger

Published by the Royal Society of Chemistry, www.rsc.org

CHAPTER 1

Synthetic Aspects of Peptide– and Protein–Polymer Conjugates in the Post-click Era

MARIA MEIßLER, SEBASTIAN WIECZOREK, NIELS TEN BRUMMELHUIS* AND HANS G. BÖRNER*

Laboratory for Organic Synthesis of Functional Systems, Department of Chemistry, Humboldt-Universität zu Berlin, Brook-Taylor-Str. 2, D-12489 Berlin, Germany
*Email: niels.ten.brummelhuis@hu-berlin.de; h.boerner@hu-berlin.de

1.1 Introduction

The synthesis of peptide– and protein–polymer conjugates offers the possibility to integrate properties of biological macromolecules into synthetic systems, thereby obtaining hybrid materials with unique functions.[1–9] The synthetic polymers within these structures provide a versatile range of properties,[4] whereas the peptide or protein domain introduces highly specific functions, ranging from enzymatic activity[10] to specific interaction capabilities or recognition,[11] and to disease modifying activities (Figure 1.1).[12] The multidisciplinary field of creating bioconjugates gives access to a variety of materials for application in materials science, biotechnology, or pharmacology, where the bioconjugates act at the interface between biology and synthetic materials.[6–9]

RSC Polymer Chemistry Series No. 22
Bio-inspired Polymers
Edited by Nico Bruns and Andreas F. M. Kilbinger

Published by the Royal Society of Chemistry, www.rsc.org

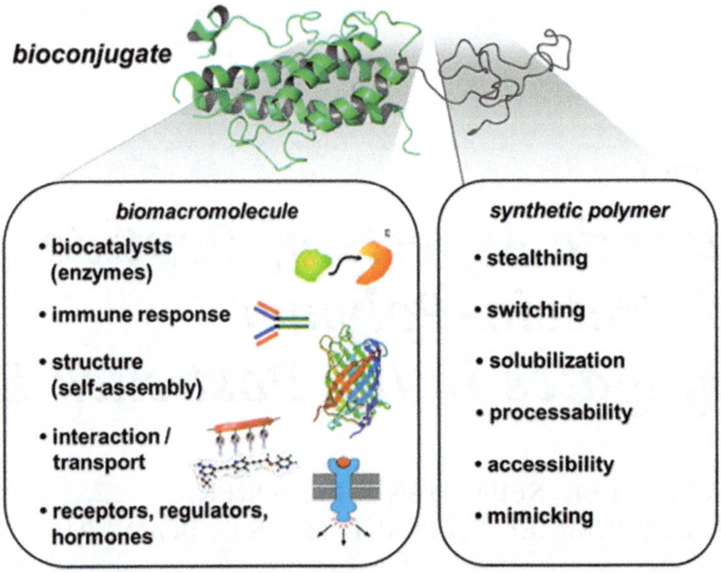

Figure 1.1 Description of the versatile properties of a biomacromolecule (PDB: 1HGU)[13] and a synthetic polymer. The combination of both structures into one hybrid material offers a wide range of functions, which enables their application in materials science, biotechnology, or pharmacology.[1–9]

First attempts to combine synthetic polymers with biomacromolecules emerged in the early 1950s, when initial reports concerning the synthesis of peptide–polymer conjugates were published by Jatzkewitz.[14,15] About 20 years later, Abuchowski, Davis, and coworkers described the attachment of a linear poly(ethylene oxide) (PEO) to bovine serum albumin (BSA) and bovine liver catalase.[16,17] Particularly noteworthy was that the extensive modification of these proteins with a synthetic polymer block did not inhibit the functionality of the biological macromolecules while simultaneously reducing their immunogenicity and increasing blood circulation times.[16,17] This widely used strategy of incorporating PEO onto proteins or peptides with the objective of improving the properties of biomacromolecules is termed "PEG*ylation*", referring to poly(ethylene glycol) or PEG.[18] Due to polymer modification, the stability of the resulting protein–polymer conjugates toward proteolytic digestion and antibody interactions can often be decreased significantly.[19,20] This enhancement in stability and solubility of the modified proteins represents an important improvement for *in vivo* applications, such as in pharmaceutical research and the development of protein-based drugs.[5,21] Ever since these pioneering works showed the potential of PEG*ylation*, and the combination of proteins or peptides with synthetic polymers in general, many research groups have started to investigate peptide–polymer conjugates as a new class of hybrid materials.[1,5,18,21]

Controlled techniques to connect the building blocks from the synthetic and the biological worlds are indispensable for the preparation of well-defined peptide– and protein–polymer conjugates.[22] In 2001, Sharpless, Kolb, and Finn introduced the concept of "*click*" chemistry, which describes the most important criteria to attach two molecules to each other in a highly selective and efficient manner.[23] A reaction defined under this term has to result in very high yields and easily isolable products, while generating only non-hazardous side products that can easily be removed afterwards. Besides the copper(I)-catalyzed cycloaddition (CuAAC) between alkynes and azides described independently by Sharpless[24] and Meldal,[25] a wide range of ligation strategies have since been found to fulfill the criteria of "*click*" reactions.[23] Of these (in addition to CuAAC), the strain-promoted azide–alkyne cycloaddition (SPAAC), Staudinger ligation, oxime formation, Michael-type additions and other thiol-ene reactions are among the most frequently cited representatives.[26] The advent of various highly efficient reactions has also proved advantageous for the preparation of peptide– and protein–polymer conjugates; these synthetic strategies offer versatile opportunities in the fields of materials science and biomedicine, where they are being applied widely.[26,27]

In conjunction with the emerging "*click*" reactions in this century, scientists have focused on the development of innovative strategies to design more complex structures.[6–9] The rich world of chemical ligation tools, which have traditionally been used for protein modification, has extensively been reviewed.[1,4,5,10,12,28–32] This book chapter will mainly provide an overview of the last decade's progress over the preparation of functional bioconjugates, also referred to as "*post-click*" methods.[8] After a brief survey of various well-established and widely used ligation techniques, we will focus on novel types of chemistry as well as chemoenzymatic approaches and biotransformations to create functional bioconjugates using enzymatically catalyzed reactions. Finally, recent advances in this field will be described to provide insight into potential future directions for the preparation of functional peptide– and protein–polymer conjugates.

1.2 General Concepts for Bioconjugation

The most widely used method to prepare bioconjugates is based on a convergent strategy – the so-called "*grafting to*" approach (Figure 1.2a).[22,29] In order to obtain well-defined structures, the peptide or protein, which contains one or more reactive groups, is reacted with a polymer bearing complementary reactive groups. The "*grafting to*" approach is applied most since the independent synthesis and characterization of both components prior to the ligation enables a facile structural and chemical analysis of the resulting bioconjugates. Some of the disadvantages compared to other methods (*vide infra*) are that two macromolecules need to be coupled, which is often slow and/or inefficient due to the hindered accessibility of the reactive groups. Additionally, such reactions can only be performed using relatively low concentrations, and separation of the starting materials from the product is often difficult. Therefore, highly efficient coupling reactions are required.[28]

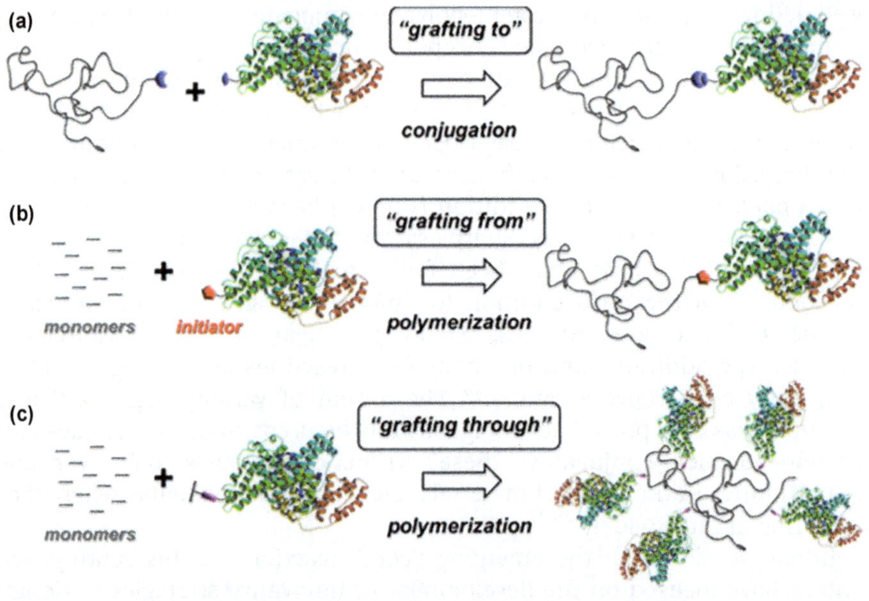

Figure 1.2 Synthetic strategies for synthesis of peptide– and protein–polymer conjugates (PDB: 3V03).[34] The "*grafting to*" approach, where a polymer bearing reactive groups is attached to a biomacromolecule (a); the "*grafting from*" strategy, where an initiating moiety is introduced to a peptide or protein, enabling the polymerization from the specific site of the biomacromolecule (b); and the "*grafting through*" method, which exploits the (co)polymerization of macromonomers containing a peptide or protein domain (c).[7]

However, the synthesis and characterization of both components independently enables a facilitated structural and chemical analysis of the resulting bioconjugates.

A range of different reactions for convergent ligation have been developed so far.[6–9] On the one hand, different amino acid side chains within the peptides and proteins can be addressed using various ligation strategies. Thiol groups of cysteine residues and primary amine groups at the N-terminus or on lysine side chains are the most common sites for attachment to polymers.[33] On the other hand, more appropriate functional groups can be introduced to improve the coupling efficiency and to enable site-specific conjugation, *e.g.* phosphine or azide residues for Staudinger ligation, aldehydes for oxime formation, or strained cyclooctynes for SPAAC.[26]

In the "*grafting from*" approach a moiety with the ability to initiate or mediate polymerization is introduced in the biomacromolecule (Figure 1.2b).[22,31,35–40] This divergent strategy often relies on controlled radical polymerization techniques, since these are highly tolerant to the presence of functional groups which are frequently found in peptides and proteins.[31] Furthermore, protein denaturation, resulting in loss of function, is unacceptable, making polymerization in aqueous media under mild conditions desirable. Controlled radical polymerization techniques such as atom transfer radical polymerization (ATRP),[36,41] reversible addition-fragmentation chain transfer (RAFT),[42–45] or nitroxide-mediated polymerization (NMP),[46] are compatible with aqueous solvents and avoid adverse side reactions.

One of the first examples following the "*grafting from*" approach was the preparation of a protein–polymer conjugate composed of BSA and poly(*N*-isopropylacrylamide) (PNIPAM).[47] The modification of the protein with a maleimide-functionalized chain transfer agent (CTA) enabled the RAFT polymerization of NIPAM, resulting in a thermoresponsive bioconjugate.

The "*grafting through*" approach implies the incorporation of a polymerizable group into the biomacromolecule, allowing for copolymerization with synthetic monomer units. This results in a bioconjugate product with a comb-like structure, where a certain number of the polymer side chains contain peptide or protein moieties (Figure 1.2c).[22,31] Compared to the concept of "*grafting to*", the strategy avoids side reactions and low coupling efficiencies due to the attachment of peptides or proteins to monomers before the polymerization process, at which stage purification is easier.[31] By comparison, the "*grafting through*" method is less modular than the "*grafting to*" strategy. In the "*grafting to*" approach, theoretically, any peptide containing a suitable reactive group can be coupled to a well-defined precursor polymer to produce different products with, *e.g.*, different degrees of functionalization or with different peptides, *etc.* For the "*grafting through*" method, optimized polymerization conditions have to be found for each new monomer and the average length will not be precisely reproducible.

The first application using the "*grafting through*" approach was the synthesis of a thermoresponsive antibody–polymer conjugate described in 1987.[48] A more recent example deals with the incorporation of fibrinogen in a protein–polymer conjugate. Pluronic F127 was end-group-modified with an acrylate moiety on one side, while the other was coupled to fibrinogen.[49] The acrylate moiety was used for UV-activated free-radical polymerization. Since many of the fibrinogen molecules were functionalized with multiple Pluronic F127 moieties, polymerization triggered cross-linking, yielding a thermoresponsive hydrogel.

Beyond these three main concepts, the inverse bioconjugation approach offers another strategy to connect peptides or proteins with synthetic polymers. Using a solid support, which is preloaded with a polymer block, the biological molecule can be assembled in a stepwise fashion through solid-phase synthesis. Mutter and coworkers first showed the attachment of PEO to a poly(styrene) resin *via* a benzyl ether linker.[50] This concept was finally developed further by Bayer and Rapp leading to a commercially available PAP resin, which is widely applied in solid-phase peptide synthesis.[51] In a similar approach, Lutz, Börner, and coworkers demonstrated the preparation of cleavable and non-cleavable soluble polystyrene supports by ATRP for the liquid-phase synthesis of peptide–polymer conjugates.[52]

1.3 Chemical Synthesis of Peptide- and Protein–Polymer Conjugates

The number of applications for peptide- and protein–polymer conjugates is constantly increasing. While classical synthetic reactions take advantage of naturally occurring amino acid side chains for coupling reactions with polymers, many innovative strategies are based on methods to attach synthetic polymers selectively, and with high efficiency, to improve yields and purity of the conjugation products.[6–9] The following sections will give an overview of the most commonly used techniques, sorted according to the different reactive functional groups. Novel bioconjugation approaches will be discussed in detail.

1.3.1 Coupling with Amines

Exploiting primary amines is attractive since amines are among the most reactive functional groups present in peptides and proteins, and are found in relatively high abundance at the surface of globular proteins.[53] Various ligation strategies to attach polymers to them have been established.[31] Most common is the modification of a synthetic macromolecule with an activated carboxylic acid group in order to address the peptide or protein N-terminus or a lysine side chain, if available.[33] Typically, polymeric derivatives of *N*-hydroxysuccinimide (NHS)[54] are used (Figure 1.3a). This reaction allows for the coupling of lysine residues in an almost quantitative manner, resulting in a stable amide linkage. For example, the NHS end group modification of PNIPAM derived from RAFT polymerization enabled the ligation to a cyclic peptide.[55] The resulting thermoresponsive peptide–polymer conjugate possesses the ability to form channels within phospholipid membranes.

However, the applicability of NHS-esters is limited due to their susceptibility to nucleophilic addition reactions, resulting in hydrolytic instability in

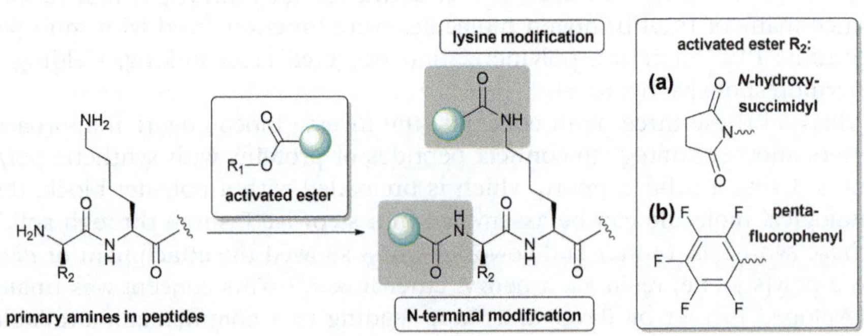

Figure 1.3 Bioconjugation of synthetic polymers by targeting primary amines within biomacromolecules resulting in amide bond formation at the N-terminus or the lysine side chain: (a) *N*-hydroxysuccinimidyl (NHS) activated ester,[54] (b) pentafluorophenyl-activated ester.[56]

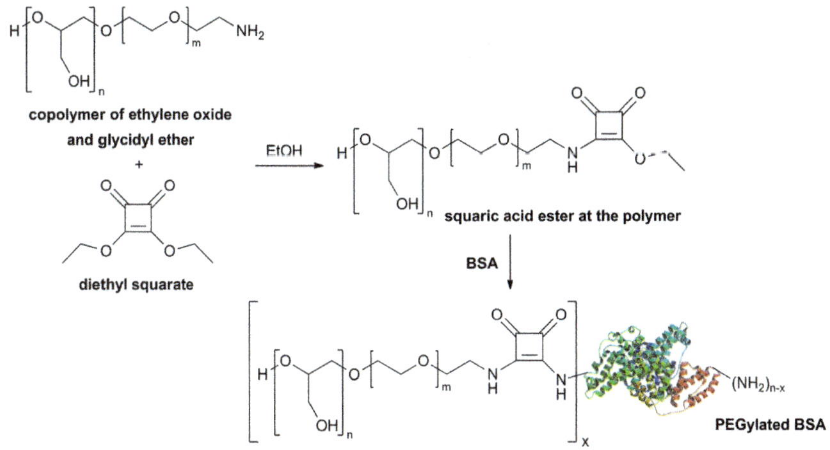

Figure 1.4 Diethyl squarate end-group functionalized PEO used for the squaric ester-mediated PEG*ylation* of BSA[58] (PDB: 3V03).[34]

aqueous media. To avoid low coupling efficiencies, the integration of pentafluorophenyl-activated carboxylic acid groups into synthetic polymers and their behavior in bioconjugation have been investigated (Figure 1.3b).[56] An appropriate modification of a collagen-like peptide was used for the site-selective conjugation of a stimulus-responsive poly(diethylene methyl ether methacrylate).[57] Coupling of two such bioconjugates resulted in the formation of triblock copolymer which showed triple-helix formation and thermoresponsiveness.

Frey and coworkers presented the squaric acid mediated PEG*ylation* of proteins.[58] Amine bearing PEO was end-group functionalized with squaric acid diethyl ester, using one of the reactive groups in the squaric acid moiety. The other was still available for the efficient functionalization of lysine residues in proteins like BSA (Figure 1.4).

Despite the efficient coupling reactions available for the functionalization of amines, their high abundance in peptides or proteins prevents selective bioconjugate formation, since a single-site modification is often not possible. Furthermore, the multiple and unspecific conjugation of synthetic polymers can result in loss of function and reduced enzymatic activity, and can potentially even induce toxicity.[33] For these reasons, it is often desirable to apply more specific and directed coupling strategies, *e.g.* using reactions in which one of the components is rarely found in peptides and proteins.

1.3.2 Coupling with Thiols

Cysteine residues are frequently addressed in biological molecules because they are rather rare in peptides and proteins. If accessible thiols are present, they can therefore allow for a site-specific polymer ligation, yielding well-defined bioconjugates.[59] Classically, thioether formation by Michael

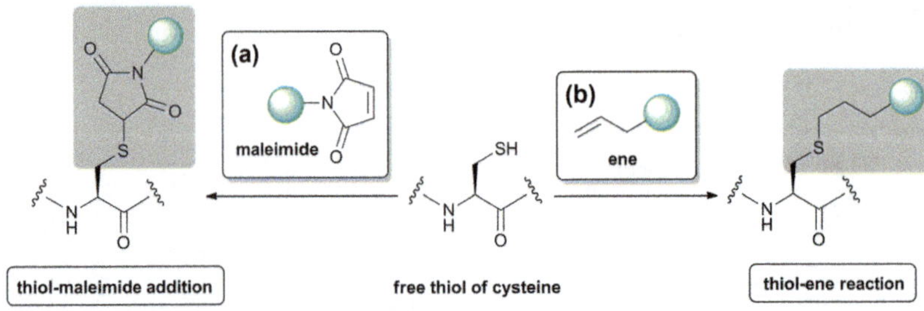

Figure 1.5 Bioconjugation of synthetic polymers by targeting thiols (cysteine residues): (a) Michael addition between thiols and maleimides,[60] and (b) thiol-ene reaction.[62]

addition of the thiol to a suitable electron-deficient alkene in the polymer blocks is used. The Michael addition reaction between thiols and maleimides constitutes the most well-known reaction, and is widely used to prepare bioconjugates of peptides and proteins (Figure 1.5a).[60]

For example, the aminolysis of the chain transfer agent end group of PNIPAM prepared by RAFT exposed thiol end groups, which can be reacted with 1,8-bis-maleimidodiethyleneglycol. Ligation of bovine serum albumin and ovalbumin to the maleimide-terminated polymer yielded protein–polymer conjugates.[61]

Due to its electrophilic properties, maleimides are susceptible to the addition of other nucleophiles besides thiols, such as primary amines. Under physiological conditions, the reaction between a thiol group and a maleimide proceeds more rapidly than with an amine, as is desired.[63] However, slight changes in the pH to alkaline conditions can already result in a shift towards more side reactions with *e.g.* lysine side chains or the N-terminus of the biomacromolecule.[64] Beyond that, higher pH-values can even effect the hydrolysis of the maleimide group to an open maleamic acid form, which is no longer reactive towards thiols.[65] After the formation of the thiol-maleimide product, the ring-opening can effect the stabilization of the resulting conjugate, which prevents a possible thiol exchange.[66]

These limitations can be avoided by using the free-radical addition of a thiol to a double bond (Figure 1.5b), which is referred to as a thiol-ene reaction[62] and can also proceed as a "*click*" reaction.[23] Initially, radical formation can be induced by heat or light.[67,68] The generated thiyl radical enters the addition reaction with the vinyl group, resulting in a carbon-centered radical. Subsequently, the desired thioether product is generated by hydrogen abstraction from other thiols, yielding further thiyl radicals. To achieve bioconjugates for *in vivo* applications, a photoinitiator is often added to generate radicals, since this procedure allows for the use of mild, relatively long wavelength (>300 nm) UV irradiation and can be performed in aqueous media under physiologic conditions.[69]

The post-polymerization modification of poly(pentafluorophenyl methacrylate) with allylamine to couple a thiol-terminated peptide domain was

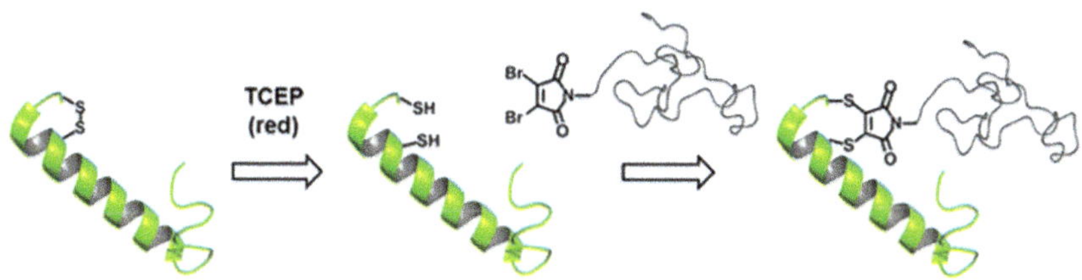

Figure 1.6 Disulfide reduction and subsequent bridging *via* addition-elimination with a dibromomaleimide to obtain a bioconjugate composed of salmon calcitonin (PDB: 2GLH)[74] and PEO.[72,73]

presented by Klok and coworkers.[70] The polymeric side chain modification was used later on for the synthesis of polyvalent peptide–polymer conjugates exhibiting HIV-1 inhibitory properties with antiviral activity depending on the bioconjugate length.[71]

In proteins, cysteine residues are not always readily accessible, since they are often involved in disulfide bridges within the complex three-dimensional biomacromolecular structures. Therefore, only a small number of cysteine residues can be used for bioconjugation reactions.[59] The ligation of polymer bound dibromomaleimides takes advantage of this circumstance, as it allows for the functionalization of disulfide moieties. Haddleton and coworkers demonstrated the applicability of this reaction for bioconjugation of salmon calcitonin (sCT).[72,73]

Therein, the Mitsunobu reaction was adapted in order to end-group functionalize a linear PEO with a dibromomaleimide moiety. This modified polymer was conjugated to sCT after reduction by tris(2-carboxyethyl)phosphine (TCEP) under mild reaction conditions, yielding disulfide rebridging conjugation (Figure 1.6).[72] The efficient ligation reaction of dibromomaleimides with cysteine side chains or disulfide bridges, respectively, enables a site specific protein modification in a fast and equimolar manner. Despite opening a disulfide bridge in a protein, the native structure might be less significantly affected since a covalent bridge between both cysteine residues is reinstalled (*cf.* Figure 1.6). Furthermore, O'Reilly and coworkers used dibromomaleimide monomers for the preparation of profluorescent polymers, which were capable of chemico-fluorescent responsiveness mediated by thiol conjugation, resulting in a switch of the fluorescence emission.[75]

1.3.3 Chemical Ligation by Oxime/Hydrazone Formation

For certain applications, the ligation reactions need to be performed *in vivo* without interfering with the given biological system. This severely limits the number of reactions that can be used due to, for example, the chemically very complex environment in biological systems and the large variety of functional groups present in a living system. In this context, the term "*bioorthogonal*" reaction was first coined by Bertozzi in 2003[76] in order to summarize efficient non-toxic chemical reactions that can be used in such systems, *e.g.*

Staudinger-based ligation techniques, oxime/hydrazone formation, or strain-promoted azide–alkyne cycloaddition reactions.[77] To be able to specifically address a particular protein target, or other moieties, these reactions use chemical groups that are not typically found in natural systems.

Ligation techniques based on carbonyl chemistry between aldehydes or ketones with nucleophilic amine derivatives, such as aminooxy or hydrazide groups, have been developed for classical bioorthogonal tools such as protein-labeling[78] and protein–polymer conjugation (Figure 1.7).[26] These types of condensation reactions were originally reported in the context of the *in vitro* assembly of a cytotoxin from non-harmful precursors through hydrazone formation.[79] Apart from the potential reversibility of the resulting imine bond, the oxime/hydrazone formation enables a highly selective route to prepare protein–polymer conjugates under physiological and additive-free conditions.[80]

In a recent example, Maynard and coworkers demonstrated the synthesis of a Boc-protected aminooxy functionalized initiator for copper(I)-mediated ATRP of NIPAM (Figure 1.8).[81] Deprotection of the polymer end groups

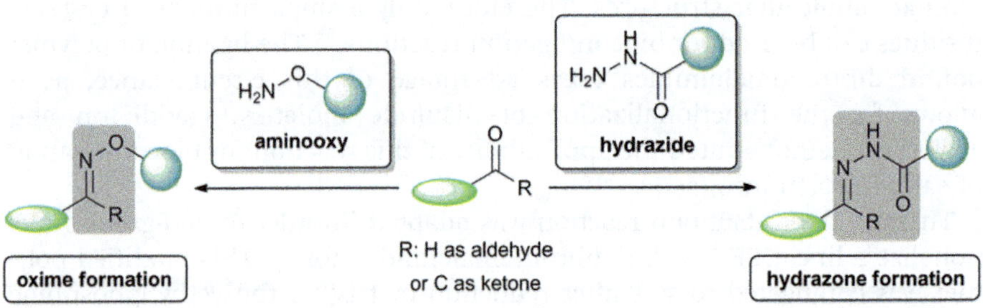

Figure 1.7 Chemical ligation of aldehyde or ketones with aminooxy or hydrazide peptide derivatives to form oximes or hydrazones, respectively.[80]

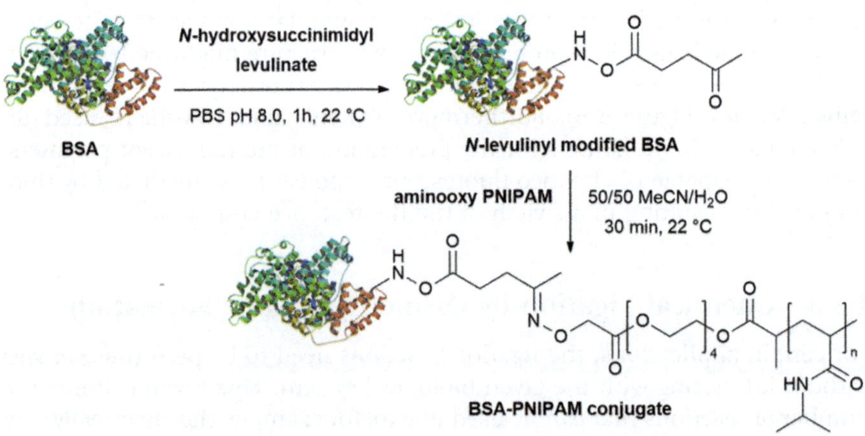

Figure 1.8 Aminooxy end-group functionalized PNIPAM synthesized by copper-catalyzed ATRP reacted with (*N*-levulinyl)-modified BSA (PDB: 3V03),[34] establishing a novel approach to design well defined protein–polymer conjugates.[34,81]

allows for bioconjugation with the aminooxy moiety. Therefore, *N*-levulinyl (a ketone containing group)-modified bovine serum albumin can be attached to the aminooxy functionalized PNIPAM rapidly and in a highly selective manner, resulting in a thermoresponsive protein–polymer conjugate.

Furthermore, oxime formation has been used to functionalize proteins; for example, N-terminal oxidation of interleukin 8 with sodium periodate[82] and biomimetic transamination of myoglobin with pyridoxal-5-phosphate,[83] followed by site-specific protein PEG*ylation* with hydroxylamine-functionalized polymer blocks. The selectivity of this reaction is limited by the fact that other amino acids within the protein can be oxidized as well, and incomplete and unselective transamination can take place.[5]

1.3.4 Staudinger Ligation

In a first *in vitro* application, the Staudinger ligation was used for the modification of an azide-functionalized glycoconjugate with a biotinylated triarylphosphine resulting in cell surface labeling.[84] The ligation technique is a modern adaptation of the classical Staudinger reaction originally published in 1919.[85] In the first and rate limiting reaction step, a phosphazide is formed under the liberation of nitrogen (Figure 1.9a).[86] The ester group in *ortho*-position to the phosphorus group is subsequently involved in an intramolecular cyclization. The cyclization is followed by hydrolysis in aqueous media, and the desired amide bond is generated with simultaneous formation of a phosphine oxide byproduct. An additional feature of this reaction is the possibility of creating a "traceless" product by changing the linker between the ester or thioester and the phosphine moiety, which

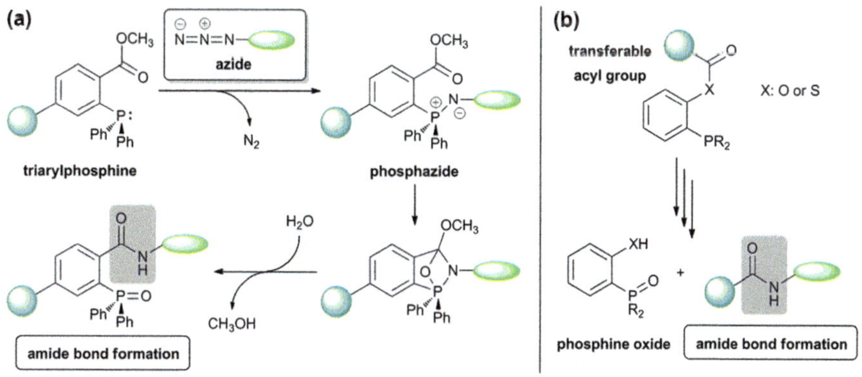

Figure 1.9 (a) Mechanism of the "classic" Staudinger ligation between a triarylphosphine and an azide resulting in the desired amide bond, simultaneously yielding the phosphine oxide byproduct. (b) Comparison with the "traceless" ligation, where the linker between the ester or thioester and the phosphine moiety is changed to allow for elimination of the phosphine oxide group by hydrolysis.[86]

results in the elimination of the phosphine oxide side product in the final hydrolysis step (Figure 1.9b).[87,88]

The ligation can be performed in aqueous media in the absence of any metal catalyst. In addition, the absence of azides and phosphines in the natural environment renders the Staudinger ligation highly suitable for bioorthogonal labeling in living systems[89,90] and for the conjugation of synthetic polymer blocks to biological macromolecules.[91] Among others, the incorporation of non-natural amino acids in protein biosynthesis,[92] the diazo transfer onto primary amines,[93] or native chemical ligation strategies[94] can be used for the introduction of the azide moiety into peptide or proteins.

An interesting example combines the previously mentioned free-radical thiol-ene "*click*" addition with the Staudinger ligation in order to obtain biodegradable peptide-functionalized poly(lactide)-*graft*-poly(ethylene oxide) copolymers (Figure 1.10).[95]

In this "*grafting through*" approach, allyl lactide and a bifunctional tri(ethylene glycol) initially react to generate a tri(ethylene glycol)-containing lactide analogue using thiol-ene chemistry. Subsequently, ring-opening

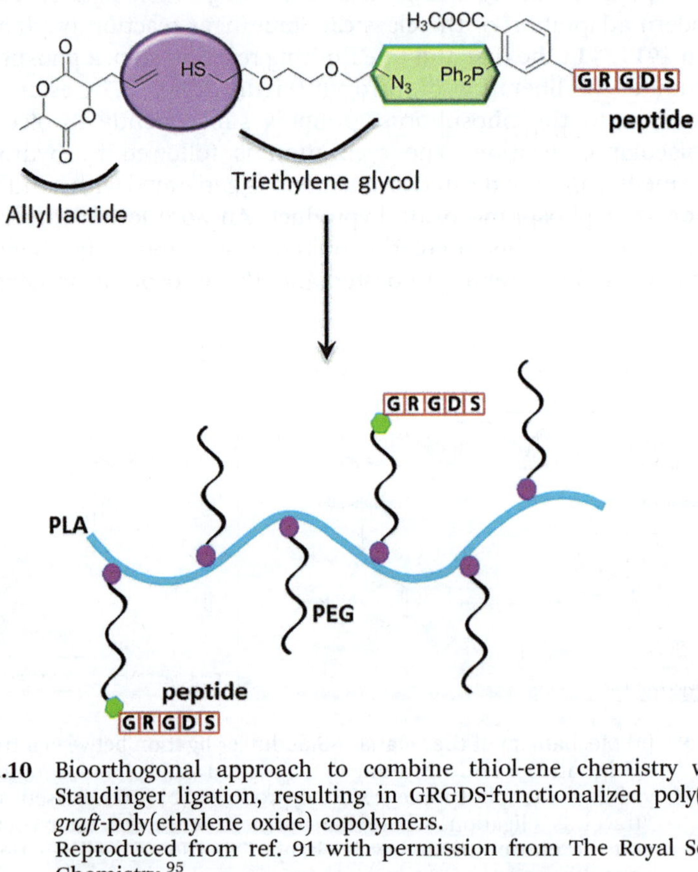

Figure 1.10 Bioorthogonal approach to combine thiol-ene chemistry with the Staudinger ligation, resulting in GRGDS-functionalized poly(lactide)-*graft*-poly(ethylene oxide) copolymers.
Reproduced from ref. 91 with permission from The Royal Society of Chemistry.[95]

polymerization induced by tin octoate yields a well-defined copolymer with a molecular weight of 6×10^3 g mol^{-1} and a dispersity index of 1.6. Finally, azide terminal groups on the copolymer tri(ethylene glycol) chains of the copolymer enable efficient post-polymerization functionalization by Staudinger ligation to install a GRGDS peptide domain.

1.3.5 Azide–Alkyne Cycloaddition

One of the organic reactions that has recently been most intensively studied to combine two macromolecules is the 1,3-dipolar cycloaddition between an azide and an alkyne resulting in a 1,2,3-triazole linker. Originally reported by Huisgen in 1961,[96] the reaction has been classified as a paradigm for the "*click*" reaction by Sharpless at the beginning of the 21st century.[23] Through additional regioselectivity using copper(I) as the catalyst (Figure 1.11),[24,25] this cycloaddition (CuAAC, copper(I)-catalyzed alkyne–azide cycloaddition) reaction became a valuable strategy to prepare peptide- and protein–polymer conjugates for a variety of applications.[8,27,97]

Again, the unique reactivity of azides (as in the previously described Staudinger ligation) often allows for almost quantitative product formation rates and mild reaction conditions.[98] Because the cycloaddition reaction is preferred over a nucleophilic substitution mechanism, biological macromolecules can be modified in a highly selective manner, whereas side reactions rarely occur, rendering protective groups redundant.[31]

Jia and coworkers took advantage of CuAAC to design an elastin-like peptide–polymer conjugate.[99] The reaction between a bifunctional, azide-terminated PEO block and an alkyne-functionalized peptide composed of two different functional domains yielded a multiblock copolymer (Figure 1.12).

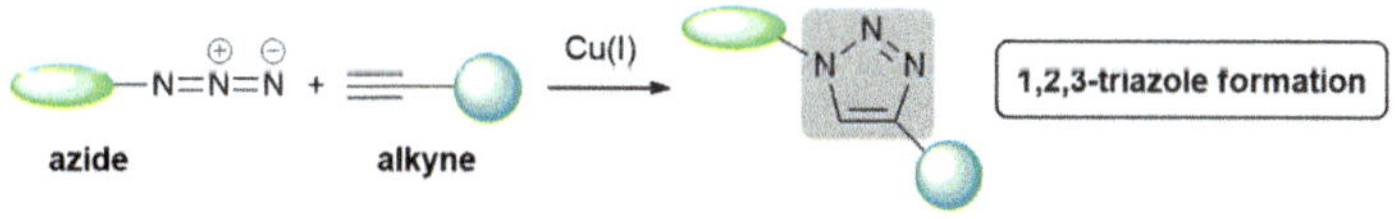

Figure 1.11 Schematic representation of the regioselective 1,3-dipolar cycloaddition between an azide and an alkyne to afford a 1,2,3-triazole-mediated by Cu(I) (CuAAC).[17,18]

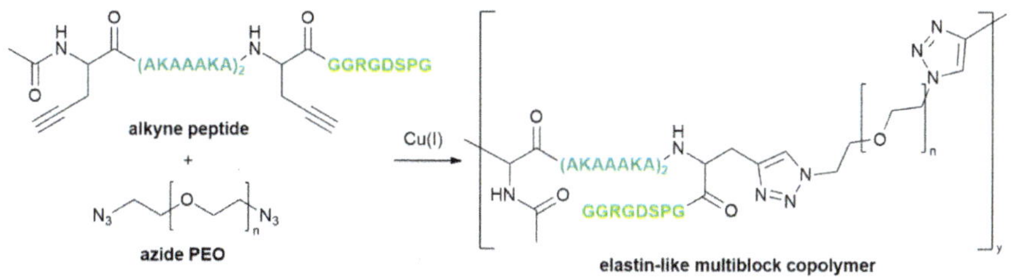

Figure 1.12 Elastin-like hybrid multiblock copolymer synthesized *via* CuAAC chemistry containing the $(AKAAAKA)_2$ sequence for cross-linking and the GGRGDSPG one to mediate cell adhesion.[99]

While the sequence $(AKAAAKA)_2$ is abundant in the cross-linking region of natural elastin, the RGD motif in GGRGDSPG mediates cell adhesion. Finally, covalent crosslinking through the lysine residues using hexamethylene diisocyanate resulted in hydrogels with elastin-like mechanical properties.

Unfortunately, utilization of copper(I) to control the regioselectivity yields several problems. Although only catalytic amounts are required, copper can induce cytotoxic effects[100] and protein denaturation,[101] which can complicate bioconjugation by CuAAC for applications *in vitro* and *in vivo*. In addition, copper salts are often rather difficult to remove from the reaction products. Nevertheless, advantages like high selectivity, mild reaction conditions, and almost quantitative product formation have proven CuAAC highly beneficial for the preparation of site-specific peptide- and protein–polymer conjugates.[102]

The drawbacks of CuAAC can be circumvented by replacing the alkyne in the copper(I)-catalyzed reaction with a ring-strained cyclooctyne. The spontaneous reaction between an azide and a cyclooctyne (OCT) was first reported in 1953 (Figure 1.13a).[103] However, until Bertozzi and coworkers took advantage of the strain-promoted azide–alkyne cycloaddition (SPAAC) with enhanced biocompatibility, this reaction had hardly been used.[104] Ever since, many different strained azophiles have been reported in connection with *e.g.* protein-labeling, cell surface modification, or *in vivo* imaging.[78] Besides the classic OCT,[104] derivatives like monofluorinated cyclooctyne (MOFO)[105] or dibenzocyclooctyne (DIBO)[106] have been developed to enhance the reactivity towards azide moieties (Figure 1.13b).

The potential of SPAAC in peptide–polymer conjugation was successfully demonstrated by Chaikof and coworkers using a cyclooctyne-functionalized peptide for surface decoration (Figure 1.14).[107] It was shown that a

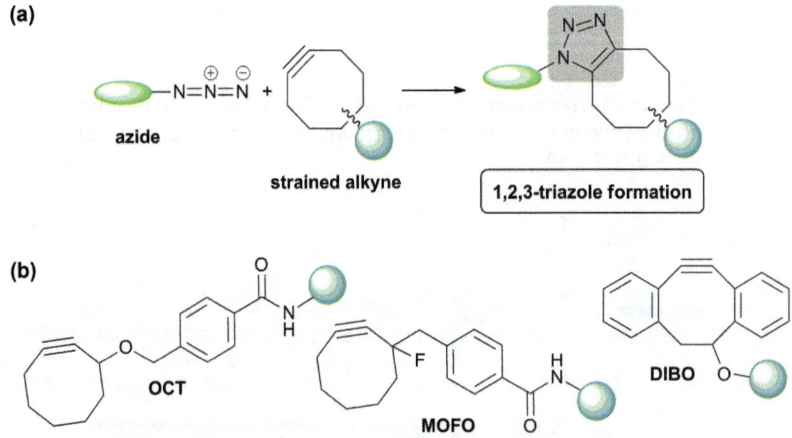

Figure 1.13 (a) Strain-promoted azide–alkyne cycloaddition (SPAAC) between an azide and cyclooctyne.[104] (b) Chemical structures of the azophiles OCT, MOFO, and DIBO, frequently used in SPAAC.[78]

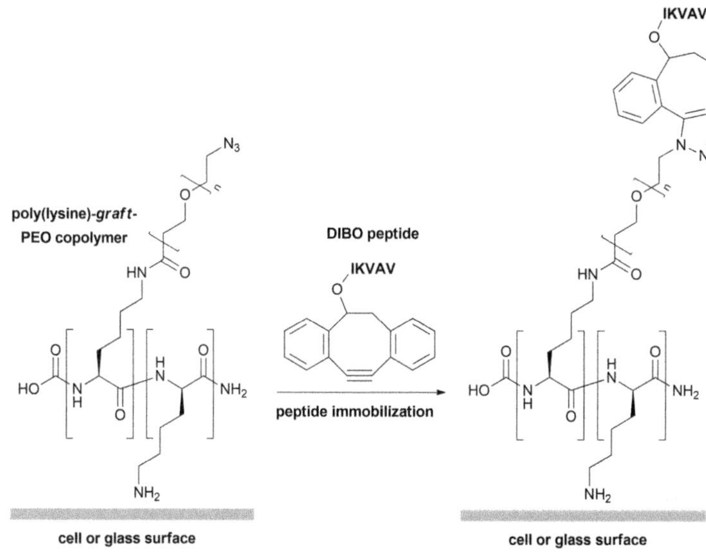

Figure 1.14 Cytocompatible poly(L-lysine)-*graft*-poly(ethylene oxide) copolymer modified with azido groups for the immobilization of a DIBO-peptide IKVAV on surfaces using SPAAC.[107]

cytocompatible poly(L-lysine)-*graft*-poly(ethylene oxide) copolymer bearing azido groups was efficiently modified with a DIBO-functionalized IKVAV peptide, yielding peptide–polymer conjugates with controlled grafting densities that could be used in peptide microarrays.[107]

Despite the advantages of site-specific bioconjugation, certain limitations of SPAAC have to be considered. Besides reaction kinetics 10–100 times slower than CuAAC,[108] hydrophobicity and instability of the highly reactive cyclooctynes[109] under physiological conditions restrict its application in living systems. For that reason, the scientific community still strives for novel approaches to design peptide- and protein–polymer conjugates in a highly selective and fast manner.

1.3.6 Diels–Alder (DA) Cycloaddition Reactions

Reactions between a diene and a dienophile yielding cyclohexene derivatives through carbon–carbon or carbon–heteroatom bond formation are referred to as Diels–Alder (DA) reactions. The concerted [4 + 2] cycloaddition was first described in 1928 by Diels and Alder,[110] and can be adapted for the efficient conjugation of peptides and proteins to synthetic polymers.[26] Despite being recognized as highly selective and additive-free, DA reactions require elevated temperatures to obtain the desired product. Furthermore, the reversibility of the reaction limits its application in materials science.[23]

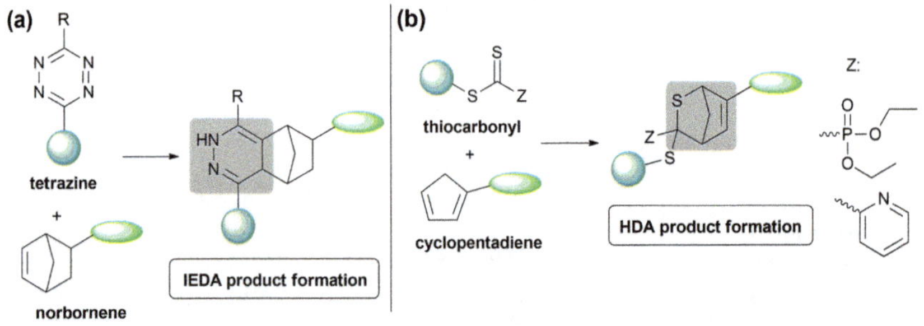

Figure 1.15 Modern variants of the Diels–Alder cycloaddition reaction: (a) Inverse-electron-demand Diels–Alder (IEDA) reaction between tetrazine and norbornene,[113] and (b) hetero Diels–Alder (HDA) reaction between thiocarbonyl and cyclopentadiene.[114]

In comparison, the inverse-electron-demand Diels–Alder (IEDA) reaction between a tetrazine and a strained alkene or alkyne followed by retro-Diels–Alder elimination of nitrogen offers a highly selective alternative, yielding stable dihydropyridazine or pyridazine products.[111] Studies on, *e.g.*, *trans*-cyclooctene have revealed rate constants up to 2×10^3 M^{-1} s^{-1}, which are much higher than those of any other "*click*"-type reactions.[112] For that reason, *trans*-cyclooctene derivatives and other strained alkenes, such as cyclobutene or norbornene derivatives (Figure 1.15a), can be used for bioconjugation purposes.[113]

An interesting example of this reaction deals with the IEDA between a multifunctional PEO-tetrazine polymer and a dinorbornene peptide.[114] In this system, a four-arm amino-functionalized PEO (20×10^3 g mol^{-1}) was modified with benzylamino tetrazine. The IEDA reaction with the cross-linking peptide KGPQGIWGQKK containing norbornene moieties triggered the formation of a hydrogel.

Another remarkable strategy is based on the addition of hetero-dienophiles, such as imines, aldehydes, or thiocarbonyl compounds, to suitable dienes (Figure 1.15b).[115] These reactions are summarized under the term hetero Diels–Alder (HDA) reactions. Barner-Kowollik and coworkers explored HDA reactions for the preparation of peptide-functionalized polymers using cellulose as the polymer scaffold (Figure 1.16).[116] As a model reaction, a thioamide-functionalized peptide (Phos-GGFPWWG) was coupled to poly(methylmethacrylate) containing a cyclopentadienyl terminal group, yielding PMMA-*b*-GGFPWWG bioconjugates. The strategy was thereafter used for the attachment of Phos-GGFPWWG to cyclopentadienyl functional cellulose. Remarkably, only small amounts of acid were required to obtain conversions near 90%.

Even azide moieties are capable of entering DA cycloaddition reactions with oxanorbornadiene derivatives, yielding triazole products. For example, the tandem [3+2]/Diels–Alder cycloreversion (CrDA) was used for the PEG*ylation* of an N-terminally azido functionalized peptide (GGRGDG).[117] Furthermore, labeling of oxanorbornadiene-modified hen egg-white lysozyme with fluorescent coumarine was demonstrated.[117] The release of furan

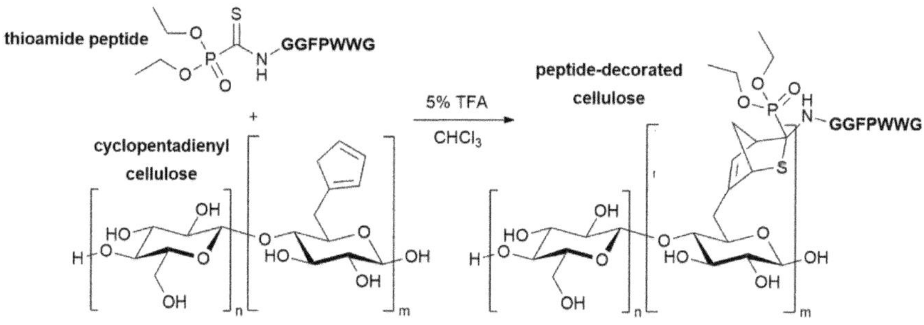

Figure 1.16 Hetero Diels–Alder (HDA) reaction between a thioamide-functionalized peptide (Phos-GGFPWWG) and cyclopentadienyl functional cellulose for the preparation of peptide-decorated cellulose.[116]

offers a versatile and efficient route to prepare peptide–polymer conjugates at moderate temperatures.

Another interesting strategy to design polymer–peptide–polymer conjugates based on the HAD reaction was demonstrated by Barner-Kowollik, Börner, and coworkers. In the first step, the formation of a peptide–PEO conjugate based on maleimide-magainin and a modified PEO bearing a tetrazole and a photoenol moiety was induced by light. Repeated irradiation in the presence of maleimide terminal polylactide lead to the formation of a hybrid triblock macromolecule.[118]

1.3.7 Chemistry with 1,2,4-Triazoline-3,5-diones (TAD)

Recent developments concerning fast and highly selective ligation methods use 1,2,4-triazoline-3,5-dione (TAD) as one of the most remarkable and versatile reagents in "*post-click*" chemistry.[8] Within the heterocyclic compound, the arrangement of the azo moiety and the electron-withdrawing carbonyls offers a significant reactivity towards a variety of functional groups such as enes or dienes.[119] Therefore, Diels–Alder and other addition-type reactions take place, resulting in the formation of stable bonds as well as reversible connections depending on the reaction partner. The reaction proceeds in the absence of a catalyst under moderate conditions, and its orthogonality makes several reactions with TAD highly interesting for the preparation of bioconjugates.[23]

Du Prez and coworkers presented a relevant approach concerning TAD-based chemistry described as a novel "*click*" and "*trans-click*" strategy (Figure 1.17a).[120] The reaction of 4-butyl-1,2,4-triazoline-3,5-dione (BuTAD) with 2,4-hexadiene-1-ol (HDEO) yielded the Diels–Alder adduct TAD–HDEO in an extremely fast, traceless, and irreversible manner. The occurrence of a retro-Diels–Alder reaction can be excluded based on reference experiments with competing 2,4-hexadiene-1,6-diol. Furthermore, an addition-type adduct (TAD–indole) was prepared using a substituted indole, *e.g.*, tryptophan in proteins. To demonstrate the "*trans-click*" reactivity,

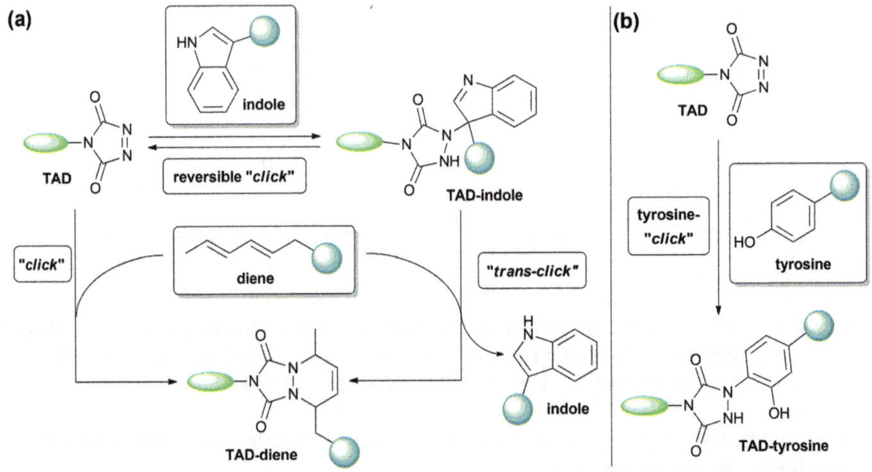

Figure 1.17 Reactivity of 1,2,4-triazoline-3,5-dione (TAD): (a) "*Click*" and "*trans-click*" reactions involving a diene and an indole derivative,[120] and (b) tyrosine-"*click*" reaction.[121,122]

HDEO was added under elevated temperatures resulting in TAD–HDEO (TAD–diene). The ability of TAD to form both reversible and non-reversible covalent linkages offers novel opportunities for polymer and materials science.[8] The potential of TAD as an extremely fast and highly selective reacting moiety has already been used in a variety of applications such as polymer end-group modification,[120] polymer–polymer conjugation,[120] the preparation of polymer networks,[120] and orthogonal protein labeling.[121]

The previously described orthogonal strategies, such as ligation by oxime/hydrazone formation,[80] Staudinger ligation,[87] or SPAAC,[121,122] *etc.*, require chemical modification of the peptide and proteins for chemoselective polymer conjugation. To circumvent this, Barbas and coworkers studied TAD derivatives possessing azide, alkyne, and ketone groups. They demonstrated the reactivity of the TAD moiety by addressing tyrosine moieties in peptides and proteins (Figure 1.17b).[123] As an example of the potential of TAD-based chemistry in the creation of peptide- and protein–polymer conjugates, TAD-functionalized PEO was used for the PEG*ylation* of chymotrypsinogen, which contains four tyrosine residues for addition-type reactions.[91] An alkyne-PEO (5×10^3 g mol^{-1}) was modified with a urazole-bearing azide moiety by CuAAC, where the urazole served as a precursor to the reactive TAD. Upon oxidation to TAD–PEO, the reaction with chymotrypsinogen resulted in addition products bearing multiple PEO chains. Taking into consideration that TAD-reactions based on Diels–Alder reactions are even more efficient than those based on addition reactions,[120] this work demonstrated that TAD chemistry can be used effectively for the development of novel bioconjugation strategies for materials science applications.

1.4 Chemoenzymatic Approaches

Strategies for the preparation of bioconjugates taking advantage of enzymes have been investigated with the objective of addressing functional groups in peptides or proteins in a highly selective manner. From a formal point of view, enzymes can be separated into two categories, by which site-specific modification of biological molecules can be realized.[6] On the one hand, a number of enzymes are capable of introducing appropriately functionalized synthetic polymers onto biomacromolecular substrates.[124] On the other hand, enzymes can be used for the modification of a single moiety.[125] Transglutaminase (TGase) and sortase (Srt) are representatives of the first-mentioned class of enzymes, whereas formylglycine transferase and lipoic acid ligase have been used for the transformation of functional groups.[6] The potential of these enzymes for bioconjugate synthesis and protein modification will be discussed in the following sections.

1.4.1 Transglutaminase (TGase) Catalyzed Ligation

Transglutaminase (TGase) enzymes are capable of catalyzing an acyl transfer between two proteins involving the glutamyl group of a glutamine (acyl donor) and a primary amine (acyl acceptor), usually the ε amino group of a lysine (Figure 1.18).[126] In the catalytic cycle, first an acyl-enzyme intermediate is formed, simultaneously releasing ammonia. The presence of an acyl acceptor leads to the transamidation product, whereas TGase is regenerated.[127] Using a PEO-amine derivative as an acyl acceptor, site-specific TGase-mediated PEG*ylation* of proteins is feasible.

In a recent example, the recombinant human growth hormone (hGH) was modified with an amine-functionalized PEO-block (20×10^3 g mol^{-1}) using microbial TGase (mTGase) to mediate the PEG*ylation* (Figure 1.19). Performed in a water/ethanol mixture, mTGase allowed the selective ligation at hGH Gln_{141}, yielding the site-specific conjugate PEO–Gln_{141}–hGH.[128]

Due to the substrate selectivity of the enzyme, one or two glutamine residues within a protein can be addressed selectively in most cases.[129,130] Taking into account that two large molecules are covalently connected, the site-specific and nearly quantitative product formation is particularly noteworthy.[128] A disadvantage of TGase-mediated and other enzyme catalyzed

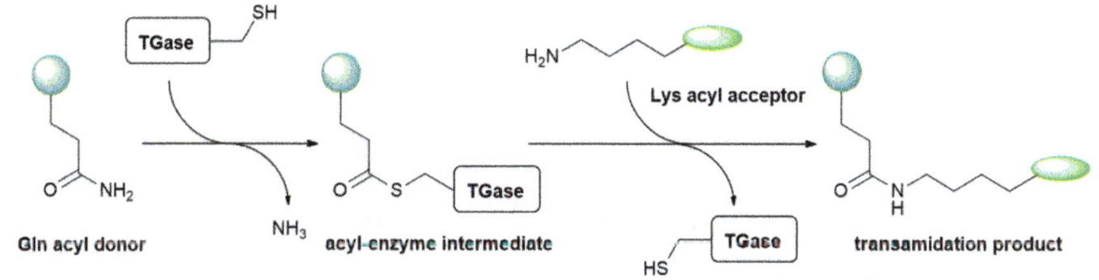

Figure 1.18 Catalytic process of the TGase-mediated transamidation reaction.[127]

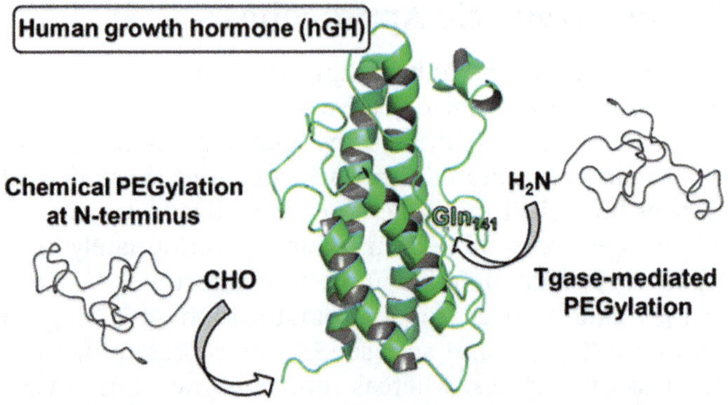

Figure 1.19 Site-specific TGase-mediated PEG*ylation* of the recombinant human growth hormone (PDB: 1HGU)[13] at hGH Gln_{141}.[128]

reactions is that product purification, in particular enzyme separation, can be rather tedious. The purification can be simplified by involving thermoresponsive polymers such as PNIPAM in protein–polymer conjugates, exploiting the lower critical solution temperature behavior of the polymer to obtain two immiscible phases.[131]

1.4.2 Sortase (Srt)-mediated Ligation

Sortases are transpeptidases, which use the linear epitope LPXT↓G as a recognition site within a target peptide or protein.[132] The cleavage of the amide bond between the threonine and glycine residue (↓) results in an acyl-enzyme intermediate. Subsequently, the thioester intermediate is attacked by a nucleophilic co-substrate comprised of an N-terminal glycine residue in order to form a native peptide bond. Sortase A (SrtA) derived from *Staphylococcus aureus* is commonly used for site-specific N- and C-terminal protein labeling,[133,134] the synthesis of protein–lipid conjugates,[135] or intramolecular peptide cyclization.[136] Furthermore, SrtA enables the site-specific PEG*ylation* of cytokine with an exogenous oligo-glycine functionalized-PEO.[137] The enzyme-based ligation strategy results in the extension of the cytokine half-life and biological activity.

Unfortunately, the transpeptidation mediated by SrtA is subject to reversibility, since the newly formed ligation product is a potential substrate for another transpeptidation cycle.[138] In order to prevent the reverse reaction and thus selectively obtain the desired product, the amide bond between the threonine and glycine residues within the recognition site can be replaced by an ester.[139]

1.4.3 Enzyme-induced Functional Group Modifications

Besides transglutaminases and sortases, a variety of enzymes are able to catalyze site-specific protein modifications in order to introduce reactive

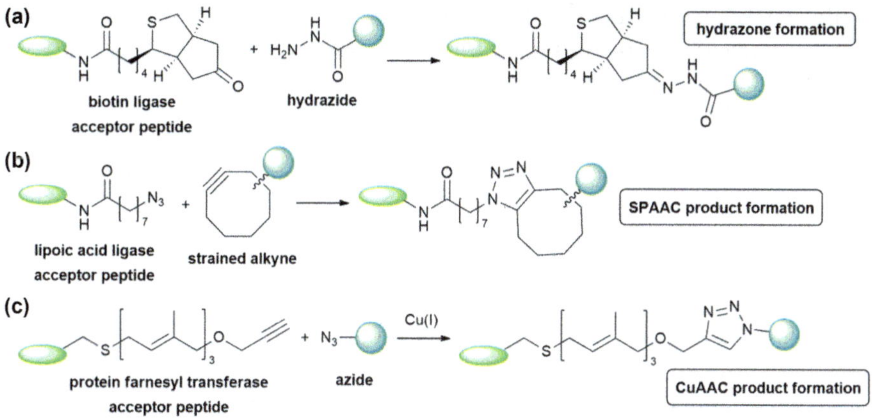

Figure 1.20 Enzymes for site-specific protein modification: (a) Biotin ligase,[141] (b) lipoic acid ligase,[142] and (c) protein farnesyltransferase.[143]

groups for further functionalization and ligation.[140] Biotin ligase, for example, catalyzes the modification of recombinant cell-surface proteins with biotin-mimicking ketones. The ketones can subsequently be used in ligation with hydrazide- or hydroxylamine-functionalized molecules (Figure 1.20a).[141] Microbial lipoic acid ligase is also used for protein labeling by attaching an alkyl–azide onto a cell-surface engineered acceptor peptide.[142] The introduction of the azide group permits the conjugation of a cyclooctyne-containing probe *via* SPAAC (Figure 1.20b).

Protein farnesyltransferase recognizes a short tetramer peptide motif and can be applied for chemoselective functionalization of proteins with synthetic azido-farnesyl analogs, followed by CuAAC for chemical labeling (Figure 1.20c).[144] Similarly, protein expression vectors can be tagged with CXPXR sequences, which act as recognition sites for formylglycine transferase.[143] The enzyme catalyzes cysteine oxidation to an aldehyde-bearing formylglycine, which enables site-specific ligation with α-nucleophilic moieties.

The variety of possibilities to address functional groups in biological molecules provided by chemoenzymatic approaches offers a wide range of opportunities for protein modification and bioconjugate synthesis with regard to pharmaceutical and materials science applications.[124]

1.5 Biotransformations

Besides the chemoenzymatic approaches described above, where functional groups are added to allow for conjugation with other macromolecules, enzymes demonstrate other possibilities to obtain peptide- and protein–polymer conjugates, which are normally difficult to obtain by common chemical synthetic strategies. Enzyme responsiveness can be used to modulate bioconjugate properties in order to create hybrid materials with more complex functions.[145] In living systems, enzymes catalyze a broad range of reactions to stimulate biological processes such as protein expression, cellular adhesion, muscle contraction, *etc.*[146] Some of these processes have been adapted for a variety of biotransformation reactions with the objective of integrating novel properties into bioconjugates. As a result, enzymatically catalyzed reactions are able to affect self-assembly,[147] to induce controlled release of, for example, therapeutic drugs,[148] or to initiate adhesion to diverse surfaces.[149,150]

Cenker and coworkers demonstrated the enzyme responsiveness of a PEO-peptide micelle system that can be proteolytically cleaved, leading to the release of non-aggregated amyloid-based peptides.[151] The peptide–polymer conjugate incorporating the *β*A*β*AKLVFF sequence was cleaved by α-chymotrypsin to produce *β*A*β*AKLVF and F–PEO. Whereas the hexamer did not aggregate into β-sheet structures, the heptapeptide *β*A*β*AKLVFF formed well-defined β-sheet ribbons, which have been associated with Alzheimer's disease.[152]

Börner and coworkers used acid phosphatase to induce microstructure formation.[153] The peptide-segment contained three phosphothreonine residues in the $(TV)_5$ peptide motif of the peptide–PEO conjugate, which prevented β-sheet formation. Enzymatic dephosphorylation of the threonine side chains therefore induced self-assembly of the bioconjugate.

A more recent example of a biotransformation of peptide–polymer conjugates combines biocombinatorial procedures with enzymatic activation processes in order to realize mussel-glue inspired adhesion.[154] Phage-display biopanning incorporating tyrosinase to transform L-tyrosine residues to L-dopa enabled the direct selection of enzymatically activable 12-mer peptide adhesion domains for aluminum oxide (Figure 1.21).

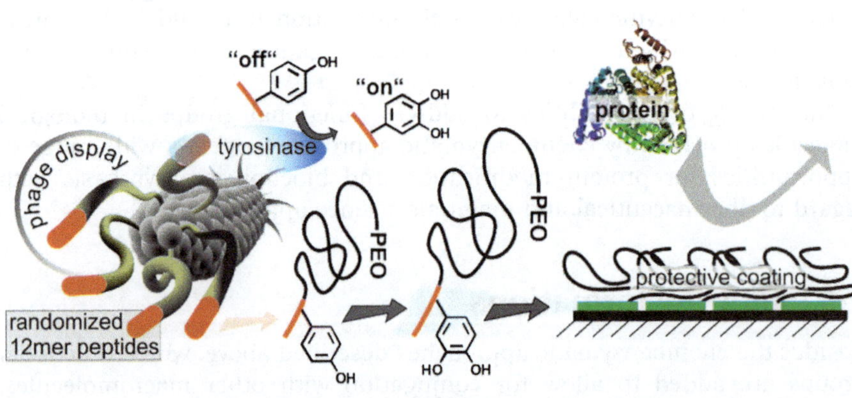

Figure 1.21 Concept to integrate tyrosinase activation processing steps into phage display biopanning to generate mussel-glue inspired adhesives and antifouling coatings[154] (PDB: 3V03).[34]
This figure was kindly provided by Dr Patrick Wilke.

After identification of activable adhesive domains, the 12-mer peptides were synthesized as peptide–PEO conjugates by the inverse bioconjugation approach. Switching of these conjugates from weak to strong binders was successfully triggered by tyrosinase. The activated systems revealed effective adsorption even under saltwater conditions. Furthermore, the coatings from enzymatically activated conjugates possessed antifouling properties and effectively suppressed interactions between blood plasma protein cocktails and bacteria with the coated surface, emphasizing the potential for medical as well as industrial applications.

1.6 Conclusions and Future Perspectives

Within the last decade, a wide range of synthetic and chemoenzymatic strategies have been developed to allow for the preparation of complex hybrid materials combining peptides or proteins with synthetic polymers.[6–9,124] Of the four main strategies for the creation of protein- or peptide–polymer conjugates, *i.e.* "*grafting from*", "*through*", "*to*", and through inverse bioconjugation, the "*grafting to*" method is the most widely used, although advances in controlled radical polymerization techniques including ATRP,[41] RAFT,[42] and NMP[46] have significantly contributed to the feasibility of the other methods. The main challenges in the preparation of well-defined conjugates are the site-specific and bioorthogonal modification of peptides and proteins and highly efficient product formation. Since the copper(I)-catalyzed azide–alkyne cycloaddition (CuAAC)[18,19] reaction was proven to be cytotoxic, a variety of ligation techniques that can be considered "*click*" reactions, such as the ligation by oxime/hydrazone formation,[80] Staudinger ligation,[87] and strain-promoted azide–alkyne cycloadditions (SPAAC),[123] have been established. Recent advances in bioconjugation by inverse-electron-demand Diels–Alder (IEDA) reactions between tetrazine and strained alkenes, for example, have provided insight into highly efficient and specific potential future reactions for the preparation of functional peptide- and protein–polymer conjugates.[115] Furthermore, the reactivity towards strained alkenes can be modulated by varying the electron withdrawing groups using nitrones, nitrile oxides, diazoalkanes, and syndones, as alternative 1,3-dipoles.[9] Equally, chemistry with 1,2,4-triazoline-3,5-dione (TAD) based on Diels–Alder reactions can be suggested for the development of novel bioconjugation strategies.[120]

Some other highly efficient cycloaddition reactions have been used for polymer end-group modification and polymer–polymer conjugation.[155] These include an additive-free version of the hetero Diels–Alder (HDA) reaction[156] or the nitrile-imine-mediated tetrazole-ene cycloaddition (NITEC), the latter only requiring photoirradiation.[157] Both techniques meet the criteria to be considered bioorthogonal reactions and can be performed under mild reaction conditions. As such, these reactions should also be applicable for the preparation of bioconjugates.

As discussed, a wide range of ligation reactions can be (potentially) used for bioconjugate synthesis,[158] and new reactions are being investigated to achieve even higher reactivity, more efficient product formation, and to circumvent side reactions. Increased reactivity and specificity are often incompatible properties though. Therefore, it is necessary to validate the applicability of each system individually. Chemoenzymatic approaches offer a great alternative to prepare peptide- and protein–polymer conjugates. Enzymes are capable of addressing functional groups occurring in peptides or proteins in a highly selective manner: most of the appropriate enzymes can be used for the modification of single (or very few) moieties in a biopolymer, either with small molecules that can be used in further ligation strategies or for the direct attachment of appropriately functionalized synthetic polymers onto biomacromolecular substrates. Though a large body of work exists on the (site-specific) modification of peptides and proteins (*e.g.* for labeling) using enzymes,[130,131] the exploitation of these reactions for the creation of well-defined functional protein- and peptide–polymer conjugates is still in its infancy.

References

1. J. C. M. van Hest, *Polym. Rev.*, 2007, **47**, 63.
2. R. K. Iha, K. L. Wooley, A. M. Nyström, D. J. Burke, M. J. Kade and C. J. Hawker, *Chem. Rev.*, 2009, **109**, 5620.
3. A. Carlsen and S. Lecommandoux, *Curr. Opin. Colloid Interface Sci.*, 2009, **14**, 329.
4. H. A. Klok, *Macromolecules*, 2009, **42**, 7990.
5. E. M. Pelegri-O'Day, E.-W. Lin and H. D. Maynard, *J. Am. Chem. Soc.*, 2014, **136**, 14323.
6. A. C. Obermeyer and B. D. Olsen, *ACS Macro Lett.*, 2015, **4**, 101.
7. I. Cobo, M. Li, B. S. Sumerlin and S. Perrier, *Nat. Mater.*, 2015, **14**, 143.
8. P. Espeel and F. E. Du Prez, *Macromolecules*, 2015, **48**, 2.
9. S. Liu, K. T. Dicker and X. Jia, *Chem. Commun.*, 2015, **51**, 5218.
10. M. A. Gauthier and H.-A. Klok, *Polym. Chem.*, 2010, **1**, 1352.
11. Q. Liu, J. Wang and B. J. Boyd, *Talanta*, 2015, **136**, 114.
12. H. G. Börner, *Prog. Polym. Sci.*, 2009, **34**, 811.
13. Image from the RCSB PDB (www.rcsb.org) of PDB ID 1HGU (L. Chantalat, N. D. Jones, F. Korber, J. Navaza, A. G. Pavlovsky (1995) The crystal structure of wild type growth hormone at 2.5 angstrom resolution *Protein Pept. Lett.*, **2**, 333–340).
14. H. Jatzkewitz, *Hoppe-Seyler's Z. Physiol. Chem.*, 1954, **297**, 149.
15. H. Jatzkewitz, *Z. Naturforsch., B: Chem. Biochem.*, 1955, **10**, 27.
16. A. Abuchowski, T. Vanes, N. C. Palczuk and F. F. Davis, *J. Biol. Chem.*, 1977, **252**, 3578.
17. A. Abuchowski, J. R. Mccoy, N. C. Palczuk, T. Vanes and F. F. Davis, *J. Biol. Chem.*, 1977, **252**, 3582.
18. N. Nischan and C. P. R. Hackenberger, *J. Org. Chem.*, 2014, **79**, 10727.

19. F. M. Veronese, *Biomaterials*, 2001, **22**, 405.
20. F. M. Veronese and G. Pasut, *Drug Discovery Today*, 2005, **10**, 1451.
21. I. W. Hamley, *Biomacromolecules*, 2014, **15**, 1543.
22. S. Perrier, S. Dehn, R. Chapman and K. A. Jolliffe, *Polym. Rev.*, 2011, **51**, 214.
23. H. C. Kolb, M. G. Finn and K. B. Sharpless, *Angew. Chem., Int. Ed.*, 2001, **40**, 2004.
24. V. V. Rostovtsev, L. G. Green, V. V. Fokin and K. B. Sharpless, *Angew. Chem., Int. Ed.*, 2002, **41**, 2596.
25. C. W. Tornoe, C. Christensen and M. Meldal, *J. Org. Chem.*, 2002, **67**, 3057.
26. W. Tang and M. L. Becker, *Chem. Soc. Rev.*, 2014, **43**, 7013.
27. W. X. Xi, T. F. Scott, C. J. Kloxin and C. N. Bowman, *Adv. Funct. Mater.*, 2014, **24**, 2572.
28. H. A. Klok and M. A. Gauthier, *Chem. Commun.*, 2008, 2591.
29. J. F. Lutz and H. G. Börner, *Prog. Polym. Sci.*, 2008, **33**, 1.
30. G. N. Grover and H. D. Maynard, *Curr. Opin. Chem. Biol.*, 2010, **14**, 818.
31. L. A. Canalle, D. W. P. M. Löwik and J. C. M. van Hest, *Chem. Soc. Rev.*, 2010, **39**, 329.
32. R. M. Broyer, G. N. Grover and H. D. Maynard, *Chem. Commun.*, 2011, **47**, 2212.
33. P. Thordarson, B. Le Droumaguet and K. Velonia, *Appl. Microbiol. Biotechnol.*, 2006, **73**, 243.
34. RCSB PDB (www.rcsb.org) of PDB ID 3V03 (K. A. Majorek, P. J. Porebski, A. Dayal, M. D. Zimmerman, K. Jablonska, A. J. Stewart, M. Chruszcz, W. Minor (2012) Structural and immunologic characterization of bovine, horse, and rabbit serum albumins, *Mol. Immunol.*, **52**, 174–182).
35. J. Nicolas, G. Mantovani and D. M. Haddleton, *Macromol. Rapid Commun.*, 2007, **28**, 1083.
36. H. Rettig, E. Krause and H. G. Börner, *Macromol. Rapid Commun.*, 2004, **25**, 1251.
37. Y. Mei, K. L. Beers, H. C. M. Byrd, D. L. Vanderhart and N. R. Washburn, *J. Am. Chem. Soc.*, 2004, **126**, 3472.
38. K. L. Heredia and H. D. Maynard, *Org. Biomol. Chem.*, 2007, **5**, 45.
39. B. S. Sumerlin, *ACS Macro Lett.*, 2012, **1**, 141.
40. S. Hansson, V. Trouillet, T. Tischer, A. S. Goldmann, A. Carlmark, C. Barner-Kowollik and E. Malmstrom, *Biomacromolecules*, 2013, **14**, 64.
41. B. S. Lele, H. Murata, K. Matyjaszewski and A. J. Russell, *Biomacromolecules*, 2005, **6**, 3380.
42. A. B. Lowe and C. L. McCormick, *Prog. Polym. Sci.*, 2007, **32**, 283.
43. J. Hentschel, M. G. J. ten Cate and H. G. Börner, *Macromolecules*, 2007, **40**, 9224.
44. C. Boyer, V. Bulmus, T. P. Davis, V. Ladmiral, J. Q. Liu and S. Perrier, *Chem. Rev.*, 2009, **109**, 5402.
45. H. G. Börner, in *Controlled/Living Radical Polymerization: Progress in RAFT, DT, NMP & OMRP*, American Chemical Society, 2009, vol. 1024, p. 265.

46. M. L. Becker, J. Q. Liu and K. L. Wooley, *Biomacromolecules*, 2005, **6**, 220.
47. P. De, M. Li, S. R. Gondi and B. S. Sumerlin, *J. Am. Chem. Soc.*, 2008, **130**, 11288.
48. N. Monji and A. S. Hoffman, *Appl. Biochem. Biotechnol.*, 1987, **14**, 107.
49. Y. Shachaf, M. Gonen-Wadmany and D. Seliktar, *Biomaterials*, 2010, **31**, 2836.
50. H. Hellermann, H. W. Lucas, J. Maul, V. N. R. Pillai and M. Mutter, *Makromol. Chem.*, 1983, **184**, 2603.
51. W. Rapp, in *Combinatorial Peptide and Nonpeptide Libraries*, Wiley-VCH Verlag GmbH, 2007, p. 425.
52. A. Meszynska, N. Badi, H. G. Börner and J. F. Lutz, *Chem. Commun.*, 2012, **48**, 3887.
53. R. F. Doolittle, in *Prediction of Protein Structure and the Principles of Protein Conformation*, ed. G. D. Fasman, Springer, New York, 1989, p. 599.
54. A. J. Lomant and G. Fairbanks, *J. Mol. Biol.*, 1976, **104**, 243.
55. M. Danial, C. M. N. Tran, K. A. Jolliffe and S. Perrier, *J. Am. Chem. Soc.*, 2014, **136**, 8018.
56. P. Theato, J. U. Kim and J. C. Lee, *Macromolecules*, 2004, **37**, 5475.
57. K. T. Wiss, O. D. Krishna, P. J. Roth, K. L. Kiick and P. Theato, *Macromolecules*, 2009, **42**, 3860.
58. C. Dingels, F. Wurm, M. Wagner, H. A. Klok and H. Frey, *Chem. – Eur. J.*, 2012, **18**, 16828.
59. M. H. Stenzel, *ACS Macro Lett.*, 2013, **2**, 14.
60. T. Tokoroyama, *Eur. J. Org. Chem.*, 2010, **2010**, 2009.
61. M. Li, P. De, H. M. Li and B. S. Sumerlin, *Polym. Chem.*, 2010, **1**, 854.
62. M. W. Jones and D. M. Haddleton, in *Thiol-X Chemistries in Polymer and Materials Science*, The Royal Society of Chemistry, 2013, p. 95.
63. M. D. Partis, D. G. Griffiths, G. C. Roberts and R. B. Beechey, *J. Protein Chem.*, 1983, **2**, 263.
64. C. F. Brewer and J. P. Riehm, *Anal. Biochem.*, 1967, **18**, 248.
65. G. T. Hermanson, in *Bioconjugate Techniques*, Academic Press (Elsevier Inc.), San Diego, 2nd edn, 2008, p. 767.
66. S. D. Fontaine, R. Reid, L. Robinson, G. W. Ashley and D. V. Santi, *Bioconjugate Chem.*, 2015, **26**, 145.
67. A. B. Lowe, *Polym. Chem.*, 2010, **1**, 17.
68. H. Schlaad, *Polymer*, 2014, **55**, 5509.
69. A. B. Lowe, *Polym. Chem.*, 2014, **5**, 4820.
70. N. K. Singha, M. I. Gibson, B. P. Koiry, M. Danial and H. A. Klok, *Biomacromolecules*, 2011, **12**, 2908.
71. M. Danial, M. J. Root and H. A. Klok, *Biomacromolecules*, 2012, **13**, 1438.
72. M. W. Jones, R. A. Strickland, F. F. Schumacher, S. Caddick, J. R. Baker, M. I. Gibson and D. M. Haddleton, *J. Am. Chem. Soc.*, 2012, **134**, 1847.
73. M. W. Jones, R. A. Strickland, F. F. Schumacher, S. Caddick, J. R. Baker, M. I. Gibson and D. M. Haddleton, *Chem. Commun.*, 2012, **48**, 4064.

74. RCSB PDB (www.rcsb.org) of PDB ID 2GLH (G. Andreotti, B. L. Mendez, P. Amodeo, M. A. Morelli, H. Nakamuta, A. Motta, (2006) Structural determinants of salmon calcitonin bioactivity: the role of the Leu-based amphipathic alpha-helix, *J. Biol. Chem.*, **281**, 24193–24203).
75. M. P. Robin and R. K. O'Reilly, *Chem. Sci.*, 2014, **5**, 2717.
76. H. C. Hang, C. Yu, D. L. Kato and C. R. Bertozzi, *Proc. Natl. Acad. Sci. U. S. A.*, 2003, **100**, 14846.
77. E. M. Sletten and C. R. Bertozzi, *Angew. Chem., Int. Ed.*, 2009, **48**, 6974.
78. M. King and A. Wagner, *Bioconjugate Chem.*, 2014, **25**, 825.
79. D. Rideout, *Science*, 1986, **233**, 561.
80. E. T. Kool, D. H. Park and P. Crisalli, *J. Am. Chem. Soc.*, 2013, **135**, 17663.
81. K. L. Heredia, Z. P. Tolstyka and H. D. Maynard, *Macromolecules*, 2007, **40**, 4772.
82. H. F. Gaertner and R. E. Offord, *Bioconjugate Chem.*, 1996, 7, 38.
83. J. M. Gilmore, R. A. Scheck, A. P. Esser-Kahn, N. S. Joshi and M. B. Francis, *Angew. Chem., Int. Ed.*, 2006, **45**, 5307.
84. E. Saxon and C. R. Bertozzi, *Science*, 2000, **287**, 2007.
85. H. Staudinger and J. Meyer, *Helv. Chim. Acta*, 1919, **2**, 619.
86. F. L. Lin, H. M. Hoyt, H. van Halbeek, R. G. Bergman and C. R. Bertozzi, *J. Am. Chem. Soc.*, 2005, **127**, 2686.
87. E. Saxon, J. I. Armstrong and C. R. Bertozzi, *Org. Lett.*, 2000, 2, 2141.
88. B. L. Nilsson, L. L. Kiessling and R. T. Raines, *Org. Lett.*, 2000, **2**, 1939.
89. M. F. Debets, J. C. M. van Hest and F. P. J. T. Rutjes, *Org. Biomol. Chem.*, 2013, **11**, 6439.
90. A. Borrmann and J. C. M. van Hest, *Chem. Sci.*, 2014, **5**, 2123.
91. B. E. I. Ramakers, J. C. M. van Hest and D. W. P. M. Lowik, *Chem. Soc. Rev.*, 2014, **43**, 2743.
92. K. L. Kiick, E. Saxon, D. A. Tirrell and C. R. Bertozzi, *Proc. Natl. Acad. Sci. U. S. A.*, 2002, **99**, 19.
93. S. F. M. van Dongen, R. L. M. Teeuwen, M. Nallani, S. S. van Berkel, J. J. L. M. Cornelissen, R. J. M. Nolte and J. C. M. van Hest, *Bioconjugate Chem.*, 2009, **20**, 20.
94. J. P. Xiao and T. J. Tolbert, *Org. Lett.*, 2009, **11**, 4144.
95. D. E. Borchmann, N. ten Brummelhuis and M. Weck, *Macromolecules*, 2013, **46**, 4426.
96. R. Huisgen, *Proc. Chem. Soc.*, 1961, 357.
97. K. Kempe, A. Krieg, C. R. Becer and U. S. Schubert, *Chem. Soc. Rev.*, 2012, **41**, 176.
98. M. F. Debets, C. W. J. van der Doelen, F. P. J. T. Rutjes and F. L. van Delft, *ChemBioChem*, 2010, **11**, 1168.
99. S. E. Grieshaber, A. J. E. Farran, S. Lin-Gibson, K. L. Kiick and X. Q. Jia, *Macromolecules*, 2009, **42**, 2532.
100. L. M. Gaetke and C. K. Chow, *Toxicology*, 2003, **189**, 147.
101. G. J. Brewer, *Chem. Res. Toxicol.*, 2010, **23**, 319.
102. J.-F. Lutz and Z. Zarafshani, *Adv. Drug Delivery Rev.*, 2008, **60**, 958.

103. A. T. Blomquist and L. H. Liu, *J. Am. Chem. Soc.*, 1953, **75**, 2153.
104. N. J. Agard, J. A. Prescher and C. R. Bertozzi, *J. Am. Chem. Soc.*, 2004, **126**, 15046.
105. N. J. Agard, J. M. Baskin, J. A. Prescher, A. Lo and C. R. Bertozzi, *ACS Chem. Biol.*, 2006, **1**, 644.
106. X. H. Ning, J. Guo, M. A. Wolfert and G. J. Boons, *Angew. Chem., Int. Ed.*, 2008, **47**, 2253.
107. V. R. Krishnamurthy, J. T. Wilson, W. X. Cui, X. Z. Song, Y. Lasanajak, R. D. Cummings and E. L. Chaikof, *Langmuir*, 2010, **26**, 7675.
108. S. I. Presolski, V. Hong, S. H. Cho and M. G. Finn, *J. Am. Chem. Soc.*, 2010, **132**, 14570.
109. J. C. Jewett and C. R. Bertozzi, *Chem. Soc. Rev.*, 2010, **39**, 1272.
110. O. Diels and K. Alder, *Justus Liebigs Ann. Chem.*, 1928, **460**, 98.
111. D. L. Boger, *Chem. Rev.*, 1986, **86**, 781.
112. M. L. Blackman, M. Royzen and J. M. Fox, *J. Am. Chem. Soc.*, 2008, **130**, 13518.
113. M. F. Debets, S. S. van Berkel, J. Dommerholt, A. J. Dirks, F. P. J. T. Rutjes and F. L. van Delft, *Acc. Chem. Res.*, 2011, **44**, 805.
114. H. Du and K. Ding, in *Comprehensive Enantioselective Organocatalysis*, Wiley-VCH Verlag GmbH & Co. KGaA, 2013, p. 1131.
115. D. L. Alge, M. A. Azagarsamy, D. F. Donohue and K. S. Anseth, *Biomacromolecules*, 2013, **14**, 949.
116. T. Tischer, A. S. Goldmann, K. Linkert, V. Trouillet, H. G. Börner and C. Barner-Kowollik, *Adv. Funct. Mater.*, 2012, **22**, 3853.
117. S. S. van Berkel, A. T. J. Dirks, M. F. Debets, F. L. van Delft, J. J. L. M. Cornelissen, R. J. M. Nolte and F. P. J. T. Rutjes, *ChemBioChem*, 2007, **8**, 1504.
118. K. Hiltebrandt, T. Pauloehrl, J. P. Blinco, K. Linkert, H. G. Börner and C. Barner-Kowollik, *Angew. Chem., Int. Ed.*, 2015, **54**, 2838.
119. I. K. Korobitsyna, A. V. Khalikova, L. L. Rodina and N. P. Shusherina, *Chem. Heterocycl. Compd.*, 1983, **19**, 117.
120. S. Billiet, K. De Bruycker, F. Driessen, H. Goossens, V. Van Speybroeck, J. M. Winne and F. E. Du Prez, *Nat. Chem.*, 2014, **6**, 815.
121. H. Ban, J. Gavrilyuk and C. F. Barbas, *J. Am. Chem. Soc.*, 2010, **132**, 1523.
122. H. Ban, M. Nagano, J. Gavrilyuk, W. Hakamata, T. Inokuma and C. F. Barbas, *Bioconjugate Chem.*, 2013, **24**, 520.
123. E. M. Sletten and C. R. Bertozzi, *Acc. Chem. Res.*, 2011, **44**, 666.
124. T. Matsumoto, T. Tanaka and A. Kondo, *Biotechnol. J.*, 2012, **7**, 1137.
125. M. Rashidian, J. K. Dozier and M. D. Distefano, *Bioconjugate Chem.*, 2013, **24**, 1277.
126. P. Strop, *Bioconjugate Chem.*, 2014, **25**, 855.
127. J. W. Keillor, C. M. Clouthier, K. Y. P. Apperley, A. Akbar and A. Mulani, *Bioorg. Chem.*, 2014, **57**, 186.
128. D. da Silva Freitas, A. Mero and G. Pasut, *Bioconjugate Chem.*, 2013, **24**, 456.

129. A. Fontana, B. Spolaore, A. Mero and F. M. Veronese, *Adv. Drug Delivery Rev.*, 2008, **60**, 13.
130. A. Mero, B. Spolaore, F. M. Veronese and A. Fontana, *Bioconjugate Chem.*, 2009, **20**, 384.
131. P. Wilke, W. L. A. Brooks, R. Kuhnle, B. Sumerlin and H. G. Börner, *ACS Symp. Ser.*, 2012, **1101**, 271.
132. M. Ritzefeld, *Chem. – Eur. J.*, 2014, **20**, 8516.
133. C. S. Theile, M. D. Witte, A. E. M. Blom, L. Kundrat, H. L. Ploegh and C. P. Guimaraes, *Nat. Protoc.*, 2013, **8**, 1800.
134. C. P. Guimaraes, M. D. Witte, C. S. Theile, G. Bozkurt, L. Kundrat, A. E. M. Blom and H. L. Ploegh, *Nat. Protoc.*, 2013, **8**, 1787.
135. J. M. Antos, G. M. Miller, G. M. Grotenbreg and H. L. Ploegh, *J. Am. Chem. Soc.*, 2008, **130**, 16338.
136. Z. M. Wu, X. Q. Guo and Z. W. Guo, *Chem. Commun.*, 2011, **47**, 9218.
137. M. W. Popp, S. K. Dougan, T. Y. Chuang, E. Spooner and H. L. Ploegh, *Proc. Natl. Acad. Sci. U. S. A.*, 2011, **108**, 3169.
138. J. M. Antos, M. W. L. Popp, R. Ernst, G. L. Chew, E. Spooner and H. L. Ploegh, *J. Biol. Chem.*, 2009, **284**, 16028.
139. D. J. Williamson, M. A. Fascione, M. E. Webb and W. B. Turnbull, *Angew. Chem., Int. Ed.*, 2012, **51**, 9377.
140. D. Rabuka, *Curr. Opin. Chem. Biol.*, 2010, **14**, 790.
141. I. Chen, M. Howarth, W. Y. Lin and A. Y. Ting, *Nat. Methods*, 2005, **2**, 99.
142. M. Fernandez-Suarez, H. Baruah, L. Martinez-Hernandez, K. T. Xie, J. M. Baskin, C. R. Bertozzi and A. Y. Ting, *Nat. Biotechnol.*, 2007, **25**, 1483.
143. D. Rabuka, J. S. Rush, G. W. deHart, P. Wu and C. R. Bertozzi, *Nat. Protoc.*, 2012, **7**, 1052.
144. J. W. Wollack, J. M. Silverman, C. J. Petzold, J. D. Mougous and M. D. Distefano, *ChemBioChem*, 2009, **10**, 2934.
145. M. Zelzer, S. J. Todd, A. R. Hirst, T. O. McDonald and R. V. Ulijn, *Biomater. Sci.*, 2013, **1**, 11.
146. A. Gora, J. Brezovsky and J. Damborsky, *Chem. Rev.*, 2013, **113**, 5871.
147. R. V. Ulijn, *J. Mater. Chem.*, 2006, **16**, 2217.
148. P.-F. Caponi and R. V. Ulijn, in *Smart Materials for Drug Delivery: Volume 1*, The Royal Society of Chemistry, 2013, vol. 1, p. 232.
149. P. Wilke and H. G. Börner, *ACS Macro Lett.*, 2012, **1**, 871.
150. S. J. Todd, D. J. Scurr, J. E. Gough, M. R. Alexander and R. V. Ulijn, *Langmuir*, 2009, **25**, 7533.
151. V. Castelletto, J. E. McKendrick, I. W. Hamley, U. Olsson and C. Cenker, *Langmuir*, 2010, **26**, 11624.
152. I. W. Hamley, *Chem. Rev.*, 2012, **112**, 5147.
153. H. Kühnle and H. G. Börner, *Angew. Chem., Int. Ed.*, 2009, **48**, 6431.
154. P. Wilke, N. Helfricht, A. Mark, G. Papastavrou, D. Faivre and H. G. Börner, *J. Am. Chem. Soc.*, 2014, **136**, 12667.
155. G. Delaittre, N. K. Guimard and C. Barner-Kowollik, *Acc. Chem. Res.*, 2015, **48**, 1296.

156. M. Glassner, G. Delaittre, M. Kaupp, J. P. Blinco and C. Barner-Kowollik, *J. Am. Chem. Soc.*, 2012, **134**, 7274.
157. M. Dietrich, G. Delaittre, J. P. Blinco, A. J. Inglis, M. Bruns and C. Barner-Kowollik, *Adv. Funct. Mater.*, 2012, **22**, 304.
158. A. S. Goldmann, M. Glassner, A. J. Inglis and C. Barner-Kowollik, *Macromol. Rapid Commun.*, 2013, **34**, 810.

CHAPTER 2

Glycopolymers

LAURA HARTMANN* AND KIRA NEUHAUS

Heinrich-Heine-University Düsseldorf, Institute for Organic Chemistry and Macromolecular Chemistry, Universitätsstr. 1, 40225 Düsseldorf, Germany
*Email: laura.hartmann@hhu.de

2.1 Introduction

When you hear 'sugar', your first thought is most likely 'sweet', but sugars are more than just part of our nutrition. Their structures – from monosaccharides to linear polysaccharides and branched oligosaccharides – are as diverse and complex as their functions. Today, we know that sugars play a key role in mediating numerous biological events, especially in cell–cell contacts. However, many open questions remain starting with the identification of biologically active sugar molecules, their structure–function correlation, and application in *e.g.* biotechnology and biomedicine. Indeed, numerous areas of biosciences are now focusing on the impact of sugars in glycochemistry, glycobiology, glycomics, or glycovirology. One of the challenges hampering the identification of bioactive sugars and the investigation of their biological function is their inherently low affinity. Nature overcomes this by taking advantage of multivalency, whereby several sugars interact at the same time or with the same or multiple biomacromolecules, *e.g.* binding receptors. A second challenge is the structural complexity of carbohydrate-based biomolecules. For example, oligosaccharides are often comprised of a diverse array of repeating carbohydrate motifs, where individual monomers are linked together in a specific order and fashion and then bound for presentation on proteins or lipid membranes. The overall monomer content,

RSC Polymer Chemistry Series No. 22
Bio-inspired Polymers
Edited by Nico Bruns and Andreas F. M. Kilbinger

Published by the Royal Society of Chemistry, www.rsc.org

linkage pattern, and presentation have been shown to be critical for their biological recognition and subsequent activity, *e.g.*, an immune response.

In order to mimic the multivalent presentation of sugars, scientists have endeavored to create numerous natural, synthetic, and bio-inspired scaffolds decorated with simple and complex sugar ligands. One important example are glycopolymers. Through classical polymerization methods, polymer chains carrying a large number of sugars in the side chain are easily accessible and can be considered glycoprotein or polysaccharide mimetics. However, glycopolymers are just one example of multivalent glycoconjugates. Other classes of glycoconjugates, including glycodendrimers, glycoparticles, glycolipids, and glyconanomaterials that make use of scaffolds such as carbon nanotubes, cyclodextrins, or nanocrystals, have also garnered significant attention from scientists. The aforementioned functionalized scaffolds have the potential to impart biological specificity and can simultaneously be used as model systems to probe carbohydrate–biomacromolecule interactions in an effort to learn more about the role of multivalency.

Why are glycopolymers particularly interesting? One reason is their straightforward synthesis and modularity, which allow researchers to access a wide range of polymeric materials in a short period of time. Another reason is the large number of sugars that can be covalently linked on a polymer chain. By controlling the degree of polymerization, the number of sugar ligands per molecule can be easily tuned. Furthermore, using advanced polymerization techniques, copolymers combining different sugar ligands, complex block copolymer structures, and different polymer architectures, such as star polymers or micellar structures, various kinds of polymer-conjugates or hybrid materials can be obtained.[1]

The first examples of the synthesis of glycopolymers date back to the 1940s. Since then, the field of glycopolymers has profited from numerous developments in polymer and carbohydrate chemistry and biology, *e.g.* the development of living and controlled radical polymerization and new insights into the biological importance of glycosylated materials. This chapter will focus on examples of linear glycopolymers (Figure 2.1), where sugar ligands are attached as side chains *via* a linker to a polymer scaffold. The different sections will give a short overview of the synthesis, characterization, and application of linear glycopolymers as glycomimetic systems. For more details on the historical context of glycopolymer synthesis, other types of glycopolymers, and applications beyond the scope of this chapter, the reader is referred to other book chapters and various review articles on the topic.[2–16]

2.2 Synthesis of Glycopolymers

In general, the approaches to synthesize glycopolymers are as diverse as the number of monomers available for their synthesis. However, two main strategies can be differentiated for attaching sugar ligands to a polymer scaffold (Figure 2.2). First, the sugar ligand can be attached to the monomer

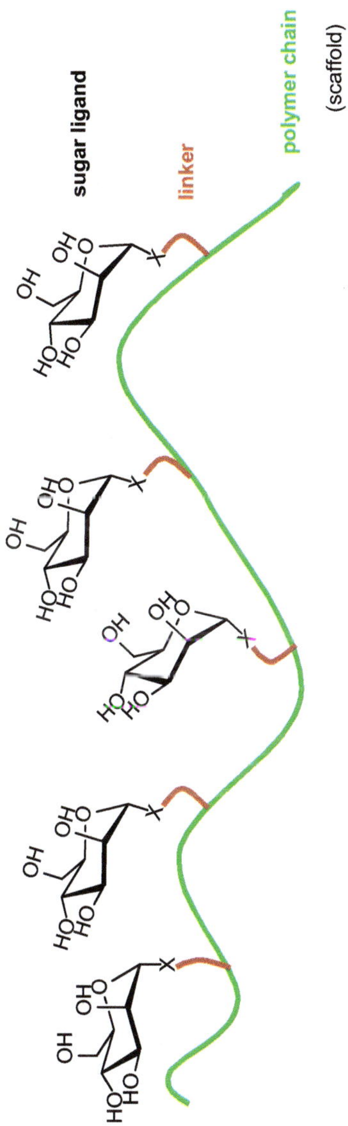

Figure 2.1 Schematic presentation of a linear glycopolymer with pendant sugar ligands in the side chain.

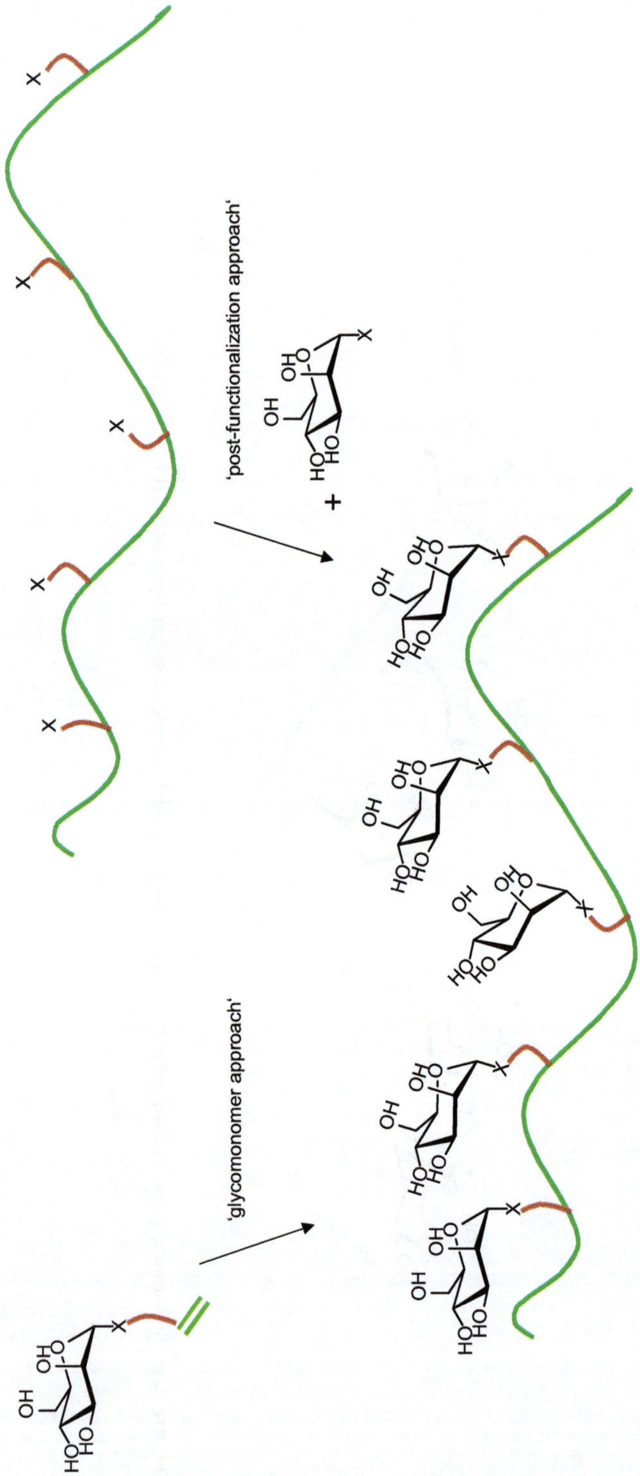

Figure 2.2 Glycomonomer *vs.* post-functionalization approaches.

prior to polymerization. This will be referred to here as the 'glycomonomer approach'. Second, the sugar ligand can be attached after polymerization *e.g.* to functional groups in the side chains. This will be referred to as the 'post-functionalization approach'. Both strategies allow for the synthesis of copolymers of various compositions. The following sections highlight key examples for both strategies.

2.2.1 Synthesis of Glycopolymers *via* Glycomonomers

Glycomonomers are generally prepared by transforming sugar ligands into polymerizable monomers by covalent attachment of the sugar ligand to a well-known monomeric unit. The following section presents examples of different monomeric units, their functionalization with different sugar ligands, and their subsequent polymerization. For more detailed information, the reader is referred to a recent book chapter by Stenzel and coworkers, which provides a detailed overview on the different strategies for glycopolymer synthesis.[3,4]

In general, the choice of monomeric unit defines which polymerization technique is applicable (*e.g.* ionic *vs.* radical polymerization), and thereby the degree of polymerization, polydispersity, and copolymerization parameters. However, it can be just as important to think about how to connect the sugar ligand to the monomeric unit. In most cases, the sugar ligand is modified at the anomeric position and a linker is introduced between the sugar ligand and the monomeric unit (Figure 2.3). It has been shown that the length and/or chemistry of the linker can influence the recognition of the glycopolymer by a protein receptor;[17] however, more studies are required to understand these effects in detail. Furthermore, the linker can impart additional properties to the glycopolymers, *e.g.* by introducing degradable moieties. It is also noteworthy that the connectivity between the sugar ligand and the monomeric unit is important, as some linkages can be intrinsically susceptible to hydrolysis as, for example, ester functionalities.

Glycomonomers containing vinyl functionalities are one of the most important classes of glycomonomers, as they can be applied in radical polymerization reactions including controlled radical polymerization. Depending on the structure of the target glycomonomer and polymer, *e.g.* the regioselective attachment of the vinyl functionality, the glycomonomer needs to carry protecting groups such as an acetyl protecting group. Furthermore, protecting groups are often used to influence the solubility of the glycopolymers, which can be important during purification.

Figure 2.3 Schematic representation of a glycomonomer.

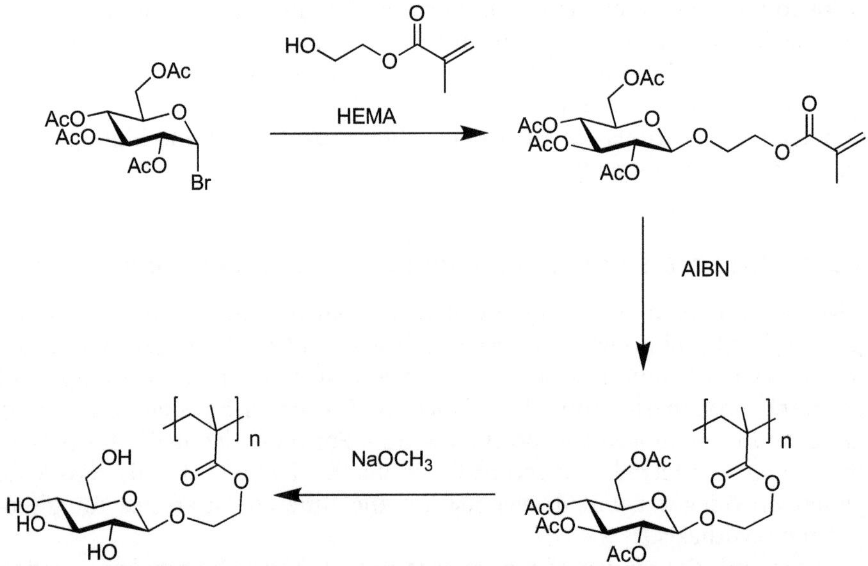

Figure 2.4 Reaction of 2,3,4,6-penta-*O*-acetyl-α-D-bromo-glucopyranose with 2-hydroxyethyl methacrylate, followed by free radical polymerization of the glycomonomer, and subsequent deprotection to obtain a glycopolymer.[18,19]

Usually, acetylated glycomonomers are easily obtained from the glycosylation of commercially available or readily accessible peracetylated carbohydrates with a nucleophilic donor such as 2-hydroxyethyl methacrylate, as shown in Figure 2.4.[18,19] The combined influence of the chemistry of the hemiacetal and the reaction conditions affect the regio- and stereoselectivity of the reaction and, in most cases, can be used to selectively generate the α or β anomer.

Another straightforward route towards protected glycomonomers uses isopropylidene protected carbohydrates. For example, in Figure 2.5, the *syn* hydroxyl groups at positions C1 and C2 and C3 and C4 are protected using acetone in the presence of an acid catalyst leaving the C6 hydroxyl group free for further functionalization. It is important to note here that the type of sugar ligand and reaction conditions determine the position of the free hydroxyl group (*e.g.* similar reaction conditions with glucose provide glucofuranose with the C3 position free).[20] Various monomeric units can now be introduced at the free hydroxyl group, *e.g.* vinyl units for radical polymerization such as methacrylate or acrylate groups.[21,22]

Non-protected glycomonomers for radical polymerization are easily obtained by the functionalization of amino sugars such as galactosamine. The high reactivity of the amine *vs.* the hydroxyl groups allows for selective functionalization with, for example, acryloyl or methacryloyl chloride, as is shown in the example for glucosamine in Figure 2.6.[23,24]

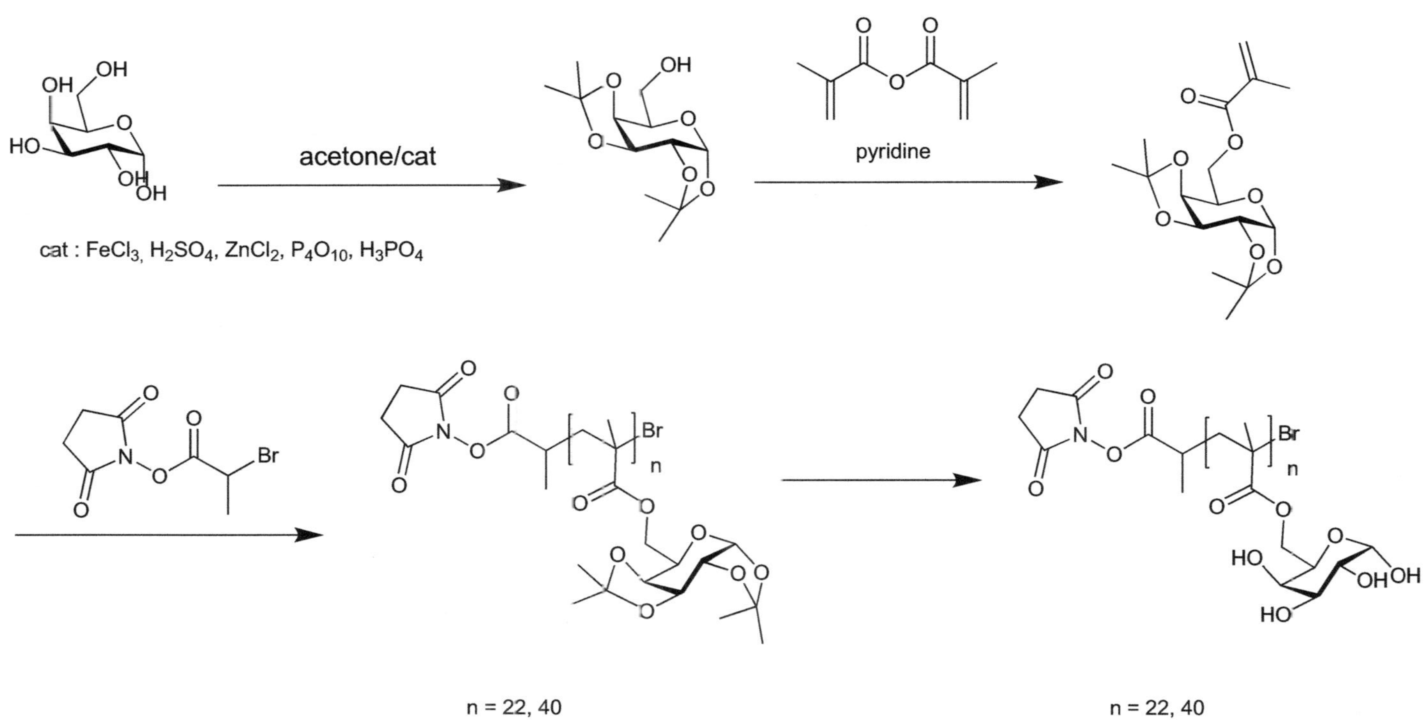

Figure 2.5 Installing isopropylidene protecting groups on galactose, followed by reaction with methacryloyl chloride to obtain a glycomonomer, subsequent ATRP, and deprotection to obtain a glycopolymer.[21,22]

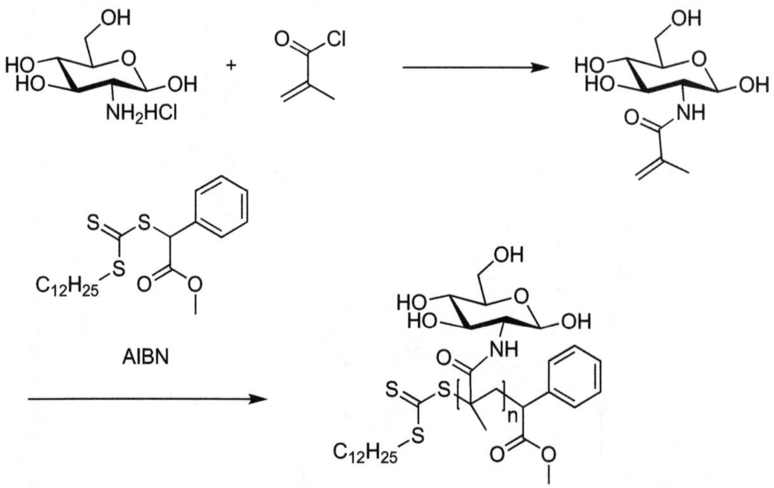

Figure 2.6 Synthesis of 2-(methacrylamido)glucopyranose from glucosamine hydrochloride and methacryloyl chloride, and subsequent polymerization *via* RAFT.[86]

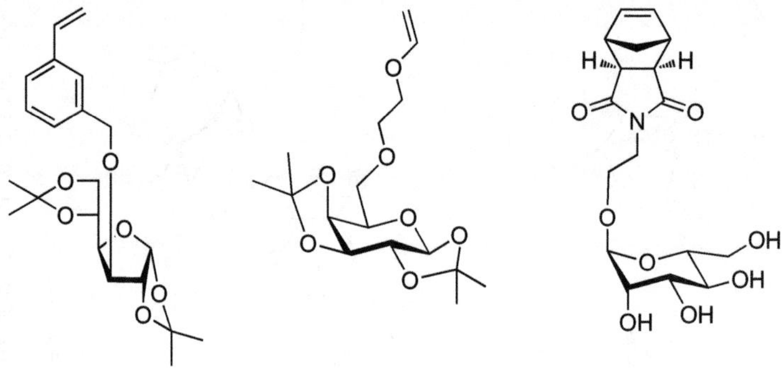

Figure 2.7 Glycomonomers for anionic[25] and cationic[26] polymerization.

Protected vinyl based glycomonomers are also suitable for ionic polymerization. Isopropylidene protected monomers are often applied in anionic polymerization,[25] while monomers for cationic polymerizations are often based on vinyl-oxy functional groups.[26] Figure 2.7 provides examples of glycomonomers suitable for ionic polymerization.

Isopropylidene-protected sugar ligands can also be transferred onto monomers for ring opening metathesis polymerization (ROMP) by reaction with imide-fused norbornene, the work pioneered by Kiessling and coworkers since the 1990s.[27–29] Figure 2.8 shows an example of the synthesis of a sulfated glycomonomer bearing a norbornene moiety and its subsequent polymerization using a first generation Grubbs catalyst.[30] Due to the increasing insolubility of the polymer formed during the reaction, an emulsion polymerization was employed leading to water soluble polymers in up to 74% yield.

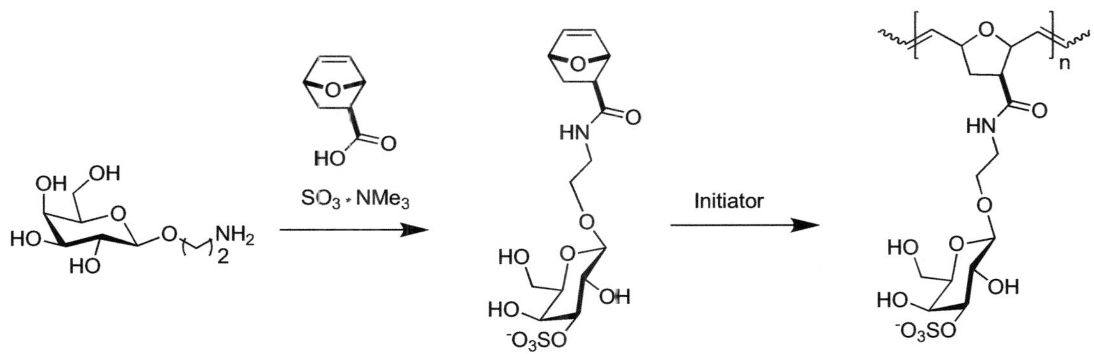

Figure 2.8 Synthesis of sulfated glycopolymers *via* glycomonomer approach based on ROMP.[30]

2.2.2 Synthesis of Glycopolymers *via* Post-polymerization Strategies

The direct polymerization of glycomonomers allows for the synthesis of fully functionalized glycopolymers with every repeating unit carrying a sugar ligand, if desired. With the use of controlled polymerization techniques, the chain length can also be controlled leading to low dispersity materials. However, there are also good reasons to create functionalizable polymer scaffolds first, with the sugar ligands appended later *via* polymer analogue reactions. This strategy is often preferred when more complex sugar ligands, *e.g.* oligosaccharides, are used. The post-polymerization strategy also allows for the straightforward synthesis of libraries of glycopolymers, as well as copolymers, presenting different sugar ligands using the same polymeric scaffold. This way, the potential influence of chain length and polydispersity can be neglected and only the composition of the sugar side chains is taken into account when, for example, analyzing the lectin binding affinity of a series of glycopolymers.

The following section highlights five examples for the synthesis of glycopolymers *via* this post-polymerization strategy.

Boyer and Davis[31] performed RAFT polymerization of pentafluorophenyl acrylate to obtain polymer scaffolds with pendant active ester groups susceptible to further functionalization with amino sugars, as shown in Figure 2.9. The high reactivity of the pentafluorophenyl side chains enabled the complete conversion, resulting in fully glycosylated polymers of different chain lengths. Furthermore, the RAFT end-group was transformed *via* a simultaneous aminolysis–thiol-ene reaction to yield biotinylated glycopolymers.

Another example of the application of active ester side-chains was introduced by Kiessling and coworkers[32] making use of a norbornene derivative containing an *N*-hydroxylsuccinimide (NHS) side chain. ROMP of the functional monomer afforded a polymer scaffold with pendant NHS groups that were then further functionalized with sugar amines (Figure 2.10). In this work, special focus was devoted to the introduction of fluorescent end labels, as they are often required for the subsequent use of glycopolymers in binding studies and cell culture studies (see Sections 2.3 and 2.4).

In general, the use of "click" chemistry has proven to be a versatile method for the glycoconjugation of polymeric scaffolds. Major advantages are the possibility to apply non-protected sugar ligands, the fast and efficient conversion to the desired glycopolymer, and the well-established synthesis of the according monomers and functionalized sugar ligands.[14]

Haddleton and coworkers[33] prepared a functionalizable polymer scaffold *via* catalytic chain transfer polymerization of trimethylsilane-protected propargyl methacrylate using bis(boron difluorodimethylglyoximate) cobalt(II) as the catalyst, as shown in Figure 2.11. After deprotection of the alkyne side chains and end-group functionalization of the terminal vinyl group, Cu(I)-catalyzed azide–alkyne cycloaddition (CuAAc) ("click" reaction) of sugar azides was used for the introduction of sugar side chains.

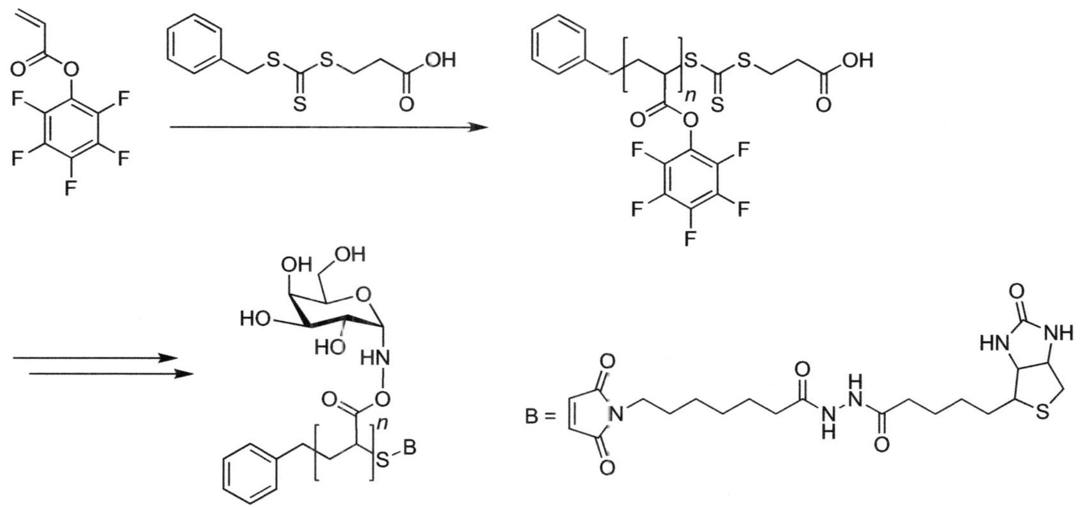

Figure 2.9 Synthesis of pentafluoro side chain polymer *via* RAFT and post-functionalization with sugar amine to obtain a glycopolymer.[31]

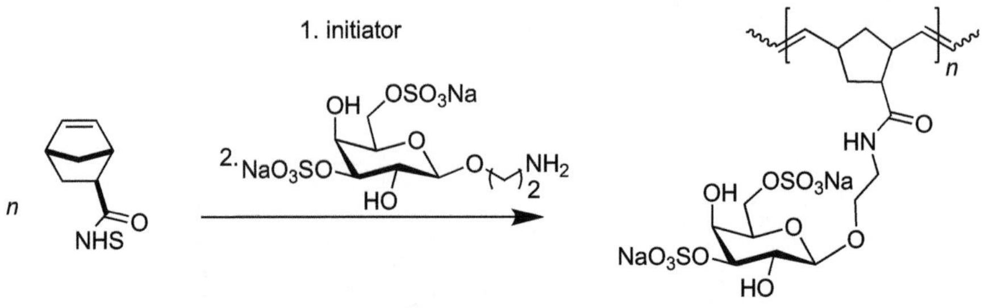

Figure 2.10 Synthesis of *N*-hydroxysuccinimid side functional polymer and post-functionalization with sugar amine to obtain final a glycopolymer.[32]

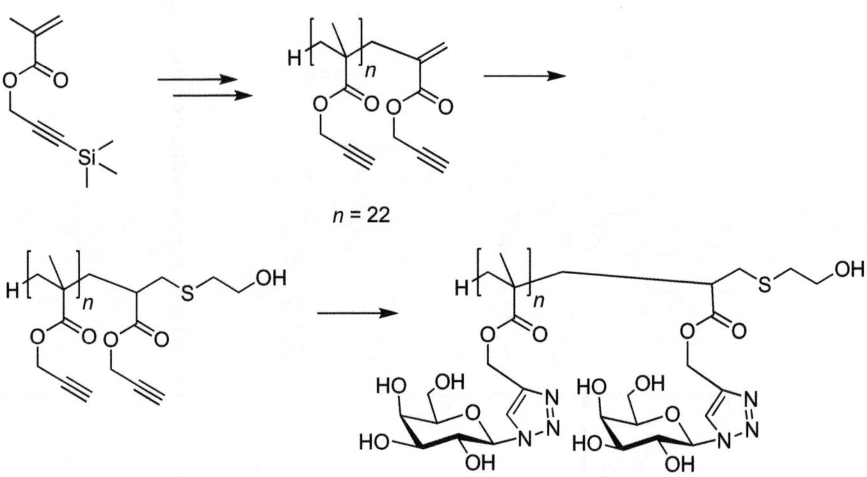

Figure 2.11 Synthesis of an alkyne side functional polymer and introduction of azido sugar ligands *via* CuAAc.[33]

Schlaad and coworkers[34] applied another "click" reaction for glycoconjugation – the thiol-ene reaction. They first synthesized a well-defined poly[2(-3-isobutenyl)-2-oxazoline] scaffold *via* cationic ring opening polymerization, which was then followed by the addition of sugar thiols such as acetylated thioglucose *via* photoaddition to the polymeric vinyl groups (Figure 2.12).

Hartmann and coworkers reported the use of CuAAC[35] and thiol-ene click chemistry[36,37] on a solid support to obtain monodisperse, sequence-controlled glycooligomers (Figure 2.13). They obtained alkyne or alkene functionalized scaffolds *via* stepwise assembly of tailor-made building blocks on a solid support using Fmoc peptide coupling chemistry. Subsequent attachment of the sugar ligands on the support allowed for the precise control of the position and distance of the sugar ligands along the polymer backbone, as well as for the combination of different sugar ligands within one glycopolymer.[38]

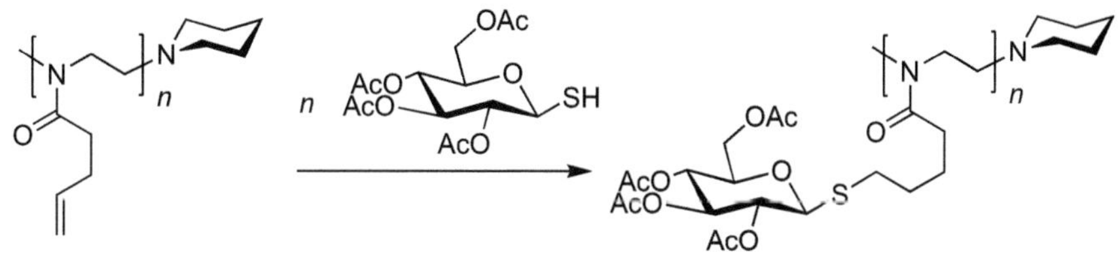

n = 63

Figure 2.12 Functionalization of alkene side functional polymer *via* thiol-ene conjugation of protected thioglucose.[34]

2.2.3 Synthesis of Glyco- and Block Copolymers

One of the advantages of using polymer scaffolds for the multivalent presentation of sugar ligands is the straightforward synthesis of complex copolymer structures. Again, several examples of a variety of successfully synthesized glycocopolymer structures have been reported in the literature to date. Here, we would like to highlight examples that fall into two important classes of glycocopolymers: First, glycopolymers combining different sugar ligands and second, glyco block copolymers combining a sugar-functionalized block with a non-sugar functionalized block, *e.g.*, a hydrophobic block that allows for the latter self-assembly of the polymer. For more examples, the reader is referred to recent reviews.[39] Noteworthy, copolymerization strategies also allow for the synthesis of different architectures of glycopolymers, *e.g.*, comb polymers or branched polymers; however, the discussion of such polymers is beyond the scope of this chapter.

2.2.3.1 Glycocopolymers for Self-assembly

The following section provides five examples of the synthesis of co- and block copolymers combining sugar ligands and hydrophobic segments within the same molecule. In general, the introduction of hydrophobic segments to a glycopolymer is used in order to induce aggregation or self-assembly of the polymer, for example, to form particles or micelles. This is particularly interesting for the application of such glycopolymers, *e.g.* in drug delivery or for surface modifications, and will be further discussed in Section 2.4.

Schubert and coworkers[40] explored the use of pentafluorostyrene and its copolymerization with styrene to obtain homo- and copolymer scaffolds for further functionalization with protected thiosugars *via* a thiol-*para*-fluoro "click" reaction (Figure 2.14). The first functionalization of the side chains resulted in about 60% conversion. Subsequent functionalization led to an increase in side chain conversion of up to 90%. The observed increase in functionalization could also be applied in a stepwise fashion to give access to heteromultivalent glycocopolymers. Selected copolymers showed self-assembly into spherical nanoparticles during nanoprecipitation with diameters ranging from 70 to 720 nm.

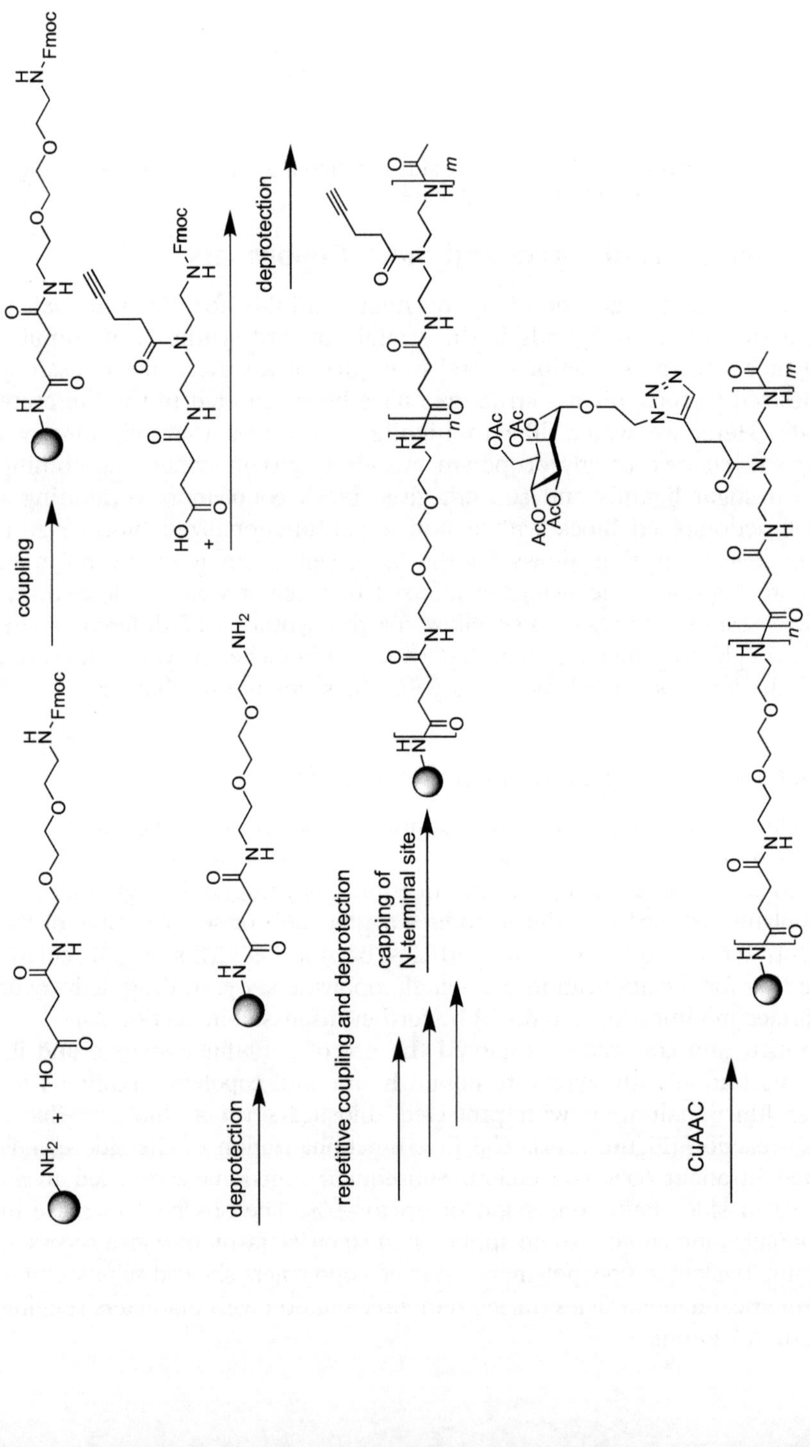

Figure 2.13 Solid phase polymer synthesis of monodisperse sequence-defined glycooligomers using CuAAc to attach sugar ligands on the polymer backbone.[35]

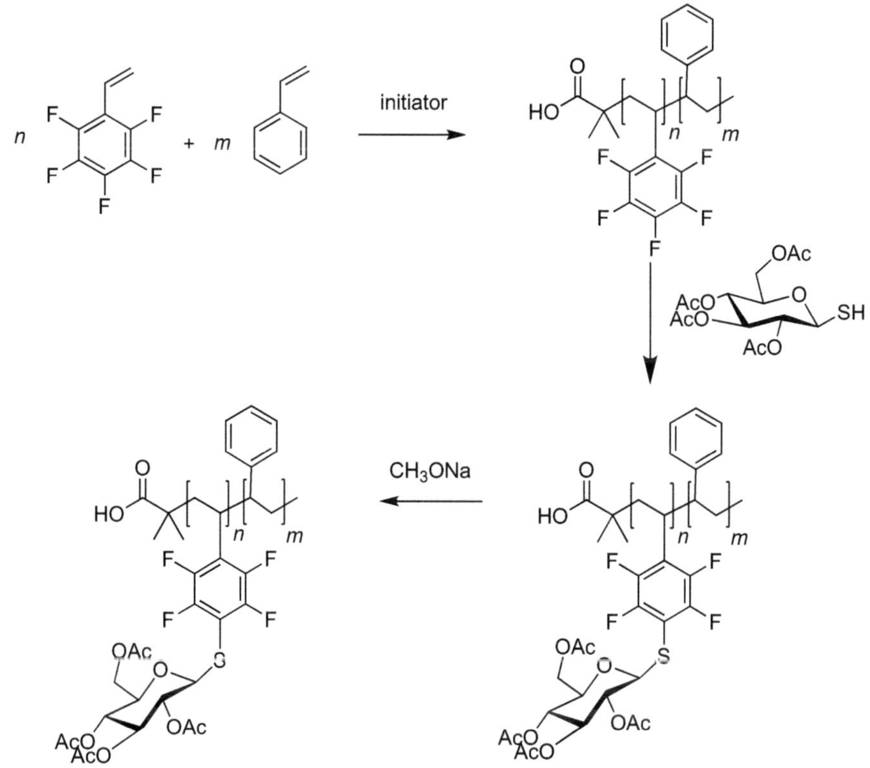

Figure 2.14 Synthesis of block copolymer presenting pentafluorostyrene blocks for conjugation with acetylated thiosugars.[40]

Hawker and coworkers[41] synthesized copolymers presenting sugar ligands and lipid side chains by copolymerization of *N*,*N*-di(octadecyl)acrylamide and 1,2,5,6-di(isopropylidene)-D-glucose-2-propenoate *via* nitroxide-mediated living free-radical polymerization (NMP). Similar to the copolymers prepared by Schubert *et al.* in the previous example, and in contrast to a glyco block copolymer that contains one hydrophilic glycoblock and one hydrophobic block, the copolymers by Hawker *et al.* are composed of several hydrophobic and hydrophilic sugar units within the same polymer (Figure 2.15). This allows for the fine tuning of the resulting surface activity, where these amphiphilic polymers form stable Langmuir films with control over the molecular area.

Pasparakis and Alexander[42] reported the synthesis of a diblock copolymer combining a poly(2-glucosyloxyethyl methacrylate) glycopolymer block with a poly(diethyleneglycol methacrylate) block *via* RAFT polymerization (Figure 2.16). The poly(diethyleneglycol methacrylate) segment, similar to poly(*N*-isopropylacrylamide) (PNIPAAm), can change from hydrophilic to hydrophobic *via* temperature induced changes in solubility. This LCST behavior leads to assembly of the block copolymers into vesicles at temperatures above 20 °C.

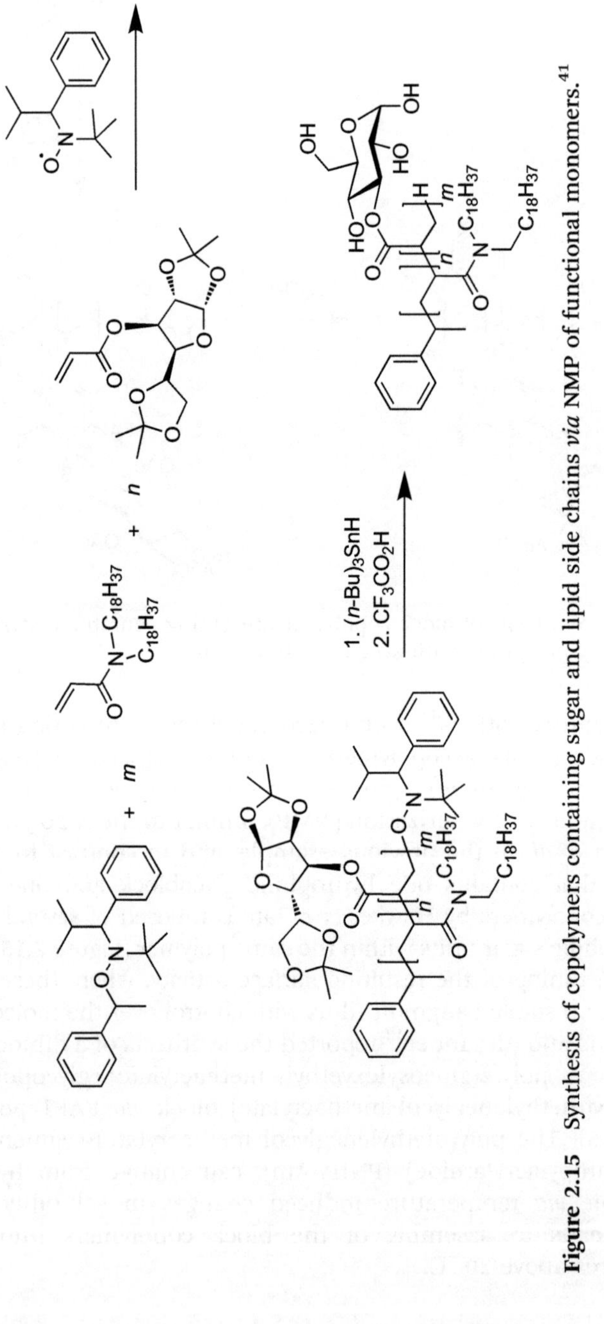

Figure 2.15 Synthesis of copolymers containing sugar and lipid side chains *via* NMP of functional monomers.[41]

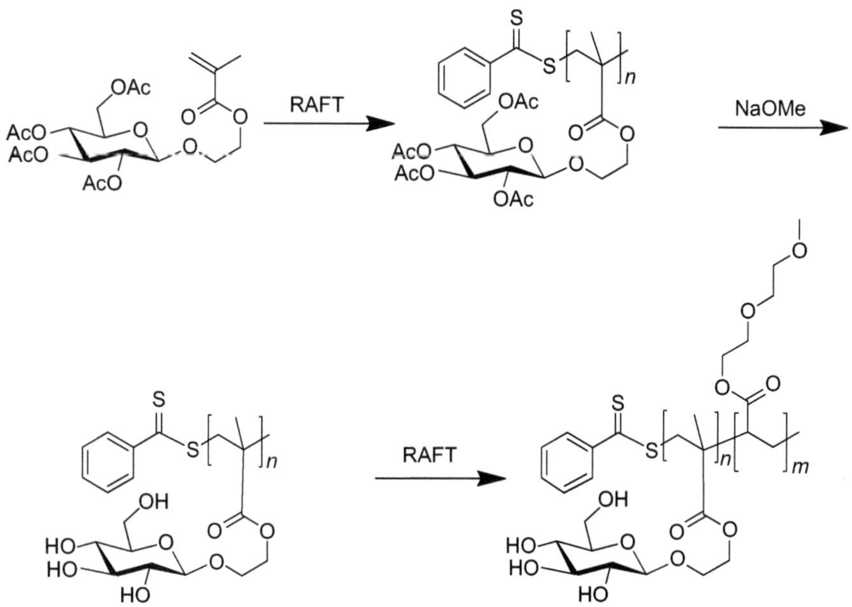

Figure 2.16 Synthesis of temperature sensitive block copolymer *via* RAFT polymerization of glycomonomer and diethyleneglycolmethacrylate.[42]

Going from diblock to triblock copolymers is possible *via* different approaches, which give access to even more complex self-assembling systems that can include additional functionalities such as biodegradability. For example, an ABA type block copolymer presenting two glycopolymer segments connected by a hydrophobic degradable caprolactone block was reported by Chen and Wulff.[43] The synthesis starts with the ring-opening polymerization of ε-caprolactone followed by ATRP, as shown in Figure 2.17.

An ABC type triblock copolymer was synthesized by Stenzel and coworkers[44] linking a glycopolymer segment with a poly(*n*-butyl acrylate) and a poly-(4-vinylpyridine) block *via* RAFT polymerization (Figure 2.18). In contrast to previous examples, where the composition of the copolymer determined its aggregation behavior to form one dominant colloidal species, here the pH sensitive self-assembly of the triblock copolymer allowed for the formulation of various aggregates such as globular, cylindrical, or Janus-type micelles.

2.2.3.2 *Heteromultivalent Glycopolymers*

Although various studies have effectively shown the potential of glycopolymers to present sugar ligands in high number and density, most of these systems only present one kind of sugar ligand, *e.g.* a mannose residue. However, in order to mimic more complex oligo- and polysaccharide structures and/or understanding the potential interplay of several sugar ligands, it is more interesting to combine different sugar ligands along the same

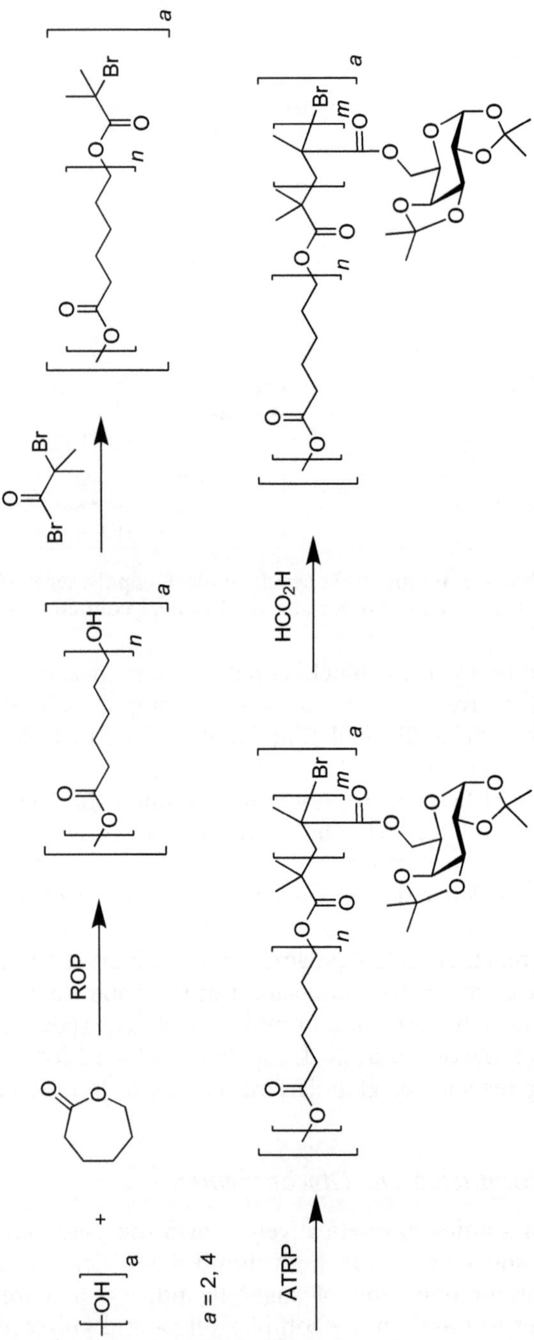

Figure 2.17 Synthesis of ABA triblock copolymer *via* ROP and ATRP using a protected glycomonomer.[43]

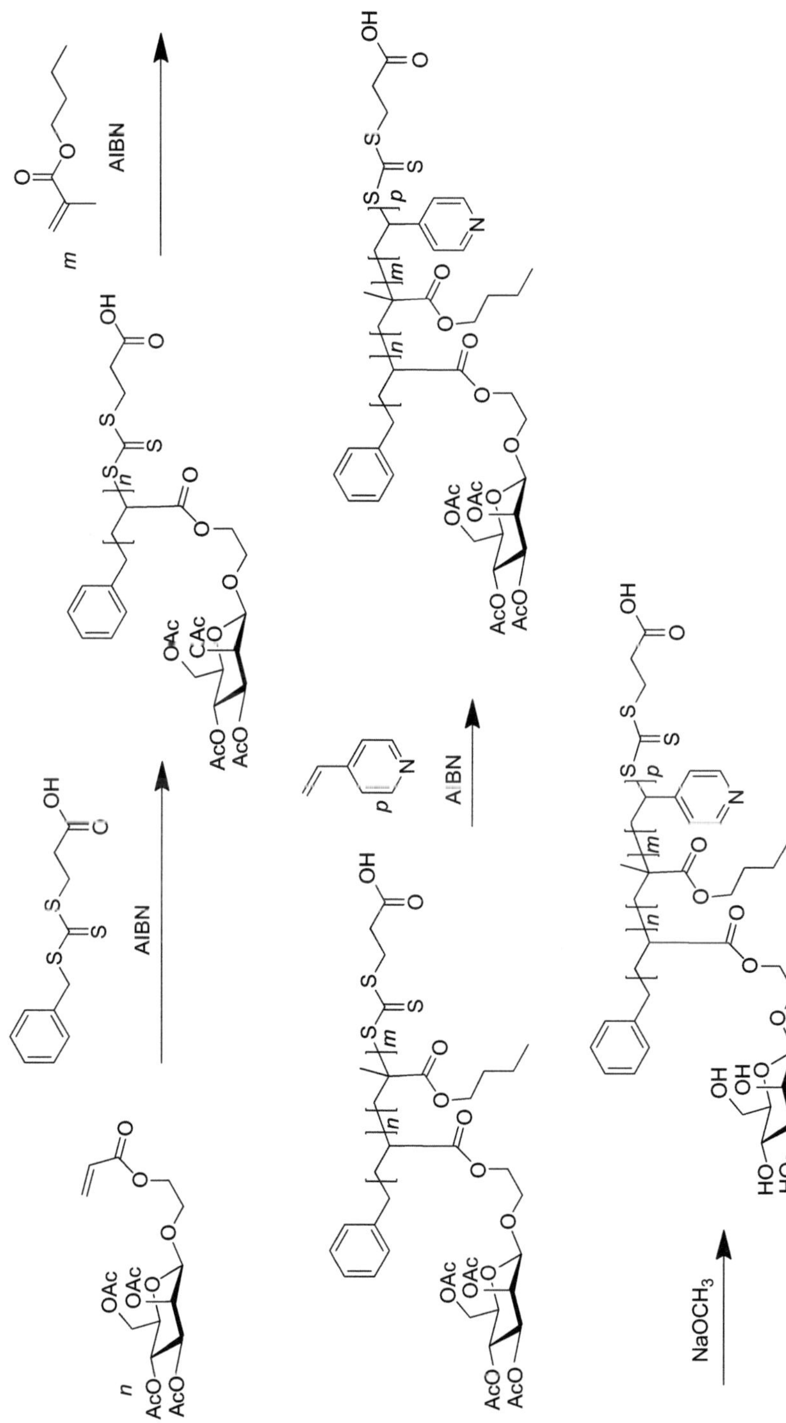

Figure 2.18 Synthesis of an ABC triblock copolymer *via* RAFT polymerization using a protected glycomonomer.[44]

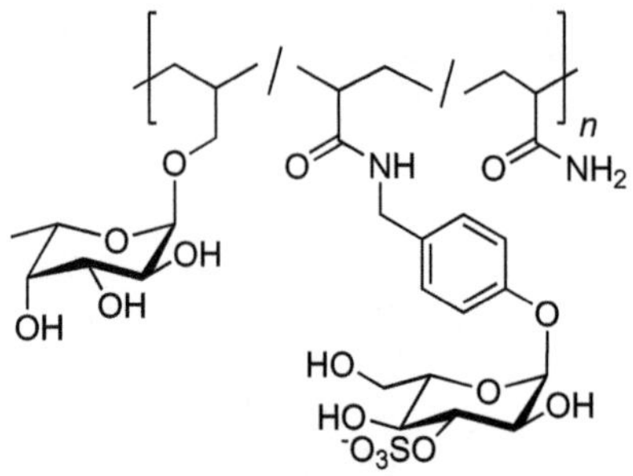

Figure 2.19 Structure of a fucosylated terpolymer used for selectin binding.[46]

polymer scaffold. Such systems are called heteromultivalent polymers. For a more detailed overview on other heteromultivalent glycomaterials, the reader is referred to a recent review by Jimenez *et al.* on multivalency in heterogeneous glycoenvironments.[45] Studies are increasingly investigating the influence of heteromultivalent glycopolymers on their resulting binding properties. Investigations include combinations of different sugar ligands as well as the combination of sugar ligands and non-sugar ligands to make use of secondary binding effects. The following section provides briefly three examples of heteromultivalent glycopolymers that are important representatives of different strategies for the combination of different sugar ligands on a polymer scaffold.

Kobayashi and coworkers[46] synthesized a copolymer *via* copolymerization of glycopolymers presenting fucoside and 3-sulfo-galactoside residues as key fragments of the sialyl Lewis^X (sLe^X) oligosaccharide ligand (Figure 2.19). They were also able to demonstrate that only the heteromultivalent glycopolymer inhibited both P- and L-selectin binding to sLe^X, while the homopolymer showed no affinity. Therefore, this work nicely represents the potential of glycopolymers to mimic complex oligo- and polysaccharide ligands by attaching smaller, more easily accessible ligand fragments to a polymer backbone.[47,48]

Haddleton and coworkers reported the synthesis of a well-defined alkyne side chain functionalized polymer that provided the straightforward introduction of different sugar ligands by a ‘coclicking’ approach (Figure 2.20).[49] One of the advantages of such a post-polymerization approach is the possible variation of sugar ligand ratios while keeping the scaffold parameters, such as chain length and polydispersity, constant. Thus, this strategy gives access to a library of glycopolymers to study the influence of epitope density through the combination of binding and non-binding sugar ligands targeting different protein receptors.[49–51]

A recent example from Haddleton and coworkers[52] combined sugar and non-sugar ligands to use secondary binding effects to enhance not only the overall affinity, but also the specificity of the glycopolymer. In the process, they developed a new 3-step tandem post-polymerization protocol based on an epoxy side-chain functionalized polymer, as shown in Figure 2.21.

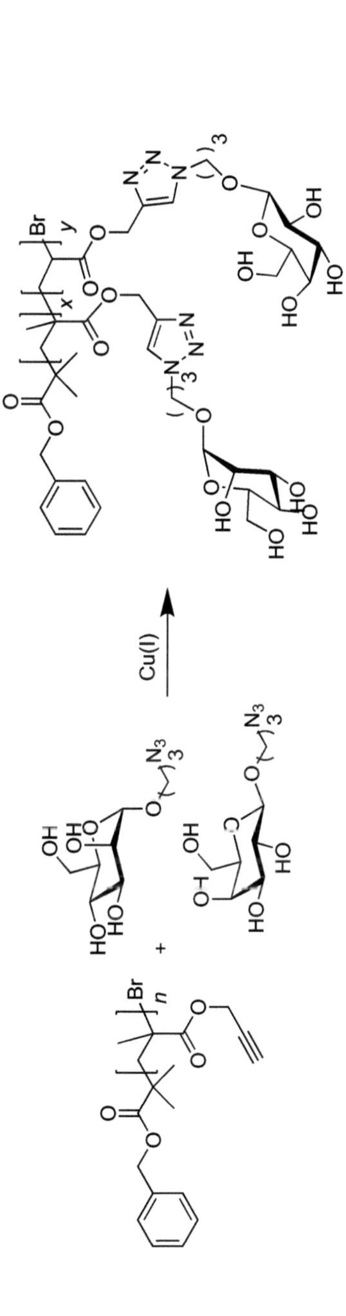

Figure 2.20 Synthesis of heteromultivalent glycopolymer *via* a 'coclicking approach' on an alkyne side chain functional polymer.[49]

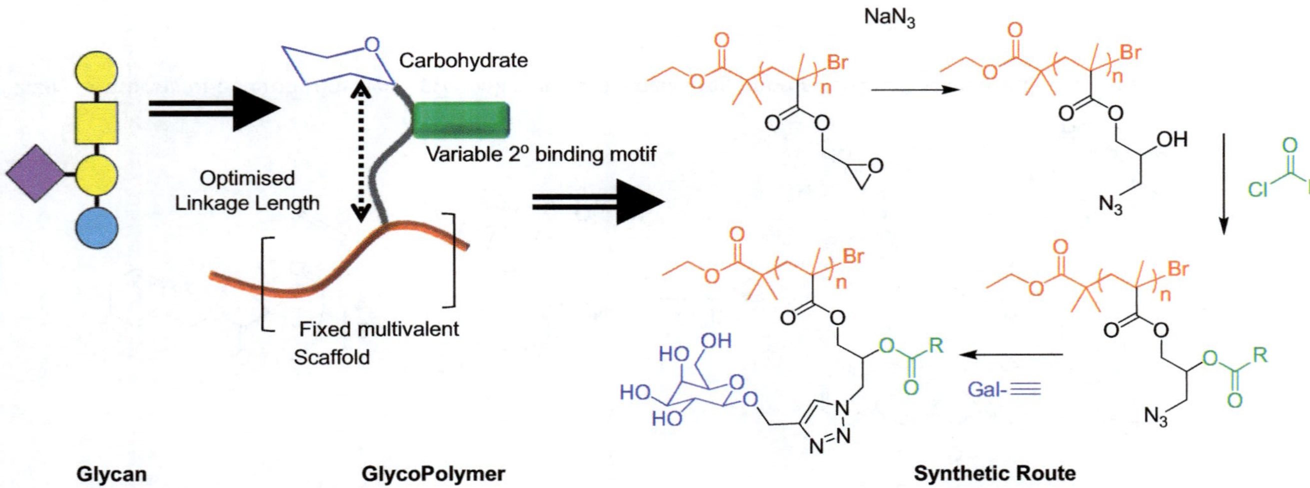

Figure 2.21 Design principle and synthetic route to derive heteromultivalent glycopolymers presenting sugar and non-sugar ligands. Reproduced from ref. 52 with permission from the Royal Society of Chemistry.

Regioselective ring opening of the azide, followed by subsequent insertion of a non-sugar ligand to the released hydroxyl group, and CuAAc between an alkyne functionalized galactose ligand and an azido side chain afforded the final glycopolymer. Screening of a library of differently functionalized polymers led to an optimized glycopolymer with a 20-fold increase in binding affinity and increased specificity towards the toxin CTx related to cholera infections.

2.3 Analyzing Glycopolymers

One important area of application for glycopolymers is their use in biochemistry, biomedicine, or biotechnology (see Section 2.4); however, today still little is known about the interplay between the chemical structure of a glycopolymer and its resulting biological properties. Therefore, the rational design of a glycopolymer is often not possible. Still, great efforts are currently being undertaken to gain deeper insight into the interactions of glycopolymers with protein receptors, usually the target for glycopolymer binding. The following section provides an overview of binding mechanisms that have been hypothesized and/or successfully demonstrated for glycopolymer binding to protein receptors. Some of the most common protein receptors used for glycopolymer binding studies, and the techniques used to analyze and quantify glycopolymer-receptor interactions are also presented.

2.3.1 Multivalent Binding of Glycopolymers

The binding of a single sugar ligand to a protein receptor is usually weak with affinities in the μM to mM regime for monosaccharides. Nature overcomes this problem by using multiple sugar ligands to target the same or several different receptor binding sites simultaneously, leading to an overall increase in binding avidity. This effect is called multivalency and, in general, it is known to play an important role in many biological interactions, *e.g.*, between carbohydrates, proteins, and other biomacromolecules. For this reason, multivalency has been the target of increased interest for a number of years. This section provides a short overview of the impact of glycopolymers as one important class of multivalent glycomaterials. For more detailed information on the theoretical aspects, thermodynamics and kinetics, or the biological relevance of multivalency, the reader is referred to other books and reviews on the topic.[13,53–59]

Indeed, it has been shown in many studies that the use of a polymer backbone presenting a low affinity sugar ligand multiple times leads to a dramatic increase in the binding avidity of the overall glycopolymer. The question is, how does the use of several low affinity ligands lead to such a high avidity binding event?

Kiessling and coworkers studied the different multivalent binding modes of glycopolymers (see Figure 2.22).[60] Due to the high number and density of sugar ligands displayed on a glycopolymer, different multivalent binding

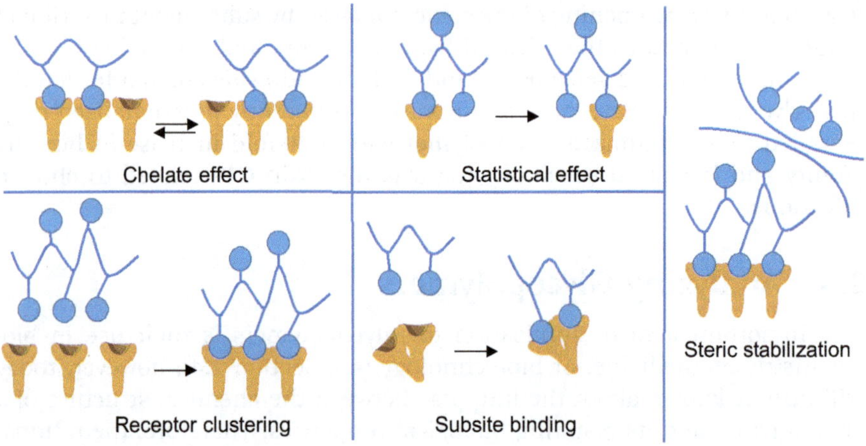

Figure 2.22 Multivalent binding modes of glycopolymers. Adapted with permission from J. E. Gestwicki, C. W. Cairo, L. E. Strong, K. A. Oetjen and L. L. Kiessling, *J. Am. Chem. Soc.*, 2002, **124**, 14922–14933.[60] Copyright (2002) American Chemical Society.

modes can come into action upon binding to, for example, a lectin receptor. First, several sugar ligands can occupy two or more binding sites of the same receptor in a chelate binding effect. Here, the entropic cost paid by the first binding event favors binding to additional binding sites that are now energetically more favorable, leading to a dramatic increase in the overall binding avidity. Similarly, the simultaneous binding to a subsite of the binding pocket, *e.g.* by another sugar ligand on the backbone, can increase the binding avidity. Due to the low affinity of a single sugar ligand, it is expected that they have a high on- and off-rate, meaning that ligands are more often free than bound. Presumably, a high local concentration of neighboring sugar ligands on the polymer backbone allows for the fast exchange of sugar ligands in the bound state, which is often referred to as statistical rebinding. Another consequence of the high number and density of sugar ligands is the potential intermolecular binding or receptor clustering. Here, several receptor molecules bind simultaneously to one glycopolymer, thereby increasing the effective avidity of the glycopolymer. Indeed, this effect is usually observed for glycopolymer binding and can be visualized by a turbidimetry assay where, upon glycopolymer–receptor cluster formation, an insoluble precipitate forms, as quantified by the increasing turbidity of the sample.[61]

As discussed in the next section, determining the binding of glycopolymers is not only relevant in terms of direct binding to the receptor, but also for the determination of the inhibitory potential of the glycopolymer. To examine this type of interaction, the receptor is pre-incubated with a ligand of known affinity, and the amount of glycopolymer required to replace said ligand is determined. In this competition study, another binding mode where the bound glycopolymer can sterically shield the binding pocket, thus minimizing the potential binding of competing ligands, can lead to an increase in the apparent binding avidity of the glycopolymer.

While, on paper, we can look at these different binding modes separately, a combination of the different binding modes is taking place during the actual binding event. Furthermore, depending on the assay applied, quantifying the binding affinity or avidity of different binding modes might be favored or disfavored, *e.g.*, receptor clustering is more likely to be observed in solution based assays, whereas steric shielding might be more pronounced in chip based assays where the receptor or ligand is covalently bound to a surface. Nevertheless, due to their ease of synthesis and structural homogeneity in comparison to most biologically derived oligo- or polysaccharide samples, glycopolymers are an important class of highly valent structures with the potential to further investigate multivalent binding modes. Such probes may also provide critical information regarding correlations between the multivalent binding modes and the resulting biological properties or functions of synthetic and natural glycoligands.

2.3.2 Binding Studies of Glycopolymers Targeting Lectins

Recently, the focus of glycopolymer research has moved from the development of new synthetic methods to derive glycopolymers for the study of their binding behavior and potential biomedical applications. As discussed in the previous section, one of the key features of glycopolymers is their increased avidity in binding carbohydrate-binding receptors such as lectins. Understanding such interactions provides a foundation for the development of glycopolymers for novel therapies, *e.g.* antiviral or antibacterial. In order to quantify binding avidity and identify high avidity glycopolymers for further *in vitro* and *in vivo* studies, different binding assays should be applied. It is important to note that a direct comparison between avidity values derived from different assays is difficult, since absolute values are highly dependent on several factors. Therefore, relative values are usually reported instead, *e.g.* relating the measured affinity to the affinity of a known weak binder such as the according monosaccharide. Such relative affinities can then be further normalized by the number of sugar ligands attached to the polymer, providing a relative or valence corrected affinity per sugar ligand.

Overall, there are several assays and techniques that allow for the determination of the affinity and avidity of a glycopolymer. The following section will highlight three examples of binding assays applied to study the lectin binding of glycopolymers, namely surface plasmon resonance (SPR), isothermal calorimetry (ITC), and soft colloidal probe reflection interference microscopy (SCP-RICM). From the technical point of view, SPR is a chip-based method where one of the components – ligand or receptor – is surface-bound. Other chip- or surface-based methods are quartz crystal microbalance (QCM) or enzyme-linked immunosorbent assay (ELISA). Alternatively, both components can be in solution, as in isothermal calorimetry (ITC) experiments, fluorescence correlation spectroscopy (FCS), saturation-transfer difference nuclear magnetic resonance (STD-NMR), or turbidimetry measurements. Finally, both ligand and receptor can be surface-bound, *e.g.* in AFM or cell adhesion studies and, as will be discussed, in the SCP-RICM assay. The choice of assay can therefore be driven by technical considerations, *e.g.* the possibility to immobilize the glycopolymer. More importantly, the applied assay can take into consideration the potential application of the glycopolymer as, for example, an inhibitor of viral adhesion, and the ability of the assay to mimic the future application as closely as possible.

A good example for the choice of assay when considering a potential application was presented by Becer *et al.* using SPR to determine the ability of a series of mannose- and galactose-functionalized copolymers to inhibit binding of the highly glycosylated gp120 envelope glycoprotein of HIV-1 to the DC-SIGN receptor. In order to investigate the potential binding mechanism and use of the glycopolymer in inhibition therapy to potentially block the HIV infection, they employed direct and competition/inhibition assays (Figure 2.23). For the direct binding assay, DC-SIGN was immobilized on a chip and glycopolymer binding was measured. K_D values were obtained by this method, giving a measure for the avidity of the glycopolymer in comparison to the natural glycoprotein gp120. In order to test the ability of the glycopolymers to compete with and potentially inhibit gp120 binding, a competition/inhibition assay was performed. Here, gp120 was immobilized on the chip surface and DC-SIGN was flown over the surface at a fixed concentration in the presence of varying concentrations of glycopolymer. If the glycopolymer competed for binding between gp120 and DC-SIGN, a reduction in the binding signal with the increasing glycopolymer concentration would be expected. These results are often fitted with a so-called Hill plot giving the half-minimum inhibitory concentration or IC_{50} values of the glycopolymer. Indeed, Becer *et al.* observed IC_{50} values in the nM range for the fully mannosylated glycopolymer, indicating the potential of this polymer to inhibit binding to DC-SIGN and thereby potentially block the DC-SIGN mediated adhesion and infection of viruses.

Another important aspect of the multivalent binding of glycopolymers is the change in thermodynamics *via* the simultaneous binding of several sugar ligands to a protein receptor. This information is not accessible *via* SPR or similar surface-based techniques. Here, ITC has to be used in order to derive the free binding energy and thereby the enthalpic and entropic contributions from ligand–receptor binding. Ambrosi *et al.*[62] demonstrated the use of ITC to determine the thermodynamics of glycopolymers bearing β-D-galactose side groups binding to the lectin peanut agglutinin (PNA). They found the optimal set-up to be a titration of the glycopolymer into a PNA solution in citrate buffer. Indeed, finding the right conditions in order to avoid protein or protein–ligand complex precipitation during the measurements, as well as to remove any possible contribution of buffer protonation effects to the enthalpic term, are crucial for a successful ITC experiment. Ambrosi *et al.* found that monomeric ligands show enthalpically driven

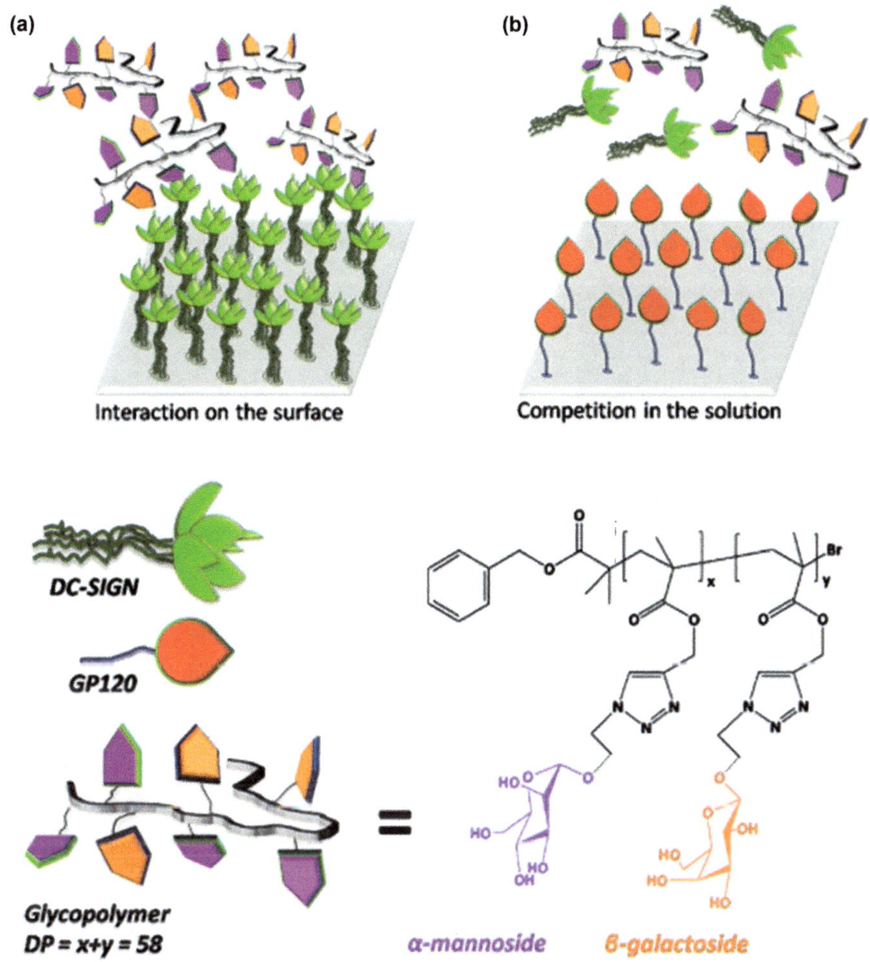

Figure 2.23 Schematic presentation of the direct binding and competition assay performed *via* SPR to access the binding of heteromultivalent glycopolymers to DC-SIGN.
Figure reprinted from ref. 51, http://pubs.acs.org/doi/abs/10.1021%2Fja1056714.

binding to PNA while, for multivalent glycopolymers, the enthalpic contribution becomes less favorable but is compensated by a favorable entropic contribution. The authors noted that, in addition to potential multivalent binding modes, hydrophobic effects where water molecules are released from the PNA binding site upon complexation with the glycopolymer may also contribute entropically.

Recently, Schmidt and coworkers[63] introduced a novel assay to determine binding affinities and IC_{50} values by mimicking more closely cell–cell contacts and their inhibition. They synthesized soft colloidal particles (SCPs) functionalized with mannose ligands and incubated these microgels on a

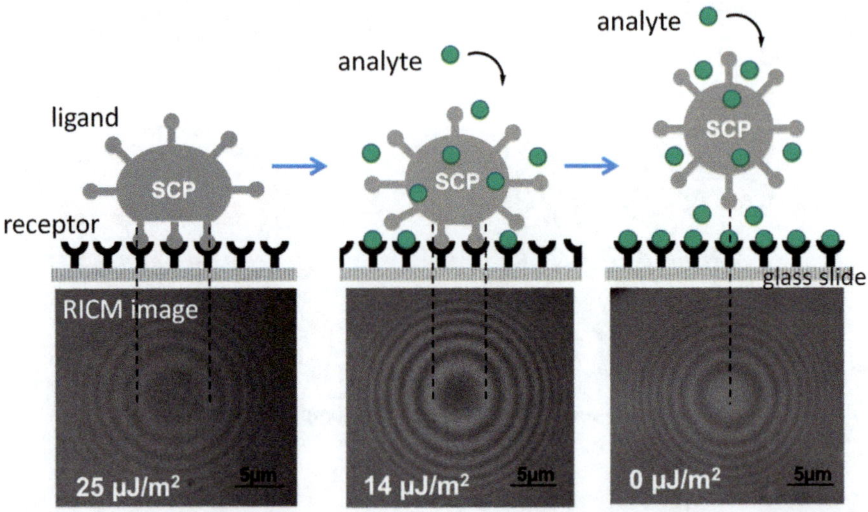

Figure 2.24 Schematic presentation of the direct, as well as inhibition-competition, assay using soft colloidal probes and their analysis *via* reflection interference contrast microscopy, adapted from Pussak *et al.*[63]

Concanavalin A (ConA) receptor functionalized glass surface. Upon ligand–receptor binding between the microgel and the glass surface, the SCPs form a distinct contact area on the glass slide that can be detected *via* reflection interference contrast microscopy (RICM). Using mathematical modeling, this contact area can be correlated with the adhesion energy. Performing this experiment in the presence of increasing concentrations of inhibitors competing with the SCP binding gives the according IC_{50} values of the tested inhibitor (Figure 2.24). This assay allows for the fast and simple detection of affinities with an optical microscope. In addition, both the ligand and receptor are bound to (soft) surfaces, thus giving new insight into the interplay of ligand and receptors in cell–cell contacts.

2.4 Biomedical and Biotechnological Applications of Glycopolymers

The use of glycopolymers in biomedical and biotechnological applications is steadily growing. The following section highlights a few examples to show some of the main applications in biomedicine, namely the use of glycopolymers for tumor targeting,[64–66] drug delivery,[67–73] as anti-pathogenic agents,[74–78] and as immune stimulants.[79–81]

Reineke and coworkers[82] established the synthesis of AB type block copolymers comprised of a glycopolymer block and a cationic block *via* aqueous RAFT polymerization of 2-deoxy-2-methacrylamido glucopyranose and the primary amine-containing *N*-(2-aminoethyl) methacrylamide. The cationic block functions as a non-viral vector to complex DNA or siRNA and

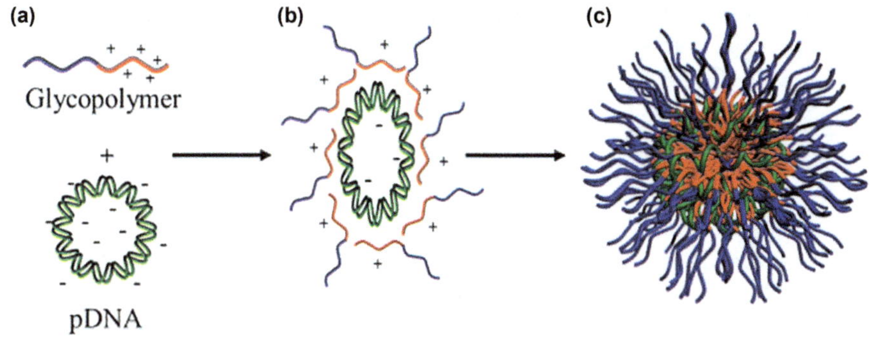

Figure 2.25 Schematic presentation of the polyplex formation of a diblock glycopolymer with DNA.
Figure reprinted from ref. 82, http://pubs.acs.org/doi/abs/10.1021%2Fbm200643c.

form polyplexes that are surrounded by a glycopolymer corona, while the glycopolymer block stabilizes the formed polyplexes against aggregation and non-specific interactions during transport (Figure 2.25). In addition, depending on the sugar ligands attached to the polymer, specific interactions with cell surface-bound receptors can be used to enhance the specific uptake of the polyplexes into *e.g.* glioblastoma cells.

Stenzel and coworkers[83] synthesized polymeric micelles presenting fucose and glucose ligands based on poly-(propargyl methacrylate)-*block*-poly(*n*-butyl acrylate) (PPMA-*b*-PBA) block copolymers and their CuAAC functionalization with 2-azidoethyl β-L-fucopyranoside ($FucEtN_3$) and 2-azideoethyl β-D-glucopyranoside ($GlcEtN_3$). It has been shown that many cancer cell lines upregulate fucose. Therefore, L-fucose could be an attractive ligand to target cancer cells. Indeed, Stenzel and coworkers found that with an increasing degree of fucosylation, the micelles show preferential and specific uptake into different cancer cell lines but not into healthy cells.

Kiessling and coworkers[84] reported an interesting example of the use of glycopolymers as immune stimulants. They synthesized a heteromultivalent glycopolymer presenting a CD22 ligand (Neu5Ac_2,6Gal_1,4Glc) and a dinitrophenyl group (DNP) as ligands for the B cell antigen receptor (BCR) to investigate B cell activation (Figure 2.26). CD22 is a known inhibitory co-receptor on the surface of B cells that reduces B cell activation. Only the simultaneous presentation of DNP as a ligand for the BCR and the binding of the sugar ligand to the CD22 receptor on the copolymer resulted in attenuated B cell activation. These results indicate that the co-localization of the two receptors initiated by binding the copolymer leads to diminished B cell activation and therefore gives rise to a new approach to design and synthesize immune stimulants based on heteromultivalent glycopolymers.

Yan *et al.*[85] synthesized glycopolymers to inhibit the binding of *E. coli* to intestinal epithelial cells as a method to reduce the inflammation caused by Crohn's disease. They attached *n*-heptyl α-D-mannose, a ligand to the FimH

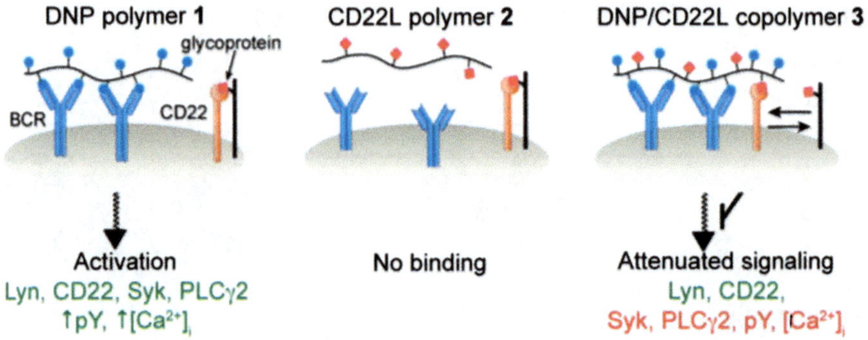

Figure 2.26 Schematic presentation of homo and heteromultivalent glycopolymers displaying DNP and the CD22 ligand and their binding to the B cell receptor and CD22 leading to the attenuation of B cell activation. Figure reprinted with permission from A. H. Courtney, E. B. Puffer, J. K. Pontrello, Z.-Q. Yang and L. L. Kiessling, *Proc. Natl. Acad. Sci. U. S. A.*, 2009, **106**, 2500–2505.[84]

receptor of *E.coli* bacteria, to a series of linear and star-shaped glycopolymers with varying chain length. Interestingly, the authors found that the enhancement in activity in terms of disrupting cell adhesion was not due to a higher affinity for the single FimH receptors, but more likely due to the ability of the polymer to induce crosslinking and aggregation of the bacteria (Figure 2.27). This again highlights the importance of different multivalent binding modes for the resulting biological activity of glycopolymers.

2.5 Conclusions

As the result of fruitful years of research, linear glycopolymers can be synthesized in many different ways, as outlined in the first part of this chapter. Today, we have straightforward access to glycopolymers of various sizes, architectures, and compositions. Therefore, the focus has now shifted from the synthesis to the application of glycopolymers in biotechnology and biomedicine. The second part of this chapter highlighted examples of the use of glycopolymers as sensors to study multivalent interactions, as well as of their application in fields such as drug delivery, cancer therapy, or as antipathogenic agents.

Indeed, there are many more examples in the literature on the synthesis and application of glycopolymers that we have not had the chance to present or acknowledge in this chapter. Furthermore, glycopolymers represent only a subclass of the larger class of glycomaterials including glycopeptides, glyconanoparticles, and glycodendrimers. Therefore, we believe that glycomaterials will continue to play an important role in the fields of biomaterials and biomedical research. We hope that this book chapter will help to promote future research in these areas while providing new researchers with a first insight into the field of linear glycopolymers.

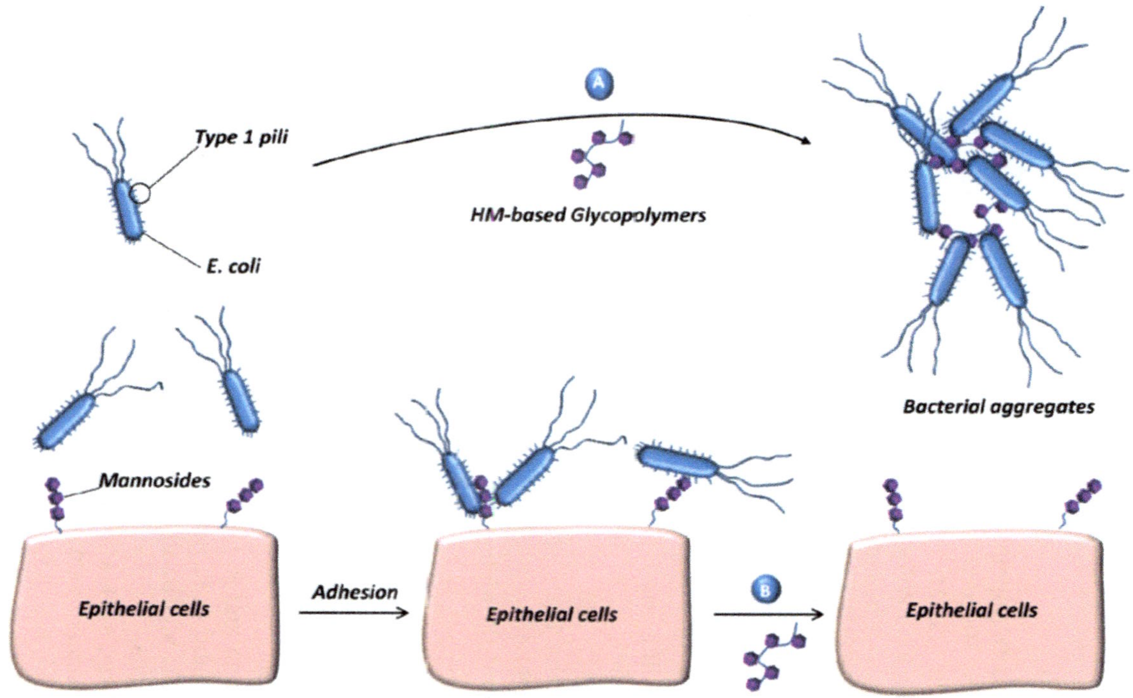

Figure 2.27 Schematic presentation of the *n*-heptyl α-D-mannose based glycopolymers and their anti-adhesive action. Reprinted with permission from X. Yan, A. Sivignon, N. Yamakawa, A. Crepet, C. Travelet, R. Borsali, T. Dumych, Z. Li, R. Bilyy, D. Deniaud, E. Feury, N. Barnich, A. Darfeuille-Michaud, S. G. Gouin, J. Bouckaert and J. Bernard, *Biomacromolecules*, 2015, **16**, 1827–1836.[85] Copyright (2015) American Chemical Society.

Acknowledgements

The authors thank the Heinrich-Heine-Universität, the Deutsche Forschungsgemeinschaft, and the Boehringer-Ingelheim-Stiftung for financial support, as well as Prof. Dr. Nicole L. Snyder for continuous support and proofreading of the manuscript.

References

1. B. Voit and D. Appelhans, *Macromol. Chem. Phys.*, 2010, **211**, 727–735.
2. C. R. Becer, *Macromol. Rapid Commum.*, 2012, **33**, 742–752.
3. K. Babiuch and M. H. Stenzel, in *Encyclopedia of Polymer Science and Technology*, John Wiley & Sons, Inc., 2002, DOI: 10.1002/0471440264.pst618.
4. S. Pearson, G. Chen and M. H. Stenzel, in *Engineered Carbohydrate-Based Materials for Biomedical Applications*, John Wiley & Sons, Inc., 2011, ch. 1, pp. 1–118, DOI: 10.1002/9780470944349.
5. G. Yilmaz and C. R. Becer, *Eur. Polym. J.*, 2013, **49**, 3046–3051.
6. Q. Zhang and D. M. Haddleton, in *Hierarchical Macromolecular Structures: 60 Years after the Staudinger Nobel Prize Ii*, ed. V. Percec, 2013, vol. 262, pp. 39–59.
7. L. L. Kiessling and J. C. Grim, *Chem. Soc. Rev.*, 2013, **42**, 4476–4491.
8. L. L. Kiessling, L. E. Strong and J. E. Gestwicki, *Annu. Rep. Med. Chem.*, 2000, **35**, 321–330.
9. D. Kushwaha, P. Dwivedi, S. K. Kuanar and V. K. Tiwari, *Curr. Org. Synth.*, 2013, **10**, 90–135.
10. V. Ladmiral, E. Melia and D. M. Haddleton, *Eur. Polym. J.*, 2004, **40**, 431–449.
11. T. K. Lindhorst, *Host-Guest Chem.*, 2002, **218**, 201–235.
12. Y. Miura, *Polym. J.*, 2012, **44**, 679–689.
13. R. Roy, in *Glycoscience Synthesis of Substrate Analogs and Mimetics*, ed. H. Driguez and J. Thiem, 1997, vol. 187, pp. 241–274.
14. S. Slavin, J. Burns, D. M. Haddleton and C. R. Becer, *Eur. Polym. J.*, 2011, **47**, 435–446.
15. V. Wittmann and R. J. Pieters, *Chem. Soc. Rev.*, 2013, **42**, 4492–4503.
16. G. Yilmaz and C. R. Becer, *Front. Bioeng. Biotechnol.*, 2014, **2**, 39.
17. J. Kumar, L. McDowall, G. Chen and M. H. Stenzel, *Polym. Chem.*, 2011, **2**, 1879–1886.
18. S. R. S. Ting, E. H. Min, P. Escale, M. Save, L. Billon and M. H. Stenzel, *Macromolecules*, 2009, **42**, 9422–9434.
19. M. Ambrosi, A. S. Batsanov, N. R. Cameron, B. G. Davis, J. A. K. Howard and R. Hunter, *J. Chem. Soc., Perkin Trans. 1*, 2002, 45–52, DOI: 10.1039/B108421F.
20. O. T. Schmidt, in *Methods in Carbohydrate Chemistry*, Academic Press, Inc., New York, 1963, DOI: 10.1002/ange.19630750720.
21. V. Ladmiral, L. Monaghan, G. Mantovani and D. M. Haddleton, *Polymer*, 2005, **46**, 8536–8545.

22. A. B. Lowe and R. Wang, *Polymer*, 2007, **48**, 2221–2230.
23. R. L. Whistler and J. L. Goatley, *J. Polym. Sci.*, 1961, **50**, 127–132.
24. R. L. Whistler, H. P. Panzer and H. J. Roberts, *J.Org. Chem.*, 1961, **26**, 1583–1588.
25. S. Loykulnant, M. Hayashi and A. Hirao, *Macromolecules*, 1998, **31**, 9121–9126.
26. F. D'Agosto, M.-T. Charreyre, F. Delolme, G. Dessalces, H. Cramail, A. Deffieux and C. Pichot, *Macromolecules*, 2002, **35**, 7911–7918.
27. M. Kanai, K. H. Mortell and L. L. Kiessling, *J. Am. Chem. Soc.*, 1997, **119**, 9931–9932.
28. K. H. Mortell, M. Gingras and L. L. Kiessling, *J. Am. Chem. Soc.*, 1994, **116**, 12053–12054.
29. K. H. Mortell, R. V. Weatherman and L. L. Kiessling, *J. Am. Chem. Soc.*, 1996, **118**, 2297–2298.
30. D. D. Manning, X. Hu, P. Beck and L. L. Kiessling, *J. Am. Chem. Soc.*, 1997, **119**, 3161–3162.
31. C. Boyer and T. P. Davis, *Chem. Commun.*, 2009, 6029–6031, DOI: 10.1039/b910296e.
32. R. M. Owen, J. E. Gestwicki, T. Young and L. L. Kiessling, *Org. Lett.*, 2002, **4**, 2293–2296.
33. L. Nurmi, J. Lindqvist, R. Randev, J. Syrett and D. M. Haddleton, *Chem. Commun.*, 2009, 2727–2729, DOI: 10.1039/B904249K.
34. A. Gress, A. Völkel and H. Schlaad, *Macromolecules*, 2007, **40**, 7928–7933.
35. D. Ponader, F. Wojcik, F. Beceren-Braun, J. Dernedde and L. Hartmann, *Biomacromolecules*, 2012, **13**, 1845–1852.
36. F. Wojcik, S. Lel, A. G. O'Brien, P. H. Seeberger and L. Hartmann, *Beilstein J. Org. Chem.*, 2013, **9**, 2395–2403.
37. F. Wojcik, A. G. O'Brien, S. Goetze, P. H. Seeberger and L. Hartmann, *Chem. – Eur. J.*, 2013, **19**, 3090–3098.
38. D. Ponader, P. Maffre, J. Aretz, D. Pussak, N. M. Ninnemann, S. Schmidt, P. H. Seeberger, C. Rademacher, G. U. Nienhaus and L. Hartmann, *J. Am. Chem. Soc.*, 2014, **136**, 2008–2016.
39. Q. Yang, in *Engineered Carbohydrate-Based Materials for Biomedical Applications*, John Wiley & Sons, Inc., 2011, ch. 2, pp. 119–141, DOI: 10.1002/9780470944349.
40. C. R. Becer, K. Babiuch, D. Pilz, S. Hornig, T. Heinze, M. Gottschaldt and U. S. Schubert, *Macromolecules*, 2009, **42**, 2387–2394.
41. H. Götz, E. Harth, S. M. Schiller, C. W. Frank, W. Knoll and C. J. Hawker, *J. Polym. Sci., Part A: Polym. Chem.*, 2002, **40**, 3379–3391.
42. G. Pasparakis and C. Alexander, *Angew. Chem., Int. Ed.*, 2008, **47**, 4847–4850.
43. Y.-M. Chen and G. Wulff, *Macromol. Rapid Commum.*, 2002, **23**, 59–63.
44. A. Dag, J. Zhao and M. H. Stenzel, *ACS Macro Lett.*, 2015, **4**, 579–583.
45. J. L. Jimenez Blanco, C. Ortiz Mellet and J. M. Garcia Fernandez, *Chem. Soc. Rev.*, 2013, **42**, 4518–4531.

46. Y. Nishida, H. Uzawa, T. Toba, K. Sasaki, H. Kondo and K. Kobayashi, *Biomacromolecules*, 2000, **1**, 68–74.
47. Y. Nishida, H. Dohi and K. Kobayashi, *Trends Glycosci. Glycotechnol.*, 2005, **17**, 59–69.
48. K. Sasaki, Y. Nishida, T. Tsurumi, H. Uzawa, H. Kondo and K. Kobayashi, *Angew. Chem., Int. Ed.*, 2002, **41**, 4463–4467.
49. V. Ladmiral, G. Mantovani, G. J. Clarkson, S. Cauet, J. L. Irwin and D. M. Haddleton, *J. Am. Chem. Soc.*, 2006, **128**, 4823–4830.
50. J. Geng, G. Mantovani, L. Tao, J. Nicolas, G. Chen, R. Wallis, D. A. Mitchell, B. R. G. Johnson, S. D. Evans and D. M. Haddleton, *J. Am. Chem. Soc.*, 2007, **129**, 15156–15163.
51. C. R. Becer, M. I. Gibson, J. Geng, R. Ilyas, R. Wallis, D. A. Mitchell and D. M. Haddleton, *J. Am. Chem. Soc.*, 2010, **132**, 15130–15132.
52. M. W. Jones, L. Otten, S. J. Richards, R. Lowery, D. J. Phillips, D. M. Haddleton and M. I. Gibson, *Chem. Sci.*, 2014, **5**, 1611–1616.
53. J. D. Badjic, A. Nelson, S. J. Cantrill, W. B. Turnbull and J. F. Stoddart, *Acc. Chem. Res.*, 2005, **38**, 723–732.
54. A. Barnard and D. K. Smith, *Angew. Chem., Int. Ed.*, 2012, **51**, 6572–6581.
55. A. Mulder, J. Huskens and D. N. Reinhoudt, *Org. Biomol. Chem.*, 2004, **2**, 3409–3424.
56. C. R. Becer and L. Hartmann, in *Glycopolymer Code: Synthesis of Glycopolymers and their Applications*, The Royal Society of Chemistry, Cambridge, UK, 2015, DOI: 10.1039/9781782622666.
57. C. Fasting, C. A. Schalley, M. Weber, O. Seitz, S. Hecht, B. Koksch, J. Dernedde, C. Graf, E.-W. Knapp and R. Haag, *Angew. Chem., Int. Ed.*, 2012, **51**, 10472–10498.
58. L. L. Kiessling and R. A. Splain, in *Annual Review of Biochemistry*, ed. R. D. Kornberg, C. R. H. Raetz, J. E. Rothman and J. W. Thorner, 2010, vol. 79, pp. 619–653.
59. V. M. Krishnamurthy, L. A. Estroff and G. M. Whitesides, in *Fragment-based Approaches in Drug Discovery*, Wiley-VCH Verlag GmbH & Co. KGaA, 2006, ch. 2, pp. 11–53, DOI: 10.1002/3527608761.
60. J. E. Gestwicki, C. W. Cairo, L. E. Strong, K. A. Oetjen and L. L. Kiessling, *J. Am. Chem. Soc.*, 2002, **124**, 14922–14933.
61. Y. Chen, M. S. Lord, A. Piloni and M. H. Stenzel, *Macromolecules*, 2015, **48**, 346–357.
62. M. Ambrosi, N. R. Cameron, B. G. Davis and S. Stolnik, *Org. Biomol. Chem.*, 2005, **3**, 1476–1480.
63. D. Pussak, D. Ponader, S. Mosca, S. V. Ruiz, L. Hartmann and S. Schmidt, *Angew. Chem. Int. Ed.*, 2013, **52**, 6084–6087.
64. C. B. Carlson, P. Mowery, R. M. Owen, E. C. Dykhuizen and L. L. Kiessling, *ACS Chem. Biol.*, 2007, **2**, 119–127.
65. J. Lu, W. Zhang, L. Yuan, W. Ma, X. Li, W. Lu, Y. Zhao and G. Chen, *Macromol. Biosci.*, 2014, **14**, 340–346.
66. R. T. C. Sheridan, J. Hudon, J. A. Hank, P. M. Sondel and L. L. Kiessling, *ChemBioChem*, 2014, **15**, 1393–1398.

67. M. Mahkam, *Drug Delivery*, 2007, **14**, 147–153.
68. S. Menon, R. M. Ongungal and S. Das, *Macromol. Chem. Phys.*, 2014, **215**, 2365–2373.
69. M. Tranter, Y. Liu, S. He, J. Gulick, X. Ren, J. Robbins, W. K. Jones and T. M. Reineke, *Mol. Ther.*, 2012, **20**, 601–608.
70. R. H. Utama, Y. Jiang, P. B. Zetterlund and M. H. Stenzel, *Biomacromolecules*, 2015, **16**, 2144–2156.
71. Y. Wang, X. Zhang, P. Yu and C. Li, *Int. J. Pharm.*, 2013, **441**, 170–180.
72. M. Ahmed and R. Narain, *Biomaterials*, 2011, **32**, 5279–5290.
73. M. Ahmed and R. Narain, *Biomaterials*, 2012, **33**, 3990–4001.
74. A. Bernardi, J. Jimenez-Barbero, A. Casnati, C. De Castro, T. Darbre, F. Fieschi, J. Finne, H. Funken, K.-E. Jaeger, M. Lahmann, T. K. Lindhorst, M. Marradi, P. Messner, A. Molinaro, P. V. Murphy, C. Nativi, S. Oscarson, S. Penades, F. Peri, R. J. Pieters, O. Renaudet, J.-L. Reymond, B. Richichi, J. Rojo, F. Sansone, C. Schaeffer, W. B. Turnbull, T. Velasco-Torrijos, S. Vidal, S. Vincent, T. Wennekes, H. Zuilhof and A. Imberty, *Chem. Soc. Rev.*, 2013, **42**, 4709–4727.
75. S.-J. Richards, M. W. Jones, M. Hunaban, D. M. Haddleton and M. I. Gibson, *Angew. Chem., Int. Ed.*, 2012, **51**, 7812–7816.
76. T. Tanaka, H. Ishitani, Y. Miura, K. Oishi, T. Takahashi, T. Suzuki, S.-i. Shoda and Y. Kimura, *ACS Macro Lett.*, 2014, **3**, 1074–1078.
77. A. Wild, K. Babiuch, M. Koenig, A. Winter, M. D. Hager, M. Gottschaldt, A. Prokop and U. S. Schubert, *Chem. Commun.*, 2012, **48**, 6357–6359.
78. L. N. Albers and W. B. Gleason, *J. Carbohydr. Chem.*, 2014, **33**, 455–470.
79. J. J. Hollenbeck, E. B. Puffer, J. K. Pontrello and L. L. Kiessling, *Biochemistry*, 2003, **42**, 8602.
80. E. B. Puffer, J. K. Pontrello, J. J. Hollenbeck, J. A. Kink and L. L. Kiessling, *ACS Chem. Biol.*, 2007, **2**, 252–262.
81. E.-H. Song, M. J. Manganiello, Y.-H. Chow, B. Ghosn, A. J. Convertine, P. S. Stayton, L. M. Schnapp and D. M. Ratner, *Biomaterials*, 2012, **33**, 6889–6897.
82. A. E. Smith, A. Sizovs, G. Grandinetti, L. Xue and T. M. Reineke, *Biomacromolecules*, 2011, **12**, 3015–3022.
83. K. Babiuch, A. Dag, J. Zhao, H. Lu and M. H. Stenzel, *Biomacromolecules*, 2015, **16**, 1948–1957.
84. A. H. Courtney, E. B. Puffer, J. K. Pontrello, Z.-Q. Yang and L. L. Kiessling, *Proc. Natl. Acad. Sci. U. S. A.*, 2009, **106**, 2500–2505.
85. X. Yan, A. Sivignon, N. Yamakawa, A. Crepet, C. Travelet, R. Borsali, T. Dumych, Z. Li, R. Bilyy, D. Deniaud, E. Feury, N. Barnich, A. Darfeuille-Michaud, S. G. Gouin, J. Bouckaert and J. Bernard, *Biomacromolecules*, 2015, **16**, 1827–1836.
86. S. R. S. Ting, E. H. Min, P. B. Zetterlund and M. H. Stenzel, *Macromolecules*, 2010, **43**, 5211–5221.

CHAPTER 3

Synthesis of Non-natural Polymers with Controlled Primary Structures

ABDELAZIZ AL OUAHABI,† RAJ KUMAR ROY† AND JEAN-FRANÇOIS LUTZ*

Precision Macromolecular Chemistry, Institut Charles Sadron, UPR22-CNRS, BP84047, 23 Rue du Loess, 67034 Strasbourg Cedex 2, France
*Email: jflutz@unistra.fr

3.1 Introduction

As discussed in all the chapters of this book, Nature is an endless well of inspiration for the design of man-made polymers and materials. The sources of bio-inspiration are very diverse and may occur at different scales, *i.e.* macroscopic, microscopic, or molecular.[1–5] At the macroscopic scale, many researchers have mimicked the superior mechanical,[6] optical,[7] self-repair,[8] stimulus-responsive,[9] or surface[10] properties of biomaterials. At the nanoscale, the self-assembly of biopolymers into hierarchical organized morphologies has been very inspirational for the design of synthetic polymer materials in recent years. For instance, the self-organization strategies of Nature have been successfully imported in the field of block copolymer self-assembly.[11–14] At the molecular scale, however, the mimicry of nature

†Both authors contributed equally to this work and are listed in alphabetical order.

RSC Polymer Chemistry Series No. 22
Bio-inspired Polymers
Edited by Nico Bruns and Andreas F. M. Kilbinger

Published by the Royal Society of Chemistry, www.rsc.org

becomes more challenging. Indeed, there is still an important gap between structurally-defined biopolymers such as nucleic acids, proteins, and polysaccharides and polydisperse synthetic polymers prepared by classical polymerization methods such as chain-growth or step-growth polymerization.[15] For instance, it is well known that nucleic acids and proteins exhibit controlled primary structures, *i.e.* tacticity and monomer sequences. As a consequence, these biopolymers can fold into precise three-dimensional structures and fulfil advanced biological functions such as information storage, signaling, biocatalysis, and selective molecular transport. In comparison, synthetic polymers exhibit in general less-controlled molecular structures. In fact, most of the broadly-studied synthetic macromolecules are homopolymers, *i.e.* polymers constructed with the same repeating monomer unit. In some homopolymers, tacticity can be controlled and therefore semi-crystalline materials can be attained.[16] However, when two or more monomers are used to build a polymer, its primary structure is, in general, poorly controlled.[17,18] Synthetic copolymers have been studied since the mid-1940s[19] and different types of copolymers, such as statistical, alternating, block, periodic, and gradient copolymers, have been developed.[20] Statistical copolymers are probably the most common class of copolymers. In such structures, comonomers are statistically distributed along the formed copolymer chains. This copolymerization strategy permits to attain properties that are in between those of the corresponding homopolymers (*i.e.* averaged properties) and is therefore a convenient approach to roughly adjust mechanical or solution properties.[16] Nevertheless, the control of the primary structure in these copolymers is, of course, very imprecise. Block copolymers constitute another important family of copolymers. They are made of two or more individual homopolymer segments that are covalently connected to another.[21] Such copolymer design permits to combine the properties of the corresponding homopolymers in a single macromolecule (*i.e.* additive properties).[16] A classic example of that is the design of amphiphilic block copolymers that display opposite affinities for a given solvent. Besides widely studied statistical and block copolymers, other class of copolymers such as alternating, periodic, or gradient copolymers have received so far less academic and industrial attention.[22] On the whole, man-made copolymers exhibit ill-defined primary structures compared to those of biopolymers. As a consequence, a large amount of research has been dedicated to the synthesis of non-natural sequence-controlled polymers over the last years.[23,24] Early contributions to the field have been made for developing peptidomimetics and foldamers, which are short oligomers that mimic the structure of proteins.[25–28] Similarly, interesting progress has been made during the last two decades in the preparation of artificial macromolecules that resemble nucleic acids.[29–33] However, although non-natural, these types of polymers remain conceptually and structurally very close to biological polymers. Quite recently, there has been a broader interest in developing sequence-controlled polymers that are inspired by Nature, but that are chemically different from proteins and nucleic acids.[24,34,35] In this context, several new concepts have been proposed for regulating comonomer sequences in synthetic polymerizations.[24] For instance, many research groups have investigated monomer sequence regulation in chain-growth and step-growth polymerizations. Precisely-controlled synthetic primary structures have also been developed using solid-phase chemistry[36] or, more broadly speaking, multi-step growth approaches. In addition, sequence-controlled polymers have been prepared using bio-inspired polymerization concepts, *e.g.* using supramolecular templates and catalytic molecular machines. These recent developments in the field of sequence-controlled polymers are discussed and critically analyzed in this chapter.

3.2 Sequence-controlled Polymers Prepared by Chain-growth Polymerization

Chain-growth polymerizations are convenient methods for preparing (co)polymers. In particular, "living" chain-growth polymerizations,[37] which include anionic,[38] cationic,[39] and controlled radical polymerizations,[40–44] are very useful to prepare polymers with control over their chain-length, molecular weight distribution, composition, and architecture. However, as these methods involve highly reactive propagating species (either ions or radicals), it is challenging to control comonomer sequences in chain-growth copolymers. Nevertheless, significant advances have been made in recent years, for example using template and kinetic strategies.[24] For instance, sequences can be controlled by selecting comonomers of different chemical reactivity. In living chain-growth polymerizations, all chains grow simultaneously and therefore the differences in reactivity of the comonomers are "written" in the microstructure of the formed copolymers.[45] In addition, the use of external parameters,[46] such as time controlled addition of reactants or physical stimuli, allows to regulate even more finely the microstructure of synthetic polymers. These new options are described in detail in the following sections.

3.2.1 Anionic Polymerization

The term 'living' polymers was first coined by Michael Szwarc in the context of the termination-free anionic polymerization of styrene.[37] Over the last decades, living polymerization techniques have been used for the preparation of well-defined polymers with controlled end-groups, composition, and architectures.[47] However, as mentioned earlier, it is difficult to control monomer sequences in chain-growth polymerizations involving multiple comonomers. Nevertheless, in some cases, comonomer cross-propagation is favored over homopolymerization, thus leading to the formation of alternating polymers. For example, diphenylethylene (DPE) is a bulky monomer that does not homopolymerize in anionic polymerization but can form alternating copolymers with suitable comonomers.[48] In a recent example, Hutchings and coworkers studied the sequence-controlled anionic copolymerization of DPE and 1,1-bis(4-*tert*-butyldimethylsiloxyphenyl) ethylene (DPE-OSi) with styrene or butadiene.[49] The comonomer sequence was controlled by tuning the experimental parameters such as the nature of the solvent. It was shown that the copolymerization of styrene and DPE formed nearly perfect alternating copolymers in THF, whereas sequence-control was lost in an apolar solvent like benzene. Furthermore, the microstructure of the copolymers was tuned by varying the type of comonomers. For instance, the copolymerization of styrene with DPE-OSi in benzene did not lead to the formation of alternating copolymers, but resulted in telechelic copolymers. In another example, the same group studied the effect of electron donating or withdrawing substituents onto the phenyl rings of DPE.[50] They showed that the one-pot terpolymerization of DPE, DPE-OSi, and styrene results in terpolymers with controlled microstructures. During copolymerization, the lower reactive monomer DPE-OSi is not incorporated in the formed polymer chains before the complete consumption of more reactive DPE. As a result, the resulting terpolymer contains first an alternating sequence of styrene and DPE, followed by a short terminal polystyrene segment containing DPE-OSi units. Following the work of Hutchings, controlled comonomer sequences were also recently prepared by Li and coworkers using substituted DPE derivatives.[51] Another strategy to synthesize sequence-controlled polymers *via* a living anionic polymerization route was recently reported by He and coworkers.[52,53] In this work, 2,4-disubstituted butadiene derivatives were used as monomers. Although butadiene has, in general, a tendency to polymerize both in 1,2- and 1,4-fashion, the use of appropriate substituents may lead to exclusive 1,4-polymerization. For instance, 2-isopropyl-4-phenyl butadiene leads to nearly quantitative 1,4 addition during anionic polymerization, thus resulting in the formation of periodic functional copolymers.[53] Ultimately, these polymers are converted into sequence-controlled polyacetylenes by oxidative dehydrogenation.

The development of sequence-controlled polymers also necessitates appropriate analytical techniques.[54] In gradient copolymers, for example, it is difficult to characterize accurately the gradient length using post-polymerization analysis. Here, *in situ* characterization methods can be very useful to monitor comonomer sequences more precisely. In some recent reports, Frey and coworkers have monitored the monomer sequence distribution in living anionic copolymerization by real-time ^{1}H nuclear magnetic resonance (NMR) spectroscopy.[55,56] This technique allows following the consumption of each monomer and their incorporation into the formed polymer chains throughout the whole copolymerization process. For instance, the copolymerization of styrenic derivatives, such as protected hydroxy-styrene, *p*-(1-ethoxy ethoxy)styrene (pEES) and 4-*tert*-butoxystyrene (*t*BuOS), was studied by NMR spectroscopy. As expected from such structurally-alike monomers, the formation of a random copolymer, with a slight deviation from the ideal copolymerization behavior, was evidenced by NMR analysis. Another copolymerization using chemically different monomers such as styrene and pEES resulted in a more-pronounced gradient copolymer, as evidenced by real-time ^{1}H NMR spectroscopy.[56] However, stringent reaction conditions, such as non-protic solvents, oxygen free conditions, and protecting functional groups, are necessary to avoid undesirable side reactions and often limit the application of anionic polymerizations for such purposes.

3.2.2 Cationic Polymerization

Sequence-controlled polymers can also be prepared by living cationic polymerization. One of the earliest examples was reported by Higashimura and Sawamoto.[57] This synthesis involved the sequential and successive addition of four different vinyl ether derivatives along with HI/ZnI_2, which generated oligomeric growing species in a living fashion. Apart from the tetramer with the desired sequence, the intermediate oligomers were also trapped and isolated by quenching the reaction with MeOH. Alternating copolymers have also been prepared by cationic polymerization. For example, as shown by Kamigaito and coworkers,[58] β-methylstyrene derivatives (*e.g.* 4-hydroxy-3-methoxyphenyl propenyl) do not homopolymerize due to the steric bulk around the carbocation but they do copolymerize with less bulky *p*-methoxystyrene. In addition to this, the cationic polymerization of *p*-methoxystyrene reacts at a much faster rate with electron rich monomers such as (4-hydroxy-3-methoxy-β-methylstyrene). As a matter of fact, the combination of these two monomers forms alternating copolymers.[58] Apart from their alternating structure, the introduction of a phenolic functionality in the formed copolymer chain is also interesting, since it is difficult to prepare such functional polymers by other methods.

Aoshima and coworkers have described the synthesis of alternating polymers by living cationic polymerization of aldehydes and vinyl ether derivatives.[59] Aromatic aldehydes are difficult to polymerize cationically but copolymerize with reactive vinyl ethers. For instance, benzaldehyde (BzA) was copolymerized with isobutyl vinyl ether (IBVE), leading to polymers having about 30% of BzA units but devoid of any BzA–BzA sequences, which ruled out the possibility of BzA homopolymerization. In a next step, a strictly alternating benzaldehyde/vinyl ether comonomer pair was studied. It was found that the controlled cationic copolymerization of the pairs *p*-methoxyBzA/IBVE and BzA/2-chloroethylVE (CEVE) resulted in alternating copolymers using the $EtSO_3H/GaCl_3$ initiating system in the presence of 1,4-dioxane as the solvent. Additionally, the cationic polymerization of aldehyde and vinyl ether led to an acid labile acetal linkage, which could be exploited to prepare sequence-controlled degradable polymers. Apart from alternating copolymers, the authors explored other controlled architectures, such as block copolymers with an acid degradable block and core degradable star polymers (Figure 3.1).[60] The precise placement of cleavable units at predetermined positions and their selective degradation was possible by this approach.

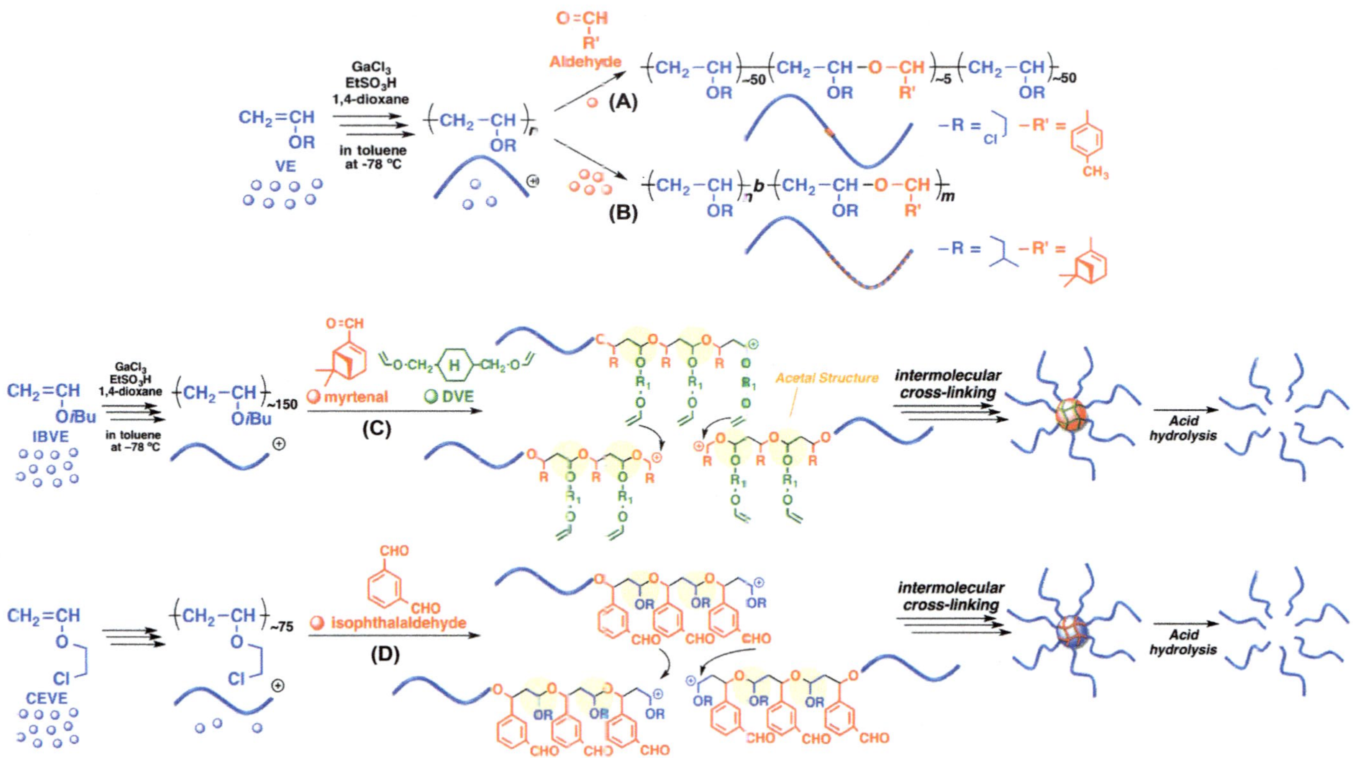

Figure 3.1 Synthesis of sequence-controlled polymers by living cationic polymerization. (A) Copolymers containing acid-degradable acetal units. (B) Block copolymers with an acid degradable segment. (C) and (D) Synthesis of star-shaped polymers with acid-degradable acetal units in the core.
Reprinted from ref. 60 with permission from the Royal Society of Chemistry.

3.2.3 Controlled Radical Polymerization

Controlled radical polymerization (CRP) techniques, such as nitroxide mediated radical polymerization (NMP),[44] atom-transfer radical polymerization (ATRP),[42,61] and reversible addition-fragmentation chain-transfer radical polymerization (RAFT),[41] have become extremely popular during the last two decades for the synthesis of controlled polymer architectures.[43] In recent years, it has been shown that these approaches are convenient for the synthesis of sequence-controlled polymers. Although this field of research is still expanding, the concepts that have been reported so far for monomer sequence regulation in CRP can be broadly classified into three main categories: (i) kinetic approaches, (ii) templated strategies, and (iii) multiblock copolymers with short block lengths. The CRP synthesis of simpler sequence-controlled polymers such as block, gradient, and alternating copolymers has been described in numerous publications, and the reader is guided to comprehensive reviews for further information on those topics.[17,42,43,62,63]

For instance, it is well-known that the radical copolymerization of equimolar amounts of a donor monomer such as styrene with an acceptor monomer such as maleic anhydride or *N*-substituted maleimides leads to the formation of nearly perfect alternating copolymers.[64–68] The concept relies on the fact that such acceptor comonomers have a low tendency to homopolymerize but can easily copolymerize with a styrene derivative. When used in a CRP process, such comonomer pairs lead to well-defined polymers with controlled chain-length, dispersity, and sequences. This situation becomes even more interesting when an excess of donor monomer is used with respect to the acceptor one (Figure 3.2).[69] In this case, kinetically-favored donor/acceptor copolymerization occurs until the acceptor monomer is consumed, followed by the homopolymerization of the excess of styrene. For CRP in which all chains are initiated at the same time and grow at nearly the same rate, these different regimes are "written" in the microstructure of the formed copolymer, thus leading to the formation of sequence-controlled regions.[70] Our group has shown that this simple behavior can be used as a versatile strategy to "write" information on synthetic polymer chains.[45,69,71–73] Interestingly, this concept is still working with very small amounts of acceptor comonomers (*e.g.* one mole percent as compared to the initiator), thus allowing the possibility to form small functional zones in polystyrene chains.[45,69] The acceptor comonomer can be used at the beginning of the polymerization, but it can also be added later during the course of the CRP of styrene.[71] Thus, by knowing the homopolymerization kinetics of styrene, one can precisely predict the placement of the *N*-substituted maleimide in the chain (Figure 3.3a). Such a time-controlled monomer addition can be performed multiple times during styrene polymerization, thus allowing the synthesis of complex microstructures (Figure 3.3a–c).[71,74] Yet, it was observed that the placement precision was more accurate at high styrene conversion because the donor/acceptor ratio

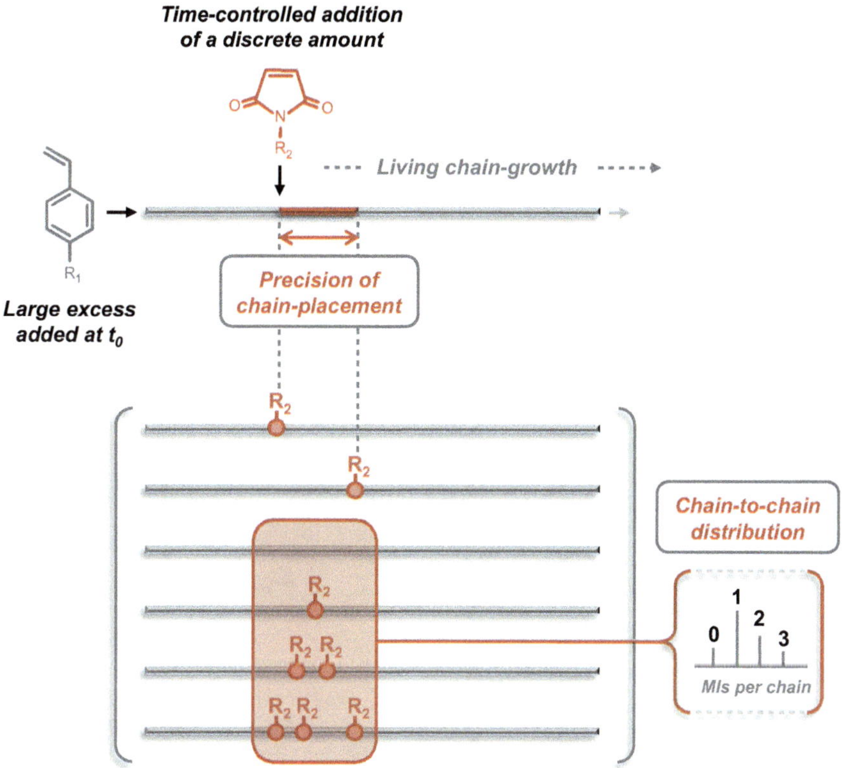

Figure 3.2 Concept for the local chain-installation of a discrete amount of an acceptor monomer during the living polymerization of a large excess of a donor monomer. The bottom part of the figure shows a simplified view of the chain-to-chain distribution of the acceptor comonomer in the final sample.
Reprinted with permission from J.-F. Lutz, *Acc. Chem. Res.*, 2013, **46**, 2696–2705.[45] Copyright (2013) American Chemical Society.

was lower. Therefore, ultra-precise incorporation can be attained by using successive feeds of donor and acceptor comonomers (Figure 3.3a).[75] Altogether, this concept is very versatile and can be applied to a wide variety of *N*-substituted maleimides.[76] For example, this platform was used for the incorporation of reactive functions,[77–79] biological moieties,[80] and highly-functional dendrons.[81] Moreover, the nature of the donor monomer can be varied, thus allowing the synthesis of functional sequence-controlled polymers such as water-soluble polymers,[82,83] polyelectrolytes,[84,85] and semi-crystalline polymers.[86] In addition, such sequence-controlled polymers can be folded into structured objects using intramolecular covalent or non-covalent chemistry.[87–92] For instance, precisely positioned covalent bridges can be obtained by intramolecular chemistry in dilute conditions.[93–95] Linear precursors, prepared by sequence-controlled ATRP of styrene with *N*-propargyl maleimide, were folded by intramolecular alkyne-based

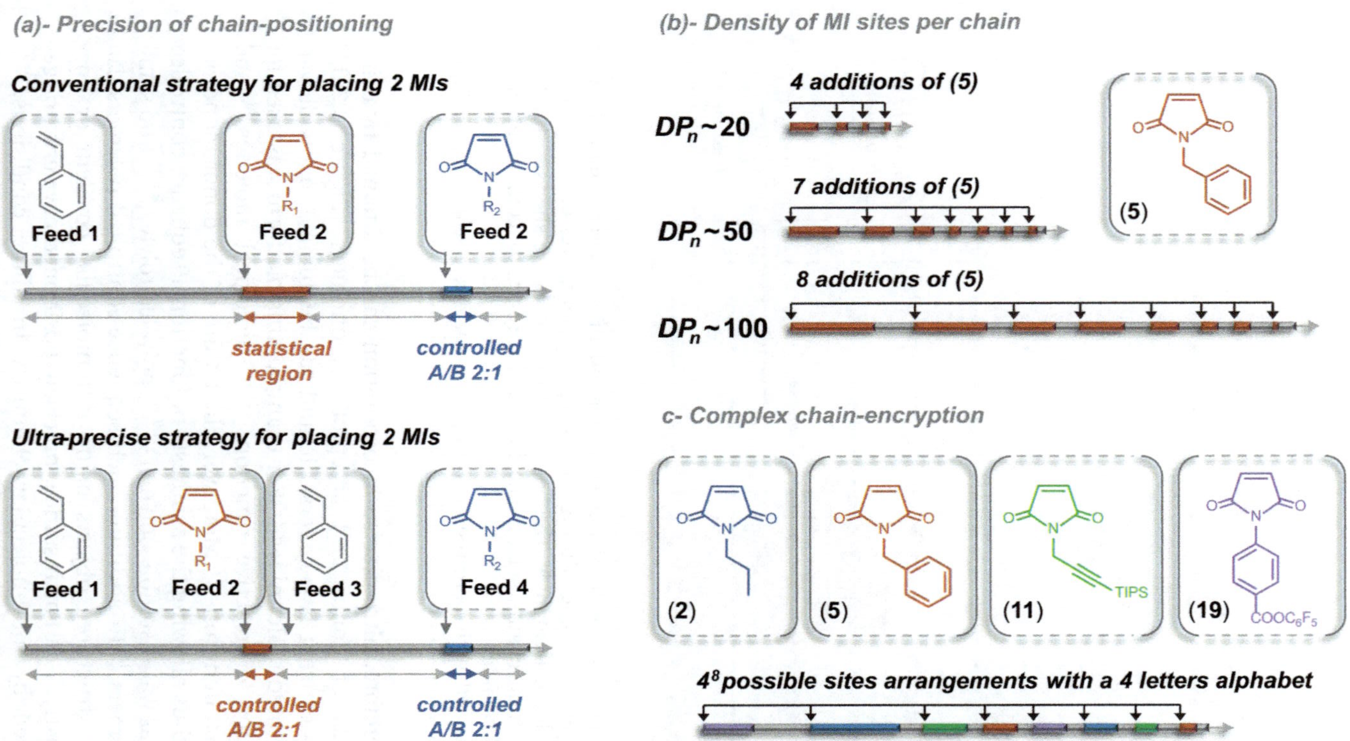

Figure 3.3 Possibilities of microstructural control using donor/acceptor sequence-controlled copolymerization: (a) chain-positioning of functional *N*-substituted maleimides using a conventional strategy and an ultra-precise strategy. Reproduced from M. Zamfir and J.-F. Lutz, *Nat Commun.*, 2012, **3**, 1138 with permission from Nature Publishing Group.[75] (b) Maximum incorporation of discrete *N*-substituted maleimide zones in polystyrene chains of different chain-length.[74] (c) Microstructural encoding using a set of 4 different *N*-substituted maleimides.[74] (b) and (c) are reproduced from ref. 74 with permission from John Wiley and Sons. Copyright © 2012 Wiley-VCH Verlag GmbH & Co. KGaA, Weinheim.

chemistry into defined topological macromolecules with different shapes.[93] This example clearly highlights the versatility of the styrene/*N*-substituted maleimides copolymerization platform for the preparation of complex polymer structures. In a more recent study, it was shown that pseudo tertiary structures can also be prepared using that chemistry.[96] In this case, short functional zones were sequentially compacted into globular domains, thus leading to the formation of "structured" random coils.

Besides the styrene/*N*-substituted maleimide approach reported by our group, other kinetic concepts have been explored in CRP for the preparation of sequence-controlled polymers. For example, Kamigaito and coworkers described a highly selective 1 : 2 sequence-regulated copolymerization of naturally occurring (+)-*d*-limonene with maleimides in a fluoroalcohol (Figure 3.4).[97] Such AAB copolymers were prepared by RAFT polymerization. Limonene does not undergo homopolymerization; however, it has a tendency to copolymerize with polar monomers such as maleic anhydride or acrylates, although the incorporation of limonene is low in the formed copolymer. However, it was observed that the copolymerizability of this monomer is enhanced by the use of a fluorinated alcohol. Limonene was replaced for hydroxyl-functionalized limonene and was radically copolymerized with cyclohexylmaleimide (CyMI) using the fluorinated solvent $PhC(CF_3)_2OH$. The resulting copolymer possesses 1 : 2 sequence-regulated (limonene to maleimide) propagation with a periodically functionalized single hydroxyl group within every three-monomer units.

Similarly, the copolymerization of a hydroxy-functionalized maleimide (*N*-2-hydroxyethylmaleimide) with limonene resulted in a periodic copolymer having two hydroxyl groups for every three-monomer units.[98] In another work, copolymers with sequence-controlled side-chains were prepared. Sequence-defined oligomers were prepared by successive atom transfer radical addition of styrene (S) or methyl acrylate (A) to an alkyl halide-possessing either an A or S unit (*i.e.* methyl α-bromopropionate or 1-phenylethylhalide). The resulting halide-ended sequence-regulated oligomer was subsequently converted into a maleimide-functionalized macromonomer and copolymerized by radical polymerization.[99]

Another important strategy that has been investigated for regulating monomer sequences in controlled radical polymerizations is the use of macromolecular templates.[100] Such strategies have been so far primarily investigated by Sawamoto and coworkers.[101] For instance, these authors proposed the use of sequence-defined templates prepared by living cationic polymerization to regulate comonomer sequences in CRP.[102] Being that aim very difficult to achieve, they first studied low molecular weight models.[102,103] For example, a dual initiator for both living cationic and living radical polymerizations was used to perform a single cationic monomer addition of a vinyl ether containing a pendant primary amine function (see the section on living cationic polymerizations for details).[102] Afterwards, this template was tested for controlled radical monoinsertion. Due to acid–amine supramolecular interactions, a single methacrylic acid monomer unit was successfully

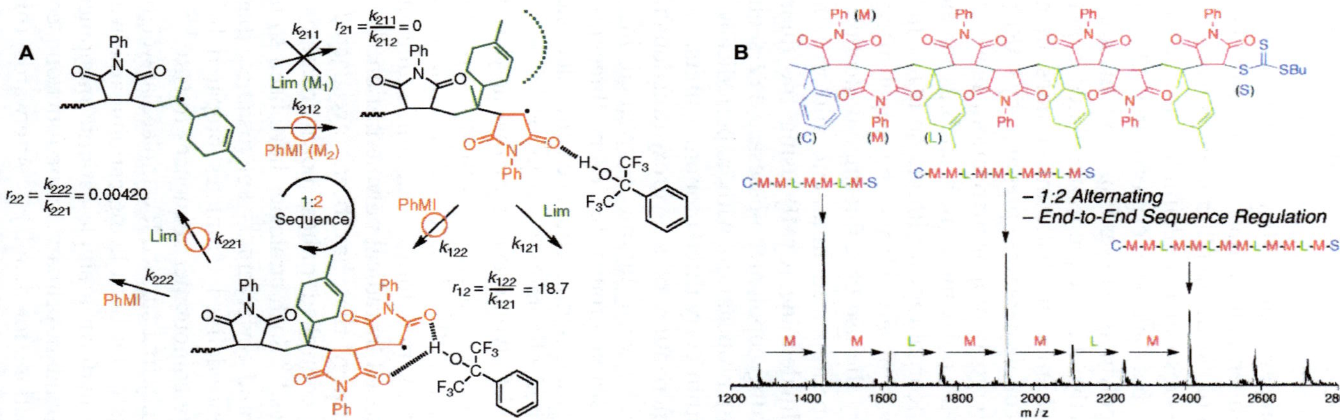

Figure 3.4 [A] AAB sequence-controlled radical copolymerization of *d*-limonene (M1) with phenylmaleimide (M2). [B] MALDI-TOF-MS spectrum of the copolymer of *d*-limonene and phenylmaleimide.
Adapted with permission from K. Satoh, M. Matsuda, K. Nagai and M. Kamigaito, *J. Am. Chem. Soc.*, 2010, **132**, 10003–10005.[97] Copyright (2010) American Chemical Society.

incorporated in the template. Although promising, this strategy is difficult to apply to longer sequence-defined templates.[104] Thus, a simpler approach employing preformed polymerizable templates was also investigated by Sawamoto and coworkers.[105,106] In this strategy, vinyl monomers were covalently or non-covalently immobilized on a cleavable template and polymerized by CRP in dilute conditions. For example, polymerizable acrylate and methacrylate moieties were attached in the 1,8-positions of naphthalene *via* ester linkages.[106] The rigid naphthalene core acts as a template and guides the acrylate/methacrylate alternating copolymerization. Although the non-templated radical copolymerization of these two monomers leads to random comonomer sequences, predominantly alternating sequences were obtained with the template. In order to increase the length of the sequence-controlled repeat unit, Sawamoto and coworkers have also investigated the complexation of metal ions with monomers.[105] For instance, two styrene derivatives and one 4-vinylpiridine were immobilized in a precise ABA sequence on a palladium atom. The cyclopolymerization of this template led to the synthesis of well-defined ABA periodic copolymers.

The CRP synthesis of multiblock copolymers, with blocks of very short average chain-length, is another interesting trend in the field of sequence-controlled polymers.[20] In general, it is relatively difficult to synthesize block copolymers longer than tetra- or penta-blocks using macroinitiator strategies.[107] Nevertheless, this situation has recently changed with the discovery of optimal CRP techniques leading to polymers with high-chain-end fidelity. For instance, Whittaker and coworkers have reported the synthesis of hexa- and octa-block copolymers by stepwise one-pot single-electron transfer living radical polymerization (SET-LRP).[108,109] Afterwards, a similar strategy was employed by Haddleton and coworkers for the synthesis of sequence-controlled multiblock glycopolymers.[110–113] Although the chain-end fidelity in SET-LRP is high, it is not sufficient to carry out a large number of block iterations as evidenced by the gradual increase of dispersity with the number of blocks. However, this was later improved with the use of photo-induced copper-mediated multiblock synthesis.[114–116] Perrier and coworkers also reported the RAFT synthesis of multiblock copolymers as long as icosablock copolymers.[117,118] In this approach, the chain-end fidelity was maximized by lowering the concentration of the radical initiator and thus reducing the probability of bimolecular terminations during the RAFT process.[119]

3.2.4 Ring-opening Polymerization

Some examples of sequence-controlled polymers prepared by ring opening polymerization of cyclic esters have been reported.[120] For instance, Thomas, Coates, and coworkers reported a very original method for the synthesis of alternating aliphatic polyesters.[121] These authors used a stereoselective yttrium-based organometallic catalyst to control comonomer sequences. It had been previously shown that the polymerization of a racemic mixture of β-butyrolactone in the presence of this catalyst led to the formation of syndiotactic polymers.[122] However, instead of using a racemic mixture of a given monomer, the authors copolymerized two different enantiopure β-lactone derivatives with opposite configurations. Therefore, this stereo-controlled process was transformed into a sequence-controlled polymerization. For example, it was demonstrated that the copolymerization of (*R*)-β-butyrolactone and (*S*)-4-ethoxymethyl-2-propiolactone led to the formation of copolymers with almost perfect alternating sequences. Furthermore, some preliminary degradation studies indicated that the control over comonomer sequences had a pronounced influence on the macroscopic properties of the formed polyesters. This concept was afterwards investigated by Carpentier, Guillaume and coworkers for the ring opening polymerization of β-malolactonates.[123] For instance, the alternating copolymerization of allyl- and benzyl-functionalized β-malolactonates was achieved. The hydrolysis of the benzyl-moiety in the final copolymers allowed the synthesis of precise polyacids.

3.2.5 Ring-opening Metathesis Polymerization

Ring opening metathesis polymerization (ROMP) is a living polymerization technique employing cycloalkenes as monomers.[124,125] In the 1990s, Cho and coworkers reported the ROMP of disubstituted cyclooctenes, wherein methyl and phenyl groups were substituted at the 4- and 6-positions respectively.[126] Interestingly, the resulting linear polymer could be considered as the alternating copolymer of butadiene with ethylene, propylene, and styrene. However, as these polymerizations are not regioselective, the resulting polymers were not truly sequence-specific. More recently, Hillmyer and coworkers reported the Grubbs catalyst mediated ROMP of 3-substituted cyclooctenes, which proceeds in a regio- and stereo-selective fashion. The monomers are connected in a head–tail fashion to form the polymer chain.[127] In particular, they have shown that ROMP of cyclooctene derivatives with substituents $R_1 \neq R_2 \neq R_3 \neq H$ and $R_1 = H$ leads to sequence-specific polymers with very high stereo- and regio-regularity.[128] Xia and Elling have also reported that the copolymerization of cyclopropene derivatives with low ring-strain containing cycloolefins proceeds in an alternating fashion since the monomers have a low tendency for homopolymerization. In general, cyclopropenes homopolymerize due to their high ring-strain; however, the presence of bulky substituents on cyclopropenes inhibits homoaddition and ensures single monomer addition to the Ru-based Grubbs catalyst. O'Reilly and coworkers reported an elegant kinetic strategy to control monomer sequences in the ROMP of norbornene derivatives.[129] In this approach, the reactivity differences between *exo*-and *endo*-norbornenes in ROMP was exploited for sequence regulation. The time-controlled addition of small amounts of fast reacting *exo*-norbornenes into the polymerization of a large excess of *endo*-norbornenes allowed the synthesis of local functional regions in the formed polymer chains.

In general, cycloalkenes exhibit ring strain, which makes them suitable monomers for ROMP. However, large cyclic monomers with low ring strain are difficult to polymerize. Recently, some approaches have been developed to polymerize such low strain cycloalkenes. One of them is entropy-driven ring opening metathesis polymerization (ED-ROMP), in which the concentration of the reaction medium is kept in such a way that the reaction favors chains over rings. In a recent study, Meyer and coworkers prepared a macrocyclic ring, having minimal sequences of lactic acid and glycolic acid, by a ring closing metathesis reaction as shown in Figure 3.5.[130] Afterwards, ED-ROMP of such low strain cyclic compounds resulted in the formation of sequence-controlled copolymers. Hawker and coworkers utilized relay metathesis polymerization as another strategy to polymerize such low ring strain macrocycles.[131] Here, close proximity of the cyclic olefin to an alkyne is the key that permits fast intramolecular cyclization, followed by rapid polymerization. This monomer design has allowed the preparation of sequence-controlled polymers with a variety of functional groups in the backbone such as ester, amide, or sulfonamide.

3.3 Sequence-controlled Polymers Prepared by Step-growth Polymerization

3.3.1 Acyclic Diene Metathesis Polymerization

Acyclic diene metathesis (ADMET) polymerization is an important technique for the preparation of precision polymers.[132] In this approach, α,ω-diene monomers are polymerized *via* a metathesis reaction, thus leading to the formation of linear unsaturated polymers that can be transformed by hydrogenation into saturated polyolefins. The ADMET process is a typical step-growth polymerization, in which high molecular weight polymers are obtained at nearly complete conversion of monomers. As a result, polymers with poorly controlled molecular weight and broad dispersity ($Đ \sim 2$) are obtained through this process. On the other hand, this method is very refined for controlling the microstructure of polymers. Indeed, by selecting dienes with defined backbones and substituents, it is possible to synthesize copolymers with precise periodic patterns. For instance, Wagener and coworkers have systematically studied the effect of both the position and nature of branching groups in branched polyethylene.[133–135] In addition, the same research group has reported the precise positioning of polar groups on polyolefin backbones.[136,137] More recently, Li and coworkers reported the ADMET synthesis of periodic vinyl copolymers containing γ-butyrolactone units in the main chain.[138,139] It was found that the periodic incorporation of rigid γ-butyrolactone motifs led to enhanced thermal properties (*i.e.* T_g and T_m) in comparison to linear polyethylene. In another example, the same group incorporated an ester functional group in the symmetric diene monomer (di(10-undecenyl) itaconate) by esterification of itaconic acid and 10-undecenol.[140] Interestingly, the reactive double bond from itaconic acid persisted under the polymerization conditions and could be modified after polymerization by Michael addition. These sequence-controlled polymers

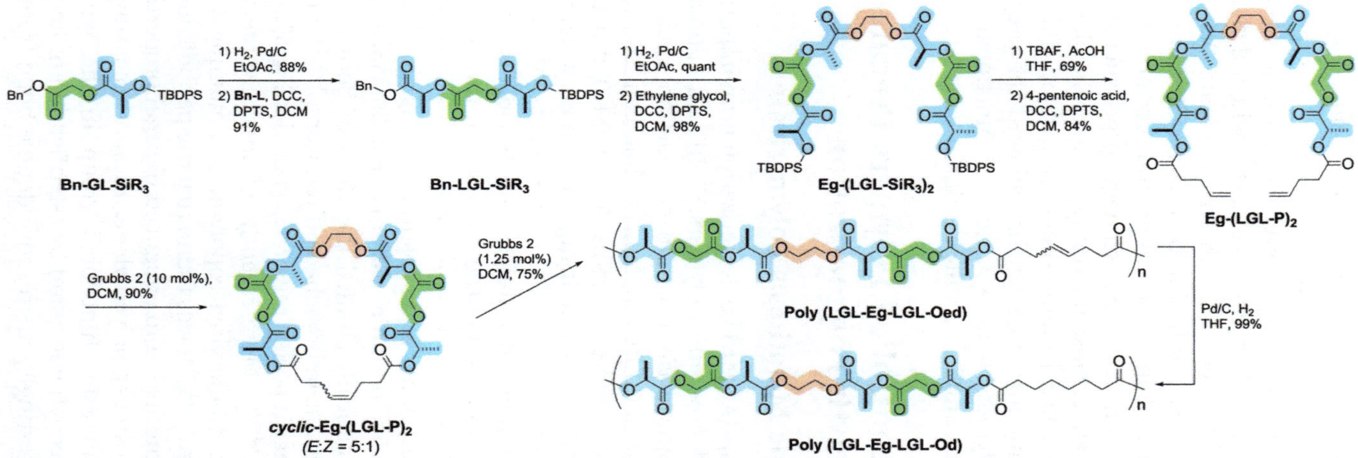

Figure 3.5 Preparation of sequence-controlled copolymers by entropy-driven ring opening metathesis polymerization. Reprinted with permission from R. M. Weiss, A. L. Short and T. Y. Meyer, *ACS Macro Lett.*, 2015, **4**, 1039–1043.[130] Copyright (2015) American Chemical Society.

were converted into self-degradable polyesters when the double bond of the itaconic esters was reacted with *sec*-butylamine *via* Michael addition reaction.

3.3.2 Click Step-growth Polymerization

Efficient click reactions such as copper-catalyzed azide–alkyne cycloaddition (CuAAC),[141] thiol-ene,[142] and thiol-yne[143] reactions are interesting chemistries for the preparation of periodic copolymers. The first example of a sequence-controlled periodic copolymer prepared by CuAAC was reported by our group.[144] As discussed in Section 3.2.3, the controlled radical copolymerization of large amounts of styrene with discrete amounts of *N*-substituted maleimides allows the synthesis of sequence-controlled polymers.[45] This chain-growth copolymerization concept can also be combined with a step-growth mechanism to synthesize periodic copolymers. Short polystyrene segments (*i.e.* DP ∼ 20) containing functional *N*-substituted maleimides in the middle of their chains were synthesized by sequence-controlled ATRP and subsequently transformed into α-alkyne-, ω-azido-heterotelechelic structures by post-polymerization modifications.[144] These reactive segments were afterwards polymerized by step-growth CuAAC to afford periodic copolymers (Figure 3.6). Besides CuAAC, some examples of periodic copolymers prepared by thiol-ene and thiol-yne chemistry have been reported. For instance, Du Prez and coworkers have reported very recently the synthesis of periodic microstructures by thiol-ene step-growth polymerization.[145] In this approach, α-thiolactone, ω-acrylate heterotelechelic polymers were first reacted with a functional amine to afford α-thiol, ω-acrylate reactive segments. The thiolactone-amine reaction permits the incorporation of a wide variety of functional substituents at the α-chain-end of the polymers. Afterwards, these functional segments were polymerized by thiol-ene step-growth chemistry. More recently, the group of Bowman has described the synthesis of nucleobase-functionalized periodic copolymers by thiol-ene polymerization.[146] Gao and coworkers also reported the synthesis of nucleobase-containing sequence-controlled polymers by click polymerization of dithiols with either diene or diyne monomers.[147]

3.3.3 Other Step-growth Approaches

Poly(lactic-*co*-glycolic acid) (PLGA) copolymers are important biodegradable polymers for biomedical applications. They are usually synthesized by ring opening polymerization of lactide and glycolide that leads to random sequences. In a series of recent papers, Meyer and coworkers have investigated a step-growth strategy to prepare sequence-controlled PLGA.[148–150] In these works, carboxylic acid, alcohol AB-type reactive monomers containing defined lactic, glycolic acid minimal sequences were synthesized and polymerized into periodic polyesters (Figure 3.7a). The resulting sequence-controlled copolymers exhibited very interesting biodegradation properties.

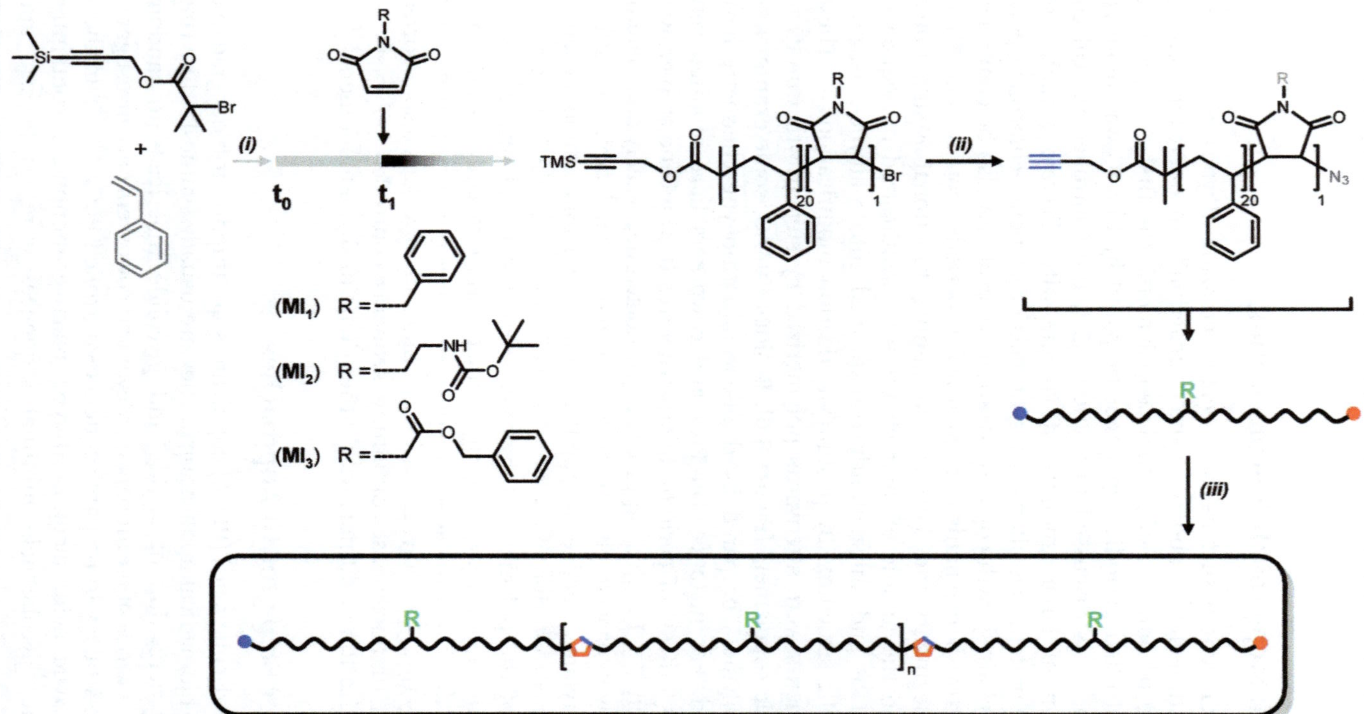

Figure 3.6 Synthesis of periodic copolymers by step-growth CuAAC of sequence-controlled α-alkyne-, ω-azido-heterotelechelic pre-polymers.
Reprinted with permission from M.-A. Berthet, Z. Zarafshani, S. Pfeifer and J.-F. Lutz, *Macromolecules*, 2010, **43**, 44–50.[144] Copyright (2010) American Chemical Society.

For instance, the *in vitro* hydrolytic degradation of PLGA with alternating sequences was found to be remarkably different from that of conventional random PLGA copolymers (Figure 3.7b), although the overall monomer composition was identical in both types of polymers.[149,151] Sequence control allows the sustained release of encapsulated guests and a more homogenous degradation profile, which are also essential for drug delivery applications.[150]

Kamigaito and coworkers reported the synthesis of sequence-regulated ABCC-vinyl copolymers by using metal-catalyzed step-growth radical polymerization.[152] This polymerization proceeds through a C–C bond forming reaction between an unconjugated C=C bond and the radical species generated through the metal-catalyzed activation of a C–Cl bond. For instance, the synthesis of periodic copolymers containing the ABCC repeating sequence vinyl-chloride-styrene-methyl acrylate-methyl acrylate was reported. It is interesting to note that the sequence-defined building-blocks utilized in this work can also be used as macromonomers in radical chain-growth polymerization to afford side-chain sequence-defined copolymers.[153] In a recent publication, Ramakrishnan and coworkers synthesized a copolymer of 1,20-eicosanediol and diethyl 2-propargyl-malonate *via* a melt-transesterification technique.[154] The primary structure of the copolymer contained periodic repetitions of polyethylene like segments, separated by ester linkages. Additionally, the copolymerization also ensured the periodic placement of reactive propargyl groups in the resulting polymer chains. For instance, a polyethylene immiscible functionality, such as an alkyl-fluoroalkyl segment, was introduced *via* post-polymerization modification.

3.3.4 Multicomponent Reactions

Multicomponent reactions (MCR) are reactions in which more than two reagents react either simultaneously or sequentially to form a single product with very high selectivity.[155] As the products of MCR contain a specific sequence of synthons (*i.e.* the order of MCR dictates the sequence in the product), such reactions are attractive for the preparation of sequence-controlled polymers. For instance, Li and coworkers have described the synthesis of periodic poly(ester amide)s *via* a step-growth polymerization based on the 3-component Passerini reaction.[156] In a following work, even more complex periodic patterns were synthesized using a combination of the Passerini reaction, thiol-ene reaction, and step-growth Passerini polymerization.[157] You and coworkers have described an efficient multicomponent polymerization based on the ring opening of thiolactone, followed by a thiol-Michael addition reaction (Figure 3.8a).[158] This reaction was used to prepare ACBCA dienes that were copolymerized with a CBC dithiol by thiol-ene step-growth polymerization, to afford ACBCA periodic copolymers. In a related work, You and coworkers used the same multicomponent reaction to prepare dialkyne ABCB motifs (Figure 3.8b) that were afterwards polymerized by

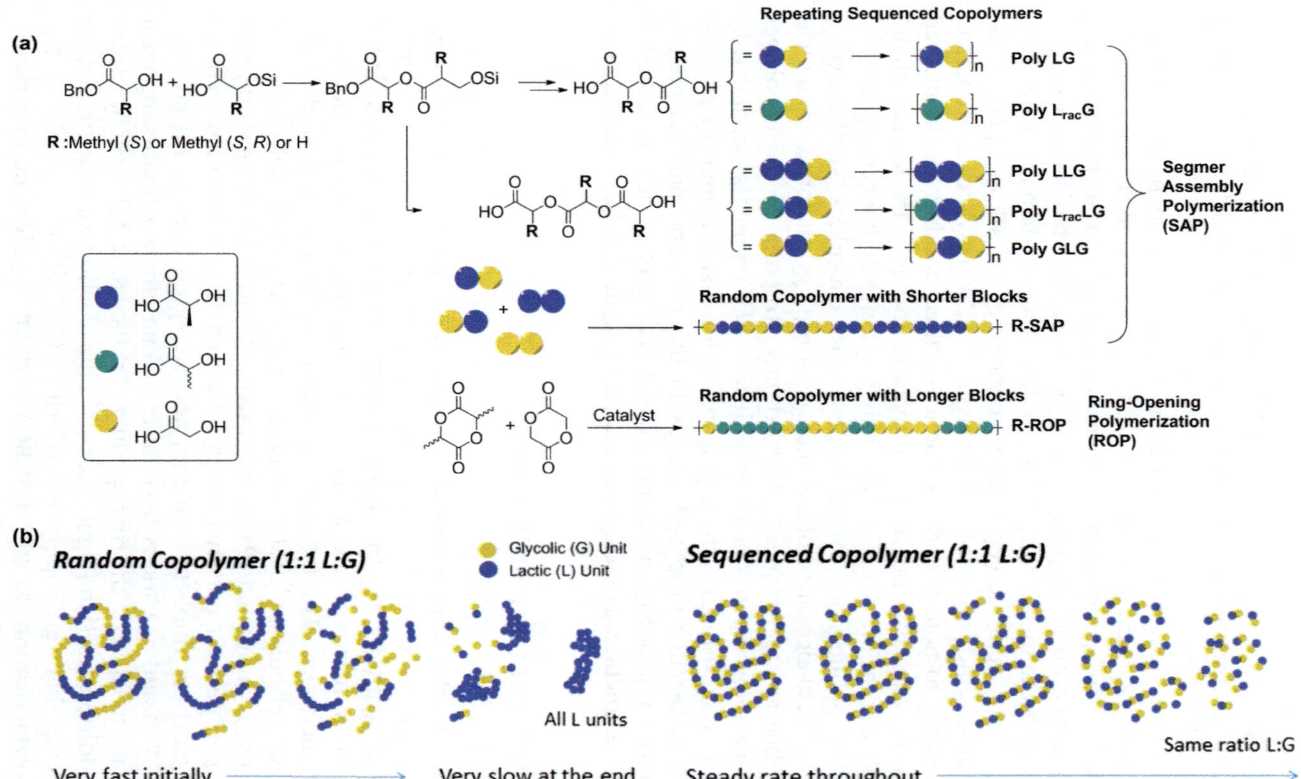

Figure 3.7 (a) Synthesis of sequence-controlled PLGA by step-growth polymerization. (b) Proposed degradation mechanism for random and sequence-controlled PLGAs.
Adapted with permission from J. Li, S. N. Rothstein, S. R. Little, H. M. Edenborn and T. Y. Meyer, *J. Am. Chem. Soc.*, 2012, **134**, 16352–16359.[150] Copyright (2012) American Chemical Society.

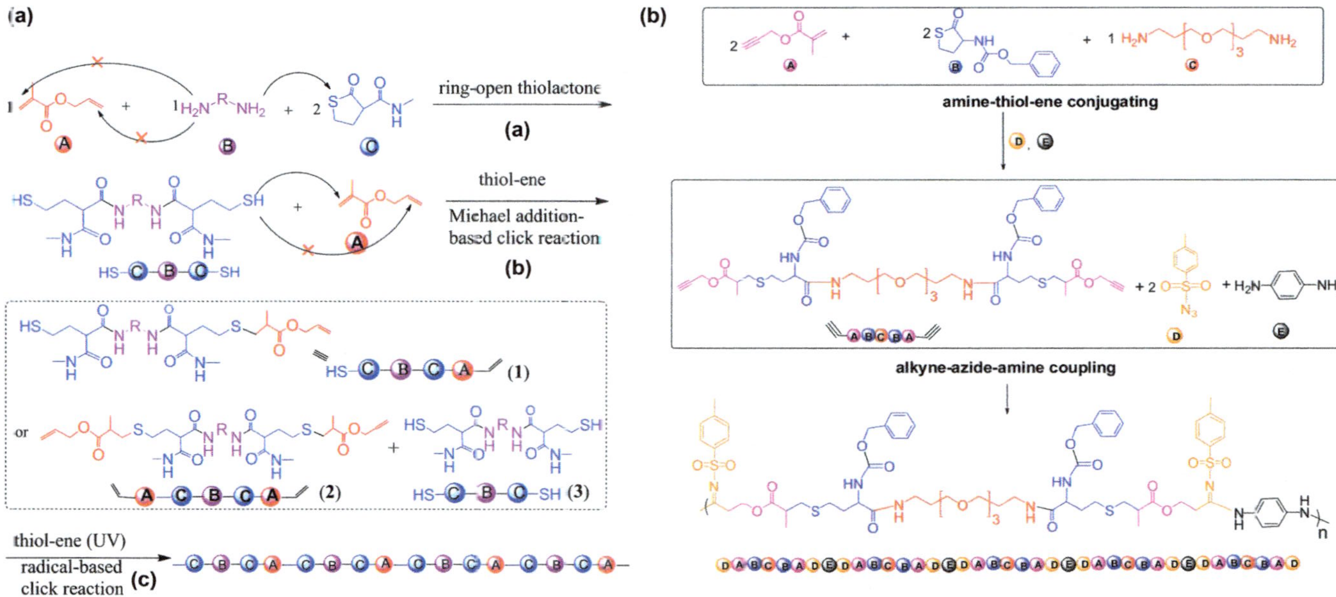

Figure 3.8 Synthesis of sequence-controlled periodic copolymers using multicomponent reactions. (a) Synthesis of periodic ACBC copolymers by a thiolactone ring-opening/thiol-Michael multicomponent reaction followed by thiol-ene step-growth polymerization.[158] Reproduced from ref. 158 with permission from the Royal Society of Chemistry. (b) Synthesis of periodic DABCBADE copolymers by a thiolactone ring-opening/thiol-Michael multicomponent reaction followed by alkyne–azide–amine step-growth polymerization. Adapted with permission from Z. Zhang, Y.-Z. You, D.-C. Wu and C.-Y. Hong, *Macromolecules*, 2015, **48**, 3414–3421.[159] Copyright (2015) American Chemical Society.

step-growth alkyne–azide–amine coupling to afford unprecedented sequence-controlled DABCBADE periodic copolymers.[159]

3.4 Sequence-controlled Polymers Prepared by Multi-step-growth Polymerization

3.4.1 Conventional Iterative Synthesis

A straightforward approach for synthesizing sequence-controlled polymers is to link monomers one-by-one (*i.e.* multi-step growth). This approach has been utilized as early as the 1940s for the solution synthesis of small peptides.[19] In such strategies, monomers bearing two different functional groups (AB-type) are used and therefore sequence-regulation is only made possible by implementing protection–deprotection steps.[24] As a consequence, such strategies are experimentally-demanding and require time to isolate the pure products. However, iterative synthesis has been considerably simplified by the introduction of crosslinked polymer supports (*i.e.* solid-phase chemistry) by Merrifield[160] and, afterwards, of other types of supports such as soluble polymer supports,[161] fluorous tags,[162] and ionic liquids.[163] In such approaches, mono-protected AB monomers are attached sequentially to the support. Thus, sequenced-defined oligomers can be easily prepared using repetitive coupling/deprotection cycles.[24] After each step, the use of a support greatly facilitates the purification (*i.e.* by using simple filtration or precipitation) and allows for considerable time savings. Historically, these approaches have been primarily developed and optimized for the synthesis of sequence-defined biological macromolecules such as peptides, oligosaccharides, and oligonucleotides. However, this strategy was afterwards extended to the synthesis of bio-hybrid polymers[164,165] and, more recently, to a wide variety of non-natural sequence-defined oligomers such as oligoureas, oligocarbamates, oligoesters, polyamines, polyamidoamines, and oligophosphates.[23] These recent developments are too broad to be comprehensively covered in the present book chapter and the reader is guided to previous reviews for additional information on the subject.[17,23,24] Nevertheless, it is worth specifying that those chemistries, which are very different from the ones employed for biopolymers synthesis, can be used to prepare sequence-defined oligomers. For example, Moore and coworkers have described the synthesis of sequence specific phenylacetylene oligomers.[166] Using silyl protected monomers and solid supported synthesis, different sequence-controlled phenyacetylene oligomers were obtained.[167] This example is one among many but illustrates the versatility of iterative synthesis for preparing sequence-controlled polymers. Alternatively, some efficient coupling chemistries, which have been optimized for biopolymer synthesis, can also be used for the preparation of non-natural sequence-defined polymers. For instance, Fmoc protocols, which have been introduced for peptide synthesis, have been employed to synthesize sequence-defined polymers containing both natural and non-natural building blocks.[168,169] Similarly, phosphoramidite chemistry, which has been developed for oligonucleotide synthesis,[170] has recently become popular for the preparation of non-natural sequence-defined polymers. For instance, Sleiman and coworkers have reported the synthesis of sequence-controlled oligonucleotide–oligomer bio-hybrids.[171] After the synthesis of short DNA sequences, oligomerization was continued by the stepwise attachment of commercially available oligonucleotide spacers as synthetic monomers, one by one, in order to obtain short oligonucleotide–polymer conjugates with hydrophilic and hydrophobic motifs. Our group has also recently reported the phosphoramidite synthesis of monodisperse non-natural information-containing macromolecules.[172] In particular, digitally-encoded polymers with high chain lengths were prepared on a DNA synthesizer.[173] Interestingly, these polymers contained ASCII-encoded text messages as shown in Figure 3.9. Using the automated phosphoramidite approach and two synthetic phosphoramidite monomers defined as 0 and 1 bits, binary-coded poly(phosphodiesters) containing 16, 24, 56, and 104 coded monomer units were easily synthesized. The use of automated synthesis with a large excess of phosphoramidite monomers, 1000 Å controlled pore glass, efficient capping, and an appropriate purification technique were the keys to this success. Thus, supported iterative synthesis remains so far the most appropriate way to access long monodisperse sequence-controlled macromolecules, as evidenced by this recent example.

3.4.2 Protecting-group-free Iterative Synthesis

As discussed in the previous section, conventional iterative synthesis requires, in general, the use of protecting groups, which complicates the experimental protocols. However, although mandatory for biopolymer synthesis, the use of protecting groups can be avoided for the synthesis of non-natural sequence-defined oligomers. As shown in Figure 3.10, various orthogonal iterative strategies, such as AA + BB, AA + BC, and AB + CD approaches, have been proposed in recent years. All these approaches have in common that they involve two successive chemoselective coupling steps employing two different types of reactive building blocks. As a consequence, the monomers can be attached to another one sequentially without requiring the use of protecting groups. All these concepts have been summarized in detail in a recent review and are therefore only briefly described in this book chapter.[174]

The so-called AA + BB approach (Figure 3.10) involves the successive ligation of bifunctional monomers AA and BB, which are used in large excess as compared to the immobilized reacting sites.[175] For instance, this method was used by Rose and coworkers for the synthesis of monodisperse nylon-type oligoamides using succinic acid anhydride and 1,6-diaminohexane as AA and BB monomers, respectively.[176] However; the use of a large excess of monomers and the activator was mandatory to avoid incomplete conversion and intermolecular coupling, which would lead to polydisperse oligomers. In the same spirit, the group of Börner described the synthesis of

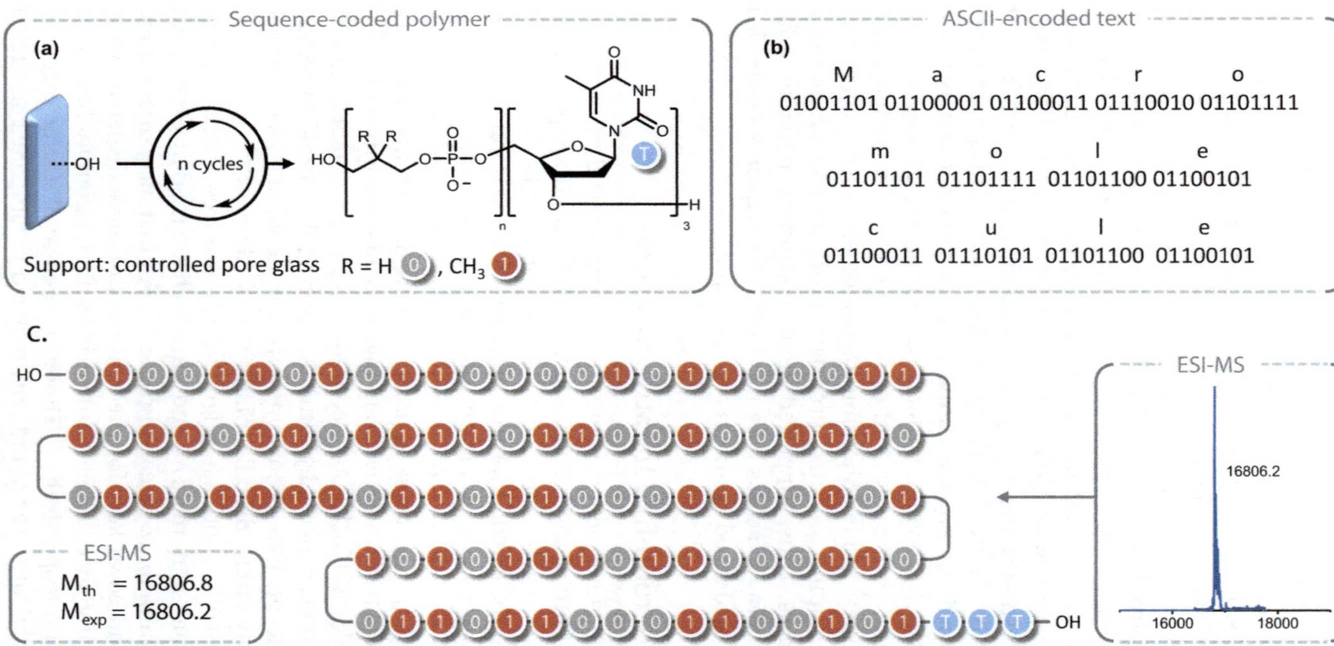

Figure 3.9 (a) Synthesis of digitally encoded polymers by automated phosphoramidite chemistry on a controlled pore glass solid support. In such strategies, two comonomers are intentionally defined as 0 and 1 bits. The monomer T is a thymine nucleotide, which is used to start the sequence and facilitates the polymer characterization. (b) Extended-ASCII encoded text that was implemented in the primary structure of a polymer using the monomer-based digital code. (c) Schematic representation of the primary structure of the corresponding sequence-coded polymer and its characterization by negative mode electrospray mass spectrometry (ESI-MS).
Adapted from A. Al Ouahabi, M. Kotera, L. Charles and J.-F. Lutz, *ACS Macro Lett.*, 2015, **4**, 1077–1080.[173] Copyright (2015) American Chemical Society.

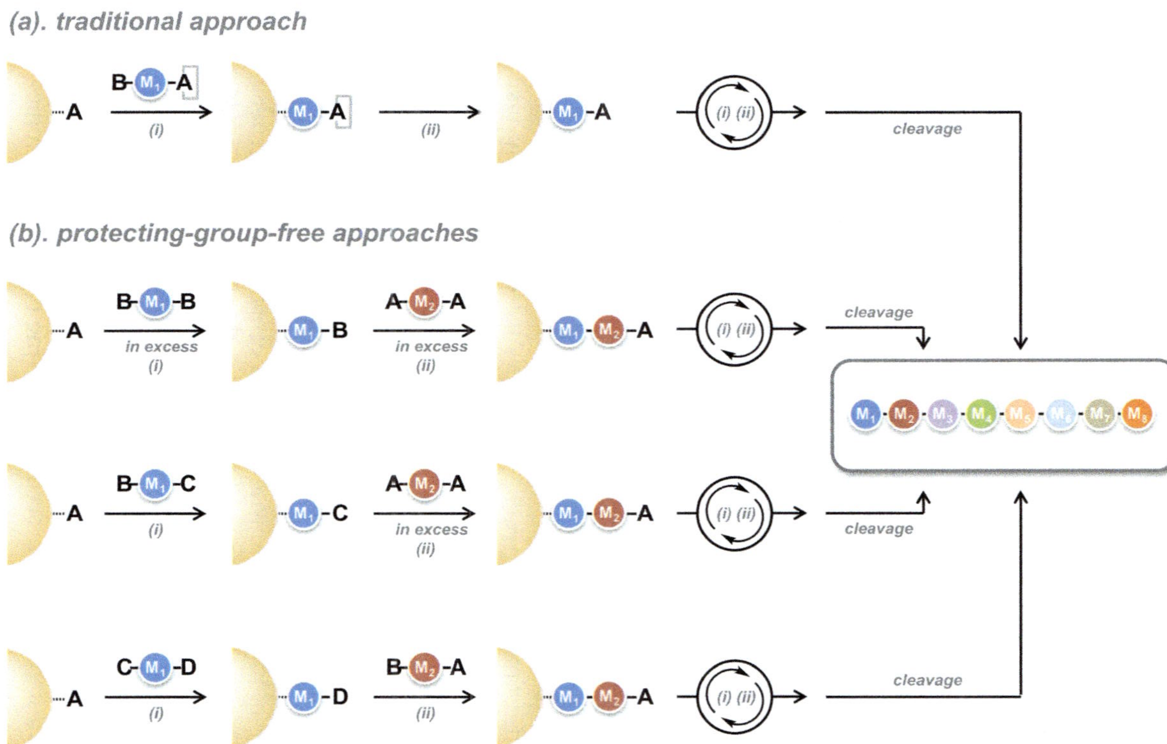

Figure 3.10 Solid-phase iterative strategies for preparing sequence-defined polymers. For clarity, all these concepts are drawn on solid-supports (orange spheres) but can also be performed on other types of support. (a) Traditional approach employing AB building-blocks. (i) Coupling step, in which function A reacts with B. The function A of the monomer is protected during that step, and (ii) deprotection. (b) Protecting-group-free approaches. (Top) AA + BB approach, in which function A reacts with B. Bifunctional monomers are used in large excess in steps (i) and (ii). (Middle) AA + BC approach. Function A reacts only with B under coupling conditions (i). Function A reacts only with C under coupling conditions (ii). Bifunctional monomer AA is used in large excess in step (ii). (Bottom) AB + CD approach. Function A reacts only with C. Function B reacts only with D. Coupling steps (i) and (ii) are chemoselective.
Reprinted from ref. 174 with permission from John Wiley and Sons. Copyright © 2015 WILEY-VCH Verlag GmbH & Co. KGaA, Weinheim.

monodisperse sequence-controlled oligo(amidoamines) using various diamines and succinic anhydride.[177–179] The AA + BC approach (Figure 3.10) is related to the AA + BB approach, but it employs three reactive functions, A, B, and C, instead of two. For example, Alabi and Porel synthesized monodisperse sequence-defined polymers using dithiols (AA) and building blocks bearing acrylamide and allyl functions (BC).[180] These two types of double bonds B and C can react orthogonally with a thiol function, either by photoinitiated thiol-ene coupling or by Michael addition. The repetition of these two coupling steps on a fluorous tag leads to the synthesis of sequence-defined oligomers. Beyond the AA + BB and AA + BC approaches that employ at least one reactant that may lead to intermolecular side reactions, the AB + CD approach is a fully chemoselective process involving four reactive functions (A, B, C, and D) and two orthogonal coupling steps (Figure 3.10). A variant of this approach, *i.e.* the so-called submonomer strategy, has been particularly used by Zuckermann and coworkers for the synthesis of sequence-defined peptoids (*i.e.* poly(*N*-substituted glycines)).[25] This strategy involves a non-symmetrical bromo-carboxylic acid bifunctional monomer and a primary amine derivative, and is performed in two-steps, *i.e.* substitution of the halogen derivative by the nucleophilic amine and acetylation of the secondary generated amine by the bromocarboxylic acid monomer using efficient coupling agents. The coupling steps are repeated until the desired sequence is reached. It is, however, important to note that the starting monomers are bearing only three reactive functions and the fourth is generated *in situ* during the addition steps. Using this non symmetrical chemoselective strategy, a wide variety of sequence-controlled peptidomimetics have been prepared over the past years.[181–183] The AB + CD approach has also been used by our group for the synthesis of sequence-defined poly(triazole amide)s.[184] In particular, the synthesis of information-containing macromolecules was achieved by this iterative strategy.[185] The concept was also recently extended to the synthesis of digitally-encoded poly(alkoxyamine amide)s.[186] Using two different dibromo-acid anhydrides defined as 0-bit and 1-bit and an amino-nitroxide building-block as a spacer, the alternating iterative amidification between the amino-spacer and acid anhydride and the radical coupling of the nitroxide with a carbon-centered radical generated by copper-activation of the alkyl bromide enable the synthesis of monodisperse sequence-coded polymers. This method is rapid and very efficient for the preparation of binary-encoded information-containing macromolecules. Moreover, the digital information stored in the polymer chains can be easily decoded by tandem mass spectrometry[187,188] but also erased by thermal degradation. Du Prez and coworkers have also recently reported an easy two-step iterative procedure involving one primary amine building-block and another building-block containing both thiolactone and vinyl functions.[189] The aminolysis of the thiolactone allows the formation of a fourth reactive function (*i.e.* a thiol), which can be engaged in thiol-ene coupling. The repetition of these steps on a solid support allows the efficient synthesis of monodisperse sequence-defined oligomers.

All the protecting-group-free approaches described in Figure 3.10 involve two different types of monomers and two successive orthogonal reactions. However, more complex multicomponent reactions can also be used to attain sequence-defined oligomers. For instance, Meier and coworkers have described in an elegant series of papers the synthesis of sequence-defined monodisperse polymers using iterative coupling steps based on multi-component reactions such as the three-component Passerini or four-component Ugi reactions.[190–192] For example, an orthogonal strategy based on successive Passerini and thiol-ene coupling steps has been reported.[190] This strategy involves three main types of building blocks, *i.e.* one bifunctional monomer bearing thiol and carboxylic acid functions, one monomer containing both alkene and aldehyde functions, and a monofunctional isocyanide monomer. The repetition of these coupling steps allows the synthesis of uniform polymers with controlled monomer sequences. This protecting-group-free iterative strategy has also been recently extended to a four-component reaction.[191] Sequence-controlled polymers were obtained in this work using successive Ugi and thiol-ene coupling steps.

3.4.3 Successive Radical Insertion

Another multi-step growth strategy explored recently by polymer chemists to achieve well-defined polymers with controlled monomer sequences is the successive monoinsertion of vinyl monomers in a radical polymerization. This goal is, however, relatively difficult to achieve due to the highly reactive nature of radicals. The first example of this type was reported by Huang and coworkers.[193] These authors studied the atom-transfer radical addition (ATRA) of allyl alcohol. This monomer cannot polymerize under ATRA conditions and is therefore only mono-inserted. However, after performing this step, if the hydroxyl methyl side chain is transformed into a carboxylic acid, which is afterwards esterified, the terminal bond becomes active for another allyl alcohol single addition. Thus, iterative cycles of single-monomer addition and side-group functionalization allows the synthesis of controlled monomer sequences. Moad and coworkers have also studied the successive insertion of monomers in a trithiocarbonate-mediated RAFT process.[194,195] For instance, the insertion of a styrene monomer unit followed by the insertion of a *N*-isopropyl acrylamide single unit has been described.[194] Although very interesting, this approach remains unfortunately limited to the formation of very short oligomers such as dimers and seems very difficult to extend to longer monomer sequences. Taking these limitations into account, Junkers and coworkers have described an efficient strategy for the preparation of monodisperse sequence-defined oligomers by either RAFT[196] of SET-LRP.[197] This approach combines a near-monoinsertion CRP process, in which one molar equivalent of monomer is used as compared to the initiator, with a preparative recycling size exclusion chromatography (SEC) purification technique. After performing the monomer insertion step, a polydisperse sample is usually obtained. This mixture is then purified by online SEC to afford the pure mono-inserted product. Afterwards, this purified oligomer is involved in the next iterative cycle. Monodisperse sequence-defined RAFT tetramers and SET-LRP pentamers have been successfully obtained *via* this approach.

3.4.4 Convergent and Divergent Strategies

As discussed in the previous two sections, monodisperse sequence-defined polymers can be efficiently prepared by monomer-by-monomer iterative coupling. However, such approaches remain time-consuming to obtain long sequences. This situation can be improved by using convergent or divergent strategies, in which the overall number of steps is reduced. For instance, convergent synthesis has been extensively used for the total synthesis of proteins.[198] Recently, these approaches have been also studied for the synthesis of monodisperse non-natural polymers. One of the simplest strategies to afford monodisperse polymers is convergent–divergent iterative exponential growth (IEG), in which mono-protected bifunctional monomers are first coupled to another. The resulting dimer is then divided in two batches, which are selectively deprotected on one terminus and coupled to each other, thus affording a tetramer, which is then involved in a similar deprotection/coupling cycle. The cycle is repeated a certain number of times until the desired length is reached. This amplification strategy was tested for several years for the synthesis of monodisperse oligomers and was, in particular, recently rejuvenated by Hawker and coworkers for the synthesis of monodisperse aliphatic polyesters.[199] For instance, the synthesis of 64-mer monodisperse poly(ε-caprolactone)up was easily achieved *via* IEG. However, this approach remained limited to the synthesis of homopolymers.

Recently, it was extended to the synthesis of monodisperse alternating and periodic copolymers by Johnson, Jamison, and coworkers.[200,201] In particular, these authors combined the IEG concept with an online flow methodology that greatly simplified the approach (Figure 3.11).[200] This Flow-IEG strategy was, for example, used for the synthesis of sequence-controlled poly(triazole ester)s containing repeating hydrophilic and hydrophobic motifs. Such polymers were prepared by exponential iterative CuAAC using building blocks with adequate protecting groups or latent reactive functions. The same methodology was afterwards applied to chiral monomers and allowed the synthesis of monodisperse periodic copolymers with controlled tacticity.[201] Barner-Kowollik and coworkers have also recently described a related divergent photochemical strategy allowing the synthesis of periodic monodisperse copolymers.[202] In that case, the iterative growth was not exponential but performed in two directions from a bifunctional core. Beyond the synthesis of periodic copolymers, convergent–divergent strategies can also be used to simplify the synthesis of sequence-coded aperiodic macromolecules. For example, our group has recently described the synthesis of digitally-encoded poly(triazole amide)s by stepwise ligation of oligomers containing coded dyads.[203] Such an approach is only partly convergent

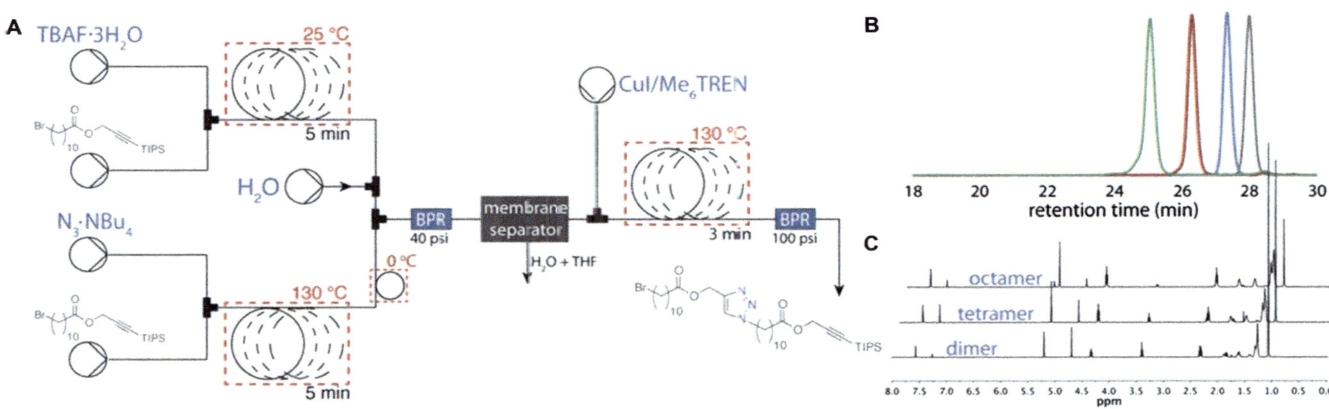

Figure 3.11 Flow iterative growth strategy reported by Johnson, Jamison, and coworkers for the synthesis of monodisperse periodic copolymers.
This figure is reprinted with permission from F. A. Leibfarth, J. A. Johnson and T. F. Jamison, *Proc. Natl. Acad. Sci.* U.S.A., 2015, **112**, 10617–10622.[200]

because it still utilizes iterative coupling steps. However, it allows a reduction by half of the number of coupling steps as compared to a traditional monomer-by-monomer approach. This concept was also expanded to the synthesis of sequence-coded poly(alkoxyamine amide)s, a very interesting new class of information-containing macromolecules.[204] To achieve that, a library of four α-acid, ω-bromo oligo(alkoxyamine amide)s containing all possible binary dyads, *i.e.* 00, 01, 10, and 11, was first synthesized and characterized. These building blocks were afterwards coupled together by stepwise iterative ligation on a solid support. Using this simple library of dyads, any desired binary sequence can be potentially synthesized using a reduced number of coupling steps.

3.5 Use of Templates and Catalytic Molecular Machines

Asides from chain-growth, step-growth, and multi-step growth strategies, sequence-defined polymers may also be prepared using polymerization concepts inspired by biological polymerizations such as replication, transcription, and translation. For instance, sequence-defined templates can be used for monomer sequence regulation in non-natural polymerizations. Alternatively, catalytic molecular machines inspired by biocatalysts such as enzymes and ribozymes have been tested for the synthesis of sequence-controlled polymers. These developments are summarized in this last section of the chapter.

3.5.1 Template-assisted Sequence-controlled Polymerization

A vast amount of literature dealing with the non-enzymatic translation of DNA sequences into sequence-defined xeno-nucleic acids, such as peptide nucleic acid (PNA), threose nucleic acid, and hexitol nucleic acid, has been described. This area of research is particularly broad and the reader is therefore referred to a specialized review for additional information on the topic.[32] In the present section, the synthesis of non-natural polymers using nucleic-acid templates is more specifically discussed.[205] Schuster and coworkers reported the DNA-templated synthesis of sequence-defined conjugated oligomers.[206,207] For example, two complementary DNA strands, in which some cytosine bases were modified by either aniline or pyrrole reactive monomers, were hybridized in order to bring these monomers at an optimal distance for sequence-defined polymerization.[207] Block and alternating sequences were synthesized using this approach. More recently, Liu and coworkers have shown that the sequences of nucleic acid templates can be translated into non-natural sequences using PNA adapters.[208] In this work, polymer building blocks without any structural relationship to oligonucleotides, such as polyethylene glycol, α-(D)-peptides, and β-peptides, were attached by a cleavable linker to programmed PNA motifs capable to recognize specific sequences of a DNA template (Figure 3.12). PNA/DNA

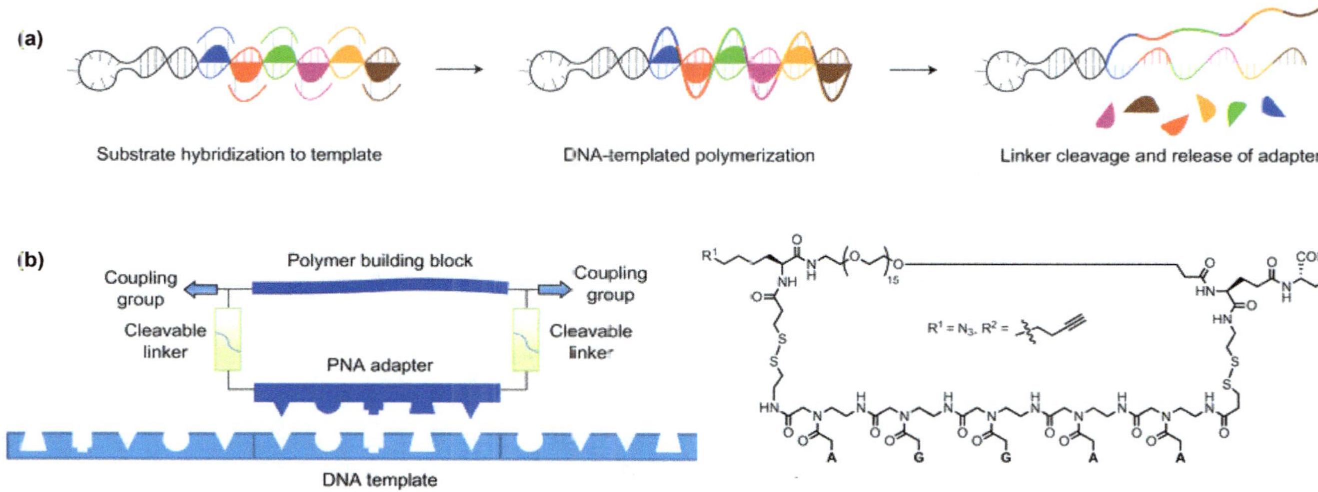

Figure 3.12 Translation of nucleic acid sequences into synthetic sequence-defined polymers using PNA adapters. (a) Hybridization of macrocyclic substrates with codons on a DNA template, organizing synthetic polymer building blocks along the template. Oligomerization of these pre-organized substrates and cleavage releases the PNA adapters and liberates the synthetic polymer product. (b) Representation of a macrocyclic substrate for the translation system in (b).
Adapted from J. Niu, R. Hili and D. R. Liu, *Nat. Chem.*, 2013, **5**, 282–292 with permission from Nature Publishing Group.[208]

hybridization allowed the alignment of the reactive non-natural building blocks in a specific sequence that could be polymerized. After polymerization of the buildings blocks, the resulting synthetic sequence-controlled polymer was cleaved from the PNA adapters.

Another interesting approach for sequence-regulation using templates has been reported by Turberfield, O'Reilly, and coworkers.[209–212] In this work, two complementary DNA single strands were functionalized at the 5′ or 3′ ends with reactive monomers such that, after DNA hybridization, the reactive groups were brought in close proximity and reacted. After reaction, the DNA complex was disrupted by strand exchange and the strand bearing the formed growing oligomer was hybridized with a new single strand containing another reactive functional monomer, thus allowing the formation of a defined non-natural sequence. This hybridization/reaction was reproduced a certain number of times in order to afford sequence-defined oligomers. For instance, as proof-of-concept, sequence-defined tetramers were synthesized by Wittig reactions between modified DNA strands containing either phosphine ylide or aldehyde functional groups.

3.5.2 Rotaxane-based Catalytic Machines

Another interesting strategy, but up to now underexplored for sequence-regulation in polymer science, is the use of molecular catalytic machines. Leigh and coworkers have recently reported the synthesis of sequence-defined oligopeptides using rotaxane-based machines that mimic ribosomal protein synthesis.[213,214] In this approach, a macrocycle bearing a thiolate group moves on a molecular track formed by an oligotriazole-amide bearing immobilized amino acids (Figure 3.13). The ring picks up and polymerizes the amino acids encountered on its way to synthesize a sequence-defined peptide, following the sequence-ordered defined by the 1D track. In particular, the thiolate initiator takes up the amino acids and brings them to a catalytic site where the peptide oligomerization takes place. After this transfer, the thiolate group is regenerated and can undergo a new cycle. This process continues until all the amino acids immobilized on the tracks are consumed and polymerized. Although only very short oligomers have been prepared so far using this approach, this elegant example highlights the relevance of molecular machines for sequence-controlled polymer synthesis.

3.6 Outlook

All the literature examples highlighted in this chapter indicate that the field of sequence-controlled polymers has significantly evolved during the last five years. First of all, several new approaches for monomer sequence regulation in synthetic polymerizations have been reported. For instance, interesting routes for controlling monomer sequences in chain-growth and step-growth polymerizations have been described, although the polymers

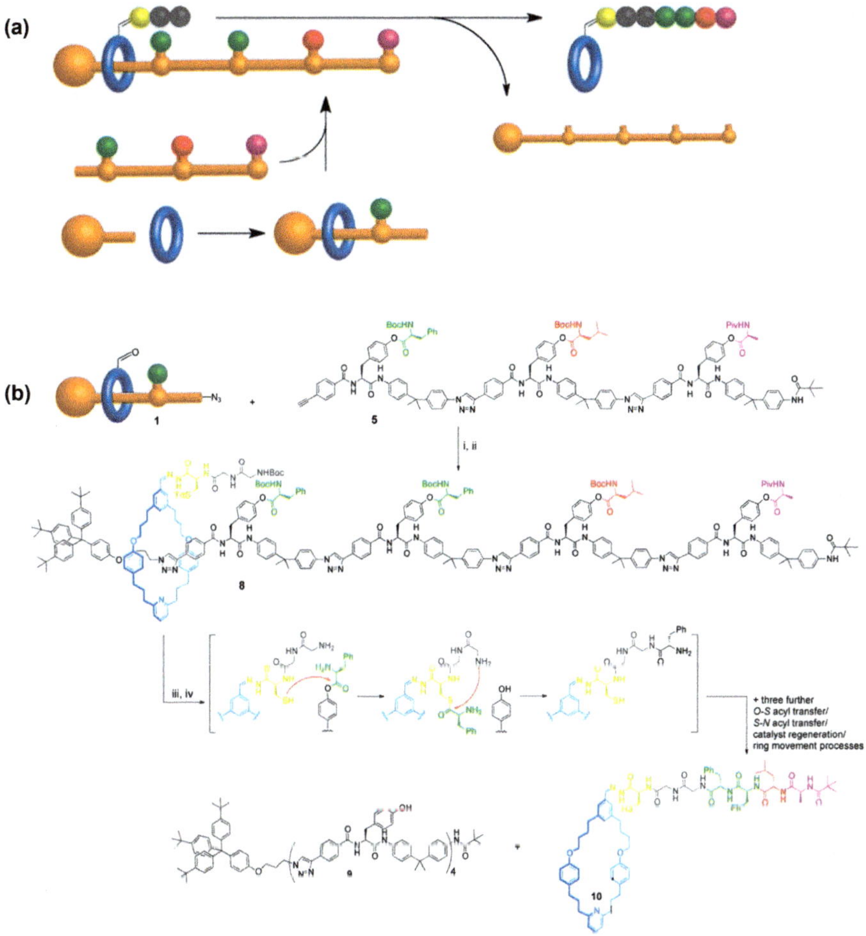

Figure 3.13 Synthesis of a sequence-defined oligopeptide using a four-barrier rotaxane molecular machine.
Adapted with permission from ref. 214.

obtained by such means are usually not perfectly molecularly defined. Furthermore, monodisperse sequence-defined polymers can be obtained using either iterative synthesis or bio-inspired concepts such as templated polymerizations or molecular machines. Perhaps even more importantly, new types of sequence-controlled polymers are now available. Beyond classical copolymers such as statistical, block, alternating, and periodic copolymers, it is now possible to synthesize unprecedented sequence-coded polymers with adjustable properties.[34,35,215,216] These recent developments open up a new range of applications for synthetic polymers in, for example, the fields of data storage, catalysis, nano-electronics, and nanomedicine.

References

1. G. M. Whitesides and B. Grzybowski, *Science*, 2002, **295**, 2418–2421.
2. A. E. Barron and R. N. Zuckerman, *Curr. Opin. Chem. Biol.*, 1999, **3**, 681–687.
3. E. Dujardin and S. Mann, *Adv. Eng. Mater.*, 2002, **4**, 461–474.
4. C. Sanchez, H. Arribart and M. M. Giraud Guille, *Nat. Mater.*, 2005, **4**, 277–288.
5. U. G. K. Wegst, H. Bai, E. Saiz, A. P. Tomsia and R. O. Ritchie, *Nat. Mater.*, 2015, **14**, 23–36.
6. S. J. Rowan, *Nat. Chem.*, 2009, **1**, 347–348.
7. Y. Zhao, Z. Xie, H. Gu, C. Zhu and Z. Gu, *Chem. Soc. Rev.*, 2012, **41**, 3297–3317.
8. P. Cordier, F. Tournilhac, C. Soulie-Ziakovic and L. Leibler, *Nature*, 2008, **451**, 977–980.
9. M. A. C. Stuart, W. T. S. Huck, J. Genzer, M. Muller, C. Ober, M. Stamm, G. B. Sukhorukov, I. Szleifer, V. V. Tsukruk, M. Urban, F. Winnik, S. Zauscher, I. Luzinov and S. Minko, *Nat. Mater.*, 2010, **9**, 101–113.
10. T.-S. Wong, S. H. Kang, S. K. Y. Tang, E. J. Smythe, B. D. Hatton, A. Grinthal and J. Aizenberg, *Nature*, 2011, **477**, 443–447.
11. J. D. Hartgerink, E. Beniash and S. I. Stupp, *Science*, 2001, **294**, 1684–1688.
12. O. Ikkala and G. ten Brinke, *Science*, 2002, **295**, 2407–2409.
13. Z. Li, E. Kesselman, Y. Talmon, M. A. Hillmyer and T. P. Lodge, *Science*, 2004, **306**, 98–101.
14. A. H. Groschel, A. Walther, T. I. Lobling, F. H. Schacher, H. Schmalz and A. H. E. Muller, *Nature*, 2013, **503**, 247–251.
15. J.-F. Lutz and H. G. Börner, *Macromol. Rapid Commun.*, 2011, **32**, 113–114.
16. N. Badi, D. Chan-Seng and J.-F. Lutz, *Macromol. Chem. Phys.*, 2013, **214**, 135–142.
17. J.-F. Lutz, *Polym. Chem.*, 2010, **1**, 55–62.
18. J.-F. Lutz, *Nat. Chem.*, 2010, **2**, 84–85.
19. J.-F. Lutz, in *Sequence-Controlled Polymers: Synthesis, Self-Assembly, and Properties*, American Chemical Society, 2014, vol. 1170, pp. 1–11.
20. J.-F. Lutz, *ACS Macro Lett.*, 2014, **3**, 1020–1023.
21. N. Hadjichristidis, S. Pispas and G. Floudas, *Block Copolymers: Synthetic Strategies, Physical Properties, and Applications*, John Wiley & Sons, Inc., Hoboken, 2003.
22. J.-F. Lutz, T. Pakula and K. Matyjaszewski, in *Advances in Controlled/Living Radical Polymerization*, American Chemical Society, 2003, vol. 854, pp. 268–282.
23. N. Badi and J.-F. Lutz, *Chem. Soc. Rev.*, 2009, **38**, 3383–3390.
24. J.-F. Lutz, M. Ouchi, D. R. Liu and M. Sawamoto, *Science*, 2013, **341**, 1238149.
25. R. N. Zuckermann, J. M. Kerr, S. B. H. Kent and W. H. Moos, *J. Am. Chem. Soc.*, 1992, **114**, 10646–10647.

26. C. Y. Cho, E. J. Moran, Cherry, J. C. Stephans, S. P. Fodor, C. L. Adams, A. Sundaram, J. W. Jacobs and P. G. Schultz, *Science*, 1993, **261**, 1303–1305.
27. S. H. Gellman, *Acc. Chem. Res.*, 1998, **31**, 173–180.
28. G. Guichard and I. Huc, *Chem. Commun.*, 2011, **47**, 5933–5941.
29. P. E. Nielsen, M. Egholm, R. H. Berg and O. Buchardt, *Science*, 1991, **254**, 1497–1500.
30. D. M. Rosenbaum and D. R. Liu, *J. Am. Chem. Soc.*, 2003, **125**, 13924–13925.
31. R. E. Kleiner, Y. Brudno, M. E. Birnbaum and D. R. Liu, *J. Am. Chem. Soc.*, 2008, **130**, 4646–4659.
32. Y. Brudno and D. R. Liu, *Chem. Biol.*, 2009, **16**, 265–276.
33. V. B. Pinheiro, A. I. Taylor, C. Cozens, M. Abramov, M. Renders, S. Zhang, J. C. Chaput, J. Wengel, S.-Y. Peak-Chew, S. H. McLaughlin, P. Herdewijn and P. Holliger, *Science*, 2012, **336**, 341–344.
34. H. M. Colquhoun and J.-F. Lutz, *Nat. Chem.*, 2014, **6**, 455–456.
35. J.-F. Lutz, *Macromolecules*, 2015, **48**, 4759–4767.
36. *Solid-Phase Synthesis: A Practical Guide*, ed. S. A. Kates and F. Albericio, CRC Press, New York, 2000.
37. M. Szwarc, *Nature*, 1956, **178**, 1168–1169.
38. N. Hadjichristidis, M. Pitsikalis, S. Pispas and H. Iatrou, *Chem. Rev.*, 2001, **101**, 3747–3792.
39. S. Aoshima and S. Kanaoka, *Chem. Rev.*, 2009, **109**, 5245–5287.
40. K. Matyjaszewski and J. Xia, *Chem. Rev.*, 2001, **101**, 2921–2990.
41. G. Moad, E. Rizzardo and S. H. Thang, *Polymer*, 2008, **49**, 1079–1131.
42. M. Ouchi, T. Terashima and M. Sawamoto, *Chem. Rev.*, 2009, **109**, 4963–5050.
43. K. Matyjaszewski, *Science*, 2011, **333**, 1104–1105.
44. J. Nicolas, Y. Guillaneuf, C. Lefay, D. Bertin, D. Gigmes and B. Charleux, *Prog. Polym. Sci.*, 2013, **38**, 63–235.
45. J.-F. Lutz, *Acc. Chem. Res.*, 2013, **46**, 2696–2705.
46. F. A. Leibfarth, K. M. Mattson, B. P. Fors, H. A. Collins and C. J. Hawker, *Angew. Chem., Int. Ed.*, 2013, **52**, 199–210.
47. A. Hirao, R. Goseki and T. Ishizone, *Macromolecules*, 2014, **47**, 1883–1905.
48. H. Yuki and Y. Okamoto, *Polym. J.*, 1970, **1**, 13–18.
49. L. R. Hutchings, P. P. Brooks, D. Parker, J. A. Mosely and S. Sevinc, *Macromolecules*, 2015, **48**, 610–628.
50. A. Natalello, J. N. Hall, E. A. L. Eccles, S. M. Kimani and L. R. Hutchings, *Macromol. Rapid Commun.*, 2011, **32**, 233–237.
51. W. Sang, H. Ma, Q. Wang, X. Hao, Y. Zheng, Y. Wang and Y. Li, *Polym. Chem.*, 2016, **7**, 219–234.
52. Y. Zhang, J. Li, X. Li and J. He, *Macromolecules*, 2014, **47**, 6260–6269.
53. J. Li and J. He, *ACS Macro Lett.*, 2015, **4**, 372–376.
54. H. Mutlu and J.-F. Lutz, *Angew. Chem., Int. Ed.*, 2014, **53**, 13010–13019.
55. A. Natalello, A. Alkan, P. von Tiedemann, F. R. Wurm and H. Frey, *ACS Macro Lett.*, 2014, **3**, 560–564.

56. A. Natalello, M. Werre, A. Alkan and H. Frey, *Macromolecules*, 2013, **46**, 8467–8471.
57. M. Minoda, M. Sawamoto and T. Higashimura, *Polym. Bull.*, 1990, **23**, 133–139.
58. K. Satoh, S. Saitoh and M. Kamigaito, *J. Am. Chem. Soc.*, 2007, **129**, 9586–9587.
59. Y. Ishido, R. Aburaki, S. Kanaoka and S. Aoshima, *Macromolecules*, 2010, **43**, 3141–3144.
60. M. Kawamura, A. Kanazawa, S. Kanaoka and S. Aoshima, *Polym. Chem.*, 2015, **6**, 4102–4108.
61. K. Matyjaszewski and N. V. Tsarevsky, *J. Am. Chem. Soc.*, 2014, **136**, 6513–6533.
62. K. Matyjaszewski, M. J. Ziegler, S. V. Arehart, D. Greszta and T. Pakula, *J. Phys. Org. Chem.*, 2000, **13**, 775–786.
63. Z. M. O. Rzaev, *Prog. Polym. Sci.*, 2000, **25**, 163–217.
64. B. Klumperman, *Polym. Chem.*, 2010, **1**, 558–562.
65. B. Klumperman, in *Sequence-Controlled Polymers: Synthesis, Self-Assembly, and Properties*, American Chemical Society, 2014, vol. 1170, pp. 213–221.
66. M. Hisano, K. Takeda, T. Takashima, Z. Jin, A. Shiibashi and A. Matsumoto, *Macromolecules*, 2013, **46**, 3314–3323.
67. M. Hisano, K. Takeda, T. Takashima, Z. Jin, A. Shiibashi and A. Matsumoto, *Macromolecules*, 2013, **46**, 7733–7744.
68. D. Yamamoto and A. Matsumoto, *Macromolecules*, 2013, **46**, 9526–9536.
69. J.-F. Lutz, B. V. K. J. Schmidt and S. Pfeifer, *Macromol. Rapid Commun.*, 2011, **32**, 127–135.
70. D. Benoit, C. J. Hawker, E. E. Huang, Z. Lin and T. P. Russell, *Macromolecules*, 2000, **33**, 1505–1507.
71. S. Pfeifer and J.-F. Lutz, *J. Am. Chem. Soc.*, 2007, **129**, 9542–9543.
72. M. Zamfir and J.-F. Lutz, in *Progress in Controlled Radical Polymerization: Materials and Applications*, American Chemical Society, 2012, vol. 1101, pp. 1–12.
73. N. Baradel, O. Shishkan, S. Srichan and J.-F. Lutz, in *Sequence-Controlled Polymers: Synthesis, Self-Assembly, and Properties*, American Chemical Society, 2014, vol. 1170, pp. 119–131.
74. D. Chan-Seng, M. Zamfir and J.-F. Lutz, *Angew. Chem., Int. Ed.*, 2012, **51**, 12254–12257.
75. M. Zamfir and J.-F. Lutz, *Nat. Commun.*, 2012, **3**, 1138.
76. S. Pfeifer and J.-F. Lutz, *Chem. – Eur. J.*, 2008, **14**, 10949–10957.
77. R. Kakuchi, M. Zamfir, J.-F. Lutz and P. Theato, *Macromol. Rapid Commun.*, 2012, **33**, 54–60.
78. H. Zhao, W. Gu, R. Kakuchi, Z. Sun, E. Sterner, T. P. Russell, E. B. Coughlin and P. Theato, *ACS Macro Lett.*, 2013, **2**, 966–969.
79. E. L. Malins, S. Amabilino, G. Yilmaz, F. H. Isikgor, B. M. Gridley and C. R. Becer, *Eur. Polym. J.*, 2015, **62**, 347–351.

80. N. Baradel, S. Fort, S. Halila, N. Badi and J.-F. Lutz, *Angew. Chem., Int. Ed.*, 2013, **52**, 2335–2339.
81. N. Baradel, O. Gok, M. Zamfir, A. Sanyal and J.-F. Lutz, *Chem. Commun.*, 2013, **49**, 7280–7282.
82. S. Srichan, D. Chan-Seng and J. F. Lutz, *ACS Macro Lett.*, 2012, **1**, 589–592.
83. S. Srichan, H. Mutlu, N. Badi and J.-F. Lutz, *Angew. Chem., Int. Ed.*, 2014, **53**, 9231–9235.
84. S. Srichan, L. Oswald, M. Zamfir and J.-F. Lutz, *Chem. Commun.*, 2012, **48**, 1517–1519.
85. S. Srichan, H. Mutlu and J.-F. Lutz, *Eur. Polym. J.*, 2015, **62**, 338–346.
86. S. Srichan, N. Kayunkid, L. Oswald, B. Lotz and J.-F. Lutz, *Macromolecules*, 2014, **47**, 1570–1577.
87. M. Ouchi, N. Badi, J.-F. Lutz and M. Sawamoto, *Nat. Chem.*, 2011, **3**, 917–924.
88. T. Terashima, T. Mes, T. F. A. De Greef, M. A. J. Gillissen, P. Besenius, A. R. A. Palmans and E. W. Meijer, *J. Am. Chem. Soc.*, 2011, **133**, 4742–4745.
89. N. Giuseppone and J.-F. Lutz, *Nature*, 2011, **473**, 40–41.
90. J.-F. Lutz, *Kobunshi*, 2013, **62**, 509–510.
91. M. Gonzalez-Burgos, A. Latorre-Sanchez and J. A. Pomposo, *Chem. Soc. Rev.*, 2015, **44**, 6122–6142.
92. Y. Liu, T. Pauloehrl, S. I. Presolski, L. Albertazzi, A. R. A. Palmans and E. W. Meijer, *J. Am. Chem. Soc.*, 2015, **137**, 13096–13105.
93. B. V. K. J. Schmidt, N. Fechler, J. Falkenhagen and J.-F. Lutz, *Nat. Chem.*, 2011, **3**, 234–238.
94. M. Zamfir, P. Theato and J.-F. Lutz, *Polym. Chem.*, 2012, **3**, 1796–1802.
95. O. Shishkan, M. Zamfir, M. A. Gauthier, H. G. Börner and J.-F. Lutz, *Chem. Commun.*, 2014, **50**, 1570–1572.
96. R. K. Roy and J.-F. Lutz, *J. Am. Chem. Soc.*, 2014, **136**, 12888–12891.
97. K. Satoh, M. Matsuda, K. Nagai and M. Kamigaito, *J. Am. Chem. Soc.*, 2010, **132**, 10003–10005.
98. M. Matsuda, K. Satoh and M. Kamigaito, *Macromolecules*, 2013, **46**, 5473–5482.
99. T. Soejima, K. Satoh and M. Kamigaito, *ACS Macro Lett.*, 2015, **4**, 745–749.
100. N. ten Brummelhuis, *Polym. Chem.*, 2015, **6**, 654–667.
101. S. Ida, M. Ouchi and M. Sawamoto, *Macromol. Rapid Commun.*, 2011, **32**, 209–214.
102. S. Ida, T. Terashima, M. Ouchi and M. Sawamoto, *J. Am. Chem. Soc.*, 2009, **131**, 10808–10809.
103. S. Ida, M. Ouchi and M. Sawamoto, *J. Am. Chem. Soc.*, 2010, **132**, 14748–14750.
104. M. Ouchi, Y. Hibi, T. Arima, D. Hayata and M. Sawamoto, in *Sequence-Controlled Polymers: Synthesis, Self-Assembly, and Properties*, American Chemical Society, 2014, vol. 1170, pp. 149–160.

105. Y. Hibi, M. Ouchi and M. Sawamoto, *Angew. Chem., Int. Ed.*, 2011, **50**, 7434–7437.
106. Y. Hibi, S. Tokuoka, T. Terashima, M. Ouchi and M. Sawamoto, *Polym. Chem.*, 2011, **2**, 341–347.
107. J. G. Kim, C. D. Cowman, A. M. LaPointe, U. Wiesner and G. W. Coates, *Macromolecules*, 2011, **44**, 1110–1113.
108. C. Boyer, A. H. Soeriyadi, P. B. Zetterlund and M. R. Whittaker, *Macromolecules*, 2011, **44**, 8028–8033.
109. A. H. Soeriyadi, C. Boyer, F. Nyström, P. B. Zetterlund and M. R. Whittaker, *J. Am. Chem. Soc.*, 2011, **133**, 11128–11131.
110. Q. Zhang, J. Collins, A. Anastasaki, R. Wallis, D. A. Mitchell, C. R. Becer and D. M. Haddleton, *Angew. Chem., Int. Ed.*, 2013, **52**, 4435–4439.
111. Z. Qiang, C. Jennifer, A. Athina, W. Russell, A. M. Daniel, C. R. Becer, W. Paul and M. H. David, in *Sequence-Controlled Polymers: Synthesis, Self-Assembly, and Properties*, American Chemical Society, 2014, vol. 1170, pp. 327–348.
112. Q. Zhang, A. Anastasaki, G.-Z. Li, A. J. Haddleton, P. Wilson and D. M. Haddleton, *Polym. Chem.*, 2014, **5**, 3876–3883.
113. F. Alsubaie, A. Anastasaki, P. Wilson and D. M. Haddleton, *Polym. Chem.*, 2015, **6**, 406–417.
114. A. Anastasaki, V. Nikolaou, G. S. Pappas, Q. Zhang, C. Wan, P. Wilson, T. P. Davis, M. R. Whittaker and D. M. Haddleton, *Chem. Sci.*, 2014, **5**, 3536–3542.
115. Y.-M. Chuang, A. Ethirajan and T. Junkers, *ACS Macro Lett.*, 2014, **3**, 732–737.
116. A. Anastasaki, V. Nikolaou, N. W. McCaul, A. Simula, J. Godfrey, C. Waldron, P. Wilson, K. Kempe and D. M. Haddleton, *Macromolecules*, 2015, **48**, 1404–1411.
117. G. Gody, T. Maschmeyer, P. B. Zetterlund and S. Perrier, *Nat. Commun.*, 2013, **4**, 2505.
118. G. Gody, T. Maschmeyer, P. B. Zetterlund and S. Perrier, *Macromolecules*, 2014, **47**, 3451–3460.
119. J. Vandenbergh and T. Junkers, *Macromolecules*, 2014, **47**, 5051–5059.
120. E. Brulé, C. Robert and C. M. Thomas, in *Sequence-Controlled Polymers: Synthesis, Self-Assembly, and Properties*, American Chemical Society, 2014, vol. 1170, pp. 349–368.
121. J. W. Kramer, D. S. Treitler, E. W. Dunn, P. M. Castro, T. Roisnel, C. M. Thomas and G. W. Coates, *J. Am. Chem. Soc.*, 2009, **131**, 16042–16044.
122. A. Amgoune, C. M. Thomas, S. Ilinca, T. Roisnel and J.-F. Carpentier, *Angew. Chem., Int. Ed.*, 2006, **45**, 2782–2784.
123. C. G. Jaffredo, Y. Chapurina, S. M. Guillaume and J.-F. Carpentier, *Angew. Chem., Int. Ed.*, 2014, **53**, 2687–2691.
124. C. W. Bielawski and R. H. Grubbs, *Prog. Polym. Sci.*, 2007, **32**, 1–29.
125. A. B. Chang, G. M. Miyake and R. H. Grubbs, in *Sequence-Controlled Polymers:* Synthesis, Self-Assembly, and Properties, American Chemical Society, 2014, vol. 1170, pp. 161–188.

126. I. Cho, *Macromol. Symp.*, 2003, **195**, 89–94.
127. J. Zhang, M. E. Matta and M. A. Hillmyer, *ACS Macro Lett.*, 2012, **1**, 1383–1387.
128. B. R. Elling and Y. Xia, *J. Am. Chem. Soc.*, 2015, **137**, 9922–9926.
129. D. Moatsou, C. F. Hansell and R. K. O'Reilly, *Chem. Sci.*, 2014, **5**, 2246–2250.
130. R. M. Weiss, A. L. Short and T. Y. Meyer, *ACS Macro Lett.*, 2015, **4**, 1039–1043.
131. W. R. Gutekunst and C. J. Hawker, *J. Am. Chem. Soc.*, 2015, **137**, 8038–8041.
132. M. D. Schulz and K. B. Wagener, *Macromol. Chem. Phys.*, 2014, **215**, 1936–1945.
133. J. C. Sworen, J. A. Smith, J. M. Berg and K. B. Wagener, *J. Am. Chem. Soc.*, 2004, **126**, 11238–11246.
134. G. Rojas, B. Inci, Y. Wei and K. B. Wagener, *J. Am. Chem. Soc.*, 2009, **131**, 17376–17386.
135. B. Inci and K. B. Wagener, *J. Am. Chem. Soc.*, 2011, **133**, 11872–11875.
136. M. E. Seitz, C. D. Chan, K. L. Opper, T. W. Baughman, K. B. Wagener and K. I. Winey, *J. Am. Chem. Soc.*, 2010, **132**, 8165–8174.
137. B. S. Aitken, C. F. Buitrago, J. D. Heffley, M. Lee, H. W. Gibson, K. I. Winey and K. B. Wagener, *Macromolecules*, 2012, **45**, 681–687.
138. Z.-L. Li, L. Li, X.-X. Deng, L.-J. Zhang, B.-T. Dong, F.-S. Du and Z.-C. Li, *Macromolecules*, 2012, **45**, 4590–4598.
139. Z.-L. Li, A. Lv, F.-S. Du and Z.-C. Li, *Macromolecules*, 2014, **47**, 5942–5951.
140. A. Lv, Z.-L. Li, F.-S. Du and Z.-C. Li, *Macromolecules*, 2014, **47**, 7707–7716.
141. J.-F. Lutz, *Angew. Chem., Int. Ed.*, 2007, **46**, 1018–1025.
142. C. E. Hoyle and C. N. Bowman, *Angew. Chem., Int. Ed.*, 2010, **49**, 1540–1573.
143. A. B. Lowe, C. E. Hoyle and C. N. Bowman, *J. Mater. Chem.*, 2010, **20**, 4745–4750.
144. M.-A. Berthet, Z. Zarafshani, S. Pfeifer and J.-F. Lutz, *Macromolecules*, 2010, **43**, 44–50.
145. F. Driessen, F. E. Du Prez and P. Espeel, *ACS Macro Lett.*, 2015, **4**, 616–619.
146. W. Xi, S. Pattanayak, C. Wang, B. Fairbanks, T. Gong, J. Wagner, C. J. Kloxin and C. N. Bowman, *Angew. Chem., Int. Ed.*, 2015, **54**, 14462–14467.
147. J. Han, Y. Zheng, B. Zhao, S. Li, Y. Zhang and C. Gao, *Sci. Rep.*, 2014, **4**, 4387.
148. R. M. Stayshich and T. Y. Meyer, *J. Am. Chem. Soc.*, 2010, **132**, 10920–10934.
149. J. Li, R. M. Stayshich and T. Y. Meyer, *J. Am. Chem. Soc.*, 2011, **133**, 6910–6913.
150. J. Li, S. N. Rothstein, S. R. Little, H. M. Edenborn and T. Y. Meyer, *J. Am. Chem. Soc.*, 2012, **134**, 16352–16359.

151. C. M. Thomas and J.-F. Lutz, *Angew. Chem., Int. Ed.*, 2011, **50**, 9244–9246.
152. K. Satoh, S. Ozawa, M. Mizutani, K. Nagai and M. Kamigaito, *Nat. Commun.*, 2010, **1**, 1–6.
153. S. Takamasa, S. Kotaro and K. Masami, in *Sequence-Controlled Polymers: Synthesis, Self-Assembly, and Properties*, American Chemical Society, 2014, vol. 1170, pp. 189–200.
154. J. Mandal, S. Krishna Prasad, D. S. S. Rao and S. Ramakrishnan, *J. Am. Chem. Soc.*, 2014, **136**, 2538–2545.
155. R. Kakuchi, *Angew. Chem., Int. Ed.*, 2014, **53**, 46–48.
156. X.-X. Deng, L. Li, Z.-L. Li, A. Lv, F.-S. Du and Z.-C. Li, *ACS Macro Lett.*, 2012, **1**, 1300–1303.
157. J. Zhang, X. X. Deng, F. S. Du and Z. C. Li, in *Sequence-Controlled Polymers: Synthesis, Self-Assembly, and Properties*, American Chemical Society, 2014, vol. 1170, pp. 223–234.
158. L. Yu, L.-H. Wang, Z.-T. Hu, Y.-Z. You, D.-C. Wu and C.-Y. Hong, *Polym. Chem.*, 2015, **6**, 1527–1532.
159. Z. Zhang, Y.-Z. You, D.-C. Wu and C.-Y. Hong, *Macromolecules*, 2015, **48**, 3414–3421.
160. R. B. Merrifield, *Angew. Chem., Int. Ed.*, 1985, **24**, 799–810.
161. D. J. Gravert and K. D. Janda, *Chem. Rev.*, 1997, **97**, 489–510.
162. W. Zhang and D. P. Curran, *Tetrahedron*, 2006, **62**, 11837–11865.
163. W. Miao and T.-H. Chan, *J. Org. Chem.*, 2005, **70**, 3251–3255.
164. J.-F. Lutz and H. G. Börner, *Prog. Polym. Sci.*, 2008, **33**, 1–39.
165. H. G. Börner, *Prog. Polym. Sci.*, 2009, **34**, 811–851.
166. J. K. Young, J. C. Nelson and J. S. Moore, *J. Am. Chem. Soc.*, 1994, **116**, 10841–10842.
167. E. L. Elliott, C. R. Ray, S. Kraft, J. R. Atkins and J. S. Moore, *J. Org. Chem.*, 2006, **71**, 5282–5290.
168. D. Chan-Seng and J.-F. Lutz, *ACS Macro Lett.*, 2014, **3**, 291–294.
169. D. Chan-Seng and J.-F. Lutz, in *Sequence-Controlled Polymers: Synthesis, Self-Assembly, and Properties*, American Chemical Society, 2014, vol. 1170, pp. 103–116.
170. S. L. Beaucage and R. P. Iyer, *Tetrahedron*, 1992, **48**, 2223–2311.
171. T. G. W. Edwardson, K. M. M. Carneiro, C. J. Serpell and H. F. Sleiman, *Angew. Chem., Int. Ed.*, 2014, **53**, 4567–4571.
172. A. Al Ouahabi, L. Charles and J.-F. Lutz, *J. Am. Chem. Soc.*, 2015, **137**, 5629–5635.
173. A. Al Ouahabi, M. Kotera, L. Charles and J.-F. Lutz, *ACS Macro Lett.*, 2015, **4**, 1077–1080.
174. T. T. Trinh, C. Laure and J.-F. Lutz, *Macromol. Chem. Phys.*, 2015, **216**, 1498–1506.
175. L. Hartmann, *Macromol. Chem. Phys.*, 2011, **212**, 8–13.
176. K. Rose and J. Vizzavona, *J. Am. Chem. Soc.*, 1999, **121**, 7034–7038.
177. L. Hartmann, E. Krause, M. Antonietti and H. G. Börner, *Biomacromolecules*, 2006, **7**, 1239–1244.

178. L. Hartmann, S. Häfele, R. Peschka-Süss, M. Antonietti and H. G. Börner, *Chem. – Eur. J.*, 2008, **14**, 2025–2033.
179. L. Hartmann and H. G. Börner, *Adv. Mater.*, 2009, **21**, 3425–3431.
180. M. Porel and C. A. Alabi, *J. Am. Chem. Soc.*, 2014, **136**, 13162–13165.
181. A. M. Rosales, R. A. Segalman and R. N. Zuckermann, *Soft Matter*, 2013, **9**, 8400–8414.
182. J. Sun and R. N. Zuckermann, *ACS Nano*, 2013, **7**, 4715–4732.
183. J. Sun, C. Proulx and R. N. Zuckermann, in *Sequence-Controlled Polymers: Synthesis, Self-Assembly, and Properties*, American Chemical Society, 2014, vol. 1170, pp. 35–53.
184. S. Pfeifer, Z. Zarafshani, N. Badi and J.-F. Lutz, *J. Am. Chem. Soc.*, 2009, **131**, 9195–9196.
185. T. T. Trinh, L. Oswald, D. Chan-Seng and J.-F. Lutz, *Macromol. Rapid Commun.*, 2014, **35**, 141–145.
186. R. K. Roy, A. Meszynska, C. Laure, L. Charles, C. Verchin and J.-F. Lutz, *Nat. Commun.*, 2015, **6**, 7237.
187. L. Charles, C. Laure, J.-F. Lutz and R. K. Roy, *Macromolecules*, 2015, **48**, 4319–4328.
188. L. Charles, C. Laure, J.-F. Lutz and R. K. Roy, *Rapid Commun. Mass Spectrom.*, 2016, **30**, 22–28.
189. P. Espeel, L. L. G. Carrette, K. Bury, S. Capenberghs, J. C. Martins, F. E. Du Prez and A. Madder, *Angew. Chem., Int. Ed.*, 2013, **52**, 13261–13264.
190. S. C. Solleder and M. A. R. Meier, *Angew. Chem., Int. Ed.*, 2014, **53**, 711–714.
191. S. C. Solleder, K. S. Wetzel and M. A. R. Meier, *Polym. Chem.*, 2015, **6**, 3201–3204.
192. S. C. Solleder, D. Zengel, K. S. Wetzel and M. A. R. Meier, *Angew. Chem., Int. Ed.*, 2016, **55**, 1204–1207.
193. X. M. Tong, B. H. Guo and Y. B. Huang, *Chem. Commun.*, 2011, **47**, 1455–1457.
194. S. Houshyar, D. J. Keddie, G. Moad, R. J. Mulder, S. Saubern and J. Tsanaktsidis, *Polym. Chem.*, 2012, **3**, 1879–1889.
195. G. Moad, C. Guerrero-Sanchez, J. J. Haven, D. J. Keddie, A. Postma, E. Rizzardo and S. H. Thang, in *Sequence-Controlled Polymers: Synthesis, Self-Assembly, and Properties*, American Chemical Society, 2014, vol. 1170, pp. 133–147.
196. J. Vandenbergh, G. Reekmans, P. Adriaensens and T. Junkers, *Chem. Commun.*, 2013, **49**, 10358–10360.
197. J. Vandenbergh, G. Reekmans, P. Adriaensens and T. Junkers, *Chem. Sci.*, 2015, **6**, 5753–5761.
198. S. B. H. Kent, *Chem. Soc. Rev.*, 2009, **38**, 338–351.
199. K. Takizawa, C. Tang and C. J. Hawker, *J. Am. Chem. Soc.*, 2008, **130**, 1718–1726.
200. F. A. Leibfarth, J. A. Johnson and T. F. Jamison, *Proc. Natl. Acad. Sci. U. S. A.*, 2015, **112**, 10617–10622.

201. J. C. Barnes, D. J. C. Ehrlich, A. X. Gao, F. A. Leibfarth, Y. Jiang, E. Zhou, T. F. Jamison and J. A. Johnson, *Nat. Chem.*, 2015, **7**, 810–815.
202. N. Zydziak, F. Feist, B. Huber, J. O. Mueller and C. Barner-Kowollik, *Chem. Commun.*, 2015, **51**, 1799–1802.
203. T. T. Trinh, L. Oswald, D. Chan-Seng, L. Charles and J.-F. Lutz, *Chem. – Eur. J.*, 2015, **21**, 11961–11965.
204. R. K. Roy, C. Laure, D. Fischer-Krauser, L. Charles and J.-F. Lutz, *Chem. Commun.*, 2015, **51**, 15677–15680.
205. X. Li, Z.-Y. J. Zhan, R. Knipe and D. G. Lynn, *J. Am. Chem. Soc.*, 2002, **124**, 746–747.
206. B. Datta and G. B. Schuster, *J. Am. Chem. Soc.*, 2008, **130**, 2965–2973.
207. W. Chen and G. B. Schuster, *J. Am. Chem. Soc.*, 2013, **135**, 4438–4449.
208. J. Niu, R. Hili and D. R. Liu, *Nat. Chem.*, 2013, **5**, 282–292.
209. M. L. McKee, P. J. Milnes, J. Bath, E. Stulz, A. J. Turberfield and R. K. O'Reilly, *Angew. Chem., Int. Ed.*, 2010, **49**, 7948–7951.
210. M. L. McKee, P. J. Milnes, J. Bath, E. Stulz, R. K. O'Reilly and A. J. Turberfield, *J. Am. Chem. Soc.*, 2012, **134**, 1446–1449.
211. P. J. Milnes, M. L. McKee, J. Bath, L. Song, E. Stulz, A. J. Turberfield and R. K. O'Reilly, *Chem. Commun.*, 2012, **48**, 5614–5616.
212. P. J. Milnes and R. K. O. Reilly, in *Sequence-Controlled Polymers: Synthesis, Self-Assembly, and Properties*, American Chemical Society, 2014, vol. 1170, pp. 71–84.
213. B. Lewandowski, G. De Bo, J. W. Ward, M. Papmeyer, S. Kuschel, M. J. Aldegunde, P. M. E. Gramlich, D. Heckmann, S. M. Goldup, D. M. D'Souza, A. E. Fernandes and D. A. Leigh, *Science*, 2013, **339**, 189–193.
214. G. De Bo, S. Kuschel, D. A. Leigh, B. Lewandowski, M. Papmeyer and J. W. Ward, *J. Am. Chem. Soc.*, 2014, **136**, 5811–5814.
215. B. N. Norris, S. Zhang, C. M. Campbell, J. T. Auletta, P. Calvo-Marzal, G. R. Hutchison and T. Y. Meyer, *Macromolecules*, 2013, **46**, 1384–1392.
216. Y. K. Ilana, S. B. Jonathon and R. H. Geoffrey, in *Sequence-Controlled Polymers: Synthesis, Self-Assembly, and Properties*, American Chemical Society, 2014, vol. 1170, pp. 379–393.

CHAPTER 4

Single-chain Nanoparticles

JUSTIN P. COLE,[a] CHRISTOPHER K. LYON[a] AND ERIK B. BERDA*[a,b]

[a] Department of Chemistry, University of New Hampshire, 23 Academic way, Durham, NH 03824, USA; [b] Material Science Program, University of New Hampshire, 23 Academic way, Durham, NH 03824, USA
*Email: Erik.Berda@unh.edu

4.1 Introduction

Nature is replete with examples of functional nanostructures capable of completing complex tasks. These structures are composed of well-defined linear polymer chains folded into precise, three-dimensional shapes. An obvious yet unmet goal of chemists is to reproduce the functionality of natural systems using synthetic polymers. This warrants a tour de force involving modern controlled polymerization techniques and post-polymerization modification strategies in combination with current theories of polymer physics. Areas such as catalysis,[1–7] sensing,[8] nanoreactors,[9] and nanomedicine[10–13] stand to benefit from advances in this research.

Biomacromolecules such as proteins derive their function from their three-dimensional shape and the precise functional group placement on the surface and interior of the structure. Also known as the tertiary structure, this shape is a result of a perfectly controlled monomer sequence, or primary structure. Synthetic polymers containing a perfect monomer sequence are inaccessible using contemporary techniques. However, recent advances in polymerization techniques allow for the fabrication of multi-block polymers with narrow molecular weight distributions[14] and materials with relatively controlled monomer sequences[15,16] by step-growth and chain-growth

RSC Polymer Chemistry Series No. 22
Bio-inspired Polymers
Edited by Nico Bruns and Andreas F. M. Kilbinger

Published by the Royal Society of Chemistry, www.rsc.org

methods. While these techniques are worth considering as a large step forward, they are not free from producing microstructural defects or broad molecular weight distributions.

Henceforth, there is a new paradigm in the synthesis of functional soft nanomaterials: the manipulation of single polymer chains, commonly referred to as single-chain technology.[17] In particular, the collapse of single polymer chains into architecturally defined nanostructures has recently garnered significant attention.[18,19] Simple in principle, as seen in Figure 4.1, these single-chain nanoparticles (SCNPs) exhibit behavior more complex than anticipated and are an intense topic of focus by a number of research groups,[20–23] including our own.

To our knowledge, the intramolecular cross-linking of linear single polymer chains was first reported by Kuhn and Balmer as early as 1962.[24] The topic was revisited in the late 60s by Rossi *et al.*,[25,26] and again both theoretically and experimentally by various researchers throughout the 80s[27,28] and 90s.[29–31] The arrival of controlled radical polymerization methods and the development of efficient post-polymerization modification reactions revitalized the interest in the topic at the turn of the century.

The purpose of this chapter is to provide overview and insight for what is currently an expanding field. Herein, we will discuss (i) the chemistry and conditions used to synthesize SCNPs, (ii) the techniques used for characterization, and (iii) their applications.

4.2 Synthesis of SCNPs

Numerous chemical reactions and interactions have been used in the synthesis of SCNPs, including covalent and non-covalent chemistries. The process typically involves the synthesis of functional polymers, followed by the exposure of a dilute polymer solution to some chemical or physical stimulus that begets cross-link formation. The reactions and structures of cross-links used in SCNP formation are summarized in Table 4.1.

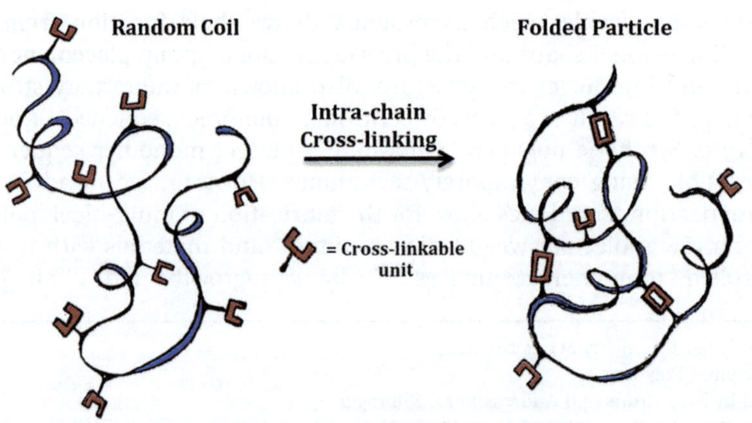

Figure 4.1 Linear polymers furnished with functional groups that form intramolecular cross-links when triggered in dilute solution.

Table 4.1 Chemistry of SCNP formation. Reproduced from ref. 19 with permission from the Royal Society of Chemistry.

Covalent chemistry			
Before cross-linking	Structure of cross-link	Type of chemistry	Ref.
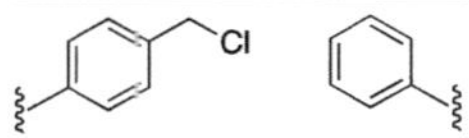	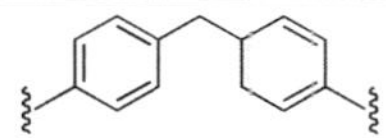	Friedel–Crafts alkylation	32
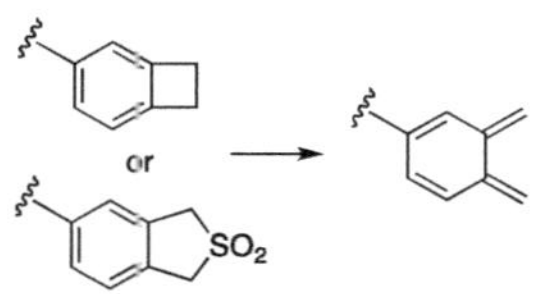	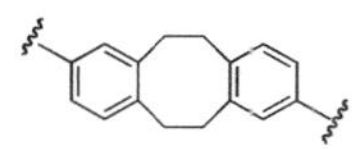	Thermal [4 + 4] cycloaddition	33–36
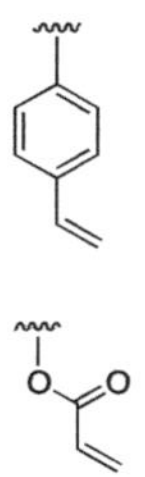	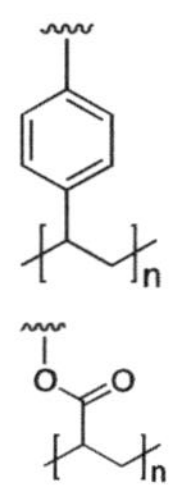	Free radical polymerization	37, 38
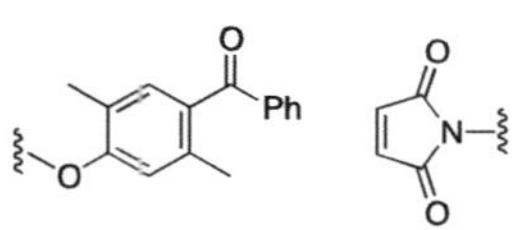	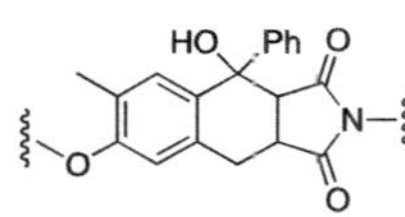	Photoinduced [4 + 2] cycloaddition	39
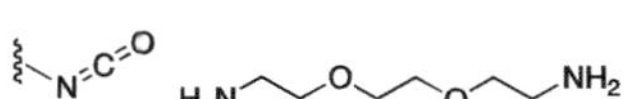	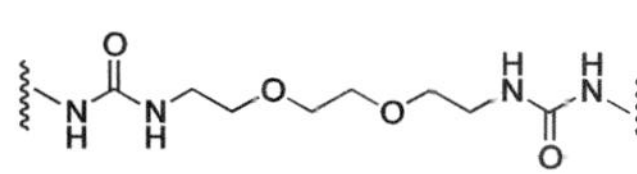	Isocyanate-amine	40
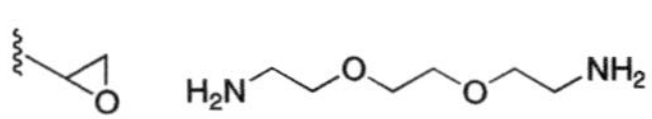	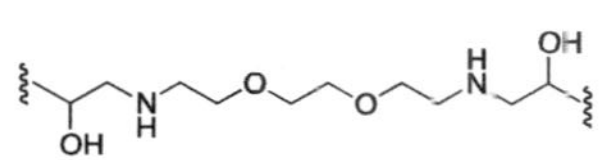	Epoxide-amine	

Table 4.1 (*Continued*)

Covalent chemistry			
Before cross-linking	Structure of cross-link	Type of chemistry	Ref.
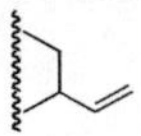	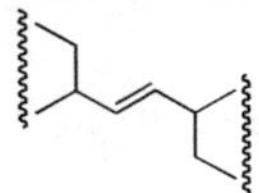	Olefin metathesis	41
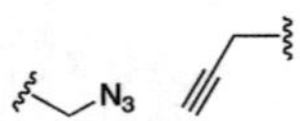	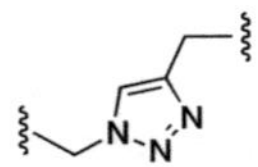	Azide–alkyne "click" chemistry	10, 42–44
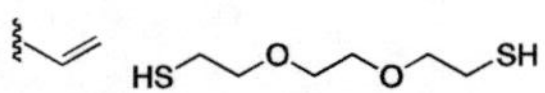	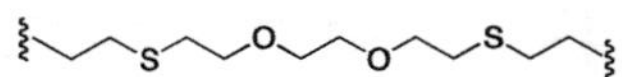	Thiol-ene "click" chemistry	45
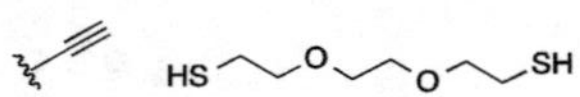	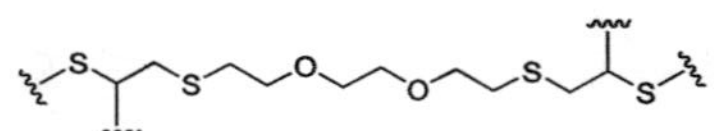	Thiol-yne "click" chemistry	46
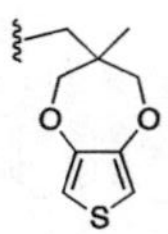	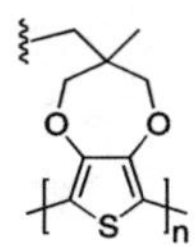	Oxidative polymerization of thiophene	47
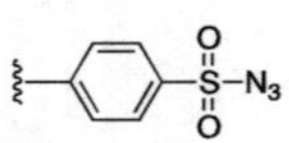	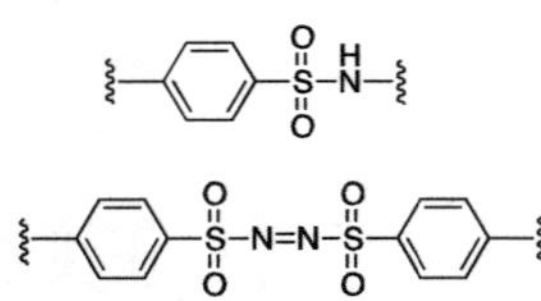	Sulfonyl nitrene insertion/coupling	48
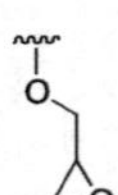	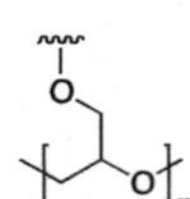	Cationic polymerization of epoxide	1
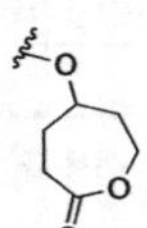	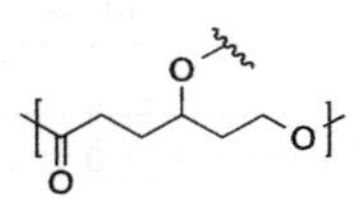	Organocatalyzed ring-opening polymerization	49

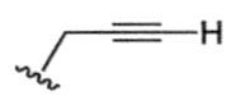	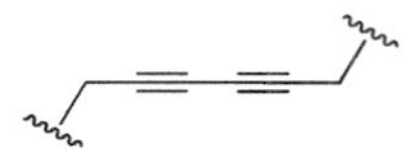	Glaser–Hay coupling	50
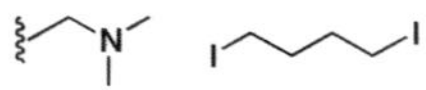	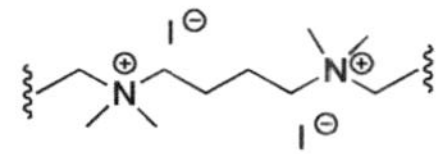	Menshutkin reaction	51
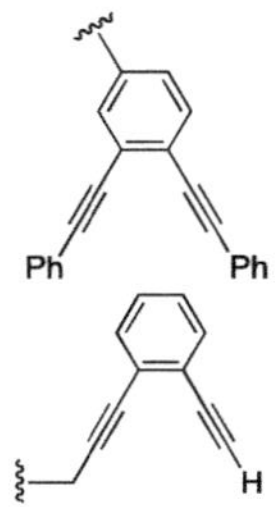	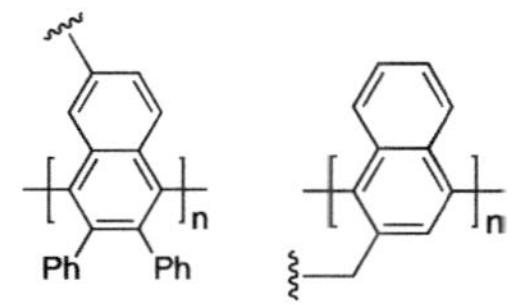	Bergman cyclization	52, 53
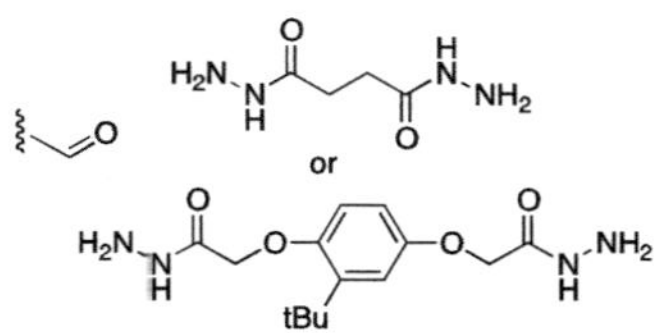	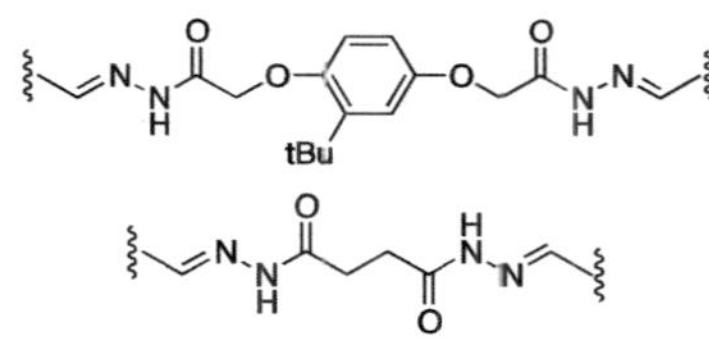	Hydrazone formation	20, 54
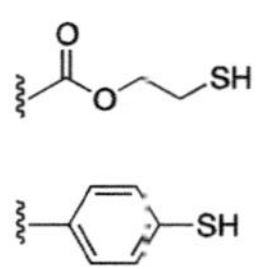	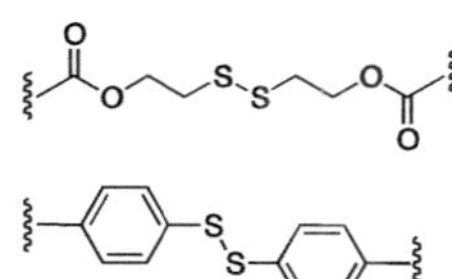	Disulfide formation	55, 56
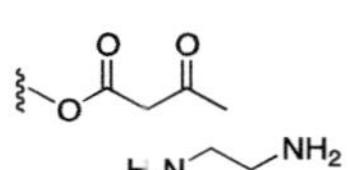	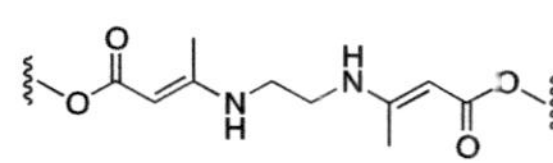	Enamine formation	57, 58
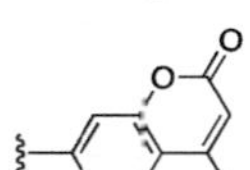	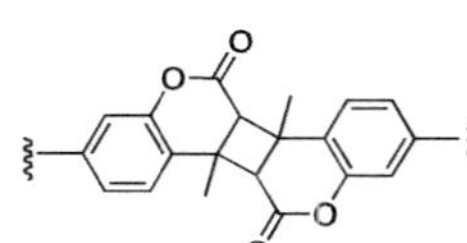	Photodimerization of coumarin	9, 59

Table 4.1 (*Continued*)

Covalent chemistry			
Before cross-linking	Structure of cross-link	Type of chemistry	Ref.
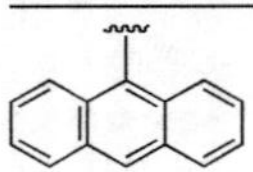	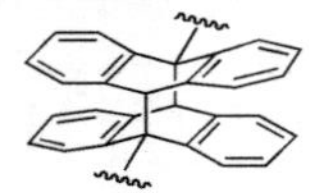	Photodimerization of anthracene	60
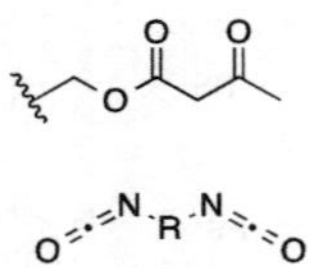	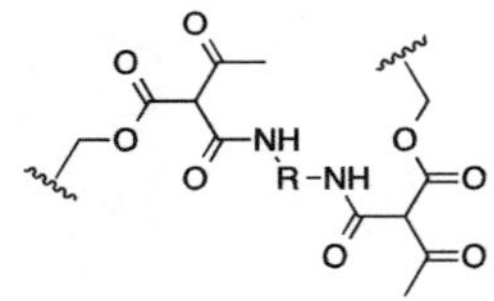	Amide formation	61
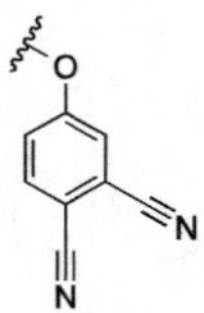 CuCl	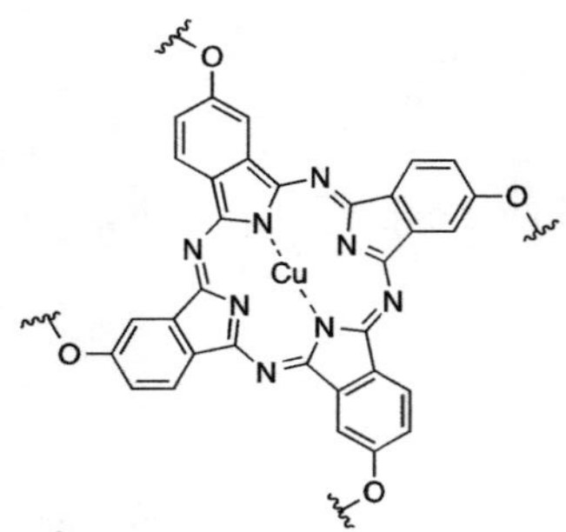	Phthalocyanine formation	62
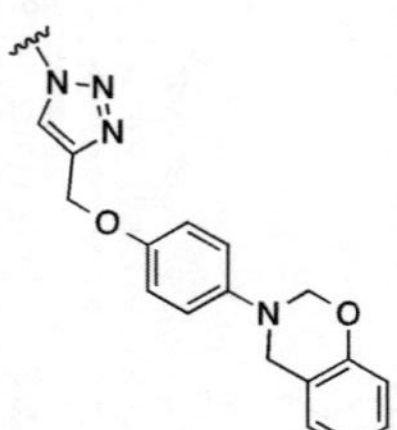	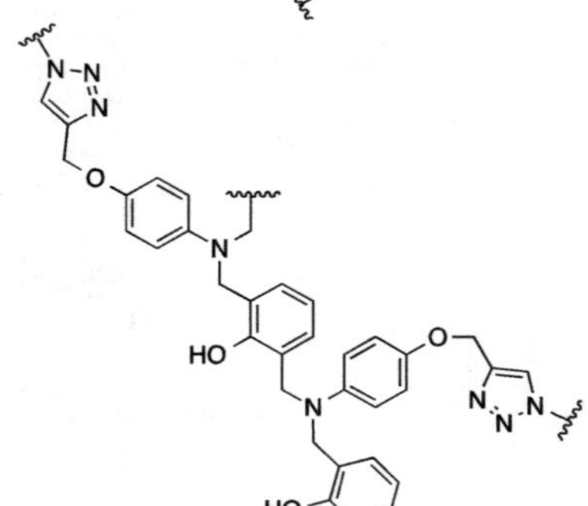	Benzoxazine crosslinking	63

Non-covalent chemistry		
Structure		Ref.
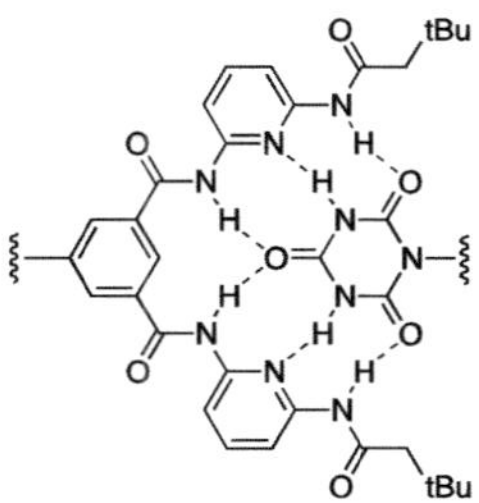	Hamilton wedge, cyanuric acid hydrogen bonding	64, 65
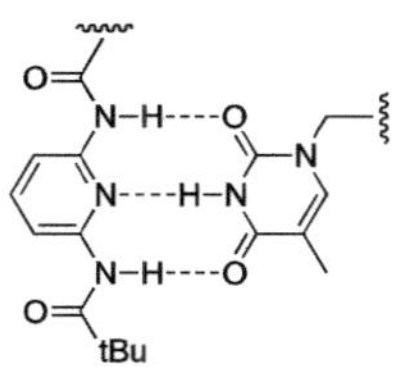	Thymine, diaminopyridine hydrogen bonding	65, 66
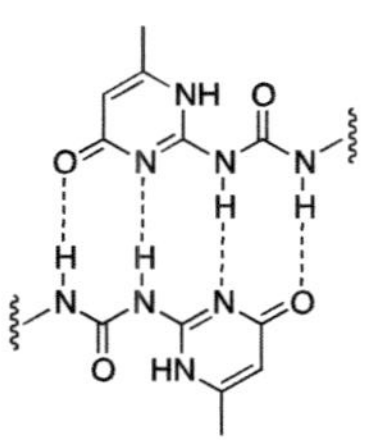	2-Ureido-4[1*H*]-pyrimidinone (UPy) hydrogen bonding	67–71

Table 4.1 (*Continued*)

Non-covalent chemistry		
Structure		Ref.
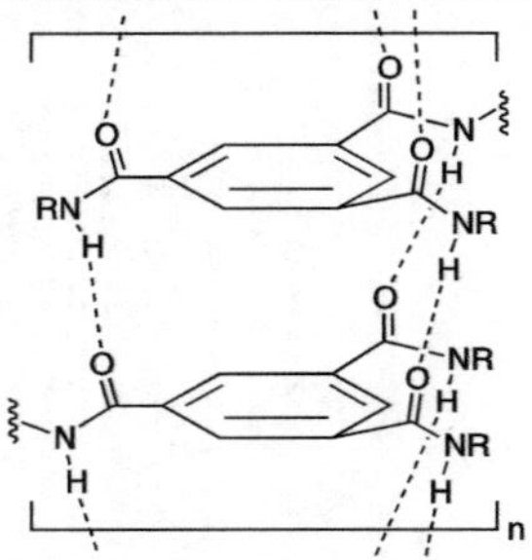	Benzene-1,3,5-tricarboxamide (BTA) hydrogen bonding	2, 3, 8, 72, 73
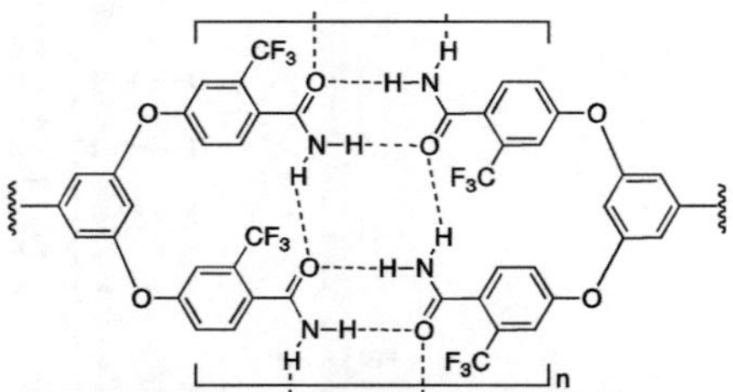	Dendritic self-complimentary hydrogen bonding	74

Metal coordination			
Before cross-linking	Structure of cross-link	Type of chemistry	Ref.
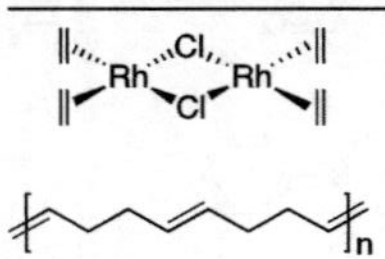	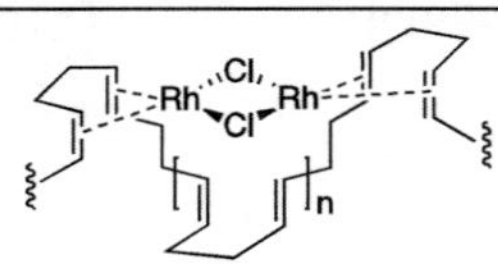	Rhodium coordination	75, 76
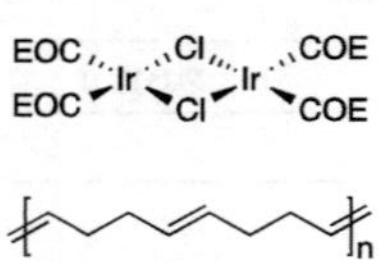	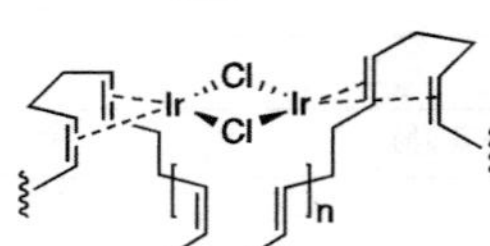	Iridium coordination	76

Ni(II)acac 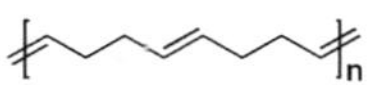	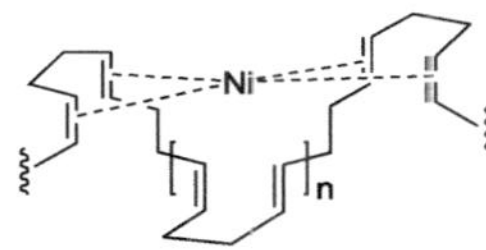	Nickel coordination	
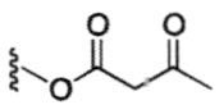 $Cu(OAc)_2$	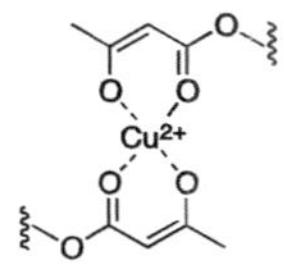		
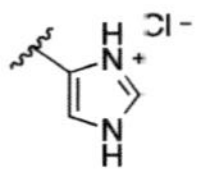 Cu(II)	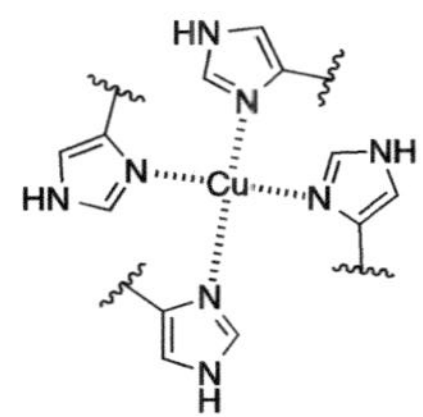	Copper coordination	5, 77
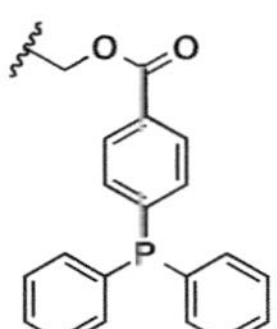 $Pd(COD)Cl_2$	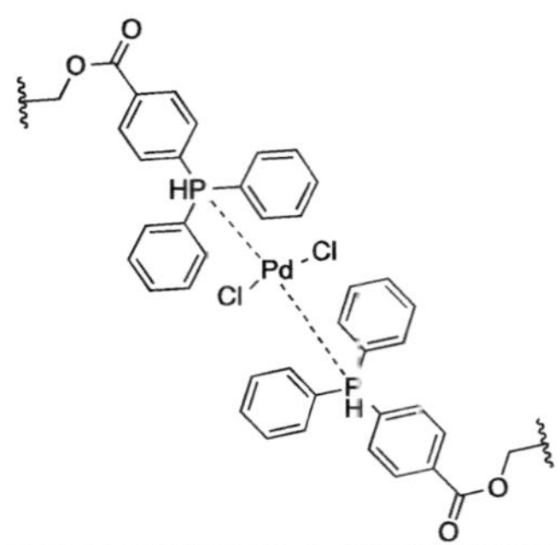	Palladium coordination	6

4.2.1 Covalent Cross-linking Reactions

The formation and subsequent [4 + 4] cycloaddition of pendant benzoquinone dimethide species was utilized by Hawker and coworkers in the synthesis of architecturally defined SCNPs (Figure 4.2).[33] Statistical copolymers of 4-vinylbenzocyclobutene were synthesized by nitroxide-mediated free radical polymerization. The nanoparticle synthesis employed a continuous addition strategy, in which a concentrated polymer solution

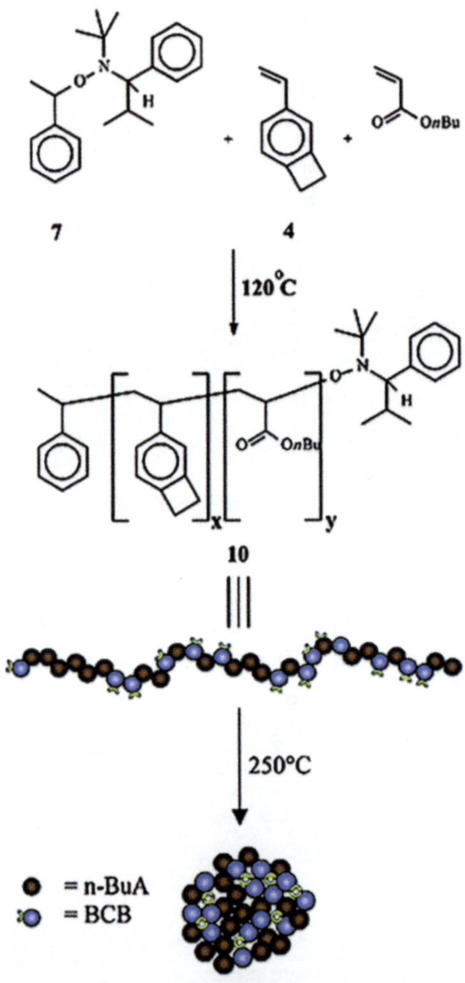

Figure 4.2 Representative scheme of the intramolecular collapse of linear polymers *via* benzocyclobutene chemistry.
Reprinted with permission from E. Harth, B. Van Horn, V. Y. Lee, D. S. Germack, C. P. Gonzales, R. D. Miller and C. J. Hawker, *J. Am. Chem. Soc.*, 2002, **124**, 8653–8660.[33] Copyright (2002) American Chemical Society.

was slowly added to a larger volume of heated solvent. At high temperatures (*ca.* 250 °C), the benzocyclobutene group undergoes a cycloreversion reaction to form benzoquinone dimethide, resulting in cross-link formation. As a result of the continuous addition method, the final polymer concentration is higher than typical ultra dilute conditions. While this strategy is effective at decreasing the solvent demand, it is still impractical to produce commercially relevant quantities of material using this method.

As a follow up, Harth and coworkers reported the synthesis of two related monomers. The first was a more synthetically accessible vinylbenzosulfone monomer that displayed similar behavior, but required fewer steps and less purification.[78] The second monomer included an electron donating group on the benzocyclobutene ring, which lowered the isomerization temperature to 150 °C, allowing for the possibility of incorporating thermally sensitive functional groups.[79]

"Click" reactions are an attractive means to synthesize SCNPs due to their high efficiency, high functional group tolerance, and mild reaction conditions.[23] Copper-mediated azide–alkyne cycloaddition,[10,42–44] thiol-ene addition,[45,80] thiol-yne addition,[81] and amine-isocyanate addition[40] click reactions have been used to synthesize cross-links and form SCNPs.

In a study by Pomposo and coworkers, SCNPs were constructed from alkyne-functionalized polymers using the copper-catalyzed Glaser–Hay alkyne coupling reaction. Interestingly, the alkyne-bearing polymer was synthesized without necessity of protection using redox-initiated RAFT polymerization.[50]

While not traditionally considered a "click" reaction, the tetrazine-norbornene reaction enjoys fast, quantitative conversions at room temperature without the need for a catalyst. O'Reilly and coworkers exploited the benefits of this reaction for SCNP synthesis.[82]

There are many well-studied photochemical reactions that are clean, high yielding, relatively fast, and require no chemical catalysts. A number of these have been examined for SCNP formation, including the photochemically triggered Diels–Alder reaction between 2,5-dimethylbenzophenone and maleimide,[39] the photo-dimerization of coumarin,[9,59] the photo-dimerization of anthracene,[60] and the photo-induced nitrile imine mediated tetrazole-ene cycloaddition.[83]

In addition to the aforementioned transformations, the photo-induced Bergman cyclization was utilized by Zhu and coworkers to form intramolecular polymeric nanoparticles.[52,53] The appropriate reactive diyne unit was incorporated into the polymer using both direct controlled radical polymerization and alkyne protection strategies. The resulting linear polymers were subjected to photochemical Bergman cyclization conditions using a continuous addition strategy, similar to the case previously discussed.[33]

A few cases of the use of nitrene chemistry to form SCNPs have been reported in the literature. In one example, Pu and coworkers took advantage of the sulfonyl azide moiety, which is capable of releasing nitrogen at high temperature to form sulfonyl nitrenes.[48] Due to the highly reactive nature of the resultant nitrene, the cross-links that are formed are not well-defined. In a similar approach by Li and coworkers, azido-functionalized polystyrene was exposed to UV light, resulting in the loss of nitrogen and formation of nitrene groups.[84] This strategy has multiple advantages: the use of photochemistry eliminates the need for high temperatures, and any unreacted azide groups can be further functionalized using the azide–alkyne "click" reaction.

Olefin metathesis is well-studied and offers a diverse catalogue of catalysts and synthetic opportunities. Coates and coworkers synthesized polymers functionalized with vinyl groups using a zinc-mediated alternating copolymerization technique. The polymers were subjected to olefin metathesis conditions using Grubbs' catalyst, resulting in the formation of intramolecular 1,2-disubstituted olefin cross-links.[41]

Amine-functionalized SCNPs were synthesized by copolymerizing Boc protected aminostyrene with chloromethylstyrene, and using the benzyl chloride units as a reactive handle to add styryl groups through a post-polymerization S_N2 reaction. The styrene groups were polymerized intramolecularly in dilute solution, followed by removal of the Boc groups. The amine-functionality of these nanoparticles was demonstrated by adding functionality *via* the Schotten–Baumann reaction.[37]

Pomposo and coworkers reported an example of SCNPs with catalytic activity.[1] Linear glycidy-functionalized polymers were collapsed *via* $B(C_6F_5)_3$ catalyzed ring-opening polymerization. The chain ends of the polymerized glycidyl units remained active, containing the catalyst in the nanoparticle. The authors found that polymers greater than 100 kDa required greater dilution in order to avoid intermolecular reactions. The catalytic aspect of this research is discussed further in the applications section of this review.

Pyun and coworkers synthesized a novel polymer containing propylenedioxy-thiophene units using ATRP, followed by SCNP formation through the oxidative polymerization of the pendant thiophene units. The ester units connecting the parent polymer to the thiophene units were then reductively cleaved, confirming the formation of polythiophene cross-links in the SCNPs.[47]

Zhao and coworkers reported the synthesis of single-chain tadpoles as shape amphiphiles. The tadpole particles were synthesized from a block copolymer precursor, containing a hydrophobic polystyrene block attached to a hydrophilic poly(2-(dimethylamino)ethyl methacrylate) (DMAEMA) block. The DMAEMA block was collapsed using a quaternization reaction of the tertiary amine with a diiodo cross-linker. The self-assembly of these tadpole-shaped structures was then studied in various polar and non-polar solvents.[51] These and other single-chain tadpoles[85–87] represent nanoscale architectures with multiple functionalities and controlled three-dimensional structure.

4.2.2 Dynamic Covalent Chemistry

Dynamic covalent chemistry is an interesting tool for the fabrication of macromolecular structures.[88] Based on the type of chemistry used, dynamic covalent bonds can be reversible in nature, kinetically fixed, or cleaved in response to their chemical environment. Conditions that affect dynamic covalent bonds include relevant chemical and physical stimuli such as pH, oxidation, or temperature.

The dynamic reaction of hydrazides with aldehydes produces hydrazones, which are reversible in the presence of acid. Fulton and coworkers reported SCNPs formed from poly(4-vinylbenzaldehyde) and a bis(hydrazide) cross-linker in the presence of catalytic trifluoroacetic acid.[20] In order to isolate the nanoparticles, the acid catalyst was quenched in order to kinetically trap the dynamic hydrazone bonds. The cross-linking density was studied as a function of the cross-linker concentration. The dynamic nature of the cross-links was confirmed by fully adorning an aldehyde-functionalized polymer with a monofunctional hydrazide, followed by isolation, redissolution, and addition of a hydrazide cross-linker along with an acid catalyst. Upon this treatment, a reduction in volume was observed, which is only possible in the case of hydrazone exchange. In a subsequent publication, the same authors reported a similar polymer system functionalized with oligo(ethylene glycol) units to add thermoresponsive behavior. At low pH, nanoparticles with hydrazone cross-links are kinetically trapped. Upon raising the pH and the temperature, the particles fall out of solution due to LCST behavior, and the dynamic nature of the cross-links results in hydrogel formation. Upon cooling, this process is reversed.[54]

Enamine bond formation operates in a similarly dynamic manner. Pomposo, Fulton, and coworkers synthesized reversible SCNPs using enamine chemistry. Initially, enamine-functionalized polymers were synthesized by the condensation of butylamine with a polymer containing beta-keto ester units. Nanoparticles were synthesized by a dynamic exchange reaction with ethylene diamine. The enamine cross-links were cleaved upon addition of phosphoric acid, and reformed with additional ethylene diamine.[57]

Disulfides are common in biological systems and are attractive for use in single-chain assembly owing to their dynamic nature and sensitivity to redox chemistry. In our laboratory, a polymer bearing anhydride units was synthesized and cross-linked in dilute solution upon addition of 4-aminophenyl disulfide. The disulfide links were reduced and oxidized reversibly using dithiothreitol (DTT) and iron(III) chloride, respectively.[55] Thayumanavan *et al.* have also taken advantage of disulfide chemistry using a different strategy.[89] Pyridyldisulfide ethyl methacrylate (PDSEMA) was used as a protected thiol monomer, with hydroxyethyl methacrylate as a comonomer. In the presence of stoichiometric DTT, the pyridinethiol moieties are removed and the remaining thiol units in the polymer readily form disulfides. This process was carried out at concentrations as high as 10 mg mL^{-1}, which is considerably higher than those in many other publications. Further addition of excess DTT results in the unravelling of the nanoparticles. The ability of the water-soluble nanoparticles to encapsulate hydrophobic guests and release them upon unfolding was also explored.

4.2.3 Non-covalent Chemistry

Non-covalent supramolecular interactions are the primary driving force behind the folding of biopolymers. Several researchers have explored non-covalent interactions such as hydrogen bonding as a means to effect SCNP formation. The use of H-bonds in SCNP systems typically requires a protection strategy to prevent aggregate formation during polymer synthesis. Several publications by Meijer and coworkers demonstrate this concept using the dimerization of 2-ureidopyrimidinone (UPy), displayed in Figure 4.3.[67–70] Once in dilute solution, the protecting groups are removed, resulting in quadruple H-bond formation and polymer collapse. These "metastable" SCNPs remain soluble when cast into a film; however, upon

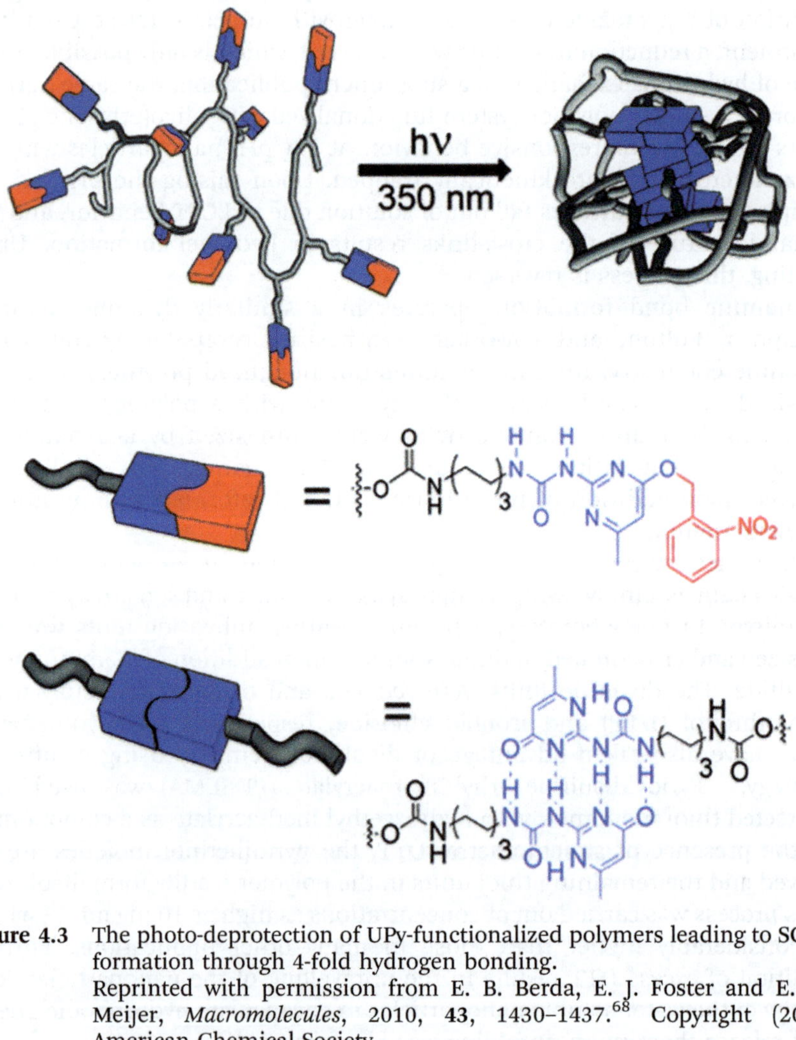

Figure 4.3 The photo-deprotection of UPy-functionalized polymers leading to SCNP formation through 4-fold hydrogen bonding.
Reprinted with permission from E. B. Berda, E. J. Foster and E. W. Meijer, *Macromolecules*, 2010, **43**, 1430–1437.[68] Copyright (2010) American Chemical Society.

heating, inter-chain H-bonds begin to form, resulting in an insoluble network.[67]

The UPy systems have been used to study several variables involved in SCNP formation. The rigidity of the polymer backbone and the molecular weight of the polymer did not have any effect on the polymers' ability to fold into SCNPs. Not surprisingly, solvent effects were found to be important – solvents with H-bond acceptors were found to prevent SCNP formation.[71]

Benzene-1,3,5-tricarboxamide (BTA) is well-known in the field of supramolecular chemistry due to its propensity to form helical assemblies *via* hydrogen bonding.[90] The BTA group has been used extensively in SCNP synthesis, both using the aforementioned protection strategy,[72] and by the direct polymerization of BTA-containing monomers in solvents capable of disrupting H-bonding.[2–4,8,91,92] Also, BTA-functionalized polymers have been synthesized using post-polymerization azide–alkyne "click" chemistry.[70]

An interesting example of single-chain folding involves the use of host–guest chemistry. Cucurbit[*n*]urils are known to form complexes with aromatic molecules. Scherman and coworkers synthesized methyl viologen- and naphthyl-functionalized polymers which formed ternary host–guest complexes when introduced to cucurbit[*n*]urils in dilute solution.[93,94] All cases were studied at particularly low polymer concentrations. At concentrations above 0.1 mg mL^{-1}, intermolecular aggregation was observed. Notably, this strategy did not require the use of any protection chemistry.

The use of metal-ligand coordination to form SCNPs is of great interest, largely for catalytic purposes, which are discussed further in the applications section of this chapter. In principle, the preparation of these SCNPs is very simple: a metal complex is added to a dilute solution of a ligand-bearing polymer, and the polymer-bound ligands undergo ligand exchange, resulting in intramolecular, metal-containing cross-links. Examples include polycyclooctadiene folded by metals such as rhodium,[75] iridium, or nickel;[7] acac-functionalized polymers containing copper(II),[5] and a palladium cross-linked triphenylphosphine-containing polymer.[6]

4.2.4 Multiple Intra-chain Interactions

Computational[95,96] and experimental investigations have concluded that the use of multiple orthogonal cross-linking interactions results in more compact SCNPs. In a study by Hosono and coworkers, UPy and BTA units were combined in a block copolymer, resulting in SCNPs with orthogonal H-bonding pairs (Figure 4.4).[70] Altintas *et al.* reported the dual-point folding of a single chain using a different set of orthogonal H-bonding dimers.[65] Another example by Chao *et al.* involved the supramolecular interaction between pendant tetraaniline units combined with permanent covalent cross-linking from a thiol-ene click reaction.[80]

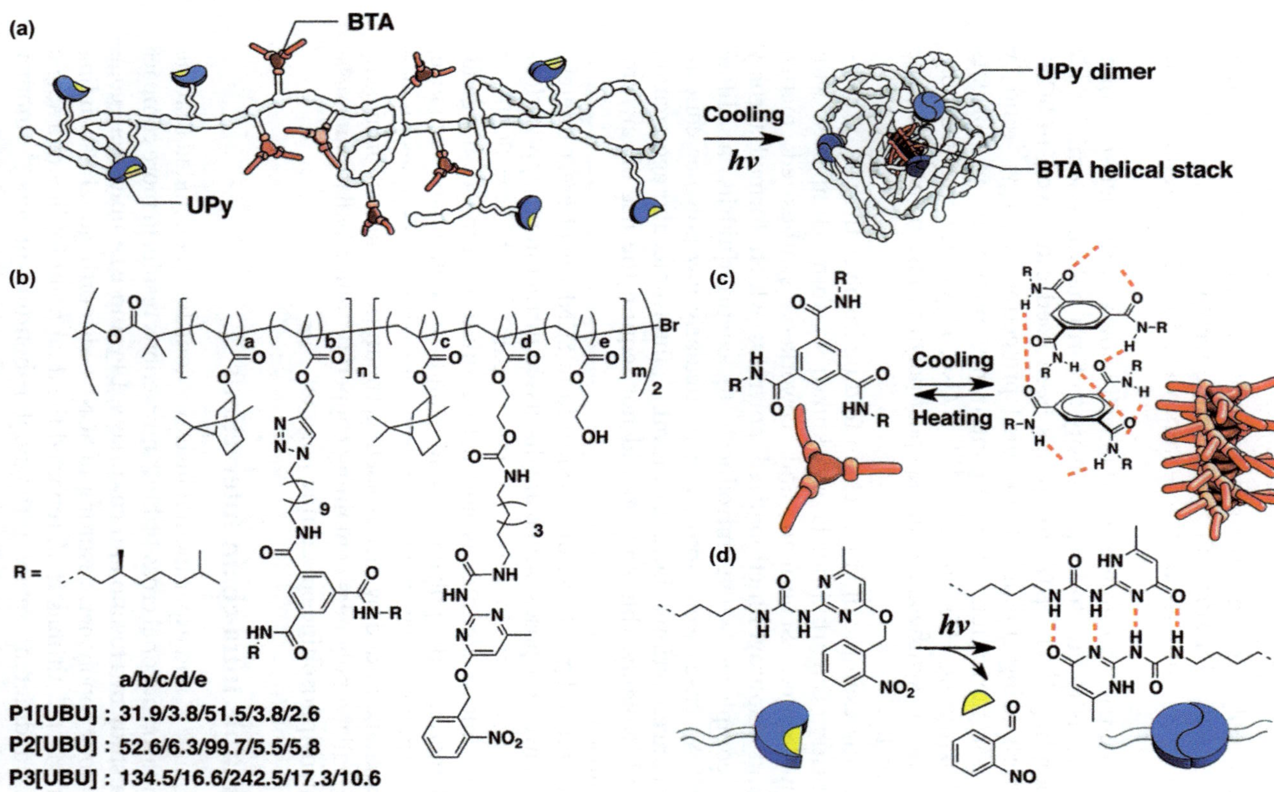

Figure 4.4 Orthogonal hydrogen bonding units in triblock copolymers. (a) Graphical representation of SCNP formation; (b) structure of block copolymer; (c) self-assembly of BTA units *via* threefold-symmetric hydrogen bonding; (d) photo-deprotection of UPy units to allow intramolecular 4-fold hydrogen bonding.
Reprinted with permission from N. Hosono, M. A. J. Gillissen, Y. Li, S. S. Sheiko, A. R. A. Palmans and E. W. Meijer, *J. Am. Chem. Soc.*, 2013, **135**, 501–510.[70] Copyright (2012) American Chemical Society.

4.2.5 Outlook

Surveying the literature, there are countless examples of the synthesis of SCNPs using a single covalent, dynamic covalent, or supramolecular interaction. Hence, the simple formation of SCNPs is relatively well-studied and not novel. Driving research in this field forward will require additional work related to the use of multiple orthogonal intra-chain interactions, clever methods of characterization, and the exploration of more complex polymer architectures, such as block copolymers, branched structures, and sequence-controlled polymers. The combination of the synthetic principles involved in single-chain technology and the state of the art in sequence controlled polymers developed by Perrier,[14] Lutz,[16] Whittaker,[97] and Hawker[98] provides possibilities that could lead to significant advances toward the fabrication of synthetic polymers capable of mimicking the structural precision found in natural folded macromolecules.

4.3 Characterization of SCNPs

As a result of the complex nature of SCNP formation, multiple analytical techniques are often required for the characterization of these materials. In this section, we will discuss the detection of the appearance or disappearance of cross-linking functional groups, as well as changes in size and morphology. In research involving SCNPs, there generally exists the opportunity for polymers to form cross-links between two or more polymer chains and form aggregates. It is therefore important to use techniques sensitive enough to detect small concentrations of these aggregates in order to prove that these nanostructures are indeed composed of single chains, as the moniker implies.

A recent report from our lab exemplified the need for the careful analysis of multiple methods of characterization for these materials. SCNPs were designed based on a poly(norbornene imide) backbone with pendant methacryloyl units and cross-linked *via* radical chemistry. GPC, multi-angle light scattering (MALS), and viscometric data all suggested efficient SCNP formation. When detailed nuclear magnetic resonance (NMR) studies were performed, it was determined that adventitious oxygen, not the methacryloyl moiety, was responsible for the change in solution conformation of this material.[99]

4.3.1 Size Exclusion Chromatography

Size exclusion chromatography (SEC) has been one of the most important tools for the characterization of SCNPs. Early reports of SCNP characterization used qualitative SEC measurements based on linear polymer standards.[32,33] As the work continued to progress, researchers began characterizing these materials with more quantitative measurements using multiple detectors such as MALS and viscometry.

The molecular weight of collapsed SCNPs cannot be compared to linear polymer standards when using SEC; however, SEC still provides vital data for characterizing SCNP formation. A decrease in the hydrodynamic radius can be observed, as well as an increase in the hydrodynamic radius for multi-chain aggregates. Harth *et al.* set out to carefully study SCNPs *via* SEC.[33] They first created a family of styrene and vinylbenzocyclobutane (BCB) copolymers and collapsed these polymers using BCB cross-linking. The linear polymers and the collapsed SCNPs were compared to polystyrene standards using SEC. It was observed that, when the polymer was collapsed, it had an increase in retention time as well as a decrease in apparent molecular weight. The BCB cross-linking reaction does not produce any side products (*vide supra*), so the decrease in apparent molecular weight can be ascribed to a decrease in the hydrodynamic volume. In addition, the authors used the change in apparent molecular weight, as well as dynamic light scattering (DLS) measurements, to calculate the decrease in hydrodynamic volume. In order to prove that the change in hydrodynamic volume could be attributed to BCB cross-linking, ^{1}H NMR studies were conducted to prove the disappearance of the BCB moiety.

Interestingly, when an SCNP collapses from the linear chain to a globular particle, a decrease in the polydispersity index ($Đ$) is often observed. Pomposo *et al.* performed a computational study that examined SCNP formation assuming theta solvent conditions.[100] Using these parameters, SCNPs could be treated as small linear polymers with a comparable hydrodynamic volume. Using the standard SEC calibration equation ($M_{app} = cM^{\beta}$), the researchers attributed this decrease in $Đ$ to the scaling factor derived from a hydrodynamic radius equation. Pomposo's work highlighted the importance of understanding the complex physics of SCNP formation using both mathematical and computational methods. This line of SCNP research is very important to understanding how these materials form, but there is still a lot of theoretical work to be done in this area.

4.3.2 Light Scattering

Light scattering was established as an analytical technique by renowned scientists such as Einstein,[101] Raman,[102] Debye,[103] and Zimm[104] in the early 20th century, and has become one of the most important means for absolute molecular weight characterization. Light scattering measurements do not rely on standards, as the molar mass is calculated based on first principles.[105] The two most common light scattering techniques for SCNP measurements are DLS and MALS. Our laboratory utilizes a MALS detector in tandem with SEC to prove that molecular weights are consistent from the parent polymer to the SCNPs.[55,60] Other groups have used DLS in a similar manner.[39,45,94]

4.3.3 Viscometry

Solution viscometry relates a particle's intrinsic viscosity to its molecular weight *via* the Mark–Houwink equation:

$$[\eta] = KM^{a}$$

where K and a are coefficients that describe the polymer's conformation as well as the interaction between polymer and solvent. These coefficients can be determined using intrinsic viscosity data from the viscometer as well as the molar mass determined by MALS. The hydrodynamic volume (V_h) and radius (R_h) can be found by viscometry as well using the Einstein–Simha relation:

$$V_h = \frac{M[\eta]}{2.5\,N_A}$$

$$R_h = \left(\frac{3V_h}{4\pi}\right)^{\frac{1}{3}}$$

where $[\eta]$ is the intrinsic viscosity, M is the molar mass, V_h is the hydrodynamic volume, N_A is Avogadro's number, and R_h is the hydrodynamic radius.

Hawker and coworkers studied SCNP formation using viscometric measurements (Figure 4.5).[40] In this experiment, they studied two parent polymers (100 kDa and 150 kDa) with pendant isocyanate moieties that were collapsed by an external diamine cross-linker. They found that the intrinsic viscosity of a particle is inversely proportional to the degree of crosslinking. The SCNPs, despite a 50% increase in molecular mass compared to their parents, both decreased in intrinsic viscosity compared to their parent polymers and the viscosities of the two nanoparticles were similar to one another. These data are consistent with Einstein's prediction that the intrinsic viscosity of a constant density sphere is independent of its molecular weight.

Perez-Baena *et al.* expanded on this work and determined the relationship between intrinsic viscosity and molecular weight of SCNPs as a function of cross-linking density.[106]

4.3.4 NMR Spectroscopy

^{1}H NMR spectroscopy is widely used in SCNP research to monitor the appearance or disappearance of signals from internal or external cross-link bond formation. Beyond this, advanced NMR techniques have proven useful for studying the coil-to-SCNP transition. Zhao *et al.* studied the ^{1}H NMR spin–spin relaxation time (T_2) of SCNPs cross-linked by the photodimerization of coumarin.[9] It was observed that increasing the degree of dimerization caused a decrease in the spin–spin relaxation time. This was attributed to sections of the chains having reduced mobility after collapse. A reduction of polymer proton signals was also observed when going from parent to SCNPs.

DOSY experiments were performed by Loinaz and coworkers in order to determine the diffusion coefficient of thermoresponsive poly(*N*-isopropylacrylamide) based SCNPs in solution.[43] The diffusion coefficient was inversely proportional to the hydrodynamic volume, meaning that the SCNPs had a greater diffusion coefficient than their parents.

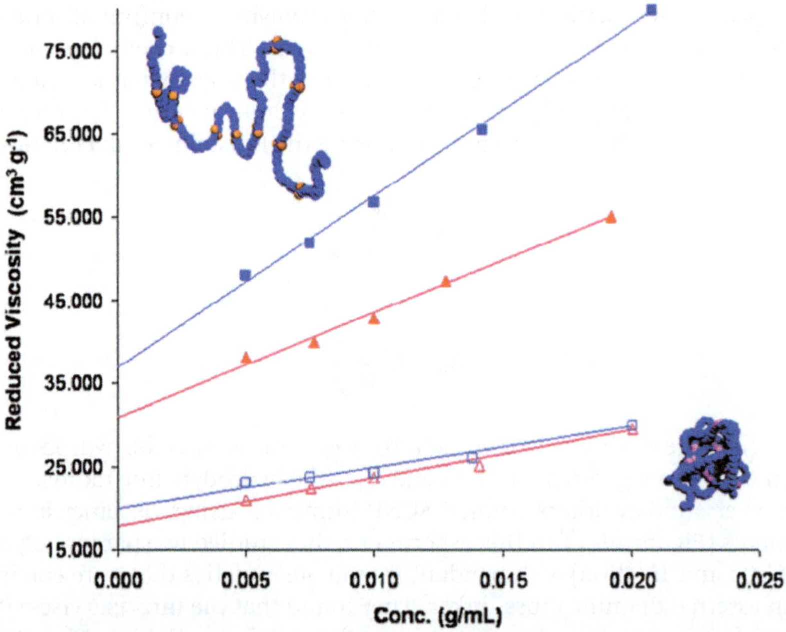

Figure 4.5 Plot of reduced viscosity *versus* concentration for control copolymers (■, 150 kDa; ▲, 100 kDa) and their analogous crosslinked nanoparticles (□, 150 kDa; △, 100 kDa) in THF. Both copolymers were prepared from starting polymers incorporating *ca.* 15 mol% NCO functional groups along the backbone.
Reprinted with permission from J. B. Beck, K. L. Killops, T. Kang, K. Sivanandan, A. Bayles, M. E. Mackay, K. L. Wooley and C. J. Hawker, *Macromolecules*, 2009, **42**, 5629–5635.[40] Copyright (2009) American Chemical Society.

4.3.5 Characterizing the Morphology of SCNPs

The morphology of SCNPs is challenging to characterize, because of the complex interactions with solvents and the interaction between polymer chains at different concentrations. Several parallels can be drawn to the field of dendrimer research and the characterization of the morphology of those materials. As discussed in a review[107] by Meijer *et al.,* dendrimers tend to have very different morphologies from what was expected based on system design. In the same way, the growing library of SCNP research suggests that these materials behave very differently in solution compared to what would be expected based on linear chains. One of the most powerful tools for SCNP morphology characterization is solution-free microscopy. Atomic force microscopy (AFM) and transmission electron microscopy (TEM) allow for in depth studies of nanoparticle size, shape, and aggregation. To determine the morphology of these materials in solution, researchers have used small-angle neutron scattering (SANS) and small-angle X-ray scattering (SAXS). The corroboration of microscopy and scattering data, along with molecular simulations, can provide a glimpse into the true nature of SCNPs. In order to observe individual SCNPs, concentrations much lower than the overlap concentration (C^*) must be used to avoid the formation of aggregates. Several studies have shown that increasing the molecular weight of the parent polymer increases the size of the particle, and that increasing the amount of cross-linking decreases the particle size.

Meijer and coworkers carried out detailed AFM studies on 2-ureidopyrimidonone (UPy) cross-linked polymers (*vide supra*).[67] AFM provided enough information to allow the authors to deduce the possible morphology of SCNPs formed from different polymer chains. A method for calculating the diameter of these particles was developed, which assumed a hemi-ellipsoidal conformation. AFM images with the diameter calculation applied showed SCNPs with a similar size distribution to that obtained from GPC data. In a subsequent study,[68] Meijer *et al.* reported monitoring the deprotection of UPy groups and the chain collapse by AFM. Variables such as concentration of sample, solvent, and choice of surface strongly influenced these data. In this study, the difference between SCNPs and aggregates was evident in the height and phase images, ensuring that the characterization applied only to unimolecular entities. Globular, single-chain particles possessed a darkened core compared to no changes in phase for multi-chain aggregates.

Meijer and coworkers also reported a study[69] on particle aggregation mediated by solvent evaporation. AFM sample preparation requires the material to be drop cast onto substrates and then have the solvent evaporated. Aggregation can come about during the final stages of evaporation, when the particles concentrate in the center of an evaporating droplet. The researchers conducted AFM studies designed to intentionally form particle aggregates by varying the vapor pressure and solvent during evaporation. It was determined that nanoparticle mobility in solution and the amount of time required for solvent evaporation are the major factors that control the SCNP morphology observed by AFM.

A recent report[108] from Hosono and coworkers sought to explore the folding of UPy and BPA containing SCNP using single-molecule force microscopy (SMFS) (Figure 4.6). One end of the polymer chain was attached to an AFM cantilever and the polymer was nano-mechanically unfolded and the force profile was recorded and analyzed. The unfolding data gave some insight into the internal structure as well as allowed the researchers to calculate the dissociation constant (k_d) for BTA-containing SCNPs.

When SCNPs are drop cast onto a level surface for AFM and TEM sample preparation, the morphology of the particle changes (Figure 4.7). Dimensional analysis can be carried out by AFM, while TEM generally shows better diameter correlation with DLS studies. Zhao *et al.*[51] and Pomposo *et al.*[1] set out to compare AFM, TEM, and DLS data for several SCNPs. TEM results were close to the hydrodynamic diameter determined by DLS, while AFM results were much larger. This size discrepancy between methods is most likely due to the broadness of the AFM tip. TEM can provide a more accurate estimate of the diameter of SCNPs in solution, but the diameter is still

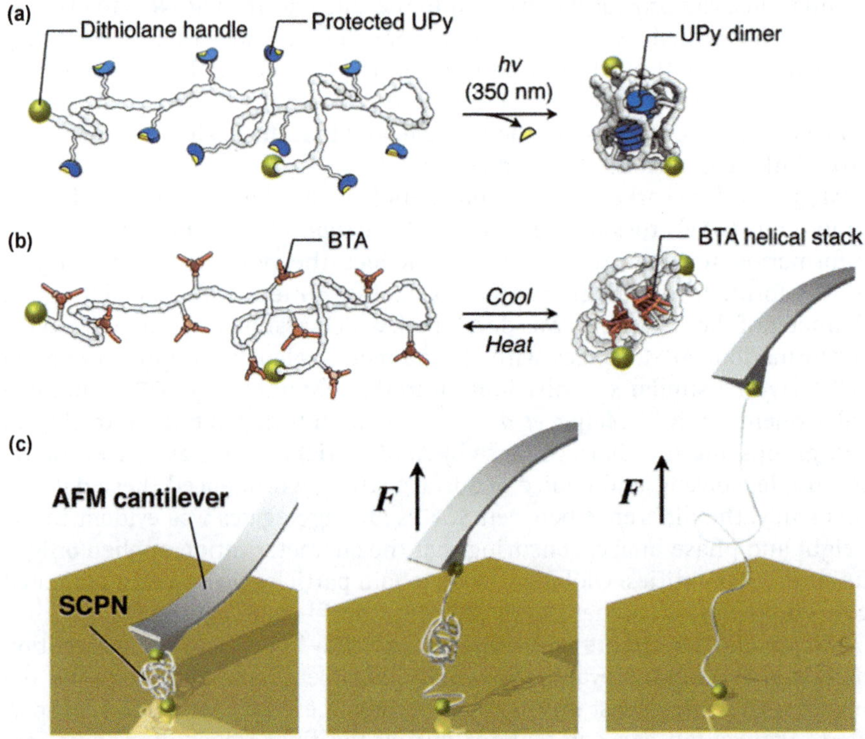

Figure 4.6 Schematic representations of folding polymer with (a) UPy modules P[UPy*n*] and (b) BTA modules P[BTA*n*], which self-assemble into dimer and helical columnar aggregates, respectively, resulting in SCNP formation. (c) Schematic illustrations of the mechanical unfolding experiment on SCNPs.
Reprinted with permission from N. Hosono, A. M. Kushner, J. Chung, A. R. A. Palmans, Z. Guan and E. W. Meijer, *J. Am. Chem. Soc.*, 2015.[108] Copyright (2015) American Chemical Society.

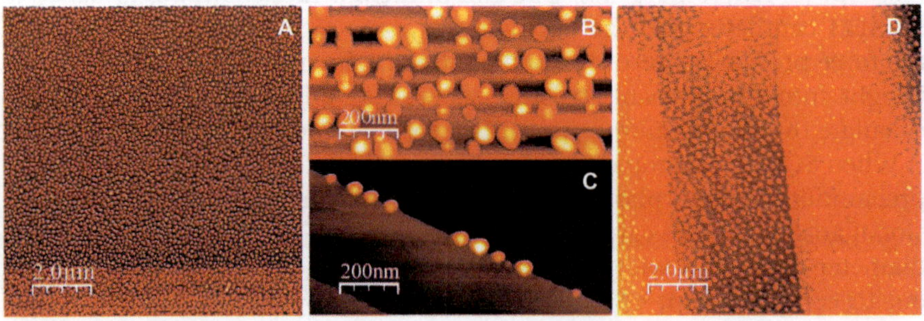

Figure 4.7 AFM images of single-chain nanoparticles. Panels A and B are on mica, while panels C and D are on graphite.
Reprinted with permission from E. B. Berda, E. J. Foster and E. W. Meijer, *Macromolecules*, 2010, **43**, 1430–1437.[68] Copyright (2009) American Chemical Society.

underestimated, as the particle is swollen in solution but more compact when dry. The variation in measurements between instruments and sample preparation highlights the need to utilize multiple methods of characterization for these materials.

Techniques such as AFM and TEM provide valuable data for the characterization of the morphology of SCNPs, but their behavior in the absence of solvent and their interaction with substrates are still not well understood. In order to study SCNPs in solution, small angle scattering techniques such as SANS and SAXS can be used for characterization.[95] In contrast to TEM and AFM, which tend to suggest compact morphologies, SAXS and SANS indicate less compact morphologies, especially in good solvent conditions. The radius of gyration and the form factor for SCNPs can also be worked out using these scattering techniques.[1,3,11,70,95,109] In addition, researchers have used molecular dynamics (MD) simulations to better understand the data provided by small angle scattering techniques.[1,11,95]

Meijer *et al.* demonstrated that SAXS could provide evidence of SCNP formation by a clear reduction in the radius of gyration from coils to nanoparticles.[70] O'Reilly and coworkers showed that larger R_g values were obtained *via* SANS when compared to data from AFM and TEM. The larger R_g values were attributed to aggregates forming at room temperature, which was backed up by data from DLS temperature studies.[82] These studies show that solvent free techniques are not representative of how SCNPs behave in solution. MD simulations suggest that, when the quality of the solvent for the polymer system is decreased, more compact morphologies result, similar to the morphologies observed in solvent free techniques.[11]

Pomposo and coworkers used MD simulations and experimental studies to compare heterofunctionalized nanoparticles and their homofunctional counterparts.[95] MD simulations suggested that heterofunctional species would be more compact than their homofunctional counterparts, and experimental SEC/MALS traces and SAXS data supported these results. SAXS data showed larger particle sizes when compared to the MD simulations, but this was attributed to bending and torsional barriers inherent in the polymer that were not taken into account in the simulations.

Voets *et al.* performed an in-depth study of SCNPs collapsed by BTA self-assembly using SANS, SAXS, and DLS in order to elucidate the folding process as a function of the chain length.[91] Intermolecular self-assembly of BTA is generally a cooperative process, but the experiments showed a lack of cooperativity in the intramolecular folding process. In a recent study by Huurne and coworkers, SAXS was used to characterize the structure of a BiPy-BTA containing poly(norbornene imide) under several different solvent conditions. Under optimal solvent conditions, individual BiPy-BTA stacks could actually be observed, affording the first SAXS visualization of the interior of an SCNP.

Data suggest that most SCNPs resemble an intrinsically disordered globular coil, but simulations of the folding process for a single polymer chain have shown that there is potential for these materials to mimic the control of many natural processes. Yoshinaga and coworkers,[110] as well as Danilov *et al.*,[111] performed Monte Carlo simulations that provided great

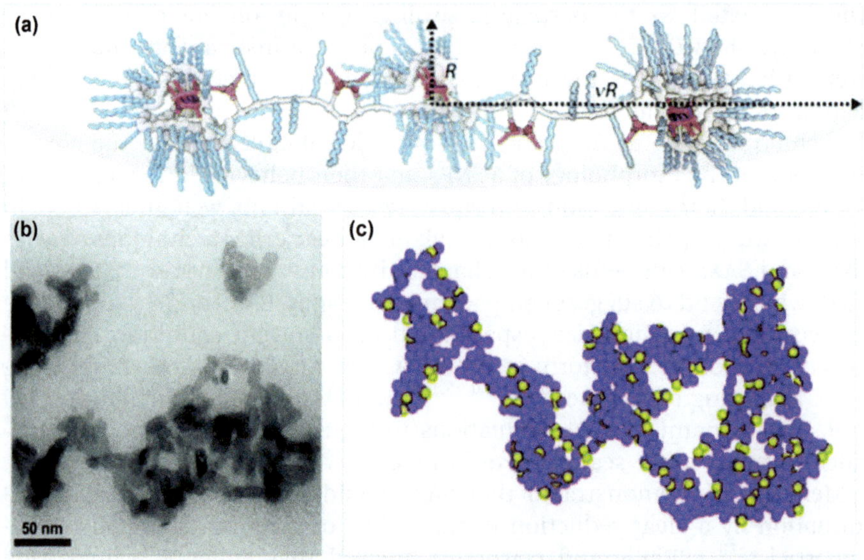

Figure 4.8 (a) Ellipsoidal structure proposed by Meijer *et al.*; (b) visualization of ellipsoidal SCNPs formed by photocrossinking of pendant anthracene groups; (c) SCNP structure obtained from MD simulations.
Reprinted with permission from (a) P. J. M. Stals, M. A. J. Gillissen, T. F. E. Paffen, T. F. A. de Greef, P. Lindner, E. W. Meijer, A. R. A. Palmans and I. K. Voets, *Macromolecules*, 2014, **47**, 2947–2954[91] and (b) A. J. Moreno, F. Lo Verso, A. Sanchez-Sanchez, A. Arbe, J. Colmenero and J. A. Pomposo, *Macromolecules*, 2013, **46**, 9748–9759.[95] Copyright (2013), (2014) American Chemical Society. (c) Reprinted from ref. 60 with permission from John Wiley and Sons. Copyright © 2013 WILEY-VCH Verlag GmbH & Co. KGaA, Weinheim.

insight to the folding process of SCNPs. The data from these studies suggest that SCNPs can adopt unique conformations similar to globular proteins, though current materials are far from this ideal.

The current literature suggests that SCNPs do not adopt a compact globular conformation. Cross-linking reactions are generally carried out in good solvents to avoid aggregation, thus cross-links are formed based on short-range interactions in the polymer chain. Ellipsoidal structures were observed *via* SANS by Meijer *et al.* (Figure 4.8a)[91] and *via* TEM in works from our laboratory (Figure 4.8b).[60] Pomposo and coworkers performed computations[95] that suggest that multiple orthogonal cross-linking chemistries will lead to a more compact globular material (Figure 4.8c), and experiments from the Meijer lab[70] and our group[80] have supported this claim.

4.4 Potential Applications

SCNPs show promise for use as catalysts,[1–5,112] sensors,[8] nanoreactors,[9] and in nanomedicine.[10–12,112,113] This class of materials is particularly suited to these applications because of their small size and the plethora of functional groups that can be incorporated. In addition, the interior of the particle can serve as a useful chemical environment because it can be hydrophobic or hydrophilic depending on solvent and polymer design. The size of the interior can be controlled by the molecular weight of the parent polymer and the amount of intra-chain crosslinking. This section will summarize recent research involving SCNP applications.

4.4.1 Catalysis

Enzymes are an amazing class of macromolecules that can catalyze reactions in aqueous media, and are therefore an important target to mimic in the lab. A large amount of research has been focused on modeling enzymes, but synthetic systems tend to be less efficient catalysts. Most of these synthetic systems have an active site that is exposed to an aqueous environment, unlike enzymes, which protect their active sites in a hydrophobic pocket.[114] Researchers have begun using an SCNP scaffold by incorporating polymer backbones into synthetic enzyme models, thereby creating a hydrophobic pocket for the model active site. This opens up interesting possibilities for tuning efficient catalytic systems by altering the polymer solubility, increasing the accessibility to a larger library of substrates, and increasing the turnover frequency (TOF). In addition to tuning the catalytic activity, the SCNP motif allows for facile recovery and regeneration of the catalyst.[115] Most homogeneous catalysts have similar solubility to the reactants and products, and these reactions are generally carried out in organic solvents. This is especially troubling when these catalysts are used on an industrial scale since separating and recovering the catalyst can create large amounts of waste. By adding a polymer support to the catalyst, the product and the catalyst can be easily separated.

Perez-Baena *et al.* reported SCNPs that contained a tris(pentafluorophenyl)boron ($B(C_6F_6)_3$) active site to mimic the function of reductase and polymerase.[1] Catalysis with this SCNPs could only be carried out in halogenated solvents, toluene, and benzene. Solvents that could form adducts with boron were unsuitable, as they quenched the catalyst. The kinetics and TOF of catalysis by these particles were influenced by size, composition, quantity, and location of the active sites in the SCNPs. The TOF was also proportional to the molecular weight and inversely proportional to the hydrodynamic radius of the SCNPs.

In an attempt to mimic an enzyme's catalytic activity in water, Terashima *et al.* reported a polymer consisting of PEG, BTA, and diphenylphosphinostyrene (SDP) polymerized by ruthenium catalyzed CRP.[3] In aqueous conditions, these polymers self assembled to form SCNPs and the SDP moieties ligated to the ruthenium catalyst in the hydrophobic cavities within the particle. The BTA units in the polymer formed α-helices that could withstand aqueous catalytic conditions (0.4 M HCOONa at 40 °C). The SDP-bound ruthenium was able to catalyze the transfer hydrogenation of cyclohexanone and acetophenone in water with a TOF comparable to that of other catalysts.[116–118] Subsequently, the authors reported that the BTA-induced self-assembly was not necessary for the catalytic system. The ligation of the SDP moiety to the ruthenium was enough to collapse the linear polymer into a folded structure containing hydrophobic pockets, where the catalysis could take place. In both reports, the SCNP-bound catalyst was more effective than the free catalyst.[2]

Huerta *et al.* demonstrated that compartmentalizing the active site within a hydrophobic pocket was critical for the catalytic activity in these systems.[4] In this report, a water-soluble methacrylate copolymer was synthesized, containing BTA units for self-assembly as well as L-proline (L-pro) units for organocatalytic activity. The L-pro catalyzed aldol condensation of cyclohexanone and *p*-nitrobenzaldehyde was studied. It was determined that higher molecular weight polymers (~38 kDa) were more active than lower molecular weight polymers (~28 kDa) and that no reaction occurred in polymers that did not contain BTA. Interestingly, catalysis by the lower molecular weight SCNPs showed no diastereoselectivity and reduced enatioselectivity. Catalytic efficiency in this system depended on the hydrophobic environment within the SCNPs, which shields the active site from H-bonding interactions and increases the effective concentration of the substrate. Controlling the size of the SCNPs in this system is essential for tuning the catalytic efficiency.

The library of reactions catalyzed by SCNPs is rapidly growing. Pomposo *et al.* reported the synthesis of a catalytic SCNP system used as a catalyst for alkyne homo-coupling.[5] These metal-containing SCNPs showed greater activity, even at a lower catalyst concentration, when compared to a free copper complex. In a recent report, Willenbacher *et al.* reported using SCNPs formed by Pd complexation to catalyze Sonagashira coupling.[6] Iridium containing SCNPs based on pCOD were used to catalyze the reduction of amides in a report by Mavila *et al.*[7] The same report also featured rhodium-containing SCNPs used to catalyze cross couplings.

He *et al.* reported SCNPs made by photo-crosslinking coumarin-containing random copolymers of poly(*N*,*N*-dimethylaminoethyl methacrylate).[9] These amino-functionalized SCNPs were then used as nanoreactors to synthesize gold nanoparticles (AuNPs). A dilute solution of the SCNPs was stirred with $HAuCl_4$ and the tertiary amines in the polymer reduced and stabilized the AuNPs. The authors found that the more compact SCNPs provided a higher effective concentration of amine and, therefore, a faster reduction of gold.

4.4.2 Nano-medicine

Perez-Baena *et al.* reported a potential MRI contrast agent based on SCNPs that contained multiple gadolinium(III) sites.[10] This material was based on a water soluble, azide functionalized acrylic polymer that was cross-linked by di-alkyne functionalized Gd(III) diethylenetriaminepentaacetic acid (DTPA).

When compared to Magnevist, a commonly used contrast agent, the conformationally locked Gd(III) centers in the SCNP system showed a 2-fold increase in relaxation time. When the Gd(III) cross-linker was tested without the SCNP scaffold, the relaxation times were four times slower. Harth and coworkers also developed a Gd-containing SCNP that fluoresced for use as an imaging agent in biological systems.[112]

Pomposo and coworkers studied poly(MMA-*r*-AEMA)-based SCNPs using SANS.[11] The polymers were crosslinked with bi- and trifunctional cross-linking units that reacted *via* Michael addition. These nanoparticles were then studied as a potential nanocarrier for vitamin B_9. The particles were loaded with vitamin B_9 and the release was monitored by UV-Vis spectroscopy. Complete release of B_9 was achieved in 5–6 hours in neutral pH water. The release exponent was calculated to be approximately 0.5, which suggests a Fickian diffusion mechanism for this SCNP based system.

Harth *et al.* reported the synthesis of a targeted drug delivery system for breast cancer that was based on polycarbonate nanosponges.[12,45] These nanoparticles were cross-linked by reacting a diamine cross-linker with pendant epoxide moieties. A targeting peptide able to recognize tumor cells that had been exposed to ionizing radiation was attached to the nanosponge with thiol-ene "click" chemistry. The nanoparticle was then loaded with the drug paclitaxel. Mouse studies on this SCNP based system showed slower tumor growth and greater paclitaxel retention times compared to systemic paclitaxel. The same group developed a similar system based on a lung cancer model, where paclitaxel and camptothecin were administered sequentially.[113]

4.4.3 Chemical Sensors

SCNPs have been deomonstrated as sensors for metal ions by Gillissen *et al.*[8] A poly(norbornene imide) polymer was cross-linked with 3,3-bis(acylamino)-2,2-bipyridine substituted benzene-1,3,5-tricarboxamides (BiPy-BTAs). The BiPy-BTA moiety served as both the metal binding site and the fluorescing agent. Because BiPy-BTA fluoresces by aggregate induced emission, the fluorescence is quenched by the presence of a metal ion binding to the BiPy site (Figure 4.9).

4.4.4 Self-assembly

One of the challenges that face SCNP research is to design a synthetic polymeric system that can act as building blocks for more complex nanomachinery. In enzymes, there is a perfectly defined primary structure that imparts the secondary and tertiary structure, which is what imparts blocking or opening the active site. While we have not achieved a perfectly defined primary structure, Barner-Kowollik *et al.* reported a crude tertiary structure imparted by the interaction of cyanuric acid-Hamilton wedge[64] and thymine-diaminopyridine.[66] This system was polymerized *via* ATRP featuring an

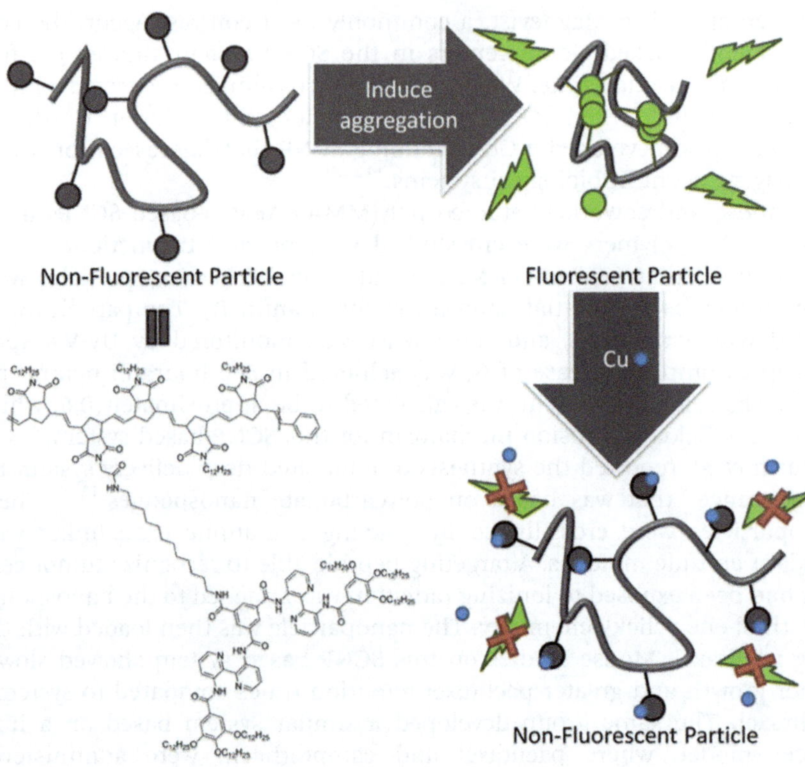

Figure 4.9 A Schematic representation of the sensing function of the BiPy-BTA functional polymers.
Reproduced from ref. 8 with permission from the Royal Society of Chemistry.

initiator that was functionalized with one of the H-bonding units. The opposing H-bonding unit was attached post-polymerization *via* "click" chemistry. In another report, the researchers created a single polymer chain with the two orthogonal H-bonding units separated by a block of polystyrene.[65] Variable temperature ^{1}H NMR spectroscopy showed the H-bonding between the cyanuric acid, Hamilton wedge, and thymine moieties by N–H chemical shift. Light scattering measurements backed up this data, showing a decrease in hydrodynamic volume of the material. The H-bonding in this system can be reversed by increasing the temperature, unfolding the polymer into a random coil.

Wen *et al.* synthesized an SCNP-shaped amphiphile based on poly(2-(dimethylamino)ethylmethacrylate)-*block*-polystyrene (PDMAEMA-*b*-PS).[51] This polymer was cross-linked *via* a Menshutkin reaction where the tertiary amine was reacted with 1,4-diiodobutane to form "tadpole" shape amphiphiles. This design features a hydrophilic SCNP head and a hydrophobic polystyrene tail. Self-assembly of these materials resulted in micelles or vesicles based on solvent choice. In a subsequent report[119] from the Zhao group, amphiphilic SCNPs were used as stabilizers for the suspension polymerization of styrene. Tao and Liu,[120] as well as Kim *et al.*,[121] have reported similar amphiphilic systems.

4.5 Summary and Outlook

Well-defined nanostructures of this size and type have the potential to make an impact in several areas of research. Because of the ease of synthesis and the plethora of possible end use applications, there is a growing interest in this class of materials. Before these nanostructures can be produced on a practical scale though, several key challenges must be overcome. First, in almost all cases, ultra-dilute conditions are needed for collapse. Second, analogous to protein characterization, in-depth structural characterization of these materials needs to be carried out. Lastly, the libraries of these structures need to be expanded from simple folded linear chains to more advanced polymer architectures. Despite these challenges, SCNPs show promise at solving real world problems and will remain a high priority in polymer research. The goal of this review was to highlight the different chemical reactions that have been utilized to synthesize SCNPs, show the most effective methods of SCNP characterization, and touch on the multitude of potential applications for this incredible class of materials.

Acknowledgements

We graciously acknowledge financial support from the University of New Hampshire, the Army Research Office for support through award W911NF-14-1-0177, and NIST for support through award 70NANB15H060.

References

1. I. Perez-Baena, F. Barroso-Bujans, U. Gasser, A. Arbe, A. J. Moreno, J. Colmenero and J. A. Pomposo, *ACS Macro Lett.*, 2013, **2**, 775–779.
2. M. Artar, T. Terashima, M. Sawamoto, E. W. Meijer and A. R. A. Palmans, *J. Polym. Sci., Part A: Polym. Chem.*, 2014, **52**, 12–20.
3. T. Terashima, T. Mes, T. F. A. De Greef, M. A. J. Gillissen, P. Besenius, A. R. A. Palmans and E. W. Meijer, *J. Am. Chem. Soc.*, 2011, **133**, 4742–4745.
4. E. Huerta, P. J. M. Stals, E. W. Meijer and A. R. A. Palmans, *Angew. Chem., Int. Ed.*, 2013, **52**, 2906–2910.
5. A. Sanchez-Sanchez, A. Arbe, J. Colmenero and J. A. Pomposo, *ACS Macro Lett.*, 2014, **3**, 439–443.
6. J. Willenbacher, O. Altintas, V. Trouillet, N. Knofel, M. J. Monteiro, P. W. Roesky and C. Barner-Kowollik, *Polym. Chem.*, 2015, **6**, 4358–4365.
7. S. Mavila, I. Rozenberg and N. G. Lemcoff, *Chem. Sci.*, 2014, **5**, 4196–4203.

8. M. A. J. Gillissen, I. K. Voets, E. W. Meijer and A. R. A. Palmans, *Polym. Chem.*, 2012, **3**, 3166–3174.
9. J. He, L. Tremblay, S. Lacelle and Y. Zhao, *Soft Matter*, 2011, **7**, 2380–2386.
10. I. Perez-Baena, I. Loinaz, D. Padro, I. Garcia, H. J. Grande and I. Odriozola, *J. Mater. Chem.*, 2010, **20**, 6916–6922.
11. A. Sanchez-Sanchez, S. Akbari, A. Etxeberria, A. Arbe, U. Gasser, A. J. Moreno, J. Colmenero and J. A. Pomposo, *ACS Macro Lett.*, 2013, **2**, 491–495.
12. R. J. Passarella, D. E. Spratt, A. E. van der Ende, J. G. Phillips, H. Wu, V. Sathiyakumar, L. Zhou, D. E. Hallahan, E. Harth and R. Diaz, *Cancer Res.*, 2010, **70**, 4550–4559.
13. G. Hariri, A. D. Edwards, T. B. Merrill, J. M. Greenbaum, A. E. van der Ende and E. Harth, *Mol. Pharmaceutics*, 2013, **11**, 265–275.
14. G. Gody, T. Maschmeyer, P. B. Zetterlund and S. Perrier, *Nat. Commun.*, 2013, **4**, 2505.
15. E. B. Berda and K. B. Wagener, *Materials Science and Technology*, Wiley-VCH Verlag GmbH & Co. KGaA, 2006.
16. J.-F. Lutz, M. Ouchi, D. R. Liu and M. Sawamoto, *Science*, 2013, **341**, 1238149.
17. M. Gonzalez-Burgos, A. Latorre-Sanchez and J. A. Pomposo, *Chem. Soc. Rev.*, 2015, **44**, 6122–6142.
18. M. Huo, N. Wang, T. Fang, M. Sun, Y. Wei and J. Yuan, *Polymer*, 2015, **66**, A11–A21.
19. C. K. Lyon, A. Prasher, A. M. Hanlon, B. T. Tuten, C. A. Tooley, P. G. Frank and E. B. Berda, *Polym. Chem.*, 2015, **6**, 181–197.
20. B. S. Murray and D. A. Fulton, *Macromolecules*, 2011, **44**, 7242–7252.
21. M. K. Aiertza, I. Odriozola, G. Cabanero, H.-J. Grande and I. Loinaz, *Cell. Mol. Life Sci.*, 2012, **69**, 337–346.
22. O. Altintas and C. Barner-Kowollik, *Macromol. Rapid Commun.*, 2012, **33**, 958–971.
23. A. Sanchez-Sanchez, I. Perez-Baena and J. A. Pomposo, *Molecules*, 2013, **18**, 3339–3355.
24. G. B. W. Kuhn, *J. Polym. Sci.*, 1962, **57**, 311.
25. P. Longi, F. Greco and U. Rossi, *Makromol. Chem.*, 1968, **116**, 113–121.
26. P. Longi, F. Greco and U. Rossi, *Makromol. Chem.*, 1969, **129**, 157–164.
27. J. E. E. B. E. Martin, *Macromolecules*, 1983, **16**, 1345–1350.
28. J. E. E. B. E. Martin, *Macromolecules*, 1983, **16**, 1350–1358.
29. V. A. Davankov and M. P. Tsyurupa, *React. Polym.*, 1990, **13**, 27–42.
30. M. P. Tsyurupa, T. A. Mrachkovsksya, L. A. Maslova, G. I. Timofeeva, L. V. Dubrovina, E. F. Titova, V. A. Davankov and V. M. Menshov, *React. Polym.*, 1993, **19**, 55–66.
31. V. A. Davankov, M. M. Ilyin, M. P. Tsyurupa, G. I. Timofeeva and L. V. Dubrovina, *Macromolecules*, 1996, **29**, 8398–8403.
32. V. A. Davankov, M. M. Ilyin, M. P. Tsyurupa, G. I. Timofeeva and L. V. Dubrovina, *Macromolecules*, 1996, **29**, 8398–8403.

33. E. Harth, B. Van Horn, V. Y. Lee, D. S. Germack, C. P. Gonzales, R. D. Miller and C. J. Hawker, *J. Am. Chem. Soc.*, 2002, **124**, 8653–8660.
34. A. Tuteja, M. E. Mackay, C. J. Hawker, B. Van Horn and D. L. Ho, *J. Polym. Sci., Part B: Polym. Phys.*, 2006, **44**, 1930–1947.
35. J. N. Dobish, S. K. Hamilton and E. Harth, *Polym. Chem.*, 2012, **3**, 857–860.
36. T. E. Dukette, M. E. Mackay, B. Van Horn, K. L. Wooley, E. Drockenmuller, M. Malkoch and C. J. Hawker, *Nano Lett.*, 2005, **5**, 1704–1709.
37. J. Jiang and S. Thayumanavan, *Macromolecules*, 2005, **38**, 5886–5891.
38. D. Mecerreyes, V. Lee, C. J. Hawker, J. L. Hedrick, A. Wursch, W. Volksen, T. Magbitang, E. Huang and R. D. Miller, *Adv. Mater.*, 2001, **13**, 204–208.
39. O. Altintas, J. Willenbacher, K. N. R. Wuest, K. K. Oehlenschlaeger, P. Krolla-Sidenstein, H. Gliemann and C. Barner-Kowollik, *Macromolecules*, 2013, **46**, 8092–8101.
40. J. B. Beck, K. L. Killops, T. Kang, K. Sivanandan, A. Bayles, M. E. Mackay, K. L. Wooley and C. J. Hawker, *Macromolecules*, 2009, **42**, 5629–5635.
41. A. E. Cherian, F. C. Sun, S. S. Sheiko and G. W. Coates, *J. Am. Chem. Soc.*, 2007, **129**, 11350–11351.
42. A. Ruiz de Luzuriaga, N. Ormategui, H. J. Grande, I. Odriozola, J. A. Pomposo and I. Loinaz, *Macromol. Rapid Commun.*, 2008, **29**, 1156–1160.
43. N. Ormategui, I. Garcia, D. Padro, G. Cabanero, H. J. Grande and I. Loinaz, *Soft Matter*, 2012, **8**, 734–740.
44. A. Ruiz de Luzuriaga, I. Perez-Baena, S. Montes, I. Loinaz, I. Odriozola, I. Garcia and J. A. Pomposo, *Macromol. Symp.*, 2010, **296**, 303–310.
45. D. M. Stevens, S. Tempelaar, A. P. Dove and E. Harth, *ACS Macro Lett.*, 2012, **1**, 915–918.
46. I. Perez-Baena, I. Asenjo-Sanz, A. Arbe, A. J. Moreno, F. Lo Verso, J. Colmenero and J. A. Pomposo, *Macromolecules*, 2014, **47**, 8270–8280.
47. P. T. Dirlam, H. J. Kim, K. J. Arrington, W. J. Chung, R. Sahoo, L. J. Hill, P. J. Costanzo, P. Theato, K. Char and J. Pyun, *Polym. Chem.*, 2013, **4**, 3765–3773.
48. X. Jiang, H. Pu and P. Wang, *Polymer*, 2011, **52**, 3597–3602.
49. E. H. H. Wong and G. G. Qiao, *Macromolecules*, 2015, **48**, 1371–1379.
50. A. Sanchez-Sanchez, I. Asenjo-Sanz, L. Buruaga and J. A. Pomposo, *Macromol. Rapid Commun.*, 2012, **33**(1262–1267), 1261–1266.
51. J. Wen, L. Yuan, Y. Yang, L. Liu and H. Zhao, *ACS Macro Lett.*, 2013, **2**, 100–106.
52. B. Zhu, G. Qian, Y. Xiao, S. Deng, M. Wang and A. Hu, *J. Polym. Sci., Part A: Polym. Chem.*, 2011, **49**, 5330–5338.
53. B. Zhu, J. Ma, Z. Li, J. Hou, X. Cheng, G. Qian, P. Liu and A. Hu, *J. Mater. Chem.*, 2011, **21**, 2679–2683.

54. D. E. Whitaker, C. S. Mahon and D. A. Fulton, *Angew. Chem., Int. Ed.*, 2013, **52**, 956–959.
55. B. T. Tuten, D. Chao, C. K. Lyon and E. B. Berda, *Polym. Chem.*, 2012, **3**, 3068–3071.
56. J.-H. Ryu, R. T. Chacko, S. Jiwpanich, S. Bickerton, R. P. Babu and S. Thayumanavan, *J. Am. Chem. Soc.*, 2010, **132**, 17227–17235.
57. A. Sanchez-Sanchez, D. A. Fulton and J. A. Pomposo, *Chem. Commun.*, 2014, **50**, 1871–1874.
58. L. Buruaga and J. A. Pomposo, *Polymers*, 2011, **3**, 1673–1683.
59. W. Fan, X. Tong, Q. Yan, S. Fu and Y. Zhao, *Chem. Commun.*, 2014, **50**, 13492–13494.
60. P. G. Frank, B. T. Tuten, A. Prasher, D. Chao and E. B. Berda, *Macromol. Rapid Commun.*, 2014, **35**, 249–253.
61. A. Sanchez-Sanchez and J. A. Pomposo, *J. Nanomater.*, 2015, **2015**, 7.
62. J. Jeong, Y.-J. Lee, B. Kim, B. Kim, K.-S. Jung and H.-J. Paik, *Polym. Chem.*, 2015, **6**, 3392–3397.
63. P. Wang, H. Pu and M. Jin, *J. Polym. Sci., Part A: Polym. Chem.*, 2011, **49**, 5133–5141.
64. O. Altintas, P. Gerstel, N. Dingenouts and C. Barner-Kowollik, *Chem. Commun.*, 2010, **46**, 6291–6293.
65. O. Altintas, E. Lejeune, P. Gerstel and C. Barner-Kowollik, *Polym. Chem.*, 2012, **3**, 640–651.
66. O. Altintas, T. Rudolph and C. Barner-Kowollik, *J. Polym. Sci., Part A: Polym. Chem.*, 2011, **49**, 2566–2576.
67. E. J. Foster, E. B. Berda and E. W. Meijer, *J. Am. Chem. Soc.*, 2009, **131**, 6964–6966.
68. E. B. Berda, E. J. Foster and E. W. Meijer, *Macromolecules*, 2010, **43**, 1430–1437.
69. E. J. Foster, E. B. Berda and E. W. Meijer, *J. Polym. Sci., Part A: Polym. Chem.*, 2011, **49**, 118–126.
70. N. Hosono, M. A. J. Gillissen, Y. Li, S. S. Sheiko, A. R. A. Palmans and E. W. Meijer, *J. Am. Chem. Soc.*, 2013, **135**, 501–510.
71. P. J. M. Stals, M. A. J. Gillissen, R. Nicolay, A. R. A. Palmans and E. W. Meijer, *Polym. Chem.*, 2013, **4**, 2584–2597.
72. T. Mes, R. van der Weegen, A. R. A. Palmans and E. W. Meijer, *Angew. Chem., Int. Ed.*, 2011, **50**, 5085–5089.
73. M. Ouchi, N. Badi, J.-F. Lutz and M. Sawamoto, *Nat. Chem.*, 2011, **3**, 917–924.
74. M. Seo, B. J. Beck, J. M. J. Paulusse, C. J. Hawker and S. Y. Kim, *Macromolecules*, 2008, **41**, 6413–6418.
75. S. Mavila, C. E. Diesendruck, S. Linde, L. Amir, R. Shikler and N. G. Lemcoff, *Angew. Chem., Int. Ed.*, 2013, **52**, 5767–5770.
76. S. Mavila, I. Rozenberg and N. G. Lemcoff, *Chem. Sci.*, 2014, **5**, 4196–4203.
77. Z. Zhu, N. Xu, Q. Yu, L. Guo, H. Cao, X. Lu and Y. Cai, *Macromol. Rapid Commun.*, 2015, **36**, 1521–1527.

78. T. A. Croce, S. K. Hamilton, M. L. Chen, H. Muchalski and E. Harth, *Macromolecules*, 2007, **40**, 6028–6031.
79. J. N. Dobish, S. K. Hamilton and E. Harth, *Polym. Chem.*, 2012, **3**, 857–860.
80. D. Chao, X. Jia, B. Tuten, C. Wang and E. B. Berda, *Chem. Commun.*, 2013, **49**, 4178–4180.
81. I. Perez-Baena, I. Asenjo-Sanz, A. Arbe, A. J. Moreno, F. Lo Verso, J. Colmenero and J. A. Pomposo, *Macromolecules*, 2014, **47**, 8270–8280.
82. C. F. Hansell, A. Lu, J. P. Patterson and R. K. O'Reilly, *Nanoscale*, 2014, **6**, 4102–4107.
83. J. Willenbacher, K. N. R. Wuest, J. O. Mueller, M. Kaupp, H.-A. Wagenknecht and C. Barner-Kowollik, *ACS Macro Lett.*, 2014, **3**, 574–579.
84. G. Li, F. Tao, L. Wang, Y. Li and R. Bai, *Polymer*, 2014, **55**, 3696–3702.
85. J. Wen, L. Yuan, Y. Yang, L. Liu and H. Zhao, *ACS Macro Lett.*, 2013, **2**, 100–106.
86. E. Harth, B. V. Horn, V. Y. Lee, D. S. Germack, C. P. Gonzales, R. D. Miller and C. J. Hawker, *J. Am. Chem. Soc.*, 2002, **124**, 8653–8660.
87. G. Njikang, G. Liu and S. A. Curda, *Macromolecules*, 2008, **41**, 5697–5702.
88. G. Gasparini, M. Dal Molin, A. Lovato and L. J. Prins, *Supramolecular Chemistry*, John Wiley & Sons, Ltd, 2012.
89. C. Song, L. Li, L. Dai and S. Thayumanavan, *Polym. Chem.*, 2015, **6**, 4828–4834.
90. S. Cantekin, T. F. A. de Greef and A. R. A. Palmans, *Chem. Soc. Rev.*, 2012, **41**, 6125–6137.
91. P. J. M. Stals, M. A. J. Gillissen, T. F. E. Paffen, T. F. A. de Greef, P. Lindner, E. W. Meijer, A. R. A. Palmans and I. K. Voets, *Macromolecules*, 2014, **47**, 2947–2954.
92. G. M. ter Huurne, M. A. J. Gillissen, A. R. A. Palmans, I. K. Voets and E. W. Meijer, *Macromolecules*, 2015, **48**, 3949–3956.
93. E. A. Appel, J. Dyson, J. del Barrio, Z. Walsh and O. A. Scherman, *Angew. Chem., Int. Ed.*, 2012, **51**(4185–4189), 4181–4111.
94. E. A. Appel, J. d. Barrio, J. Dyson, L. Isaacs and O. A. Scherman, *Chem. Sci.*, 2012, **3**, 2278–2281.
95. A. J. Moreno, F. Lo Verso, A. Sanchez-Sanchez, A. Arbe, J. Colmenero and J. A. Pomposo, *Macromolecules*, 2013, **46**, 9748–9759.
96. F. Lo Verso, J. A. Pomposo, J. Colmenero and A. J. Moreno, *Soft Matter*, 2014, **10**, 4813–4821.
97. A. Anastasaki, V. Nikolaou, G. S. Pappas, Q. Zhang, C. Wan, P. Wilson, T. P. Davis, M. R. Whittaker and D. M. Haddleton, *Chem. Sci.*, 2014, **5**, 3536–3542.
98. W. R. Gutekunst and C. J. Hawker, *J. Am. Chem. Soc.*, 2015, **137**, 8038–8041.
99. J. P. Cole, J. J. Lessard, C. K. Lyon, B. T. Tuten and E. B. Berda, *Polym. Chem.*, 2015, **4**, 5555–5559.

100. J. A. Pomposo, I. Perez-Baena, L. Buruaga, A. Alegria, A. J. Moreno and J. Colmenero, *Macromolecules*, 2011, **44**, 8644–8649.
101. A. Einstein, *Ann. Phys.*, 1910, **338**, 1275–1298.
102. C. V. Raman, *Nature*, 1928, **121**, 619.
103. P. Debye, *J. Appl. Phys.*, 1944, **15**, 338–342.
104. B. H. Zimm, *J. Chem. Phys.*, 1945, **13**, 141–145.
105. P. J. Wyatt, *Anal. Chim. Acta*, 1993, **272**, 1–40.
106. I. Perez-Baena, A. J. Moreno, J. Colmenero and J. A. Pomposo, *Soft Matter*, 2014, **10**, 9454–9459.
107. A. W. Bosman, H. M. Janssen and E. W. Meijer, *Chem. Rev.*, 1999, **99**, 1665–1688.
108. N. Hosono, A. M. Kushner, J. Chung, A. R. A. Palmans, Z. Guan and E. W. Meijer, *J. Am. Chem. Soc.*, 2015, **135**, 501–510.
109. P. Lundberg, N. A. Lynd, Y. Zhang, X. Zeng, D. V. Krogstad, T. Paffen, M. Malkoch, A. M. Nystroem and C. J. Hawker, *Soft Matter*, 2013, **9**, 82–89.
110. K. Yoshikawa and N. Yoshinaga, *J. Phys.: Condens. Matter*, 2005, **17**, S2817–S2823.
111. D. Danilov, C. Barner-Kowollik and W. Wenzel, *Chem. Commun.*, 2015, **51**, 6002–6005.
112. C. T. Adkins, J. N. Dobish, S. Brown and E. Harth, *ACS Macro Lett.*, 2013, **2**, 710–714.
113. G. Hariri, A. D. Edwards, T. B. Merrill, J. M. Greenbaum, A. E. van der Ende and E. Harth, *Mol. Pharmaceutics*, 2014, **11**, 265–275.
114. K. H. Shaughnessy, *Chem. Rev.*, 2009, **109**, 643–710.
115. C. A. McNamara, M. J. Dixon and M. Bradley, *Chem. Rev.*, 2002, **102**, 3275–3299.
116. X. Li, X. Wu, W. Chen, F. E. Hancock, F. King and J. Xiao, *Org. Lett.*, 2004, **6**, 3321–3324.
117. X. Wu and J. Xiao, *Chem. Commun.*, 2007, 2449–2466.
118. J. Liu, Y. Zhou, Y. Wu, X. Li and A. S. C. Chan, *Tetrahedron: Asymmetry*, 2008, **19**, 832–837.
119. Y. Zhang and H. Zhao, *Polymer*, 2015, **64**, 277–284.
120. J. Tao and G. Liu, *Macromolecules*, 1997, **30**, 2408–2411.
121. Y. Kim, J. Pyun, J. M. J. Frechet, C. J. Hawker and C. W. Frank, *Langmuir*, 2005, **21**, 10444–10458.

CHAPTER 5

Polymeric Tubular Structures

SAMANTHA DONINELLI, MICHAEL BADOUX AND ANDREAS F. M. KILBINGER*

Department of Chemistry, University of Fribourg, Chemin du Musée 9, CH-1700 Fribourg, Switzerland
*Email: andreas.kilbinger@unifr.ch

5.1 Introduction – Bio-inspiration

Tubular structures are plentiful in nature. They range from macroscopic tubes such as blood vessels, the trachea, oesophagus, or the intestine to tubes on the nanoscopic scale. Just as on the macroscopic scale, the function of nanoscopic tubes is typically associated with controlling mass transport between otherwise separated spaces. On the cellular level, this can be exemplified by *gap junctions*, which consist of hexagonal trans-membrane tubes that connect neighboring cells for mass exchange. These are important for intercellular communication or the nourishment of cells that are further away from the blood supply.

One of the most famous examples are probably the ion channels, such as the Cl^- channel or the K^+,[1] Na^+, or Ca^{2+} channels that allow ion flux across a membrane through water-filled pores, or the aquaporins that allow the passage of water molecules across biological membranes.[2] α-Hemolysin,[3] a bacterial toxin, is another example of a channel-forming protein assembly that can form pores in the membrane of a cell through which ions can leak and eventually kill the cell. Another example are the porins OmpF[4] and PhoE, tubular protein assemblies in the outer membranes of Gram-negative bacteria that allow the transport of nutrients and waste.

RSC Polymer Chemistry Series No. 22
Bio-inspired Polymers
Edited by Nico Bruns and Andreas F. M. Kilbinger

Published by the Royal Society of Chemistry, www.rsc.org

Helical virus capsids, such as the tobacco mosaic virus (TMV),[5] rely on tube-like protein assemblies that carry the genetic material of the virus. The inner and outer surface of tube-like TMV[6] and the inside of spherical bacteriophage MS2[7] can be chemically modified.

This incomplete list of naturally occurring nanoscopic tubes illustrates the fact that nature relies mostly on protein self-assembly to build stable tubular nanostructures for mass transport.

Nanotubes represent spatially defined nano-containers with chemically defined inside and outside properties and up to two openings for mass exchange with the surrounding environment. Such functional geometric objects have attracted the attention of synthetic chemists for many years and many different synthetic approaches have been exploited to create such structures. Most approaches can be roughly divided into covalent and non-covalent[8] synthetic strategies. The latter can be formed from, for example, stacked macrocycles[9–12] or *via* self-assembly of low molecular weight amphiphiles,[13,14] peptides,[15–20] proteins,[21,22] or deoxyribonucleic acids (DNA).[23–26] Several excellent articles have been published that review the field of nanotube synthesis up to the year 2005.[27–29]

Here, we will only focus on synthetic strategies that use polymeric materials. The first approach described is the synthetic chemist's approach to tubular structures. Here, a single polymeric chain is used to fold into a tubular shape. This is achieved *via* the formation of hollow tube-like helices. The length, *i.e.* the molecular weight of the polymer chain involved, therefore directly affects the length of the tube-like structure formed.

The second approach described is the chemical engineer's approach to tube like structures. As in the previous case, polymers are used to form the tubular shapes; however, it is their bulk material property, rather than a molecular property, that is exploited. Tube-like structures on the nanometer up to the micrometer scale can be obtained using either templates or self-assembly methods.

Today's know-how in polymer science allows the synthesis of materials that self-assemble into complex structures. There is a great interest in helical structures or tubes and many techniques have been developed to achieve such structures. A variety of polymers that can be employed to build helical structures allow their use in many different applications like ion channels, high performance conducting polymers, drug delivery, nanocomposites, chemical catalysis, hydrogen storage, or waste water treatment.[30–34]

One approach to the self-assembly of polymers into tube-like structures is the use of block copolymers. They enable the formation of structures with amphiphilic behavior, which tend to assemble into helices or vesicles.

It is also possible to apply engineering techniques to natural polymers like DNA. Indeed, DNA nanotubes are attractive because the structural, energetic, and dynamic control accessible with so-called DNA tiles is unsurpassed. Potential uses of DNA nanotubes are inspired by nanofilaments and nanotubes in living cells.[35]

5.2 Tubes Based on Single Polymer Chains

5.2.1 Polyaramides

One of the pioneers working on helical aramide structures was Hamilton, who in the 1990s published the first papers on the synthesis of helical secondary structures based on oligoanthranilamide derivatives.[36–38] Oligoanthranilamides were not the only example that could form helices, other oligomers were soon exploited as well. Huc and co-workers published several studies on the formation of helical structures based on oligopyridinedicarboxamide derivatives[39,40] and on quinoline-derived δ-amino acids.[41,42]

In the first experiments that Huc and co-workers carried out to predict the stability of a helix, they used an oligomer composed of an alternating structure of 2,6-diaminopyridine and 2,6-pyridinedicarboxylic acid. In this paper, Huc demonstrated that the interaction between CO–NH and NH–pyridine prevents the rotation around these two bonds. Due to these hydrogen bonds, the preferred shape adopted was a helical conformation.[39] It was then demonstrated that the protonation/deprotonation of the *N*-pyridine moiety of both monomers affected the conformation of the helix. The protonation of 2,6-diaminopyridine **1** by trifluoroacetic acid (TFA) gave a linear conformation (formed by the intercalation of **2** and **3**) due to the rotation of the CO–NH bond (Figure 5.1). If, instead of TFA, methanesulfonic acid (MsOH) was added to the mixture, the protonation of **3** induced a new helical conformation.[40] These findings were later confirmed by Lehn.[43]

One of the first studies in crescent shaped oligoamides was published by Gong.[44] In this paper, Gong and co-workers explained the importance of the three-centred hydrogen-bond which rigidified the obtained structures. They also showed the importance of having two amide linkages in the *meta*-position, as this allowed the formation of a crescent shaped oligomer.[44] The diameter of the cavity depended on the type of monomer. An oligomer composed of 5-amino-2,4-dihydroxybenzoic acid, for example, gave an inner diameter of around 10 Å. In order to change the inner diameter, it was necessary to modify the ratio of the two monomers (*meta* and *para*), or

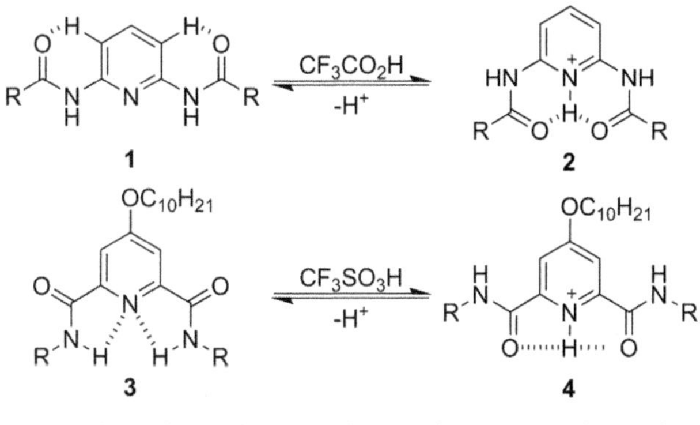

Figure 5.1 Protonation of **1** and **2** and changes in their conformation.

change the type of building blocks and use monomers with different distances between the hydrogen-bonds that were formed.[45] The structure that was formed was stable both in non-polar and polar solvents.[46] Shorter oligomers folded into helices without any problems; however, the steric hindrance of polymers or longer oligomers in the helical conformation caused increasing difficulties during the synthesis. To avoid this problem, 2,4-dimethoxybenzyl (DMB) groups were introduced to replace the amide hydrogen atoms. This avoided the intramolecular hydrogen bonding and resulted in a less rigid backbone. This strategy significantly improved the synthetic procedure.[47]

At the end, five different forces were shown to be important for the formation of a helix: the hydrogen bonds that rigidify the backbone, the *trans* configuration of the amide bond, the rigidity of the monomer, the *meta* and *para* position of the substituents, and the π–π stacking between two benzene rings close to each other in a helix.[45,48]

In 2003, Tang and co-workers[49] published an example of a narrow helix composed of 4-ethynylbenzoyl-L-vaniline monomers that were polymerized by an organorhodium catalyst. The helical conformation was confirmed by the presence of an intense Cotton effect (the Cotton effect describes a characteristic change in the optical rotatory dispersion and also the circular dichroism close to the absorption maximum of a chromophore), folding/unfolding, and the tunability of the conformation of the helix with different stimuli such as the pH, solvent, and temperature.[49] Meijer and co-workers synthesized one of the first polymeric helices with a hollow cavity.[50] The polymer was composed of two different monomers, one was a diamine (**6**) and the other one a diisocyanate (**5**), using a condensation reaction to give poly-ureidophthalimide (**7**) (Figure 5.2). The imide nitrogen carried a chiral substituent that was exploited to investigate the helix formation *via* circular dichroism spectroscopy.

Oligomers/polymers of three different lengths were obtained: **7a** contained around 30 monomers, **7b** around 7 monomers, and polymer **7c** around 3.5 monomers.

The Cotton effect in tetrahydrofuran (THF) for polymer **7a** was very intense, meaning that this copolymer folded into a chiral and ordered structure. For the other two polymers, **7b** and **7c**, the Cotton effect was rather small, meaning that they did not fold into a large helical structure (Figure 5.3). Upon heating (55 °C in THF), 55% of the ordering of the polymer **7a** was lost. This effect was observed only in THF.

Scanning tunneling microscopy (STM) measurements of oligo-ureidophthalimide (prepared from oligo(*p*-phenylenevinylene)s **8** and **9**, Figure 5.4)[51] were used to confirm that 6–8 monomers were necessary for one turn (Figure 5.5).[50,51]

To avoid the formation of macrocycles during polymerization, Gong and co-workers proposed a different strategy using a more flexible aromatic polyamide (aramide) backbone.[52] An amide side chain was introduced into the structure to facilitate the formation of the helix with

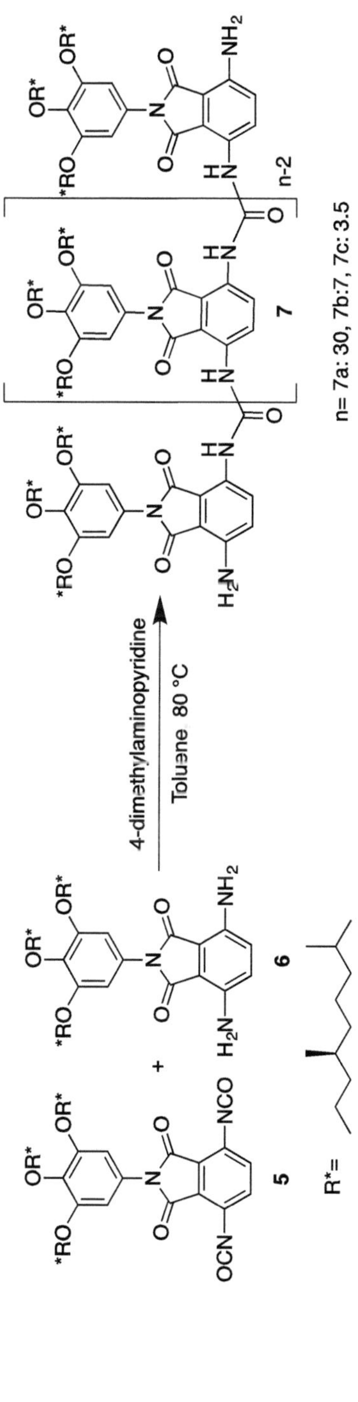

Figure 5.2 Synthesis of chiral poly-ureidophtalimide 7.

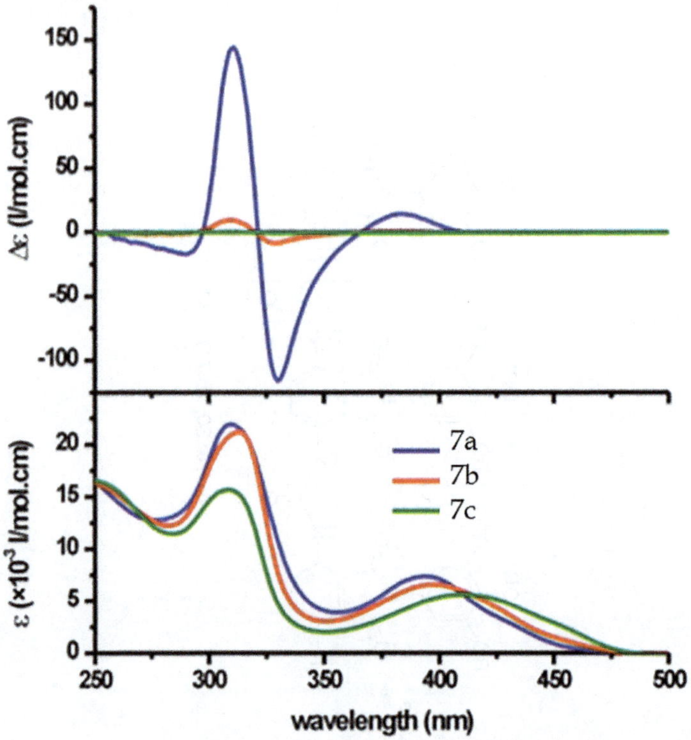

Figure 5.3 Top: circular dichroism (CD) spectra where **7a** displays a strong Cotton effect; bottom: UV-vis of poly-ureidophthalimides in THF. Reproduced from ref. 50 with permission from the Royal Society of Chemistry.

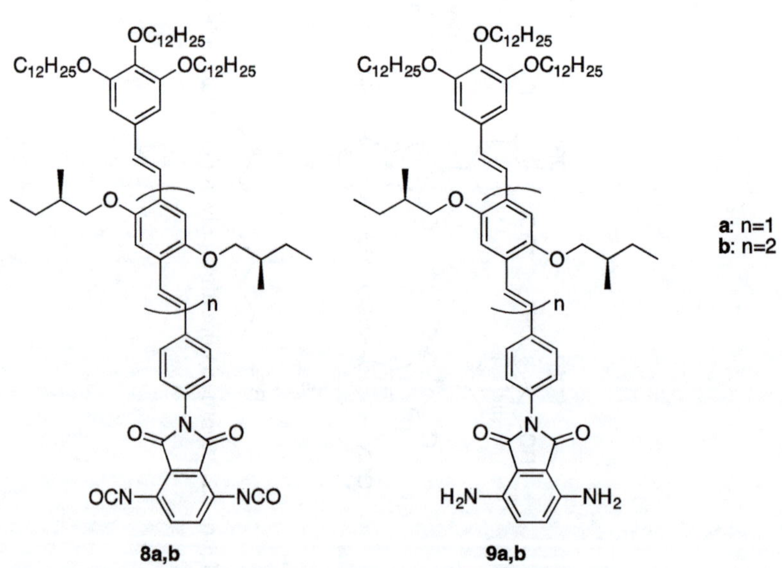

Figure 5.4 Monomers for poly-ureidophthalimide synthesis.

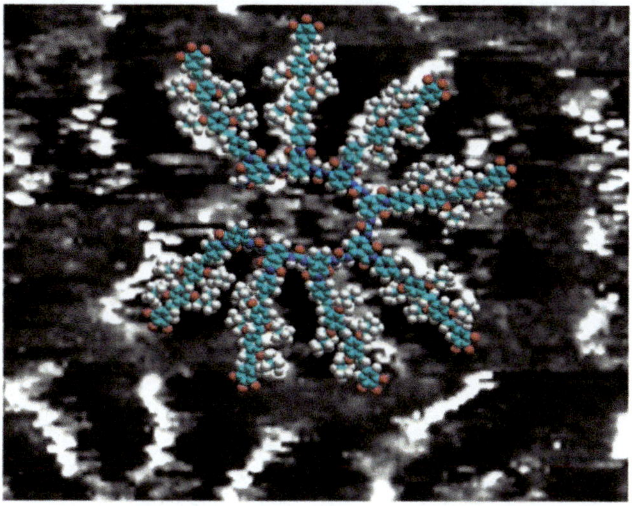

Figure 5.5 STM image with a superimposed molecular model of an oligomer composed of **8b** and **9b** with a length of 1–5 units at the 1-phenyloctane/graphite interface.
Reproduced with permission from R. W. Sinkeldam, F. J. M. Hoeben, M. J. Pouderoijen, I. De Cat, J. Zhang, S. Furukawa, S. De Feyter, J. A. J. M. Vekemans and E. W. Meijer, *J. Am. Chem. Soc.*, 2006, **128**, 16113–16121.[51] Copyright (2006) American Chemical Society.

additional H-bonds.[53–55] The polymer was prepared from di-acid chloride and di-amino monomers **10** and **11** (Figure 5.6).

Three different fractions with three different lengths of polymer **12** were obtained by a series of precipitations using poor and good solvents. In the end, polymer **12a** had an M_n of 35 100 g mol^{-1} corresponding to 18 turns, **12b** had an M_n of 18 600 g mol^{-1} (9 turns), and **12c** an M_n of 10 200 g mol^{-1} (5 turns). The low PDI (1.02–1.03) indicated that each fraction was pure. All polymers were soluble in CH_2Cl_2, $CHCl_3$, THF, dimethylformamide (DMF), and dimethylacetamide (DMA). The CD-spectra of **12a–c** showed a strong positive Cotton effect at 270 nm and a strong negative Cotton effect at 300 nm, which confirmed the formation of a helix with an excess of one handedness. The CD signal at 300 nm was recorded as a function of the temperature. A sigmoidal-shaped curve was obtained, which could indicate cooperativity between intramolecular non-covalent forces (H-bonds, π–π stacking) to form the helical structure. Similar behavior was shown also for proteins and peptides.[52]

In 2010, Li, Guan, and co-workers began a series of studies based on helical polymers. The first investigation addressed the efficiency of the synthesis of helical polymers.[56,57] The monomer that was used in this study was a methoxy-substituted *m*-amino benzoic acid (**13**). In a first test, the polymer was synthesized in *N*-methyl pyrrolidone (NMP)/pyridine in the presence of LiCl (to avoid intramolecular hydrogen-bond formation); however,

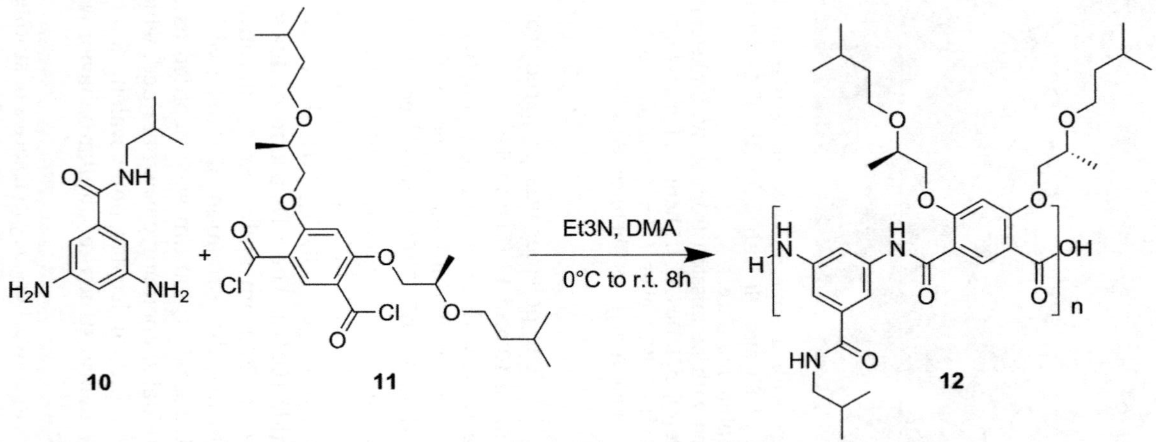

10 **11** **12**

Figure 5.6 Synthesis of polymer **12** with different lengths, **12a**: 56 Å, **12b**: 30 Å, and **12c**: 16 Å.

the *O*-demethylation of the methoxy group occurred as a side reaction. The chloride anion was found to be responsible for the nucleophilic attack and thus an extra *O*-methylation step was added to the synthesis to obtain polymer **14** (Figure 5.7).

Another solution was to HO-deprotonate the polymer to allow the formation of an intramolecular H-bond. The demethylated polymer had an M_n of 3750 g mol^{-1} and a PDI of 1.72. The deprotonated polymer, on the other hand, had an M_n of 2070 g mol^{-1} and a PDI of 1.52, and the fully *O*-methylated polymer **14**, an M_n of 2430 g mol^{-1} and a PDI of 1.47. This was due to the fact that polymer **14** and the deprotonated polymer adopted a compact helical conformation that led to an underestimated molecular weight by GPC. To confirm this theory, a chiral acid was added to the structure and a CD-spectrometry analysis in DMSO was performed. The two polymers showed a Cotton effect that corresponded to a helical conformation.[56]

Another interaction that has allowed the formation of a helix was the π–π stacking between benzene rings and the solvophobicity of the polymer. To investigate the spontaneous formation of a helical conformation in different solvents, Li and co-workers synthesized a polymer (**15**) based on 1,8-naphthalimide-3,6-diamine and an isophthalic acid precursor (Figure 5.8). Both monomers carried tetraethylene glycol chains to increase the solubility.[58]

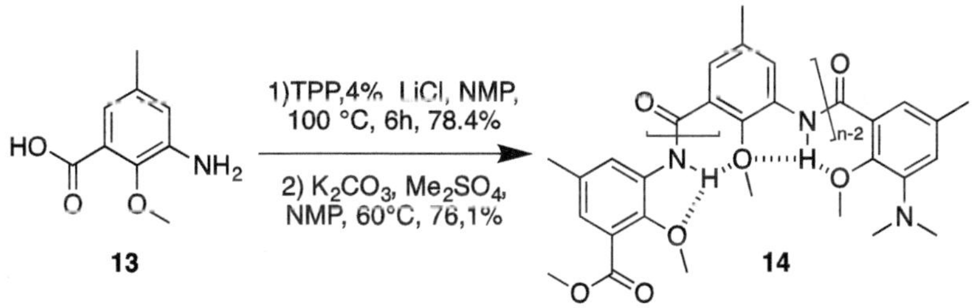

Figure 5.7 Two-step synthesis of polymer **14**.

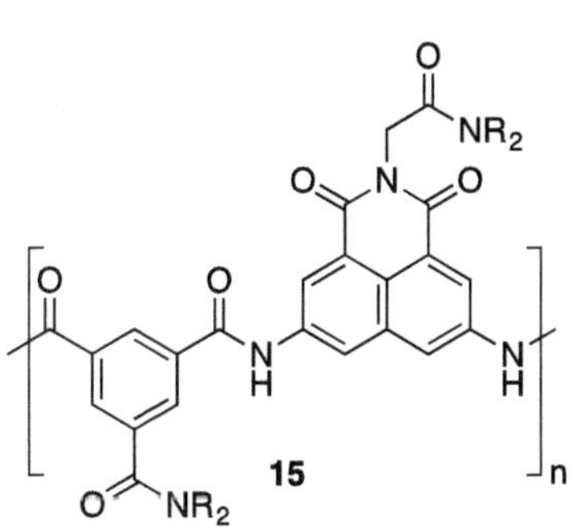

Figure 5.8 Amphiphilic polymer **15**.

This polymer (**15**) folded into a helix in both organic solvents with different polarity and also in water, as proven by fluorescence spectroscopy. Monomer NI (1,8-naphthalimide-3,6-diamine) showed fluorescence emission at 420–450 nm and the emission of an excimer band at 540–560 nm. The excimer emission was stronger than the monomer emission in all solvents except trifluoroethanol, which means that the helix conformation was energetically favored. Trifluoroethanol interacted as a strong H-bond donor with the intramolecular H-bonding of the polymer, decreasing the stability of the helical conformation.

For polar solvents (water and methanol), the main driving force for folding was solvophobicity as, in the folded conformation, the contact surface of the aromatic amide chains with the solvent was minimized. On the other hand, for solvents such as chloroform and dichloromethane, the main driving force for folding was intramolecular H-bond formation.[58]

A more recent report on helical polyaramides was published by Schulze *et al.*, which was based on the formation of large tubular helices based on tri(benzamide)s.[59] The construction of this structure was a combination of sequence-controlled organic synthesis and polymer synthesis. This structure (**16**) was composed of a *para*-substituted monomer and a *meta*-substituted monomer. This kind of combination allowed the formation of a three-center hydrogen bonded helix. A *p*-methoxybenzyl (PMB) *N*-protecting group was added on one or two amide bonds to avoid macrocyclization. These PMB-groups led to an *E*-conformation that suppressed hydrogen bond formation. After the deprotection of the PMB-amides with an acid, a conformational change from *cis* to *trans* was induced, and the formation of hydrogen bonding was observed (Figure 5.9).

It was observed that the di-PMB protected trimer readily formed a macrocycle under polymerization conditions, as the oligomer adopted a conformation in which the two ends of the oligomer were close to each other. To avoid this problem, a mono PMB-protected trimer was synthetized in order to maintain a longer distance between the reactive centers. After PMB deprotection, the polymer was no longer soluble in chlorinated solvents.

As an indication towards the formation of a helix, GPC elugrams before and after PMB-deprotection and CD spectra were recorded. The large shift present in the GPC elugrams towards higher apparent molecular weights (shorter elution times) probably indicated the rigidification of the helix upon PMB deprotection. The CD spectra of polymer **17** showed a small CD signal in the presence of a chiral solvent, indicating the formation of a chiral arrangement of the polymer backbone chromophores, consistent with the formation of a helix. Based on a molecular mechanics model, an inner diameter of the helix of 4 nm was calculated.[59]

5.2.2 Phenylene Helices

Some of the pioneers working on phenylene helices were Moore and co-workers, who demonstrated the dependence between structure and solvent

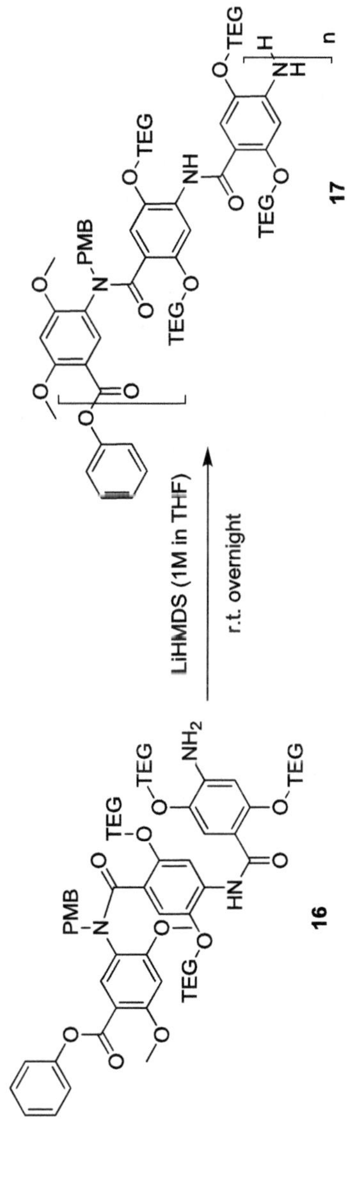

Figure 5.9 Polycondensation of the mono-PMB-protected trimer 16.

for phenyleneethylene oligomers. In solvents such as chloroform, the oligomer had a random structure while in "poor" solvents such as acetonitrile, the molecule consisting of at least 6–8 monomers had an ordered structure.[33,60,61] The solubility of the oligomer was also affected by the polarity of the side chain.[62–64] The ability to form π-stacked structures was inhibited with a benzyl ether or phenyl ether side chain, even in DMSO.[65] It was found that adding a chiral binaphthol to the compound imparted a bias in the helical conformation.[66] The problem with chiral binaphtol was that the helix lost its conformational stability. To avoid this problem, a small chiral perturbation, like a methyl group in the polyethylene glycol side chain, was added to the oligomer.[67,68] Another method to stabilize the helix was to introduce a single H-bond β-turn mimic into the oligomer. This bond rendered the helix more tolerant to denaturation by solvents.[69] Another method to impart a bias in the helical conformation was achieved with a (+)-tartaric acid derivative, which was added to the tethered oligomers to obtain a larger twist sense than with binaphthol.[70] A different strategy to fold oligo(*m*-phenyleneethynylenes) in apolar solvents was to synthesize a rigidified backbone exploiting H-bonding between neighboring side-chains.[71]

Based on the work shown above, Hecht *et al.* prepared organic nanotubes with intramolecular cross-linking of the backbone of the helix. The idea (Figure 5.10) was to polymerize a functional monomer that would fold into a helical arrangement with the cross-linking group (grey) and the solvophilic group (magenta) outside the helix. This way the helical shape could be stabilized by covalent cross-linking.

This polymer was formed using a diiodocinnamate monomer that afforded the poly(*m*-phenyleneethynylene) polymer with a tetraethylene gylcol solvophilic side chain. Before [2 + 2] photodimerization, the polymer was allowed to fold into a helix in a polar solvent. When the cross-linking was carried out, the helix became more stable and denaturation in chloroform was no longer observed.[72] In order to have efficient [2 + 2] photodimerization, short irradiation times were applied. Shorter irradiation times were

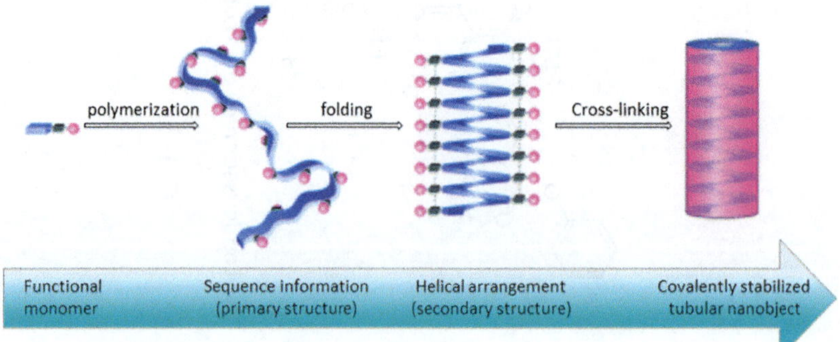

Figure 5.10 Formation of organic nanotubes.
Reproduced from ref. 72 with permission from John Wiley and Sons.

found to be better than longer irradiation times to promote intermolecular cross-linking.[73] Another type of helix obtained by Hecht were amphiphilic azobenzene polymer derivatives (Figure 5.11).[74]

The azobenzene units were connected to each other by acetylene (**18**) or diacetylene (**19**) linkages. All monomers carried an oligo(ethylene glycol) side chain to impart an amphiphilic character and thus fold into a helix in polar solvents, as proved by solvent titration.

UV-vis spectra showed a bathochromic shift for both polymers (changing the solvent from MeCN to $CHCl_3$). This was due to solvent denaturation occurring in $CHCl_3$. The transition from the *cisoid* form to the *transoid* form was also confirmed by CD spectroscopy. In MeCN, both polymers showed a Cotton effect. For PAzoE, the predominant conformation was a right-handed helix while, for PAzoB, a left-handed conformation was observed. In less polar $CHCl_3$, the Cotton effects disappeared, meaning that in this solvent the polymers adopted a random coil conformation. A substantial difference between the two polymers was that **18** did not show aggregation at low concentrations, while **19** exhibited a strong aggregation tendency. This was due to the introduction of an additional electron-withdrawing acetylene group that favored π,π-stacking.[74]

To obtain a polymer instead of an oligomer, Moore and co-workers studied the nucleation–elongation mechanism to obtain a polymer. In a closed-system, the driving force for polyimine formation in solution was the gain in energy with the folded structure. This was shown to provide a sufficient driving force to push the equilibrium to obtain high polymers. To optimize this folding mechanism, solvent quality and/or temperature could be changed and control over the molecular weight of the resulting polymer was obtained. In nucleation–elongation polymerization, the nucleation is energetically disfavored (oligomerization) but the elongation is favored (polymerization). This leads to the predominant presence of either high molecular weight polymers or monomer species and only some oligomers.[75–78]

One example of nucleation–elongation polymerization was published in 2006 by Moore and co-workers.[79] Three different *m*-phenyleneethynylene oligomers (Figure 5.12) were synthesized and their coordination polymerization studied after addition of a palladium complex. Oligomers **20** and **22** polymerized under these conditions and formed helices. On the other hand, hexamer **21** formed a macrocycle that aggregated into a columnar polymer. This clearly shows that a small change in the structure of the starting oligomers can dramatically change the mechanism of polymerization and other features like the solubility and the stability of the helix.[79]

Investigations to determine the influence of the side chain, the length of the polymer, and the functional groups on the folding procedure were carried out by Bruce and co-workers.[80] The copolymer (Figure 5.13) that was used to make this analysis was poly(*m*-phenyleneethynylene) (*m*-PPE) with different side chains. For polymer **24**, two different lengths were obtained after purification: a long chain length polymer (**24-L**, M_w of 12 591 g mol^{-1}) and a short chain length polymer (**24-S**, M_w of 8376 g mol^{-1}).

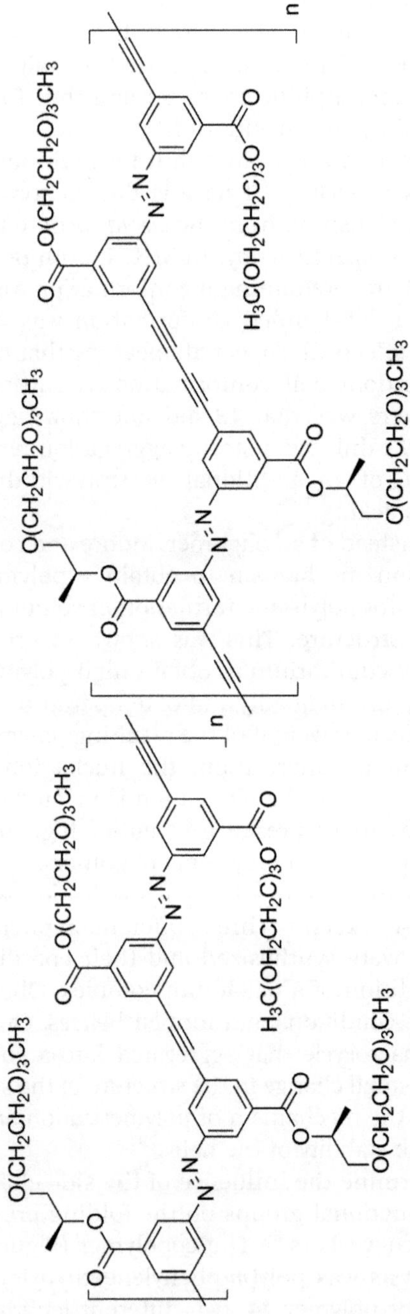

PAzoE, 18

PAzoB, 19

Figure 5.11 Polymer PAzoE and polymer PAzoB.

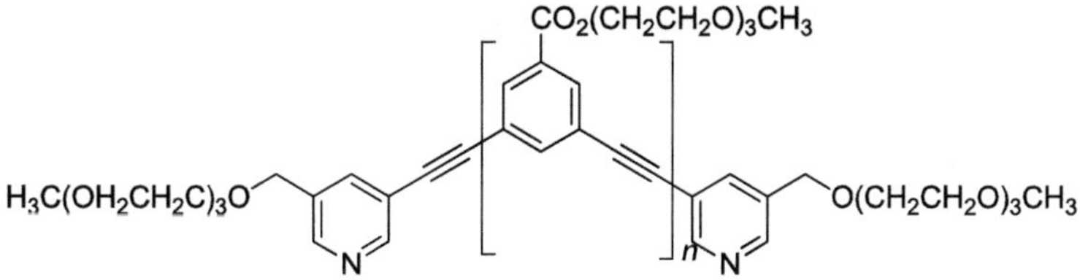

20: $n = 2$
21: $n = 4$
22: $n = 6$

Figure 5.12 Three different *m*-phenyleneethynylene oligomers.

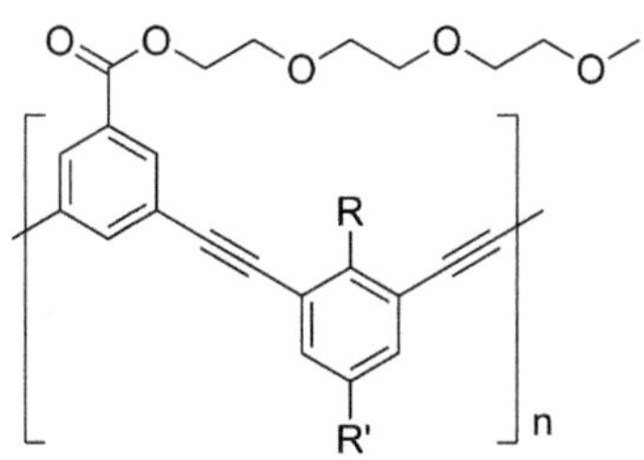

23: R= $-OCH_3$, R'= $-CH_2NHCOOC(CH_3)_3$
24: R= -H, R'= $-CH_2NH_3Cl$

Figure 5.13 *m*-PPE derivative polymers.

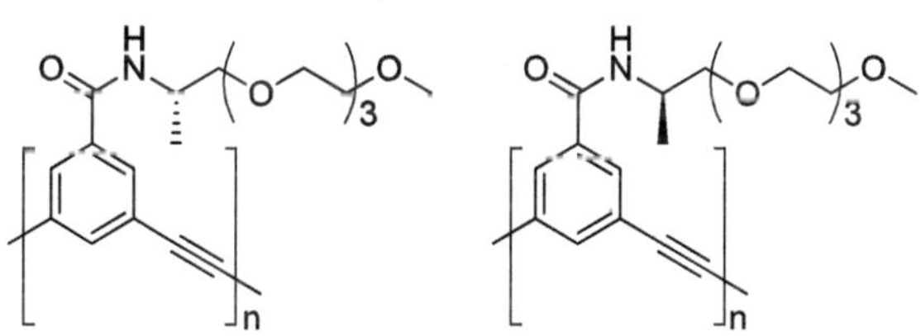

25 **26**

Figure 5.14 Poly(*m*-phenyleneethynylene)s (*S*)-PPEa (**25**) and (*R*)-PPEa (**26**).

Due to the presence of these different side chains, the solubility was increased. Both polymers were soluble in water and formed a stable helix. **24-L** was also soluble and formed stable helical structures in different alcohols and in acetonitrile. It was found that a minimum number of seven monomers was needed to obtain a helical structure but, in order to obtain a very stable helix, almost 20 monomers were required. Finally, the methoxy group of **23** positively influenced the length of the polymer obtained but decreased the ability of the polymer to fold into a helical structure in different solvents (like acetonitrile).[80]

In 2012, Hecht and co-workers published the first absolute helical sense of a poly(*m*-phenyleneethynylene) (PPE) structure.[55] To obtain this, two different monomers were synthesized. The side chains were based on derivatives of L-alanine for (*S*)-PPEa (**25**) or D-alanine for (*R*)-PPEa (**26**) (Figure 5.14).

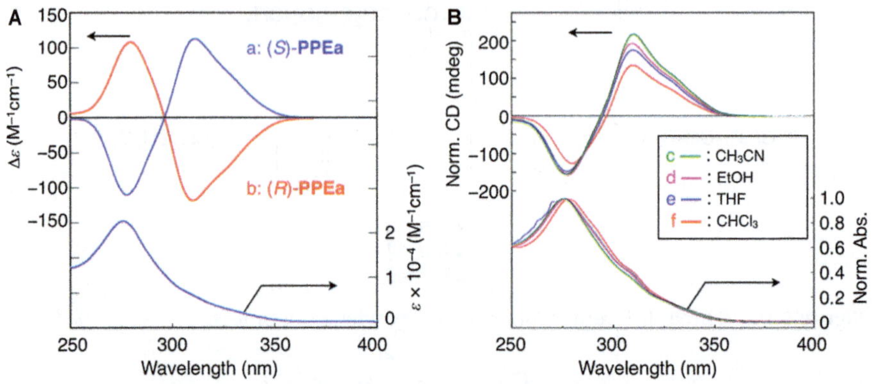

Figure 5.15 (A) CD spectra and absorption spectra of PPEs in $CHCl_3$ (0.5 mM) at 25 °C. (B) Normalized CD and absorption spectra of (*S*)-PPEa in different solvents at 25 °C (relative to the maximum absorption). Reproduced with permission from M. Banno, T. Yamaguchi, K. Nagai, C. Kaiser, S. Hecht and E. Yashima, *J. Am. Chem. Soc.*, 2012, **134**, 8718–8728.[55] Copyright (2012) American Chemical Society.

The CD spectra ($CHCl_3$) of **25** and **26** (Figure 5.15A) showed consistent Cotton effects, negative for (*S*)-PPEa and positive for (*R*)-PPEa, which revealed that both polymers were of one-handedness. This preference was probably due to the covalently attached chiral alanine side chain. (S)-PPEa showed an increase in helical sense preference with the increasing solvent polarity (Figure 5.15B). In this case, the addition of trifluoroethanol (TFE) resulted in the loss of the Cotton effect, which was regained by heating the mixture up to 55 °C. Thermal breaking of the TFE-amide bonds was responsible for this effect.

The wide-angle X-ray diffraction (WAXD) study revealed four equatorial reflections: (*R*)-PPEa 27.22, 15.52, and 10.20 Å; (*S*)-PPEa 27.29, 15.89, 13.62, and 10.37 Å; a 2D hexagonal lattice of $a = 31.43$ (*R*) and 31.51 Å (*S*), and meridional reflections at 3.56 (*S*) and 3.55 (*R*) Å that correspond to π–π stacking. As shown in Figure 5.16, the polymer was stabilized by six pairs of intramolecular H-bonds between the side chains.

In order to obtain high-resolution atomic force microscopy (AFM) images, a monolayer of polymer was crystallized on the surface of graphite. Figure 5.17 shows periodic oblique stripes in one direction for (*R*)-PPEs and in the other direction for (*S*)-PPEa; these stripes were twisted by +52° for (*S*)-PPEa and −56° for (*R*)-PPEa.

The first example of "whiskers" composed of helicene-like helical structures was published by Akagi *et al.*, who used poly(*m*-phenylene) derivatives (Figure 5.18) with racemic or chiral side chains.[81]

Polarized optical micrographs (POM) showed that, in order to fold into a whisker, the monomers needed to have a side chain length of $m = 7$ to 9 and a methyl branched structure, so that polymers **28**, **29**, and **30** formed the desired structures (Figure 5.19).

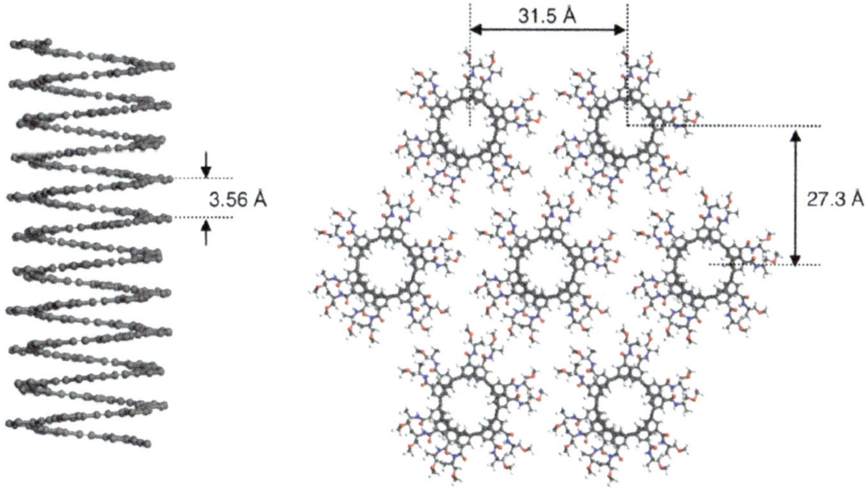

Figure 5.16 (*S*)-PPEa structure based on the XRD analysis and molecular mechanics calculations.
Reproduced with permission from M. Banno, T. Yamaguchi, K. Nagai, C. Kaiser, S. Hecht and E. Yashima, *J. Am. Chem. Soc.*, 2012, **134**, 8718–8728.[55] Copyright (2012) American Chemical Society.

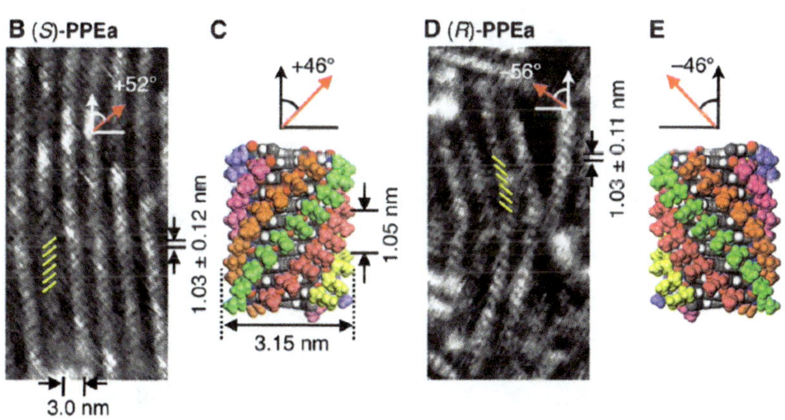

Figure 5.17 High resolution AFM images of (*S*)-PPEa (B) and (*R*)-PPEa (D).
Reproduced with permission from M. Banno, T. Yamaguchi, K. Nagai, C. Kaiser, S. Hecht and E. Yashima, *J. Am. Chem. Soc.*, 2012, **134**, 8718–8728.[55] Copyright (2012) American Chemical Society.

The chiral polymers ((*S*)-**30** and (*R*)-**30**) showed circularly polarized fluorescence because the backbone of the helix was conjugated. They also showed a linear polarized fluorescence after macroscopic alignment. X-ray diffraction (XRD) analysis showed that the whiskers were composed of hexagonally packed columns of self-assembled helices, uniaxially packed parallel to the long axis of the whiskers.

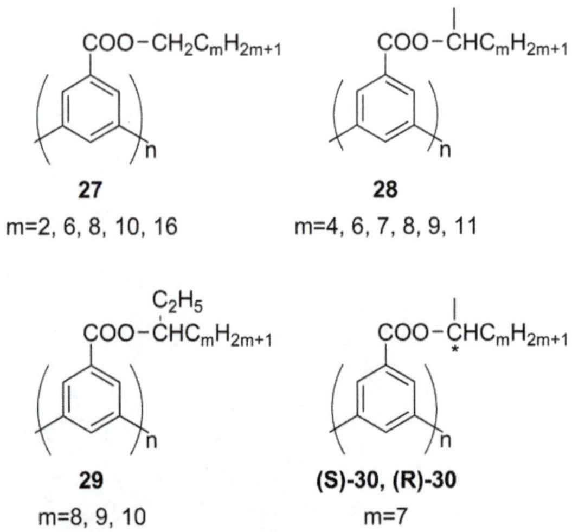

Figure 5.18 Five different poly(*m*-phenylene) polymers.

Figure 5.19 (a) POM and (b) circularly polarized optical micrograph of (*R*)-**30** in solid state.
Reproduced with permission from K. Suda and K. Akagi, *Macromolecules*, 2011, **44**, 9473–9488.[81] Copyright (2011) American Chemical Society.

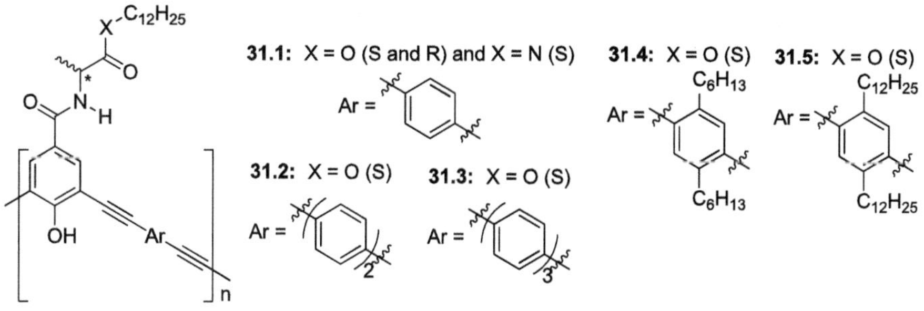

Figure 5.20 Sonogashira–Hagihara coupling polymerization of monomers **31**.

Several other chiral polymers were synthesized by Sonogashira–Hagihara coupling polymerization. These structures were composed of optically active poly(*m*-phenyleneethynylene–aryleneethynylene) with varying arylene groups (Figure 5.20).

Polymers **31.1**, **31.2** and **31.3** (Figure 5.20) containing one or more aryl groups formed stable helix conformations; on the other hand, polymers **31.4** and **31.5** with substituted aryl groups did not fold into a helix, as determined by CD-spectroscopy.

The intensity of the CD signal depended on the number of phenylene repeating units, so polymer **31.3**, containing three phenylene units, exhibited the largest CD signal. Due to intramolecular π-stacking between the terphenylene units, polymer **31.3** exhibited the most stabilized helical structure. The different intensity in the CD spectra between **31.1**, with two amide groups, and **31.1**, with an ester group, was due to the fact that the amides formed intramolecular hydrogen bonds. This led to a predominance of the one-handed helix, due to the chiral group present in the side chain. To determine the inner diameter of the 12-mer **31.1** (X = O, S), some molecular mechanics calculations were performed, and an inner diameter of 25 Å and a helical pitch of 3.8 Å were found.[82]

5.2.3 Other Helical Polymers

In addition to the aramide foldamers and phenylene helices, several other monomers such as pyridine–pyrimidine, triazole, and dihydrazine have been used to prepare tubular helical structures. In 1997, Lehn published one of the first articles about pyridine–pyrimidine oligomers that folded into a helical conformation due to intramolecular interactions.[83,84] With this oligomer, Lehn and co-workers investigated the formation of a helix that contained 12 monomers in a hexagonal structure for one turn; the inner diameter was around 8 Å and the outside diameter 25 Å.[85] With different lengths of alternating pyridine–pyrimidine (py–pym) oligomers, they showed that the presence of a chiral center was not necessary to obtain a helix. Furthermore, the barrier for the inversion of helical handedness was not dependent on the length of the oligomers. It was proposed that this structure could be used as a helicity codon.[86] Another example of a helicity codon was a pyrimidine-hydrazone oligomer. One further advantage of this structure was that imine condensation was used to form the molecule.[87] Lehn and co-workers modified their previous structure[83,85–87] and, instead of a pyridine–pyrimidine polymer, a 1,8-naphthyridine-pyrimidine (napy-pym) polymer was used. This change caused an increase in size of the helical cavity: 2 Å for the py-pym helix and 3.5 Å for the napy–pym helix.[86,88]

In some cases, an effector was needed to obtain a helical conformation. In the case of oligo-isophtalamides, the effector was a cyanuric acid template that, with the help of a solvophobic effect and stacking interactions, allowed the formation of a helical structure.[89] Other molecules like ions or a change in pH could also induce structural switching in the pyridine–pyrimidine system.[90] For example, Ag^+ could generate a double-helical entity and a change of pH could interconvert a shorter single helix into a longer double helix.[91]

Based on their work on oligomeric helices, Lehn and co-workers synthesized a dihydrazino-pyridine polymer *via* polycondensation. The hydrazone bond between pyridine groups was the basis to obtain a helical foldamer. For a molecular weight of around 9000 g mol^{-1}, the height of the helix was *ca.* 42 Å and it was composed of 12 turns. The inner diameter was around 20 Å with a pitch of 3.5 Å. This polymer was a good base structure to attach functional residues (in this case a methoxy group).[92] The hydrazone bond was also used in an amphiphilic polymer composed of di(aldehyde) and di(acylhydrazine) monomers, these monomers were mixed in an acidic aqueous solution to form the polycondensate. The resulting polymer was studied using different physical methods (size exclusion chromatography with multi angle laser light scattering detection (SEC-MALLS), small angle neutron scattering (SANS), transmission electron microscopy (TEM), matrix assisted laser desorption ionization time of flight (MALDI-TOF) mass spectrometry, electronic absorption, and emission spectrometry). The results confirmed that the polymers adopted a helical conformation. This structure was ideal to accommodate the π-stacking between turns with the hydrophilic groups pointing to the exterior of the structure. The appearance of a dramatic hypochromic effect verified this kind of folding when the polymer was dissolved in DMF.[93]

Pu and Zhang reported a helix based on an optically active polybinaphthyl polymer.[94] This polymer was synthesized with an acid-promoted intramolecular cyclization of arylalkynes (Figure 5.21). Due to the low rotation freedom, the structures folded predominately into helices. Semiempirical AM1 calculations predicted an internal diameter of 8 Å.

Fluorescence spectroscopy showed a large red shift after the reaction to obtain **33**. This shift indicated the increase of conjugation in the structure due to the formation of a planar and fused polyaromatic system.

CD spectra of polymers **34** and **35** showed the presence of a Cotton effect for polymer **35** caused by the exciton coupling in the helical units. This result was obtained also for polymer **32**. Different molecular weight samples were synthetized of polymer **35**. UV-vis spectra of these different helices were

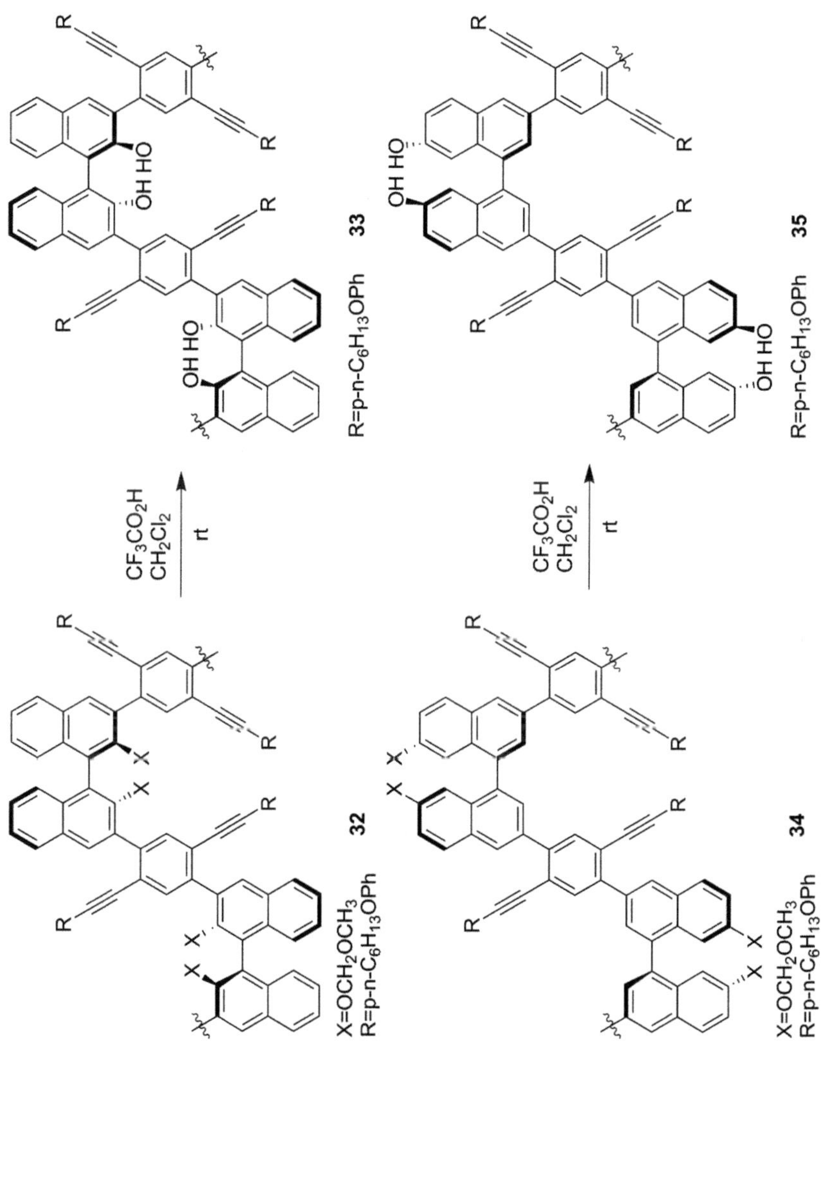

Figure 5.21 Synthesis of the Minor-Groove 33 and Major-Groove 35 helical polymers.

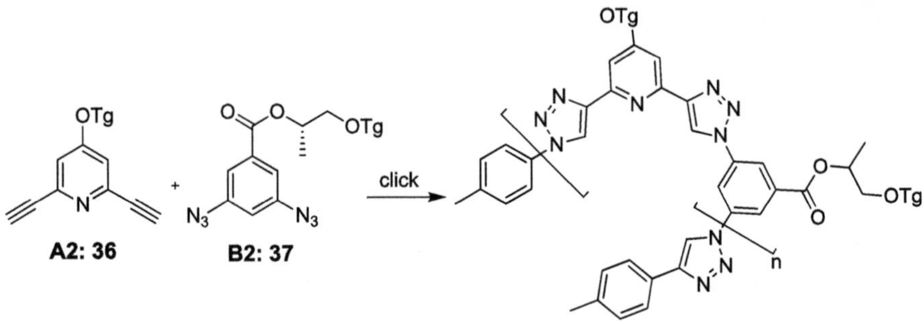

Figure 5.22 $A_2 + B_2$ step-growth polymerization.

obtained and the absorbance increased with the increasing molecular weight. The same was observed for the CD spectra where the signal intensity increased with the increasing molecular weight.[94]

Hecht and co-workers developed another class of polymers based on alkyne–azide click chemistry. The synthesis of these new polymers was an $A_2 + B_2$ step-growth polymerization, with A_2 corresponding to the 2,6-diethynylpyridine monomer and B_2 to the 3,5-diazidobenzoate monomer that reacted together under Cu-catalyzed 1,3-dipolar cycloaddition reaction conditions (Figure 5.22).

GPC results showed an underestimated M_n compared to long discrete oligomers composed of the same monomers, indicating that polymer **38** had a more compact structure and a smaller hydrodynamic radius. The BTP (2,6-bis(1,2,3-triazol-4-yl)pyridine) part of the polymer was also useful to obtain a helical foldamer both in polar and apolar solvents due to its large degree of pre-organization.[95] The same polymer was used in 2008 by Hecht and co-workers to investigate the helicity inversion induced by achiral halide ion guests.[96] Increasing the amount of water in the solvent resulted in increasing absorption at short-wavelengths in the CD spectra. This was attributed to a solvophobic driving force. In a mixture of 75% water/acetonitrile, some switchable subunits where incorporated: a simple protonation of the helix induced a helicity inversion and the helix changed from the *anti,anti*-conformation to an extended *syn,syn*-conformation. CD spectra showed that addition of KF only increased the CD signal, while addition of HCl caused an inversion of the sense of the helix. The isodichroic point at 248 nm confirmed that there was only an inversion of the same helical conformation and no other structural changes.[96]

In 2011, Meille, Porri, and co-workers characterized the crystalline form of a vinyl-type polynorbornene (PNB).[97] Fully saturated vinyl polymers or oligomers with a 2,3-*exo*-disyndiotatic (*ds*) structure were obtained using an $AlEt_2Cl$-$TiCl_4$ catalyst with a ratio monomer/$TiCl_4$ of 18 or 250, respectively. A fractionation method was used for the purification. These polymers formed a molecular tube-shape that contained some different types of small molecules like toluene, as shown in Figure 5.23 (red circles). This characteristic was confirmed also by thermogravimetry and NMR and vibrational spectroscopy.[97]

One of the most recent reports in the area of tubular helices was published in 2015 by Li and co-workers, who synthetized and reported three different aryl hydrazide polymers bearing different numbers of chiral pentaethylene glycol chains. These three polymers (Figure 5.24) were tested in different solvents (water, chloroform, dichloromethane, methanol, and others) to see the folding structure and shape dependence on the solvent. In water, these polymers showed a Cotton effect indicating that they had folded into a helix. The main driving force, however, toward folding into a helix in water was not the cross-layer intramolecular H-bonds but the hydrophobicity of the polymer. This was confirmed by the decrease of the CD signal in methanol. If the driving force was intramolecular H-bond formation, the signal should have increased due to the stabilization of these interactions.[98]

Figure 5.23 Top view of the crystal structure model of tubular *ds*PNB-toluene. Reproduced with permission from A. Buono, A. Famulari, S. V. Meille, G. Ricci and L. Porri, *Macromolecules*, 2011, **44**, 3681–3684.[97] Copyright (2011) American Chemical Society.

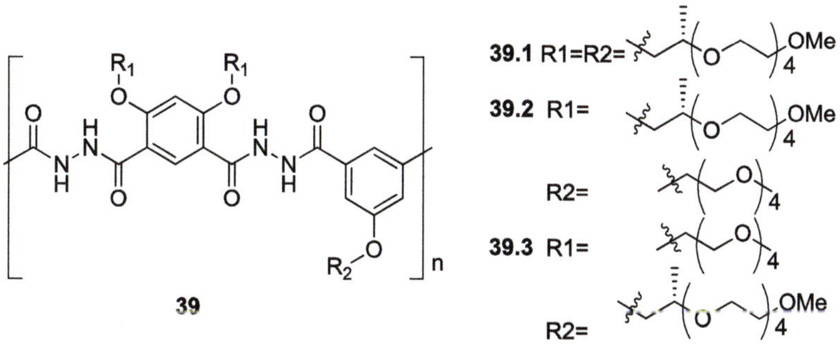

Figure 5.24 Aryl hydrazide polymers **39.1**, **39.2**, and **39.3**.

Another recent article was published by Takata and co-workers.[99] The polymer was composed of oligothiophene monomers (pink) as a planarity-switchable spacer and a C_2-chiral spirobifluorene monomer (green). Under redox conditions, the polymer switched from a one-handed helical shape to a coil-shaped rigid helical foldamer (Figure 5.25).

Polymer P1 (**40**) had an M_w of 11 000 g mol^{-1} and a PDI of 2.5, and P2 (**41**) an M_w of 72 000 g mol^{-1} and a PDI of 2.4. All these results were obtained by size exclusion chromatography (SEC) (Figure 5.26).

To compare the results obtained with these polymers, some model compounds were synthesized (Figure 5.27).

UV-vis spectroscopy showed shorter maximum absorption wavelengths for P2 (**41**), M2 (**43**), and D2 (**45**) than for P1 (**40**), M1 (**42**), and D1 (**44**). This was caused by the quarterthiophene unit twisted by the thermally induced steric interaction between the spirobifluorene and the hexyl group. Both polymers showed also a broader signal, meaning that they adopted a non-planar

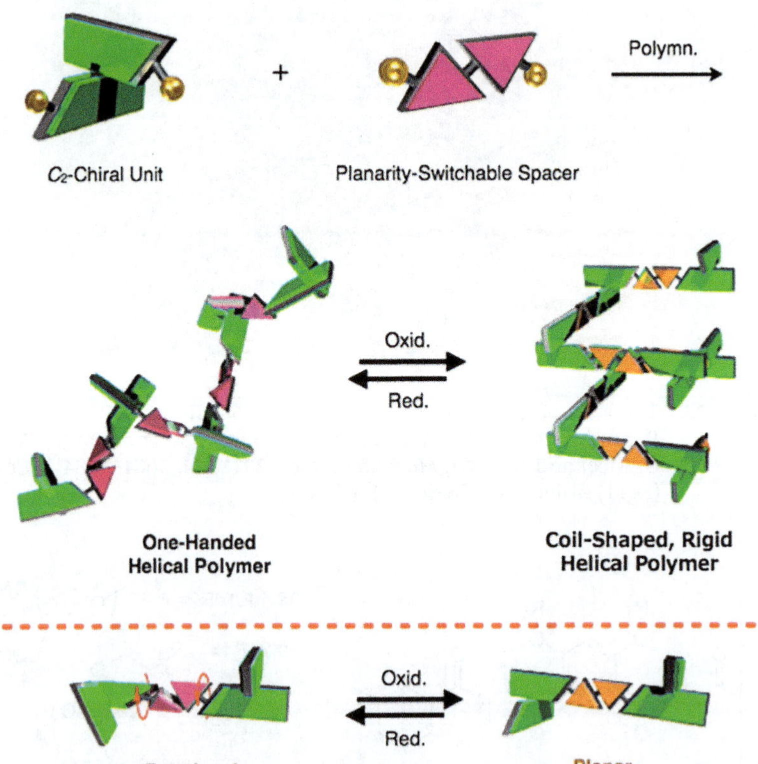

Figure 5.25 Mechanism of folding of polymer **40**.
Reproduced with permission from H. Okuda, Y. Koyama, S. Uchida, T. Michinobu, H. Sogawa and T. Takata, *ACS Macro Lett.*, 2015, **4**, 462–466.[99] Copyright (2015) American Chemical Society.

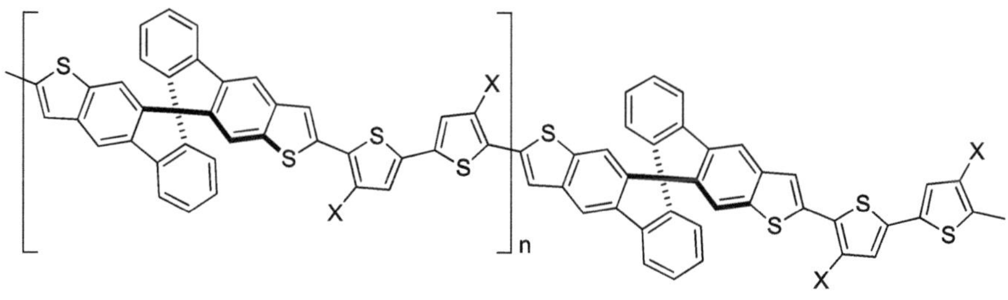

40: P1: X = H
41: P2: X = *n*-C_6H_{13}

Figure 5.26 Structure of polymers **40** and **41**.

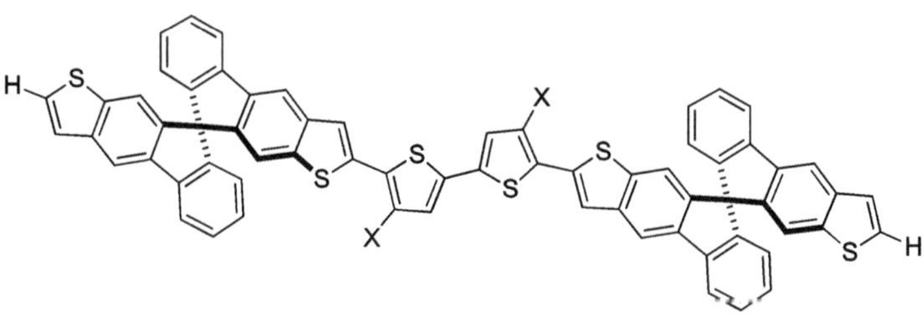

42: M1: X = H
43: M2: X = *n*-C_6H_{13}

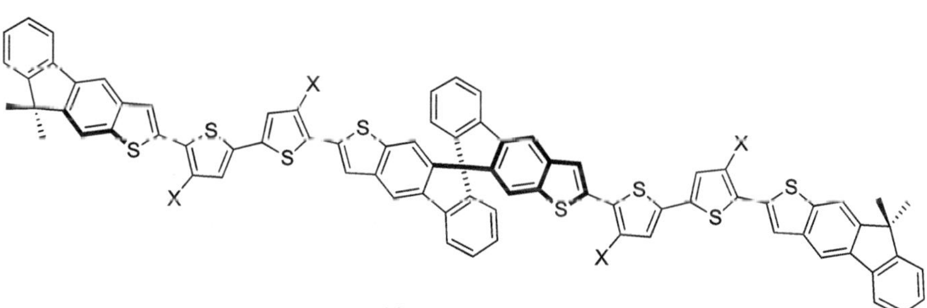

44: D1: X = H
45: D2: X = *n*-C_6H_{13}

Figure 5.27 Structure of monomers **42** and **43** and dimers **44** and **45**.

structure at the quarterthiophene stage (Figure 5.28). This non-planar structure could be attributed to the steric interactions between the polymeric main chains, as well as to the entropy-driven conformational change of the chain to adopt an extended structure.

The CD spectra showed a positive Cotton effect in the π–π^* transition region of the quarterthiophene for all the molecules. For both polymers, the effect was remarkably higher than for the monomer and dimer models, which confirmed the one-handed helical shape. After oxidation with $FeCl_3$, the Cotton effect was present at 700 nm due to the generation of a radical

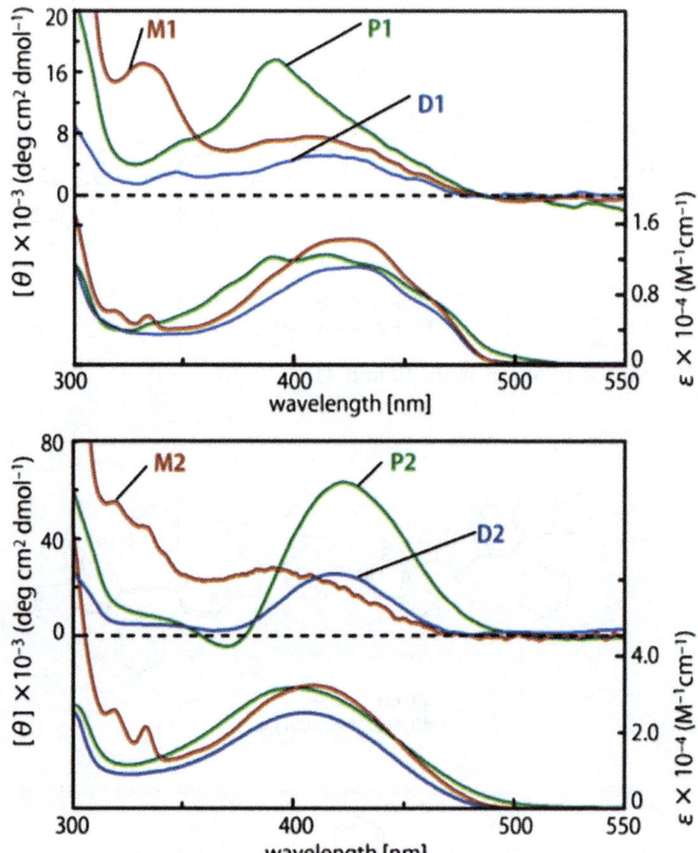

Figure 5.28 CD spectra for polymers P1 (**40**), P2 (**41**), monomers M1 (**42**), M2 (**43**), and dimers D1 (**44**), D2 (**45**).
Reproduced with permission from H. Okuda, Y. Koyama, S. Uchida, T. Michinobu, H. Sogawa and T. Takata, *ACS Macro Lett.*, 2015, **4**, 462–466.[99] Copyright (2015) American Chemical Society.

cation polymer. To restore the original spectrum, sodium bisulfite was added as a reducing agent. This work reported the first polymer in which the rigidity of the structure was controlled by a redox reaction.[99]

A nice study using STM images as evidence for helix formation of a polymer was carried out by Dong, Liu, *et al.*[100] A 2,5-bis(2-pyridyl)-1,3,4-oxadiazole was polymerized (**46**) (Figure 5.29), deposited on a graphite surface and STM images were recorded, as shown in Figure 5.30.

This helix showed a well-organized structure in which all the helices were packed in a parallel manner. Figure 5.30(f) shows a schematic representation of the helix. All the measurements reported were also confirmed by other analytical methods (XRD, CD). This polymer existed as a single helix or a double helix, as shown in Figure 5.30(e).[100]

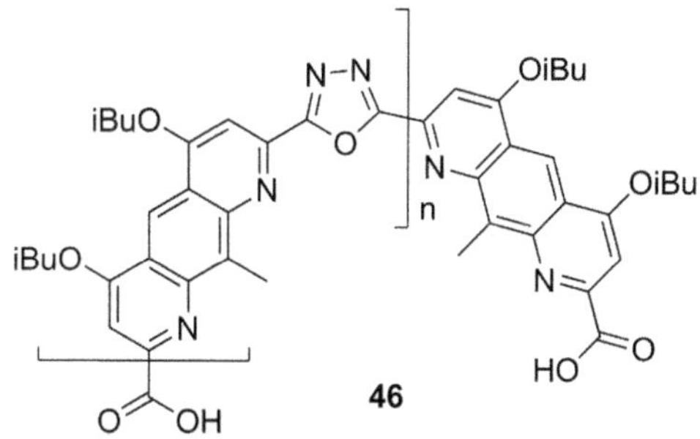

Figure 5.29 Poly(2,5-bis(2-pyridyl)-1,3,4-oxadiazole) 46.

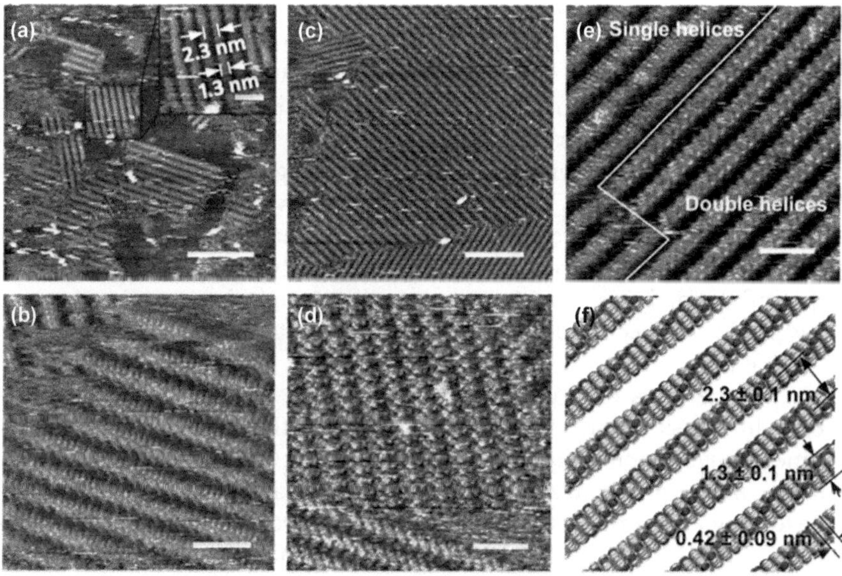

Figure 5.30 (a) STM image and partial enlarged view of polymer **46** (0.10 g L^{-1}); scale bars: 20 nm, 5 nm. (b) High-resolution STM image of **46** (0.10 g L^{-1}); scale bar: 5 nm. (c) STM image of **46** (1.0 g L^{-1}); scale bar: 20 nm. (d) High-resolution STM image of **46** (1.0 g L^{-1}); scale bar: 5 nm. (e) High-resolution STM image: single helices (upper left) and double helices (lower right); scale bar: 5 nm. (f) Schematic illustration of a regenerated arrangement of single-helical backbones with right-handedness.
Reproduced from ref. 100 with permission from John Wiley and Sons. Copyright © 2015 WILEY-VCH Verlag GmbH & Co. KGaA, Weinheim.

5.2.4 Helical Polymers with Host–Guest Interactions

Host–guest complexes in foldamer structures are based on the combination of different non-convalent interactions, which force the polymer to fold into a helical structure around the guest. In the past years, many oligomers have been synthesized with the aim to complex with guests. Neutral (polar,

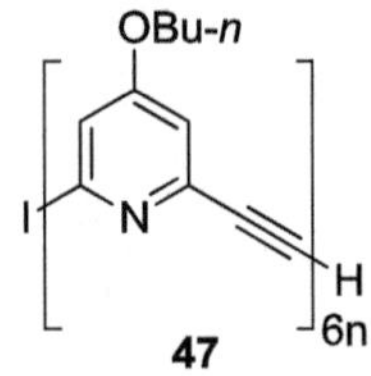

47

Figure 5.31 Poly(*m*-ethynylpyridine) **47**.

non-polar molecules, and saccharides), cationic (metal ions, ammonium), and anionic guests have been investigated.[101]

In 2004, Inouye and co-workers synthesized a poly(*meta*-ethynylpyridine) that exists as a random coil in non-polar solvents (Figure 5.31).

Addition of a saccharide into the dichloromethane solution of **47** triggered this polymer to adopt a helical conformation with a chiral sense that was transferred from the bound saccharides. The polymer was synthesized *via* Sonogashira acetylene coupling between two hexameric (*meta*-ethynylpyridine)s. The polymer had an M_n of 2200 g mol^{-1}. The induced CD signal (ICD) appeared only when *n*-octyl glycosides were added to the solution, which meant that saccharides had induced the helicity and chirality of polymer **47** *via* hydrogen bonding interactions between the saccharide-OH groups and the pyridine nitrogen atoms. Only β-glucoside showed an intense ICD signal due to saccharide hydroxy groups that fitted perfectly in the inner void of the helix.

This polymer was also useful to extract native saccharides from polar solvents into less polar solvents. The cavity of the helix contains many pyridine nitrogen atoms that render the inside of the tubular structure very polar and thus a good environment to solubilize saccharides.[102]

The folding of several derivatives of polymer **47** were investigated.[103–105] In 2011, polymers **48** and **49** bearing a metal coordination site were reported. They fold around an alkyl glycoside guest and complex a Cu(II) center forming a helical structure. The insertion of Cu(II) (Figure 5.32) into the solution changed the intensity of the ICD signal, so the helical structure was stabilized or destabilized by the cross-linking with Cu(II).[106]

CD spectra revealed the influence of the molecular size, metal cation recognition, and temperature.[107]

In 2014, Li, Hou, and co-workers synthesized a hydrazide polymer that mimics the ion transport of Gramicidin A.[108] This polymer carried phenylalanine side chains, which allowed the polymer to be inserted into a lipid bilayer at low concentrations. Patch clamp experiments showed that this polymer could transport some alkali-metal ions such as Na^+, K^+, and also NH_4^+ across the membrane. Patch-current traces up to 15 s showed that the structure was stable within the bilayer. As the NH_4^+ selectivity was higher than that for K^+, the inner pore was estimated to be smaller than 1 nm.[57,108]

Another field where helical polymers have been used is the Space-Charge-Limited Current (SCLC) hole mobility. In 2015, a discotic polycyclic aromatic

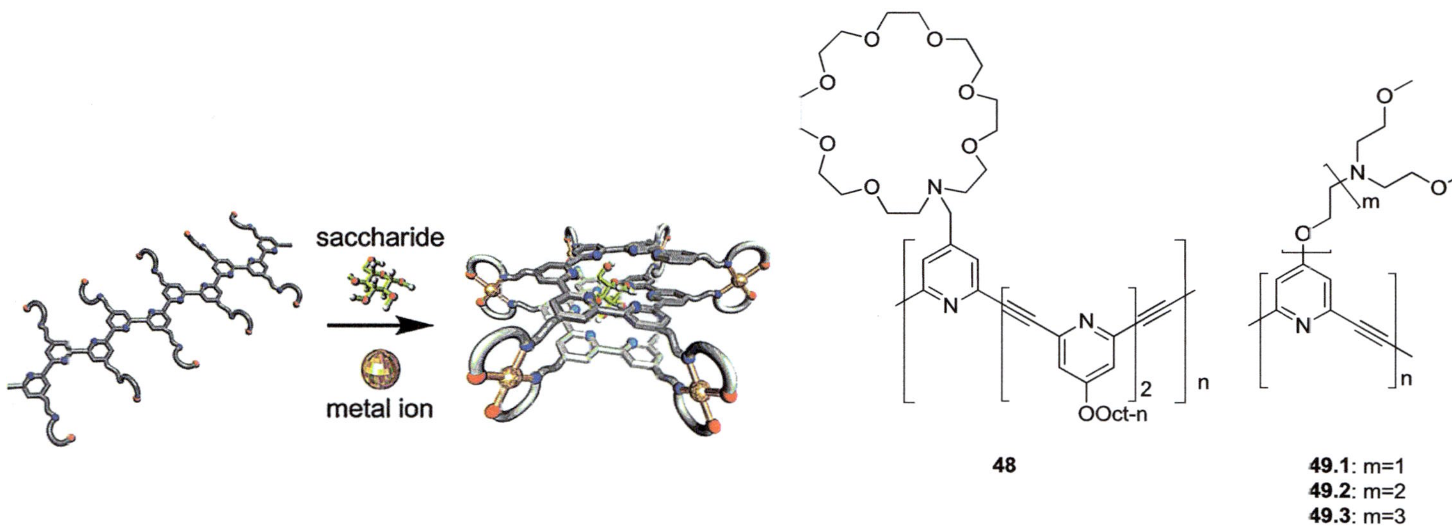

Figure 5.32 Stabilization of helical polymers **48** and **49** by coordination of metal ions. Reproduced from ref. 106 with permission from the Royal Society of Chemistry.

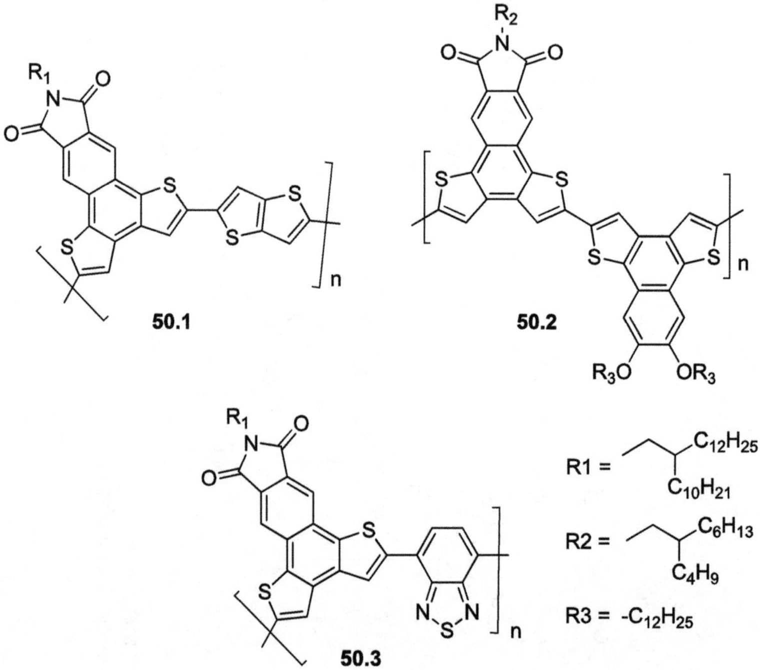

Figure 5.33 Imide-functionalized naphtha[1,2-*b*:4,3-*b*′]dithiophene (INDT) polymers.

hydrocarbon-based imide-functionalized naphtha[1,2-*b*:4,3-*b*′]dithiophene (INDT) was used as the building block to synthesize a low-band-gap donor–acceptor conjugate polymer (Figure 5.33).[109]

Fluorescence spectra of polymer **50.2** and polymer **50.3** showed a broad peak emission in both good (chloroform, THF) and poor (hexane and acetonitrile) solvents. **50.2** exhibited a maximum emission peak at 530 nm (in chloroform), which was red-shifted compared to the monomeric emission of polymer **50.1**. This showed that polymer **50.2** was highly aggregated through intrachain aggregation. When the solvent was changed from chloroform to acetonitrile, a red-shift was observed for all polymers. For **50.1** and **50.2**, this displacement was in accordance with the presence of intrachain folding. Polymer **50.3** was red-shifted by 85 nm, which means that interchain π–π stacking dominates this polymer in poor solvents. In conclusion, polymer **50.1** adopted a random-coil conformation in good solvents and showed interchain aggregation in poor solvents. Polymer **50.2** adopted a folded conformation in all solvents and **50.3** adopted a folded conformation in good solvents and interchain aggregation in poor solvents.

The highest SCLC hole mobility was measured for polymer **50.2** because of its high nanocrystallinity (2.77×10^{-2} cm^2 V^{-1} s^{-1}). For polymer **50.1**, the value was 2.20×10^{-6} cm^2 V^{-1} s^{-1}, and was 8.79×10^{-5} cm^2 V^{-1} s^{-1} for polymer **50.3**. Blending between polymer **50.2** and $PC_{71}BM$

(a C_{70} fullerene derivative) increased the crystallinity and thereby also the SCLC hole mobility (from 2.77×10^{-2} cm^2 V^{-1} s^{-1} to 1.29×10^{-1} cm^2 V^{-1} s^{-1}), one of the highest values ever reported.[109]

5.3 Engineered Polymer Nanotubes

5.3.1 Block Copolymer Self-assembly

Amphiphilic block copolymers containing poly(styrene) (PS) as a tail and charged helical poly(isocyanide) (PI) as the head group synthesized by Nolte and co-workers in the late 1990s were shown to self-assemble into micelles, vesicles, and bilayer aggregates.[110,111] The length of the poly(isocyanide) group, the interactions of the charged head group, and the pH were the keys to control the morphology of the final structure.

The structure of the monomers induced the direction of the twist in the formed helix. The two derivatized amino acids used in the head group (red in Figure 5.34) showed a relatively strong Cotton effect in the circular dichroism (CD) analysis. From this, it was concluded that PIAA contained an excess of P helices and PIAH, an excess of M helices.[112] The resulting superhelix PS-*b*-PIAA had a left-handed twist (Figure 5.35), whereas the building blocks were right-handed helices. Conversely, PIAH had left-handed building blocks and assembled into right-handed superhelices.

The group of Yan developed another method for self-assembly using an amphiphilic hyperbranched multiarm copolymer.[113] After addition to a selective solvent like acetone, macroscopic tubes with centimeter-scale length and millimeter-scale diameter were formed. The star polymer was composed of a hyperbranched poly(3-ethyl-3-oxetanemethanol) (HBPO) core and hydrophilic poly(ethylene glycol) (PEO) arms and self-assembled into a multi-walled tube.

^{1}H NMR studies showed that the surface of the observed tubes (Figure 5.36, right) was covered by densely and well-ordered PEO arms. In the selective solvent acetone, the hydrophobic interaction of the core (blue in Figure 5.36) led to spinodal microphase separation. This phenomenon induced macroscopic molecular self-assembly. One molecular stacking model that was proposed for the tube walls was the formation of juxtaposed sandwich-like microfilms consisting of condensed HBPO cores in the inside and PEO arms on the outside. These microfilms then piled in a layer by layer fashion held together by hydrogen bonds between the PEO arms, driving further the self-assembly towards multi-walled macrotubes.

Nanostructures like hollow tubes have attracted much attention mainly to meet the demand for smaller electronic devices. Tubes can be obtained using the "sculpturing" method.[114–117] Liu and co-workers synthesized hollow nanotubes using triblock copolymers.[118] The triblock copolymer poly(isoprene)-*block*-poly(2-cinnamoylethyl methacrylate)-*block*-poly(*tert*-butyl acrylate) (PI-*b*-PCEMA-*b*-P*t*BA) containing 130 isoprene, 130 CEMA, and 800 *t*BA units formed self-assembled cylindrical micelles in methanol.

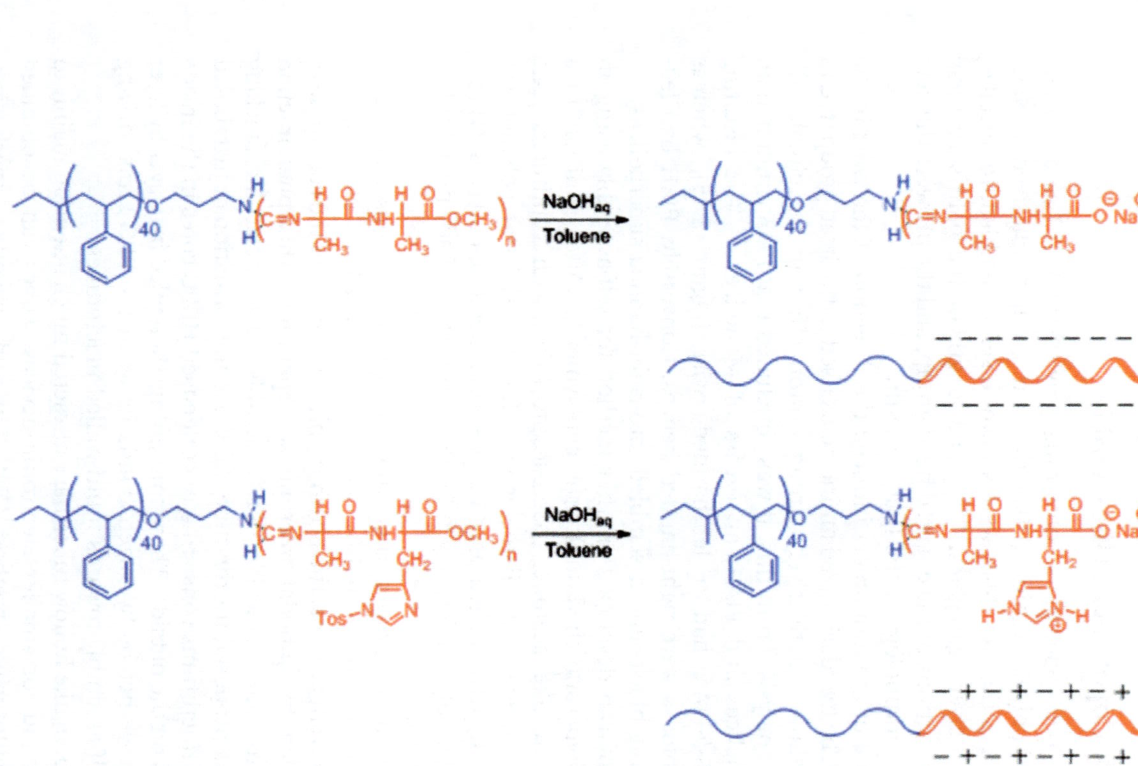

Figure 5.34 Deprotection of the block copolymers: (top) PS-*b*-poly(isocyano-L-alanine-L-alanine) (PS-*b*-PIAA) to negatively charged superamphiphiles and (bottom) PS-*b*-poly(isocyano-L-alanine-L-histidine) (PS-*b*-PIAH) to zwitterionic superamphiphiles.

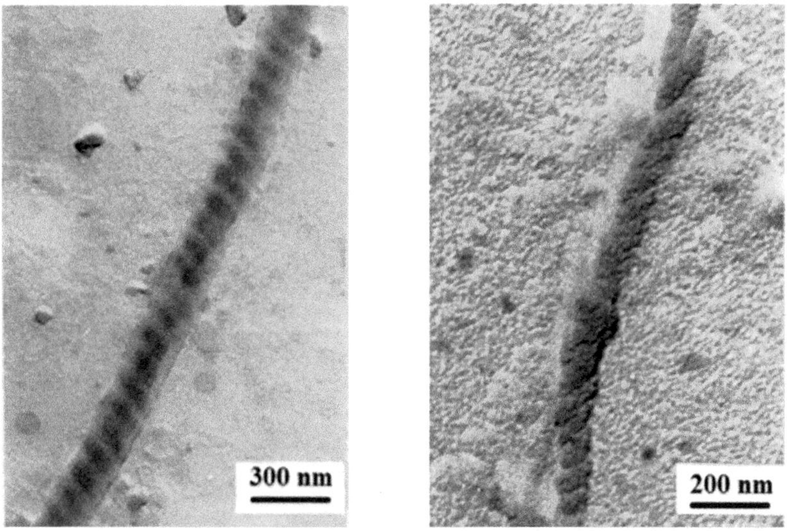

Figure 5.35 TEM images of superhelices in sodium acetate buffer of pH 5.6 (0.2 mM). Left, left-handed superhelix formed by PS-*b*-PIAA. Right, right-handed superhelix formed by PS-*b*-PIAH.
Reproduced with permission from ref. 110. Copyright 1998, The American Association for the Advancement of Science.

The micelles consisted of a PI core, a PCEMA shell, and a P*t*BA corona, which could be interlocked due to the photo-crosslinking of the PCEMA shell. Finally, degradation of the PI core of the cross-linked micelles gave a nanotube with a diameter of around 65 nm (Figure 5.37).

Using the nano "sculpturing" method, Rzayev *et al.* prepared organic nanotubes by single molecule templating of core–shell bottlebrush copolymers.[119] They synthesized the bottlebrush copolymer using a combination of living radical and ring-opening polymerization techniques. The target structure had a degradable polylactide (PLA) core (red in Figure 5.38) grafted on a poly(glycidyl methacrylate) backbone (black) containing a polyethylene oxide brush as a "stopper" on one side (green). On the other side of the PLA brush, poly(4-(3-butenyl)styrene was used as a cross-linkable shell (blue), which finally allowed the formation of a cross-linked single opened nanotube.

Nanotubes can also be obtained by direct cationic copoylmerization of divinylbenzene (DVB) and styrene (ST). The group of Larock demonstrated the direct and simple method for the preparation of nanoscale fibers and tubes.[120] They polymerized DVB with ST as a co-monomer in different ratios. In the case of DVB alone, SEM micrographs (Figure 5.39) indicated that the material was composed of clusters of fine fibers. Adding 30 wt% of styrene did not induce any changes in the obtained structures. Starting from ratios of 45 wt% of styrene and 55 wt% of DVB resulted in polymers with good morphological control. TEM images showed that the polymer formed hollow

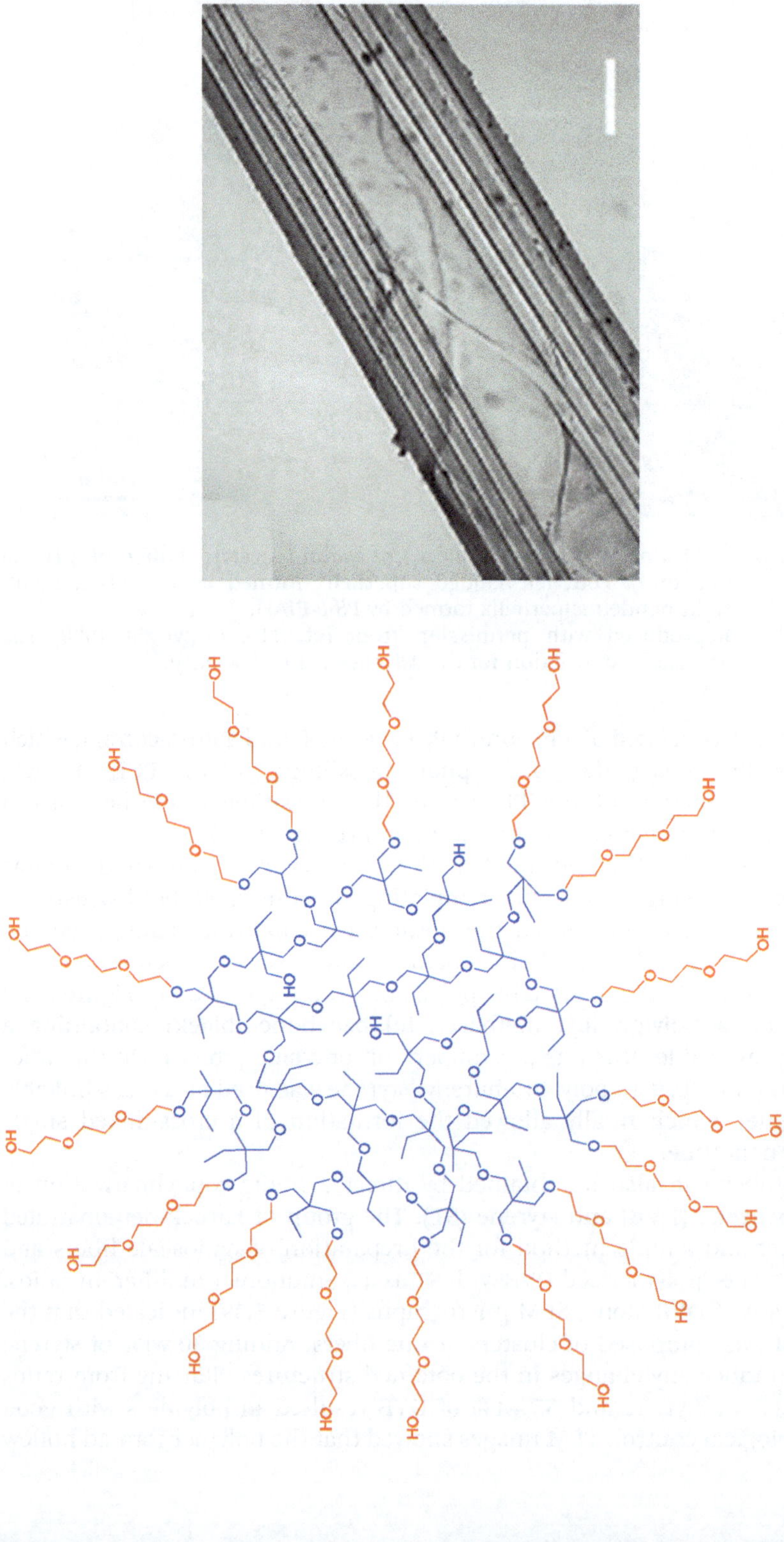

Figure 5.36 Left, sketch of the HBPO-*star*-PEO multi-arm copolymer. Blue, HBPO core; red, PEO arms. Right, optical micro-spectrometry images of a penta-walled self-assembled tube in acetone (scale bar: 300 μm). Reproduced with permission from ref. 113. Copyright 2004, The American Association for the Advancement of Science.

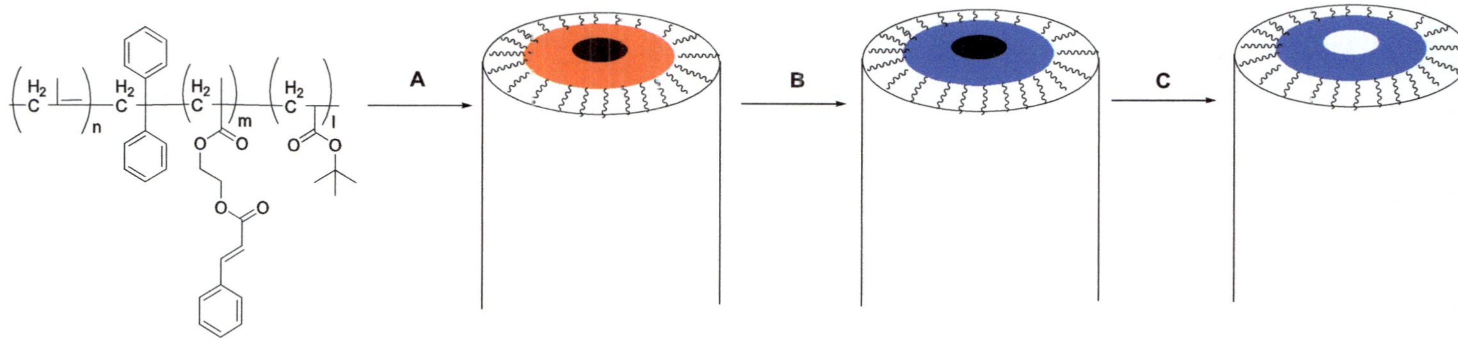

Figure 5.37 Synthesis of the block copolymer nanotube. (A) Self-assembly of the triblock copolymer PI-*b*-PCEMA-*b*-P*t*BA into cylindrical micelles with PI core (black), PCEMA shell (red)- and P*t*BA corona. (B) Photo-crosslinking of PCEMA shell (red to blue). (C) Degradation of PI core to obtain a hollow nanotube.

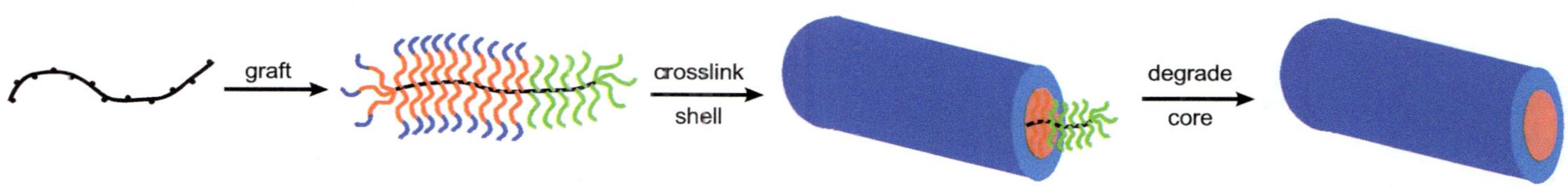

Figure 5.38 Synthetic pathway to well-defined organic nanotubes. A diblock copolymer bottlebrush was synthesized containing one block with inner red and outer blue segments and a second block serving as a sacrificial "stopper" in green.

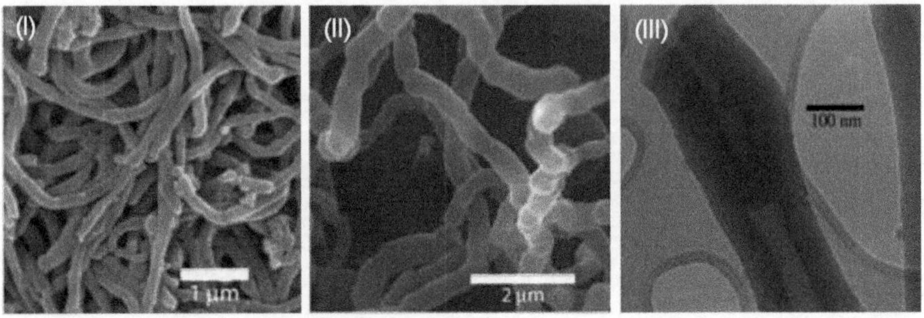

Figure 5.39 SEM (I and II) and TEM (III) images of the obtained polymers from the polymerization of (I) DVB 100%, (II) DVB with 30 wt% of ST, and (III) DVB with 45 wt% of ST.
Reproduced from ref. 120 with permission from the Royal Society of Chemistry.

tubes with some blockage at the end and in the middle of the tubes. Using 50 wt% of ST gave the same result. However, the random nature of this polymerization process rendered the shape and thickness of resulting structures unpredictable.

Lin and co-workers showed that the use of chiral block copolymers poly(L-lactide)-*b*-polystryrene (PS-*b*-PLLA) led to the formation of a hexagonally packed nanohelical microstructure of PLLA in a PS matrix.[121] Moreover, tubular nanostructures were obtained by hydrolysis of the degradable PLLA core from the core–shell cylinder microstructures. In nature, self-assembly is a tool to structure molecules. Different levels of organization can be achieved using the interplay of secondary forces, such as hydrophobic, hydrogen bonding, steric and electrostatic interactions. In the work of Lin, it was the combination of chirality and the immiscibility of the constituent blocks that were responsible for the nano organization. The authors speculated that the core–shell cylindrical microstructure of the self-assembled PS-*b*-PLLA block copolymer was induced by the formation of specific configurations of the chiral PLLA chains.

There are two accepted routes for the formation of tubular structures. Recently, ultrathin nanotubes with tunable surface functionalities were synthesized in the group of Zhou and Yan.[122] They used an alternating copolymer, which had already shown great potential in photoelectronic applications, to prepare hollow tubes with diameters of around 30 nm.[123,124] The amphiphilic alternating copolymer was prepared through an epoxy-thiol click reaction, which allowed facile insertion of functional dithiols or amines (Figure 5.40). A positively charged polymer could be obtained using tertiary amines. Negative charges could be realized using carboxy groups. Polymers decorated with immobilized peptides were also prepared. The dropwise addition of water into the solution of the copolymers in DMSO induced the self-assembly into ultrathin nanotubes. The route to the formation of hollow nanotubes was calculated with dissipative particle dynamics (DPD) simulations. Those calculations show the formation of nanosheets as the first

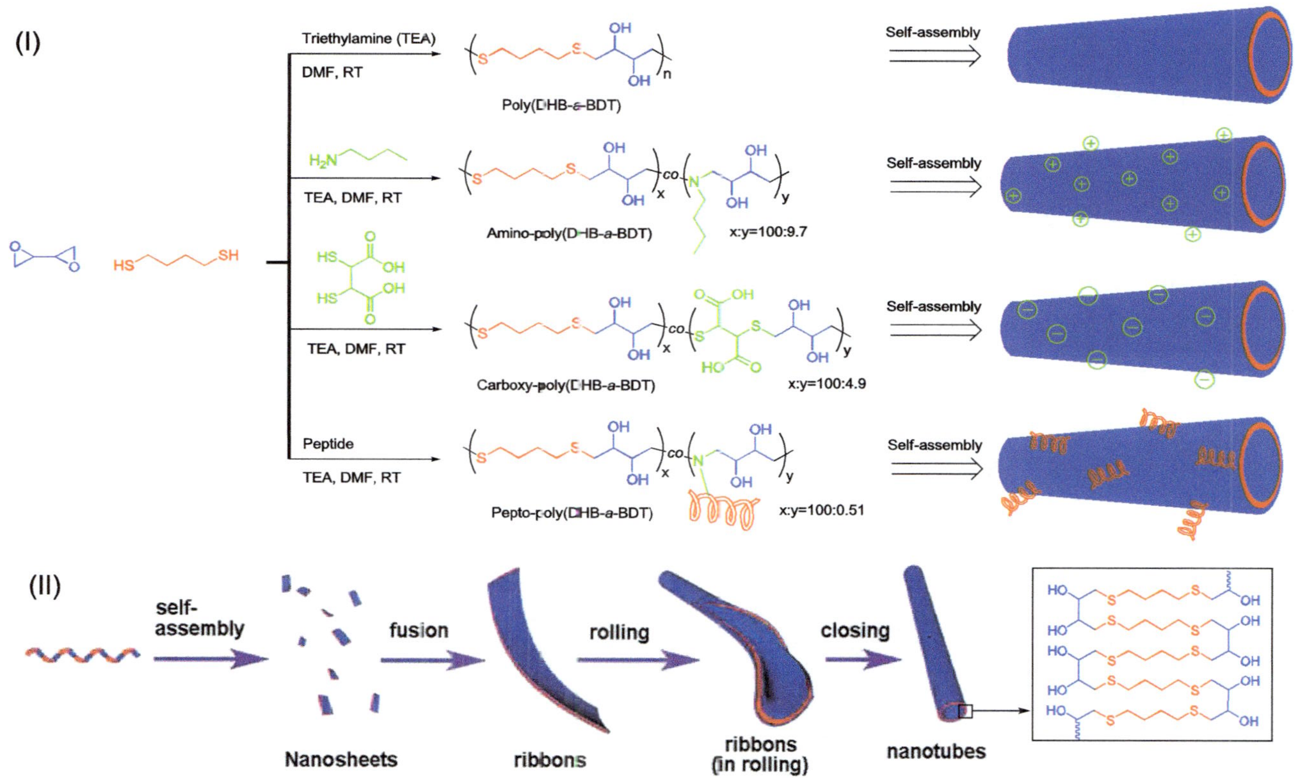

Figure 5.40 Synthesis and self-assembly of poly(2,3-dihydroxy butylene-*alt*-butylene di-thioether) poly(DHB-*alt*-BDT)s. (I) Synthetic pathway and functionalization. (II) Proposed self-assembly route and cross-section view of the membrane (supported by DPD calculations).
Reproduced from ref. 122 with permission from John Wiley and Sons. Copyright © 2015 WILEY-VCH Verlag GmbH & Co. KGaA, Weinheim.

step, the next step is the fusion to ribbons, which finally twist into an open helix that can close to form nanotubes. Evidence for the formation of hollow nanotubes was provided by TEM and AFM images. Measurements showed that the collapsed nanotubes had single membranes with thickness of 1.2 nm, which is quite rare for polymeric nanotubes, and further supported that they were rolled-up nanotubes rather than nanofibers.

As described previously, the self-assembly of block copolymers or alternating copolymers can lead to the formation of tubes through different mechanisms. Amphiphilic block copolymers have shown to be powerful polymers for the creation of helical or tubular structures. For example block copolymers of poly(ethylene glycol) (PEG) and poly(lactide) (PLA) form micelles in aqueous solution and have been widely investigated.[125–129] However, the high chain flexibility of block copolymers creates limitations upon the formation of nanotubes. Manners *et al.* described the facile but unexpected formation of organometallic self-assembled nanotubes from poly(ferrocenyldimethylsilane)-*b*-poly(dimethylsiloxane) (PFS-*b*-PDMS) in hexane.[130,131] Hedrick and co-workers demonstrated that hydrogen bonding motifs in the interface between block copolymers not only stabilize the self-assembly but also modify the morphology of the micelles.[132] In their work, they compared the self-assembly process of an amphiphilic block copolymer (PEG-*b*-PLLA) and an amphiphilic block copolymer connected by a shape persistent benzamide (BA) and hydrogen bonding (urea (U) or thiourea (TU)) sections (PEG-TU/U-BA-PLLA). As expected, the block copolymer PEG-*b*-PLLA self-assembled into spherical micelles about 20–40 nm in diameter (Figure 5.41).

In the case of a block copolymer containing a hydrogen-bonding segment and a rigid benzamide in between, the mechanism for self-assembly is different. Indeed, both urea-benzamide and thiourea-benzamide motifs

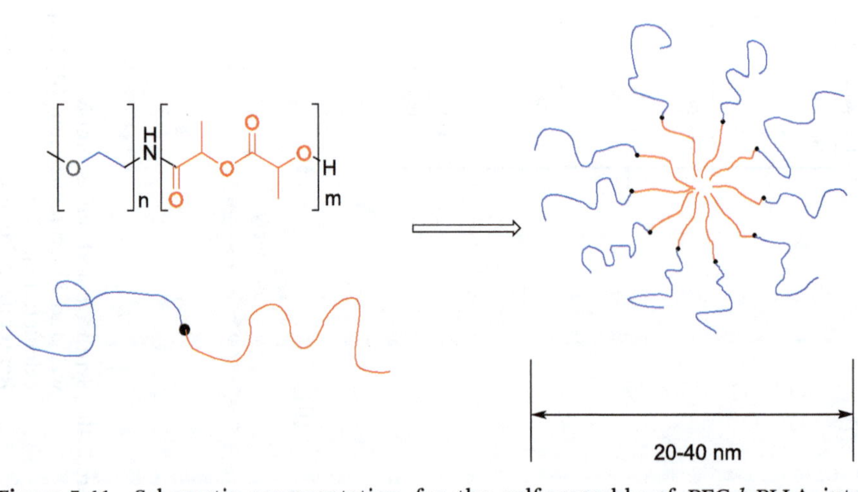

Figure 5.41 Schematic representation for the self-assembly of PEG-*b*-PLLA into micelles.

offer simple building blocks for anisotropic interactions. Instead of spherical structures, they tend to form nanotubes. A proposed mechanism for the formation of hollow tubes from block copolymers PEG-TU-BA-PLLA and PEG-U-BA-PLLA is the anisotropic growth of a bilayered tape (Figure 5.42).[133]

This approach to anisotropic block copolymer assemblies can be compared to molecular nanotubes constituted of glycolipids and other amphiphiles.[14,134] These complex molecular structures are dependent on factors like chirality, size of the head group, or molecular packing. Those examples can explain the second route to the formation of nanotubes. Therefore, the orientation within the aggregating membrane is responsible for the rolling-up into a tubular structure.

The group of Li synthesized block copolymer aggregates with temperature-controlled morphologies.[135] In their work, they assumed that the arrangement of the membrane was neither random nor isotropic, but due to the smectic organization of the liquid crystalline hydrophobic part. Indeed, the block copolymer consisted of an amphiphilic smectic poly(ethylene glycol)-*b*-poly(cholesteryl acryloxy ethyl carbonate) (PEG-*b*-PAChol) and was nanoprecipitated by solvent exchange (Figure 5.43).

These smectic polymers showed in water a variety of temperature controlled micellar aggregates including nanofibers, ellipsoidal or faceted vesicles, nanotubes, and spherical aggregates. The smectic liquid crystal phase probably serves as physical cross-linking to maintain the stability of the assemblies. All those morphologies could be observed by TEM and AFM. At low temperature (5 °C), nanofibers are obtained. As the temperature increases, nanofibers start to co-exist with vesicles (15 °C) and the percentage of vesicles increases with the increasing temperature (25 °C). At 40 °C, long tubular vesicles (nanotubes) can be observed. Finally, at 55 °C, the morphology changes again to form spherical aggregates. Because this chapter deals with tubular structures, we will focus mainly on the nanotubes obtained at 40 °C. As cryo-TEM is sensitive to the electronic density and therefore highlights only the hydrophobic part of the block copolymer PEG-*b*-PAChol, the size of the liquid crystalline part could be measured. The thickness of the membrane was 24 to 29 nm (*e* in Figure 5.44), the width from 61 to 88 nm (*d*), and the length of the tube from 100 nm to several microns (*L*).

More interestingly, these nanotubes are thermally stable after their formation even upon heating to 55 °C, which is an important feature for potential applications in drug delivery and materials science.

Another template-free method for the formation of self-assembled tubular structures has been shown by Jiang and co-workers.[136] They used polymeric Janus particles prepared from mixed-shell micelles (MSMs) that self-assembled into tubular superstructures and nanosheets. The micelles were prepared by non-covalent cross-linking of poly(acrylic acid) (PAA) blocks by addition of 1,2-propanediamine (PDA) in DMF (A on Figure 5.45). The two block copolymers were composed of hydrophilic poly(ethylene

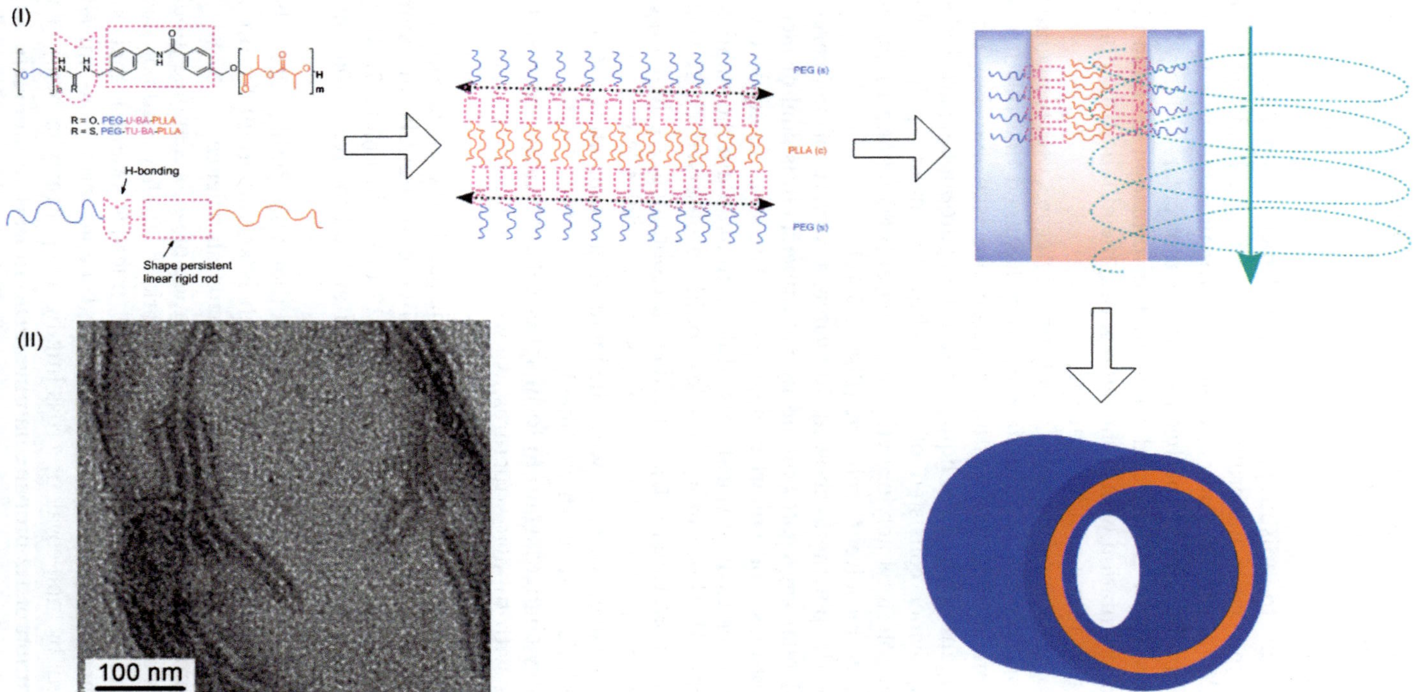

Figure 5.42 (I) Schematic representation of self-assembly and tight rolling-up of block copolymers containing rigid hydrogen bonding motifs PEG-TU-BA-PLLA and PEG-U-BA-PLLA. (II) TEM image of the PEG-TU-BA-PLLA nanostructure.
Reproduced from ref. 132 with permission from John Wiley and Sons. Copyright © 2009 WILEY-VCH Verlag GmbH & Co. KGaA, Weinheim.

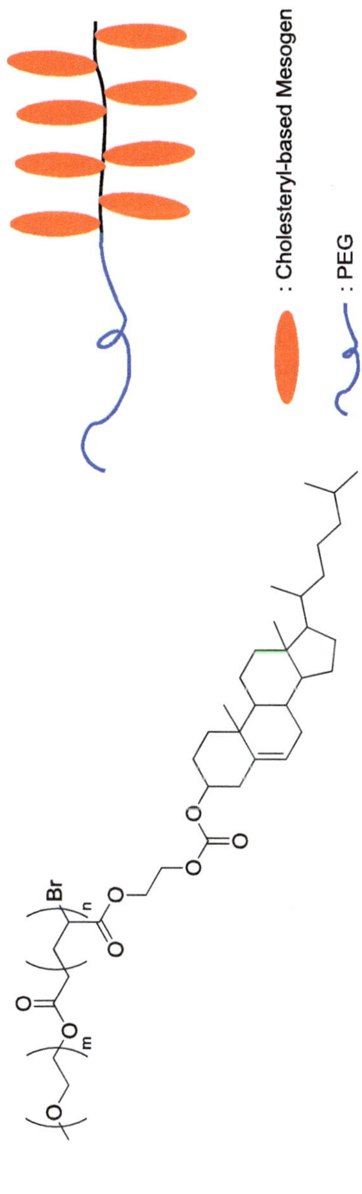

Figure 5.43 Chemical structure of the smectic amphiphile block copolymer PEG-*b*-PAChol.

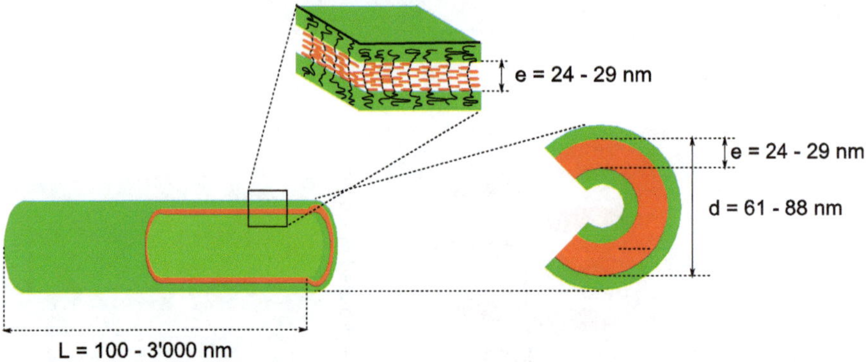

Figure 5.44 Structural model of the self-assembled nanotubes formed by PEG-*b*-PAChol in water at 40 °C.

glycol)-*b*-poly(acrylic acid) (PEG-*b*-PAA) and a second hydrophobic poly(2-vinyl naphthalene)-*b*-poly(acrylic acid) (P2VN-*b*-PAA) block. The particles constituted of P2VN-*b*-PAA and PEG-*b*-PAA block copolymers with non-covalently cross-linked PAA cores were then dialyzed against water (pH 7) (B). P2VN collapsed into separated micro-domains. By decreasing the pH to 3.1, intramolecular complexation between PEG and PAA occurred, resulting in asymmetric intramicellar phase separation between the PEO/PAA complex and P2VN (C). This phase separation resulted in the formation of amphiphilic Janus particles with P2VN hydrophobic domains formed by the aggregation of all the P2VN microdomains. Those aggregates could be observed by TEM. Numerous Janus particles did co-exist with the formed tubes, as can be observed on the TEM image in Figure 5.45.

The self-assembled nanotubes had lengths from 0.85 to 1.1 μm and widths around 0.2–0.4 μm. Some TEM images showed contrast, which demonstrates that the tubes were constructed from small particles and that they had a hollow interior. This work demonstrated that polymeric Janus particles behave like amphiphilic block copolymers or lipids. They are capable of forming layered structures with a relatively reduced curvature, such as vesicles or in this case nanotubes, even though the exact mechanism is still under investigation.

As amphiphilic block copolymers tend easily to form vesicles rather than tubes, Helmerson *et al.* reported another strategy for the formation of long polymer nanotubes.[137] They started with polymersomes, which are cross-linked vesicles of amphiphilic block copolymers filled with water. Pulling on the membrane of those polymersomes with a micropipette or with optical tweezers led to the formation of water containing nanotubes stretched from the vesicles (Figure 5.46). However, this process required the application of a lot of force. To overcome this problem, they incorporated a detergent-like triblock copolymer in the formation of the polymersomes, with the effect that the membranes were rendered more fluid.

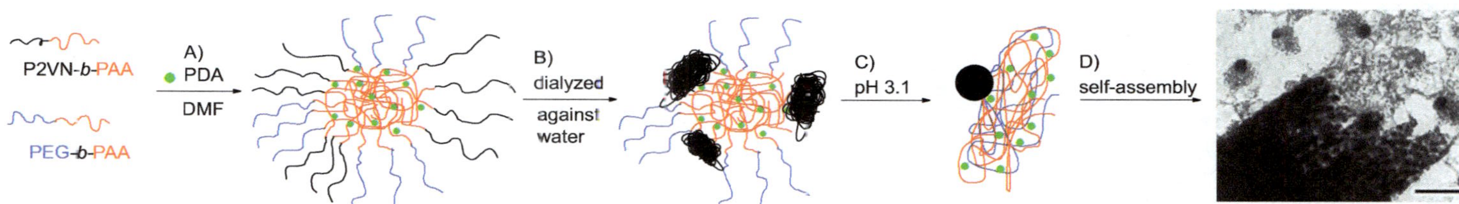

Figure 5.45 Schematic illustration of the processes for the preparation of MSMs in DMF (A) and in water (B), and the intramicelle complexation to form Janus nanoparticles (C). (D) TEM image of the aggregates (scale bar is 100 nm).
Reproduced from ref. 136 with permission from John Wiley and Sons. Copyright © 2008 WILEY-VCH Verlag GmbH & Co. KGaA, Weinheim.

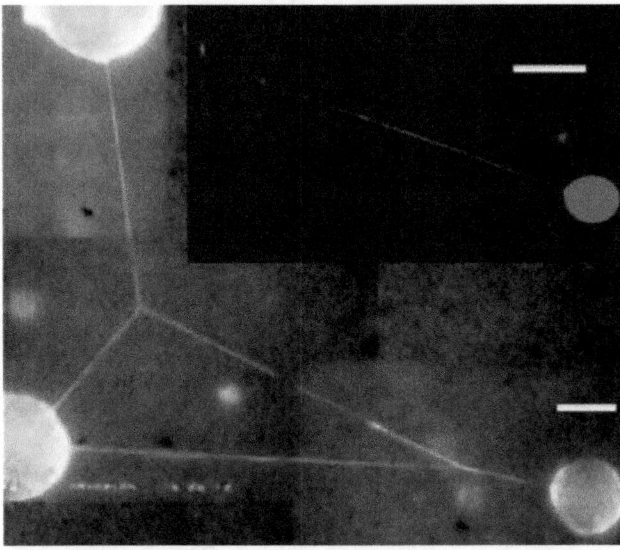

Figure 5.46 Composite image from video fluorescence microscopy of a network of polymer nanotubes and polymersomes containing the membrane dye DiO-C16, assembled using optical tweezers (scale bar is 10 μm). (Inset) Scanning confocal microscopy image of a nanotube pulled from a polymersome encapsulating sulforhodamine B dye in buffer. Variations in the intensity are due to movement of the nanotube during the scan (scale bar 10 μm).
Reproduced with permission from ref. 137. Copyright 2006, National Academy of Sciences, USA.

Poly(butadiene)-*b*-poly(ethylene oxide) PBD(1800)-*b*-PEO(900) was self-assembled into polymersomes using an electroformation technique.[138] TEM images allowed the measurement of the inner diameter of the tubes, which was around 70 nm. Moreover, with the use of water-soluble sulforhodamine B in the buffer inside the polymersomes, the authors could show that the nanotubes pulled from the polymersomes also contained the water buffer, which is important for transport applications. Indeed, those tubes can be used as vehicles for transporting biological material.

We have reported examples for the formation of tubular structures with the help of block copolymers, in most cases with amphiphilic blocks or alternating copolymers. They have all been engineered in such a way that they self-assemble into vesicles, fibers, or tubes. The mechanism for the formation of those tubes, with hollow insides in some cases, was described through two possible routes, which shows how powerful block copolymers can be. Furthermore, the field of applications is very broad: material sciences, catalysis, smaller electronic devices, strong nanocomposites, sensors, ion transport, and many others. Creating helical structures or tubes is also interesting in order to mimic nature and to try to understand the way they are created in biological systems.

5.3.2 DNA Origami

DNA is an interesting polymer for self-assembly that can be programmable due to the control over the coding sequence of a DNA strand. Indeed, tiles made out of two or three double-stranded DNA helices can be designed to assemble and form complex two-dimensional structures or DNA tubes. Winfree and co-workers synthesized DNA nanotubes with tiles consisting of two parts.[23] The central core was composed of five pseudo knotted strands that formed two double helical domains held rigidly by a pair of crossover points (Figure 5.47 I). They used two different DNA sequences with two different cores (RE blue and SE yellow). Those cores had four single-stranded sticky ends to allow binding with other tiles. They could be chosen in order to program the construction of particular structures. Those sticky ends could be tuned to couple only to themselves or in order to form diagonally striped (small "d") or perpendicularly striped lattices (small "p").

After annealing of the different sets of tiles from 95 °C to room temperature at a rate of one degree per minute, the samples were analyzed by AFM (Figure 5.48) and fluorescence microscopy. For fluorescence studies, one DNA strand was labeled with fluorescein. Various sets of tiles showed to self-assemble into DNA tubes with different local tile arrangements, symmetries, and molecular structures. However, in all cases, the formed tubes were very similar.

The formation of such tubes was reported to be due to the symmetry of the tiles used, even if the detailed geometry of the tubes and the reasons for their formation remains uncharacterized.

Similarly, Turberfield *et al.* showed the self-assembly of tiles into sheets with a perpendicular lattice that curled and closed on themselves to form tubes.[25] They used a system of tiles made of two crossed-over double-stranded DNA containing four single-strand sticky ends programmed to arrange perpendicularly (such as in Figure 5.47(V)). The set of two tiles was annealed starting from 96 °C to room temperature using a temperature linear gradient over 96 h. Negative staining with the 173.6 kDa streptavidin antibody allowed TEM imaging (Figure 5.49). There was a pronounced increase in electron transparency at the edges of the observed ribbons, which is consistent with the maxima in projected thickness of the tube membrane. It was also possible to observe the alignment of the streptavidin labels. Those labels could be perpendicular to the tube direction or forming zigzag lines, which indicated the way sheets of tiles curled and closed.

DNA tiles consisting of three double helices crossed-over together could also be programmed to form 2D lattice sheets that curled into uniform nanotubes. LaBean and co-workers reported DNA nanotube self-assembly due to the addition of thiol groups on a double-stranded DNA stem protruding from one tile.[26] As the previously cited work, the tiles were programmed to assemble into perpendicular lattices and close (Figure 5.47(V)). The curling of the 2D sheets was induced by the formation of disulfide bonds between neighboring B tiles. Those contained two extra double-stranded DNA stems,

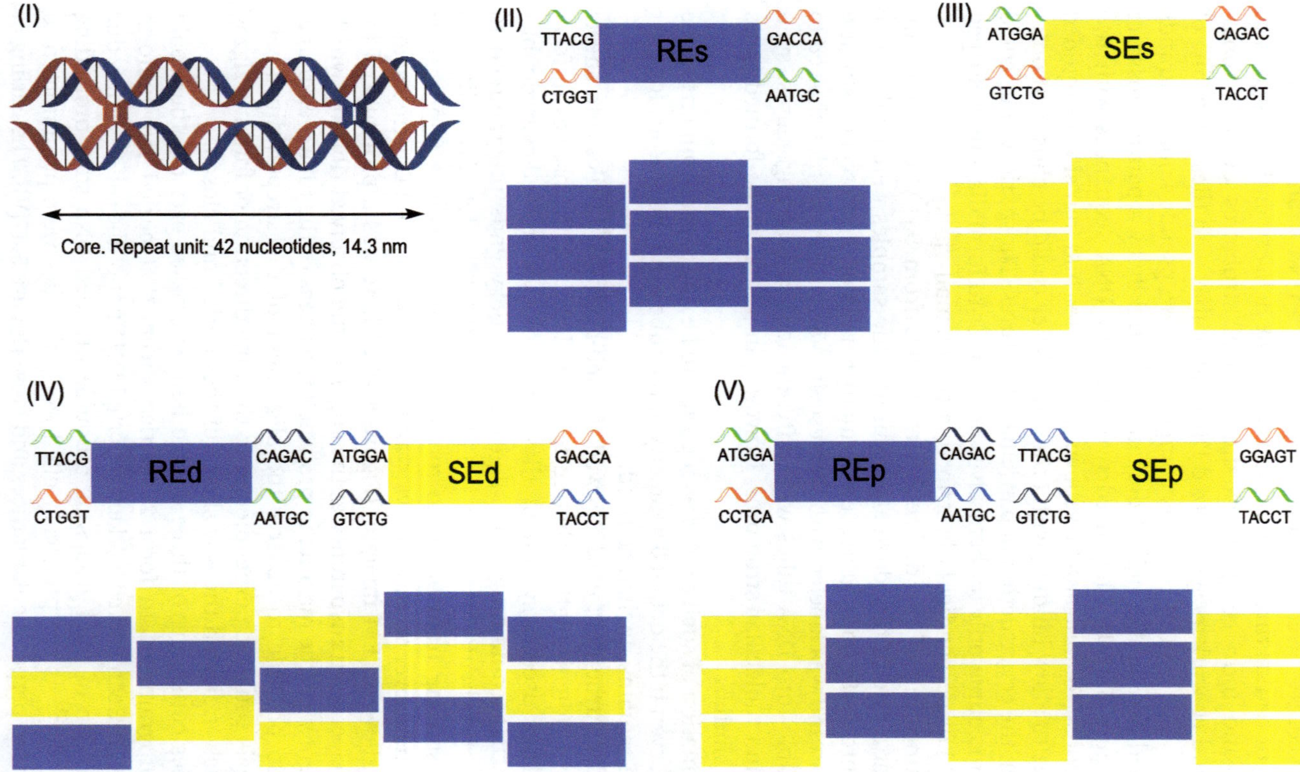

Figure 5.47 Design and modeling of DNA nanotubes. (I) Representation of two crossed-over double-strand DNAs forming the "core" of a tile, shown as a rectangle. (II) RE tile, which can bind only to itself (the colors of the terminal single stains show complementary sequences). (III) SE tile, which can bind to itself. Two cores form (IV) a diagonal lattice and (V) a perpendicular lattice.

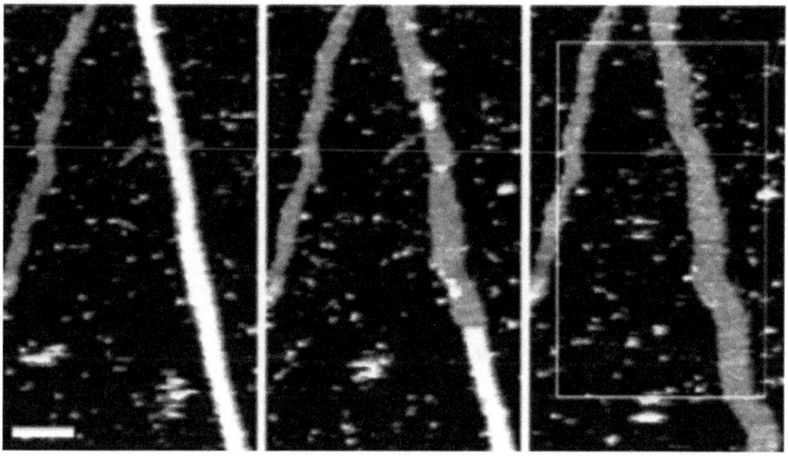

Figure 5.48 A time series of AFM images (scale bar is 100 nm). Left: a thick filament (closed tube) on the right of the image. The height of the closed tube suggests it is squashed flat. Center: the mentioned tube opens progressively from one end. Right: the fully opened tube resembles the lattice on the left.
Reproduced with permission from P. W. K. Rothemund, A. Ekani-Nkodo, N. Papadakis, A. Kumar, D. K. Fygenson and E. Winfree, *J. Am. Chem. Soc.*, 2004, **126**, 16344–16352.[23] Copyright (2004) American Chemical Society.

which formed junctions with the central helix of the tile as they projected out of the tile plane (Figure 5.50).

The nanotubes could be seeded with silver following already reported methods to produce metallized nanowires.[137,139] These nanowires demonstrated higher conductivity than previously reported ds-DNA-templated silver nanowires. However, the conductivity for ds-DNA-templated palladium nanowires and self-assembled DNA tile-templated nanowires was slightly higher.[25,140] The use of extra double-strand DNA stems showed to be efficient for the induction of folding of self-assembled 2D sheets into DNA nanotubes. Another method described in the work of Majima and Seeman also led to the formation of DNA nanotubes.[141] They used a tetra-substituted porphyrin ring that linked to one type of used tile (B tile) and ensured the folding of the 2D DNA sheet into a tube (Figure 5.51). The tiles were programmed to assemble in a perpendicular lattice, as in the previously described work (Figure 5.47(V)). On one type tile (B tile), an extra single-strand DNA was sequenced to bind to the strand attached on the porphyrin ring. The length between the end of the DNA strand on the porphyrin ring and the center of the porphyrin was 7–8 nm, which allowed the alignment of four B tiles from the DNA sheet side by side (16 nm in total). In order to show that the porphyrin template was the reason for the formation of tubes, A and B tiles alone were annealed and gave essentially 2D DNA arrays, similar to the DNA arrays previously described.[142,143] When the annealing of A and B tiles

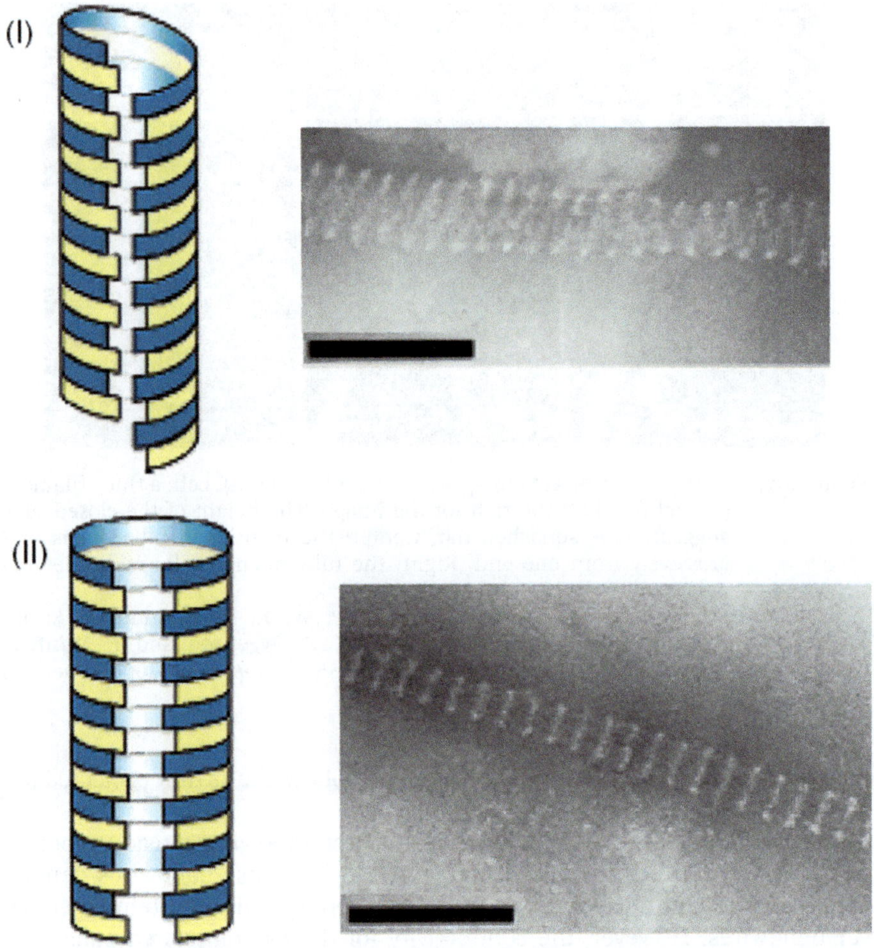

Figure 5.49 Proposed schemes of curling sheets and TEM images of formed tubes with (I) nested helical structure or (II) alternating structures (scale bar is 200 nm).
Reproduced with permission from J. C. Mitchell, J. R. Harris, J. Malo, J. Bath and A. J. Turberfield, *J. Am. Chem. Soc.*, 2004, **126**, 16342–16343.[25] Copyright (2004) American Chemical Society.

was processed with the addition of 1/4th of an equivalent of the tetra-substituted porphyrin ring, no 2D arrays were observed but fiber-like structures appeared. Analysis of the fibers with AFM showed uniform widths (approx. 55 nm) and heights (5.2–5.6 nm). However, the location of the DNA-porphyrin connectors in the DNA tubes remained unclear. The authors explained that the AFM images give the impression that the DNA-pophyrin connectors are on the outside of the tubular structures, even though, from the thickness, they could not exclude the possibility that the connecting rings are on the inside.

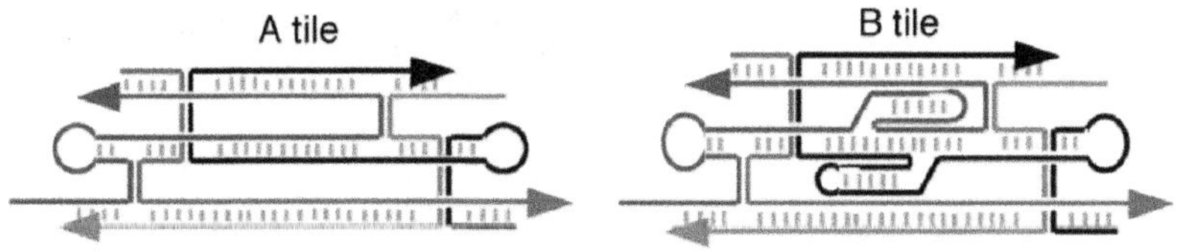

Figure 5.50 Description of the two tiles used in the construction of DNA nanotubes. Arrowheads are showing sticky single-strand ends in different shades of gray. Two extra stem-loops on the central helix of tile B are supposed to protrude, one inside and the other outside of the plane.
Reproduced with permission from ref. 26. Copyright 2004, National Academy of Sciences, USA.

As shown, DNA tiles exhibit fully programmable sequences that can be used to organize into nano-helical structures or ribbons. Furthermore, tiles can also exhibit addressable surfaces in order to organize proteins, nanoparticles, or carbon nanotubes.[144] However, reports on larger than 1 μm 2D arrays formed with DNA tiles are rare. The many difficulties encountered during the preparation of larger DNA tiles suggest that they behave somewhat differently from small DNA tiles. The group of Liu used long single-stranded DNA from a phage (M13mp18, 7200 bases genome) coupled with hundreds of short complementary DNA stands called staples.[145] They designed a rectangular shape origami from the phage genome (7056 nucleotides) folded with 168 staples. The folded array consisted of parallel crossed-over double-stranded helices containing dihedral angles of 120° between the helices, but intended to relieve the deformation so that the global twisting of the structure was minimized. This is called "zigzag DNA origami" (Figure 5.52). To those rectangular arrays, two sets of linkers can be added, the first set is meant to be attached from the top left to the bottom right of the tile (1–3 connection on Figure 5.51) and inversely, with the second set, the top right with the bottom left (2–4 connection). Two methods of addition lead to the formation of tubes. The first one is the simultaneous addition of both sets, while the second method is a stepwise approach. Stair-like arrays are formed in one direction using one set of linkers, after which the second set is added. Both methods yield similar tubes with lengths from 1 to 3 μm. The formation of tubes may be thermodynamically preferred because folding into tubes minimizes the number of unpaired DNA strands.

As demonstrated in the work described above, the use of short or longer programmable DNA strands for the creation of tiles has shown to be a powerful method to gain control over the self-assembly into nano and in some cases micrometer scale DNA tubular structures.

5.3.3 Metal-organic Nanotubes (MONTs)

Another field of chemistry that has demonstrated good control over the formation of tubes is organometallic chemistry. Indeed, metal-organic frameworks (MOFs) offer an attractive field for the creation of tubular

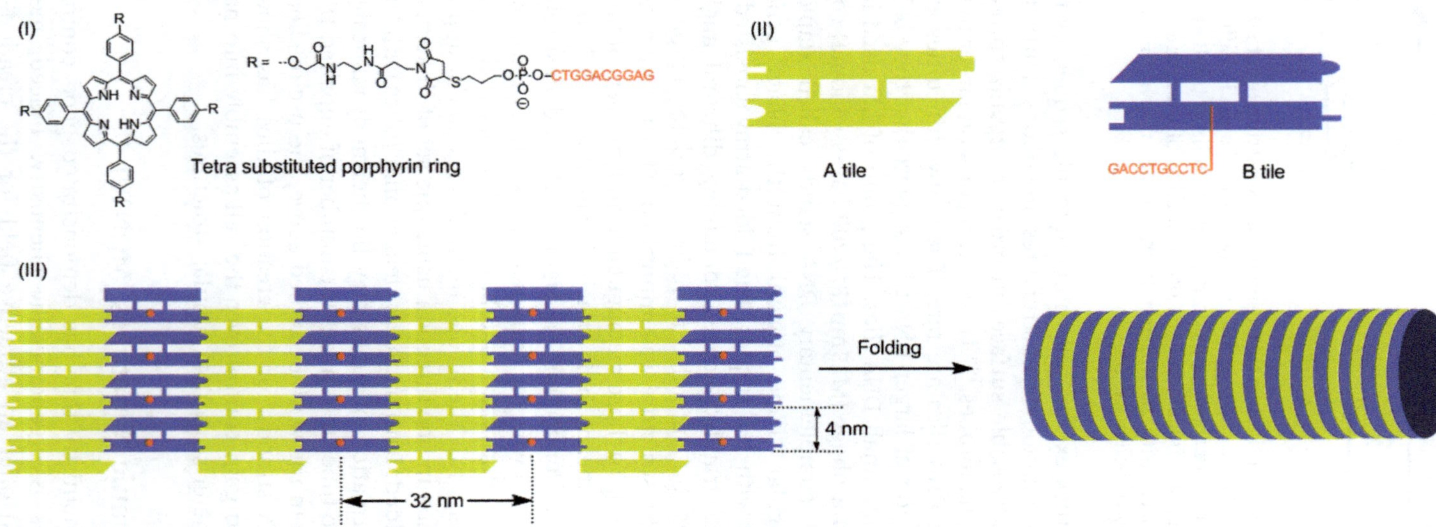

Figure 5.51 (I) The DNA-porphyrin connector with his single-strand DNA (red), and (II) the two tiles A and B, which contain the complementary single-strand DNA. (III) The perpendicular lattice array formed after annealing of A and B tiles and the resulting DNA nanotube folded from the array in the presence of the porphyrin connector.

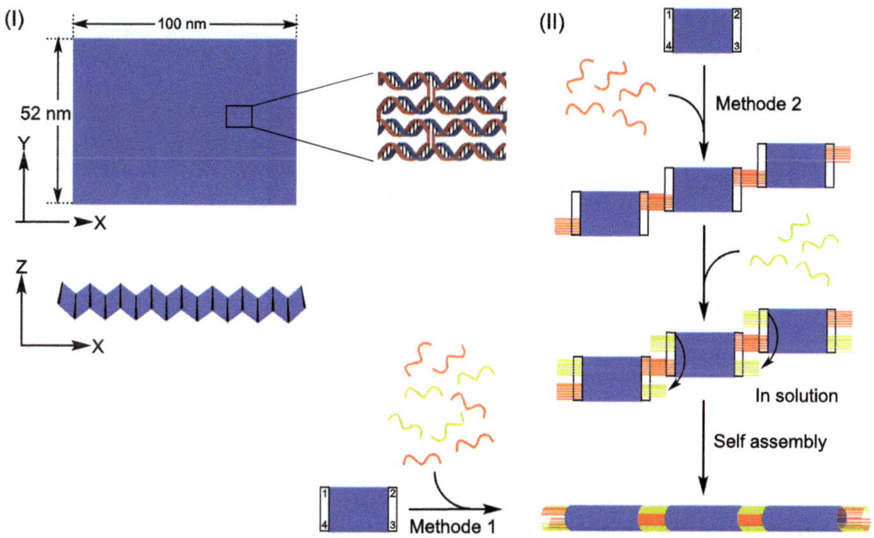

Figure 5.52 (I) Schematic drawing of the rectangular shape origami with zoom-in view on the right and side views illustrating the "zigzag" 120° dihedral angle between double-stranded helices. (II) Schematic pathway to the formation of tubular DNA structures following method 1, in which both sets of linkers are added simultaneously, or following the stepwise addition method 2, creating 1-3 connections followed by 2-4 linkers (or inversely). Arrows show the way stair-like DNA ribbons self-assembled into tubes.

structures. MOFs are tunable and well-defined structures possessing large surface areas and specific adsorption sites. In particular, single walled metal-organic nanotubes (SWMONTs) are interesting for the creation of hollow nanotubular structures; however, SWMONTs remain largely unexplored.[146]

Single-walled metal-organic nanotubes can be achieved using several methods. One method reported in the group of Sun used two different ligands with one coordinating metal center.[147] Ether was slowly diffused into a mixture containing $Zn(NO_3)_2 \cdot 6H_2O$, 5-amino-2,4,6-triiodoisophtalic acid (H_2ATIBDC), and 4,4′-bipyridine (bpy) in DMF/EtOH/H_2O. They obtained a square consisting of Zn ions with four bpy ligands aligned along one axis with ATIBDC ligands connecting those square-shaped rings together forming a SWMONT. X-ray structure determination also revealed clusters of H_2O trapped in the $Zn_8(bpy)_8(ATIBDC)_4$ complex (Figure 5.53).

Interestingly, TGA studies showed that those tubes remained in the crystalline state up to temperatures around 340 °C after the loss of coordinated water molecules. This is an interesting property for potential applications. A report from Lu and co-workers showed the formation of SWMONTs using a bidentate ligand possessing a bending angle of 120°.[148] Indeed, the appropriate choice of an organic ligand with specific functional groups and geometry is a major factor in the formation of SWMONTs. In this work, the

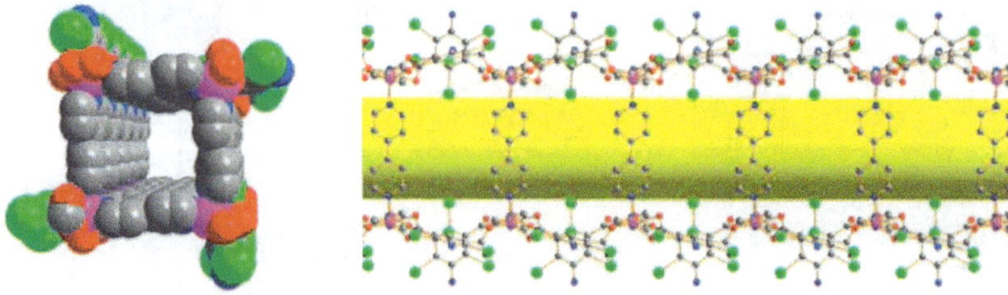

Figure 5.53 Nanotube of $[Zn(ATIBDC)(bpy)]\cdot 3H_2O$ along the *z*-axis (left) and showing the tube interior (right).
Reproduced from ref. 146 with permission from the Royal Society of Chemistry.

authors chose 4-amino-3-[(pyridin-4-methylene)amino]benzoic acid (Hapab) as the ligand coupled with cadmium perchlorate and MOH (M = Cs, K or Na). After a solvent diffusion process and through a single-step self-assembly, they could observe tubular structures with a total diameter of 3.2 nm and an inner hole diameter of 1.4 nm (Figure 5.54).

In the work of Kitagawa *et al.*, they used the square-shaped building block $[Pt(en)(bpy)]_4(NO_3)_8\cdot 5H_2O$ (en: ethylenediamine, bpy: 4,4′-bipyridine).[149] In this building block, each Pt(en) unit is coordinated in a square-planar fashion with two bpy ligands forming a square-shaped motif with 2 nm side. This square-shaped complex was oxidized using I_2 and formed a 1D halogen-bridged transition-metal covalent chain. This single-walled metal-organic nanotube $[Pt(en)(bpy)I]_4(NO_3)_8\cdot 16H_2O$ was examined to determine the sorption properties (Figure 5.55). High uptakes of H_2O, methanol, and ethanol were observed at room temperature, whereas N_2 and CO_2 were not adsorbed. Gas adsorption is one of the numerous interesting properties of metal-organic nanotubes.

The group of van der Boom reported the formation of amorphous and hollow tubes using a palladium precursor mixed with a multidentate ligand.[150] They focused on a systematic molecular approach for the formation of nanostructures varying the ligand geometry and the number of coordinating sites. However, only one type of ligand gave rise to the formation of flexible metal-organic nanotubes. A solution of $[PdCl_2(PhCN)_2]$ was mixed with a suspension of a flexible multidentate ligand (Figure 5.56 left) in toluene at 95 °C for three days in a pressured tube. The replacement of the benzonitrile ligand of palladium with a multidentate ligand led to the self-assembly into SWMONTs but also sheets, spheres, and their aggregates. Scanning electron microscopy measurements showed SWMONTs with a diameter of 70–150 nm and with uniform wall thickness about 25 nm. Nanoprobe X-ray energy dispersive spectrometry (EDS) indicated a Pd/N ratio of 0.5 : 1 for the SWMONTs, which suggests that two pyridine moieties were bound to one palladium center. The geometry around the metal center of such complexes is generally square planar and should therefore assemble into sheets. It is the rolling/folding process following the sheet development

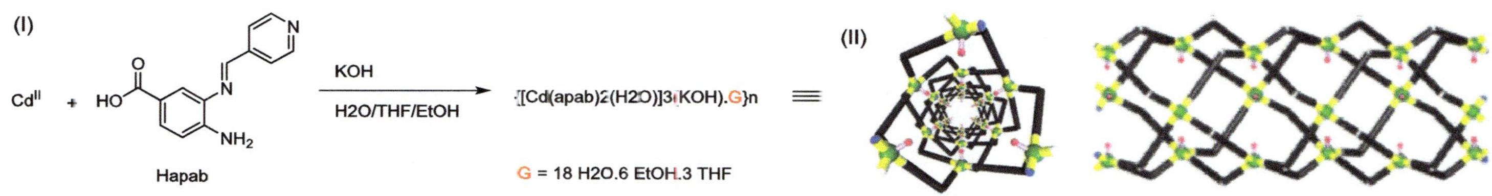

Figure 5.54 (I) Synthesis of single-walled metal-organic nanotubes by a one-step self-organization process. (II) Schematic representation of the SWMONT seen from the top and from the side with the cadmium metal core in green coordinated with the 120° bent Hapab ligand in brown.
Reproduced from ref. 148 with permission from John Wiley and Sons. Copyright © 2009 WILEY-VCH Verlag GmbH & Co. KGaA, Weinheim.

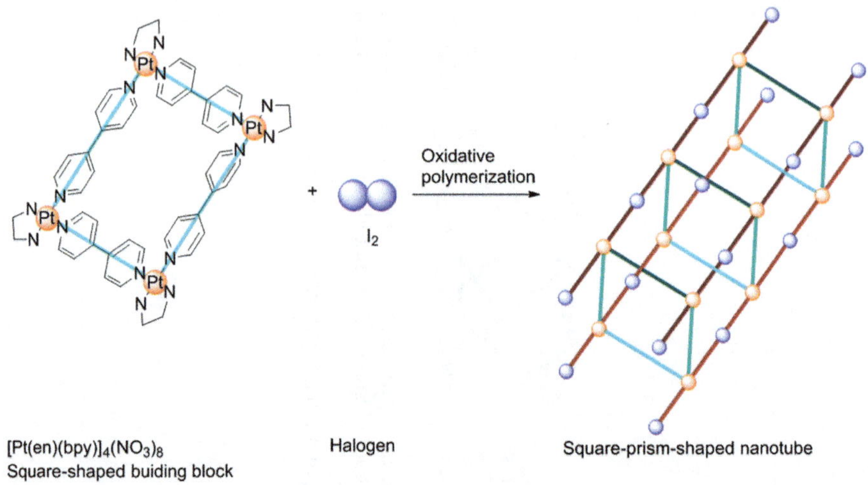

Figure 5.55 Schematic illustration of the fabrication of a single-walled metal-organic nanotube (SWMONT) using oxidative polymerization of a square-shaped building block.

that leads to the formation of tubular structures. This mechanism is supported by the presence of misfolded sheets in the observed probes and also by computational studies.[151]

There are several other reports on metal-organic frameworks that deal with adsorption of gases or liquids like water. MOFs provide a promising approach to nanoporous materials. However, among all the reported structures with pores or cages, only a few are 1D tubular-like structures.[152] In the work of Wang and co-workers, they showed the formation of double-walled metal-organic nanotubes from nonuple and triple helices.[153] The authors speculated that the use of flexible organic linkers H_2bpz (3,3′,5,5′-tetramethyl-4,4′-bipyrazole) and H_2bpdc (4,4′-biphenyldicarboxylic acid) was the key to the generation of helical chains. Those two ligands were mixed with cobalt complex $Co(CH_3COO)_2 \cdot 4H_2O$ in a mixture of DMF and acetonitrile; the whole suspension was heated at 160 °C for 72 h and cooled slowly to give purple crystals. The crystalline complex belonging to the tetragonal space group $I4_1/a$ was composed of $[Co(bpdc)_{1.5}(Hbpz)] \cdot DMF \cdot CH_3CN \cdot H_2O$ and two independent Co atoms connected to 1.5 bpdc^{2-} ligands with bis-monodentate (bpdc-I) and bis-bidentate (bpdc-II). The two Co centers adopted the same tetrahedral geometry but had different environments. Those clusters, composed of two cobalt atoms complexed with one bpdc and another half of a bpdc ligand, were interconnected with the Hbpz ligands to give a square building block (Figure 5.57).

This square molecular building block (MBB) was further connected with eight other MBBs through four bpdc-I and eight bpdc-II linkers to form a 3D framework. This framework exhibited multiple helical features along the *c* axis. Hpbz^{-} and bpdc-I were connected together to Co clusters alternately and formed three homochiral helical chains. One turn of such a chain gives

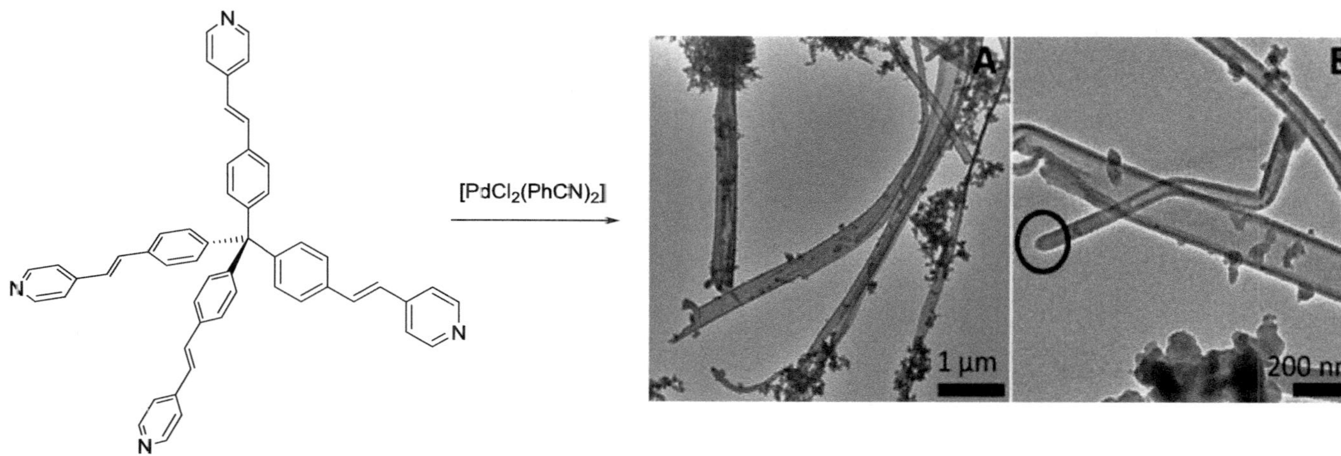

Figure 5.56 Molecular structure of the multidentate ligand (left) and TEM images of the formed SWMONTs upon heating. Example of an end-caped nanotube is shown on B.
Reproduced from ref. 150 with permission from John Wiley and Sons. Copyright © 2011 WILEY-VCH Verlag GmbH & Co. KGaA, Weinheim.

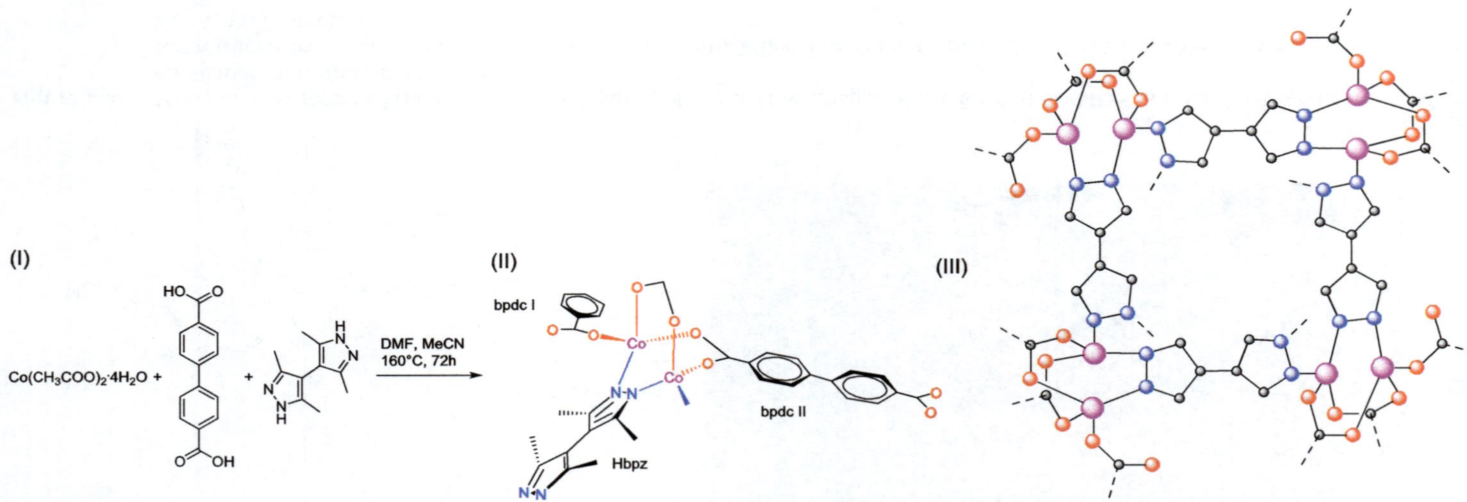

Figure 5.57 (I) Schematic representation for the preparation of the $[Co(bpdc)_{1.5}(Hbpz)]\cdot DMF\cdot CH_3CN\cdot H_2O$ complex, and (II) the coordination environment of both cobalt atoms. (III) Four cobalt clusters interconnected with four $Hbpz^-$ ligands give a saddle ring as the molecular building block.

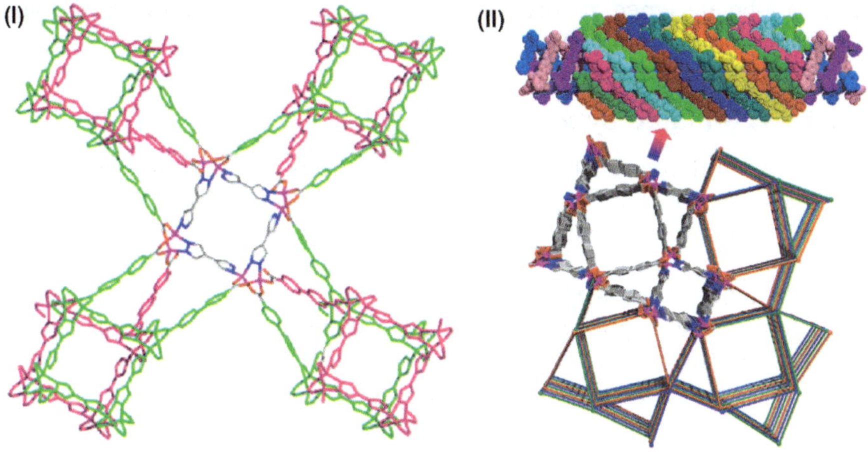

Figure 5.58 (I) Schematic picture of the molecular building block (MBB) connected with eight other MBBs *via* four bpdc-I and eight bpdc-II linkers. (II) Representation of the 3D framework consisting of double-walled MONTs generating twelvefold helical motifs.
Reproduced from ref. 153 with permission from the Royal Society of Chemistry.

a pitch of 76.85 Å, equaling nine times the unit cell on the *c*-axis, which surpasses the longest pitch observed in MOFs (63.68 Å).[154] Furthermore, there was one helical chain every unit cell on the axis leading to a nonuple helical feature (helix-I), which represents the highest-numbered association of helices in MONTs.[152,155–157] At the same time, three homochiral helices (helix-II) with opposite chirality associated to form a threefold helical motif inside the helix-I. Helix-I and helix-II together formed a twelvefold $(9+3)$ helical motif that looked like a double-walled MONT. To simplify the overall structure, the Co_2 clusters can be seen as 5 connected nodes (Figure 5.58).

Moreover, this double-walled MONT showed good CO_2 uptake and moderate sorption for H_2. However, the adsorption enthalpy value (Q_{st}) for the uptake of CO_2 was inferior to that of MOFs containing amine functionalities.[158]

As described in the previous reports, metal-organic frameworks have shown to be promising porous materials and offer an interesting alternative to carbon nanotubes, for example. However, there are only a few reports describing the construction of MONTs. Nevertheless, efforts towards the engineering of such helical containing structures have shown that the choice of ligands is primordial. Nanotubular metal-organic compounds have therefore been prepared using different types of linkers for the metals. The use of defined-length organic ligands, rigid and semi-rigid ligands, charged linkers, flexible tripodal linkers, as well as zwitterions, have shown to produce nanotubular structures with various sizes and internal connections.[159–164] Lu and co-workers described those examples in detail and therefore opened a wide domain of research in metal-organic frameworks that can be designed in order to create new tubular architectures and topologies and, subsequently, new potential applications.[146]

5.3.4 Templated Synthesis of Polymeric Tubes

Nanometer and micrometer sized tubes have been prepared from polymeric materials using suitably shaped templates to ensure the final shape.[165–167] These templates can typically be divided into several categories. Hard templates consist of thin porous membranes, often alumina[168–170] or track-etched[171] polycarbonate where the nanometer-sized pores are partially back-filled (wetted) with polymeric material.[172] Removal of the template then yields the polymer tubes. The outer tube diameter is defined by the template pore sizes while the inner (open) diameter of the tubes can often be controlled by the amount of polymer deposited into the template pores. The length of the tubes typically corresponds to the thickness of the porous membrane.

The opposite approach has also been applied for the preparation of polymer tubes, *i.e.* template fibers of defined width and length have been enclosed in polymeric materials with subsequent removal of the fiber template.

Both of the above templated methods are often referred to as *hard template* methods, as template dimensions are well-defined, pre-formed, and hence robust. In contrast to the former are the so-called *soft templates*. There, self-assembly processes are responsible for the formation of spatially defined geometric structures (the soft template) around which the polymer tube is then formed. Reverse rod-shaped micelles are a typical example around which polymeric tubes can be formed. Removal of the soft template releases the nanotubular structure.

Another very popular method is the *template-free* method. This method has mainly been used to prepare poly(aniline) tubes. It is assumed that organic acids such as β-naphthalene sulfonic acid and others initially form a salt with aniline which then self-assembles into the soft-template aiding in poly(aniline) tube formation. It could therefore also be classified under the soft-template heading. The following classification will be by type of template.

5.3.4.1 Hard Templates – Porous Membranes

Most of the work that has been carried out using porous membranes as hard templates was done on porous alumina (anodic aluminium oxide, AAO) or track-etched polycarbonate membranes. The general principle of nanotube formation relies on a continuous polymer film being formed on the inside of the pores that remains as a tubular structure once the template is removed. There are several possibilities as to how this polymer film can be formed inside the pore. A film can be obtained by wetting the pore with a polymer melt or a polymer solution. Alternatively, the polymerization of monomers inside the pore can be carried out. This can be achieved by chemical vapor deposition or by chemical or electrochemical oxidation of the monomers. Multilayers can be achieved using layer-by-layer deposition of polyelectrolytes. The outer diameter of the tube is a direct reflection of the membrane pore diameter, and the wall-thickness and membrane pore diameter together determine the tube inner diameter.

Nanotubes by wetting of pores

The simple wetting of nano sized pores (alumina and a porous polyester membrane) by polymer solutions was first introduced by Cepak *et al.*[173] It was observed that the smaller pores in the polyester membrane yielded solid nanofibrils, whereas the larger pores gave nanotubes. The nanotube diameter corresponded to the pore diameter (*ca.* 200 nm) and the nanotube length to the full length of the template pore (*ca.* 50 μm). A large variety of polymers was already included in this first report: PS, poly(2,6-dimethyl-1,4-phenylene oxide), poly(vinylidene fluoride), poly(methyl methacrylate) (PMMA), poly(bisphenol A carbonate), and poly(lactic acid) (PLA). It is interesting to note here that the pore wetting technique, as it uses solutions of pre-formed polymers, can be applied to chain-growth and step-growth polymers alike. Several reports described the wetting of porous AAO with semiconducting polymer solutions. Nanotube formation from poly(9,9-dioctylfluorene),[174,175] a fluorene-anthracene copolymer,[176] hole conducting poly(*N*-vinylcarbazole) (Figure 5.59),[177] or ferroelectric vinylidene fluoride/trifluoroethylene copolymers[178] was described.

Steinhart *et al.* demonstrated that both polymer melt and solution could be employed in the wetting of ordered porous alumina to obtain nanotubes.[180] This was exemplified among others by an ordered array of poly(tetrafluoroethylene) (PTFE) nanotubes prepared from the melt (Figure 5.59). It could be shown that wetting from the melt or a solution had profound consequences for the crystallinity of poly(vinylidene difluoride) (PVDF) nanotube walls.[181] The nanotube thickness or the inner diameter of the nanotube could be tuned by varying the concentration of the polymer solution.[182–184]

When preparing nanotubes from the melt, a critical temperature was observed for poly(propylene) (PP) and low-density poly(ethylene) (LDPE). Above this temperature, nanotubes were formed in alumina pores whereas solid nanowires were observed below the critical temperature.[179] Nanotubes with ultrahigh aspect ratio were prepared from PMMA with wall thickness of 50 nm, tube diameter of 232 nm, and a total height of 60 μm. The AAO was degassed and evacuated before being wetted with the polymer melt. The AAO template was removed using dilute H_3PO_4 rather than NaOH to prevent bundling of the free nanotubes.[185] Incorporation of perylene into PMMA[186] or palladium salts into PLA[187] gave AAO templated nanotubes with chemiluminescent properties or PLA tubes with incorporated Pd(0) crystallites after annealing. Similarly, Huang *et al.* first filled AAO pores with silver nanoparticles (AgNPs) followed by melt wetting with poly(ethylene) to form nanotubes. This yielded AgNPs that were adsorbed onto the inner walls and the open end of the nanotubes.[188]

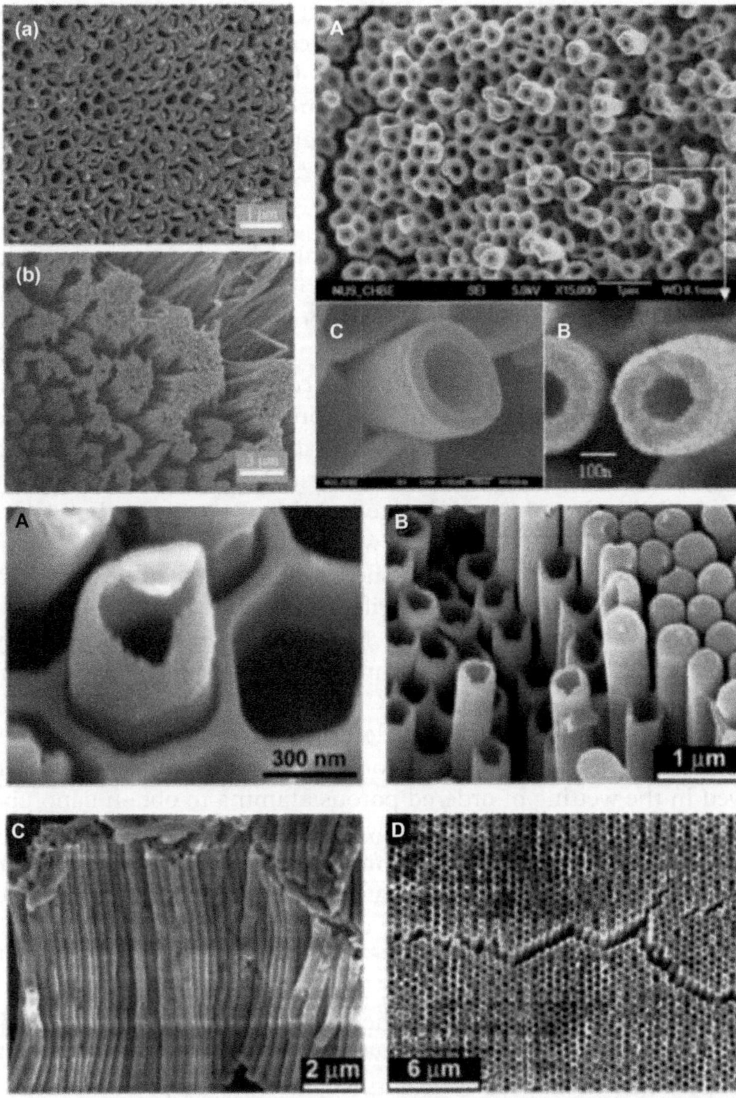

Figure 5.59 Left: scanning electron micrograph showing poly(*N*-vinylcarbazole) nanotubes: (a) in a porous AAO template, (b) after removal of the template. Reproduced from ref. 177 with permission from John Wiley and Sons. Copyright © 2007 WILEY-VCH Verlag GmbH & Co. KGaA, Weinheim. Center: (A) Field emission scanning electron micrograph of polypropylene nanotubes, (B) magnification of A, (C) polyamide 6 nanotubes. Reproduced from ref. 179 with permission from John Wiley and Sons. Copyright © 2007 WILEY-VCH Verlag GmbH & Co. KGaA, Weinheim. Right: Scanning electron micrograph of melt-wetted nanotubes. (A) poly(styrene), (B) ordered array of poly(styrene) tubes, (C) array of aligned poly(tetrafluoroethylene) tubes, and (D) poly(methyl methacrylate) tubes. Reproduced with permission from ref. 180. Copyright 2002, The American Association for the Advancement of Science.

An improved nanotube preparation procedure was reported by Jin *et al.*, who used spin-coating of polymer solutions to coat the inside of AAO pores. Even multilayer nanotubes were prepared in this manner.[189] Another variation of the standard preparation was shown by Chen *et al.*, who placed porous AAO substrates on spin-coated polymers under solvent annealing conditions. Depending on the type of solvent, tubes or rod-like structures were obtained.[190] Melt wetting porous AAO films from spin-coated PS films could also be achieved using microwave irradiation.[191] Spin-coating of a photoresist onto an AAO membrane allowed Li *et al.* to selectively form polyamide-6 nanotubes only in spatially defined areas where the pores were opened up after being exposed to ultraviolet light.[192] The wetting of pores of a 60 μm thick AAO membrane with an epoxy based negative photoresist (SU8) followed by exposure to UV light was reported by Huby *et al.* Dissolution of the AAO template using NaOH released the nanotubes. After dispersion in isopropanol, these were integrated into a photonic integrated chip and the optical propagation in nanotubes was investigated.[193,194]

Nanotubes via *polymerization inside the pores*

Almost a decade older than the wetting process of membrane pores is a technique whereby monomers are polymerized within the pores to form tubes. The first report by Martin *et al.* used a track-etched polycarbonate membrane with 0.5 μm diameter linear pores. The pyrrole monomer solution and the chemical oxidant, an aqueous $FeCl_3$ solution, diffused into the pores from opposite directions yielding poly(pyrrole) nanotubes.[195] The same report also described the electrochemical synthesis of poly(pyrrole) nanotubes using a polycarbonate membrane mounted onto a Pt-disk electrode. The length of the nanotubes obtained was identical to the membrane thickness (8 μm).

This type of nanotube preparation *via* polymerization inside membrane pores has become particularly popular for the preparation of electronically conductive polymer nanotubes. Several excellent reviews have been published on the subject.[196–201] Here, we will mainly focus on synthetic aspects of nanotube preparation as well as some of their applications.

One of the first polymers used for nanotube synthesis was poly(aniline), which can be prepared from aniline using chemical oxidants or potentiostatic electro polymerization. Parthasarathy *et al.* reported that poly(aniline) tubes formed in a polycarbonate template exhibited higher conductivities than the bulk material. They could show *via* polarized infrared spectroscopy that the polymer chains were oriented perpendicular to the nanotube axis and attributed the higher conductivity to a higher degree of order.[202] Later, Xiong *et al.* found that poly(aniline) prepared *via* chemical oxidation (ammonium peroxydisulphate) in porous AAO showed a hexagonal crystal form, in contrast to the pseudo-rhombic crystal form of bulk poly(aniline).[203] Delvaux *et al.* showed that chemical and electrochemical oxidation could lead to poly(aniline) nanotube formation in polycarbonate membranes.[204] Cao *et al.* exploited the poly(aniline) nanotube synthesis to prepare cobalt nanowires inside the nanotubes. A chemical oxidation process using ammonium meta vanadate and porous AAO was used for the oxidative polymerization of aniline.[205] It was reported that the morphology and order of the nanotubes were influenced by the synthesis method and the pore size of the template.[206]

Poly(pyrrole) is another polymer that can easily be synthesized *via* oxidative (chemical or electrochemical) polymerization. De Vito *et al.* investigated the formation of poly(pyrrole) nanotubes from polycarbonate membrane pores using a chemical oxidation ($FeCl_3$) process. In particular, improved colloidal stability in the presence of the anionic surfactant diamyl sulfosuccinate was reported.[207] Several syntheses of poly(pyrrole) nanotubes using chemical and electrochemical oxidation have been reported.[208,209] It was found that the conductivity of such nanotubes increased dramatically with the decreasing nanotube outside diameter. While the above syntheses typically rely on pyrrole solutions for the polymerization, Jang *et al.* reported a vapor deposition polymerization synthesis. There, the porous AAO was wetted with a ferric chloride solution, dried and exposed to pyrrole vapor to form poly(pyrrole) inside the AAO pores. Dissolution of the template using NaOH gave poly(pyrrole) nanotubes. These were subsequently carbonized at 2000 °C to yield carbon nanotubes.[210] Poly(pyrrole) nanotubes were decorated with gold crystals on the outside[211] or within[212,213] the nanotube walls. For the latter, the switching transition from a high conducting to a low conducting state, which was observed for regular poly(pyrrole) nanotubes at liquid nitrogen temperature, could be shifted to 100 K.

The synthetic methods described for poly(pyrrole) nanotubes were also applied to the synthesis of other polymers such as poly(thiophene),[214–216] poly(3-methylthiophene),[217] or poly(3,4-ethylenedioxythiophene).[218–223] Microtubular heterojunctions of poly(phenylene)–poly(thiophene) were electrochemically deposited onto porous AAO *via* oxidation of benzene and thiophene from solutions in boron trifluoride-diethyl ether.[224]

Kim *et al.* showed that poly(phenylene vinylene) tubes could be prepared starting from α,α′-dichloro-*p*-xylene. In a quartz tube, the monomer was vaporized, pyrolysed, and subsequently deposited as a precursor polymer inside the pores of a poly(carbonate) or alumina membrane. Thermal treatment (270 °C) yielded poly(phenylene vinylene) tubes inside the porous substrates.[225,226]

Bio-inspired nanotube applications

A first step towards the use of such spatially confined nanotubes in a biological context was taken by Parthasarathy *et al.*[227] As shown in Figure 5.60 a microporous polycarbonate membrane was first sputtered with a *ca.* 50 nm layer of gold that was used to form a poly(pyrrole) layer across one side of the membrane by electropolymerization, thereby closing the pores with a porous polymer.

After chemical polymerization of the nanotubes, the capped tubes were filled with enzymes by vacuum filtration. Glucose oxidase, catalase, subtilisin, trypsin, and alcohol dehydrogenase were investigated. The open end

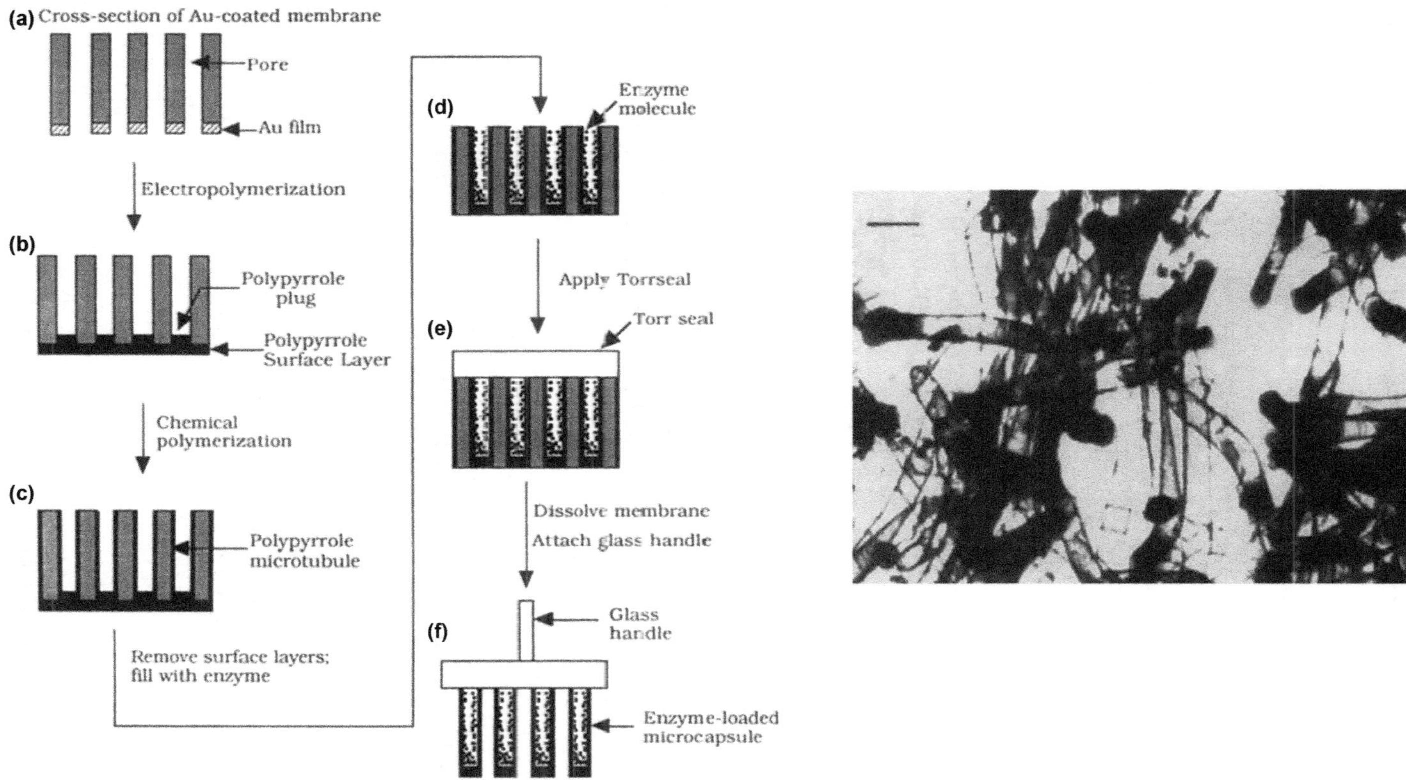

Figure 5.60 Left: preparation of enzyme filled poly(pyrrole) nanotubes sealed on one side. Right: scanning electron micrograph showing microcapsules that were not attached to the epoxy glue surface.
Reproduced with permission from ref. 227. Copyright 1994, Nature Publishing group.

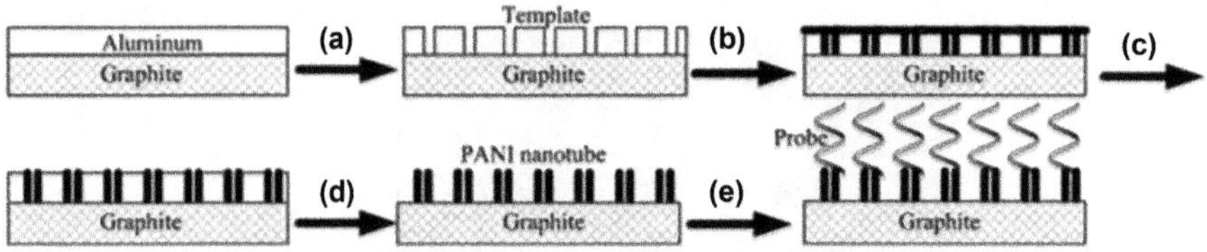

Figure 5.61 Construction of DNA modified poly(aniline) nanotubes on a graphite electrode.
Reproduced with permission from H. Chang, Y. Yuan, N. Shi and Y. Guan, *Anal. Chem.*, 2007, **79**, 5111–5115.[231] Copyright (2007) American Chemical Society.

of the tubes was sealed with an epoxy glue and the polycarbonate template removed with dichloromethane. The immobilized enzymes showed enzymatic activity as the porous poly(pyrrole) layer across one end of the tubes allowed transport of small substrate molecules, such as H_2O_2, or larger ones, such as nicotinamide–adenine dinucleotide (NAD), to pass through freely.

An amperometric biosensor based on a poly(3,4-ethylenedioxythiophene) nanotube was reported by Kros *et al.*[228] The monomer 3,4-ethylenedioxythiophene was polymerized inside a track-etched polycarbonate membrane using ferric chloride as oxidant. The enzyme glucose oxidase (GOx) was adsorbed onto the inner walls of the nanotubes and the membrane device mounted as the working electrode in a three electrode cell setup. A glucose sensitivity of 25 nA mM^{-1} was achieved. A similar sensor was originally prepared from poly(pyrrole) nanotubes[229] but poly(3,4-ethylenedioxythiophene) nanotubes were found to be less susceptible to degradation under physiological conditions.[230]

An ultrasensitive DNA sensor was constructed from graphite supported and DNA functionalized poly(aniline) nanotubes (Figure 5.61).[231] An aluminum layer was first deposited onto a graphite electrode, anodized to form a porous AAO, and aniline was chemically polymerized to form nanotubes. The closed nanotubes were polished and the alumina template removed with H_2SO_4. Covalent attachment of an oligonucleotide (5′-HOOC-CTC ACC GCC TGC CCA CTC ATT-3′) *via* carbodiimide coupling of the carboxylic acid on the 5′-end gave DNA functionalized nanotubes.

Differential pulse voltammograms allowed the detection of complementary oligonucleotides at 1 fM concentration. The differentiation of one-nucleotide mismatched oligonucleotides was also possible.

Another method of depositing polymer layers onto the inside walls of template pores is the layer-by-layer (LbL) technique,[232] using alternating layers of oppositely charged polyelectrolytes. The first report with porous templates for LbL deposition by Ai *et al.* used solutions of poly(allylamine hydrochloride) and poly(styrenesulfonic acid).[233] Both solutions were pumped through porous AAO membranes in an alternating fashion. Flexible nanotubes with wall thicknesses of 50–80 nm and lengths up to 60 μm were obtained. Nanotubes built from poly(acrylic acid), poly(allylamine hydrochloride), and poly(ethyleneimine) were able to incorporate positively charged gold nanoparticles. Similarly, negatively charged CdTe nanoparticles could be incorporated into tubes made of poly(ethyleneimine), poly(styrenesulfonic acid), and poly(allylamine hydrochloride).[234] Other combinations of polymers for LbL nanotube wall formation were investigated, such as poly(pyrrole)/poly(allylamine hydrochloride)[235] and poly(4-vinylpyridine)/poly(acrylic acid).[236] Covalent bond formation between individual layers was reported by Tian *et al.* using poly(ethylene imine) as the nucleophilic layer and 3,4,9,10-perylenetetracarboxylicanhydride as the alternating electrophile. In this case, highly fluorescent nanotubes were prepared in porous AAO.[237] Nanotubes based on just one layer of poly(acrylic acid) and one layer of poly(propyleneimine) dendrimers of different generations were synthesized by Savariar *et al.* Stable nanotube structures could be isolated from a poly(carbonate) membrane even for low generation dendrimers.[238]

A large step towards bio-functional nanotubes was taken by the same research group. An alumina membrane was first coated with two alternating layers of poly(ethylene imine) and poly(styrenesulfonic acid). Cytochrome c, with a positively charged surface at pH 7, was then alternated five times with poly(styrenesulfonic acid) in a LbL nanotube synthesis. Alternatively, cytochrome c and glutaraldehyde were employed to build up five covalent bilayers. Dissolution of the alumina template with 5% phosphoric acid yielded the protein-containing nanotubes.[239]

The first report where proteins were used to form nanotubes within porous templates employed a covalent LbL strategy with alternating layers of either glucose oxidase or hemoglobin and glutaraldehyde. Enzymatic activity was observed to increase with the increasing wall thickness of the glucose oxidase tubes.[240] In another report, an artificial hemoprotein consisting of human serum albumin (HSA) and synthetic heme was used to construct nanotubes. As the isoelectric point of HSA is low (pH 4.8), both negatively charged and positively charged protein solutions could be prepared, from which the LbL construction of the tubes was achieved.[241]

The group of Komatsu was the first to report protein nanotubes using a LbL technique where negatively charged proteins (myoglobin, HSA, ferritin) alternated with positively charged poly(L-arginine).[242] Thermal decomposition of the ferritin-poly(L-arginine) tubes gave α-Fe_2O_3 nanotubes.[243] Similarly, nanotubes built in a LbL fashion from an HSA-gold nanoparticle mixture and poly(L-arginine) within the pores of a poly(carbonate) membrane yielded gold nanotubes after calcination.[244] HSA-poly(L-arginine) tubes (Figure 5.62), to which avidin was electrostatically bound as the last layer of the tube wall, were capable of capturing biotin-labelled fluorescein or biotin-functionalized fluorescent nanoparticles.[245,246] Larger structures can also be selectively bound to the inside of the nanotubes, as shown by Komatsu *et al.* In this report, icosahedral hepatitis B virus particles were selectively captured inside LbL-HSA-poly(L-arginine) nanotubes that were decorated with the anti-hepatitis B antigen (anti-HBsAg) antibody (HBsAb).[247]

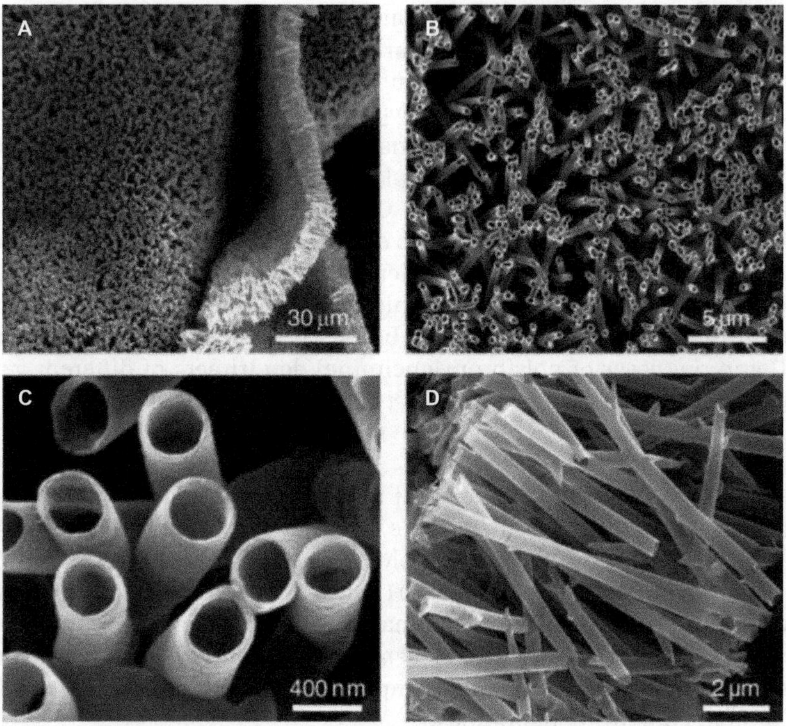

Figure 5.62 SEM images of (HAS-poly(L-arginine))$_3$ nanotubes prepared using a 400 nm diameter porous poly(carbonate) template.
Reproduced with permission from X. Qu and T. Komatsu, *ACS Nano*, 2010, **4**, 563–573.[245] Copyright (2010) American Chemical Society.

5.3.4.2 Hard Templates – Nanofibers

In the same way the inside of hollow tubes can serve as templates for tubes (see Section 5.3.4.1), solid degradable polymeric fibers can be coated on the outside to form tubes after removal of the solid fiber template. In contrast to the porous template approach, here, the degradable fiber template defines the inner diameter of the tube, whereas, in principle, the outer diameter can vary. Bognitzki *et al.* were the first to report the electrospinning of degradable poly(L-lactide) into fibers of 0.05–3.5 μm diameter and greater 100 μm in length. Chemical vapor deposition of 1,4-quinodimethane (prepared *via* pyrolysis of [2.2]paracyclophane) coated the poly(L-lactide) fibers in poly(*p*-xylylene) films of 50 nm to several micrometer thickness.[248] Subsequent annealing of the core–shell tubes at 250 °C degraded the poly(L-lactide) core and yielded poly(*p*-xylylene) tubes. The same report also describes the dip-coating of the electrospun fibers with a soluble polyimide precursor (poly(*N*,*N*-(*p*,*p*′-oxodiphenylene)pyromellitamic acid)), followed by annealing at 150–285 °C to yield the inert polyimide and degrade the inner core at the same time. Poly(xylylene) tubes with inner diameters smaller than

10 nm were obtained using an improved electrospinning method.[249] Using poly(L-lactide)-stabilized gold nanoparticles for fiber formation, followed by chemical vapor deposition of poly(*p*-xylylene) gave core–shell structures in which gold nanoparticles were embedded in the solid core. Thermal removal of poly(L-lactide) produced poly(xylylene) tubes filled with gold nanoparticle clusters. Further heating (1050 °C) decomposed all organic matter and yielded smooth gold nanowires.[250]

A combination of porous and inner core templates was used by Mayya *et al.* Nickel nanorods were first prepared using a porous alumina template. LbL deposition of poly(diallyldimethylammonium chloride) and poly(sodium-4-styrenesulfonate) coated the Ni-rods in a polymer layer before dissolution of nickel with dilute HCl gave the final polymeric tubes.[251] This approach offers many degrees of control over the tube inner diameter, length, and tube thickness.

Vertically aligned Si/SiO_2 nanowires were surface functionalized with an ATRP initiator, from which methyl methacrylate was polymerized (M_n *ca.* 217 000, PDI = 1.07). The Si tubes were removed using XeF2 to give polymer nanotubes supported on SiO_2.[252]

An unusual templating effect was observed by Zhu *et al.*, who reacted hexachlorocyclotriphosphazene with 4,4′-sulfonyldiphenol in a cross-linking polymerization in the presence of triethylamine. The precipitating triethylamine hydrochloride formed nanoscopic rod-like crystals around which the cross-linked polymer was formed. Removing the crystalline core with water gave tubes with inner diameters of 10–20 nm.[253]

Electrospun poly(L-lactide) was also used as the core material for a chemical oxidation polymerization of aniline to form poly(aniline) on the outside of the fibers. Heating up to 310 °C removed the polymer core and gave large poly(aniline) tubes with an average diameter of 890 nm and a tube thickness of 70 nm.[254]

The group of Martin investigated the use of poly(3,4-ethylenedioxythiophene) electrochemically polymerized onto electrospun poly(L-lactide) deposited on a neural microelectrode.[255] In a series of investigations, it was found that poly(pyrrole) generally lowered the impedance of the electrode less than poly(3,4-ethylenedioxythiophene).[256–258]

5.3.4.3 Soft Templates – Reverse Cylindrical Micelles

An elegant way of preparing nanotubes *via* oxidative polymerization has been shown by Jang *et al.* using rod-shaped reverse micelles.[259] Surfactant sodium bis(2-ethylhexyl) sulfosuccinate was added to a hexane/aqueous $FeCl_3$ mixture. The surfactant formed a reverse rod-like micelle containing Fe^{3+} cations in its core.[260] Adding pyrrole monomer to this mixture resulted in oxidative polymerization on the outside of the rod-like reverse micelles thus forming nanotubes. Poly(pyrrole) nanotubes with a diameter of *ca.* 98 nm and lengths of more than 5 μm were obtained. Carbonization of iron-containing poly(pyrrole) nanotubes yielded γ-Fe_2O_3 impregnated carbon nanotubes.[261]

In a similar experiment, Zhang *et al.* prepared poly(3,4-ethylenedioxythiophene) tubes using the same surfactant in hexane.[262] Nanotubes with diameters in the range of 300–800 nm and lengths greater 10 μm were obtained. It was reported that tube formation was particularly sensitive to the stirring speed, with low speeds resulting in longer tubes. The electrical response for both ammonia and UV light was observed to be strongly dependent on the diameter of the tube.[263]

5.3.4.4 Soft Templates – the "Template Free" Method

The so-called template free method was discovered accidentally by the group of Wan.[264] In this method, typically, a solution of aniline is mixed with a dopant such as β-naphthalene sulfonic acid, camphorsulfonic acid (or no dopant at all) before the oxidant ammonium persulfate is added to the solution. The mixture is kept stirring at room temperature and the precipitated tubes are washed with water. While this method does not offer the same amount of control over the morphology and the inner and outer diameter of the tubes compared to the hard template methods, its key feature is simplicity and ease of preparation.[265]

It was proposed that the monomer/dopant salts form micelles which then serve as the soft-template for the formation of tubular structures.[266] For the dopant-free or simplified method in which only aniline and ammonium persulfate are mixed, the initial formation of spherical micelles was observed by freeze-fracture transmission electron microscopy from which tubular structures were obtained.[267]

The most common monomer used in the "template-free" method is aniline. Acidic dopants that have been added to the polymerization mixture range from β-naphthalene sulfonic acid,[268] (+) and (−)10-camphorsulfonic acid,[269] hydrogensulfated fullerenol,[270] a PAMAM dendrimer carrying 24 bis-sulfonated naphthyl end-groups,[270] azobenzene sulfonic acid,[271] inorganic acids such as HCl, H_2SO_4, H_3PO_4, HBF_4,[272] or organic carboxylic acids.[273] Poly(pyrrole) tubes could also be prepared by the "template-free" method using β-naphthalene sulfonic acid.[274,275]

The "template-free" method was also used to prepare a poly(aniline) nanotube using poly(methyl vinyl ether-*alt*-maleic acid) as acidic dopant.[276] It was found that the reaction temperature had a significant effect on the tube thickness and length, while reaction time seemed to have little effect. The external diameter of the tube could be influenced *via* the aniline to dopant polymer ratio. Such poly(methyl vinyl ether-*alt*-maleic acid) containing poly(aniline) tubes were deposited onto electrode surfaces and subsequently covalently modified using functional oligonucleotides. Such a setup could be used as an oligonucleotide sensor. Potential pulse amperometry allowed the detection of complementary oligonucleotides at a concentration of 10^{-10} mol L^{-1}.

5.3.5 Other Methods to Form Polymeric Tubes

A very different approach towards nanotube formation was taken by Guo *et al.*, who used the Langmuir–Blodgett technique to form monolayers of poly(2-methoxy-5-(*n*-hexadecyloxy)-*p*-phenylene vinylene). Compression of these monolayers beyond the collapse point resulted in the spontaneous formation of nanotubes with wall thicknesses of *ca.* 20 nm and diameters of *ca.* 130 nm. A 20 nm wall thickness corresponds to *ca.* eight stacked monolayers. Larger nanotubes with 40 nm wall thickness were also observed.[277]

Chen *et al.* reported the use of a silver(I) coordinated organogel as a template for nanotube formation. This approach is similar to the fiber templating approach (see above) albeit on a smaller, *i.e.* molecular length, scale. Tripyridin-3-yl-benzene-1,3,5-tricarboxylate and silver salts formed a coordinated organogel with exceptional thermal stability. The cross-linking monomer *N*,*N*′-methylenebisacrylamide could coordinate to the silver ions of the gel fibers and form a polymer shell in the presence of the radical initiator benzoyl peroxide. Decomposition of the gel–fiber template using ammonia yielded polymer nanotubes with inner diameters of *ca.* 40 nm and wall thicknesses of *ca.* 29 nm.[278]

Poly(*o*-toluidine) nanofibers were prepared and subsequently swollen in ethanol. Evaporation of the solvent under ambient conditions gave closed nanotubes as reported by Han *et al.*[279]

5.4 Summary

The different ways to prepare polymeric tubes range from the molecular synthesis of single molecule polymeric tubes to engineering approaches exploiting the bulk properties of polymers. The dimensions of tubes accessible are also a direct reflection of the approach taken. Single molecule tubes can exhibit inner diameters starting from several Angstroms going up to several nanometers. In fact, most polymeric helices reported are tight helices without an inner void at all, and are therefore not covered in this chapter on tubular structures. Engineered tubes, on the other hand, typically exhibit inner diameters ranging from tens of nanometers up to the micrometer regime. As far as bio-inspiration is concerned, the different approaches to prepare tubular polymeric structures are also bio-inspired in different ways: the single polymer chain folding approach mimics nature by folding the chains in similar ways nature does, driven by hydrophobic effects or exploiting hydrogen bonds and weaker interactions to stabilize the helical geometry. There, the focus is mainly placed on the synthesis, rather than applying the structures to biological functions. Approaches like the folding of DNA-tiles are clearly bio-inspired in that they exploit the naturally occurring polymer and its complementary and highly selective binding capabilities for the construction of "unnatural" new structures. Finally, templated syntheses of polymeric tubes are relatively straightforward and thus, the focus in several reports has shifted from pure synthesis to applications and, in particular, applications with a biological/medical impact. With synthetic advances being made in all areas of polymeric nanotube fabrication, we expect the focus to eventually shift more and more towards the exploitation of these geometrically confined structures. As far as the inner diameters of synthetic polymeric tubes are concerned, the synthetic polymer community is aiming at ever-larger structures *via* bottom-up syntheses and engineering approaches are preparing smaller and smaller features with top-down approaches. We expect to see more and more synthetic strategies that address the interesting area of diameters in the 1–10 nm range.

Acknowledgements

SD and AFMK would like to acknowledge the Swiss National Science Foundation, MB and AFMK would like to acknowledge the National Center of Competence in Research (NCCR) "Bio Inspired Materials" for financial support.

References

1. R. MacKinnon, *Angew. Chem., Int. Ed.*, 2004, **43**, 4265–4277.
2. P. Agre, *Angew. Chem., Int. Ed.*, 2004, **43**, 4278–4290.
3. S. Bhakdi and J. Tranum-Jensen, *Microbiol. Rev.*, 1991, **55**, 733–751.
4. S. W. Cowan, R. M. Garavito, J. N. Jansonius, J. A. Jenkins, R. Karlsson, N. König, E. F. Pai, R. A. Pauptit, P. J. Rizkallah, J. P. Rosenbusch, G. Rummel and T. Schirmer, *Structure*, 1995, **3**, 1041–1050.
5. A. Klug, *Angew. Chem., Int. Ed. Engl.*, 1983, **22**, 565–582.
6. T. L. Schlick, Z. Ding, E. W. Kovacs and M. B. Francis, *J. Am. Chem. Soc.*, 2005, **127**, 3718–3723.
7. J. M. Hooker, E. W. Kovacs and M. B. Francis, *J. Am. Chem. Soc.*, 2004, **126**, 3718–3719.
8. D. T. Bong, T. D. Clark, J. R. Granja and M. R. Ghadiri, *Angew. Chem., Int. Ed.*, 2001, **40**, 988–1011.
9. S. Höger, *J. Polym. Sci., Part A: Polym. Chem.*, 1999, **37**, 2685–2698.
10. D. Zhao and J. S. Moore, *Chem. Commun.*, 2003, 807–818.
11. C. Grave and A. D. Schlüter, *Eur. J. Org. Chem.*, 2002, **2002**, 3075–3098.
12. J. Couet and M. Biesalski, *Soft Matter*, 2006, **2**, 1005–1014.
13. T. G. Barclay, K. Constantopoulos and J. Matisons, *Chem. Rev.*, 2014, **114**, 10217–10291.
14. T. Shimizu, M. Masuda and H. Minamikawa, *Chem. Rev.*, 2005, **105**, 1401–1443.
15. I. W. Hamley, *Angew. Chem., Int. Ed.*, 2014, **53**, 6866–6881.
16. P. D. U. Koert, *Phys. Chem. Chem. Phys.*, 2005, **7**, 1501–1506.
17. D. Bong and M. Ghadiri, *Angew. Chem., Int. Ed.*, 2001, **40**, 2163–2166.

18. J. Buriak and M. Ghadiri, *Mater. Sci. Eng., C*, 1997, **4**, 207–212.
19. W. Horne, C. Stout and M. Ghadiri, *J. Am. Chem. Soc.*, 2003, **125**, 9372–9376.
20. J. Hartgerink, T. Clark and M. Ghadiri, *Chem. – Eur. J.*, 1998, **4**, 1367–1372.
21. J. G. Heddle, *Nanotechnol., Sci. Appl.*, 2008, **1**, 67–78.
22. J. F. Graveland-Bikker, I. A. T. Schaap, C. F. Schmidt and C. G. de Kruif, *Nano Lett.*, 2006, **6**, 616–621.
23. P. W. K. Rothemund, A. Ekani-Nkodo, N. Papadakis, A. Kumar, D. K. Fygenson and E. Winfree, *J. Am. Chem. Soc.*, 2004, **126**, 16344–16352.
24. O. I. Wilner, R. Orbach, A. Henning, C. Teller, O. Yehezkeli, M. Mertig, D. Harries and I. Willner, *Nat. Commun.*, 2011, **2**, 540.
25. J. C. Mitchell, J. R. Harris, J. Malo, J. Bath and A. J. Turberfield, *J. Am. Chem. Soc.*, 2004, **126**, 16342–16343.
26. D. Liu, S. H. Park, J. H. Reif and T. H. LaBean, *Proc. Natl. Acad. Sci. U. S. A.*, 2004, **101**, 717–722.
27. M. A. B. Block, C. Kaiser, A. Khan and S. Hecht, in *Functional Molecular Nanostructures*, Springer Berlin Heidelberg, pp. 89–150.
28. M. A. Balbo Block and S. Hecht, *Angew. Chem., Int. Ed.*, 2005, **44**, 6986–6989.
29. T. Shimizu, *J. Polym. Sci., Part A: Polym. Chem.*, 2008, **46**, 2601–2611.
30. S. Stupp, V. LeBonheur, K. Walker, L. Li, K. Huggins, M. Keser and A. Amstutz, *Science*, 1997, **276**, 384–389.
31. G. Wong, J. X. Tang, A. Lin, Y. L. Li, P. A. Janmey and C. R. Safinya, *Science*, 2000, **288**, 2035–2039.
32. B. N. Thomas, C. R. Safinya, R. J. Plano and N. A. Clark, *Science*, 1995, **267**, 1635–1638.
33. J. C. Nelson, J. G. Saven, J. S. Moore and P. G. Wolynes, *Science*, 1997, **277**, 1793–1796.
34. J. M. Schnur, *Science*, 1993, **262**, 1669–1676.
35. H. Yan, S. H. Park, G. Finkelstein, J. H. Reif and T. H. LaBean, *Science*, 2003, **301**, 1882–1884.
36. Y. Hamuro, S. J. Geib and A. D. Hamilton, *Angew. Chem., Int. Ed.*, 1994, **33**, 446–448.
37. Y. Hamuro, S. J. Geib and A. D. Hamilton, *J Am. Chem. Soc.*, 1996, **118**, 7529–7541.
38. Y. Hamuro, S. J. Geib and A. D. Hamilton, *J. Am. Chem. Soc.*, 1997, **119**, 10587–10593.
39. V. Berl, I. Huc, R. Khoury and J. Lehn, *Chem. – Eur J.*, 2001, **7**, 2798–2809.
40. C. Dolain, V. Maurizot and I. Huc, *Angew. Chem., Int. Ed.*, 2003, **42**, 2738–2740.
41. H. Jiang, J. Leger and I. Huc, *J. Am. Chem. Soc.*, 2003, **125**, 3448–3449.
42. H. Jiang, J. Leger, C. Dolain, P. Guionneau and I. Huc, *Tetrahedron*, 2003, **59**, 8365–8374.

43. E. Kolomiets, V. Berl, I. Odriozola, A.-M. Stadler, N. Kyritsakas and J.-M. Lehn, *Chem. Commun.*, 2003, 2868–2869.
44. J. Zhu, R. Parra, H. Zeng, E. Skrzypczak-Jankun, X. Zeng and B. Gong, *J. Am. Chem. Soc.*, 2000, **122**, 4219–4220.
45. B. Gong, *Chem. - Eur. J.*, 2001, **7**, 4336–4342.
46. B. Gong, H. Zeng, J. Zhu, L. Yua, Y. Han, S. Cheng, M. Furukawa, R. D. Parra, A. Y. Kovalevsky, J. L. Mills, E. Skrzypczak-Jankun, S. Martinovic, R. D. Smith, C. Zheng, T. Szyperski and X. C. Zeng, *Proc. Natl. Acad. Sci. U. S. A.*, 2002, **99**, 11583–11588.
47. A. Zhang, J. S. Ferguson, K. Yamato, C. Zheng and B. Gong, *Org. Lett.*, 2006, **8**, 5117–5120.
48. L. Yuan, H. Zeng, K. Yamato, A. Sanford, W. Feng, H. Atreya, D. Sukumaran, T. Szyperski and B. Gong, *J. Am. Chem. Soc.*, 2004, **126**, 16528–16537.
49. K. K. L. Cheuk, J. W. Y. Lam, L. M. Lai, Y. Dong and B. Z. Tang, *Macromolecules*, 2003, **36**, 9752–9762.
50. J. J. van Gorp, J. A. J. M. Vekemans and E. W. Meijer, *Chem. Commun.*, 2004, 60–61.
51. R. W. Sinkeldam, F. J. M. Hoeben, M. J. Pouderoijen, I. De Cat, J. Zhang, S. Furukawa, S. De Feyter, J. A. J. M. Vekemans and E. W. Meijer, *J. Am. Chem. Soc.*, 2006, **128**, 16113–16121.
52. J. Cao, M. Kline, Z. Chen, B. Luan, M. Lv, W. Zhang, C. Lian, Q. Wang, Q. Huang, X. Wei, J. Deng, J. Zhu and B. Gong, *Chem. Commun.*, 2012, **48**, 11112–11114.
53. R. Liu, M. Shiotsuki, T. Masuda and F. Sanda, *Macromolecules*, 2009, **42**, 6115–6122.
54. X. Zhao and K. S. Schanze, *Langmuir*, 2006, **22**, 4856–4862.
55. M. Banno, T. Yamaguchi, K. Nagai, C. Kaiser, S. Hecht and E. Yashima, *J. Am. Chem. Soc.*, 2012, **134**, 8718–8728.
56. Y.-X. Lu, Z.-M. Shi, Z.-T. Li and Z. Guan, *Chem. Commun.*, 2010, **46**, 9019–9021.
57. D.-W. Zhang, X. Zhao and Z.-T. Li, *Acc. Chem. Res.*, 2014, **47**, 1961–1970.
58. P. Zhang, L. Zhang, H. Wang, D.-W. Zhang and Z.-T. Li, *Polym. Chem.*, 2015, **6**, 2955–2961.
59. M. Schulze and A. F. M. Kilbinger, *J. Polym. Sci., Part A: Polym. Chem.*, 2016, **54**, 1731–1741.
60. R. B. Prince, T. Okada and J. S. Moore, *Angew. Chem., Int. Ed.*, 1999, **38**, 233–236.
61. R. B. Prince, J. G. Saven, P. G. Wolynes and J. S. Moore, *J. Am. Chem. Soc.*, 1999, **121**, 3114–3121.
62. Luc Brunsveld, R. B. Prince, A. E. W. MeijerJeffrey and S. Moore, *Org. Lett.*, 2000, **2**, 1525–1528.
63. D. J. Hill and J. S. Moore, *Proc. Natl. Acad. Sci. U. S. A.*, 2002, **99**, 5053–5057.
64. M. T. Stone and J. S. Moore, *Org. Lett.*, 2004, **6**, 469–472.
65. S. Lahiri, J. L. Thompson and J. S. Moore, *J. Am. Chem. Soc.*, 2000, **122**, 11315–11319.

66. M. S. Gin, T. Yokozawa, R. B. Prince and J. S. Moore, *J. Am. Chem. Soc.*, 1999, **121**, 2643–2644.
67. R. B. Prince, L. Brunsveld, E. W. Meijer and J. S. Moore, *Angew. Chem., Int. Ed.*, 2000, **39**, 228–230.
68. R. B. Prince, J. S. Moore, L. Brunsveld and E. W. Meijer, *Chem. – Eur. J.*, 2001, **7**, 4150–4154.
69. J. M. Cary and J. S. Moore, *Org. Lett.*, 2002, **4**, 4663–4666.
70. M. S. Gin and J. S. Moore, *Org. Lett.*, 2000, **2**, 135–138.
71. X. Yang, A. L. Brown, M. Furukawa, S. Li, W. E. Gardinier, E. J. Bukowski, F. V. Bright, C. Zheng, X. C. Zeng and B. Gong, *Chem. Commun.*, 2003, 56–57.
72. S. Hecht and A. Khan, *Angew. Chem., Int. Ed.*, 2003, **42**, 6021–6024.
73. A. Khan and S. Hecht, *Synth. Methods*, 2004, **147**, 37–42.
74. Z. Yu and S. Hecht, *J. Polym. Sci., Part A: Polym. Chem.*, 2015, **53**, 313–318.
75. D. Zhao and J. S. Moore, *Macromolecules*, 2003, **36**, 2712–2720.
76. Keunchan Oh, A. Kyu-Sung Jeong and J. S. Moore, *J. Org. Chem.*, 2003, **68**, 8397–8403.
77. D. Zhao and J. S. Moore, *Org. Biomol. Chem.*, 2003, **1**, 3471–3491.
78. D. H. Zhao and J. S. Moore, *J. Am. Chem. Soc.*, 2003, **125**, 16294–16299.
79. J. W. Wackerly and J. S. Moore, *Macromolecules*, 2006, **39**, 7269–7276.
80. H. H. Nguyen, J. H. McAliley and D. A. Bruce, *J. Polym. Sci., Part A: Polym. Chem.*, 2012, **50**, 2019–2028.
81. K. Suda and K. Akagi, *Macromolecules*, 2011, **44**, 9473–9488.
82. H. Sogawa, M. Shiotsuki, T. Hirao, T. Haino and F. Sanda, *Macromolecules*, 2013, **46**, 8161–8170.
83. G. S. Hanan, J.-M. Lehn, N. Kyritsakas and J. Fischer, *J. Chem. Soc., Chem. Commun.*, 1995, 765–766.
84. D. M. Bassani, J.-M. Lehn, G. Baum and D. Fenske, *Angew. Chem., Int. Ed.*, 1997, **36**, 1845–1847.
85. L. A. Cuccia, J.-M. Lehn, J. C. Homo and M. Schmutz, *Angew. Chem., Int. Ed.*, 2000, **39**, 233–237.
86. M. Ohkita, J.-M. Lehn, G. Baum and D. Fenske, *Chem. – Eur. J.*, 1999, **5**, 3471–3481.
87. K. M. Gardinier, R. G. Khoury and J.-M. Lehn, *Chem. – Eur. J.*, 2000, **6**, 4124–4131.
88. A. Petitjean, L. A. Cuccia, J.-M. Lehn, H. Nierengarten and M. Schmutz, *Angew. Chem., Int. Ed.*, 2002, **41**, 1195–1198.
89. V. Berl, M. Krische, I. Huc, J. Lehn and M. Schmutz, *Chem. – Eur. J.*, 2000, **6**, 1938–1946.
90. M. Barboiu and J.-M. Lehn, *Proc. Natl. Acad. Sci. U. S. A.*, 2002, **99**, 5201–5206.
91. M. Barboiu, G. Vaughan, N. Kyritsakas and J.-M. Lehn, *Chem. – Eur. J.*, 2003, **9**, 763–769.
92. J. L. Schmitt and J.-M. Lehn, *Helv. Chim. Acta*, 2003, **86**, 3417–3426.
93. J. F. Folmer-Andersen, E. Buhler, S.-J. Candau, S. Joulie, M. Schmutz and J.-M. Lehn, *Polym. Int.*, 2010, **59**, 1477–1491.

94. H.-C. Zhang and L. Pu, *Macromolecules*, 2004, **37**, 2695–2702.
95. R. M. Meudtner and S. Hecht, *Macromol. Rapid Commun.*, 2008, **29**, 347–351.
96. R. M. Meudtner and S. Hecht, *Angew. Chem., Int. Ed.*, 2008, **47**, 4926–4930.
97. A. Buono, A. Famulari, S. V. Meille, G. Ricci and L. Porri, *Macromolecules*, 2011, **44**, 3681–3684.
98. R. Guo, L. Zhang, H. Wang, D.-W. Zhang and Z.-T. Li, *Polym. Chem.*, 2015, **6**, 2382–2385.
99. H. Okuda, Y. Koyama, S. Uchida, T. Michinobu, H. Sogawa and T. Takata, *ACS Macro Lett.*, 2015, **4**, 462–466.
100. J. Zhu, Z. Dong, S. Lei, L. Cao, B. Yang, W. Li, Y. Zhang, J. Liu and J. Shen, *Angew. Chem., Int. Ed.*, 2015, **54**, 3097–3101.
101. H. Juwarker, J.-M. Suk and K.-S. Jeong, *Chem. Soc. Rev.*, 2009, **38**, 3316–3325.
102. M. Inouye, M. Waki and H. Abe, *J. Am. Chem. Soc.*, 2004, **126**, 2022–2027.
103. M. Waki, H. Abe and M. Inouye, *Chem. – Eur. J.*, 2006, **12**, 7839–7847.
104. M. Waki, H. Abe and M. Inouye, *Angew. Chem., Int. Ed.*, 2007, **46**, 3059–3061.
105. H. Abe, S. Takashima, T. Yamamoto and M. Inouye, *Chem. Commun.*, 2009, 2121–2123.
106. S. Takashima, H. Abe and M. Inouye, *Chem. Commun.*, 2011, **47**, 7455–7457.
107. H. Abe, K. Okada, H. Makida and M. Inouye, *Org. Biomol. Chem.*, 2012, **10**, 6930–6936.
108. P. Xin, P. Zhu, P. Su, J.-L. Hou and Z.-T. Li, *J. Am. Chem. Soc.*, 2014, **136**, 13078–13081.
109. Y. Li, T. Dutta, N. Gerasimchuk, S. Wu, K. Shetye, L. Jin, R. Wang, D.-M. Zhu and Z. Peng, *ACS Appl. Mater. Interfaces*, 2015, **7**, 9372–9384.
110. J. Cornelissen, M. Fischer, N. Sommerdijk and R. Nolte, *Science*, 1998, **280**, 1427–1430.
111. D. M. Vriezema, A. Kros, R. de Gelder, J. Cornelissen, A. E. Rowan and R. Nolte, *Macromolecules*, 2004, **37**, 4736–4739.
112. J. M. van der Eijk, R. Nolte, W. Drenth and A. M. F. Hezemans, *Macromolecules*, 1980, **13**, 1391–1397.
113. D. Yan, Y. Zhou and J. Hou, *Science*, 2004, **303**, 65–67.
114. G. J. Liu and J. F. Ding, *Adv. Mater.*, 1998, **10**, 69–71.
115. G. J. Liu, *Adv. Mater.*, 1997, **9**, 437–439.
116. J. S. Lee, A. Hirao and S. Nakahama, *Macromolecules*, 1989, **22**, 2602–2606.
117. H. Y. Huang, E. E. Remsen, T. Kowalewski and K. L. Wooley, *J. Am. Chem. Soc.*, 1999, **121**, 3805–3806.
118. S. Stewart and G. Liu, *Angew. Chem., Int. Ed.*, 2000, **39**, 340–344.
119. K. Huang and J. Rzayev, *J. Am. Chem. Soc.*, 2009, **131**, 6880–6885.
120. T. Kesharwani, J. S. Valenstein, B. G. Trewyn, F. Li, V. S. Y. Lin and R. C. Larock, *Polym. Chem.*, 2010, **1**, 1427–1429.

121. R. M. Ho, C. K. Chen, Y. W. Chiang, B. T. Ko and C. C. Lin, *Adv. Mater.*, 2006, **18**, 2355–2358.
122. J. Chen, C. Yu, Z. Shi, S. Yu, Z. Lu, W. Jiang, M. Zhang, W. He, Y. Zhou and D. Yan, *Angew. Chem., Int. Ed.*, 2015, **54**, 3621–3625.
123. N. Blouin, A. Michaud, D. Gendron, S. Wakim, E. Blair, R. Neagu-Plesu, M. Belletête, G. Durocher, Y. Tao and M. Leclerc, *J. Am. Chem. Soc.*, 2008, **130**, 732–742.
124. E. Zhou, S. Yamakawa, K. Tajima, C. Yang and K. Hashimoto, *Chem. Mater.*, 2009, **21**, 4055–4061.
125. A. S. Sawhney, C. P. Pathak and J. A. Hubbell, *Macromolecules*, 1993, **26**, 581–587.
126. K. J. Zhu, X. Z. Lin and S. L. Yang, *J. Appl. Polym. Sci.*, 1990, **39**, 1–9.
127. C. Scholz, M. Iijima, Y. Nagasaki and K. Kataoka, *Macromolecules*, 1995, **28**, 7295–7297.
128. K. E. Uhrich, S. M. Cannizzaro, R. S. Langer and K. M. Shakesheff, *Chem. Rev.*, 1999, **99**, 3181–3198.
129. N. Kang, M. E. Perron, R. E. Prud'homme, Y. B. Zhang, G. Gaucher and J. C. Leroux, *Nano Lett.*, 2005, **5**, 315–319.
130. J. Raez, R. Barjovanu, J. A. Massey, M. A. Winnik and I. Manners, *Angew. Chem., Int. Ed.*, 2000, **39**, 3862–3865.
131. K. Yu and A. Eisenberg, *Macromolecules*, 1998, **31**, 3509–3518.
132. S. H. Kim, F. Nederberg, R. Jakobs, J. P. K. Tan, K. Fukushima, A. Nelson, E. W. Meijer, Y. Y. Yang and J. L. Hedrick, *Angew. Chem., Int. Ed.*, 2009, **48**, 4508–4512.
133. J. P. Hill, W. S. Jin, A. Kosaka, T. Fukushima, H. Ichihara, T. Shimomura, K. Ito, T. Hashizume, N. Ishii and T. Aida, *Science*, 2004, **304**, 1481–1483.
134. G. John, M. Masuda, Y. Okada, K. Yase and T. Shimizu, *Adv. Mater.*, 2001, **13**, 715–718.
135. L. Jia, D. Lévy, D. Durand, M. Impéror-Clerc, A. Cao and M.-H. Li, *Soft Matter*, 2011, **7**, 7395–7403.
136. L. Cheng, G. Zhang, L. Zhu, D. Chen and M. Jiang, *Angew. Chem., Int. Ed.*, 2008, **47**, 10171–10174.
137. J. E. Reiner, J. M. Wells, R. B. Kishore, C. Pfefferkorn and K. Helmerson, *Proc. Natl. Acad. Sci. U. S. A.*, 2006, **103**, 1173–1177.
138. B. M. Discher, Y. Y. Won, D. S. Ege, J. Lee, F. S. Bates, D. E. Discher and D. A. Hammer, *Science*, 1999, **284**, 1143–1146.
139. E. Braun, Y. Eichen, U. Sivan and G. Ben-Yoseph, *Nature*, 1998, **391**, 775–778.
140. J. Richter, M. Mertig, W. Pompe, I. Monch and H. K. Schackert, *Appl. Phys. Lett.*, 2001, **78**, 536–538.
141. M. Endo, N. C. Seeman and T. Majima, *Angew. Chem., Int. Ed.*, 2005, **44**, 6074–6077.
142. E. Winfree, F. R. Liu, L. A. Wenzler and N. C. Seeman, *Nature*, 1998, **394**, 539–544.

143. F. R. Liu, R. J. Sha and N. C. Seeman, *J. Am. Chem. Soc.*, 1999, **121**, 917–922.
144. P. W. K. Rothemund, *Nature*, 2006, **440**, 297–302.
145. Z. Li, M. Liu, L. Wang, J. Nangreave, H. Yan and Y. Liu, *J. Am. Chem. Soc.*, 2010, **132**, 13545–13552.
146. P. Thanasekaran, T.-T. Luo, C.-H. Lee and K.-L. Lu, *J. Mater. Chem.*, 2011, **21**, 13140–13149.
147. F. Dai, H. He and D. Sun, *J. Am. Chem. Soc.*, 2008, **130**, 14064–14065.
148. T.-T. Luo, H. C. Wu, Y. C. Jao, S. M. Huang, T. W. Tseng, Y. S. Wen, G. H. Lee, S. M. Peng and K.-L. Lu, *Angew. Chem., Int. Ed.*, 2009, **48**, 9461–9464.
149. K. Otsubo, Y. Wakabayashi, J. Ohara, S. Yamamoto, H. Matsuzaki, H. Okamoto, K. Nitta, T. Uruga and H. Kitagawa, *Nat. Mater.*, 2011, **10**, 291–295.
150. R. Kaminker, R. Popovitz Biro and M. E. van der Boom, *Angew. Chem., Int. Ed.*, 2011, **50**, 3224–3226.
151. R. Kaminker, L. Motiei, A. Gulino, I. Fragalà, L. J. W. Shimon, G. Evmenenko, P. Dutta, M. A. Iron and M. E. van der Boom, *J. Am. Chem., Soc.*, 2010, **132**, 14554–14561.
152. H. Deng, S. Grunder, K. E. Cordova, C. Valente, H. Furukawa, M. Hmadeh, F. Gandara, A. C. Whalley, Z. Liu, S. Asahina, H. Kazumori, M. O'Keeffe, O. Terasaki, J. F. Stoddart and O. M. Yaghi, *Science*, 2012, **336**, 1018–1023.
153. L. Hou, L.-N. Jia, W.-J. Shi, L.-Y. Du, J. Li, Y.-Y. Wang and Q.-Z. Shi, *Dalton Trans.*, 2013, **42**, 6306–6309.
154. X.-R. Hao, X.-L. Wang, C. Qin, Z.-M. Su, E.-B. Wang, Y.-Q. Lan and K.-Z. Shao, *Chem. Commun.*, 2007, 4620–4622.
155. Y. Cui, S. J. Lee and W. B. Lin, *J. Am. Chem. Soc.*, 2003, **125**, 6014–6015.
156. M. A. Braverman, J. H. Nettleman, R. M. Supkowski and R. L. LaDuca, *Inorg. Chem.*, 2009, **48**, 4918–4926.
157. Z.-Z. Lu, R. Zhang, Y.-Z. Li, Z.-J. Guo and H.-G. Zheng, *Chem. Commun.*, 2011, **47**, 2919–2921.
158. K. Sumida, D. L. Rogow, J. A. Mason, T. M. McDonald, E. D. Bloch, Z. R. Herm, T.-H. Bae and J. R. Long, *Chem. Rev.*, 2012, **112**, 724–781.
159. M. C. Hong, Y. J. Zhao, W. P. Su, R. Cao, M. Fujita, Z. Y. Zhou and A. Chan, *Angew. Chem., Int. Ed.*, 2000, **39**, 2468–2470.
160. S.-B. Ren, X.-L. Yang, J. Zhang, Y.-Z. Li, Y.-X. Zheng, H.-B. Du and X.-Z. You, *CrystEngComm*, 2009, **11**, 246–248.
161. C. Y. Su, M. D. Smith and H. C. zur Loye, *Angew. Chem., Int. Ed.*, 2003, **42**, 4085–4089.
162. T. Bataille, F. Costantino, P. Lorenzo-Luis, S. Midollini and A. Orlandini, *Inorg. Chim. A-Art. Lett.*, 2008, **361**, 9–15.
163. J. Fan, H. F. Zhu, T. A. Okamura, W. Y. Sun, W. X. Tang and N. Ueyama, *Inorg. Chem.*, 2003, **42**, 158–162.
164. Z. F. Fei, D. B. Zhao, T. J. Geldbach, R. Scopelliti, P. J. Dyson, S. Antonijevic and G. Bodenhausen, *Angew. Chem., Int. Ed.*, 2005, **44**, 5720–5725.

165. C. R. Martin, *Adv. Mater.*, 1991, **3**, 457–459.
166. C. R. Martin, *Science*, 1994, **266**, 1961–1966.
167. J. C. Hulteen and C. R. Martin, *J. Mater. Chem.*, 1997, **7**, 1075–1087.
168. H. Masuda and K. Fukuda, *Science*, 1995, **268**, 1466–1468.
169. K. Nielsch, J. Choi, K. Schwirn, R. B. Wehrspohn and U. Gösele, *Nano Lett.*, 2002, **2**, 677–680.
170. A. P. Li, F. Müller, A. Birner, K. Nielsch and U. Gösele, *J. Appl. Phys.*, 1998, **84**, 6023.
171. S. K. Chakarvarti, *Radiat. Meas.*, 2009, **44**, 1085–1092.
172. M. Steinhart, R. B. Wehrspohn, U. Gösele and J. H. Wendorff, *Angew. Chem., Int. Ed.*, 2004, **43**, 1334–1344.
173. V. M. Cepak and C. R. Martin, *Chem. Mater.*, 1999, **11**, 1363–1367.
174. D. O'Carroll, D. Iacopino and G. Redmond, *Adv. Mater.*, 2009, **21**, 1160–1165.
175. S. Moynihan, D. Lacopino, D. O'Carroll, P. Lovera and G. Redmond, *Chem. Mater.*, 2008, **20**, 996–1003.
176. D. Iacopino and G. Redmond, *Chem. Commun.*, 2011, **47**, 9170–9172.
177. S. Moynihan, D. Iacopino, D. O'Carroll, H. Doyle, D. A. Tanner and G. Redmond, *Adv. Mater.*, 2007, **19**, 2474–2479.
178. C. C. Wang, Q. D. Shen, S. C. Tang, Q. Wu, H. M. Bao, C. Z. Yang and X. Q. Jiang, *Macromol. Rapid Commun.*, 2008, **29**, 724–728.
179. X. She, G. Song, J. Li, P. Han, S. Yang, W. Shulong and Z. Peng, *Polym. J.*, 2006, **38**, 639–642.
180. M. Steinhart, J. H. Wendorff, A. Greiner, R. B. Wehrspohn, K. Nielsch, J. Schilling, J. Choi and U. Gösele, *Science*, 2002, **296**, 1997.
181. M. Steinhart, S. Senz, R. B. Wehrspohn, U. Gösele and J. H. Wendorff, *Macromolecules*, 2003, **36**, 3646–3651.
182. S. Schlitt, A. Greiner and J. H. Wendorff, *Macromolecules*, 2008, **41**, 3228–3234.
183. S. Dougherty and J. Liang, *Nanotechnology*, 2009, **20**, 295301.
184. G. Song, X. She, Z. Fu and J. Li, *J. Mater. Res.*, 2004, **19**, 3324–3328.
185. G. Chen, S. A. Soper and R. L. Mccarley, *Langmuir*, 2007, **23**, 11777–11781.
186. P. Lovera, K. Reynolds and G. Redmond, *Phys. Status Solidi A*, 2009, **206**, 2240–2244.
187. M. Steinhart, Z. Jia, A. K. Schaper, R. B. Wehrspohn, U. Gösele and J. H. Wendorff, *Adv. Mater.*, 2003, **15**, 706–709.
188. L. B. Huang, Y. Zhou, S. T. Han, Y. Yan, L. Zhou, W. Chen, P. Zhou, X. Chen and V. A. L. Roy, *Small*, 2014, **10**, 4645–4650.
189. S. Jin, Y. Lee, S.-M. Jeon, B. H. Sohn, W.-S. Chae and J.-K. Lee, *J. Mater. Chem.*, 2012, **22**, 23368–23373.
190. J.-T. Chen, C.-W. Lee, M.-H. Chi and I.-C. Yao, *Macromol. Rapid Commun.*, 2013, **34**, 348–354.
191. C.-W. Chang, M.-H. Chi, C.-W. Chu, H.-W. Ko, Y.-H. Tu, C.-C. Tsai and J.-T. Chen, *RSC Adv.*, 2015, **5**, 27443–27448.
192. X. Li, P. Han, G. Song, Z. Peng, Y. Yu, S. Feng, C. Zhou and M. Li, *Mater. Lett.*, 2015, **141**, 157–160.

193. N. Huby, J.-L. Duvail, D. Duval, D. Pluchon and B. Bêche, *Appl. Phys. Lett.*, 2011, **99**, 113302.
194. N. Huby, J. Bigeon, G. Danion, J.-L. Duvail, F. Gouttefangeas, L. Joanny and B. Bêche, *Micromachines*, 2015, **6**, 544–553.
195. C. R. Martin, L. S. Van Dyke, Z. Cai and W. Liang, *J. Am. Chem. Soc.*, 1990, **112**, 8976–8977.
196. C. R. Martin, *Acc. Chem. Res.*, 1995, **28**, 61–68.
197. S. Il Cho and S. B. Lee, *Acc. Chem. Res.*, 2008, **41**, 699–707.
198. Y.-Z. Long, M.-M. Li, C. Gu, M. Wan, J.-L. Duvail, Z. Liu and Z. Fan, *Prog. Polym. Sci.*, 2011, **36**, 1415–1442.
199. D. Zhang and Y. Wang, *J. Mater. Eng.: B*, 2006, **134**, 9–19.
200. A. Kros, R. J. M. Nolte and N. A. J. M. Sommerdijk, *Adv. Mater.*, 2002, **14**, 1779–1782.
201. L. Xia, Z. Wei and M. Wan, *J. Colloid Interface Sci.*, 2010, **341**, 1–11.
202. R. V. Parthasarathy and C. R. Martin, *Chem. Mater.*, 1994, **6**, 1627–1632.
203. S. Xiong, Q. Wang and H. Xia, *Mater. Res. Bull.*, 2004, **39**, 1569–1580.
204. M. Delvaux, J. Duchet, P.-Y. Stavaux, R. Legras and S. Demoustier-Champagne, *Synth. Methods*, 2000, **113**, 275–280.
205. H. Q. Cao, Z. Xu, H. Sang, D. Sheng and C. Y. Tie, *Adv. Mater.*, 2001, **13**, 121–123.
206. S. M. Yang, K. H. Chen and Y. F. Yang, *Synth. Methods*, 2005, **152**, 65–68.
207. S. De Vito and C. R. Martin, *Chem. Mater.*, 1998, **10**, 1738–1741.
208. J. Duchet, R. Legras and S. Demoustier-Champagne, *Synth. Methods*, 1998, **98**, 113–122.
209. S. Demoustier-Champagne and P.-Y. Stavaux, *Chem. Mater.*, 1999, **11**, 829–834.
210. J. Jang and J. H. Oh, *Chem. Commun.*, 2004, 882–883.
211. Y. He, J. Yuan and G. Shi, *J. Mater. Chem.*, 2005, **15**, 859–862.
212. I. Sarkar, A. Sarma, M. K. Sanyal, S. Thieß and W. Drube, *J. Appl. Phys.*, 2013, **114**, 163707.
213. A. Sarma, M. K. Sanyal, A. Rahman and B. Satpati, *J. Appl. Phys.*, 2012, **112**, 044304.
214. M. Fu, Y. Zhu, R. Tan and G. Shi, *Adv. Mater.*, 2001, **13**, 1874–1877.
215. N. Fujimoto, T. Nakahodo and H. Fujihara, *Chem. Lett.*, 2013, **42**, 1394–1396.
216. K. Sato, T. Nakahodo and H. Fujihara, *Chem. Commun.*, 2011, **47**, 10067–10069.
217. Y. K. Hong, D. H. Park, S. K. Park, H. Song, D. C. Kim, J. Kim, Y. H. Han, O. K. Park, B. C. Lee and J. Joo, *Adv. Funct. Mater.*, 2009, **19**, 567–572.
218. J. L. Duvail, P. Rétho, V. Fernandez, G. Louarn, P. Molinié and O. Chauvet, *J. Phys. Chem. B*, 2004, **108**, 18552–18556.
219. M. G. Han and S. H. Foulger, *Chem. Commun.*, 2005, 3092–3094.
220. S. I. Cho, W. J. Kwon, S. J. Choi, P. Kim, S. A. Park, J. Kim, S. J. Son, R. Xiao, S. H. Kim and S. B. Lee, *Adv. Mater.*, 2005, **17**, 171–175.
221. B. H. Kim, D. H. Park, J. Joo, S. G. Yu and S. H. Lee, *Synth. Methods*, 2005, **150**, 279–284.

222. J. Joo, B. H. Kim, D. H. Park, H. S. Kim, D. S. Seo, J. H. Shim, S. J. Lee, K. S. Ryu, K. Kim, J. I. Jin, T. J. Lee and C. J. Lee, *Synth. Methods*, 2005, **153**, 313–316.
223. J. Joo, K. T. Park, B. H. Kim, M. S. Kim, S. Y. Lee, C. K. Jeong, J. K. Lee, D. H. Park, W. K. Yi, S. H. Lee and K. S. Ryu, *Synth. Methods*, 2003, **135-136**, 7–9.
224. M. Fu, F. Chen, J. Zhang and G. Shi, *J. Mater. Chem.*, 2002, **12**, 2331–2333.
225. K. Kim and J.-I. Jin, *Nano Lett.*, 2001, **1**, 631–636.
226. K. Kim, B. H. Kim, S. H. Joo, J. S. Park, J. Joo and J. I. Jin, *Adv. Mater.*, 2005, **17**, 464–468.
227. R. V. Parthasarathy and C. R. Martin, *Nature*, 1994, **369**, 298–301.
228. A. Kros, S. W. F. M. van Hövell, N. A. J. M. Sommerdijk and R. J. M. Nolte, *Adv. Mater.*, 2001, **13**, 1555–1557.
229. C. G. J. Koopal, B. de Ruiter and R. J. M. Nolte, *J. Chem. Soc. Chem. Commun.*, 1991, 1691–1692.
230. A. Kros, N. A. J. M. Sommerdijk and R. J. M. Nolte, *Sens. Actuators, B*, 2005, **106**, 289–295.
231. H. Chang, Y. Yuan, N. Shi and Y. Guan, *Anal. Chem.*, 2007, **79**, 5111–5115.
232. G. Decher, *Science*, 1997, **277**, 1232–1237.
233. S. Ai, G. Lu, Q. He and J. Li, *J. Am. Chem. Soc.*, 2003, **125**, 11140–11141.
234. Z. Liang, A. S. Susha, A. Yu and F. Caruso, *Adv. Mater.*, 2003, **15**, 1849–1853.
235. S. Ai, Q. He, C. Tao, S. Zheng and J. Li, *Macromol. Rapid Commun.*, 2005, **26**, 1965–1969.
236. Y. Tian, Q. He, Y. Cui, C. Tao and J. Li, *Chem. – Eur. J.*, 2006, **12**, 4808–4812.
237. Y. Tian, Q. He, C. Tao and J. B. Li, *Langmuir*, 2006, **22**, 360–362.
238. E. N. Savariar, M. M. Sochat, A. Klaikherd and S. Thayumanavan, *Angew. Chem., Int. Ed.*, 2009, **48**, 110–114.
239. Y. Tian, Q. He, Y. Cui and J. Li, *Biomacromolecules*, 2006, **7**, 2539–2542.
240. S. F. Hou, J. H. Wang and C. R. Martin, *Nano Lett.*, 2005, **5**, 231–234.
241. G. Lu, T. Komatsu and E. Tsuchida, *Chem. Commun.*, 2007, 2980–2982.
242. X. Qu, G. Lu, E. Tsuchida and T. Komatsu, *Chem. – Eur. J.*, 2008, **14**, 10303–10308.
243. X. Qu, N. Kobayashi and T. Komatsu, *ACS Nano*, 2010, **4**, 1732–1738.
244. S. Goto, Y. Amano, M. Akiyama, C. Boettcher and T. Komatsu, *Langmuir*, 2013, **29**, 14293–14300.
245. X. Qu and T. Komatsu, *ACS Nano*, 2010, **4**, 563–573.
246. R. Kato and T. Komatsu, *Chem. Lett.*, 2011, **40**, 1338–1339.
247. T. Komatsu, X. Qu, H. Ihara, M. Fujihara, H. Azuma and H. Ikeda, *J. Am. Chem. Soc.*, 2011, **133**, 3246–3248.
248. M. Bognitzki, H. Hou, M. Ishaque, T. Frese, M. Hellwig, C. Schwarte, A. Schaper, J. H. Wendorff and A. Greiner, *Adv. Mater.*, 2000, **12**, 637–640.
249. H. Q. Hou, Z. Jun, A. Reuning, A. Schaper, J. H. Wendorff and A. Greiner, *Macromolecules*, 2002, **35**, 2429–2431.

250. F. Mitschang, M. Langner, H. Vieker, A. Beyer and A. Greiner, *Macromol. Rapid Commun.*, 2015, **36**, 304–310.
251. K. S. Mayya, D. I. Gittins, A. M. Dibaj and F. Caruso, *Nano Lett.*, 2001, **1**, 727–730.
252. M. J. Mulvihill, B. L. Rupert, R. R. He, A. Hochbaum, J. Arnold and P. D. Yang, *J. Am. Chem. Soc.*, 2005, **127**, 16040–16041.
253. L. Zhu, Y. Xu, W. Yuan, J. Xi, X. Huang, X. Tang and S. Zheng, *Adv. Mater.*, 2006, **18**, 2997–3000.
254. H. Dong, S. Prasad, V. Nyame and W. E. Jones, *Chem. Mater.*, 2004, **16**, 371–373.
255. M. R. Abidian, D. H. Kim and D. C. Martin, *Adv. Mater.*, 2006, **18**, 405.
256. M. R. Abidian and D. C. Martin, *Biomaterials*, 2008, **29**, 1273–1283.
257. M. R. Abidian, J. M. Corey, D. R. Kipke and D. C. Martin, *Small*, 2010, **6**, 421–429.
258. M. R. Abidian, K. A. Ludwig, T. C. Marzullo, D. C. Martin and D. R. Kipke, *Adv. Mater.*, 2009, **21**, 3764–3770.
259. J. Jang and H. Yoon, *Chem. Commun.*, 2003, 720–721.
260. J. Jang and H. Yoon, *Langmuir*, 2005, **21**, 11484–11489.
261. J. Jang and H. Yoon, *Adv. Mater.*, 2003, **15**, 2088–2091.
262. X. Y. Zhang, J. S. Lee, G. S. Lee, D. K. Cha, M. J. Kim, D. J. Yang and S. K. Manohar, *Macromolecules*, 2006, **39**, 470–472.
263. H. Yoon, M. Chang and J. Jang, *J. Phys. Chem. B*, 2006, **110**, 14074–14077.
264. M. Wan, *Adv. Mater.*, 2008, **20**, 2926–2932.
265. M. Wan, *Macromol. Rapid Commun.*, 2009, **30**, 963–975.
266. Z. X. Wei, Z. M. Zhang and M. X. Wan, *Langmuir*, 2002, **18**, 917–921.
267. H. Ding, J. Shen, M. Wan and Z. Chen, *Macromol. Chem. Phys.*, 2008, **209**, 864–871.
268. M. Wan and J. Li, *J. Polym. Sci., Part A: Polym. Chem.*, 2000, **38**, 2359–2364.
269. L. Zhang and M. Wan, *Thin Solid Films*, 2005, **477**, 24–31.
270. H. Q. Qiu, M. Wan, B. M. Matthews and L. Dai, *Macromolecules*, 2001, **34**, 675–677.
271. K. Huang and M. Wan, *Chem. Mater.*, 2002, **14**, 3486–3492.
272. Z. M. Zhang, Z. X. Wei and M. X. Wan, *Macromolecules*, 2002, **35**, 5937–5942.
273. L. Zhang, Y. Long, Z. Chen and M. Wan, *Adv. Funct. Mater.*, 2004, **14**, 693–698.
274. J. Liu and M. Wan, *J. Mater. Chem.*, 2001, **11**, 404–407.
275. Y. Yang and M. Wan, *J. Mater. Chem.*, 2001, **11**, 2022–2027.
276. L. Zhang, H. Peng, P. A. Kilmartin, C. Soeller and J. Travas Sejdic, *Electroanalysis*, 2007, **19**, 870–875.
277. L. Guo, Z. Wu and Y. Liang, *Chem. Commun.*, 2004, 1664–1665.
278. K. Chen, L. Tang, Y. Xia and Y. Wang, *Langmuir*, 2008, **24**, 13838–13841.
279. J. Han, L. Wang and R. Guo, *Macromol. Rapid Commun.*, 2011, **32**, 729–735.

CHAPTER 6

Bio-inspired Polymer Membranes

VIKTORIA MIKHALEVICH, CHRISTINA ZELMER, ANJA CAR, CORNELIA PALIVAN* AND WOLFGANG MEIER*

University of Basel, Department of Chemistry, Klingelbergstrasse 80, CH-4056 Basel, Switzerland
*Email: cornelia.palivan@unibas.ch; wolfgang.meier@unibas.ch

6.1 Introduction

Molecular self-assembly is a self-driven process essential to various systems in nature (*e.g.* formation of cell membranes, protein folding, DNA hybridization, *etc.*). The internal organization of self-assembled systems is governed by inter- (van der Waals, dipole–dipole, and hydrogen bonding) and intra-molecular (hydrophobic) forces that drive the molecules of the assembly into a low energy stable state.[1] In particular, biomembranes are phospholipid bilayers with embedded proteins responsible for biological communication and molecular transport that form enclosed compartments separating different biological environments.[2,3] Phospholipids provide structural stability for the membranes, and high lateral fluidity, which enable the functionality of embedded membrane proteins. One of the most important functions of a biomembrane is selective permeability, which allows selected moieties to cross the barrier. Although biomembranes are composed of similar phospholipids, their functionalities are mostly determined by the specificity of their associated proteins: there are over 300 types of ion channels in a living cell.[4]

RSC Polymer Chemistry Series No. 22
Bio-inspired Polymers
Edited by Nico Bruns and Andreas F. M. Kilbinger

Published by the Royal Society of Chemistry, www.rsc.org

The structural and functional complexity of biomembranes has challenged researchers to develop simpler artificial models to mimic their properties. Amphiphilic block copolymers are of particular interest, because of the dual environmental affinity that is associated with covalently bound hydrophobic and hydrophilic blocks. These strive to minimize their contact, and therefore drive self-assembly into assemblies with different architectures.[5] Based on their chemical specificity, as for example the hydrophilic-to hydrophobic ratio, amphiphilic block copolymers can self-assemble in dilute aqueous solutions into micelles, vesicles, tubes, wire-like structures, nanoparticles, or planar membranes at water–air interfaces.[6,7] Synthetic membranes have greater mechanical stability than phospholipids because of the higher molecular weight (M_w) of amphiphilic block copolymers, and thus are thicker and stiffer than lipid bilayers.

Polymer vesicles, the so called polymersomes, are of significant interest, because of their unique architecture, which involves an enclosed membrane with a central aqueous compartment, similar to that of biocompartments.[8] In addition, the chemical nature of amphiphilic block copolymers allows their functionalization with specific molecules at the surface, and insertion of molecules inside the membrane. Therefore, polymersomes provide three different regions for combination with active molecules: the aqueous cavity (for encapsulating hydrophilic molecules), the membrane (for insertion of hydrophobic molecules), and the boundary of the membrane with the environment (for functionalization with specific molecular groups to serve as targets or for immobilization on solid supports).[9] The internal cavity of polymersomes serves as a container for payloads, either for acting as a drug delivery system, or as a nano-space for reactions in so-called nanoreactors. A key parameter for *in situ* catalytic reactions is the permeability of the polymersome membranes to allow exchange of substrates/products with the environment. Therefore, selection of the amphiphilic copolymer and the type of the corresponding supramolecular assembly has to match both the selected biomolecule and the intended application.

In contrast to polymersomes, there are various models of planar membranes: monolayers at the water–air interface, free-standing bilayers, and solid-supported membranes. The functionality of proteins in natural membranes strongly depends on their mobility in the matrix, and this is thus an essential prerequisite for artificial membranes to mimic the dynamic environment of biomembranes in order to serve as templates for biomolecules.[10] Therefore, the building blocks forming a bio-inspired membrane need to possess high flexibility to compensate the hydrophobic mismatch between the size of the biomolecules, and the membrane thickness.[11] Furthermore, a variety of membrane properties (thickness, polarity, and surface charge) have to be considered for the successful insertion/attachment of biomolecules. Decoration of polymer membranes with biomolecules, either on their surfaces or inside the bilayers, can be achieved by various approaches, such as physical adsorption, insertion, and covalent binding.[12] Compared to physical immobilization of biomolecules on

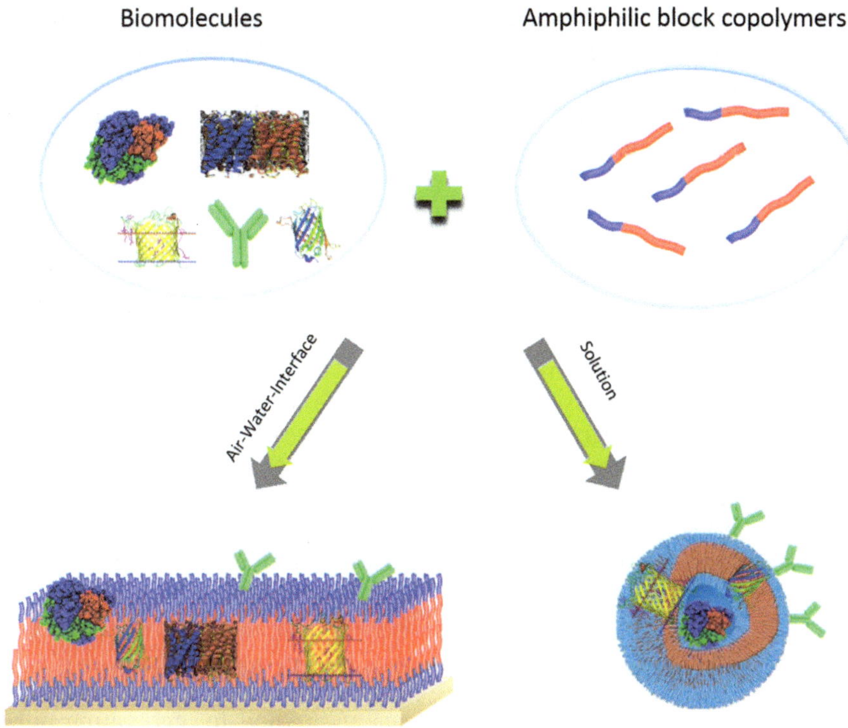

Figure 6.1 Conceptual overview of bio-inspired polymer vesicles and planar polymer membranes.

synthetic membranes, covalent attachment of molecules is more stable and can be used for targeting approaches, or for immobilization purposes.[13] Surface modification of planar membranes is a focus for the development of active surfaces for biosensing and diagnostic purposes.[14] The aim of this chapter is to introduce biomimetic strategies for producing synthetic membranes (polymersomes and planar membranes) that can be combined with biomolecules for developing novel functional materials/systems. Figure 6.1 shows the design of such polymer membranes.

6.2 Properties of Copolymers that Form Bio-inspired Membranes

Well-defined amphiphilic block copolymers can be prepared by various controlled polymerization reactions, such as atom transfer radical polymerization (ATRP), reversible addition-fragmentation chain transfer (RAFT), and nitroxide mediated polymerization (NMP).[15,16] The amphiphilic copolymers most frequently used in bio-inspired strategies consist of hydrophilic blocks, such as poly(acrylic acid) (PAA), poly(ethylene oxide) (PEO), or poly(2-methyl oxazoline) (PMOXA), and a hydrophobic block, such as polystyrene (PS), poly(butadiene) (PB), or poly(dimethylsiloxane) (PDMS).[17,18] The chemical nature of the blocks is a key factor for inducing the properties required from biomimetic approaches into artificial membranes.[19] Both the individual characteristics of each block and the properties of the final copolymer have to be tailored to generate supramolecular assemblies that are able to host biomolecules. For example, a charged surface on the polymer membrane is required for attaching biomolecules with a complementary charge by electrostatic interactions.[20] Variation in block copolymer composition and M_w influence the architecture and size of the supramolecular assembly and the membrane thickness, whilst the glass transition and crystallinity of the hydrophobic block influence the flexibility and fluidity of the membrane.[10]

Stimulus-responsiveness of the membrane is a specific property that is important in drug delivery systems: in the presence of stimuli (*e.g.* the acidic environment of a tumor) the membrane is disrupted or changes its architecture (*e.g.* from vesicle to micelle).[21] Responsiveness of hydrophobic blocks can be chemically achieved for various stimuli, such as pH, redox conditions, enzyme-degradation, light, temperature, and changes in a magnetic field.[22]

6.3 Bio-inspired Polymersomes (3D Membranes)

If the copolymer is water soluble, the easiest method for generating polymersomes is direct dissolution of the dry polymer in aqueous media.[8] However, this method cannot be used for insoluble water amphiphilic block-copolymers, for which a solvent-free or a solvent displacement method must be used.[23] With the solvent-free method, a homogeneous amphiphilic polymer film is first deposited on a solid surface, and then subsequently hydrated in an aqueous buffer solution to form supramolecular assemblies upon desorption from the support. This is the method of choice when working under organic solvent-free conditions is mandatory, as for example for the combination with biomolecules.[24] When applying the solvent displacement method, the copolymer is first dissolved in an appropriate solvent, and then slowly mixed with water: the reduced solubility of the copolymer drives the self-assembly process and formation of supramolecular assemblies, such as polymersomes.[25] The organic solvent is eventually extracted, but complete removal of solvent traces is not possible and traces of organic solvents could affect sensitive biomolecules.

Whilst the solvent-free and solvent-displacement methods yield polymersomes with diameters ranging from nanometers up to a few micrometers, giant vesicles of tens of micrometers in diameter can be formed by electroformation or copolymer stabilized double emulsions methods. Electroformation of giant vesicles is triggered by an a.c. field that is applied across an amphiphilic polymer film to enhance water diffusion and film hydration.[26] The double emulsion method consists on the formation of a leaflet of the inner membrane using one copolymer type as a stabilizer for a water-in-oil emulsion. Then, by passing water droplets through a second polymer at the oil–water interface, a complete membrane is formed. Subsequent extraction of the organic phase from the double emulsion results in the formation of polymersomes. Multilamellar polymersomes, with membranes consisting of the same or different types of amphiphilic block-copolymers, have been obtained by the 'water-in-oil-in-water' (W/O/W) double emulsion method, and it has recently been shown that even monodisperse polymersomes can be formed from double emulsions using microfluidic fabrication devices (Figure 6.2).[27,28]

The use of polymersomes for biological applications requires specific characteristics, such as shape, size, and size distribution, together with appropriate membrane properties (*e.g.* flexibility and fluidity). In order to analyze the size, shape, and size distribution of polymersomes, a combination of microscopy techniques and light scattering is used. Optical microscopy is used to visualize polymersomes with diameters >200 nm, whilst transmission electron microscopy (TEM) is used to produce high-resolution images of smaller polymersomes.[25] However, because of the low electron density in the scaffold of the vesicle sample, staining is required for contrast enhancement of TEM micrographs.[29] Light scattering methods rely on the dependence of the intensity of the scattered light on the aggregate size and morphology.[30] Static light scattering (SLS) is used to obtain the radius of gyration (R_g) of polymersomes, whilst dynamic light scatting (DLS) allows the hydrodynamic radius (R_h) to be measured. If $\rho = R_g/R_h$ is equal or very close to one, the self-assembled structures are hollow spherical objects (polymersomes) but, if other structures are present, the values of ρ are lower.[31] Another technique for studying the size, shape, and size distribution of polymersomes is atomic force microscopy (AFM), which measures attractive and repulsive forces between a sensor (the so-called cantilever) and a sample. By scanning across a surface on which polymersomes are present, a force-map is constructed, which provides evidence of the surface topology of the polymersomes. Furthermore, AFM yields information about the intrinsic elastic properties of polymersomes, *e.g.*, membrane flexibility.[32,33] Lateral diffusion inside polymersome membranes of polymer chains, which is associated with membrane fluidity, can be studied by fluorescence correlation spectroscopy (FCS) using fluorescent molecules as probes for membrane dynamics[10] or by electron paramagnetic resonance (EPR) with spin labels as probes for the local dynamics inside membranes.[34]

6.3.1 Biomolecule Surface-functionalized Vesicles

Surface functionalization of polymersomes with biological ligands, such as sugars, antibodies, enzymes, or proteins, greatly increases their recognition by targeted interaction sites. Thus, biomolecule surface-functionalized polymersomes have attracted much attention for the development of targeted drug-delivery systems.[35] There are various approaches for attaching or exposing biomolecules on the hydrophilic outer layer of amphiphilic

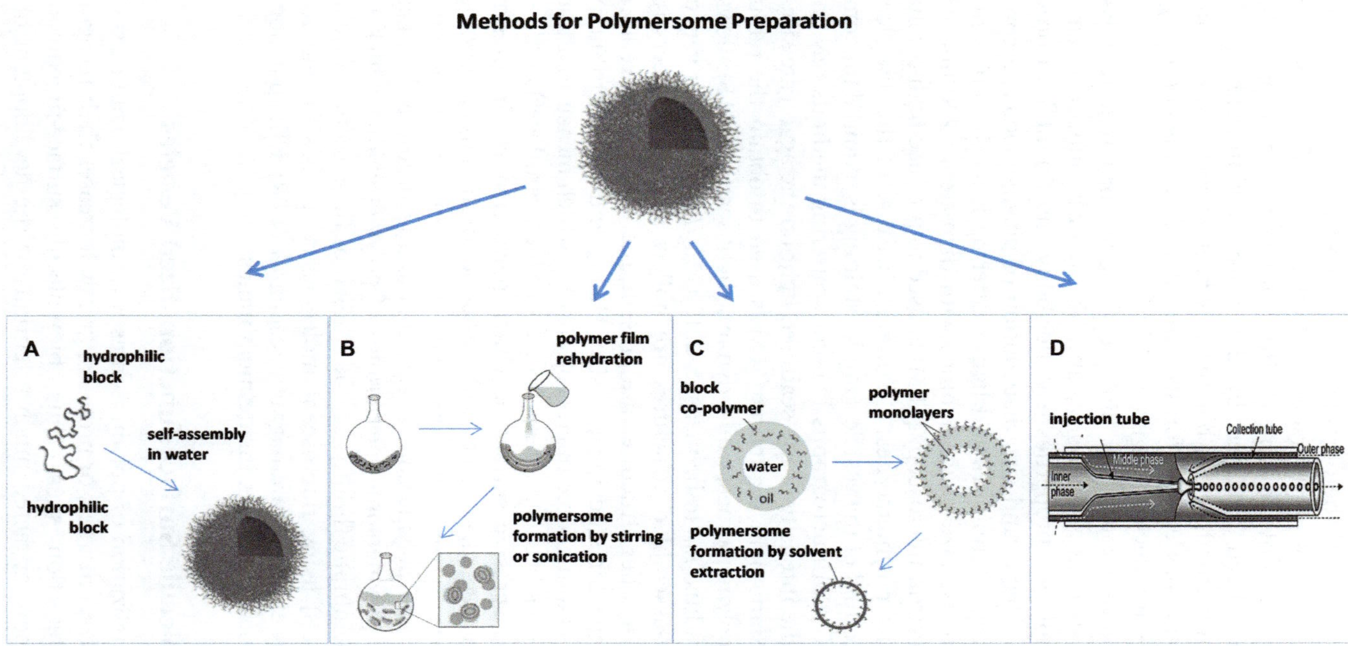

Figure 6.2 Commonly used methods for preparing polymersomes *via* (A) the direct dissolution method, (B) the solvent free method, (C) the solvent displacement method, and (D) the double emulsion method for giant unilamellar vesicles. Reproduced with permission from H. C. Shum, J.-W. Kim and D. A. Weitz, *J. Am. Chem. Soc.*, 2008, **130**, 9543–9549. Copyright (2008) American Chemical Society.[28]

polymer membranes based on modifications of the membrane either prior or post vesicle formation.[9]

One approach prior to vesicle formation is to use block copolymers, which have as end-groups either biological molecules or specific ligands that favor the binding of biomolecules, and then to allow them to self-assemble to form vesicles (Figure 6.3A). The advantage of this approach is that pre-functionalized polymers can be analyzed and characterized in their non-assembled state. The surface-density of the functional molecules can be controlled and adjusted to the desired proportion by mixing different amounts of functionalized and non-functionalized copolymers. However, it also has to be recognized that the biological entity may alter the self-assembly process of the amphiphilic block-copolymer. Various examples of bio-inspired polymersomes have been reported, as for example tetra(*p*-phenylene)-*block*-PEG_{12}-α-D-mannopyrannose polymersomes, which showed enhanced binding affinity to the carbohydrate recognizing plant lectin Concanavalin A, and specific binding to the pili of the ORN 178 *E. coli* strain.[36,37]

Polymersomes of poly(butadiene)-*block*-poly(ethylene oxide) (PB-*b*-PEO) with metal-trisNTA groups exposed at their surface have been successfully used for specific binding of various His-tagged proteins, and the results indicated that the binding affinity could be modulated by changing the metal ion.[38,39]

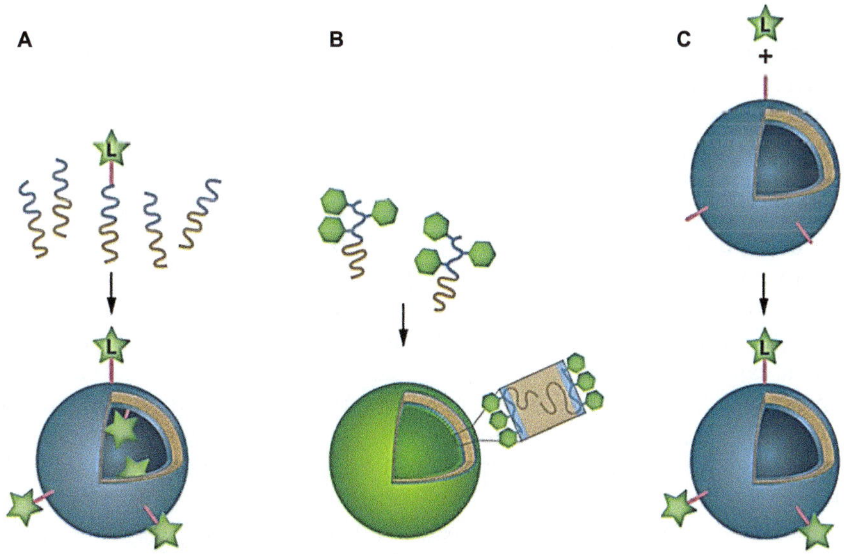

Figure 6.3 Routes for surface modification of polymersomes with biofunctionalities by (A) end-group functionalization of block copolymers before self-assembly, (B) biomolecule conjugated hybrid block copolymers used for self-assembly, and (C) surface conjugation of biomolecules to pre-formed vesicles.
(Figure reproduced with permission from ref. 9.)

Another strategy for functionalizing polymeric membranes is to use a copolymer where one of the hydrophilic blocks is intrinsically biofunctional (Figure 6.3B). Such hybrid block copolymers can be synthesized, *e.g.*, by conjugating linear carbohydrates to a synthetic homopolymer. For example, heparin decorated vesicles, formed by connecting heparin as the hydrophilic block to a hydrophobic block of PDMS effectively inhibited the malaria transferring parasite *P. falciparum* from invading host red blood cells.[40] Peptides are also frequently used as hydrophilic blocks to modify polymers, and such peptide-based vesicles represent a new class of chimaeric polymersomes, named pepsomes. Depending on the polypeptide used, pepsomes can be highly responsive to external stimuli. For example, asymmetric poly(ethylene glycol)-*block*-poly(L-leucine)-*block*-poly(L-glutamic acid) (PEG-*b*-PLeu-*b*-PGA) triblock copolymers are stimulus-responsive to pH changes: upon protonation in an acidic environment, the PGA block undergoes a phase transition from hydrophilic randomly coiled to a hydrophobic α-helix structure. The pH response of PEG-*b*-PLeu-*b*-PGA based pepsomes has been used to release doxorubicin hydrochloride (DOX · HCl) inside low-pH tumor cells, resulting in high antitumor activity in RAW 264.7 and MCF-7/ADR cell lines.[41] Thermoresponsive polymersomes have been formed using poly(*N*-isopropylacrylamide) (PNIPAM) as one of the blocks, which exhibit their lowest critical solution temperature (LCST) at 37 °C and undergo a phase transition from hydrophilic to hydrophobic. PNIPAM conjugation with the hydrophilic protein amFP497 results in an amphiphilic polymer–protein bioconjugate that spontaneously self-assembles to pepsomes above 37 °C. During pepsome formation, both the PE545 protein and DOX · HCl are effectively encapsulated, and then selectively released when the system is cooled below the LCST of PNIPAM.[42] DNA–polymer hybrids have also been generated to self-assemble into vesicles. Through the co-assembly of pore-forming proteins with an amphiphilic sequence of a short nucleotide connected to a hydrophobic polymer, an encapsulated enzyme was activated within a polymersome cavity as the reconstituted protein allowed the necessary substrate exchange across the protein pore.[43] In addition, a specific nucleotide sequence on a polymersome surface has been used to bind to immunoglobulin E (IgE) protein, an allergy biomarker, resulting in an excellent candidate for target-therapy.[44]

A second strategy for obtaining polymersomes with biofunctional surfaces is to chemically conjugate biological entities to pre-formed polymersomes (Figure 6.3C). This method is advantageous because the biological molecules do not interfere with the self-assembly process of the amphiphilic polymers. Major drawbacks of the method are the need for the chemical conjugation method chosen for ligand-surface binding to be viable in water, and the problem that the functional conjugation groups may be hidden in the assembled structures, and therefore inaccessible for the attachment of biomolecules. A conjugation method frequently applied for biomolecule coupling is the biotin–streptavidin conjugation approach, in which various different biotin-bearing ligands can be bound to the polymersome membrane.[45–48] By exploiting biotin–streptavidin binding, vesicles of biotin end-group-functionalized PMOXA-*b*-PDMS-*b*-PMOXA have been bound to the scavenger receptor A1 on the surface of macrophages. In the transgenic scavenger receptor A1-expressing COS-7 cells, the vesicles were taken up after cell surface binding.[47] A more efficient way to covalently conjugate targeting ligands to the surfaces of polymersomes is the Huisgen 1,3-dipolar cycloaddition of azides and alkynes, which results in stable and irreversible thiazol ring binding. The click reaction was applied to conjugate alkyne-bearing biotin and the green fluorescent protein (GFP) to the surface of polymersomes. These polymersomes pre-formed from poly(styrene)-*block*-poly(acrylic acid) block copolymer (PS-*b*-PAA) exposed azide functionalities on their external surface.[49] Azide-alkyne click chemistry was used to conjugate peptides that mimic fibronectin to polymersomes of PEO-*b*-PB because fibronectin binds selectively to colon cancer cells and therefore, the hybrid polymersomes could have potential as a targeted drug delivery system.[50] Other conjugation methods have been adapted for polymersome surface modification, *e.g.* maleimide-thiol click reactions,[51] or the reaction of *N*-hydroxy sucinimidyl ester with amine groups.[52]

6.3.2 Reconstitution of Membrane Proteins into Polymer Membranes

Another bio-inspired approach is to design polymersomes as enclosed reaction compartments for the development of nanoreactors, nanodevices, or artificial organelles, in which active compounds are not only protected from the environment, but also allowed to act *in situ*.[53] For such function, membrane permeability is of crucial importance, since it allows the exchange of substrates/products with the environment of the polymersomes. Various methods have been reported to generate polymersomes with permeable membranes: (i) polymers forming intrinsically porous membranes,[54] (ii) polymer membranes that are permeable to ions as *e.g.* specific oxygen species,[55] (iii) pore formation in pH responsive polymer membranes by chemical treatment,[56] (iv) polymer membrane permeabilization by UV-irradiation, and (v) biopores or membrane proteins inserted into polymer membranes.[57–59]

All of these approaches lead to permeable nano-compartments; however, biometric membranes are only those resulting from insertion of trans-membrane proteins or biopores.[60–64] The small pore-forming polypeptide gramicidin (gA), with a length of 2.6 nm, has been successfully inserted into PMOXA-*b*-PDMS-*b*-PMOXA membranes with thicknesses ranging from 9.2 nm up to 13.2 nm by addition of gA to pre-formed polymersomes (Figure 6.4).[65] The significant hydrophobic mismatch between the gA size and the membrane thickness was overcome by the flexibility of the membrane, which supported successful insertion of the biopore in a similar manner to that reported for other membrane proteins.[10] The spontaneous gA insertion enabled the translocation of H^+, Na^+, and K^+ through the

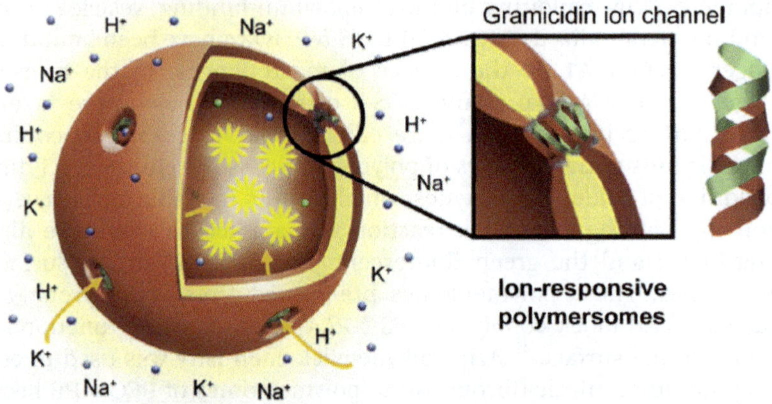

Figure 6.4 Bio-inspired polymeric vesicle with reconstituted gA biopores. In contrast to lipid based vesicles, gA exhibits increased hydrophobic mismatch with the polymeric membrane (9–13 nm thickness) whilst preserving its ion exchange function.
(Reproduced from M. Lomora *et al.*, Polymersomes with engineered ion selective permeability as stimuli-responsive nanocompartments with preserved architecture, *Biomaterials*, **53**, 406–414, Copyright 2015 with permission from Elsevier.[65,67])

membrane, whilst preserving the structural integrity of the polymersomes.[65] In addition, the flexibility of PMOXA-*b*-PDMS-*b*-PMOXA membranes allowed the successful insertion of ionomycin, an ionophore of only 1.5 nm size, which mediated Ca^{2+} transport through the membrane even though its thickness was >7 times greater than the length of the ionophore.[66,67]

Reconstitution of membrane proteins inside polymer membranes increases the complexity of the task, because many membrane proteins are very sensitive to environmental changes and denature rapidly with the subsequent loss of biological activity. Thus, the reconstitution process has to be customized for each individual membrane protein. Usually, prior to reconstitution into block copolymer membranes, membrane proteins are inserted into lipid bilayers, where the hydrophobic mismatch is very low (lipid bilayers have a thickness ranging from 3–5 nm). This is mediated through the addition of detergents, which stabilize water-insoluble proteins in aqueous solution and mediate the protein insertion.[68,69] After reconstitution, the detergents are removed *via* dialysis or using detergent adsorbing resins, so-called biobeads.

To adapt the detergent-assisted reconstitution method from liposomes to polymersomes, it is necessary to create conditions for insertion when there is a high hydrophobic mismatch between the size of the membrane proteins and the thickness of the membrane.[70] At low detergent concentrations, polymersomes and detergent micelles co-exist, whilst at higher detergent concentrations, the polymer membrane is disrupted to release the surface tension.[71] Thus, conditions for the insertion of membrane proteins into polymer membranes should slightly destabilize the polymersome membrane and favor the insertion, while avoiding the complete disruption of the membrane.[63,64,72] Copolymers that are sufficiently flexible can overcome the hydrophobic imbalance with respect to proteins by bending the hydrophobic region around the membrane protein down to the size of the protein. This has been demonstrated on PDMS-containing amphiphilic block copolymer membranes with membrane thicknesses between 9 nm to 13 nm, in which different membrane proteins with hydrophobic mismatches between 3.1 nm up to 7.1 nm were successfully reconstituted.[10] In addition, flexible synthetic membranes are able to replace the necessary lipid environment when delipidated membrane proteins are inserted. Membrane thickness and copolymer flexibility are the key limiting factors for protein reconstitution, as demonstrated for model gramicidin inserted into polymer membranes.[65] The highly sensitive membrane protein, Complex I, was delipidated and then inserted into PMOXA-*b*-PDMS-*b*-PMOXA polymersomes by membrane destabilization mediated by detergent. Upon insertion, Complex I acted as an electron transfer mediator from the surrounding environment into the polymersome membrane.[73] Depending on the hydrophilic-to-hydrophobic ratio, the delipidated Complex I could be activated upon insertion in the synthetic membrane, the key molecular parameter being the size of the hydrophobic domain.

Another approach towards the formation of biomimetic membranes is based on spontaneous protein insertion into synthetic membranes during the polymersome self-assembly process. The first membrane protein ever reconstituted in a polymer membrane was the highly stable porin outer membrane protein F (OmpF), which was integrated into PMOXA-*b*-PDMS-*b*-PMOXA based polymersomes and resulted in membrane permeability for molecules up to 600 Da.[64] In a similar manner, the light-driven transmembrane proton pump α-helical model protein bacteriorhodopsin (BR), together with the F_0F_1-ATP synthase membrane protein, have been successfully integrated into poly(2-ethyl-2-oxazoline)-*block*-poly(dimethylsiloxane)-*block*-poly(2-ethyl-2-oxazoline) (PEtOz-*b*-PDMS-*b*-PEtOz) polymersome membranes. In this artificial proteo-organelle, BR created a light-driven proton gradient across the polymer membrane. The decrease in pH in the polymersome cavity subsequently triggered F_0F_1-ATP synthase to biosynthesize ATP.[74,75] In order to create biomimetic membranes, the orientation of the membrane proteins is essential when they are not passive transport channels. Two approaches that favor the desired orientation inside membranes are: (i) the use of asymmetric membranes, and (ii) insertion in pre-formed polymersomes with membranes slightly destabilized by addition of detergents (see above). For example, Aquaporin 0 insertion has been directed by using an asymmetric poly(ethylene oxide)-*block*-poly(dimethylsiloxane)-*block*-poly(2-methyl-2-oxazoline) (PEO-*b*-PDMS-*b*-PMOXA) block copolymer scaffold,[76] whilst Complex I has been inserted with the desired orientation in preformed polymersomes.[73] These examples show that protein orientation, along with functional reconstitution, can be promoted by combining polymer chemistries to tailor block copolymers with engineered bio-inspired systems. However, the study of polymer membrane properties and their effect on protein reconstitution is still at an early stage of research, and further investigations are essential.

6.3.3 Bio-inspired Block Copolymer/Lipid Hybrid Vesicles

Besides purely block copolymer based bio-inspired vesicles, there are recent examples of hybrid systems generated from mixed block copolymers and lipid membranes. Such systems combine the advantages of both synthetic polymers (chemical diversity, high membrane stability, and low permeability) and natural or synthetically modified phospholipids (biocompatibility and biofunctionality).[77] For example, the HER2/neu receptor antibody adjacent to folic acid functionalized polymer/lipid hybrid vesicles based on PEO-*b*-PB and hydrogenated soy phosphatidylcholine (HSPC) have a significantly higher tumor targeting efficiency than either the purely polymer based HER2/neu receptor antibody or the folic acid functionalized vesicles (Figure 6.5).[78]

However, polymers and lipids are often immiscible because of differences in entropy and enthalpy and therefore, the membranes formed may contain discrete lipid domains. The effect of membrane structure on molecular recognition has been studied for mixed and phase-separated hybrid vesicles formed from polyisobutylene-block-polyethyleneoxide copolymer (PIB_{87}-b-PEO_{17}) and the ganglioside M1 receptor bearing 1,2-dipalmitoyl-sn-glycero-3-phosphatidylcholine (DPPC). Ganglioside M1 was recognized by the bacterial protein cholera toxin B regardless of the membrane organization.[79] Overall, ongoing studies show that it is still difficult to control

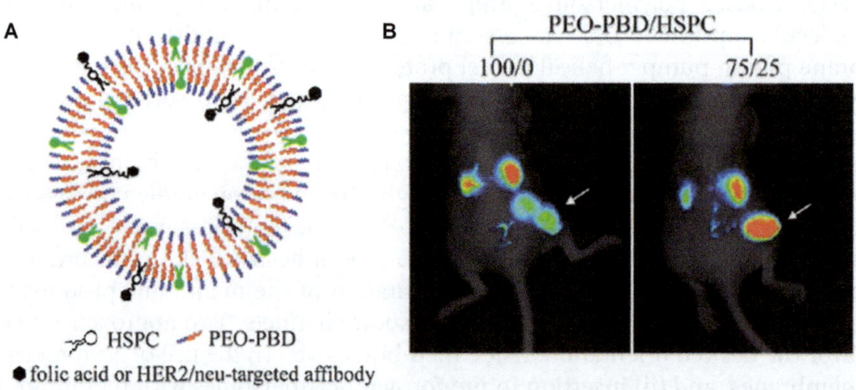

Figure 6.5 (A) Schematic representation of a polymer–lipid hybrid vesicle, surface functionalized with HER2/neu receptor. (B) Polymer–lipid vesicles from PEO-PBD/HSPC blends show significantly enhanced tumor targeting activity compared to the block copolymer alone.
(Figure reproduced with permission from Z. Cheng, D. R. Elias, N. P. Kamat, E. D. Johnston, A. Poloukhtine, V. Popik, D. A. Hammer and A. Tsourkas, *Bioconjugate Chem.*, 2011, **22**, 2021–2029. Copyright (2011) American Chemical Society.[78])

precisely the molecular composition and nanoscopic membrane structure of block copolymer/lipid hybrid vesicles.[77] A major challenge for future studies will be to precisely modulate the molecular texture on the membrane, since this may lead to new insights into how membrane nano-composition and hence membrane nano-mechanics translate into biological functionality.

6.3.4 Protein-polymer Nanoreactors

Inspired by nature, where chemical reactions take place inside biocompartments, synthetic nanoreactors have been developed to facilitate chemical reactions in confined spaces that protect the active compounds involved in the reactions from environmental conditions, as for example proteolytic attack. Compartmentalization is generally achieved by encapsulating and entrapping biomolecules in the polymersome structure (inner cavity, membrane) during formation.[57,64] Various nanoreactores have been designed either as models for confined reactions or for translational applications.[35] For example, laccase encapsulated in poly(*N*-vinylpyrrolidone)-*block*-poly(dimethylsiloxane)-*block*-poly(*N*-vinylpyrrolidone) (PNVP-*b*-PDMS-*b*-PNVP) polymersomes produced reactive oxygen species, which diffused through the semipermeable membrane,[80] whilst hemoglobin encapsulated in PMOXA-*b*-PDMS-*b*-PMOXA compartments generated nanoreactors with dual functionality: detoxification of peroxynitrites and storage of oxygen.[81]

A further step was very recently realized through the design of triggered nanoreactors, which were able to start *in situ* enzymatic reactions in the presence of stimuli in their environment.[82] Such nanoreactors were based on the simultaneous encapsulation of an enzyme and insertion in the polymersome membrane of a chemically engineered channel porin, which had a pH-sensitive cap. Upon a pH change in the environment of the nanoreactor, the molecular cap was released, and the membrane allowed transport of substrates/products, which then started the *in situ* enzymatic reaction (Figure 6.6).

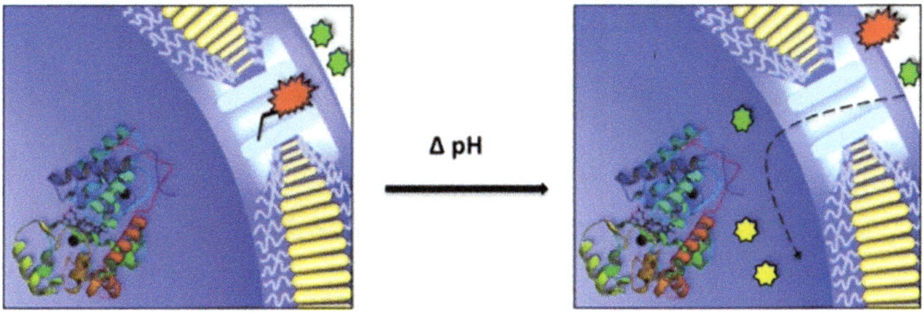

Figure 6.6 Nanoreactor with pH dependent activity: the down regulation in pH triggers the cap cleavage from the transmembrane protein OmpF, which opens up the channel in the protein and enables a substrate/product exchange 'on demand'.
(Figure reproduced with permission from T. Einfalt, R. Goers, I. A. Dinu, A. Najer, M. Spulber, O. Onaca-Fischer and C. G. Palivan, *Nano Lett.*, 2015, **15**, 7596–7603. Copyright (2015) American Chemical Society.[82])

Using the same concept of compartmentalized chemical conversion, polymersomes have been used as compartments for *in situ* cascade reactions by encapsulating one enzyme in a polymersome, which provided the substrate for another enzyme outside (or *vice versa*), co-encapsulating two enzymes, or encapsulating them in separate polymersomes.[54,56,83]

Similar to co-encapsulation of two different enzymes, a three step cascade reaction has been realized in a single polymersome by loading glucose oxidase (GOx) into the polymersome cavity, *Candida antarctica* lipase B (CalB) into the hydrophobic part of the amphiphilic membrane, and horseradish peroxidase (HRP) on the outer surface.[84] In this particular arrangement, monoacetylated glucose added to the environment of the nanoreactor was deacetylated by CalB, and further metabolized to gluconolactone and H_2O_2 by GOx. The hydrogen peroxide molecules migrated out of the polymersome and subsequently activated HRP to convert ABTS into $ABTS^+$ (Figure 6.7):

An elegant approach was to create multicompartments required for more complex systems.[85,86] By encapsulating polystyrene-*block*-poly(3-(isocyano-l-alanyl-aminoethyl)thiophene) (PS-*b*-PIAT) nanoreactors in PB-*b*-PEO polymersomes, a fully active multicompartment system was introduced.[17,87] However, these examples are still model systems and are much simpler than cells or other biocompartments.

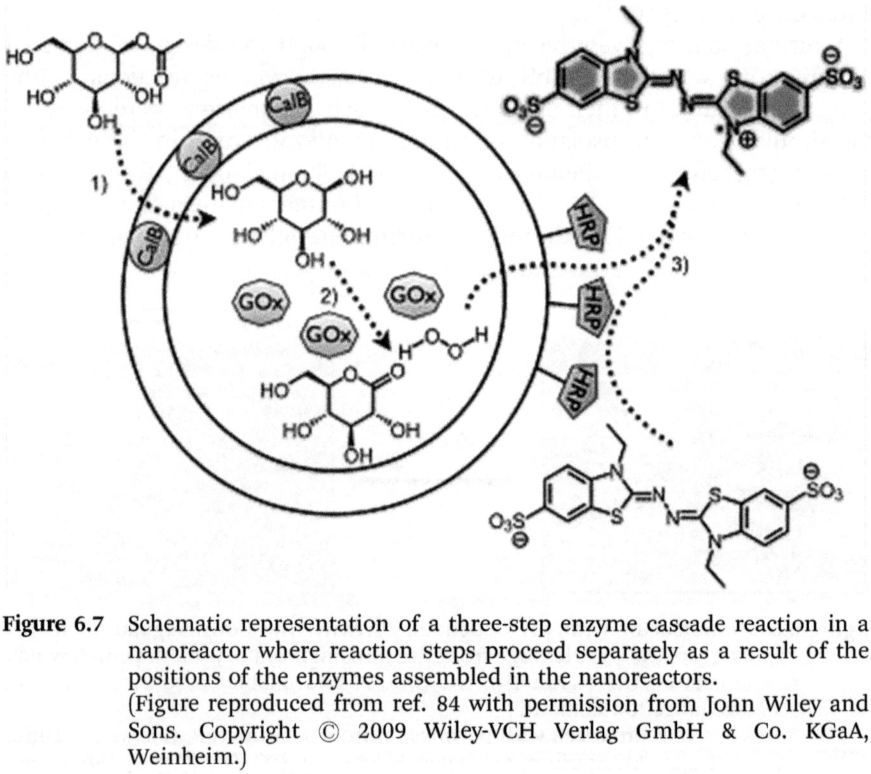

Figure 6.7 Schematic representation of a three-step enzyme cascade reaction in a nanoreactor where reaction steps proceed separately as a result of the positions of the enzymes assembled in the nanoreactors.
(Figure reproduced from ref. 84 with permission from John Wiley and Sons. Copyright © 2009 Wiley-VCH Verlag GmbH & Co. KGaA, Weinheim.)

6.4 Bio-inspired Planar Polymer Membranes (2D Membranes)

Various planar membrane models have been developed, either for fundamental studies or for translational applications: monolayers at the air–water interface, freestanding films in solution, solid supported membranes, and membranes on a porous solid support. Planar biomimetic membranes based on amphiphilic block copolymers are important artificial systems often used to mimic natural membranes. Their advantages, compared to artificial lipid membranes, are their improved stability and the possibility of chemically tailoring their structures.[7] The simplest model of such a planar membrane is a monolayer at the air–water interface, formed when amphiphilic molecules are spread on water. As cell membrane models, it is more common to use free-standing membranes in which both sides of the membrane are accessible to water or buffer, and thus a bilayer is formed.[88] The disadvantage of these two membrane models is the lack of stability, which can be overcome by the development of a solid supported membrane model. Characterization of such planar membranes can be challenging and several techniques, such as AFM, quartz crystal microbalance (QCM), infrared (IR) spectroscopy, confocal laser scan microscopy (CLSM), electrophoretic mobility, surface plasmon resonance (SPR), contact angle, ellipsometry, electrochemical impedance spectroscopy (EIS), patch clamp, or X-ray electron spectroscopy (XPS) have been used to characterize their properties.[89]

The imaging of polymer films at the air–water interface is possible by Brewster-angle microscopy (BAM), but only with a limited resolution. Imaging membranes on solid supports is achieved through AFM, which gives information about the surface topography and the mechanical properties of the membrane. Generally, as a result of the high resolution of AFM, imaging of surfaces at the nanoscale is possible. Several AFMs also possess additional functionalities such as enabling simultaneous imaging and force analysis, high-speed, or conductivity measurements. If the polymer is labeled with a fluorescent dye, CLSM can be used to visualize the polymer film on a transparent surface. However, when characterization of biomolecules combined with synthetic membranes is required, these imaging techniques need to be combined with other techniques, such as QCM, electric conductance, FT-IR, or Raman spectroscopy.

EIS or a patch clamp can be used to characterize the reconstituted proteins, since the conductivity of the polymer membrane changes after incorporation of the proteins into the membranes. Consequently, the patch clamp technique is even suitable for investigating monomolecular resolution on transmembrane channels. EIS measurements can be used to characterize the electrochemical properties of polymer membranes, with the advantage of being able to provide both qualitative and quantitative information.

SPR and ellipsometry are used to characterize the thickness and optical properties of membranes, whilst QCM provides information on the attachment of (bio)molecules to the membrane and their distribution on the surface. SPR is based on changes in the refractive index and is widely used as a label-free detection technique for studying molecular interactions, such as ligand binding to membrane proteins.[90] QCM measures the change in frequency of a quartz crystal resonator and can be used to produce both qualitative and quantitative information on the protein in the membrane.[91] Various IR techniques (FT-IT, ATR-IR, and PM-IRAS) provide information about sample composition, and whether or not biomolecules are attached to the surface.

6.4.1 Monolayer at the Water–Air Interface and Free-standing Membranes

When spread over water or buffer solution, lipids and block copolymers can self-assemble as monolayer films at the water–air interface (Figure 6.8A), whilst bilayers are often formed when they are spread over small apertures in aqueous solution (Figure 6.8B). Freestanding monolayer films based on PMOXA-*b*-PDMS-*b*-PMOXA triblock copolymers have been reported to allow incorporation of OmpF.[92]

In another example of freestanding block copolymer membranes, α-haemolysin (αHL), OmpG, and alamethicin have been successfully incorporated into PMOXA-*b*-PDMS-*b*-PMOXA membranes.[93]

The disadvantage of free standing membranes is their lack of stability, even when crosslinked using layer-by-layer techniques (LbL) or hydrogels as substrates.[94–96] Therefore, solid supports are more suitable for stabilizing membranes, and allow a better characterization of mono- and bilayer membranes.

6.4.2 Solid Supported Membranes

Lipid-bilayer membranes on solid substrates are often used as cell-surface models connecting biological and artificial materials. They can be placed either directly on solids or on ultrathin polymer supports, such as brushes or hydrogels,[97,98] which mimic the extracellular matrix. A similar approach has been applied to polymer membranes with the advantage of tunable thickness, easier chemical modifications to allow stimulus responsiveness, or the attachment of active molecules by incorporation of reactive end groups. In addition, incorporated proteins have lower interactions with the support because of the increased membrane thickness, and therefore behave as in a natural environment.[99]

There are various methods for producing polymer membranes on surfaces, divided into "grafting-from" and "grafting-to" methods. In the "grafting-from" method (Figure 6.9), initiator molecules must first be

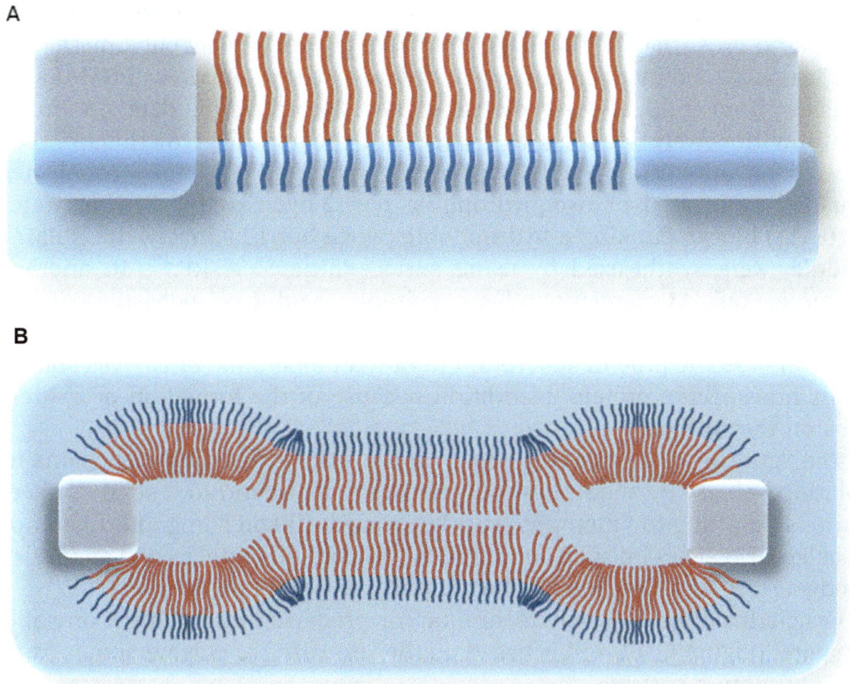

Figure 6.8 Schematic representation of non-supported membranes. (A) A monolayer at the air–water interface, and (B) a freestanding planar film.

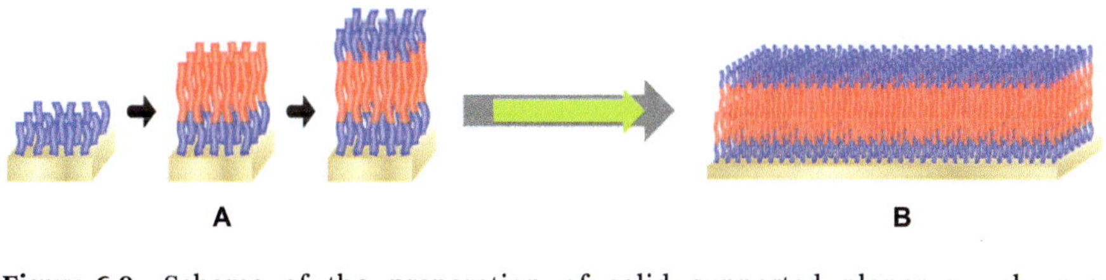

Figure 6.9 Scheme of the preparation of solid supported planar membranes. (A) Surface initiated polymerization (SIP), and (B) solid supported polymer brush.
(Reproduced from ref. 7 with permission from the Royal Society of Chemistry.)

immobilized on the surface. Then, the monomers are polymerized directly on the surface. This method usually provides polymer brushes, which are covalently attached to the surface. Polymer brushes are defined as polymer chains that are surface-confined by being tethered by one end to a surface or an interface. A high density of polymer chains is forced, because the chains are able to stretch away from the surface.[100]

Various brush systems have been used for cell adhesion and biosensing studies, but only a few have been used for insertion of biomolecules (enzymes or membrane proteins). One of the first examples of such brushes was based on surface-initiated ATRP for the growth of poly(butyl methacrylate)-*co*-poly(2,2-dimethylaminoethyl methacrylate) (PBMA-*co*-PDMAEMA) brushes from gold and silicon substrates. Experimental data on brush conformation were identical to those from theoretical predictions.[101] For biosensing applications, gold-supported amphiphilic triblock copolymer brushes, composed of two hydrophilic poly(2-hydroxyethyl methacrylate) (PHEMA) blocks flanking a hydrophobic poly(*n*-butyl methacrylate) (PBMA) middle, were synthesized by using surface-initiated ATRP.[102] In another example, poly(11-mercaptoundecyl sulfonic acid)-*block*-poly(sulfobetaine methacrylate) (PSA-*b*-PSBMA) brushes were synthesized as an agent to prevent non-specific protein adsorption in human blood: the brush was able to resist non-specific protein adsorption because of the formation of zwitterions on the surface.[103]

The "grafting-to" method is based on the synthesis of polymer chains in solution, which are then physically or covalently bound to a solid surface. Various "grafting-to" methods have been used: (i) spreading and fusion of micelles or vesicles on solid supports, (ii) transfer of monolayers (Langmuir–Blodgett (LB) and Langmuir–Schaefer (LS)), and (iii) LbL methods.[104–107]

Vesicle fusion is a simple technique for preparing planar films on solid supports (Figure 6.10), and is based on the spreading of polymersomes from solution onto a surface.

The formation of membranes by vesicle fusion is dependent on several parameters, such as the composition and size polydispersity of the vesicles, critical osmotic pressure, surface charge, and roughness of the solid substrate. Additionally, environmental conditions, such as ionic strength and solution pH play a crucial role.[7] A big disadvantage of this method is the

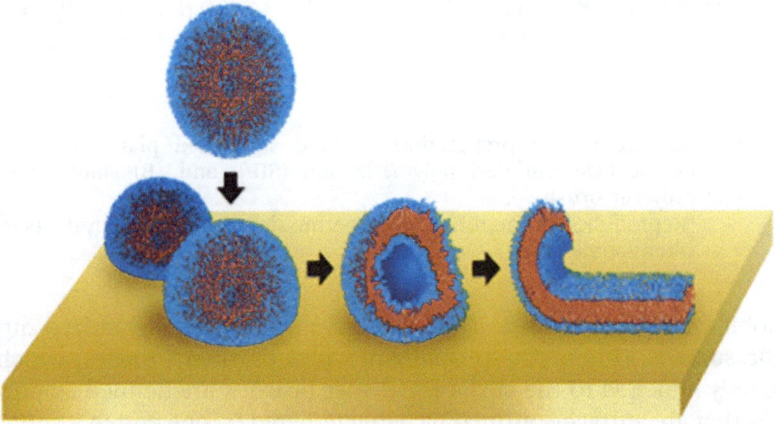

Figure 6.10 Scheme for the preparation of planar membranes by vesicle fusion. (Reproduced from ref. 7 with permission from the Royal Society of Chemistry.)

inhomogeneity of the planar membrane which, together with the low reproducibility of the obtained membranes, explains why there are only a few reports based solely on membranes prepared by vesicle fusion on a solid surface.[105,106]

Whereas the vesicle fusion method produces polymer membranes with high numbers of defects, the monolayer transfer methods are more easily controlled and allow the homogeneous deposition of polymer films. With these techniques, it is also possible to develop multilayer structures even with mixed layer composition.[7]

LB films (Figure 6.11A) are produced in two steps: (i) by spreading a thin layer of an amphiphilic copolymer on the surface of a liquid (water or buffer), followed by (ii) adsorption of the layer on a solid support, which is dipped or immersed in the liquid and then removed. At low concentrations,

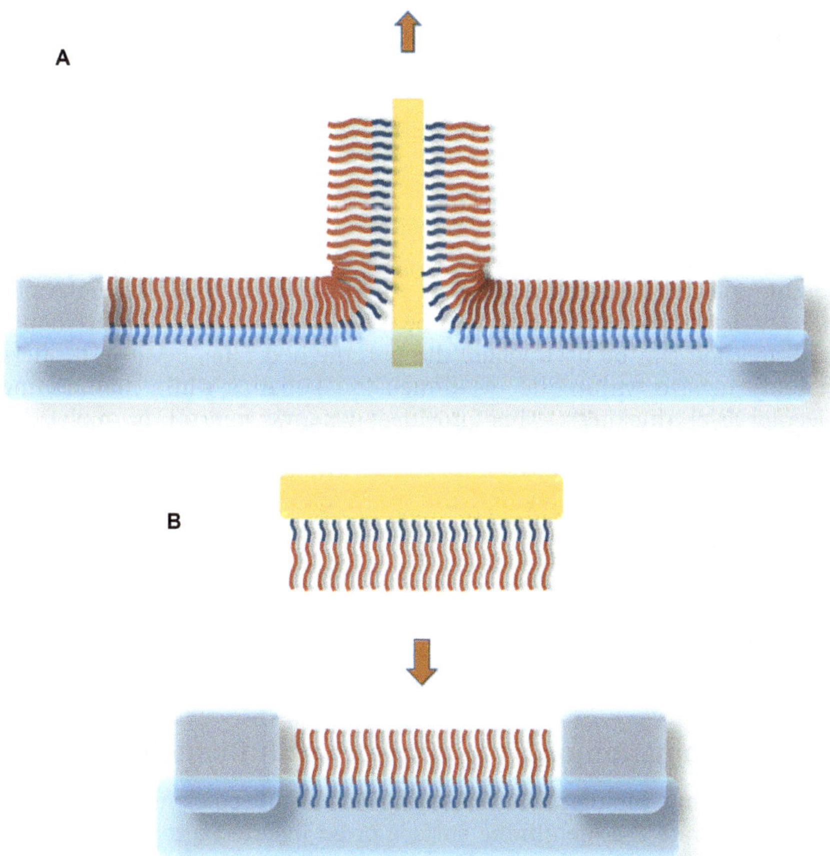

Figure 6.11 Schemes for the preparation of polymer membranes using the (A) LB and (B) LS techniques.

amphiphilic molecules occupy only the surface of the liquid but, with increasing concentrations, the surface becomes completely covered with copolymer molecules. Any additional molecules arrange themselves into micelles, starting at a concentration called critical micelle concentration (cmc). As long as the cmc is not exceeded, the amphiphilic molecules assemble at the air–water interface as a thin monolayer, which can be transferred onto a solid support. Deposition on a solid substrate is performed by dipping or immersing the solid substrate at a constant velocity. The monolayer is adsorbed homogeneously on the support and then taken up from the liquid. Depending on the desired morphology of the polymer membrane, the procedure can be repeated to achieve mono-, bi-, or multilayer polymer membranes. It depends on the film composition and the hydrophilicity of the solid substrate whether immersion or emersion is used. With LB, thin, homogeneous films with defined thicknesses can be obtained. LS films are prepared in a similar manner to LB films, but the dipping or emersion of the solid substrate is horizontal and not vertical (Figure 6.11B).

Whilst the LB method can only be performed with amphiphilic molecules, the LbL method is a cyclical process in which different charged polymers are successively used to generate the membrane: one polymer layer with a specific charge is adsorbed onto a substrate, then, after washing, an oppositely charged polymer is adsorbed on top of the first layer, and so on.[108] The advantage of the LbL method compared with the LB method is that LbL assembly does not require special conditions or equipment: LbL membranes can be easily prepared by using only beakers/tweezers. On the other hand, the LB method can be used where the LbL method fails, because it offers variability in size within the two-dimensional planes while maintaining molecular-level thickness. This provides dynamic control of molecular functions, and can be used for tuning the molecular recognition sites of molecules.[109]

The advantage of the "grafting to" methods is that it is possible to obtain better control over the ordering and packing of the macromolecules and, therefore, over the architecture of the polymer layers. In contrast, the advantage of the "grafting-from" method is that it provides good control over the brush thickness and the grafting density is high, whereas with the "grafting-to" method it can be difficult to achieve high grafting densities.

6.4.3 Combination of 2D Membranes with Biomolecules

In order to mimic biological cell membranes, biomolecules have been used either to decorate the membrane surface or be inserted directly inside the membrane. Various enzymes, such as laccase, GOx, or HRP, have been reported for decorating planar membranes.[91,110,111] These enzymes were either chemically linked to the membrane by reactions such as esterification and amidation, or just physically adsorbed. For example, physically adsorbed laccase on copolymer membranes displayed a higher activity than free laccase under similar concentration conditions because of the protection offered by the polymer hydrophilic domain.[91]

Transmembrane proteins have been incorporated into artificial polymer membranes for fundamental studies of their structures within synthetic membranes and for the selective transport of molecules. Such membrane proteins can be inserted in synthetic solid-supported membranes by: (i) destabilization of the membrane with an electric current, which favors protein insertion, and (ii) addition of a detergent followed by its slow removal with biobeads, which drives protein insertion. Destabilization of the membrane with an electric current has been used for water soluble biomolecules, as for example the bacterial membrane polypeptide αHL,[112] whilst methods using biobeads have been used for insoluble membrane proteins, such as the potassium channel of the bacterium *Mesorhizobium loti*, MloK1.[104] Incorporation of αHL in an amphiphilic PB-*b*-PEO solid-supported membrane obtained by the LB-LS method was directly monitored by conductance measurements. MloK1 was solubilized by addition of a detergent and then reconstituted in amphiphilic PDMS-*b*-PMOXA solid-supported membranes by removing the detergent with Bio-Beads (Figure 6.12).[104]

Both reconstitution methods pose challenges, because it is possible that the biomolecules are not only reconstituted in the membrane in a transverse active form, but with another orientation, which results in non-functional membranes. In addition, for most applications, incorporation into solid supported membranes has the drawback that molecules passing through the protein remain in the small space between the membrane and the solid support,[112] although the use of porous supports based on organic (porous polycarbonate films) or inorganic (porous alumina) materials may solve this problem. The resulting pore-solid-supported membranes (Figure 6.13) combine the mechanical stability of solid-supported membranes with the advantage of free-standing membranes over the pores. Various studies, such as conformational changes in membrane proteins drawn by gradients, cargo transport, and external forces, can only be performed with this type of lipid membranes.[113]

One of the first reported examples of a pore-solid support was the insertion of Aquaporin Z (AqpZ) into PMOXA-*b*-PDMS-*b*-PMOXA membranes to produce a membrane highly permeable to water, but not to other solutes.[114] This work was based on a previous report by Meier and coworkers that showed the insertion of AqpZ into polymersome membranes.[63] In similar studies, AqpZ was incorporated into polymer membranes on various supports, such as porous alumina and cellulose acetate, to produce biomimetic water channels.[115,116]

6.4.4 Hybrid Polymer–Lipid Membranes

Biological membranes not only contain lipids and proteins, but also carbohydrates, cholesterol patches, and other components, which are often

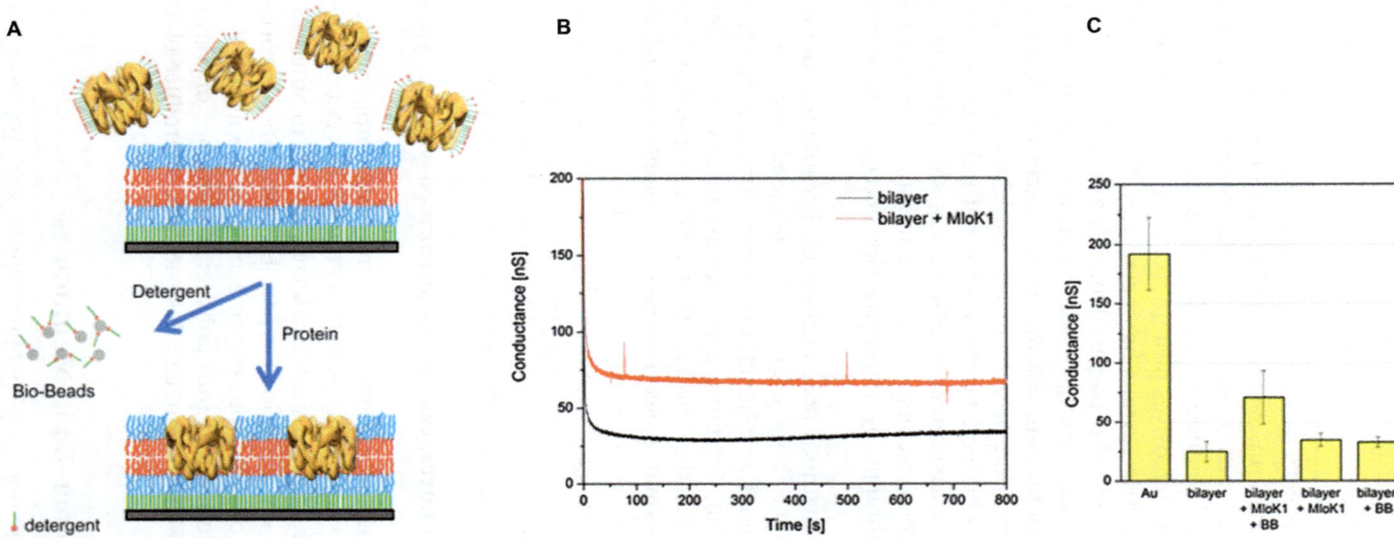

Figure 6.12 (A) Schematic representation of membrane protein insertion into a solid-supported polymer membrane using Bio-Beads. (B) Conductance of solid-supported polymer bilayer without (black line) and with incorporated MloK1 (red line) with time. (C) Conductance measured at a constant applied voltage of 40 mV (Au – gold substrate, BB – Bio-Beads). (Reproduced from J. Kowal *et al.*, Functional surface engineering by nucleotide-modulated potassium channel insertion into polymer membranes attached to solid supports, *Biomaterials*, **35**, 7286–7294. Copyright 2015 with permission from Elsevier.[104])

Figure 6.13 Schematic representation of the incorporation of membrane proteins in a porous-solid supported polymeric membrane.

important for the stability and transport of molecules. One approach for the production of membranes with triggered properties is the development of hybrid systems comprising lipids and amphiphilic block copolymers, where the polymers can increase the stability, and lipid patches can help the reconstitution of membrane proteins.

Hybrid lipid–polymer membranes are interesting to study in combination with the insertion of proteins. As an example, PDMS-*b*-PMOXA copolymers of different lengths were mixed with various lipids to study phase separation and insertion of the membrane protein Mlok1. By modulating the composition of polymer–lipid mixtures, the properties of the films changed the protein distribution between the polymer and lipid phases.[117] Thus, the distribution of proteins can be controlled according to the composition of the hybrid polymer–lipid membrane.

6.5 Immobilized Vesicles

To immobilize polymersomes on solid supports (Figure 6.14A), vesicles have to be sufficiently stable not to rupture when they interact with the surface. Polymersomes and liposomes have been immobilized on surfaces by various approaches: (i) physical interaction mediated by electrostatic forces,[118] (ii) covalent interaction based on chemical linkers,[119] and (iii) covalent interactions based on molecular recognition units, such as complementary DNA strands,[120] peptides,[121] and protein–ligand pairs.[122] An example of immobilized vesicles showed polymersomes based on two different Pluronic triblock copolymers, which were stabilized by a permanent interpenetrating polymer network and immobilized onto a glass or mica surface. For this, they were first stabilized by a permanent interpenetrating network, and then later immobilized on the surface.[123] The successful immobilization of nanoreactors on a glass substrate was achieved using the receptor–ligand pair biotin–streptavidin (Figure 6.14B). Dephosphorylation of the fluorogenic substrate ELF 97 by acid phosphatase was chosen as the model reaction inside the immobilized nanoreactors.[46]

The immobilization of nanoreactors based on protein-polymersomes on solid supports allows the engineering of active surfaces, which have the advantage of the functionality provided by the enzymatic reaction inside the nanoreactors, whilst the support provides a straightforward translational application. Various methods are available for the investigation of both planar membranes and immobilized vesicles, including AFM, QCM, IR spectroscopy, CLSM, and electrophoretic mobility. In addition, small angle X-ray scattering (SAXS) and neutron scattering (NS) can be used for immobilized vesicles. For SAXS, the sample is irradiated with X-rays and the detected elastically scattered radiation contains information about the shape and size of the molecular assemblies. NS has the advantage of deeper penetration because of the electrical neutrality of neutrons.

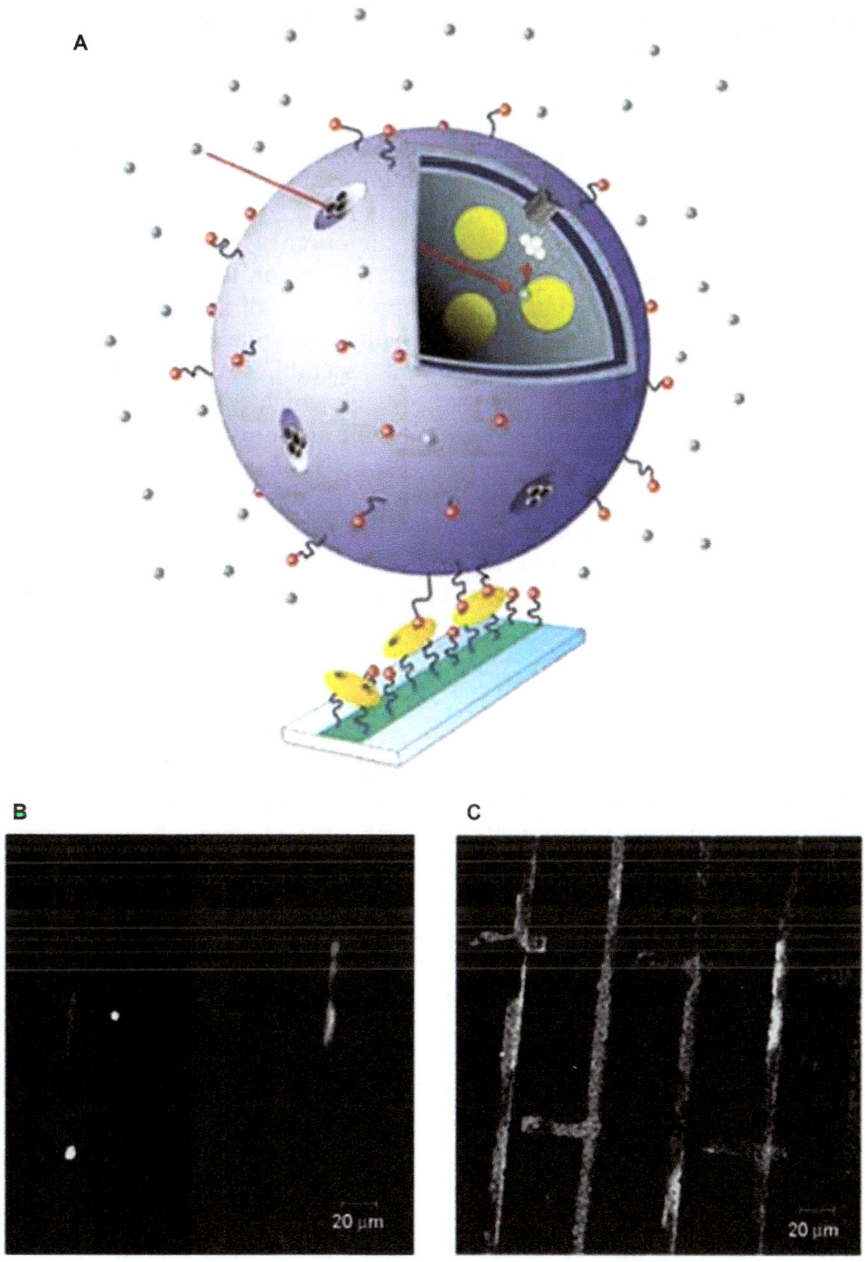

Figure 6.14 (A) Scheme of an immobilized vesicle on a surface. LSM images of structurally immobilized nanoreactors (B) before and (C) after incubation (20 minutes) with ELF 97 substrate.
(Reproduced from ref. 46 with permission from John Wiley and Sons. Copyright © 2009 Wiley-VCH Verlag GmbH & Co. KGaA, Weinheim.)

6.6 Applications of Bio-inspired Polymer Membranes

Investigations on the interactions between biomolecules and membranes and understanding their properties at the molecular level are crucial for various fields of research, such as therapeutic and diagnostic strategies, or the development of new separation techniques.

Although exploration of bio-inspired membranes is advancing rapidly, applications are still at an early stage of development. In order to accelerate the development of bio- and medical applications of such systems, a few crucial points need to be addressed: (i) to obtain a broader knowledge of the fabrication of complex protein-membrane assemblies with preserved functionality, (ii) to develop stable hybrid systems which can be scaled-up, and (iii) to tailor biodegradable and biocompatible amphiphilic copolymers with stimulus–response properties. Thus, the diversity of such platforms requires interdisciplinary knowledge of materials and surface science, strong analytical facilities, expertise in molecular biology, and a deep understanding of biological processes. The following section presents a few relevant examples of bio-inspired polymer vesicles and membranes for some potential applications.

6.6.1 Polymersomes

Polymersomes are ideal candidates for a large variety of applications, such as (i) drug delivery-containers,[124] (ii) compartments for contrast agents,[125] (iii) nanoreactors to serve as confined nano-spaces for transport phenomena and chemical transformation,[81,126–130] and (iv) artificial organelles or simple mimics of cells.[131]

Depending on the type (anticancer, anti-infection, anti-inflammatory, *etc.*) and the route of application (intravenous injection, local injection, oral application, *etc.*), polymersomes have to fulfil a complex scenario of requirements in order to be applied as drug delivery systems: no/limited toxicity, biodegradability, sufficient blood circulation time (for systemic applications), ability to reach and be taken up by target cells, and release their therapeutic cargo 'on demand'. For example, cancer therapies have mainly focused on systemic delivery of polymersomes, which can passively target tumor regions by the enhanced permeation and retention effect known as passive targeting of solid tumors.[132] Active targeting represents a more demanding utilization of polymersomes in nano-medicine, where surface modification with ligands/biomolecules offers receptor specific targeting of certain cell types. One example is the application of antibody decorated polymersomes to overcome the blood–brain barrier, which represents a crucial obstacle to the successful delivery of drugs into brains.[133] Polymersomes functionalized with the ganglioside GM1 targeting peptide and prion-targeting peptides have been developed as a potential

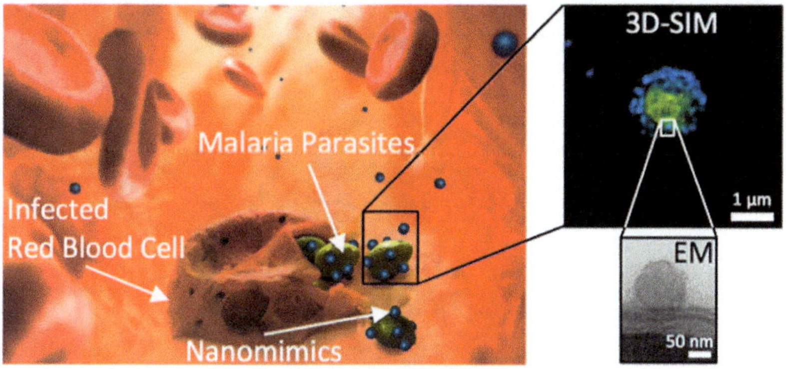

Figure 6.15 Schematic representation of merozoites bound with nanomimics after egress from red blood cells (left). Projections of merozoites (blue) blocked by sulforhodamine B (SRB)-filled nanomimics observed with structured illumination microscopy (3D-SIM) and electron microscopy (EM) (right).
(Reproduced with permission from A. Najer, D. Wu, A. Bieri, F. Brand, C. G. Palivan, H.-P. Beck and W. Meier, *ACS Nano*, 2014, **8**, 12560–12571. Copyright (2014) American Chemical Society.[40])

treatment of Alzheimer's and Parkinson's diseases.[134,135] Interestingly, nanomimics based on polymersomes that present receptors required for parasite attachment to host cells can efficiently interrupt the reproductive cycle of malaria parasites in human red blood cells. These nanomimics were able to bind to the parasite surface after egress from red blood cells, and block subsequent invasion processes (Figure 6.15).[40,136]

Such functionalized polymersomes exhibited efficacies of more than two orders of magnitude higher than the soluble form of the receptor as a result of multivalent interactions of several receptors on one nanomimic with the infective parasite.

A completely different approach is used when nanoreactors are employed. In this case, the active compounds act inside of the polymersome cavity and transport occurs through the channel proteins. For example, nanoreactors have been designed to generate reactive oxygen species (ROS) "on demand" in order to serve for photodynamic therapy. A photosensitizer conjugated with bovine serum albumin (BSA) was encapsulated inside polymersomes. Upon irradiation with light of an appropriate wavelength, the photosensitizer produced reactive oxygen species. Inside HeLa cells, the nanoreactor acted as a Trojan horse, solely inducing cell toxicity upon irradiation and only in the region where the nanoreactors were present.[128] In another example, nanoreactors were used for local controlled synthesis and release of antibiotics as an efficient system to locally inhibit bacterial growth "on demand". The encapsulated enzyme, penicillin acylase, produced antibiotics that inhibited bacterial growth for up to 7 days (Figure 6.16).[137] Such system has great potential in supporting antibacterial strategies in implant devices.[119]

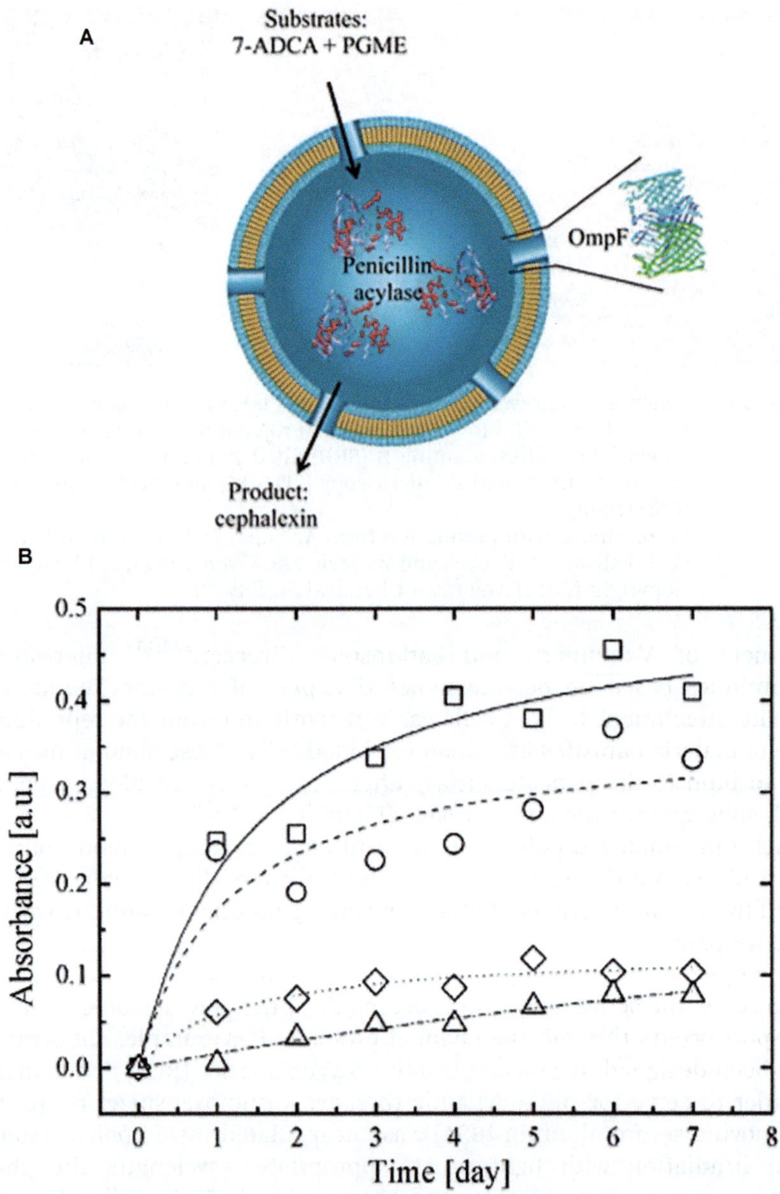

Figure 6.16 (A) Schematic presentation of a nanoreactor loaded with penicillin acylase. (B) Catalytic activity of penicillin acylase in for (◇) the free enzyme, (△) the enzyme encapsulated in non-permeabilized polymersomes, (□) nanoreactors generated at 8 °C, and (○) nanoreactors generated at RT.
(Reproduced from ref. 137 with permission from the Royal Society of Chemistry.)

The most innovative approach investigated in the last few years has been to mimic cellular organelles by designing artificial organelles based on polymersomes containing active compounds (proteins, enzymes) that act upon cellular uptake. For example, artificial peroxisomes have been developed to detoxify superoxide radicals and H_2O_2 after endocytosis.[138] Enzymes (Cu/Zn-superoxide dismutase and lactoperoxidase/catalase) were encapsulated in tandem inside polymersomes with the membrane permeabilized by insertion of OmpF. Cu/Zn-superoxide dismutase detoxified the superoxide radicals with the formation of H_2O_2, which in the second step was degraded by lactoperoxidase/catalase to water and O_2.

6.6.2 Planar Membranes

Highly sensitive biosensors are emerging systems that promise significant advances in diagnostics by combining polymer membranes and biomolecules. Polymer-based biosensors can be applied to the selective detection of molecules in solution and at surfaces. Two different types of sensor involving polymer membranes and biomolecules have mainly been used for medical purposes, namely glucose sensors and sensors based on specific enzymatic reactions. Glucose sensors, which are extremely important in diabetes diagnostics, are based on enzymatic oxidation mediated by glucose. The use of biologically suitable polymer brushes on electrodes enables the covalent attachment of glucose oxidase to provide high glucose sensing efficiency.[110] Enzyme catalysis produces a specific compound in the presence of the molecule to be detected, *e.g.* hydrogen peroxide, which is then oxidized at a platinum surface and detected amperometrically.[139] Apart from diagnostic systems, biomimetic planar membranes can be applied to facilitate complex processes at interfaces. For example, laccase adsorbed on polymer membranes can act as a sensing device for the enzymatic oxidation of phenol, *p*-chlorophenol, and aniline.[140] The concept of inserting functional membrane channel proteins into polymer membranes is a promising tool for various sophisticated applications, such as single molecule techniques, drug screening, and trace analysis. Reconstitution of membrane proteins into PEG (with fatty acid moieties) brush-supported lipid membranes enables the probing of diffusion and interactions of isolated membrane proteins.[141] This example demonstrates the applicability of such platforms for single molecule detection. Aquaporins, as biological water transport proteins, exhibit excellent water transport properties which can be preserved in polymer matrixes.[142] Therefore, the concept of using aquaporins in the design of hybrid artificial membranes for water treatment is advancing quickly, and various companies are trying to realize the principle in commercial products.[143,144] Furthermore, nanoporous cellulose acetate membrane substrates have also been used in an approach to perform nanofiltration with artificial membranes formed through the rupture of triblock copolymer (ABA) vesicles and UV polymerization.[115]

6.7 Conclusions and Perspectives

Polymer (3D) assemblies (polymersomes), and planar (2D) membranes based on self-assembly of amphiphilic block copolymers (or a combination of copolymers and lipids) can be combined with active molecules in a bio-inspired approach to obtain hybrid systems with specific properties that enable further development of multifunctional systems. Although amphiphilic polymeric nanostructures provide an elegant solution to some of the greatest challenges in drug delivery[145] (*i.e.* enhanced solubility of hydrophobic drugs, co-delivery of both hydrophilic and hydrophobic drugs, reduction of toxicity by shielding drugs inside the inner cavity, improved circulation time, controlled release, targeting, *etc.*), they are still in their early stage of development, especially for the synthesis of novel polymeric structures with improved biocompatibility and biodegradability for *in vivo* applications.[146] However, the focus of biomimetic platforms based on compartments such as polymersomes is not limited to drug delivery, and the design of nanoreactors and artificial organelles provides a novel strategy for developing multifunctional systems that act inside cells in a more controlled manner than systems based on the release of active compounds. In addition, other applications, especially in diagnostics or sensing, are profiting from the combination of biomolecules and synthetic compartments/planar membranes. For instance, biosensors based on enzyme–polymer membranes possess high sensitivity, specificity, speed, and accuracy. Furthermore, the exquisite selectivity and unique transport properties of membrane proteins can be used in various translational applications ranging from technological to environmental utilizations. Lipid membranes are being replaced with artificial block copolymer membranes in order to improve the mechanical stability and to form defect-free membranes. The creation of novel multifunctional hybrid biomolecule–polymer assemblies based on amphiphilic block copolymers is now a focus for research and application in the biomimetic field.

Abbreviations

αHL	α-hemolysin
ABTS	2,2′-azinobis(3-ethylbenzothiazoline-6-sulfonic acid)
AFM	atomic force microscopy
ATR IR	attenuated total reflectance infrared
ATRP	atom transfer radical polymerization
AqpZ	Aquaporin Z
BAM	Brewster-angle microscopy
BR	bacteriorhodopsin
BSA	bovine serum albumin
CalB	*Candida antarctica* lipase B
CLSM	confocal laser scan microscopy
Cmc	critical micelle concentration

DPPC	1,2-dipalmitoyl-sn-glycero-3-phosphatidylcholine
DLS	dynamic light scattering
EIS	electrochemical impedance spectroscopy
EM	electron microscopy
EPR	electron paramagnetic resonance
FCS	fluorescence correlation spectroscopy
FT-IR	Fourier transform infrared spectroscopy
gA	gramicidin
GFP	green fluorescent protein
GOx	glucose oxidase
HRP	horseradish peroxidase
HSPC	hydrogenated soy phosphatidylcholine
IgE	immunoglobulin E
IR	infrared
Leu	L-leucine
LB	Langmuir–Blodgett
LbL	layer-by-layer
LCST	lowest critical solution temperature
LS	Langmuir–Schaefer
MloK1	*Mesorhizobium loti* potassium channel
M_w	molecular weight
NMP	nitroxide mediated polymerization
NS	neutron scattering
OmpF	outer membrane protein F
OmpG	outer membrane protein G
PM-IRAS	polarization-modulation infrared reflection absorption spectroscopy
PAA	poly(acrylic acid)
PB	poly(butadiene)
PBMA	poly(*n*-butyl methacrylate)
PDMAEMA	poly(2,2-dimethylaminoethyl methacrylate)
PDMS	poly(dimethylsiloxane)
PEG	poly(ethylene glycol)
PEO	poly(ethylene oxide)
PEtOz	poly(2-ethyl-2-oxazoline)
PGA	poly(L-glutamic acid)
PHEMA	poly(2-hydroxyethyl methacrylate)
PIAT	poly(3-(isocyano-L-alanyl-aminoethyl)thiophene)
PIB	polyisobutylene
PMOXA	poly(2-methyl oxazoline)
PNIPAM	poly(*N*-isopropylacrylamide)
PNVP	poly(*N*-vinylpyrrolidone)
PS	poly(styrene)
PSA	poly(11-mercaptoundecyl sulfonic acid)
PSBMA	poly(sulfobetaine methacrylate)
R_g	radius of gyration

R_h	hydrodynamic radius
RAFT	reversible addition-fragmentation chain transfer
ROS	reactive oxygen species
SAXS	small angle X-ray scattering
SIP	surface initiated polymerization
SLS	static light scattering
SPR	surface plasmon resonance
SRB	sulforhodamine B
TEM	transmission electron microscopy
QCM	quartz crystal microbalance
W/O/W	water-in-oil-in-water
XPS	X-ray electron spectroscopy

Acknowledgements

This work was supported by the Swiss National Science Foundation, the Swiss Nanoscience Institute, and by the NCCR Molecular Systems Engineering, and this is gratefully acknowledged. The authors would like to thank Dr Bernard Goodman for proofreading the manuscript.

References

1. J. P. Hill, L. K. Shrestha, S. Ishihara, Q. Ji and K. Ariga, *Molecules*, 2014, **19**, 8589–8609.
2. D. M. Engelman, *Nature*, 2005, **438**, 578–580.
3. G. van Meer, D. R. Voelker and G. W. Feigenson, *Nat. Rev. Mol. Cell Biol.*, 2008, **9**, 112–124.
4. N. Tabassum and A. Feroz, *J. Appl. Pharm. Sci.*, 2011, **1**, 20–25.
5. K. Letchford and H. Burt, *Eur. J. Pharm. Biopharm.*, 2007, **65**, 259–269.
6. H.-A. Klok and S. Lecommandoux, *Adv. Mater.*, 2001, **13**, 1217–1229.
7. J. Kowal, X. Zhang, I. A. Dinu, C. G. Palivan and W. Meier, *ACS Macro Lett.*, 2014, **3**, 59–63.
8. C. LoPresti, H. Lomas, M. Massignani, T. Smart and G. Battaglia, *J. Mater. Chem.*, 2009, **19**, 3576–3590.
9. S. Egli, H. Schlaad, N. Bruns and W. Meier, *Polymers*, 2011, **3**, 252–280.
10. F. Itel, A. Najer, C. G. Palivan and W. Meier, *Nano Lett.*, 2015, **15**, 3871–3878.
11. F. Itel, M. Chami, A. Najer, S. Lorcher, D. Wu, I. A. Dinu and W. Meier, *Macromolecules*, 2014, **47**, 7588–7596.
12. P. Baumann, P. Tanner, O. Onaca and C. G. Palivan, *Polymers*, 2011, **3**, 173–192.
13. B. Krishnamoorthy, V. Karanam, V. R. Chellan, K. Siram, T. S. Natarajan and M. Gregory, *J. Drug Targeting*, 2014, **22**, 469–477.
14. Y.-R. Kim, S. Jung, H. Ryu, Y.-E. Yoo, S. M. Kim and T.-J. Jeon, *Sensors*, 2012, **12**, 9530–9550.
15. K. Matyjaszewski and J. Spanswick, *Mater. Today*, 2005, **8**, 26–33.

16. W. A. Braunecker and K. Matyjaszewski, *Prog. Polym. Sci.*, 2007, **32**, 93–146.
17. M. Marguet, C. Bonduelle and S. Lecommandoux, *Chem. Soc. Rev.*, 2013, **42**, 512–529.
18. X. Zhang, P. Tanner, A. Graff, C. G. Palivan and W. Meier, *J. Polym. Sci., Part A: Polym. Chem.*, 2012, **50**, 2293–2318.
19. O. S. Bello, K. A. Adegoke and R. O. Oyewole, *IOSR J. Appl. Chem.*, 2013, **5**(22–35), 14.
20. S. Cosnier, *Biosens. Bioelectron.*, 1999, **14**, 443–456.
21. H. Oh, V. Javvaji, N. A. Yaraghi, L. Abezgauz, D. Danino and S. R. Raghavan, *Soft Matter*, 2013, **9**, 11576–11584.
22. F. Meng, Z. Zhong and J. Feijen, *Biomacromolecules*, 2009, **10**, 197–209.
23. F. Caruso, *Modern Techniques for Nano- and Microreactors/-Reactions*, Springer, 2010, pp. 133–136.
24. L. Wang, L. Chierico, D. Little, N. Patikarnmonthon, Z. Yang, M. Azzouz, J. Madsen, S. P. Armes and G. Battaglia, *Angew. Chem., Int. Ed.*, 2012, **51**, 11122–11125.
25. K. Kita-Tokarczyk, J. Grumelard, T. Haefele and W. Meier, *Polymer*, 2005, **46**, 3540–3563.
26. B. M. Discher, Y.-Y. Won, D. S. Ege, J. C. Lee, F. S. Bates, D. E. Discher and D. A. Hammer, *Science*, 1999, **284**, 1143–1146.
27. E. Lorenceau, A. S. Utada, D. R. Link, G. Cristobal, M. Joanicot and D. A. Weitz, *Langmuir*, 2005, **21**, 9183–9186.
28. H. C. Shum, J.-W. Kim and D. A. Weitz, *J. Am. Chem. Soc.*, 2008, **130**, 9543–9549.
29. J. Ding and G. Liu, *Macromolecules*, 1997, **30**, 655–657.
30. R. M. Murphy, *Curr. Opin. Biotechnol.*, 1997, **8**, 25–30.
31. O. Stauch, R. Schubert, G. Savin and W. Burchard, *Biomacromolecules*, 2002, **3**, 565–578.
32. M. Regenbrecht, S. Akari, S. Förster and H. Möhwald, *J. Phys. Chem. B*, 1999, **103**, 6669–6675.
33. F. Ahmed, A. Hategan, D. E. Discher and B. M. Discher, *Langmuir*, 2003, **19**, 6505–6511.
34. D. Wu, M. Spulber, F. Itel, M. Chami, T. Pfohl, C. G. Palivan and W. Meier, *Macromolecules*, 2014, **47**, 5060–5069.
35. C. G. Palivan, O. Fischer-Onaca, M. Delcea, F. Itel and W. Meier, *Chem. Soc. Rev.*, 2012, **41**, 2800–2823.
36. B.-S. Kim, W.-Y. Yang, J.-H. Ryu, Y.-S. Yoo and M. Lee, *Chem. Commun.*, 2005, 2035–2037.
37. B.-S. Kim, D.-J. Hong, J. Bae and M. Lee, *J. Am. Chem. Soc.*, 2005, **127**, 16333–16337.
38. R. Nehring, C. G. Palivan, O. Casse, P. Tanner, J. Tüxen and W. Meier, *Langmuir*, 2008, **25**, 1122–1130.
39. P. Tanner, M. Ezhevskaya, R. Nehring, S. Van Doorslaer, W. Meier and C. Palivan, *J. Phys. Chem. B*, 2012, **116**, 10113–10124.

40. A. Najer, D. Wu, A. Bieri, F. Brand, C. G. Palivan, H.-P. Beck and W. Meier, *ACS Nano*, 2014, **8**, 12560–12571.
41. P. Chen, M. Qiu, C. Deng, F. Meng, J. Zhang, R. Cheng and Z. Zhong, *Biomacromolecules*, 2015, **16**, 1322–1330.
42. C. K. Wong, A. J. Laos, A. H. Soeriyadi, J. Wiedenmann, P. M. Curmi, J. J. Gooding, C. P. Marquis, M. H. Stenzel and P. Thordarson, *Angew. Chem.*, 2015, **127**, 5407–5412.
43. N. Cottenye, M.-I. Syga, S. Nosov, A. H. E. Muller, L. Ploux and C. Vebert-Nardin, *Chem. Commun.*, 2012, **48**, 2615–2617.
44. D. Kedracki, P. Maroni, H. Schlaad and C. Vebert-Nardin, *Adv. Funct. Mater.*, 2014, **24**, 1133–1139.
45. J. J. Lin, J. A. Silas, H. Bermudez, V. T. Milam, F. S. Bates and D. A. Hammer, *Langmuir*, 2004, **20**, 5493–5500.
46. M. Grzelakowski, O. Onaca, P. Rigler, M. Kumar and W. Meier, *Small*, 2009, **5**, 2545–2548.
47. P. Brož, S. M. Benito, C. Saw, P. Burger, H. Heider, M. Pfisterer, S. Marsch, W. Meier and P. Hunziker, *J. Controlled Release*, 2005, **102**, 475–488.
48. P. Broz, N. Ben-Haim, M. Grzelakowski, S. Marsch, W. Meier and P. Hunziker, *J. Cardiovasc. Pharmacol.*, 2008, **51**, 246–252.
49. J. A. Opsteen, R. P. Brinkhuis, R. L. Teeuwen, D. W. Löwik and J. C. van Hest, *Chem. Commun.*, 2007, 3136–3138.
50. T. O. Pangburn, F. S. Bates and E. Kokkoli, *Soft Matter*, 2012, **8**, 4449–4461.
51. Z. Pang, H. Gao, Y. Yu, J. Chen, L. Guo, J. Ren, Z. Wen, J. Su and X. Jiang, *Int. J. Pharm.*, 2011, **415**, 284–292.
52. H. De Oliveira, J. Thevenot and S. Lecommandoux, *Wiley Interdiscip. Rev.: Nanomed. Nanobiotechnol.*, 2012, **4**, 525–546.
53. P. Tanner, S. Egli, V. Balasubramanian, O. Onaca, C. G. Palivan and W. Meier, *FEBS Lett.*, 2011, **585**, 1699–1706.
54. S. M. Kuiper, M. Nallani, D. M. Vriezema, J. J. Cornelissen, J. C. van Hest, R. J. Nolte and A. E. Rowan, *Org. Biomol. Chem.*, 2008, **6**, 4315–4318.
55. D. R. Arifin and A. F. Palmer, *Biomacromolecules*, 2005, **6**, 2172–2181.
56. D. Gräfe, J. Gaitzsch, D. Appelhans and B. Voit, *Nanoscale*, 2014, **6**, 10752–10761.
57. C. Nardin, S. Thoeni, J. Widmer, M. Winterhalter and W. Meier, *Chem. Commun.*, 2000, 1433–1434.
58. M. Lomora, F. Itel, I. A. Dinu and C. G. Palivan, *Phys. Chem. Chem. Phys.*, 2015, **17**, 15538–15546.
59. M. V. Dinu, M. Spulber, K. Renggli, D. Wu, C. A. Monnier, A. Petri-Fink and N. Bruns, *Macromol. Rapid Commun.*, 2015, **36**, 507–514.
60. F. Junge, B. Schneider, S. Reckel, D. Schwarz, V. Dötsch and F. Bernhard, *Cell. Mol. Life Sci.*, 2008, **65**, 1729–1755.
61. T. Rabilloud, *Electrophoresis*, 2009, **30**, 174–S180.
62. A. M. Seddon, P. Curnow and P. J. Booth, *Biochim. Biophys. Acta, Biomembr.*, 2004, **1666**, 105–117.

63. M. Kumar, M. Grzelakowski, J. Zilles, M. Clark and W. Meier, *Proc. Natl. Acad. Sci. U. S. A.*, 2007, **104**, 20719–20724.
64. C. Nardin, J. Widmer, M. Winterhalter and W. Meier, *Eur. Phys. J. E: Soft Matter Biol. Phys.*, 2001, **4**, 403–410.
65. M. Lomora, M. Garni, F. Itel, P. Tanner, M. Spulber and C. G. Palivan, *Biomaterials*, 2015, **53**, 406–414.
66. M. Lomora, F. Itel, I. A. Dinu and C. G. Palivan, *Phys. Chem. Chem. Phys.*, 2015, **17**, 15538–15546.
67. M. Lomora, I. A. Dinu, F. Itel, S. Rigo, M. Spulber and C. G. Palivan, *Macromol. Rapid Commun.*, 2015, **36**, 1929–1934.
68. M. T. Paternostre, M. Roux and J. L. Rigaud, *Biochemistry*, 1988, **27**, 2668–2677.
69. M. Ollivon, S. Lesieur, C. Grabielle-Madelmont and M. T. Paternostre, *Biochim. Biophys. Acta, Biomembr.*, 2000, **1508**, 34–50.
70. J.-F. Le Meins, O. Sandre and S. Lecommandoux, *Eur. Phys. J. E: Soft Matter Biol. Phys.*, 2011, **34**, 1–17.
71. V. Pata, F. Ahmed, D. E. Discher and N. Dan, *Langmuir*, 2004, **20**, 3888–3893.
72. G. Srinivas, D. E. Discher and M. L. Klein, *Nano Lett.*, 2005, **5**, 2343–2349.
73. A. Graff, C. Fraysse-Ailhas, C. G. Palivan, M. Grzelakowski, T. Friedrich, C. Vebert, G. Gescheidt and W. Meier, *Macromol. Chem. Phys.*, 2010, **211**, 229–238.
74. H.-J. Choi and C. D. Montemagno, *IEEE Trans. Nanotechnol.*, 2007, **6**, 171–176.
75. H.-J. Choi and C. D. Montemagno, *Nano Lett.*, 2005, **5**, 2538–2542.
76. R. Stoenescu, A. Graff and W. Meier, *Macromol. Biosci.*, 2004, **4**, 930–935.
77. J.-F. Le Meins, C. Schatz, S. Lecommandoux and O. Sandre, *Mater. Today*, 2013, **16**, 397–402.
78. Z. Cheng, D. R. Elias, N. P. Kamat, E. D. Johnston, A. Poloukhtine, V. Popik, D. A. Hammer and A. Tsourkas, *Bioconjugate Chem.*, 2011, **22**, 2021–2029.
79. M. Schulz, S. Werner, K. Bacia and W. H. Binder, *Angew. Chem., Int. Ed.*, 2013, **52**, 1829–1833.
80. M. Spulber, P. Baumann, S. S. Saxer, U. Pieles, W. Meier and N. Bruns, *Biomacromolecules*, 2014, **15**, 1469–1475.
81. D. Dobrunz, A. C. Toma, P. Tanner, T. Pfohl and C. G. Palivan, *Langmuir*, 2012, **28**, 15889–15899.
82. T. Einfalt, R. Goers, I. A. Dinu, A. Najer, M. Spulber, O. Onaca-Fischer and C. G. Palivan, *Nano Lett.*, 2015, **15**, 7596–7603.
83. P. Tanner, O. Onaca, V. Balasubramanian, W. Meier and C. G. Palivan, *Chem. – Eur. J.*, 2011, **17**, 4552–4560.
84. S. F. van Dongen, M. Nallani, J. J. Cornelissen, R. J. Nolte and J. van Hest, *Chem. – Eur. J.*, 2009, **15**, 1107–1114.
85. S.-H. Kim, J. W. Kim, D.-H. Kim, S.-H. Han and D. A. Weitz, *Microfluid. Nanofluid.*, 2013, **14**, 509–514.

86. H. C. Shum, Y. j. Zhao, S. H. Kim and D. A. Weitz, *Angew. Chem.*, 2011, **123**, 1686–1689.
87. R. J. Peters, M. Marguet, S. Marais, M. W. Fraaije, J. van Hest and S. Lecommandoux, *Angew. Chem.*, 2014, **126**, 150–154.
88. R. S. Ries, H. Choi, R. Blunck, F. Bezanilla and J. R. Heath, *J. Phys. Chem. B*, 2004, **108**, 16040–16049.
89. A. Jagoda, J. Kowal, M. Delcea, C. G. Palivan and W. Meier, *Biomaterials Surface Science*, Wiley-VCH, 2013, pp. 159–203.
90. S. G. Patching, *Biochim., Biophys. Acta, Biomembr.*, 2014, **1838**, 43–55.
91. C. Draghici, J. Kowal, A. Darjan, W. Meier and C. G. Palivan, *Langmuir*, 2014, **30**, 11660–11669.
92. C. Nardin, M. Winterhalter and W. Meier, *Langmuir*, 2000, **16**, 7708–7712.
93. D. Wong, T.-J. Jeon and J. Schmidt, *Nanotechnology*, 2006, **17**, 3710.
94. F. Mallwitz and A. Laschewsky, *Adv. Mater.*, 2005, **17**, 1296–1299.
95. M. E. Buck and D. M. Lynn, *Langmuir*, 2010, **26**, 16134–16140.
96. T.-J. Jeon, N. Malmstadt, J. L. Poulos and J. J. Schmidt, *Biointerphases*, 2008, **3**, FA96–FA100.
97. A. Kibrom, R. F. Roskamp, U. Jonas, B. Menges, W. Knoll, H. Paulsen and R. L. Naumann, *Soft Matter*, 2011, **7**, 237–246.
98. G. W. de Groot, S. Demarche, M. G. Santonicola, L. Tiefenauer and G. J. Vancso, *Nanoscale*, 2014, **6**, 2228–2237.
99. M. Tanaka and E. Sackmann, *Nature*, 2005, **437**, 656–663.
100. B. Zhao and W. J. Brittain, *Prog. Polym. Sci.*, 2000, **25**, 677–710.
101. E. Rakhmatullina, T. Braun, T. Kaufmann, H. Spillmann, V. Malinova and W. Meier, *Macromol. Chem. Phys.*, 2007, **208**, 1283–1293.
102. E. Rakhmatullina, A. Mantion, T. Bürgi, V. Malinova and W. Meier, *J. Polym. Sci., Part A: Polym. Chem.*, 2009, **47**, 1–13.
103. Y. Chang, Y. J. Shih, C. J. Lai, H. H. Kung and S. Jiang, *Adv. Funct. Mater.*, 2013, **23**, 1100–1110.
104. J. Ł. Kowal, J. K. Kowal, D. Wu, H. Stahlberg, C. G. Palivan and W. P. Meier, *Biomaterials*, 2014, **35**, 7286–7294.
105. E. Rakhmatullina and W. Meier, *Langmuir*, 2008, **24**, 6254–6261.
106. J. Dorn, S. Belegrinou, M. Kreiter, E. K. Sinner and W. Meier, *Macromol. Biosci.*, 2011, **11**, 514–525.
107. N. Joseph, P. Ahmadiannamini, R. Hoogenboom and I. F. Vankelecom, *Polym. Chem.*, 2014, **5**, 1817–1831.
108. J. J. Richardson, M. Björnmalm and F. Caruso, *Science*, 2015, **348**, aaa2491.
109. K. Ariga, Y. Yamauchi, T. Mori and J. P. Hill, *Adv. Mater.*, 2013, **25**, 6477–6512.
110. M. E. Welch, T. Doublet, C. Bernard, G. G. Malliaras and C. K. Ober, *J. Polym. Sci., Part A: Polym. Chem.*, 2015, **53**, 372–377.
111. S. M. Lane, Z. Kuang, J. Yom, S. Arifuzzaman, J. Genzer, B. Farmer, R. Naik and R. A. Vaia, *Biomacromolecules*, 2011, **12**, 1822–1830.
112. X. Zhang, W. Fu, C. G. Palivan and W. Meier, *Sci. Rep.*, 2013, **3**, 2196.

113. T. D. Lazzara, C. Carnarius, M. Kocun, A. Janshoff and C. Steinem, *ACS Nano*, 2011, **5**, 6935–6944.
114. H. Wang, T. S. Chung, Y. W. Tong, K. Jeyaseelan, A. Armugam, Z. Chen, M. Hong and W. Meier, *Small*, 2012, **8**, 1185–1190.
115. P. S. Zhong, T.-S. Chung, K. Jeyaseelan and A. Armugam, *J. Membr. Sci.*, 2012, **407**, 27–33.
116. P. H. Duong, T.-S. Chung, K. Jeyaseelan, A. Armugam, Z. Chen, J. Yang and M. Hong, *J. Membr. Sci.*, 2012, **409**, 34–43.
117. J. Kowal, D. Wu, V. Mikhalevich, C. G. Palivan and W. Meier, *Langmuir*, 2015, **31**, 4868–4877.
118. L. Zhang, L. Hong, Y. Yu, S. C. Bae and S. Granick, *J. Am. Chem. Soc.*, 2006, **128**, 9026–9027.
119. K. Langowska, J. Kowal, C. G. Palivan and W. Meier, *J. Mater. Chem. B*, 2014, **2**, 4684–4693.
120. B. van Lengerich, R. J. Rawle and S. G. Boxer, *Langmuir*, 2010, **26**, 8666–8672.
121. J. Voskuhl, C. Wendeln, F. Versluis, E. C. Fritz, O. Roling, H. Zope, C. Schulz, S. Rinnen, H. F. Arlinghaus and B. J. Ravoo, *Angew. Chem., Int. Ed.*, 2012, **51**, 12616–12620.
122. P. Vermette, H. J. Griesser, P. Kambouris and L. Meagher, *Biomacromolecules*, 2004, **5**, 1496–1502.
123. F. Li, T. Ketelaar, M. A. Cohen Stuart, E. J. Sudhölter, F. A. Leermakers and A. T. Marcelis, *Langmuir*, 2008, **24**, 76–82.
124. G. Gunkel-Grabole, S. Sigg, M. Lomora, S. Lorcher, C. G. Palivan and W. P. Meier, *Biomater. Sci.*, 2015, **3**, 25–40.
125. P. Broz, N. Ben-Haim, F. Santini, S. Marsch, K. Scheffler, W. Meier and P. Hunziker, *Eur. J. Nanomed.*, 2009, **2**, 43–48.
126. K. Renggli, P. Baumann, K. Langowska, O. Onaca, N. Bruns and W. Meier, *Adv. Funct. Mater.*, 2011, **21**, 1241–1259.
127. F. Axthelm, O. Casse, W. H. Koppenol, T. Nauser, W. Meier and C. G. Palivan, *J. Phys. Chem. B*, 2008, **112**, 8211–8217.
128. P. Baumann, V. Balasubramanian, O. Onaca-Fischer, A. Sienkiewicz and C. G. Palivan, *Nanoscale*, 2013, **5**, 217–224.
129. V. Balasubramanian, O. Onaca, M. Ezhevskaya, S. Van Doorslaer, B. Sivasankaran and C. G. Palivan, *Soft Matter*, 2011, **7**, 5595–5603.
130. O. Onaca, D. W. Hughes, V. Balasubramanian, M. Grzelakowski, W. Meier and C. G. Palivan, *Macromol. Biosci.*, 2010, **10**, 531–538.
131. N. Ben-Haim, P. Broz, S. Marsch, W. Meier and P. Hunziker, *Nano Lett.*, 2008, **8**, 1368–1373.
132. H. Maeda, H. Nakamura and J. Fang, *Adv. Drug Delivery Rev.*, 2013, **65**, 71–79.
133. L.-H. Dieu, D. Wu, C. G. Palivan, V. Balasubramanian and J. Huwyler, *Eur. J. Pharm. Biopharm.*, 2014, **88**, 316–324.
134. K. Stojanov, J. V. Georgieva, R. P. Brinkhuis, J. C. van Hest, F. P. Rutjes, R. A. J. O. Dierckx, E. F. J. de Vries and I. S. Zuhorn, *Mol. Pharmaceutics*, 2012, **9**, 1620–1627.

135. J. V. Georgieva, R. P. Brinkhuis, K. Stojanov, C. A. G. M. Weijers, H. Zuilhof, F. P. J. T. Rutjes, D. Hoekstra, J. C. M. van Hest and I. S. Zuhorn, *Angew. Chem., Int. Ed.*, 2012, **51**, 8339–8342, 8339/8331–8339/8316.
136. A. Najer, S. Thamboo, J. T. Duskey, C. G. Palivan, H.-P. Beck and W. Meier, *Macromol. Rapid Commun.*, 2015, **36**, 1923–1928.
137. K. Langowska, C. G. Palivan and W. Meier, *Chem. Commun.*, 2013, **49**, 128–130.
138. P. Tanner, V. Balasubramanian and C. G. Palivan, *Nano Lett.*, 2013, **13**, 2875–2883.
139. N. Dale, S. Hatz, F. Tian and E. Llaudet, *Trends Biotechnol.*, 2005, **23**, 420–428.
140. G. Bayramoglu and M. Y. Arica, *Mater. Sci. Eng., C*, 2009, **29**, 1990–1997.
141. F. Roder, S. Waichman, D. Paterok, R. Schubert, C. Richter, B. Liedberg and J. Piehler, *Anal. Chem.*, 2011, **83**, 6792–6799.
142. M. Kumar, J. E. Habel, Y.-X. Shen, W. P. Meier and T. Walz, *J. Am. Chem. Soc.*, 2012, **134**, 18631–18637.
143. http://www.aquaporin.dk/, 02.02.2016, 10:38.
144. http://www.sterlitech.com/membrane-process-development/flat-sheet-membranes/forward-osmosis-membranes.html, 02.02.2016,10:40.
145. A. Rösler, G. W. Vandermeulen and H.-A. Klok, *Adv. Drug Delivery Rev.*, 2012, **64**, 270–279.
146. W. B. Liechty, D. R. Kryscio, B. V. Slaughter and N. A. Peppas, *Annu. Rev. Chem. Biomol. Eng.*, 2010, **1**, 149–173.

CHAPTER 7

Polymeric Ionic Liquids with Micelle-like Topologies and Functions

BENJAMIN KERSCHER,[a] FABIAN SCHÜLER,[a] ANNA-KATHARINA EVERS[a] AND ROLF MÜLHAUPT*[a,b]

[a] Freiburg Materials Research Center (FMF) and Institute for Macromolecular Chemistry, Albert-Ludwigs-University of Freiburg, Stefan-Meier-Str. 31, D-79104 Freiburg, Germany; [b] Freiburg Centre for Interactive Materials and Bioinspired Technologies (FIT), Georges-Köhler-Allee 105, D-79110 Freiburg, Germany

*Email: rolfmuelhaupt@web.de

7.1 Introduction

Micellar structures are prominent features of biological systems and account for many of the biological functions essential for life. Typically, the balance of hydrophilicity and hydrophobicity drives the spontaneous self-assembly of amphiphiles and governs the formation of micelles. In nature, the most prominent amphiphiles comprise lipids consisting of hydrophobic alkyl tails and hydrophilic ionic head groups. The micelles formed from lipids and other naturally occurring amphiphiles represent important functional structures in biology and play fundamental roles in metabolism. In the gastrointestinal tract, mixed micelles of bile salts and other lipids are involved in absorption, emulsification, and conversion of fats, oils, vitamins, and cholesterol.[1,2] The self-assembly of phospholipids creates bilayers of cell

RSC Polymer Chemistry Series No. 22
Bio-inspired Polymers
Edited by Nico Bruns and Andreas F. M. Kilbinger

Published by the Royal Society of Chemistry, www.rsc.org

membranes and vesicles as compartments, which enable controlled transport, storage, and even biochemical reaction cascades in their confined microenvironments.[3–5] Complex casein micelles in mammalian milk prevent amyloidosis and pathological calcification.[6]

Today, a large variety of both natural and synthetic amphiphilic molecules are employed as surfactants, which are indispensable in daily life and industry. Owing to their tendency to form micellar aggregates and to assemble at interfaces, they enable solubilization and dispersion of various substances in different media, production of stable emulsions of otherwise immiscible fluids, reduction of surface tension, and adjustment of flow characteristics. In addition to ionic and non-ionic low-molecular-weight surfactants, synthetic amphiphilic polymers and block copolymers are tailored to serve the need for dispersants, emulsifiers, antifoaming agents, and viscosity modifiers. They are important ingredients in detergents, cosmetic products, and pharmaceutical preparations, as well as common additives for lubricants, fuel, and oil.[7–11] For example, polyethylene glycols, esterified with fatty acids or etherified with fatty alcohols, are widely applied in shampoos and lotions.[9] The prominent poly(ethylene oxide)-*block*-poly-(propylene oxide)-*block*-poly(ethylene oxide) triblock copolymers are used in pharmaceutics and in cosmetics as well.[7,8] Polyisobutylene succinimides are added to engine oil to prevent aggregation of carbonaceous particles.[11] Going well beyond such conventional polymeric amphiphiles, designed block copolymers bearing specific functional groups and polymer segments afford the unprecedented control of nanostructures and functions including stimulus–response. Block copolymer self-assembly, producing micelles of various geometries, vesicles, and other micro- and nanostructures, is exploited in molecular system engineering and medical applications such as controlled drug delivery, gene transfection, and tissue engineering.[7,12–15] Hollow polymersomes, prepared by the assembly of tailor-made amphiphilic block copolymers, represent unique compartments for encapsulation, protection, and transportation of active substances, diagnostic agents, genes, or proteins, thus exhibiting high potential in nanomedicine and *in vivo* imaging.[16–18] Moreover, enzyme-filled polymersomes function as nanoreactors hosting reaction cascades, which can be triggered by external stimuli such as changes in pH.[19,20] In a promising advance, polymersomes based on a poly(2-methyloxazoline)-*block*-poly(dimethylsiloxane)-*block*-poly(2-methyloxazoline) triblock copolymer have been used for encapsulating the enzyme penicillin acylase.[21] The resulting nanoreactors enabled the local synthesis and release of antibiotics and effectively inhibited bacterial growth.

7.2 From Supramolecular Assemblies to Micelle-like Macromolecules

Micelles formed from low-molecular-weight amphiphiles or amphiphilic block copolymers, respectively, represent supramolecular structures, which are held together by intermolecular interactions. In the absence of covalent bonds, the resulting assemblies are highly sensitive to changes in their environment.[13,22,23] Hence, the shape, geometry, and stability of micelles strongly depend on the pH, concentration, added salts, temperature, and shear. However, for many applications, especially in industrial environments, it is highly desirable to employ robust micellar structures possessing a stable topology, which is independent of the above mentioned parameters. To create robust and stable micelles, one has to switch from supramolecular assemblies of amphiphiles to micelle-like macromolecules, often referred to as unimolecular micelles, which have molecular architectures resembling a single micelle. As all segments of such a micelle-like polymer are part of one macromolecule and connected *via* covalent bonds, it cannot readily disintegrate into individual molecules as observed for supramolecular micellar assemblies.

Although unimolecular micelles do not exist in nature, it is an important challenge in bio-inspired polymer research to create such molecular architectures in the lab. While linear polymers change their conformation when exposed to different solvents, temperature, and shear, highly branched polymers, also known as cascade macromolecules, are well known to be robust since their stable tree-like conformations are much less sensitive to changes in their environment. Hence, cascade macromolecules are highly attractive candidates to serve as the core of micelle-like polymers, which are readily obtained by facile end-group modification reactions. As illustrated in Figure 7.1, there exist two different types of cascade macromolecules as core building blocks for micelle-like polymers: the perfectly branched dendrimers and the randomly branched macromolecules, referred to as hyperbranched polymers.

The preparation of dendrimers requires tedious multi-step syntheses in order to achieve the precise control of branching and to minimize structural defects.[24] The first example of such molecules was reported in 1978 by Vögtle *et al.*, who utilized the reaction of diamines with acrylonitrile, the subsequent hydrogenation of the nitrile functions, and the iteration of this sequence to generate cascade-like molecular architectures.[25] In the following

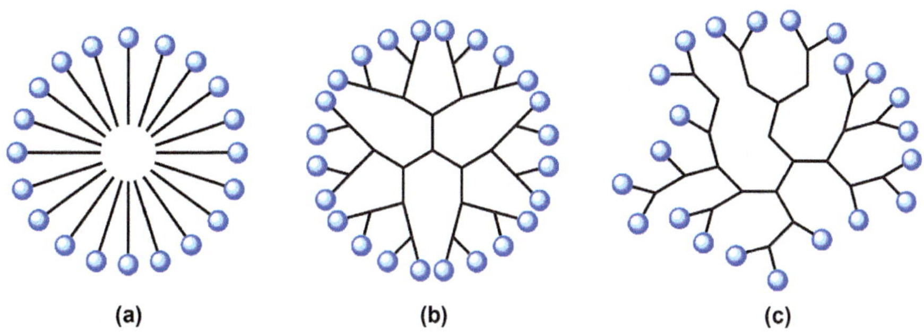

Figure 7.1 A spherical micelle formed by self-assembly of linear amphiphilic molecules (a), and unimolecular micelles on the basis of a dendrimer (b) and a hyperbranched polymer (c).

years, intense research afforded several other dendrimers, among them the prominent polyamidoamines (PAMAM), which became known as "Starburst" dendrimers.[26–29] Applying similar synthetic strategies, a great variety of tailored dendrimers are available today.[24] For example, exploiting Vögtle's dendrimer synthesis, the repeated branching sequence of Michael addition reactions of acrylonitrile to amines and the subsequent catalytic hydrogenation of the nitrile groups afforded polypropyleneimine (PPI) dendrimers in an industrially viable process without encountering a massive excess of reagents and waste.[30,31] Although dendrimers are readily converted into micelle-like macromolecules by end-group functionalization, the synthetic effort to obtain such perfectly branched architectures remains high, rendering dendrimers expensive and thus limiting their industrial use.

During the last decades, hyperbranched polymers have attracted increasing attention in academia and industry. Although hyperbranched polymers are randomly branched and have broad molar mass distributions, they exhibit properties which are very similar to those of dendrimers. However, unlike dendrimers, they are readily available by means of a one-step multi-branching polymerization reaction when using multifunctional monomers or monomer combinations. In the early 1940s it was Flory who theoretically described the formation of randomly branched polymers by means of polymerization reactions in which at least trifunctional monomers are present.[32–34] Despite these early pioneering advances and the industrial development of branched polymers such as phenolics and alkyd resins as prepolymers for applications in thermoset resin chemistry, intense research on the synthesis and physics of randomly branched polymers having much higher degree of branching did not start until the 1980s. In this decade, for example, Kricheldorf *et al.* reported branched poly(hydroxybenzoate)s, which were obtained by employing 3,5-(bistrimethylsiloxy) benzoyl chloride as comonomer in polycondensation reactions.[35] Other routes to highly branched polymers were explored by Tomalia and coworkers, who investigated the multibranching polymerization of a 2-(carboxyalkyl)-2-oxazoline.[36] At DuPont, Kim and Webster succeeded in the preparation of highly rigid hyperbranched polyphenylenes by means of Suzuki coupling reactions using 3,5-dibromophenylboronic acid.[37,38] Since then, a great variety of multifunctional monomers has been polymerized by step and chain growth polymerization to produce hyperbranched polymers. Exploiting multibranching polycondensation reactions, a considerable number of different hyperbranched polyesters,[39,40] polyamides,[41] and polyesteramides[42] has been prepared. For example, Hult and coworkers synthesized highly branched aliphatic polyesters from dihydroxyl-functional carboxylic acids, leading to an industrial process commercialized by the Perstorp Company.[43] Moreover, the ring-opening polymerization of hydroxyl-functional cyclic ethers such as 3-ethyl-3-hydroxymethyloxetane[44–46] and glycidol[46,47] afforded hyperbranched polyethers, especially when the monomers were slowly added to multifunctional polyols.[47,48] Since the 1980s, significant progress has been made with respect to the controlled synthesis of highly branched macromolecules with tailored degrees of branching, molecular architecture, molar mass, and end groups.[49,50] Today, several hyperbranched polymers are commercially available in simple one-step polymerization processes, among them the Boltorn polyester polyols (Perstorp), the Hybrane polyesteramides (DSM), and the Lupasol polyethyleneimines (BASF SE). Compared to linear polymers, the high degree of branching and multifunctionality accounts for unique features such as excellent solubility, low viscosity, high density of functional end-groups, and the existence of intramolecular cavities, which are attractive with respect to host–guest chemistry. Typical applications of hyperbranched polymers include rheology modifiers, curing agents in coatings, pigment dispersing agents, printing inks, and processing aids for thermoplastics.[42,50,51]

The facile end-group modification of highly branched polymers is highly attractive with respect to tailoring micelle-like polymeric amphiphiles with core–shell or onion-like topologies. Such micelle-like polymers can serve as molecular containers for the encapsulation and transport of guest molecules, rendering them useful for applications such as controlled drug delivery and catalysis. While cavities in the core can host different substances, the outer shell governs the solubility behavior, which can be adjusted to various media of different polarity. Unlike micelles and micellar supramolecular assemblies of block copolymers, micelle-like macromolecules are robust and do not suffer from disintegration accompanied by loss of their function when exposed to changes in external conditions such as temperature, shear forces, pH, concentration, and salt addition. As illustrated in Figure 7.2, going beyond host–guest chemistry, micelle-like polymeric

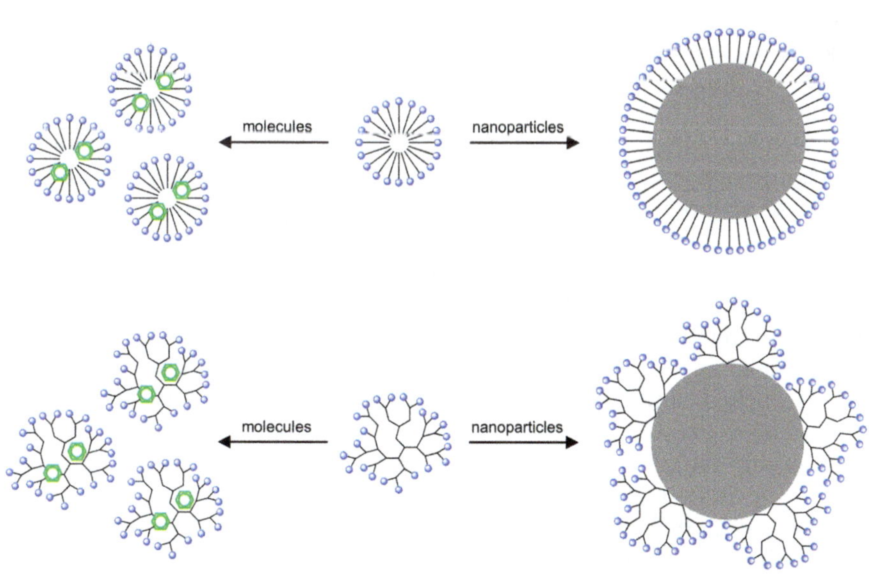

Figure 7.2 Encapsulation of small guest molecules and dispersion of nanoparticles by amphiphilic molecules with linear structure (top) and highly branched polymers with micelle-like topology (bottom).

amphiphiles are highly efficient dispersing agents, which are capable of dispersing a great variety of fluids and solids in different media by assembling at their interfaces. In contrast to the encapsulation of fairly small guest molecules in core cavities, the size of dispersed materials is much larger and varies between the nano- and millimeter scales. This facile formation of dispersions in the presence of micelle-like macromolecules is accounted for by conformational changes occurring at the polymer/particle interface. A prime requirement is high chain mobility, *i.e.* a low glass transition temperature, which is typical for polymeric ionic liquids (*cf.* Section 7.3). Micelle-like polymeric dispersing agents are of particular interest for the controlled synthesis of micro- and nanoparticles in reaction compartments formed by their self-assembly at the particle interfaces. Moreover, this *in situ* surface modification and the resulting steric stabilization enable the production of stable particle dispersions in fluids with greatly varied polarities and viscosities.

Several groups have explored the synthesis of highly branched core–shell polyelectrolytes and non-ionic amphiphiles as a versatile route to macromolecules with micelle-like topologies and designed polarities. For instance, hyperbranched polyphenylenes, as well as aliphatic cascade polymers, were rendered water-soluble by creating hydrophilic ionic carboxylate shells in the periphery of the non-polar polymer cores.[38,52] The resulting amphiphilic core–shell polymers, which have been shown to act as hosts for hydrophobic guest molecules, represent the first proof-of-concept of the controlled formation of highly branched polyelectrolytes with micelle-like molecular architecture and function. In another strategy, water-soluble micelle-like macromolecules were obtained by equipping dendritic polymers with a non-ionic shell of hydrophilic polymer chains such as polyethylene glycol (PEG).[53–55] Similar to the aforementioned polyelectrolytes, such non-ionic micelle-like polymers are attractive with respect to their application as water-soluble nanocarriers. Inspired by reverse micelles, highly branched hydrophilic polymers were equipped with a hydrophobic shell in order to solubilize water-soluble guest molecules in non-polar media. Prominent examples are hyperbranched polyglycerols and polyethyleneimines, which were rendered organophilic by reacting their end groups with long-chain fatty acids.[56–58] The reverse-micelle-like non-ionic polymers were successfully employed for the encapsulation and transportation of water-soluble dyes from an aqueous phase into non-polar solvents. In sharp contrast to the hyperbranched amphiphiles, the corresponding linear macromolecules showed no or drastically reduced phase transfer activity.[57,58] This represents clear experimental evidence that the amphiphilic core–shell-type architecture is crucial for the achievement of effective molecular encapsulation and transport, thus underlining the value of micelle-like (as well as reverse-micelle-like) hyperbranched polymers for the envisioned use in drug-delivery applications. Going well beyond the transportation of guest molecules, the capability of dendritic amphiphiles to encapsulate metal complexes was exploited to create molecular nanoreactors useful in catalysis.[59,60] Moreover, amphiphilic core–shell polymers can be employed to produce and support a great variety of metal nanoparticles. On one hand, such micelle-like polymers enable the solubilization of the respective metal salt precursor even in non-polar organic solvents. On the other hand, they govern metal particle formation during reduction of the metal salt and provide efficient steric stabilization, thus preventing aggregation and achieving the facile dispersion of metal nanoparticles tailored for application in catalysis.[61–64] In fact, owing to their remarkable capability for encapsulation, stabilization, transport, and release of organic and inorganic substances, the production of nano-sized dendritic polymers with well-defined architectures and micelle-like functions continues to be of great interest in academia and industry. Recent and ongoing research mainly concentrates on the development of novel drug-delivery systems that show high drug efficacy, possess a sufficient loading capacity, allow for controlled transport and release, are biocompatible, and exhibit a cost effectiveness that does not limit the prospect of commercialization.[65–70] Taking advantage of the high number of easily accessible functional groups, sophisticated tailor-made dendritic polymers even represent promising candidates to act as pharmaceutically active substances themselves. For example, Haag and coworkers developed micelle-like hyperbranched polyglycerols comprising a shell of sulfate ions, which exhibited an excellent anti-inflammatory effect without impairing the cell viability.[71–73]

7.3 Nanostructured and Micelle-like Polymeric Ionic Liquids

The unique property profile of ionic liquids (ILs), which are commonly defined as organic salts that are liquid at temperatures below 100 °C or even at room temperature, renders them attractive candidates for manifold applications. Due to their negligible vapor pressure, excellent solvating properties, low flammability, frequently high thermal stability, ionic conductivity, and usually broad electrochemical window, ILs are widely applied as solvents, in catalysis, and as electrolytes in electrochemical devices.[74–79]

Since the late 1990s, when Ohno *et al.* started their pioneering work on the synthesis and polymerization of different IL monomers,[80–82] the macromolecular analogues of ILs have attracted much attention. In such materials, which are referred to as poly(ionic liquid)s or polymeric ionic liquids (PILs), the IL moieties are attached to a polymeric backbone. Representing a special class of polyelectrolytes, PILs combine the typical features of ILs with intrinsic polymer properties, such as low toxicity, mechanical stability, shape durability, and especially facile processability.[83,84] Although PILs have much lower glass transition temperatures compared to the majority of conventional polyelectrolytes, they are usually solid at room temperature,[85] thus substantially reducing the risk of leakage typical for low-molecular-weight ILs.

During the last decades, by exploiting the free radical polymerization of IL monomers bearing vinyl, acrylate, methacrylate, or styrene groups, a large variety of linear PIL homopolymers and cross-linked PIL networks has been prepared.[83,85] By subsequent exchange of the low-molecular-weight counter ions, the cation–anion combination is readily varied in order to adjust the PILs' properties, such as its glass transition temperature, ionic conductivity, thermal stability, and solubility behavior.[86,87] In Figure 7.3, selected examples of linear PIL structures are shown. The majority of the PIL family is of the polycation type, with the most prominent representatives being polymers bearing imidazolium cations such as the poly(3-alkyl-1-vinylimidazolium)-type PILs. However, several polyanionic PILs, *e.g.* based on poly(styrene sulfonate), have been synthesized as well.

Combining IL properties with macromolecular architectures opens manifold potential applications. Tailored PILs have been designed for use as solid electrolytes in electrochemical devices such as lithium-ion batteries.[88–90] By mixing with low-molecular-weight ILs and lithium salts, the ionic conductivity can be drastically increased without a loss in mechanical stability. Apart from ion conducting materials, additional proposed and already tested PIL applications include selective CO_2 adsorption,[91,92] their use as sorbent coatings in chromatography,[93] catalysis,[94,95] and the fabrication of optoelectronic devices.[96]

There is intense research with respect to the production of micro- or nanostructured PILs and PIL/nanomaterial hybrid systems, not only to enhance the performance in the abovementioned applications, but also in order to extend the scope of PILs. Utilizing techniques such as reversible addition fragmentation chain transfer (RAFT) polymerization, well-defined PIL block copolymers, which are prone to self-assemble into micro- and nanophase-separated structures, have been synthesized in view of their potential use in gas separation[97] or energy storage and conversion.[98–100] The polymerization of IL monomers bearing mesogenic groups has been exploited to prepare nanostructured liquid-crystalline PILs with anisotropic ion-conductivity[101] or electroactive properties.[102] By means of heterogeneous polymerization reactions, micron-sized monodisperse PIL particles[103] and PIL latexes,[104] among them even liposome-like PIL nanoparticles possessing uni- or multilamellar inner structures and forming nanoworm mesostructures at elevated concentrations,[105,106] have been obtained. In fact, tailor-made PIL micro- and nanoparticles exhibited an impressing performance when applied as sensors,[107] recyclable catalysts,[108,109] or binders in cathodes for lithium-ion batteries.[110] Going beyond producing PILs with a particulate shape, in an alternative strategy, the development of functional PILs aims at materials containing interconnected pores or channels, respectively, such as micro-structured gels or mesoporous films. For example, PIL films with a 3D-ordered macropore architecture, prepared by the colloidal crystal templating method, have been introduced as multifunctional materials that can serve as tunable photonic crystals, anion-directed molecular gating systems, surfaces with tunable wettability, and stable

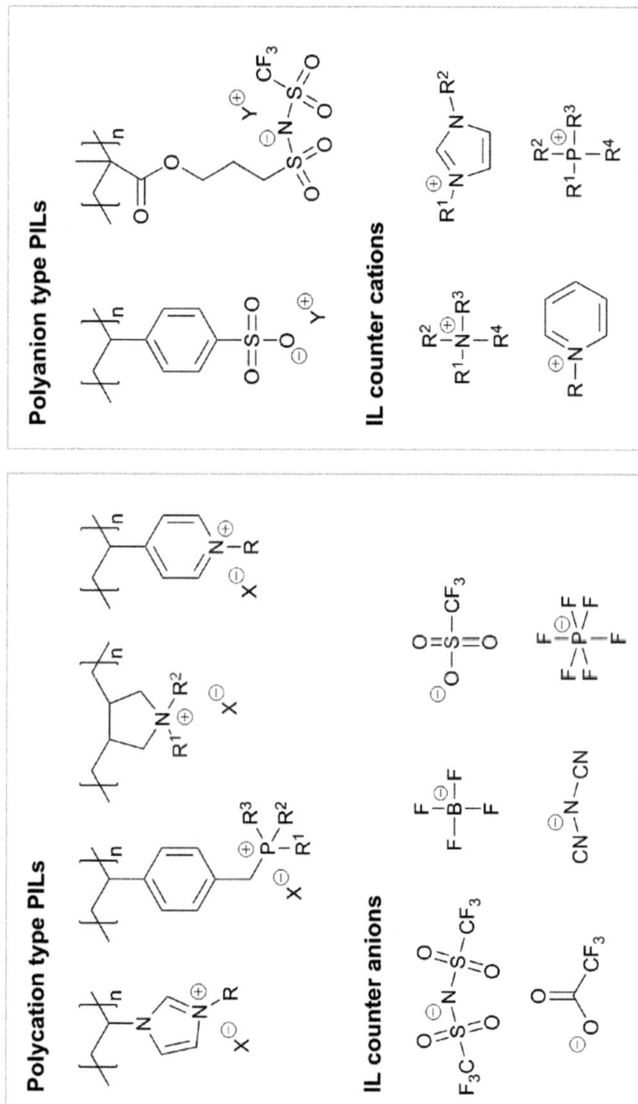

Figure 7.3 Selected examples of linear PILs.

electro-optic switches.[111] In a recent advance, Yuan and coworkers produced mesoporous membrane actuators with a gradient morphology by treating blends of poly(1-vinylimidazolium)- or poly(4-vinylpyridinium)-based PILs and carboxylic multi-acids with aqueous ammonia.[112,113] When exposed to vapors of organic solvents or changing humidity, these PIL membranes responded in terms of a fast and reversible bending movement.

The capability of PILs to function as dispersing agents for nano-objects, which has been demonstrated in phase transfer experiments,[114,115] has been exploited in the stabilization of catalytically active metal nanoparticles[116,117] and in the preparation of dispersions of conductive polymers.[118,119] Moreover, combining PILs with nano-sized foreign materials affords functional nanocomposites and hybrid structures. By polymerizing IL monomers in the presence of carbon nanotubes, mechanically reinforced bucky gels with switchable polarity[120] and highly electro-conductive bucky plastics[121] have been produced. The combination of PILs and graphene nanomaterials has been employed in the fabrication of electrodes for high-performance supercapacitors,[122] membranes with excellent ionic conductivity and mechanical properties,[123] and electrochemical sensors with low detection limits, quick response, and good selectivity.[124] Recently, new PIL composite materials comprised of $La_2O_2CO_3$ nanoparticles and poly[(*p*-vinylbenzyl)-trimethylammonium hexafluorophosphate] have been shown to be promising materials for chemoresistive CO_2 sensing under ambient conditions.[125] Representing a more sophisticated example of nanostructured PIL hybrid systems, PIL nanospheres have been coated with an inorganic mesoporous silica shell to result in nano-objects that were applied as highly efficient amphiphilic nanoreactors for conversion of hydrophobic substrates in an aqueous environment.[126]

In addition to linear PILs and cross-linked PIL networks, a large variety of PILs with branched molecular architectures has been prepared. Recent progress in the field of brush-like, star-shaped, hyperbranched, and dendrimer polyelectrolytes and PILs has been summarized in a current review by Tsukruk *et al.*[127] In fact, there exist several examples of highly branched macromolecules bearing IL moieties in the polymer backbone. In a quite simple approach, quaternization of randomly branched polyethyleneimines or polypropyleneimine dendrimers by reacting them with alkyl halides readily yielded PILs containing ammonium cations both in the branched backbone and at the chain ends.[128,129] Moreover, by means of poly(*N*-alkylation) of tailor-made multifunctional monomers, hyperbranched polymers with a skeleton comprising pyridinium or imidazolium segments can be obtained.[130,131] However, as compared to structures in which the IL groups are evenly distributed within the branched backbone, compartmentalized dendritic PILs with the IL functionalities mainly located in the periphery are much more interesting with respect to designing micelle-like PILs. Such PILs with core–shell architecture combine the features of IL groups with the inherent properties of polymeric materials and the stable micelle-like topology known from conventional dendritic amphiphiles (*cf.* Section 7.2). This is highly advantageous compared to the supramolecular micelles formed by self-assembly of amphiphilic linear PIL block copolymers.

Of course, the perfect dendrimer structures can serve as cores for micelle-like PILs. For instance, this was demonstrated for PAMAM-type dendrimers with a polyhedral oligomeric silsesquioxane (POSS) focal unit and propyl-imidazolium salt end groups.[132] Partial methylation of polypropyleneimine dendrimers can yield PIL structures in which mainly the peripheral end groups are converted into ammonium cations, while the internal tertiary amines remain unaffected.[129] In contrast, acylation of the end groups of dendritic polypropyleneimines and subsequent quaternization of the residual amine functions affords core–shell-type dendrimer PILs with the IL groups located in the core, representing useful structures for application in gene transfection.[133] However, from an economic point of view, it is much better to use hyperbranched polymers as the core of micelle-like PILs due to their easy one-step preparation. By esterification of hyperbranched polyglycerol (PG) with ω-bromoacyl chlorides and simultaneous reaction with 1,2-dimethylimidazole or pyridine, Schwab and Mecking designed PILs possessing a shell of imidazolium or pyridinium bromide groups.[134,135] After anion exchange with *p*-(diphenylphosphino)benzene sulfonate, the imidazolium-functionalized PILs served as macroligands to form rhodium–polyelectrolyte complexes, which were then employed as recoverable hydroformylation catalysts. Tosylation of the hydroxyl groups of hyperbranched PG and subsequent alkylation of *N*-methylimidazole yielded methylimidazolium tosylate moieties and thus afforded another type of PG-based core–shell PILs, which showed lower critical solution temperature (LCST)-type liquid–liquid and liquid–solid phase transitions in mixtures of methanol and chloroform.[136] The same cationic PIL, albeit equipped with other counter anions, was employed to produce novel polymer electrolytes.[137] Hyperbranched polyoxetanes comprising a shell of PEG chains with terminal methylimidazolium cations represent another example of polyether-based PILs with micelle-like topology.[138] After anion exchange with an anionic dye, these PILs self-assembled into pH-indicative vesicles in water. Starting from commercial hyperbranched Boltorn polyester, a series of PILs bearing long-chain alkyl quaternary dimethylammonium chloride end groups was prepared.[139] Although the solubility in water decreased with the increasing alkyl chain length from *n*-octyl to *n*-octadecyl, surprisingly all of these PILs were reported to have higher water-solubility compared to hexadecyl trimethyl ammonium bromide. Some of these polyelectrolytes proved to be efficient accelerators for the alkaline hydrolysis of poly(ethylene terephthalate) (PET) fabrics. However, the ammonium chloride functionalities strongly indicate a relatively low thermal stability, which may limit the usability of these PILs in advanced applications. Boltorn polyesters have also been used as cores for polyanionic micelle-like PILs.[140] After equipping the polyester polyols with carboxylic acid or sulfonic acid end groups, neutralization with 1-methylimidazole afforded the corresponding protic PILs bearing carboxylate and sulfonate anions, respectively. As for the carboxylate

PIL, an aprotic analogue with 1,3-dimethylimidazolium counter cations has been prepared as well. Branched polythiophenes with methylimidazolium bromide moieties have been introduced as novel polymeric materials that combine the characteristics of dendritic architecture, conjugated polymers, and IL groups.[141] The IL-functionalization accounted for the good solubility in a room-temperature IL and did not significantly alter the optical absorption properties compared to the parent polythiophene, thus opening possible applications of such polymers in dye-sensitized solar cells or electrochromic devices.

7.4 Compartmentalized Onion-like Polymeric Ionic Liquids

In recent years, a versatile family of compartmentalized PILs with onion-like topology has been developed.[142] These micelle-like PILs consist of a hyperbranched polyether core, an ionic inner shell of covalently attached imidazolium cations and counter anions, and a non-polar outer shell of *n*-alkyl chains in its periphery (see Figure 7.4). PIL synthesis is carried out by post-modification of hyperbranched poly(3-ethyl-3-hydroxymethyloxetane) (PEHO), which is readily accessible by means of cationic ring-opening polymerization of the oxetane monomer and usually possesses a number-average molar mass of 1800–2400 $g\,mol^{-1}$, a polydispersity of well below 2, and a degree of branching of approximately 50%. Tosylation of the hydroxyl functionalities and subsequent reaction with *N*-alkylimidazoles bearing alkyl chains of variable length, namely methyl, *n*-butyl, *n*-octyl, *n*-dodecyl, or *n*-octadecyl, affords the hyperbranched polycations with tosylate counter anions. If desired, the tosylate anions can be exchanged for a large variety of other anions by means of anion exchange reactions (herein, these PILs are abbreviated as PEHO-C*n*ImX, whereas *n* indicates the alkyl chain length and X the type of counter anion; *e.g.*, *n*-butyl-substituted PIL containing tosylate anions is referred to as PEHO-C4ImOTs).[142–144]

Variation of both the alkyl chain length and counter anion enables the easy adjustment of the polarity and PIL properties.[142] Increasing the length of the alkyl substituent in PEHO-C*n*ImOTs from methyl to *n*-octadecyl completely inverts the solubility behavior: while PEHO-C1ImOTs is highly water-soluble, amphiphilic PEHO-C18ImOTs, a reverse unimolecular micelle, is soluble in non-polar organic solvents such as toluene or benzene, but not at all in water. Exchanging the tosylate anions in methyl- or butyl-substituted PILs for 4-dodecylbenzenesulfonate or typical IL anions is another way of shifting the solubility. Moreover, by introducing anions such as bis(trifluoromethylsulfonyl)imide, significant lowering of the glass transition temperature of the PILs to temperatures well below 0 °C has been achieved.[142] The compartmentalized micelle-like structure with facile polarity design, combined with the extraordinary chemical robustness and high thermal stability up to temperatures well above 300 °C, an additional

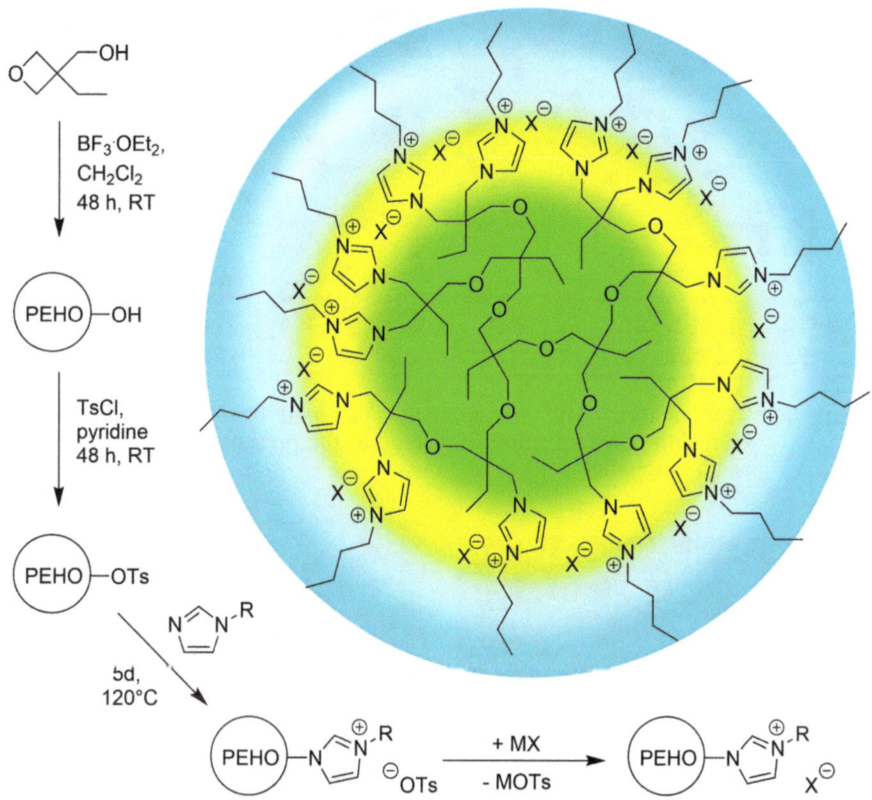

Figure 7.4 Synthetic route to hyperbranched PEHO-based PILs and illustration of the compartmentalized onion-like structure by reference to *n*-butyl-substituted PEHO-C4ImX; Ts = *p*-toluenesulfonyl.
Reproduced from ref. 142 with permission from John Wiley and Sons. Copyright © 2013 WILEY-VCH Verlag GmbH & Co. KGaA, Weinheim.

key feature of the PEHO-based PILs, renders them attractive candidates for use in various applications, which will be illustrated in the following paragraphs with the aid of selected examples.

The hyperbranched PILs represent effective stabilizers in heterogeneous polymerization reactions. When employing water-soluble PEHO-C1ImOTs as the emulsifier in the 2,2′-azobis(2-methylpropionitrile) (AIBN) initiated mini-emulsion polymerization of styrene, stable polystyrene (PS) latexes were obtained.[145] Characterization by means of transmission electron microscopy (TEM) indicated the presence of spherical PS nanoparticles (see Figure 7.5). Further studies revealed that, by modifying parameters such as the PIL/monomer ratio and the duration of the ultra-sonication treatment prior to the polymerization, the mean diameter of the obtained PS particles can be varied from 40 nm to 110 nm. In contrast to PEHO-C1ImOTs, the use of the low-molecular-weight ILs 1,3-dimethylimidazolium tosylate or 1-butyl-3-methylimidazolium tosylate did not result in the formation of stable latexes.

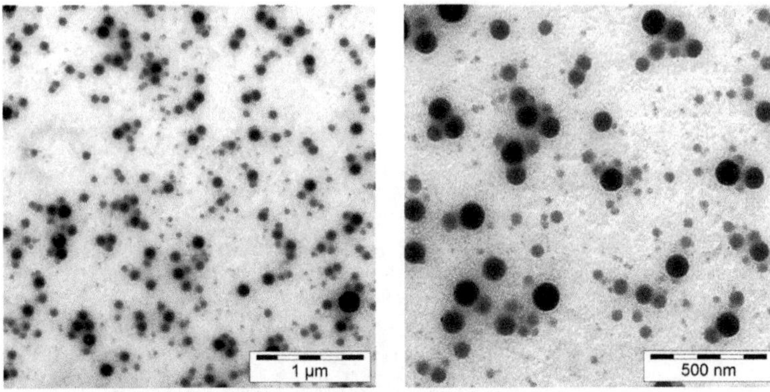

Figure 7.5 TEM images of PS nanoparticles prepared by AIBN initiated mini-emulsion polymerization of styrene in the presence of PEHO-C1ImOTs as an emulsifier.

Figure 7.6 SEM images of PS microparticles prepared by AIBN initiated dispersion polymerization of styrene in ethanol in the presence of PEHO-C18ImOTs as a stabilizer.

As a function of the polymerization process and the type of hyperbranched PIL used as the stabilizer, the size of the PS particles can be varied from the nanoscale regime to the micrometer one. Dispersion polymerization of styrene in ethanol in the presence of long-chain substituted PEHO-C18ImOTs affords stable dispersions of PS microparticles. Scanning electron microscopy (SEM) images indicated the formation of particles with a mean diameter of 3.1 μm and a narrow size distribution (see Figure 7.6). Reference experiments conducted in the presence of non-ionic *n*-octadecanoyl-modified PEHO, instead of PEHO-C18ImOTs, did not yield PS dispersions and thus emphasize the importance of the amphiphilic PIL structure for an effective stabilization of the PS microparticles. In fact, the PIL-stabilized PS micro- and nanoparticle dispersions represent interesting intermediates in the production of micro-structured films and nanostructured coatings, respectively.

In addition to the preparation of organic (nano)particles, PEHO-PILs are useful stabilizers in the synthesis and dispersion of various metal nanoparticles. These hyperbranched PILs provide excellent electrosteric stabilization and the variable polarity design enables to choose those ones with the most suitable substitution patterns with respect to the desired solvent. Octyl-substituted PEHO-C8ImOTs has been employed to produce stable dispersions of spherical silver and gold nanoparticles in *N,N*-dimethylacetamide by reduction of $AgNO_3$ and $HAuCl_4$, respectively, by $NaBH_4$.[143] In the absence of the micelle-like PIL or when using an analogue low-molecular-weight IL, no stable metal dispersions were obtained. Remarkably, in contrast to PEHO-C8ImOTs, the use of an analogous PIL having a semi-fluorinated outer shell instead of hydrocarbon chains resulted in the formation of both spherical and flaky silver nanoparticles as a function of the PIL concentration.[143] In the presence of organophilic PEHO-C18ImOTs, the thermolysis of appropriate metal carbonyl precursors carried out in non-polar solvents with a high boiling point, such as 1-octanol or diphenylmethane, afforded dispersions of iron, cobalt, and rhodium nanoparticles. Employing the same PIL as a stabilizer in the reduction of Pd(II) acetate by either ascorbic acid or hydrogen in toluene solution produced small palladium nanoparticles (Pd-NPs) with a mean diameter of between 4 nm and 6 nm.[145] By modifying the chemical structure of the hyperbranched stabilizer, even smaller particles can be obtained. In an octadecyl-substituted PIL that still contains residual hydroxyl functions, which can be easily prepared by tosylation of only 50% of the hydroxyl groups of PEHO prior to its further reaction with 1-*n*-octadecylimidazole, the hydroxymethyl groups can act as reducing agents themselves (*cf.* work by Mecking *et al.*[61]). When dissolving this bifunctional PIL and Pd(II) acetate in toluene and heating at 60 °C, Pd-NPs with a mean size of 3 nm were formed. In samples with a high PIL/Pd ratio, no particles were found at all, which indicated the formation of Pd clusters that were so small that they could not be detected by means of TEM. As evidenced by TEM analysis, the hyperbranched PILs can effectively prevent nanoparticle agglomeration. In Figure 7.7, TEM images of selected PIL-stabilized nano-metal dispersions are shown.

The PIL-stabilized Pd-NP dispersions represent highly active catalysts in the liquid phase hydrogenation of various olefins under mild conditions. In the presence of a PEHO-C18ImOTs/Pd-NP catalyst, exposure to hydrogen at normal pressure and room temperature rapidly converted styrene into ethylbenzene.[145] After only 20 minutes, quantitative conversion was achieved. Notably, as evidenced by a reference experiment performed under the same conditions, the catalytic activity of the Pd-NPs towards styrene hydrogenation was as high as that of a commercial palladium catalyst supported on activated carbon (Pd/C). For both PIL-stabilized Pd-NPs and for Pd/C, the turnover frequency and activity after 10 minutes were 530 h^{-1} and 6100 mmol product (g Pd)$^{-1}$ h^{-1}, respectively.[145] The PIL/Pd-NP catalyzed hydrogenation of other substrates, such as cyclohexene, 1-hexene, or

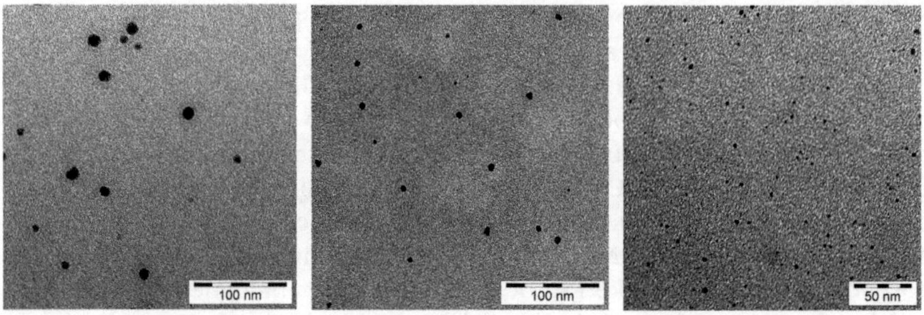

Figure 7.7 TEM images of PIL-stabilized metal nanoparticles (NPs): Co-NPs prepared by thermolysis of $Co_2(CO)_8$ in octanol in the presence of PEHO-C18ImOTs (left); Pd-NPs prepared by reduction of Pd(II) acetate by ascorbic acid in toluene in the presence of PEHO-C18ImOTs (middle); and Pd-NPs prepared by reduction of Pd(II) acetate by residual hydroxymethyl groups of partially functionalized PEHO-C18ImOTs in toluene (right).

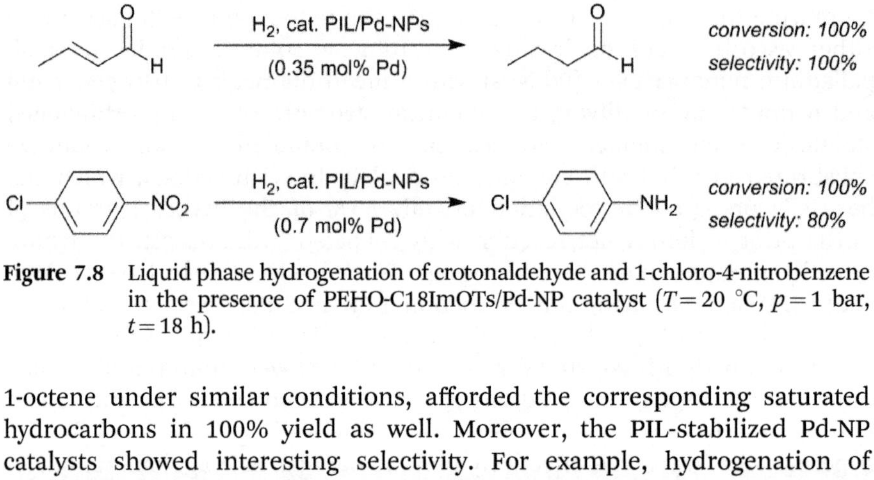

Figure 7.8 Liquid phase hydrogenation of crotonaldehyde and 1-chloro-4-nitrobenzene in the presence of PEHO-C18ImOTs/Pd-NP catalyst ($T = 20$ °C, $p = 1$ bar, $t = 18$ h).

1-octene under similar conditions, afforded the corresponding saturated hydrocarbons in 100% yield as well. Moreover, the PIL-stabilized Pd-NP catalysts showed interesting selectivity. For example, hydrogenation of crotonaldehyde did not affect the C=O double bond and exclusively yielded butyraldehyde. 1-Chloro-4-nitrobenzene was converted into *p*-chloroaniline with a selectivity of 80%, indicating that side or consecutive reactions such as the dechlorination to aniline are substantially suppressed in the presence of the PIL/Pd-NP catalyst (see Figure 7.8). Dehalogenation is a well-known problem in the preparation of halogenoanilines from halogenonitrobenzenes, as the amino substitution in the aromatic system labilizes the halogen.[146]

Due to their amphiphilic structure, the PEHO-based PILs represent effective phase transfer agents for transportation of hydrophilic molecules into non-polar media.[142,144] A series of experiments, where PILs with an outer shell of *n*-butyl, *n*-dodecyl, and *n*-octadecyl substituents were used for transferring the water-soluble dye Congo red from an aqueous phase into chloroform, demonstrated the transport efficacy of the PILs.[145] After extracting the aqueous phase with solutions of the PILs in chloroform at different concentrations, the absorption of the organic phase at 521 nm was determined as a measure for the amount of dye molecules that had been transferred. As can be seen from Figure 7.9, both PEHO-C12ImOTs and PEHO-C18ImOTs showed high transport efficacies even at low PIL concentration. This can be attributed to the pronounced reverse-micelle-type structure. In contrast, short-chain substituted PEHO-C4ImOTs was much less effective, and the shape of the curve indicates the formation of aggregates, rather than transportation in unimolecular micelles (*cf.* work by Zou *et al.*[147]). When taking into account the different molar masses of the PIL molecules related to the different lengths of the alkyl chains, it is obvious that PEHO-C18ImOTs exhibits the best transport properties.

The phase transfer activity of the amphiphilic hyperbranched PILs is not limited to polar molecules. PEHO-C18ImOTs enables the transportation of various nanomaterials as well. For example, magnetite nanoparticles and functionalized graphene nanosheets were readily transferred from water into toluene by PEHO-C18ImOTs, yielding highly stable dispersions.[142,144,148] However, a biphasic phase transfer approach is not necessarily required to produce stable nanomaterial dispersions in non-polar media. When directly mixing graphene with PEHO-C18ImOTs in toluene, the PIL effected dispersion and effective stabilization of the nanosheets.[142] In contrast, without the PIL, rapid sedimentation was observed (see Figure 7.10). This feature renders PEHO-C18ImOTs an attractive dispersing agent for applications where absolutely anhydrous dispersions are needed.

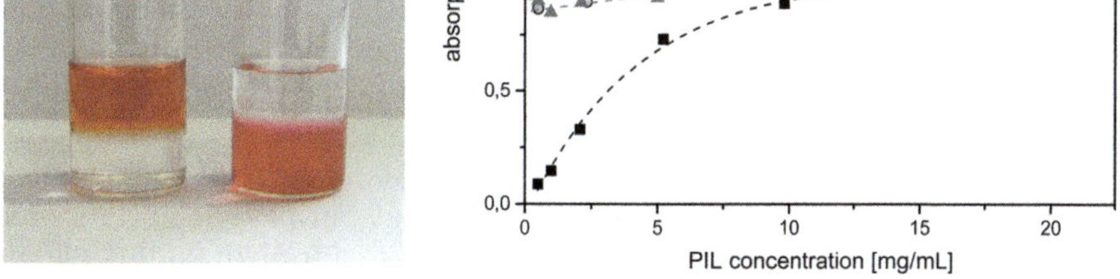

Figure 7.9 Phase transfer of Congo red from water into chloroform mediated by hyperbranched PILs: Congo red in water/chloroform without (a) and in the presence of (b) PEHO-C18ImOTs (left); absorption of the chloroform phase after transfer of the dye for PILs with different alkyl substituents as a function of the PIL concentration (right).

Figure 7.10 Response of a PIL-stabilized ferrofluid prepared by phase transfer of magnetite nanoparticles from water into toluene by PEHO-C18ImOTs to a permanent magnet (left) and graphene nanosheets in toluene in the absence (a) and presence (b) of PEHO-C18ImOTs, 1 hour after mixing and ultra-sonication treatment (right).

The capability of the hyperbranched PEHO-alkylimidazolium tosylates to function as transporters for polar molecules and nanoparticles can be easily expanded from solvents to thermoplastics. Due to their excellent thermal stability, chemical robustness, and the ability to withstand high shear forces, the PILs can be used in common polymer melt compounding processes, and adjustment of the alkyl chain length enables matching of the polarity with various polymers.

When melt blending methyl-substituted PEHO-C1ImOTs or *n*-butyl substituted PEHO-C4ImOTs with poly(styrene-*co*-acrylonitrile) (SAN) at 210 °C, the PILs formed uniformly distributed nanodroplets in the SAN matrix.[144,145] For PIL concentrations of up to 10 wt%, the diameters of the PIL phases were between 50 nm and 200 nm, as revealed by TEM analysis, and hence, the blends were fully transparent. Notably, the PIL additives did not affect the glass transition temperature of SAN at all, whereas the low-molecular-weight IL 1-butyl-3-methylimidazolium tosylate used for comparison severely plasticized the matrix polymer. When melt compounding SAN together with SiO_2 or TiO_2 nanoparticles in the presence of PEHO-C1ImOTs, the inorganic nanoparticles accumulated in the PIL nanophases, which emphasizes the scope of these micelle-like PILs to be employed in the fabrication of novel compartmentalized materials.

Increasing the alkyl chain length of the PILs to *n*-octadecyl enables their use as additives for non-polar polyolefins such as polypropylene (PP). Melt blending of PEHO-C18ImOTs and PP at 200 °C afforded PIL/PP blends comprising nanoscale PIL phases of spherical shape. The good compatibility of organophilic PEHO-C18ImOTs and PP was reflected by the uniform distribution and the small average size of the nanophases, which was 120 nm at 0.25 wt% PEHO-C18ImOTs content and slightly increased to 160 nm at a PIL concentration of 2.0 wt%.[142,144] When loading PEHO-C18ImOTs with the

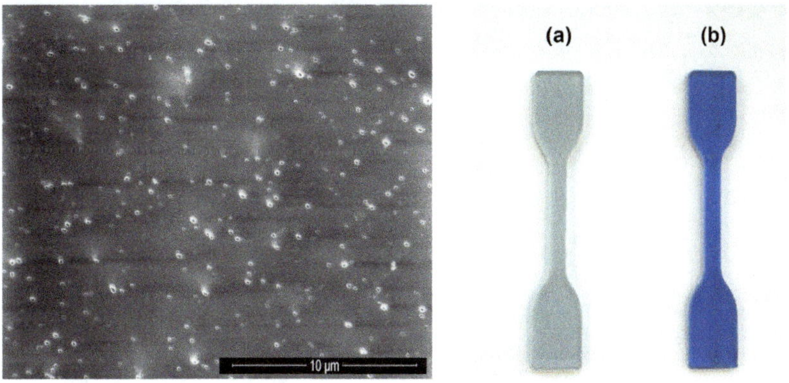

Figure 7.11 SEM image of a PIL/PP blend comprising 2.0 wt% PEHO-C18ImOTs (left); PP compounds containing 0.11 wt% E133 prepared without PIL (a) and prepared with the aid of 0.50 wt% PEHO-C18ImOTs (b), respectively (right).
Reproduced from ref. 142 with permission from John Wiley and Sons. Copyright © 2013 WILEY-VCH Verlag GmbH & Co. KGaA, Weinheim.

water-soluble food colorant E133 prior to melt compounding, the PIL effected a bright shining, brilliant coloration of the PP due to solubilization of the dye molecules in the PIL nanocompartments, which were so small that they did not prevent light transmission (see Figure 7.11). In contrast, without PIL, very poor dispersion of the polar dye was observed. Not only the incompatible PIL PEHO-C4ImOTs, which formed much larger phases in PP compared to PEHO-C18ImOTs, but also the low-molecular-weight IL 1-methyl-3-octadecylimidazolium tosylate failed to produce a brightly colored material, as scattering of light and poor dispersion of the dye, respectively, severely impaired the color effect.[142,144]

In an ethylene–octene copolymer, a highly gas-permeable polyolefin elastomer (POE), PEHO-C18ImOTs formed phases that were slightly larger, as in the case of PP, but still did not exceed a mean diameter of 200 nm for concentrations of up to 0.5 wt% PIL.[148] When using these PIL nanophases for solublization of the water-soluble pH indicator methyl red sodium salt in the non-polar POE during melt processing, an intensely and homogeneously colored material was obtained. Exposure of the dye-loaded PIL/POE blend to vapors of hydrochloric acid or aqueous ammonia induced a fast and reversible change of color, which was visible to the naked eye after a few seconds and became more and more obvious after several minutes. In Figure 7.12, this change of color is depicted in terms of photographs and UV/VIS spectra, recorded after 20 minutes of exposure to the respective vapors. For comparison, the blend placed over neutral water is shown, too. This simple example of a material responsive to acidic and basic vapors illustrates the potential of such amphiphilic hyperbranched PILs to be utilized in the fabrication of smart materials and functional systems.

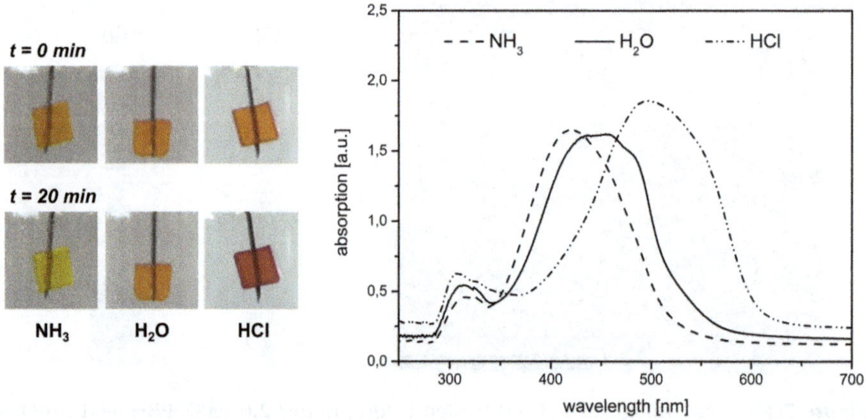

Figure 7.12 Change of color of an ethylene–octene copolymer blend comprising 0.25 wt% PEHO-C18ImOTs and 0.04 wt% methyl red sodium salt: photographs before exposure and after 20 minutes of exposure to vapors of aqueous ammonia, neutral water, and hydrochloric acid (left); UV/VIS spectra recorded after 20 minutes of exposure to the respective vapors (right).

Exploiting the ability of PEHO-C18ImOTs to effectively prevent agglomeration of functionalized graphene nanosheets in non-polar media, this PIL has been advantageously employed in the production of PS/graphene nanocomposites.[142,144] Following a solution blending approach, mixing of the components in toluene and precipitation afforded melt-processable nanocomposites comprising well-dispersed but interconnected graphene sheets. Comparison with a reference sample prepared under identical conditions with the same amount of graphene but without PIL revealed that the PIL effected a 20-fold increase in the electrical conductivity.[142] Obviously, the PIL improves significantly the dispersion of the graphene nanosheets in the PS matrix without preventing percolation. This feature of PEHO-C18ImOTs enormously broadens the scope of these hyperbranched PILs with respect to the production of advanced functional nanocomposites.

7.5 Conclusions

Inspired by naturally occurring micellar structures and micelle-forming synthetic surfactants and block copolymers, the development of macromolecules having an architecture that resembles one single micelle goes well beyond the scope of micelles formed by supramolecular self-assembly. Micelle-like macromolecules represent an intriguing new class of polymeric nanomaterials with attractive prospects for manifold applications. In contrast to micelles formed by self-assembly of myriads of individual molecules, such unimolecular micelle-like structures exhibit a stable conformation and there is no risk of disintegration upon exposure to shear, changes in temperature or pH, and salt addition.

In order to mimic micelles at the macromolecular level, both perfectly branched dendrimers and randomly branched polymers can be employed. In view of their dense topology, stable conformation, and multifunctionality, these highly branched macromolecules qualify in an ideal way as core molecules to produce a wide range of micelle-like polyelectrolytes and non-ionic amphiphiles by creating a shell of complementary polarity by means of versatile end-group conversion. Owing to their multifunctionality and their capability for the encapsulation of guest molecules and particles, there is intense research aiming at the synthesis of tailor-made macromolecules with sophisticated core–shell architectures. Today, particularly in view of their potential use in biomedical applications, the research focus in the field of highly branched amphiphiles mainly concentrates on the development of novel biocompatible transport systems that respond to cellular environments and interact with bioactive components, thus enabling controlled delivery and triggered release of active substances.

As a new generation of micelle-like polymers, micelle-like polymeric ionic liquids (PILs) are attracting considerable attention, since they combine the typical features of low-molecular-weight ILs and micelles with intrinsic properties of polymeric materials, such as low leakage levels, robustness, low toxicity, and facile processing. Going well beyond micro and nanostructure formation in linear PILs, tailoring of highly branched PILs affords amphiphilic and compartmentalized macromolecules with micelle-like core–shell topologies. In order to qualify for highly diversified applications, ranging from dispersing agents to nanoreactor compartments and polymer nanocomposites, such micelle-like PILs need to be chemically robust, exhibit high thermal stability, and should allow for the facile adjustment of their polarity and properties. Hyperbranched polyoxetanes bearing alkylimidazolium moieties in their periphery represent a versatile micelle- and onion-like PIL family that fulfills all of these requirements. They represent effective emulsifiers and dispersants in heterogeneous polymerization reactions. When employed in the synthesis of metal nanoparticles, highly stable and catalytically active metal nanoparticle dispersions in organic media are obtained. The long-chain-alkyl substituted reverse-micelle-type PILs are effective carriers for transportation of water-soluble molecules and even nanomaterials from an aqueous phase into non-polar solvents. By means of simple melt compounding, uniformly dispersed PIL nanodroplets are formed in thermoplastics such as SAN and even non-polar polyolefins. These PIL nanophases represent valuable compartments for hosting various molecules and nanomaterials. For instance, water-soluble dye molecules are transported into hydrophobic PP. Moreover, PIL-mediated dispersion of graphene nanosheets in PS substantially increases the electrical conductivity compared to ordinary PS/graphene composites.

This family of polyoxetane-based PILs with micelle-like topology illustrates the attractive potential of highly branched amphiphilic PILs for a great variety of applications. Particularly, in view of potential industrial applications ranging from printing inks to catalyst supports and nanocomposites, owing to their availability in a one-step process, the use of hyperbranched polymers as core molecules for facile molecular tailoring of micelle-like PILs is advantageous with respect to dendrimers, which require costly and tedious multi-step syntheses. The development of hyperbranched micelle-like PILs holds great prospects for designing advanced materials and systems, among them membranes, sensors, energy storage and conversion devices, recyclable catalysts, polymer nanocomposites, and responsive materials and surfaces.

Acknowledgements

The authors thank Gil Markus Scheuermann, Institute for Macromolecular Chemistry, Albert-Ludwigs-University of Freiburg, and Mario Stein and Jörg Wieland, Institute for Organic Chemistry, Albert-Ludwigs-University of Freiburg, for the evaluation of the catalytic activity of the PIL-stabilized palladium nanoparticle dispersions in hydrogenation reactions.

References

1. G. R. Thompson, *J. Clin. Pathol., Suppl. (R. Coll. Pathol.)*, 1971, **5**, 85.
2. H. I. Friedman and B. Nylund, *Am. J. Clin. Nutr.*, 1980, **33**, 1108.
3. W. Dowhan, in *Advances in Lipobiology*, ed. R. W. Gross, Elsevier, 1997, vol. 2, pp. 79–107.
4. A. R. Thiam, R. V. Farese Jr. and T. C. Walther, *Nat. Rev. Mol. Cell Biol.*, 2013, **14**, 775.
5. D. Segré, D. Ben-Eli, D. W. Deamer and D. Lancet, *Origins Life Evol. Biospheres*, 2001, **31**, 119.
6. C. Holt, J. A. Carver, H. Ecroyd and D. C. Thorn, *J. Dairy Sci.*, 2013, **96**, 6127.
7. F. H. Schacher, P. A. Rupar and I. Manners, *Angew. Chem., Int. Ed.*, 2012, **51**, 7898.
8. V. M. Nace, *Nonionic Surfactants: Polyoxyalkylene Block Copolymers*, Marcel Dekker, New York, 1996.
9. C. Dingels, M. Schömer and H. Frey, *Chem. Unserer Zeit*, 2011, **45**, 338.
10. U. Zoller, *Handbook of Detergents, Part E: Applications*, CRC Press, Boca Raton, 2008.
11. Y. Shen and J. Duhamel, *Langmuir*, 2008, **24**, 10665.
12. M. Antonietti and S. Förster, *Adv. Mater.*, 2003, **15**, 1323.
13. G. Riess, *Prog. Polym. Sci.*, 2003, **28**, 1107.
14. Z. L. Tyrrell, Y. Shen and M. Radosz, *Prog. Polym. Sci.*, 2010, **35**, 1128.
15. X.-B. Xiong, A. Falamarzian, S. M. Garg and A. Lavasanifar, *J. Controlled Release*, 2011, **155**, 248.
16. R. P. Brinkhuis, F. P. J. T. Rutjes and J. C. M. van Hest, *Polym. Chem.*, 2011, **2**, 1449.
17. J. S. Lee and J. Feijen, *J. Controlled Release*, 2012, **161**, 473.
18. R. Bleul, R. Thiermann and M. Maskos, *Macromolecules*, 2015, **48**, 7396.

19. D. M. Vriezema, P. M. L. Garcia, N. Sancho Oltra, N. S. Hatzakis, S. M. Kuiper, R. J. M. Nolte, A. E. Rowan and J. C. M. van Hest, *Angew. Chem., Int. Ed.*, 2007, **46**, 7378.
20. D. Gräfe, J. Gaitzsch, D. Appelhans and B. Voit, *Nanoscale*, 2014, **6**, 10752.
21. K. Langowska, C. G. Palivan and W. Meier, *Chem. Commun.*, 2013, **49**, 128.
22. D. F. Evans and B. W. Ninham, *J. Phys. Chem.*, 1986, **90**, 226.
23. M. Desai, N. J. Jain, R. Sharma and P. Bahadur, *J. Surfactants Deterg.*, 2000, **3**, 193.
24. F. Vögtle, G. Richardt and N. Werner, *Dendritische Moleküle: Konzepte, Synthesen, Eigenschaften, Anwendungen*, Teubner Verlag, Wiesbaden, 2007.
25. E. Buhleier, W. Wehner and F. Vögtle, *Synthesis*, 1978, **1978**, 155.
26. R. G. Denkewalter, J. Kolc and W. J. Lukasavage, *U.S. Pat.* 4289872, 1981.
27. D. A. Tomalia, H. Baker, J. Dewald, M. Hall, G. Kallos, S. Martin, J. Roeck, J. Ryder and P. Smith, *Polym. J.*, 1985, **17**, 117.
28. G. R. Newkome, Z.-q. Yao, G. R. Baker and V. K. Gupta, *J. Org. Chem.*, 1985, **50**, 2003.
29. C. Hawker and J. M. J. Fréchet, *J. Chem. Soc., Chem. Commun.*, 1990, 1010.
30. C. Wörner and R. Mülhaupt, *Angew. Chem. Int. Ed. Engl.*, 1993, **32**, 1306.
31. E. M. M. de Brabander-van den Berg and E. W. Meijer, *Angew. Chem. Int. Ed. Engl.*, 1993, **32**, 1308.
32. P. J. Flory, *J. Am. Chem. Soc.*, 1941, **63**, 3083.
33. P. J. Flory, *J. Am. Chem. Soc.*, 1941, **63**, 3091.
34. P. J. Flory, *J. Am. Chem. Soc.*, 1941, **63**, 3096.
35. H. R. Kricheldorf, Q.-Z. Zang and G. Schwarz, *Polymer*, 1982, **23**, 1821.
36. P. A. Gunatillake, G. Odian and D. A. Tomalia, *Macromolecules*, 1988, **21**, 1556.
37. Y. H. Kim and O. W. Webster, *Polym. Prepr. (Am. Chem. Soc., Div. Polym. Chem.)*, 1988, **29**, 310.
38. Y. H. Kim and O. W. Webster, *J. Am. Chem. Soc.*, 1990, **112**, 4592.
39. M. G. McKee, S. Unal, G. L. Wilkes and T. E. Long, *Prog. Polym. Sci.*, 2005, **30**, 507.
40. B. Voit, *J. Polym. Sci., Part A: Polym. Chem.*, 2005, **43**, 2679.
41. M. Scholl, Z. Kadlecova and H.-A. Klok, *Prog. Polym. Sci.*, 2009, **34**, 24.
42. P. Froehling, *J. Polym. Sci., Part A: Polym. Chem.*, 2004, **42**, 3110.
43. E. Malmström, M. Johansson and A. Hult, *Macromolecules*, 1995, **28**, 1698.
44. M. Bednarek, T. Biedron, J. Helinski, K. Kaluzynski, P. Kubisa and S. Penczek, *Macromol. Rapid Commun.*, 1999, **20**, 369.
45. H. Magnusson, E. Malmström and A. Hult, *Macromol. Rapid Commun.*, 1999, **20**, 453.
46. M. Schömer, C. Schüll and H. Frey, *J. Polym. Sci., Part A: Polym. Chem.*, 2013, **51**, 995.

47. A. Sunder, R. Hanselmann, H. Frey and R. Mülhaupt, *Macromolecules*, 1999, **32**, 4240.
48. R. Hanselmann, D. Hölter and H. Frey, *Macromolecules*, 1998, **31**, 3790.
49. B. I. Voit and A. Lederer, *Chem. Rev.*, 2009, **109**, 5924.
50. Y. Zheng, S. Li, Z. Weng and C. Gao, *Chem. Soc. Rev.*, 2015, **44**, 4091.
51. C. Gao and D. Yan, *Prog. Polym. Sci.*, 2004, **29**, 183.
52. G. R. Newkome, C. N. Moorefield, G. R. Baker, M. J. Saunders and S. H. Grossman, *Angew. Chem.*, 1991, **103**, 1207.
53. M. W. P. L. Baars, R. Kleppinger, M. H. J. Koch, S.-L. Yeu and E. W. Meijer, *Angew. Chem., Int. Ed.*, 2000, **39**, 1285.
54. S. Xu, Y. Luo and R. Haag, *Macromol. Rapid Commun.*, 2008, **29**, 171.
55. C. Li, D. Li, C. Qiu and W. Hou, *J. Polym. Res.*, 2013, **20**, 204.
56. A. Sunder, M. Krämer, R. Hanselmann, R. Mülhaupt and H. Frey, *Angew. Chem., Int. Ed.*, 1999, **38**, 3552.
57. S.-E. Stiriba, H. Kautz and H. Frey, *J. Am. Chem. Soc.*, 2002, **124**, 9698.
58. M. Krämer, M. Kopaczynska, S. Krause and R. Haag, *J. Polym. Sci., Part A: Polym. Chem.*, 2007, **45**, 2287.
59. M. Q. Slagt, S.-E. Stiriba, R. J. M. Klein Gebbink, H. Kautz, H. Frey and G. van Koten, *Macromolecules*, 2002, **35**, 5734.
60. T. Mizugaki, Y. Miyauchi, M. Murata, K. Ebitani and K. Kaneda, *Chem. Lett.*, 2005, **34**, 286.
61. S. Mecking, R. Thomann, H. Frey and A. Sunder, *Macromolecules*, 2000, **33**, 3958.
62. M. Ooe, M. Murata, T. Mizugaki, K. Ebitani and K. Kaneda, *Nano Lett.*, 2002, **2**, 999.
63. U. Schlotterbeck, C. Aymonier, R. Thomann, H. Hofmeister, M. Tromp, W. Richtering and S. Mecking, *Adv. Funct. Mater.*, 2004, **14**, 999.
64. C. Deraedt, L. Salmon, L. Etienne, J. Ruiz and D. Astruc, *Chem. Commun.*, 2013, **49**, 8169.
65. M. Irfan and M. Seiler, *Ind. Eng. Chem. Res.*, 2010, **49**, 1169.
66. S. Gupta, R. Tyagi, V. S. Parmar, S. K. Sharma and R. Haag, *Polymer*, 2012, **53**, 3053.
67. X. Zhang, X. Zhang, P. Yu, Y. Han, Y. Li and C. Li, *J. Pharm. Sci.*, 2013, **102**, 145.
68. X. Zeng, W. Tao, Z. Wang, X. Zhang, H. Zhu, Y. Wu, Y. Gao, K. Liu, Y. Jiang, L. Huang, L. Mei and S.-S. Feng, *Part. Part. Syst. Charact.*, 2015, **32**, 112.
69. E. Mohammadifar, A. Nemati Kharat and M. Adeli, *J. Mater. Chem. B*, 2015, **3**, 3896.
70. I. N. Kurniasih, J. Keilitz and R. Haag, *Chem. Soc. Rev.*, 2015, **44**, 4145.
71. J. Dernedde, A. Rausch, M. Weinhart, S. Enders, R. Tauber, K. Licha, M. Schirner, U. Zügel, A. von Bonin and R. Haag, *Proc. Natl. Acad. Sci. U. S. A.*, 2010, **107**, 19679.
72. R. Haag, J. Dernedde, R. Tauber, B. Gesche, S. Enders and H. Weinhart, *WO Pat.* 2008/015015 A2, 2008.

73. D. Maysinger, D. Gröger, A. Lake, K. Licha, M. Weinhart, P. K.-Y. Chang, R. Mulvey, R. Haag and R. A. McKinney, *Biomacromolecules*, 2015, **16**, 3073.
74. N. V. Plechkova and K. R. Seddon, *Chem. Soc. Rev.*, 2008, **37**, 123.
75. P. Hapiot and C. Lagrost, *Chem. Rev.*, 2008, **108**, 2238.
76. P. Wasserscheid and W. Keim, *Angew. Chem., Int. Ed.*, 2000, **39**, 3772.
77. J. P. Hallett and T. Welton, *Chem. Rev.*, 2011, **111**, 3508.
78. R. Hagiwara and J. S. Lee, *Electrochemistry*, 2007, **75**, 23.
79. D. R. MacFarlane, M. Forsyth, P. C. Howlett, J. M. Pringle, J. Sun, G. Annat, W. Neil and E. I. Izgorodina, *Acc. Chem. Res.*, 2007, **40**, 1165.
80. H. Ohno and K. Ito, *Chem. Lett.*, 1998, **27**, 751.
81. H. Ohno, *Electrochim. Acta*, 2001, **46**, 1407.
82. H. Ohno, M. Yoshizawa and W. Ogihara, *Electrochim. Acta*, 2004, **50**, 255.
83. D. Mecerreyes, *Prog. Polym. Sci.*, 2011, **36**, 1629.
84. J. Yuan and M. Antonietti, *Polymer*, 2011, **52**, 1469.
85. J. Yuan, D. Mecerreyes and M. Antonietti, *Prog. Polym. Sci.*, 2013, **38**, 1009.
86. R. Marcilla, J. A. Blazquez, R. Fernandez, H. Grande, J. A. Pomposo and D. Mecerreyes, *Macromol. Chem. Phys.*, 2005, **206**, 299.
87. R. L. Weber, Y. Ye, S. M. Banik, Y. A. Elabd, M. A. Hickner and M. K. Mahanthappa, *J. Polym. Sci., Part B: Polym. Phys.*, 2011, **49**, 1287.
88. A.-L. Pont, R. Marcilla, I. De Meatza, H. Grande and D. Mecerreyes, *J. Power Sources*, 2009, **188**, 558.
89. M. Döbbelin, I. Azcune, M. Bedu, A. Ruiz de Luzuriaga, A. Genua, V. Jovanovski, G. Cabañero and I. Odriozola, *Chem. Mater.*, 2012, **24**, 1583.
90. J. H. Lee, J. S. Lee, J.-W. Lee, S. M. Hong and C. M. Koo, *Eur. Polym. J.*, 2013, **49**, 1017.
91. J. Tang, H. Tang, W. Sun, H. Plancher, M. Radosz and Y. Shen, *Chem. Commun.*, 2005, 3325.
92. S. Supasitmongkol and P. Styring, *Energy Environ. Sci.*, 2010, **3**, 1961.
93. J. López-Darias, J. L. Anderson, V. Pino and A. M. Afonso, *Anal. Bioanal. Chem.*, 2011, **401**, 2965.
94. A. Wolfson, I. F. J. Vankelecom and P. A. Jacobs, *Tetrahedron Lett.*, 2003, **44**, 1195.
95. D. Liu and E. Y.-X. Chen, *Biomass Bioenergy*, 2013, **48**, 181.
96. T. Y. Kim, M. Suh, S. J. Kwon, T. H. Lee, J. E. Kim, Y. J. Lee, J. H. Kim, M. P. Hong and K. S. Suh, *Macromol. Rapid Commun.*, 2009, **30**, 1477.
97. Y. Gu and T. P. Lodge, *Macromolecules*, 2011, **44**, 1732.
98. Y. Ye, J.-H. Choi, K. I. Winey and Y. A. Elabd, *Macromolecules*, 2012, **45**, 7027.
99. J.-H. Choi, Y. Ye, Y. A. Elabd and K. I. Winey, *Macromolecules*, 2013, **46**, 5290.
100. K. M. Meek and Y. A. Elabd, *J. Mater. Chem. A*, 2015, **3**, 24187.

101. K. Hoshino, M. Yoshio, T. Mukai, K. Kishimoto, H. Ohno and T. Kato, *J. Polym. Sci., Part A: Polym. Chem.*, 2003, **41**, 3486.
102. S. Lee, G. A. Becht, B. Lee, C. T. Burns and M. A. Firestone, *Adv. Funct. Mater.*, 2010, **20**, 2063.
103. M. Tokuda, H. Minami, Y. Mizuta and T. Yamagami, *Macromol. Rapid Commun.*, 2012, **33**, 1130.
104. J. Yuan and M. Antonietti, *Macromolecules*, 2011, **44**, 744.
105. J. Yuan, S. Soll, M. Drechsler, A. H. E. Müller and M. Antonietti, *J. Am. Chem. Soc.*, 2011, **133**, 17556.
106. M. Koebe, M. Drechsler, J. Weber and J. Yuan, *Macromol. Rapid Commun.*, 2012, **33**, 646.
107. R. Marcilla, M. Sanchez-Paniagua, B. Lopez-Ruiz, E. Lopez-Cabarcos, E. Ochoteco, H. Grande and D. Mecerreyes, *J. Polym. Sci., Part A: Polym. Chem.*, 2006, **44**, 3958.
108. Y. Xiong, Y. Wang, H. Wang and R. Wang, *Polym. Chem.*, 2011, **2**, 2306.
109. X. Liang, H. Xiao and C. Qi, *Fuel Process. Technol.*, 2013, **110**, 109.
110. J. Yuan, S. Prescher, K. Sakaushi and M. Antonietti, *J. Mater. Chem. A*, 2015, **3**, 7229.
111. J. Huang, C.-a. Tao, Q. An, W. Zhang, Y. Wu, X. Li, D. Shen and G. Li, *Chem. Commun.*, 2010, **46**, 967.
112. Q. Zhao, J. W. C. Dunlop, X. Qiu, F. Huang, Z. Zhang, J. Heyda, J. Dzubiella, M. Antonietti and J. Yuan, *Nat. Commun.*, 2014, **5**, 4293.
113. K. Täuber, B. Lepenies and J. Yuan, *Polym. Chem.*, 2015, **6**, 4855.
114. R. Marcilla, M. L. Curri, P. D. Cozzoli, M. T. Martínez, I. Loinaz, H. Grande, J. A. Pomposo and D. Mecerreyes, *Small*, 2006, **2**, 507.
115. T. Y. Kim, H. W. Lee, J. E. Kim and K. S. Suh, *ACS Nano*, 2010, **4**, 1612.
116. X.-d. Mu, J.-q. Meng, Z.-C. Li and Y. Kou, *J. Am. Chem. Soc.*, 2005, **127**, 9694.
117. K. T. Prabhu Charan, N. Pothanagandhi, K. Vijayakrishna, A. Sivaramakrishna, D. Mecerreyes and B. Sreedhar, *Eur. Polym. J.*, 2014, **60**, 114.
118. R. Marcilla, E. Ochoteco, C. Pozo-Gonzalo, H. Grande, J. A. Pomposo and D. Mecerreyes, *Macromol. Rapid Commun.*, 2005, **26**, 1122.
119. M. Salsamendi, J. Abad, R. Marcilla, C. Pozo-Gonzalo, A. Urbina, H. Grande, J. Colchero and D. Mecerreyes, *Polym. Adv. Technol.*, 2011, **22**, 1665.
120. S. H. Hong, T. T. Tung, L. K. Huyen Trang, T. Y. Kim and K. S. Suh, *Colloid Polym. Sci.*, 2010, **288**, 1013.
121. T. Fukushima, A. Kosaka, Y. Yamamoto, T. Aimiya, S. Notazawa, T. Takigawa, T. Inabe and T. Aida, *Small*, 2006, **2**, 554.
122. T. Y. Kim, H. W. Lee, M. Stoller, D. R. Dreyer, C. W. Bielawski, R. S. Ruoff and K. S. Suh, *ACS Nano*, 2011, **5**, 436.
123. Y.-S. Ye, C.-Y. Tseng, W.-C. Shen, J.-S. Wang, K.-J. Chen, M.-Y. Cheng, J. Rick, Y.-J. Huang, F.-C. Chang and B.-J. Hwang, *J. Mater. Chem.*, 2011, **21**, 10448.
124. H. Chen and G. Zhao, *J. Solid State Electrochem.*, 2012, **16**, 3289.

125. C. Willa, J. Yuan, M. Niederberger and D. Koziej, *Adv. Funct. Mater.*, 2015, **25**, 2537.
126. Y. Yang, M. Ambrogi, H. Kirmse, Y. Men, M. Antonietti and J. Yuan, *Chem. Mater.*, 2015, **27**, 127.
127. W. Xu, P. A. Ledin, V. V. Shevchenko and V. V. Tsukruk, *ACS Appl. Mater. Interfaces*, 2015, **7**, 12570.
128. P. Yu, Xu, T. Yi Guo, C. Hui Zhao, I. Broadwell, Q. Gen Zhang and Q. L. Liu, *J. Appl. Polym. Sci.*, 2013, **128**, 3853.
129. J. L. Kreider and W. T. Ford, *J. Polym. Sci., Part A: Polym. Chem.*, 2001, **39**, 821.
130. S. Monmoton, H. Lefebvre, F. Costa-Torro and A. Fradet, *Macromol. Chem. Phys.*, 2008, **209**, 2382.
131. A. S. Amarasekara, B. Callis and B. Wiredu, *Polym. Bull.*, 2012, **68**, 901.
132. K. Naka, R. Shinke, M. Yamada, F. Djouadi Belkada, Y. Aijo, Y. Irie, S. Ram Shankar, K. Sai Smaran, N. Matsumi, S. Tomita and S. Sakurai, *Polym. J.*, 2014, **46**, 42.
133. F. Tack, H. M. Janssen, E. W. Meijer, M. M. F. Janicot and M. E. Brewster, *WO Pat.* 2007/006700 A1, 2007.
134. E. Schwab and S. Mecking, *J. Polym. Sci., Part A: Polym. Chem.*, 2005, **43**, 4609.
135. E. Schwab and S. Mecking, *Organometallics*, 2005, **24**, 3758.
136. M. Tamaki, T. Taguchi, Y. Kitajyo, K. Takahashi, R. Sakai, T. Kakuchi and T. Satoh, *J. Polym. Sci., Part A: Polym. Chem.*, 2009, **47**, 7032.
137. T. Zheng, S. Ren, Q. Zhou, Q. Li, L. Zhang, H. Li and Y. Lin, *J. Polym. Res.*, 2014, **21**, 361.
138. Y. Fan, D. Zhang, J. Wang, H. Jin, Y. Zhou and D. Yan, *Chem. Commun.*, 2015, **51**, 7234.
139. Y. Du, Q. Chen, L. Shen, Y. Xing and J. Dai, *J. Appl. Polym. Sci.*, 2011, **121**, 2927.
140. V. V. Shevchenko, A. V. Stryutsky, N. S. Klymenko, M. A. Gumenna, A. A. Fomenko, V. N. Bliznyuk, V. V. Trachevsky, V. V. Davydenko and V. V. Tsukruk, *Polymer*, 2014, **55**, 3349.
141. T. V. Richter, C. Bühler and S. Ludwigs, *J. Am. Chem. Soc.*, 2012, **134**, 43.
142. F. Schüler, B. Kerscher, F. Beckert, R. Thomann and R. Mülhaupt, *Angew. Chem., Int. Ed.*, 2013, **52**, 455.
143. K. Schadt, B. Kerscher, R. Thomann and R. Mülhaupt, *Macromolecules*, 2013, **46**, 4799.
144. B. Kerscher, F. Schüler, A.-K. Appel, K. Schadt and R. Mülhaupt, in *Hierarchical Macromolecular Structures: 60 Years after the Staudinger Nobel Prize II*, ed. V. Percec, Springer International Publishing, 2013, vol. 262, pp. 431–446.
145. F. Schüler, PhD thesis, Albert-Ludwigs-University of Freiburg, 2009.
146. R. Baltzly and A. P. Phillips, *J. Am. Chem. Soc.*, 1946, **68**, 261.
147. J. Zou, Y. Zhao and W. Shi, *J. Phys. Chem. B*, 2006, **110**, 2638.
148. B. Kerscher, PhD thesis, Albert-Ludwigs-University of Freiburg, 2013.

CHAPTER 8

Biological and Bio-inspired Heterogeneous Composites: From Resilient Palm Trees to Stretchable Electronics

MARKUS RÜGGEBERG,[a,b] ANDRÉ R. STUDART*[c] AND INGO BURGERT*[a,b]

[a] Institute for Building Materials, ETH Zürich, Stefano-Franscini-Platz 3, 8093 Zürich, Switzerland; [b] Applied Wood Materials, EMPA – Swiss Federal Laboratories for Materials Science and Technology, Überlandstrasse 129, 8600 Dübendorf, Switzerland; [c] Complex Materials, Department of Materials, ETH Zürich, Vladimir-Prelog-Weg 5, 8093 Zürich, Switzerland
*Email: andre.studart@mat.ethz.ch; iburgert@ethz.ch

8.1 Introduction

It has been shown in various studies that biological materials can reach impressive architectures and mechanical properties by smart hierarchical assembly and interface design.[1] Yet, examples of successful transfer of the principles underlying such natural architectures into bio-inspired materials are still rather scarce. Designing materials that combine stiffness, toughness, and strength, as well as reach reasonable dimensions as one can find in nature, is still one of the major challenges in materials science and engineering.[2,3]

RSC Polymer Chemistry Series No. 22
Bio-inspired Polymers
Edited by Nico Bruns and Andreas F. M. Kilbinger

Published by the Royal Society of Chemistry, www.rsc.org

In fact, there are many possible biological role models, *e.g.* wood,[4–6] bone,[7] antlers,[8] spider silk,[9] byssal threads,[10] and nacre,[11] with astonishing mechanical performances. They combine stiffness, strength and toughness in an optimized fashion for the organism while, at the same time, most of them are lightweight materials. This is remarkable in view of the common assembly of biological materials made up of only a small number of elements and a rather limited number of different building blocks. However, linking the outstanding mechanical properties to the specific structure of the organisms has quite often proven to be challenging due to the complexity of the structure itself. Studies on various biological materials with very different structures and functions have coincided in identifying the material heterogeneity smartly embedded in the structural hierarchy as well as the interface design as key parameters for their remarkable performance.[12,13] While homogeneous materials are often either stiff but brittle or ductile but soft, a certain level of heterogeneity together with a balanced assembly of components with different intrinsic properties can result in desired combined features such as optimized stiffness and toughness.[14] Hierarchical structuring, in turn, tremendously enlarges the possibilities and range of adaptation of mechanical properties for a given set of components. However, the resulting mechanical properties of such composite materials cannot only be tuned through the choice, distribution, and orientation of the single building blocks, but also by the design of the interface between the components.[15] Thus, a well-balanced combination of heterogeneity, hierarchy, and interfacial design allows for tailoring and optimizing the mechanical behavior.

A deep understanding of the underlying principles of such concerted composites and their extraction is the basis for a successful biomimetic transfer[12] for the design of a new class of synthetic, bio-inspired heterogeneous materials with outstanding property combinations. In the following, we will present examples of the underlying principles of hierarchy, structural adaptation, and interfacial design of biological materials focussing on the plant kingdom. The common natural building blocks of palms and trees are introduced and a short general comparison of material adaptation between both systems is presented. On this basis, we particularly focus on the formation of heterogeneous composites in palm tissue structures, as they show an eminent material adaptation strategy. Thereafter, synthetic bio-inspired materials, designed on the basis of the aforementioned biological principles, will be exemplarily presented. Polymer-based composites with hierarchical reinforcement and heterogeneous mechanical properties are used as substrates to enable the fabrication of a highly stretchable electronic device. Taking these two examples of a natural and a synthetic heterogeneous composite, special attention will be put on the incorporation of gradients in these materials and its proven or potential benefits for the specific loading conditions. A final discussion at the end of this review compiles a brief comparison of natural heterogeneous composite materials and their bio-inspired counterparts, in order to evaluate the current level of biomimetic transfer and thereby critically discuss the concept of gradual transitions in interfacial design.

8.2 The Natural Building Blocks of Plants

Generally, plant cell walls represent a natural fiber-reinforced nanocomposite made up of stiff cellulose fibrils of high aspect ratio, embedded in pliant matrix polymers.[16–18] These basic building blocks are quite similar in terms of structure and composition for all plants (and thus trees and palms); however, the volume fractions of the different polymers and the orientation of the cellulose microfibrils can vary considerably, depending on the different types of tissue or cell wall layers.[19] In trees and palms, the cell walls consist of a middle lamella made up of pectin and lignin that glues the cells together. A thin primary cell wall consisting mainly of cellulose, hemicelluloses, and pectin allows for cell expansion during the differentiation of cells. After the cells have reached their final size and shape, a thick and rigid secondary cell wall made up of cellulose, hemicelluloses, and lignin is formed, which will be the focus in the following.

By being able to control the composition of the basic building blocks at the level of the natural fiber-reinforced cell walls, plants possess powerful tools for material adaptation towards function. In principle, plants can utilize at least three potential ways of tuning the mechanical properties at the secondary cell wall level (Figure 8.1). The first is simply based on cell wall material accumulation, by means of the formation of thinner or thicker cell walls, which affects tissue density and thereby stiffness and strength (Figure 8.1a). The second strategy is related to the orientation of the stiff cellulose fibrils (Figure 8.1b). Within the secondary cell wall, the cellulose fibrils run parallel to each other, their inclination to the longitudinal axis of the cell is termed microfibril angle. Due to the very high tensile stiffness of cellulose compared to that of hemicelluloses and lignin,[20,21] the cellulose fibrils are the main load bearing elements within the cell wall. Consequently, their orientation within the cell wall layers has a

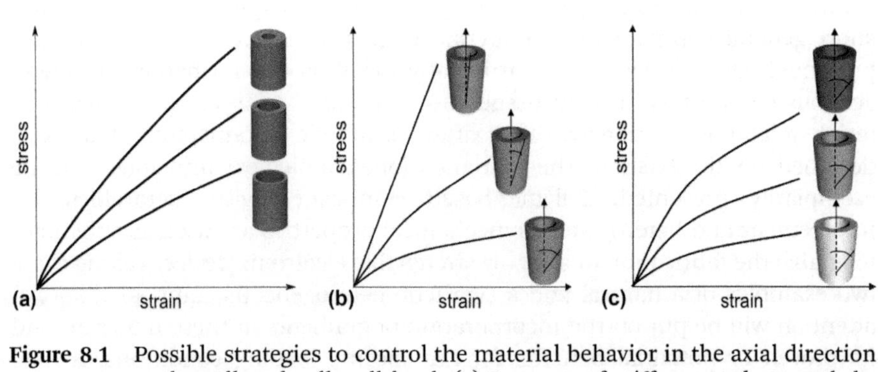

Figure 8.1 Possible strategies to control the material behavior in the axial direction at the cell and cell wall level. (a) Increase of stiffness and strength by thicker cell walls (tissue density). (b) Change of cellulose microfibril angle. (c) Alteration of matrix stiffness for a given high microfibril angle. Adapted from Eder *et al.*,[26] 2009, New Zealand Forest Research Institute Limited.

prevalent influence on the mechanical properties (*e.g.* stiffness and toughness), and numerous studies have proven that the adaptation of this parameter at the micrometer scale serves for adapting the mechanical properties of cell walls and tissues to the requirements at the macroscopic scale.[22–25] A small microfibril angle with cellulose fibrils oriented almost parallel to the cell axis results in a stiff but rather brittle material behavior, whereas a large microfibril angle renders the plant material less stiff but tough. Hence, the orientation of the cellulose fibril in the cell wall is a key parameter for plants, and in particular trees, to control and tune their mechanical behavior at the cell wall level. The third way of adapting the mechanical properties at the cell wall level is based on the specific interplay of the stiff cellulose fibrils with the soft matrix components. While this influence is rather minor for small microfibril angles, it becomes more dominant when the cellulose fibrils are more tilted and the microfibril angle is higher (Figure 8.1c). Under these conditions, the axial stiffness of the material is strongly influenced by the shear stiffness of the matrix components and, hence, alterations in the lignin content can be utilized to tune the plant's material response.

8.3 Palms as Role Models for Biological Heterogeneous Composites

It is a sight to behold when strong and rigid palm trunks bend in tropical thunderstorms. Likewise, young and slender trees have the flexibility to be strongly distorted without catastrophic failure; however, they lose this ability when growing thicker during aging. Although both palms and young trees show a similar behavior, being governed by an optimized combination of stiffness and toughness, the structures of these two plant systems are entirely different at most levels of the structural hierarchy.[26] This is particularly due to the different growth patterns of dicot trees and monocot palms. Trees possess so-called secondary growth, which leads to a subsequent increase of the diameter of the trunk. Palms do not show secondary thickening growth, and only a moderate increase of the trunk diameter during aging becomes visible.[27] This causes strong mechanical implications, since trees can increase the second moment of area by increasing the trunk diameter to compensate for increasing bending moments, a geometric adaptability which is very limited in palm species. As a consequence, palms need to almost exclusively rely on a material adaption strategy while growing taller, which takes place by ongoing and lifelong cell and tissue alterations, which is hardly the case in the wood of trees. Hence, a major difference of palms and trees at the cell level is that most of the wood cells die after their final formation, whereas most palm cells stay alive and are able to undergo further structural and chemical alterations. These growth constraints and the ensuing advanced material adaptability, which is presented in the following across different levels of structural hierarchy, make palms particularly interesting biological role models to be studied with regard to heterogeneous composites for biomimetic transfer.

8.3.1 Hierarchical Structure and Mechanics of Palms

The palm stem represents a fibre-reinforced composite material at several hierarchical levels. At the macroscopic tissue scale, a complex 3D-interwoven network of fibrous strands, so called vascular bundles, is embedded in a soft matrix of globular, parenchymatous tissue (Figure 8.2a).[28] A vascular bundle consists of conducting elements and a fiber cap, with the latter one reaching diameters of up to two millimeters (Figure 8.2d). The fiber cap is by far denser and stiffer than the surrounding parenchymatous tissue and thus, density and stiffness of the trunk are closely related to the distribution of the vascular bundles in the trunk.[29] Consequently, different local volume fractions of vascular bundles in the trunk result in gradients in density and thus stiffness in the radial direction (Figure 8.2b and c) and height. The gradient in stiffness in the radial direction is especially pronounced and can span up to three orders of magnitude, correlating with a gradient in density. Thus, the distribution of (physiologically costly) fibrous reinforcing material within the palm trunk is optimized for bending stiffness, as it is preferably placed at the bottom and the periphery of the trunk where bending moments and stresses are maximal.

As described above, palms lack secondary thickening growth and only a very moderate increase of the trunk diameter during aging becomes visible, which is based on intrusive growth of fiber cells, cell division, and growth of the parenchymatous cells in between the vascular bundles. Even though the increase of the second moment of area is very limited, palms do need to adapt to increasing bending moments caused by height growth. The analysis of individual trunks with different heights belonging to one species reveals a correlation between stiffness at breast height and the height of the individual.[31] Assuming a reasonable correlation between the height and the age of the individual, palms appear to be mechanically highly dynamic systems, which adapt to increasing mechanical demands by increasing the density (Figure 8.2b) and thus stiffness of their tissues. Anatomical investigations have revealed that, since the cells are still living, a successive deposition of additional wall layers occurs to the cell walls of fibers within the vascular bundles,[31] as well as to the cell walls of the parenchymatous tissue in between the vascular bundles. As shown schematically in Figure 8.2g for a fiber, the cell wall consists of a multitude of secondary cell wall layers, to which new layers are added from the inside by the living cell until the entire cell cross-section can be filled with the cell wall. Thus, at the integral level of hierarchy, heterogeneity and fiber reinforcement are efficiently used to *ab initio* optimize the trunk structure to the mechanical requirements, whereas changes in the mechanical requirements within the lifetime of the palms are compensated by increasing the density of the tissue through the deposition of additional cell wall material in the individual cells.[29,31]

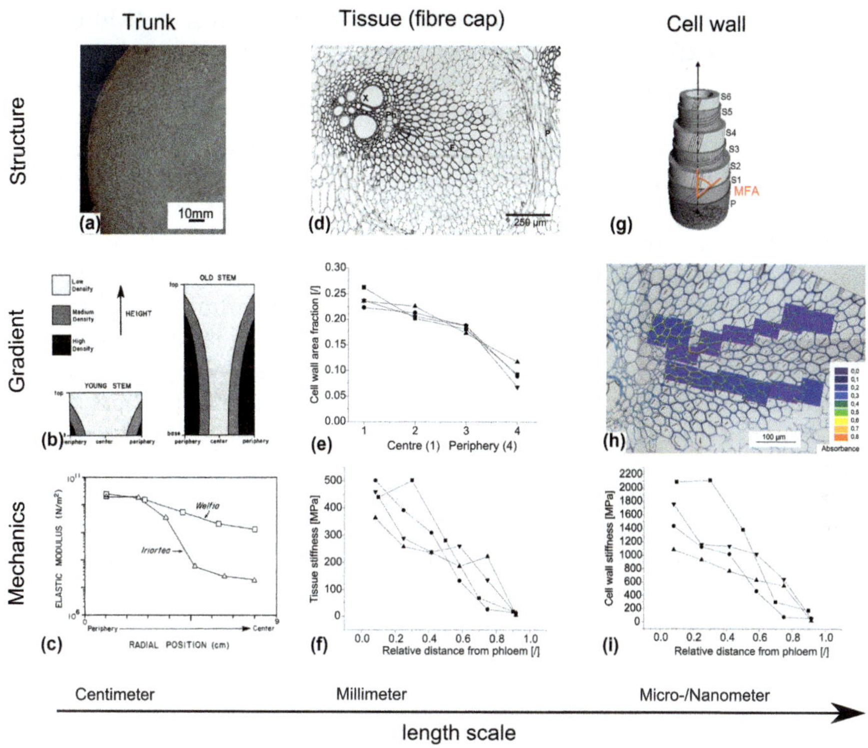

Figure 8.2 Structure and mechanics of palms from the macroscopic to the nanometer scale. (a) Distribution of vascular bundles. (b) Density distribution and its dynamics during aging. (c) Stiffness as a function of the radial position for the palm species *Iriartea* and *Welfia*. (d) Vascular bundle with fiber cap (F) and surrounding parenchymatous tissue (P), X = xylem, Ph = phloem. (e) Gradient in the cell wall area fraction as a measure of the density across individual fiber caps. (f) Tissue stiffness across individual fiber caps. (g) Multi-lamellar structure of cell walls of fiber cells of bamboo and palms, P: primary cell wall, S1–S6: secondary cell wall layers, successively deposited during aging, MFA = microfibril angle, indicated within the single cell wall layers, the orientation in S1 is taken from measurements on *W. robusta*. (h) Gradient in lignification across the fiber cap; (i) calculated cell wall stiffness based on (f) and cell wall area fraction. (b) and (c) are reproduced with permission from Rich,[29] 1987, The University of Chicago; (d)–(i) are reproduced from Rüggeberg *et al.*,[30] 2008, The Royal Society; (g) is adapted from Eder *et al.*,[26] 2009, New Zealand Forest Research Institute Limited.

8.3.2 Controlled Local Composition and Reinforcement Orientation

A crucial factor for the control of the mechanical performance by vascular bundle distribution in the stem is the mechanical interplay between the stiff bundles and the pliant surrounding parenchyma tissue. Interestingly, next to the gradients observed on the level of the integral trunk, palms like

Washingtonia robusta also show a gradient in stiffness across individual fiber caps (Figure 8.2f).[30] This gradient correlates with a gradient in the cell wall area fraction, which can be taken as a measure of the density (Figure 8.2e) and corresponds to the mechanical adaptation strategy presented in Figure 8.1a. Additionally, the gradient in stiffness persists also for calculations on cell wall stiffness (Figure 8.2i), indicating further adaptation of the mechanical properties at the level of the composition of the cell wall building blocks. At the ultrastructural level, the fiber cells of the vascular bundles exhibit a high microfibril angle throughout the entire fiber cap (data not shown), which rules out an adaptation based on changes in cellulose fibril orientation (which is the major adaptation strategy in trees (Figure 8.1b)). However, chemical analysis of the matrix composition reveals a gradient in lignification across the fiber caps (Figure 8.2h). Fibers close to the center of a vascular bundle show a by far higher degree of lignification than fibers at the periphery of the fiber cap in close vicinity to the soft parenchymatous cells. A higher degree of lignification can lead to a higher stiffness and strength of the matrix of the cell composite. Hence, in accordance with the mechanical adaptation strategy presented in Figure 8.1c, the gradient in stiffness at the cell wall level is set up by a specific interplay of the stiff cellulose fibrils with the soft matrix components.[30,32,33] In particular for high microfibril angles, the shear stiffness of the matrix components is a crucial factor and alterations in lignin content can have strong impact on the axial stiffness of the material.

As illustrated in Figure 8.2, palms are capable of highly efficient adaptation at various levels of structural hierarchy based on the building blocks that form heterogeneous composites. This makes the stem a highly resilient system of optimized natural functionality and, at the same time, an interesting biological role model for the design of bio-inspired heterogeneous composites.

8.4 Bio-inspired Heterogeneous Composites

Heterogeneous composites with a graded architecture have been fabricated in the past as a means to alleviate stress concentrations arising from the combination of mechanically dissimilar materials. This has led to the development of a broad class of composites known as functionally graded materials (FGMs).[34–36] The numerous examples of unique heterogeneous architectures in nature have recently motivated the further development of synthetic FGMs inspired by natural design principles.

Here, we provide examples of artificial heterogeneous composites that illustrate the level of control over the orientation and local concentration of building blocks that has been achieved in synthetic systems. A comparative analysis between such artificial materials and the rich multiscale architecture of plants provides insights on the microstructural control achievable using current manufacturing technologies and on the challenges and opportunities for future development of heterogeneous composites inspired by biological design principles.

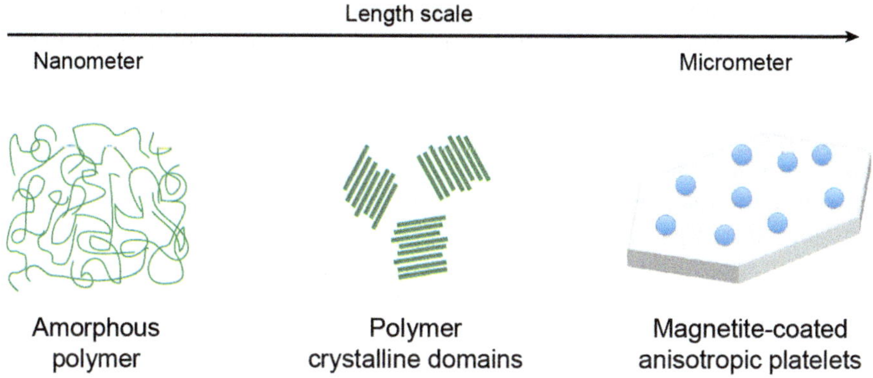

Figure 8.3 Examples of synthetic building blocks used to fabricate polymer-based heterogeneous composites.

8.4.1 The Synthetic Building Blocks

Alike in plant systems, building blocks spanning over multiple length scales have been used to create heterogeneous architectures (Figure 8.3). Since crystalline domains of polymers can be stiffer than an amorphous organic phase, these building blocks are often combined in semi-crystalline polymers or phase-separated copolymers to enable tuning of the elastic modulus of the material, similarly to the lignin–hemicellulose mixtures found in plants. At a coarser length scale, anisotropic stiff particles analogous to plant cellulose fibrils can be used to further reinforce the polymer phase in specific orientations within the composite.

Despite these analogies, the processing routes currently exploited to fabricate bio-inspired heterogeneous composites differ markedly from the complex biological processes used to assemble multi-scale building blocks in plants. In the following, we discuss selected examples of synthetic processing tools that have been developed to gain control over the local composition and orientation of such building blocks. Similarly to the gradient profiles found within plants at multiple length scales, graded heterogeneous architectures are also utilized in synthetic systems to achieve special functionalities that would not be accessible using uniform materials. This is illustrated later in this chapter with the manufacturing of graded polymer substrates that combine the large global and small local deformations required in stretchable electronic applications.

8.4.2 Controlled Local Composition

Analogous to the reinforcing effect of lignin and cellulose fibrils within the cell walls of trees or palms, polymer crystalline domains and inorganic anisotropic particles have been used as reinforcing elements of a soft amorphous phase to generate synthetic polymer-based composites with graded architectures. Building blocks that can impart a strong reinforcing effect are advantageous for the design and fabrication of composites with extreme gradients in mechanical properties. One such example is depicted in Figure 8.4, which shows the elastic modulus range achievable in a synthetic system comprising a soft polyurethane matrix that is hierarchically reinforced with polyurethane crystalline domains, laponite nanoplatelets, and alumina microplatelets at progressively larger length scales.[37] By changing the relative concentration of such reinforcing building blocks, it is possible to create individual composite films with elastic modulus spanning over more than four orders of magnitude (Figure 8.4a). In this system, the fraction of crystalline hard domains relative to the amorphous soft domains is controlled by changing the ratio of hydrophilic and hydrophobic segments in the synthesis of polyurethane. The hydrophilic domains can be selectively reinforced by laponite platelets, which preferentially segregate within the crystalline structure of the polymer. To further increase the elastic modulus of the composites, alumina microplatelets can be incorporated at a larger length scale. In this case, reinforcement occurs through stress transfer from the surrounding matrix to the stiff platelets.

The preparation of composites with extreme mechanical gradients is possible by combining individual films reinforced to different extents into a single component using established joining technologies, such as thermal or solvent welding. Effective joining relies on the use of thermoplastic matrices that allow for diffusion and interpenetration of macromolecules at the interface between the initial films during the welding process. An example of a bulk composite part obtained through such a joining process is shown in Figure 8.4b. The local elastic modulus across the height of this composite part changes in a stepwise fashion from 40 MPa (M0040 layer) to 7 GPa (M7000 layer). The cross-section of the resulting graded composite shows that the interface between the original films is hardly visible in scanning electron microscopy images, indicating the effectiveness of the welding process in promoting intermolecular diffusion and penetration of macromolecules at the interface. As will be discussed later, such molecular interpenetration and entanglement at the interface is crucial to ensure efficient stress transfer throughout the graded composites and thus avoid undesired delamination failure under mechanical loads.

8.4.3 Controlled Reinforcement Orientation

In addition to local variations in the concentration of reinforcing building blocks, processing tools have also been proposed to enable the control of the orientation of anisotropic particles within synthetic composites. Orientation control is often achieved by aligning particles suspended in a fluid followed by consolidation of this suspension through a chemical reaction, evaporation, or solidification process. The most straightforward approach to align anisotropic particles in such a suspension is to apply shear during processing, which can be induced using, for example, a doctor blade in tape casting or a nozzle of defined geometry in 3D printing. Although shear is

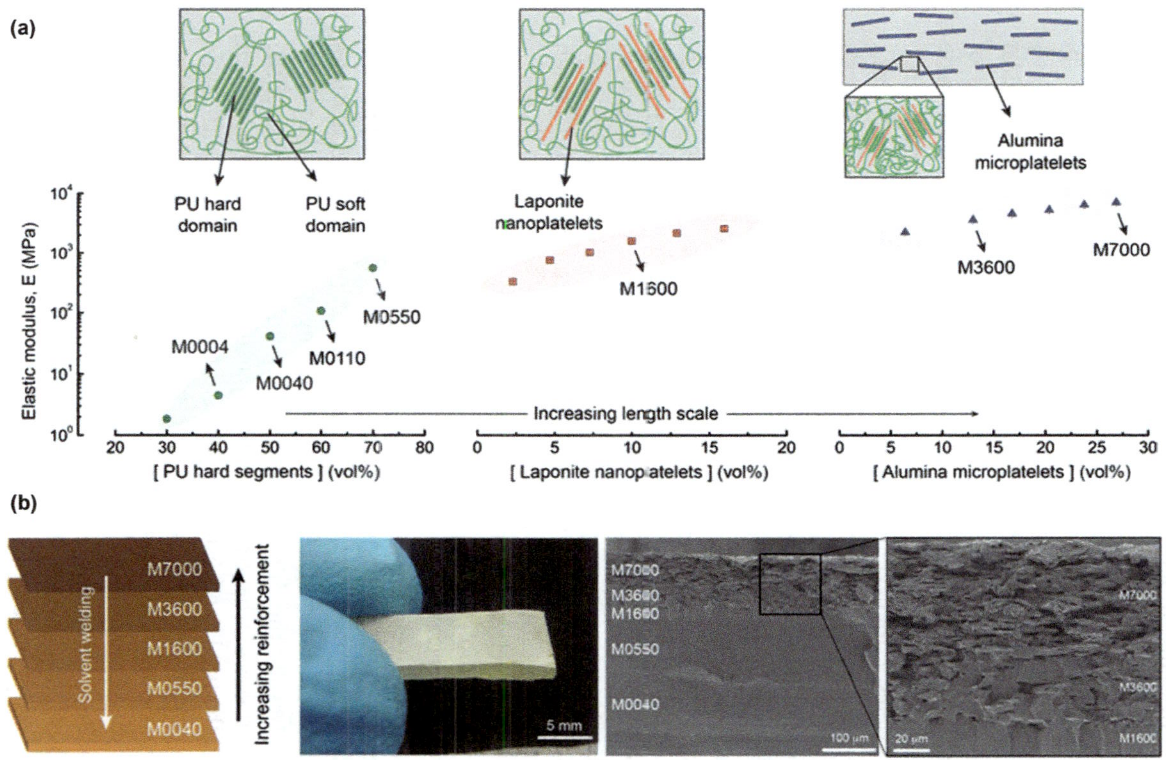

Figure 8.4 Example of heterogeneous polymer-based composites exhibiting a through-thickness gradient in the local composition and elastic properties. (a) Elastic modulus of polyurethane-based composites reinforced with different volume fractions of hard segments (left), laponite nanoplatelets (center), and alumina microplatelets (right) at increasing hierarchical levels. (b) Heterogeneous composite with extreme elastic gradient in the out-of-plane direction prepared through solvent welding of individual layers with different reinforcement levels. The nomenclature used here indicates the elastic modulus E in MPa of individual layers. Example: M3600 refers to an individual layer with $E = 3600$ MPa.
Adapted with permission from Libanori *et al.*,[38] 2012, Macmillan Publishers Limited.

easy to apply using rather simple equipment, it does not allow for deliberate control over the orientation of particles in specific pre-defined directions. Biological architectures, including for example the arrangement of cellulose microfibrils within the plant cell wall or the periodic orientation of inorganic particles found in many mineralized tissues, require other alignment strategies beyond simple shearing.

An elegant approach to align anisotropic particles in deliberate orientations in a fluid is to use external electric or magnetic fields.[39,40] The application of magnetic fields for particle orientation control has been recently exploited to generate bio-inspired architectures with remarkable mechanical properties (Figure 8.5).[41] This method is applicable to different types of anisotropic particles, ranging from short fibers to platelets, which are made magnetically responsive through the adsorption of superparamagnetic iron oxide nanoparticles (SPIONs) on the particle surface (Figure 8.5a, left). Provided that the long-axis particle size is within the typical range 2–50 μm, magnetic alignment is possible using surprisingly low external fields on the order of 1–10 mT (Figure 8.5a, right). Smaller and larger particles require higher alignment fields due to the predominant effect of thermal energy and gravity, respectively, on the motion of the suspended particles (Figure 8.5a, middle). Because of such ultra-high magnetic response, SPION concentrations as low as 0.01–0.1 vol% are needed to make the anisotropic particles controllable with an external field. Coating of the anisotropic particles with SPIONs can be accomplished by simple electrostatic adsorption in water followed by drying, delivering SPIONs in the form of emulsions or through the *in situ* synthesis of the oxide coating (Figure 8.5a).[41–43]

The ability to magnetically control the orientation of anisotropic particles suspended in a fluid followed by the consolidation of the fluid to fix such alignment provides an exciting pathway for the microstructural design of bio-inspired composites. In an illustrative example of the potential of this approach, periodic particle orientation patterns have been created by imposing a periodically changing magnetic field on a suspension of magnetized particles that was consolidated through a slowly advancing particle-jamming front (Figure 8.5b).[44] Despite the different shape, size, and chemical nature of the building blocks involved, the gradual shift in particle orientation throughout the structure of the final composite resembles the change in microfibril angle across the cell wall layers in plants (Figure 8.5b). The orientation of anisotropic particles in a polymer matrix strongly affects the mechanical properties of the resulting composites. When aligned in the loading direction, these particles enhance the composite elastic modulus and fracture strength. If oriented in the out-of-plane direction with respect to the surface, a significant increase in the hardness and wear resistance of the composite is observed (Figure 8.5c). Thus, the alignment of anisotropic particles in such a periodic fashion reduces the global anisotropy of the composite by providing mechanical reinforcement in several different directions, and might also be an effective mechanism to increase the fracture strength and toughness of the composite through crack arresting mechanisms.[45]

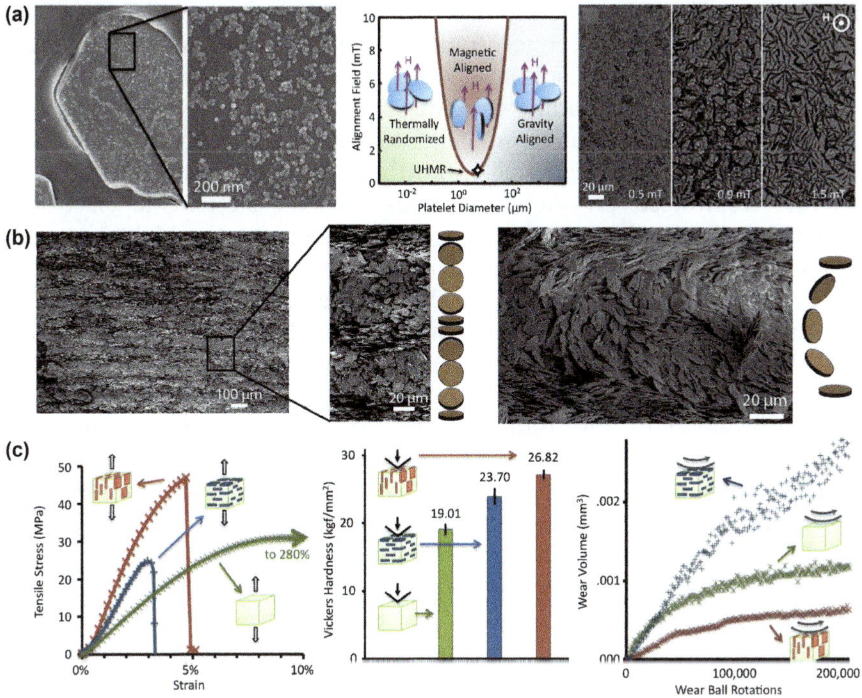

Figure 8.5 Magnetically responsive alumina platelets, their organization in periodic architectures, and their use for tailored reinforcement of polymer-based composites. (a, left) Example of ultra-high magnetically responsive (UHMR) alumina platelets coated with 0.5 vol% of iron oxide nanoparticles (SPIONs) to enable magnetic control. (a, middle) The theoretical minimum magnetic field required for alignment of the alumina platelets coated with SPIONs. (a, right) Out-of-plane alignment of the coated alumina platelets in a fluid in the presence of a uniform magnetic field of approximately 1 mT. (b) Magnetic assembly of platelets into periodic reinforcement architectures using the Magnetically-Assisted Slip Casting (MASC) process. (c) Tensile properties, hardness, and wear behavior of polymer-based composites reinforced with magnetized alumina platelets aligned in deliberate orientations.
Figures in (a) and (c) have been adapted with permission from Erb *et al.*,[41] 2012, The American Association for the Advancement of Science. Images in (b) have been adapted with permission from Le Ferrand *et al.*,[44] 2015, Macmillan Publishers Limited.

8.4.4 Functional Devices Based on Bio-inspired Heterogeneous Composites

Plants utilize their ability to control the local composition and the microfibril orientation within the cell walls to accommodate differential strains within their composite architecture. This is evidenced for example by changes in the microfibril orientation in trees depending on the specific function of the tissue,[25] as well as in the graded change in elastic modulus observed in fiber caps within the trunk of a palm tree (Figure 8.2). By gradually adjusting the concentration of lignin at a high microfibril angle across the fiber cap, the plant may reduce stress concentrations between the stiff fiber and the surrounding softer tissue.

Similarly, the formation of gradients to alleviate high local mechanical stresses has also been exploited in synthetic composites to fabricate, for instance, reinforced polymer-based substrates for stretchable electronic devices. In this application, the end goal is to combine the high electronic performance of stiff semiconductor circuitry with the rubbery nature of soft polymer substrates to fabricate stretchable electronic displays, wearable electronic devices, and tissue-compatible biomedical implants. The challenge lies in preventing premature mechanical failure of the device by reducing the high stress concentrations expected at the interface between the stiff semiconductor thin film and the underlying stretchable polymer substrate.

One possible solution to the problem is to utilize a graded polymer-based composite, referred to here as a patch, to provide a smooth mechanical coupling between the semiconductor layer and the rubbery substrate. To demonstrate this idea, polyurethanes hierarchically reinforced with nano and microplatelets (Figure 8.4) were used to create graded patches through casting of individual films following by solvent welding. As in the fiber cap of the palm tree, the gradient is built such that the stiffer part of the patch is in contact with the rigid electronic layer and the softer part is connected to the underlying compliant substrate. Finite element simulations of such a graded patch design in comparison to other homogenous controls revealed that this is indeed an effective approach to prevent early failure of the device by keeping low strains on the top surface of the patch next to the stiff semiconductor layer (ε_{top}) and by allowing for high deformations and thus lower stresses at the bottom of the patch (σ_{bott}) that is attached to the underlying stretchy substrate (Figure 8.6). In these simulations, the distributions of stresses and strains across the height of patch–substrate constructs were calculated when a global tensile strain of 25% was applied to the underlying substrate (Figure 8.6a and b). For the graded configuration, the low compressive strains at the patch top surface (ε_{top}) ensures no early mechanical fracture of the semiconductor layer, whereas the lower stresses at the bottom of the patch (σ_{bott}) allows for high stretchability of the underlying substrate without premature delamination at the patch–substrate interface (Figure 8.6c). These simulation results were experimentally confirmed by performing delamination tests and local strain measurements on fabricated graded patches and control specimens. In a proof-of-concept experimental demonstration, a light emitting diode (LED) was placed on top of the fabricated graded patch and was shown to remain functional even after the application of a global strain as high as 150% to the underlying substrate (Figure 8.6d). By contrast, tests with a control homogeneous specimen (S-0040) showed failure of the LED at global strains within the range 5–10%.

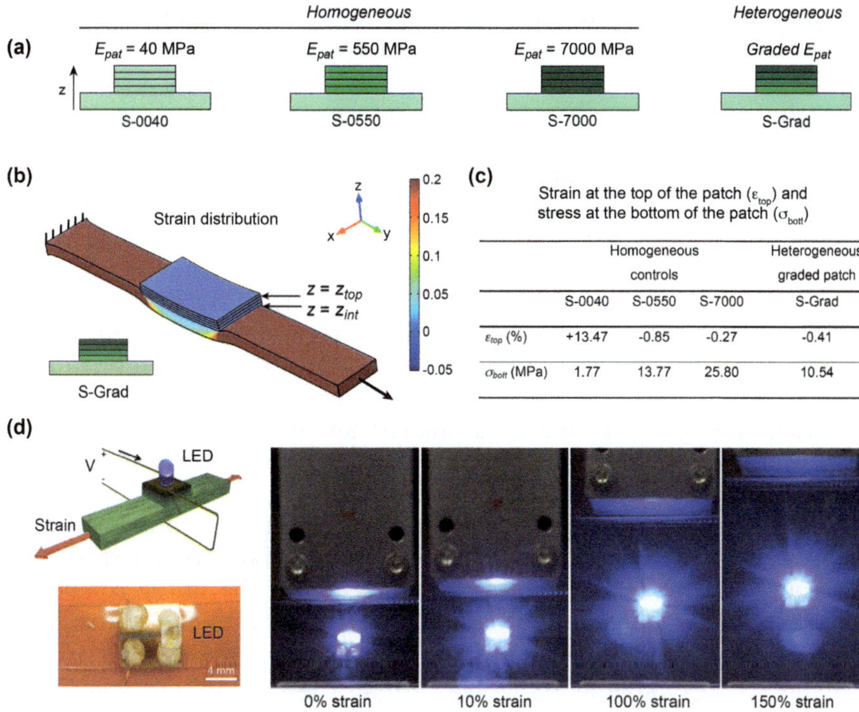

Figure 8.6 Finite element simulations and proof-of-concept demonstration of functional graded composites for stretchable electronic applications.[46] (a) Homogeneous and heterogeneous patches used as model configurations in the finite element simulations. E_{pat} is the elastic modulus of the patch. (b) Simulated strain distribution throughout the graded patch and underlying substrate for an applied global strain of 25%. (c) Strain at the top surface of the patch (ε_{top}) and stress at the bottom of the patch (σ_{bott}) obtained from simulations of the graded construct as compared to different homogeneous controls. (d) Proof-of-concept demonstration of the high stretchability enabled by the graded patch without compromising the functional response of a light emitting diode (LED). Adapted with permission from Libanori *et al.*,[38] 2012, Macmillan Publishers Limited.

8.5 Discussion

The last decades have shown a tremendous increase of research activities in the fields of biomimetics and bio-inspired materials. This increase is directly coupled to advances in analytical and fabrication technologies that enable a deeper understanding of the complexity of biological materials and their underlying structure–function relationships down to the nanometer scale, but also the synthesis of such complex materials and structures.[47] Thus, the underlying structure–function relationships of an increasing number of biological materials with peculiar, unique and often astonishing mechanical behavior have been unraveled[1] and the number of bio-inspired heterogeneous composites being synthesized is steadily rising.

Here, we present the palm tree as a natural composite material and show the synthesis of bio-inspired heterogeneous composites, both incorporating gradual transitions in their mechanical properties. The palm tree exhibits a precise spatial control of the chemical composition, orientation of reinforcements, and density from the nanometer scale up to the entire trunk, which governs the stiffness distribution in the trunk and within the individual fiber caps, and enables the formation of gradual transitions in the mechanical properties at both small and large length scales. In the synthetic heterogeneous composite, the precise spatial control of the chemical composition, distribution of reinforcements, and their orientation were achieved as well. Abrasion resistance, hardness, and extensibility can be locally manipulated and thus tailored for improved functionality. By joining multiple layers, gradual transitions in structural and mechanical properties were achieved. The benefits of such gradient materials were demonstrated by the specific example of connecting a brittle (electronic) component to a highly stretchable support. A gradual variation of stiffness in the connecting material enabled the protection of the brittle component from excessive strains while maintaining the extensibility of the whole device.

In nature, such gradual transitions bridging pronounced mismatches in mechanical properties have not only been found in plants, but in a multitude of other biological materials. This design principle appears for example in the byssus threads of molluscs,[48] as well as at the dentin–enamel junction of teeth.[49] In synthetic materials, extensibility or thermal expansion mismatches may exist, for instance, between the bulk material and protective coatings or between an electronic component and a flexible support (Figure 8.6). By incorporating gradual transitions, peak stresses are lowered as the load is distributed over a larger interfacial volume.[50]

However, it is also obvious, that gradients are not the only possible interfacial design in natural composite materials. The layered structure of the hybrid spicules of glass sponges[51] and the brick and mortar architecture of nacre[52] are just two important examples of biological materials showing high stiffness and toughness despite the incorporation of weak, non-graded interfaces. Experimental and theoretical studies have shown that the incorporation of weak interfaces results in a tremendous increase in the toughness of the composite compared to the brittle behavior of the pure material without the interface.[11] Optimum combinations of stiffness ratio, thickness of the interfacial layer, and their distance (wavelength) maximize the toughness.[45]

Thus, remarkably and appearing to be counterintuitive at first hand, at least two very different interfacial designs can lead to increased toughness of a material, left aside designs that result in optimization towards other criteria.[15] The fundamental difference in the two design strategies mentioned here is the way potentially excessive loads are endured. Whereas gradual transitions aim at lowering stress concentrations and thus at avoiding fracture, the incorporation of periodic structures leads to a tough system tolerating damage specifically at the weak interfaces. This design principle leads to the absorption of large amounts of energy without catastrophic failure and hence toughness is predominantly reached by damage tolerance. The existence of these different strategies is the immediate consequence of the difference in structure and mechanical behavior of the materials themselves. If only one interface between two components with a mismatch in (mechanical, thermal or other) properties exists, such as the dentin–enamel junction in teeth or coatings on bulk materials, a failure at this interface would be catastrophic for the entire system. The gradual transition lowers the possibility of such catastrophic failure by lowering the stress concentration. In contrast to that, the incorporation of multiple (weak) interfaces in entirely brittle materials, such as in nacre or in glass sponge helps to stop cracks when loaded excessively and thus prevents catastrophic failure by unstable crack growth. Biological composites impressively demonstrate that the mechanical behavior of the building blocks, the structure of the composite, and the interfacial design represent a powerful functional entity to tune the global mechanical response of the material. Gradual transitions are beneficial in some cases, but would lead to inferior behavior compared to weak interfaces in other cases.

The beneficial aspects of gradual transitions may also be discussed for fiber-reinforced composite materials. Potentially hazardous stress concentrations at the interface between stiff fibers and the soft matrix can be lowered, and failure by delamination can be avoided using a graded interfacial design. The discovery of gradual transitions in stiffness between the stiff fiber caps and the soft surrounding parenchymatous tissue in palm trees seems to support this hypothesis. However, the transfer to technical fiber-reinforced composites has to be made with care. Stress concentrations in fiber-reinforced composite materials appear, according to the shear-lag model, only at the fiber ends.[53] Most technical fiber-reinforced composites contain practically continuous fibers, which makes the benefits of gradual transitions questionable. In contrast, the incorporation of multiple interfaces may even promote abrupt transitions and weak interfaces for acting as multiple crack stoppers in case of impact loads. Hence, the optimum design of the interface depends on the specific structure of these fiber-reinforced composites. For the palm, such stress concentrations might nevertheless exist, as the single fiber cells are relatively short and connected by the middle lamella, which is comparatively weak. However, entirely resolving the specific loading scenario and the local stress distribution at the interface of fiber cells and the surrounding soft tissue is a very difficult task and has not been addressed yet. Thus, the role model of the palm tree remarkably shows the benefits and the underlying difficulties of biomimetic research.

8.6 Remaining Challenges and Outlook

With the help of advanced analytical and fabrication tools, mankind has just started to deeply understand and to mimic the hierarchical, intricate structure of biological materials. A number of successful transfers of principles and developed bio-inspired materials with new functionalities have proven the great potential of this biomimetic approach towards the development of new materials. With the advent of 3D printing, the remaining challenge of upscaling synthetic heterogeneous composites with hierarchical structure and precise spatial control of the local composition, structure, and functional properties will be tackled. Biological materials are grown from bottom to top (additive manufacturing), which is the radical opposite from the top-down manufacturing process (subtractive manufacturing) for most of the large-scale conventional engineering materials. The fabrication of complex composites has been demonstrated on small scales up to the centimeter range, but remains time consuming and expensive on larger scales. 3D printing will enable the additive manufacturing of composite materials with a high level of complexity, including controlled hierarchy, local chemistry, and orientation of the building blocks. In this paradigm-shifting manufacturing approach, upscaling will only be limited by the size of the printer and not by the synthetic process and time. First developments in this direction have recently been reported[44,54–59] and are expected to give further impulse to the field of biomimetic materials research and technology.

Acknowledgements

A.R.S. thanks the financial support of the Swiss National Center of Competence in Research (NCCR) Bio-Inspired Materials. I.B. thanks BAFU and Lignum, Switzerland for financial support of the Wood Materials Science group.

References

1. J. W. C. Dunlop and P. Fratzl, in *Annual Review of Materials Research*, ed. D. R. R. M. Z. F. Clarke, editon edn, 2010, vol. 40, pp. 1–24.
2. E. Munch, M. E. Launey, D. H. Alsem, E. Saiz, A. P. Tomsia and R. O. Ritchie, *Science*, 2008, **322**, 1516–1520.
3. A. R. Studart, *Adv. Mater.*, 2012, **24**, 5024–5044.
4. I. Burgert, in *Materials Design Inspired by Nature: Function through Inner Architecture*, ed. P. Fratzl, J. W. C. Dunlop and R. Weinkamer, editon edn, 2013, pp. 128–150.
5. T. Speck and I. Burgert, in *Annual Review of Materials Research*, ed. D. R. Clarke and P. Fratzl, editon edn, 2011, vol. 41, pp. 169–193.

6. P. T. Martone, M. Boller, I. Burgert, J. Dumais, J. Edwards, K. Mach, N. Rowe, M. Rüggeberg, R. Seidel and T. Speck, *Integr. Comp. Biol.*, 2010, **50**, 888–907.
7. H. Peterlik, P. Roschger, K. Klaushofer and P. Fratzl, *Nat. Mater.*, 2006, **5**, 52–55.
8. H. S. Gupta, S. Krauss, M. Kerschnitzki, A. Karunaratne, J. W. C. Dunlop, A. H. Barber, P. Boesecke, S. S. Funari and P. Fratzl, *J. Mech. Behav. Biomed. Mater.*, 2013, **28**, 366–382.
9. F. Vollrath and D. T. Edmonds, *Nature*, 1989, **340**, 305–307.
10. M. J. Harrington, A. Masic, N. Holten-Andersen, J. H. Waite and P. Fratzl, *Science*, 2010, **328**, 216–220.
11. A. P. Jackson, J. F. V. Vincent and R. M. Turner, *Proc. R. Soc. B*, 1988, **234**, 415–440.
12. P. Fratzl, *J. R. Soc., Interface*, 2007, **4**, 637–642.
13. P. Fratzl and R. Weinkamer, *Prog. Mater. Sci.*, 2007, **52**, 1263–1334.
14. P. Fratzl, H. S. Gupta, F. D. Fischer and O. Kolednik, *Adv. Mater.*, 2007, **19**, 2657–2661.
15. J. W. C. Dunlop, R. Weinkamer and P. Fratzl, *Mater. Today*, 2011, **14**, 70–78.
16. J. Fahlen and L. Salmen, *Biomacromolecules*, 2005, **6**, 433–438.
17. D. G. Hepworth and J. F. V. Vincent, *Ann. Bot.*, 1998, **81**, 761–770.
18. L. Salmen and A. M. Olsson, *J. Pulp Pap. Sci.*, 1998, **24**, 99–103.
19. D. Fengel and G. Wegener, *Wood – Chemistry, Ultrastructure, Reaction*, Walter de Gruyter, Berlin, New York, 1984.
20. A. Bergander and L. Salmen, *J. Mater. Sci.*, 2002, **37**, 151–156.
21. L. Salmen, *C. R. Biol.*, 2004, **327**, 873–880.
22. I. D. Cave, *Wood Sci. Technol.*, 1969, **3**, 40–48.
23. L. Goswami, J. W. C. Dunlop, K. Jungnikl, M. Eder, N. Gierlinger, C. Coutand, G. Jeronimidis, P. Fratzl and I. Burgert, *Plant J.*, 2008, **56**, 531–538.
24. M. Labronici and H. Ishida, *Compos. Interfaces*, 1998, **5**, 87–116.
25. H. Lichtenegger, A. Reiterer, S. E. Stanzl-Tschegg and P. Fratzl, *J. Struct. Biol.*, 1999, **128**, 257–269.
26. M. Eder, M. Rüggeberg and I. Burgert, *N. Z. J. For. Sci.*, 2009, **39**, 115–124.
27. P. M. Rich, K. Helenurm, D. Kearns, S. R. Morse, M. W. Palmer and L. Short, *Bull. Torrey Bot. Club*, 1986, **113**, 241–246.
28. P. B. Tomlinson, *The Structural Biology of Palms*, Clarendon Press, Oxford, 1990.
29. P. M. Rich, *Bot. Gaz.*, 1987, **148**, 42–50.
30. M. Rüggeberg, T. Speck, O. Paris, C. Lapierre, B. Pollet, G. Koch and I. Burgert, *Proc. R. Soc. B*, 2008, **275**, 2221–2229.
31. P. M. Rich, *Am. J. Bot.*, 1987, **74**, 792–802.
32. D. Raabe, C. Sachs and P. Romano, *Acta Mater.*, 2005, **53**, 4281–4292.
33. M. Rüggeberg, T. Speck and I. Burgert, *New Phytol.*, 2009, **182**, 443–450.
34. A. Mortensen and S. Suresh, *Int. Mater. Rev.*, 1995, **40**, 239–265.

35. M. T. Tilbrook, R. J. Moon and M. Hoffman, *Compos. Sci. Technol.*, 2005, **65**, 201–220.
36. U. Schulz, M. Peters, F. W. Bach and G. Tegeder, *Mater. Sci. Eng., A*, 2003, **362**, 61–80.
37. R. Libanori, F. H. L. Munch, D. M. Montenegro and A. R. Studart, *Compos. Sci. Technol.*, 2012, **72**, 435–445.
38. R. Libanori, R. M. Erb, A. Reiser, H. Le Ferrand, M. J. Sueess, R. Spolenak and A. R. Studart, *Nat. Commun.*, 2012, **3**, 1265.
39. M. A. Correa-Duarte, M. Grzelczak, V. Salgueirino-Maceira, M. Giersig, L. M. Liz-Marzan, M. Farle, K. Sierazdki and R. Diaz, *J. Phys. Chem. B*, 2005, **109**, 19060–19063.
40. C. Park and R. E. Robertson, *Dent. Mater.*, 1998, **14**, 385–393.
41. R. M. Erb, R. Libanori, N. Rothfuchs and A. R. Studart, *Science*, 2012, **335**, 199–204.
42. M. R. Sommer, R. M. Erb and A. R. Studart, *ACS Appl. Mater. Interfaces*, 2012, **4**, 5086–5091.
43. T. P. Niebel, F. J. Heiligtag, J. Kind, M. Zanini, A. Lauria, M. Niederberger and A. R. Studart, *RSC Adv.*, 2014, **4**, 62483–62491.
44. H. Le Ferrand, F. Bouville, T. P. Niebel and A. R. Studart, *Nat. Mater.*, 2015, **14**, 1172–1179.
45. O. Kolednik, J. Predan, F. D. Fischer and P. Fratzl, *Adv. Funct. Mater.*, 2011, **21**, 3634–3641.
46. R. Libanori, R. M. Erb, A. Reiser, H. Le Ferrand, M. J. Süess, R. Spolenak and A. R. Studart, *Nat. Commun.*, 2012, **3**, 1265.
47. U. G. K. Wegst, H. Bai, E. Saiz, A. P. Tomsia and R. O. Ritchie, *Nat. Mater.*, 2015, **14**, 23–36.
48. M. J. Harrington and J. H. Waite, *Adv. Mater.*, 2009, **21**, 440–444.
49. P. Zaslansky, A. A. Friesem and S. Weiner, *J. Struct. Biol.*, 2006, **153**, 188–199.
50. S. Suresh, *Science*, 2001, **292**, 2447–2451.
51. J. Aizenberg, J. C. Weaver, M. S. Thanawala, V. C. Sundar, D. E. Morse and P. Fratzl, *Science*, 2005, **309**, 275–278.
52. J. D. Currey, *Proc. R. Soc. B*, 1977, **196**, 443–463.
53. D. Hull and T. W. Clyne, *An Introduction to Composite Materials*, 2nd edn, Cambridge University Press, Cambridge, 1996.
54. B. G. Compton and J. A. Lewis, *Adv. Mater.*, 2014, **26**, 5930–5935.
55. L. S. Dimas, G. H. Bratzel, I. Eylon and M. J. Buehler, *Adv. Funct. Mater.*, 2013, **23**, 4629–4638.
56. L. Guiducci, J. C. Weaver, Y. J. M. Brechet, P. Fratzl and J. W. C. Dunlop, *Adv. Mater. Interfaces*, 2015, **2**, 1500011.
57. D. Kokkinis, M. Schaffner and A. R. Studart, *Nat. Commun.*, 2015, **6**, 8643.
58. J. J. Martin, B. E. Fiore and R. M. Erb, *Nat. Commun.*, 2015, **6**, 86411.
59. A. R. Studart, *Chem. Soc. Rev.*, 2016, **45**, 359–376.

CHAPTER 9

Translating Mussel Adhesion: Four Uncertainties about the Interface

J. HERBERT WAITE

Marine Science Institute, Lagoon Road, University of California Santa Barbara, Santa Barbara, CA 93106, USA
Email: herbert.waite@lifesci.ucsb.edu

9.1 Introduction

The attachment of marine mussels to submerged solid surfaces has been the focus of numerous studies over the last twenty years, motivated both by fouling concerns and the technical limitations of adhesion in the presence of moisture, salt, corrosion, and biofilms. Several recent reviews with different perspectives have appeared on this subject.[1–3] Mussel attachment is mediated by a byssus, a peculiar brush-like holdfast that consists of hundreds of disc tipped threads made one thread at a time (Figure 9.1). Byssal attachment provides stability and, in the turbulent intertidal zone, an attribute of even greater survival value – tenacity. The factors contributing to byssal tenacity have been investigated at multiple length scales: at the centimeter scale, the radial distribution of byssal threads[4] and the stiffness gradient of threads;[5] at the millimeter scale, the spatulate shape of plaques;[6] at the micrometer scale, the porous "cellular" structure of the plaque;[7] and at the nanometer scale, the interfacial chemistry at the plaque–substrate interface[8] (Figure 9.1A–D). This overview dwells on the interfacial chemistry of plaque adhesion and, more particularly, on four critical uncertainties

RSC Polymer Chemistry Series No. 22
Bio-inspired Polymers
Edited by Nico Bruns and Andreas F. M. Kilbinger

Published by the Royal Society of Chemistry, www.rsc.org

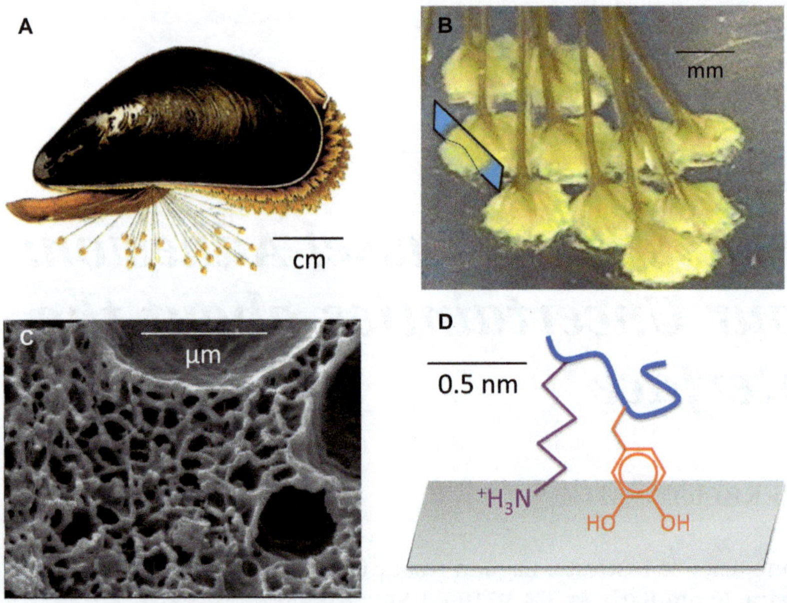

Figure 9.1 The multiple length scales of mussel adhesion. (A) Centimeter length scale pertains to the radial distribution of threads around an individual mussel, *Mytilus edulis*. (B) The foot is shown extended in the process of making a new thread. (C) The millimeter scale refers to the spatulate geometry of threads and plaques. (D) The micrometer scale is seen in a cross-sectional slice through a plaque, as in B, and reveals an extensive porosity. The nanometer scale is best captured by a portion of a single adhesive protein adsorbed to a target surface featuring the two most abundant side chains: lysine and Dopa.

about the interface pertaining to surface chemistry, hydration, pH, and redox chemistry. The interfacial emphasis is an eclectic one in that the interface is no more important than the other length scales but, in many respects, is more amenable to experimental exploration.

9.2 Are Interfacial Films Cleared Away?

Surfaces mean everything in adhesion technology. That being said, much uncertainty reigns in the realm of surfaces, especially surfaces in natural environments. The only universal agreement is that surfaces are unlike the bulk. Not knowing the surface chemistry of an adherend frustrates understanding the interfacial chemistry and the design of engineered coatings and adhesives. Mussels attach to a variety of surfaces in seawater. Many years of study have been devoted to investigating chemical changes that occur on clean, freshly prepared surfaces submerged in natural seawater.[9,10] The progression of changes is location- and surface-specific but generally describable as follows: (1) an oxide layer forms nearly spontaneously, which can grow in thickness over time. This is especially common on metals such as aluminum, steel, and titanium. (2) The surface becomes hydrated. Hydration means the formation of structured water, 4–5 water molecules thick. (3) Deposition of a conditioning film. This is formed from dissolved low molecular weight organics such as fatty acids, amino acids, phenolics, *etc.*, which are derived by metabolism or degradation of natural materials and highly dependent on location. (4) Finally, there is biofilm deposition. Biofilms are made by microbes, which settle on surfaces in order to feed on the conditioning film. Biofilms are exopolymers made by microbes as a scaffolding matrix for their growth and duplication, but often become food for yet another community of organisms. Biofilm exopolymers are often exemplified by alginates, but their chemistry is actually much more diverse.[11] If there is continuous progression of growth in layers 3–4, one can imagine surface films developing into thick slimes; as it happens. More realistically though, and particularly on intertidal surfaces, there is a steady state between film build-up and film degradation[12] leading to a surface patchwork[13,14] that ever changes and upon which hydrated, corroded, conditioned, and biofilmed patches all occur.

More important to the present overview is what mussels do about the surfaces they encounter before attaching to them. If they clean surfaces down to the bare hydrated mineral of the rock, for example, then they only have to get past the hydration layer to stick. If not, then they may be adhering to the conditioning or biofilms and, if any of these are weak links in the system, adhesion will only be as strong as the weakest link.

Although numerous studies have investigated the ability of biofilms to recruit or inhibit settlement by mussels or other epifauna,[15] there are none that examine what mussels do about films after they alight on a surface. Ongoing experiments with *M. californianus* in our laboratory suggest that, besides killing microbes, mussels do nearly nothing.[16] First, mussels use the tactile abilities of their feet to select patches of surface that are less filmed; surfaces with dense biofilms are usually avoided by mussels, who, on finding such surfaces, crawl elsewhere. Second, confocal examination of plaque interfaces stained with fluorescent dyes (Syto9 and propidium iodide) that distinguish living and dead microbes has revealed that the interfacial microbes are dead. This intriguing result suggests that, with or without biofilms, microbes are killed during plaque formation. Two possible biocide candidates are the acidic pH of protein deposition (discussed below) and/or co-secreted cationic peptides. Most mfps are strongly cationic, and cationic peptides often have antimicrobial activity (cat peptides).[17]

One fascinating experiment to explore the mussels' capacity to clean filmed surfaces was done with a gold film vacuum-evaporated onto soda glass.[18] The film was visible but so weak that it was smudged by the slightest finger contact.[18] The glass slides with gold films presented to mussels were not cleaned or altered in the least, even when the films were patterned as in Figure 9.2. Plaques were deposited over the intact film, even though the film was the weakest link and, under tension, plaques completely delaminated at the gold–glass interface (Figure 9.2).[16]

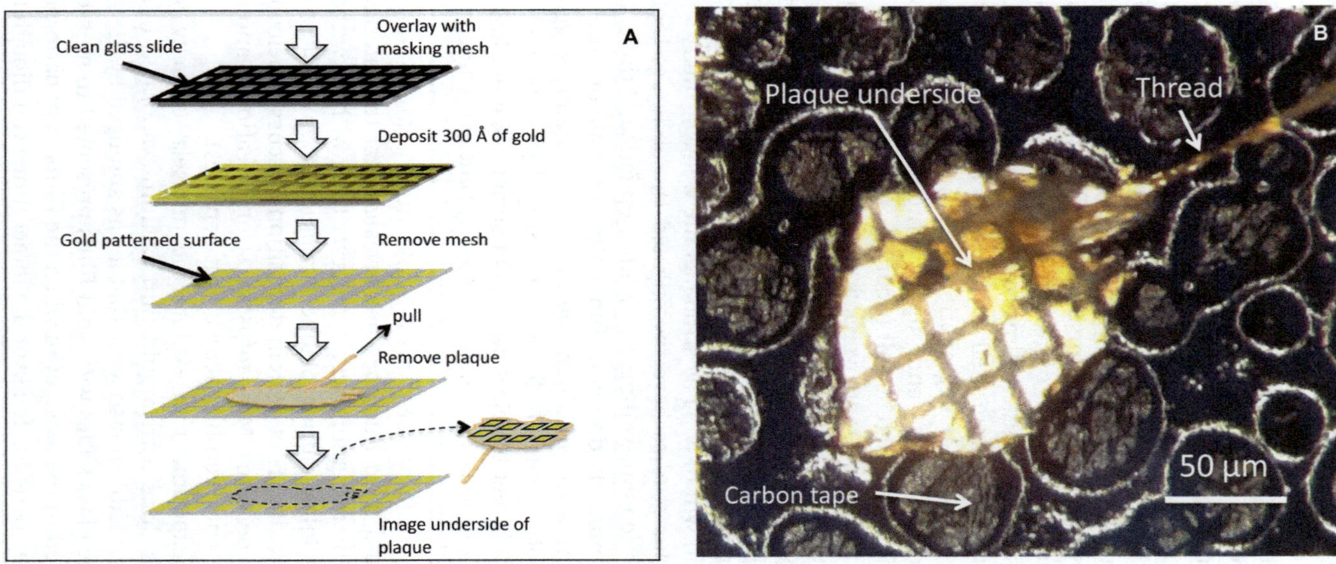

Figure 9.2 Mussel adhesion to a weak boundary layer. (A) A 300 Å thick film of gold was sputtered through a masking mesh onto a glass microscope slide and presented to a mussel for plaque attachment. (B) Mussels attached plaques to the film without first clearing the gold or disrupting the masking pattern. Each plaque (viewed from underside) delaminated from the slide at the interface between the gold and glass surfaces. The grey islands surrounding the tape are from the mounting tape. Unpublished data by N. Martinez Rodriguez and J. H. Waite, 2014.

In summary, although mussels avoid thick biofilms, upon committing to plaque deposition on a certain surface, they do little to change films besides killing the resident microbes. If a weak film such as sputtered gold is presented, plaque adhesion is undermined, but this is not necessarily the outcome with all biofilms, some of which are attached by near-covalent bounds to underlying mineral surfaces.[19] If the adhesive proteins infiltrate a strongly attached exopolymer network to create a second polymer network (as they do in *Dosidicus gigas*, in which a chitinous scaffold is infiltrated by coacervates of adhesive histidine rich-proteins[20]), the outcome could be quite robust.

9.3 Do Mussels Displace Surface Water?

All surfaces in water become hydrated. Hydration involves one or more layers of surface-associated water, which are more structured than bulk water and in exchange with it. Hydration usually compromises adhesion by hindering the intimacy between functional groups on the surface and the polymer, but does not necessarily prevent practical bonding. Dental fillings, for example, rarely achieve chemical intimacy with the wet tooth surface but can be held in place for years by mechanical interlocking.[21] Several studies have established that mussel adhesive proteins, such as mfp-3 and mfp-5, bind chemically to surfaces, which begs the question, do the proteins penetrate the hydration layer and, if so, how? In investigations relying on either atomic force microscopy (AFM) or the surface forces apparatus (SFA), the force applied during the approach of the two surfaces may be adequate to squeeze aside hydration in the same way that adhesive tape can bind to wet glass with enough applied pressure. However, other techniques have convincingly confirmed that surface dehydration by mfp-3s can be chemically achieved under essentially compression-free conditions.[22]

One technique, Overhauser Dynamic Nuclear Polarization (ODNP), is based on the well-known chemical shift of water in NMR spectra. Ordinarily, the liquid water signal intensity is low; however, intensity can be magnified 1000-fold by addition of a nitroxide spin label such as TEMPO.[22] Precession of the unpaired electron in TEMPO at the Larmor frequency results in Nuclear Overhauser-mediated polarization of the protons in water. These get polarized within $\sim$15 Å of the spin labels and then relax with a relaxation time determined by the local diffusivity, *i.e.* in bulk water, the diffusivity is high and so relaxation is rapid; by contrast, in hydration layers, relaxation takes $\sim$10-fold longer than in bulk water. Next, the trick is to covalently tether spin labels to surfaces of interest and measure how relaxation rates in hydration layers change as adhesive proteins approach and locally dehydrate the surfaces.

For proteins adsorbing to spin-labeled polystyrene (a hydrophobic surface), the relaxation times do increase significantly relative to the controls, but only for one protein, mfp-3s, which has a relaxation time of $\tau \sim 750$ ps.[22]

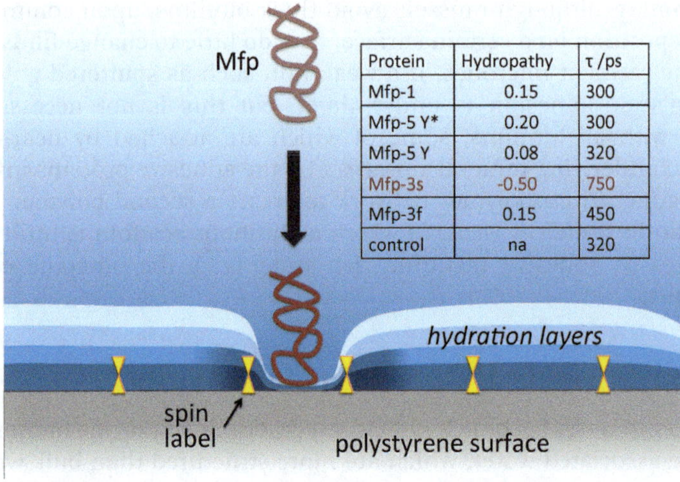

Protein	Hydropathy	τ /ps
Mfp-1	0.15	300
Mfp-5 Y*	0.20	300
Mfp-5 Y	0.08	320
Mfp-3s	-0.50	750
Mfp-3f	0.15	450
control	na	320

Figure 9.3 Mfp-mediated surface water displacement from polystyrene (PS). PS beads were labeled with tethered nitroxide spin labels. Upon excitation during electron spin resonance, polarization was transferred to the surrounding water molecules within 15 Å. Polarized water molecules present relaxation correlation times (τ in Table inset) that reflect the local environment. Longer times denote more confinement and lower diffusivity. Only one adhesive protein, mfp-3s, is capable of increasing the surface water relaxation time, presumably by adsorbing directly onto the PS surface. Mfp-3s is notable for having a high hydrophobicity (inset table), as indicated by the negative hydropathy value, which denotes a normalized per average free energy of transfer from water to the non-polar solvent. Data from ref. 22.

Other adhesive proteins tested including mfp-5, mfp-3f, and mfp-1 were found to present values marginally or not different from those of the controls at $\tau \sim 300$ ps (Figure 9.3). The presence/absence of 3,4-dihydroxyphenylalanine (Dopa) did not affect relaxation. The decreased relaxation rate induced by mfp-3s suggested that surface water becomes cramped (more ordered) upon mfp-3 binding to the polystyrene surface[22] (Figure 9.3). The broadened electron paramagnetic resonance line-widths of the spin label indicated that it too became more constrained at the surface. This very promising new methodology for investigating water at surfaces is still limited by the difficulty of engineering stable tethered spin labels on surfaces other than polystyrene, a hydrophobic surface.

A less technically complicated approach for tracking the surface dehydration ability of proteins and polymers is by attenuated total reflectance Fourier transform infrared spectroscopy (ATR-FTIR). Accordingly, a solution of a surface binding solute is pumped into a flow cell with a Zn–Se crystal sensor, where it is irradiated with IR light and reflected to a detector. The surface chemistry and area of the sensor are specified and the surface area increased using different types of nanobeads (50 nm in diameter).

Time-lapsed adsorption of solute to the target surfaces is monitored by absorbance changes in the infrared spectra. Peptide analogs of mfp-3s (GYDGYNWPYGYNGYRYGWNKGWNGY where Y can be converted to Dopa Y*) with and without Dopa were tested at concentrations of ~0.2 mM on titania and hydroxyapatite nanobeads - both of which have polar hydrated surfaces.[23] The results were as striking as intriguing. In the wavenumber range of 3000–4000 cm^{-1}, where the OH stretching and bending modes of water occur, the absorbance increased for the peptide without Dopa whereas, with Dopa, the absorbance mode was negative on both minerals but greatest for titania (Figure 9.4). We attributed the negative absorbance to surface water displacement.

Ostensibly, the results and interpretations of the ODNP and ATR-FTIR analyses were contradictory. Whereas ODNP concluded that aromatic hydrophobicity, not Dopa, is important to surface dehydration, ATR-FTIR instead found that, for the same sequences, only adsorption of the Dopa homolog removed the surface water. Future studies will need to reconcile the two. As ODNP has only been done on a hydrophobic surface and ATR-FTIR

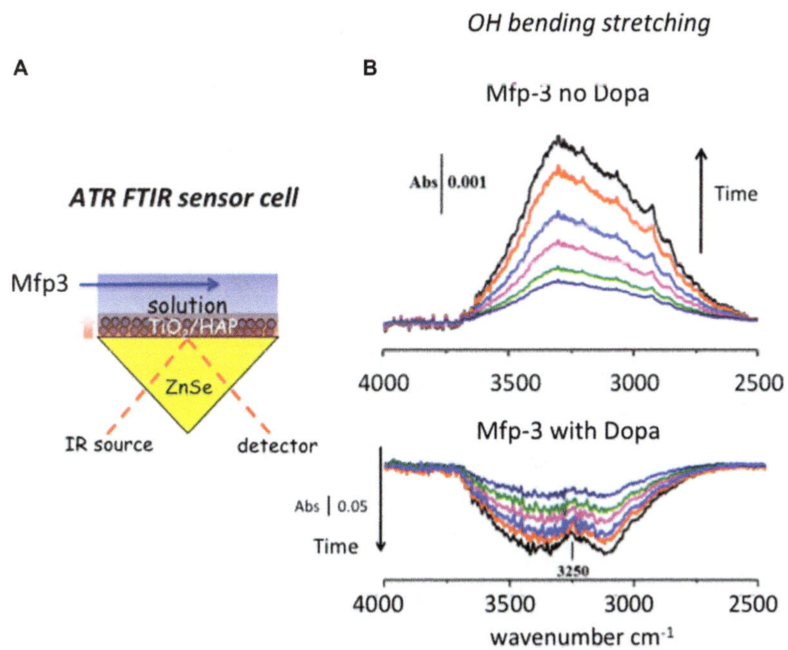

Figure 9.4 Surface water displacement from titania. Hydration layers are present on all solid surfaces immersed in water and represent an impediment to energetic bonding. (A) ATR-FTIR spectroscopy equipped with a Zn–Se sensor flow cell was used in a time-lapse manner to track the abundance of surface water on titania *via* the –OH bending and stretching modes during the adsorption of two mfp-3 inspired peptides. (B) Spectra of mfp-3 with tyrosine, but no Dopa, and mfp-3 with tyrosine converted to Dopa. Data from ref. 23.

only with hydrophilic surfaces, perhaps the answer is simple: a hydrophobic domain is needed to evict water from a hydrophobic surface, and catechols are required to remove water from mineral surfaces. With respect to the first, this is known as the "hydrophobic effect", but the mechanism of catechol-mediated dehydration remains unknown.

Aluminosilicate minerals (*e.g.* clays) exhibit yet another mechanism of surface hydration at salt concentrations ≥0.15 M: they are coated by hydrated potassium ions.[24] On mica, catechol-functionalized oligomers are unable to penetrate the hydrated K^+ layer to bind the underlying oxides, however, tandem catechol-lysine pairings evict K^+ ions and their associated water for strong surface binding.[25] Apparently, the positive charge on lysine is used to repel the hydrated potassium ions, thereby creating a footprint for the bidentate catechol (Figure 9.5). Perhaps, this explains why intertidal marine mussels evolved with such a high incidence of tandem Dopa-Lys and Dopa-Arg pairs in their adhesive proteins mfp-3 and mfp-5.[1] In summary, catecholic functionalities mediate effective adsorption to polar surfaces in water; at higher salinities, a tandem of catechol and cation (Lys or Arg) motifs are

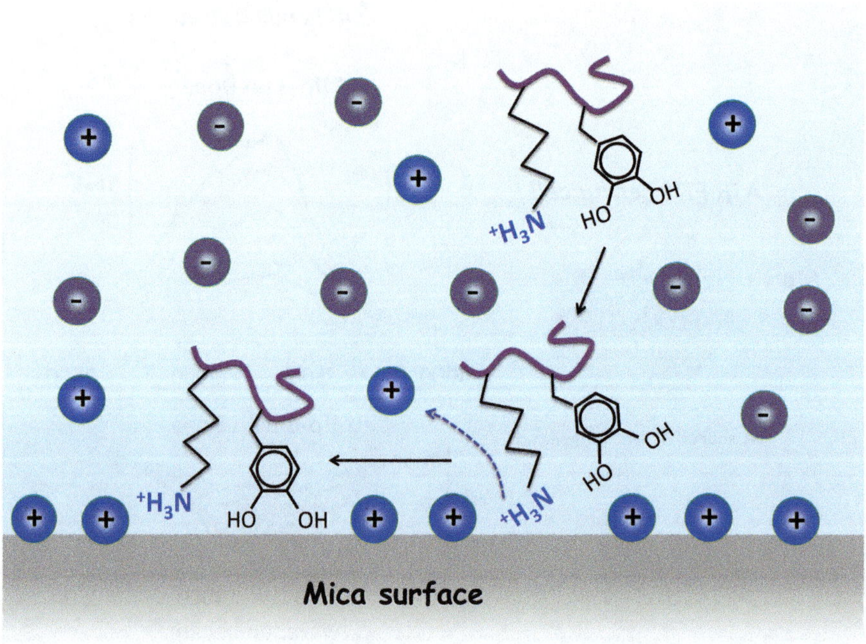

Figure 9.5 Displacement of hydrated cations from mica surfaces. Alumino-silicate minerals, including mica, adsorb K^+ (blue balls) at aluminate sites. The potassium effectively blocks catechol from forming bidentate H-bonds and coordination complexes with the underlying oxide. Tandem placement of charged amines, such as lysine, next to the catechol moieties effectively repels K^+, thus allowing the catechol groups to bond to the surface. (Interpreted from ref. 25.)

required for cation adsorbing surfaces. On non-polar surfaces, aromatic hydrophobicity dehydrates best.

9.4 Is the pH of Adhesive Deposition the Same as Seawater pH?

pH is defined as the negative logarithm of α, the chemical activity of protons but, at low concentrations is often approximated as $-\log\,[H^+]$. Marine mussels are surrounded by seawater buffered at pH $\sim$8, although ocean acidification is gradually changing this with significant consequences for marine life, including mussel attachment.[26] It would be reasonable to assume that the formation of byssus has been adapted to the properties of seawater including the pH, but this is only partly true. In actuality, the initial deposition of byssal proteins requires bizarre pH conditions that can only be compared with those of the stomach lining, and this is possible only because the foot forms a sealed, cup-like isolation chamber over the target surface from which seawater is replaced by a natural fluid. After many failed attempts, the interfacial pH during the deposition of plaque proteins was successfully measured using a surface-tethered pH sensitive fluorescent dye.[27] The surface was mica coated with a lipid bilayer. Oregon Green 488, a fluorescein derivative (pK_a of 4.5), was covalently attached to the exposed head groups of the bilayer (Figure 9.6A) and conditioned in seawater before presentation to juvenile mussels crawling about in a petri dish. In the pH range pK_a -2 to pK_a $+2$, the fluorescence was roughly sigmoidal with the pH, enabling the interpolation of the interfacial pH from the fluorescence intensity. Mussel attachment to the fluorochrome-labeled surface was followed by inverted confocal microscopy. The pH of deposition under the foot was briefly held at between pH 2–3.5, as indicated by a non-fluorescent dark spot (Figure 9.6B, top panel) before returning gradually to seawater pH. In an apparently independently evolved fiber formation process, spider silk is also acidified just prior to entering the spinnerets.[28]

The acidic pH of deposition is interconnected with many other system attributes. These are worth noting with the caveat that others may exist: the acidic secretion (a) etches the mineral surfaces, (b) is a biocide for many surface microbes, (c) maximizes mfp solubility, (d) is needed for mfp-coacervate formation, (e) inhibits Dopa and thiol oxidation, and (f) spotlights pH as an important switch for processing. Having said this, important details such as the identity and concentration of the acid, as well as the mechanism of production, remain unknown.

The importance of acidic interfacial pH during plaque formation was anticipated by the adhesion of individual purified mfps, as measured by a surface forces apparatus. Asymmetric deposition of monomolecular mfp films on mica resulted in adhesion energies that showed a strong pH-dependence for all mfps. For example, for the interfacial proteins mfp-3f and mfp-5, the greatest adhesion always occurred at pH 2.5–3, and decreased

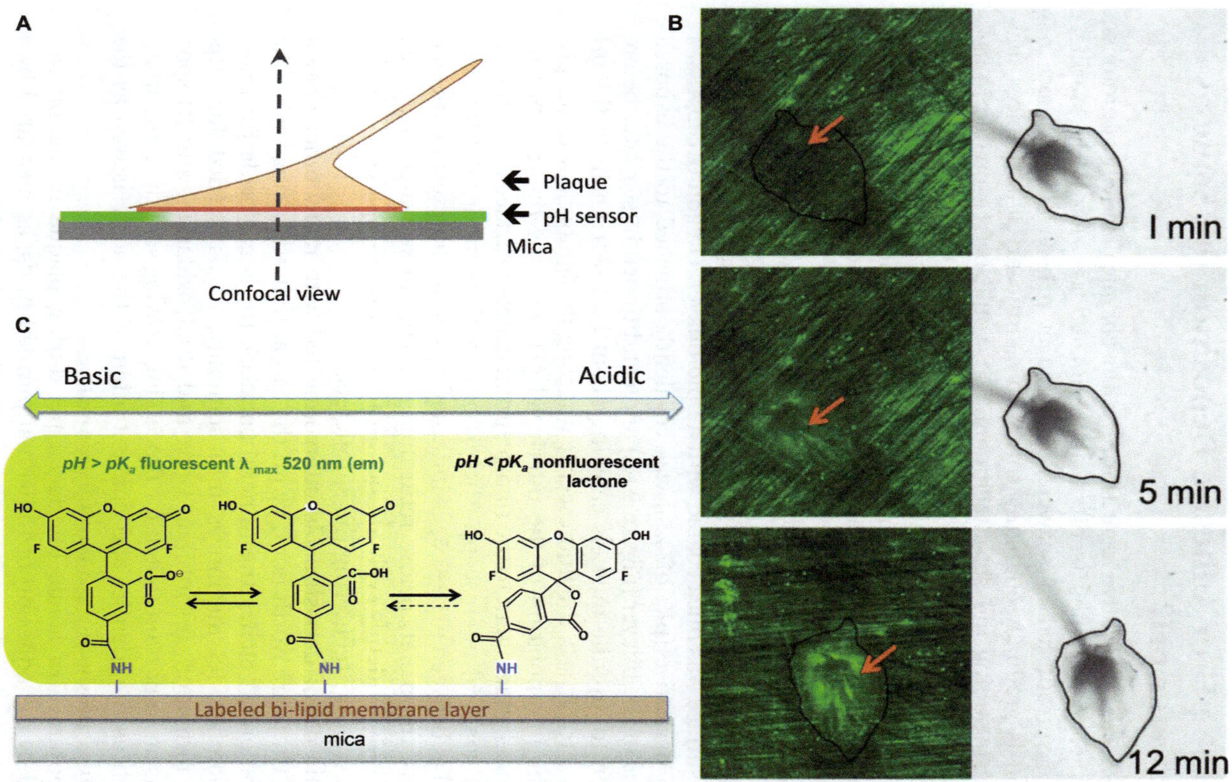

Figure 9.6 The pH under the foot during adhesive protein deposition by mussels. (A) Side view of a plaque attached to pH-sensitive fluorochrome (Oregon green 488) tethered to a bilayer on mica. The confocal view is denoted by the arrow. (B) Fluorescence intensity at three different times post protein deposition (1, 5, and 10 min). (C) Fluorescence intensity and the equivalent pH plotted as a function of time, post-deposition.
Data from ref. 27.

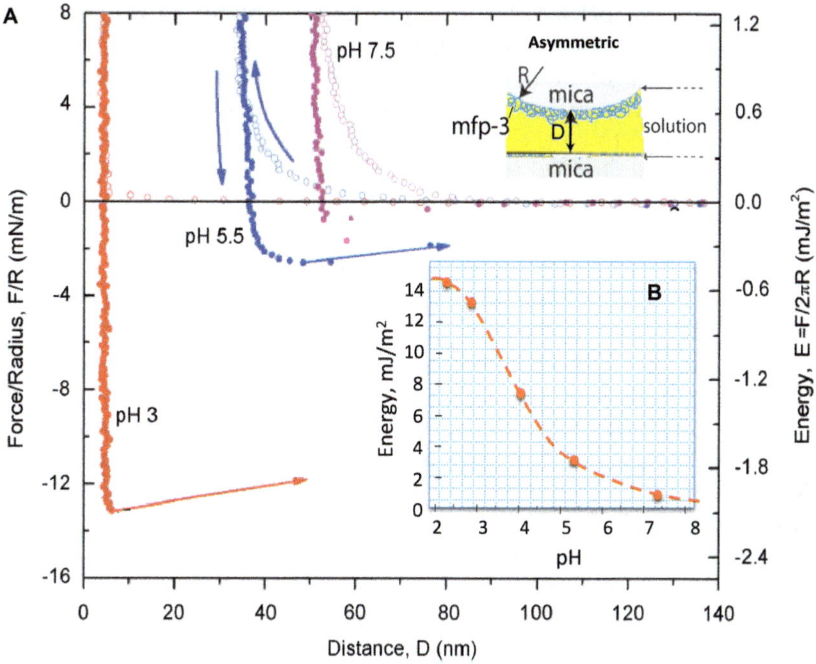

Figure 9.7 The pH dependent adhesion of mussel adhesive proteins. (A) In the asymmetric configuration (Inset, upper right) of mfp-3 and mfp-5 on mica in the surface forces apparatus, both the adhesion force and energy decrease exponentially with the increasing pH. (B) The same trend for mfp-5 (inset, lower right), but plotting only the energy corresponding to the force minima against the pH.
Data from ref. 29.

exponentially to zero with the increasing pH[29] (Figure 9.7). At first, this trend was not thought to be biologically relevant, as surely the loss of mfp adhesion at seawater pH was inconsistent with natural mussel adhesion. Further analysis combining both SFA results with chemical analysis revealed that the apparent loss of adhesion with the increasing pH was coupled to a chemical change in the proteins that reduced their adhesion to mica. The use of a buffer, such as borate at pH 8, for example, which retards the autooxidation of Dopa by forming an oxidation resistant boronate derivative, achieved the same adhesion as that at acidic pH.[30] This was interpreted to mean that deposition at acidic pH is a precaution against premature oxidation and that, once Dopa (reduced) residues are adsorbed to polar surfaces, they are effectively protected against oxidation. Further protection of the non-adsorbed Dopa groups may not be necessary, indeed it may not be desired if they are needed for cohesion through covalent crosslinking. This will be discussed in the last section in more detail.

Scrutiny of the adhesion of mfp-5 to mica revealed a further intriguing inconsistency between the solution and interfacial chemistries. That is, if adhesion loss at pH 5.5, for example, is entirely due to Dopa oxidation (Figure 9.7), then why is it that Dopa residues in mfp-3f in bulk solution at pH 5.5 are stable against oxidation? Could it be that the solution pH is not representative of the surface pH? Although surface and solution pH values are often assumed to be the same, numerous studies have shown that they are often different. Indeed, interfacial pH arises from the pK_a and charge densities of ionizable groups of both surface and absorbed molecules. Silica surface pH was estimated to be two units lower than that in solution.[31,32] Using thioctic acid functionalized self-assembled monolayers (SAMs) on gold, the interfacial pH was nearly three units below the solution pH and similar to the pK_a of the free acid.[33] The discrepancy was caused by the surface charge, the interfacial pH, and a high surface charge density.

A similar re-evaluation of pH has occurred when modeling the properties of pores or confined spaces in hydrogels made of different polyelectrolytes. Accordingly, Longo *et al.*[34] predicted that the prevailing pH in the pores of a polycationic hydrogel will tend to be higher than that of the ambient bulk solution, whereas in its polyanionic counterpart, it will be lower. Applied to mussel adhesion, the interfacial mfps have alkaline isoelectric points (up to pI 10.5) with high cationic charge densities. If, upon adsorbing to the mica surface, these create a mesh, then predicting a compartmental pH to be higher than the solution pH is quite plausible. At an interfacial pH that is two units higher than that of the buffer (pH 5.5), *e.g.*, at pH 7.5, one would indeed expect a greater Dopa oxidation to occur. Why mussels create such high interfacial pH remains a mystery.

9.5 Is Interfacial Redox the Same as Seawater Redox?

In biology, pH and redox are inextricably linked. However, given the range of effects that redox has on the chemistry of mfps, a separate treatment is presented here. Redox refers to the electron transfer between different components, and redox potentials reflect the ease or difficulty of electron transfer. For example, there are about 2.5 μmoles O_2 per mL of seawater at 20 °C. The reduction of O_2 to water requires the transfer of $2e^-$ and 2 H^+ from somewhere to each oxygen atom to produce one molecule of water. As a half-reaction, the favorability of this reaction is indicated by the positive E_o' at standard conditions (pH 7, 1 atm, 1 M concentration, 25 °C, SHE):

$$\text{Half-reaction 1: } \tfrac{1}{2}\,O_2 + 2e^- + H^+ \rightarrow H_2O \quad E_o' = +0.8\ \text{V} \tag{9.1}$$

In other words, the transfer of electrons to oxygen is very favorable in general, but this is not useful information without the corresponding potential of electron transfer from a donor compound.

Recapitulating the "inverted cup" model of the foot (Figure 9.8), the mussel foot makes an isolated mini-reaction chamber underwater prior to secreting acidified adhesive proteins or mfps onto the target surface.

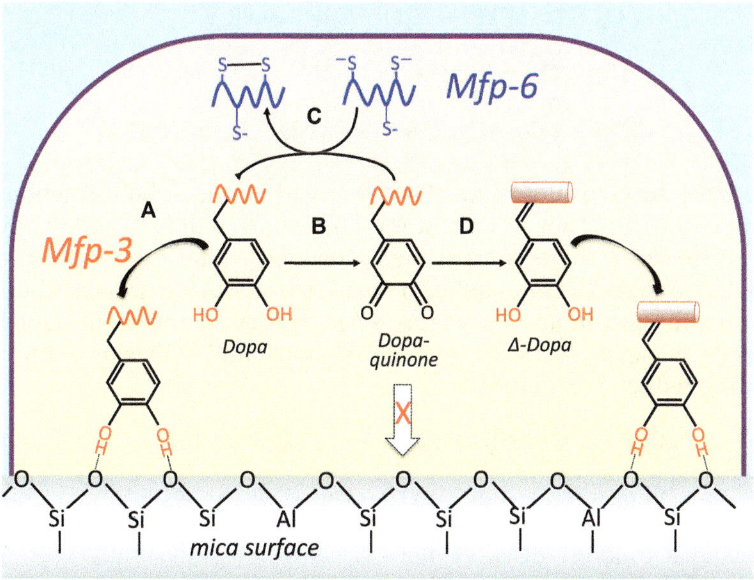

Figure 9.8 A model of interfacial redox processes during the initial deposition of plaque. Among the first proteins to be detected during plaque formation by the foot are mfp-3, mfp-5 (not shown), and mfp-6. The first two are adhesive on mica, whereas mfp-6 is not. The proposed sequence of events is: (A) mfp-3 is secreted into an acidified environment defined by the cup-like shape of the foot (see Figure 9.1A) and is bound to mica by its Dopa residues. (B) As the pH rises, more Dopa undergoes oxidation to DopaQuinone, rendering it non-adhesive. (C) DopaQuinone is regenerated as Dopa by mfp-6 mediated reduction, but results in the accumulation of oxidized mfp-6. (D) As the reducing power of mfp-6 wanes, Dopa undergoes oxidation again but, at a higher pH, DopaQuinone tautomerizes to vinyl catechol, which has potential as a reducing surrogate for mfp-6, a cross-linker, and an inducer of new secondary structures.
Data from ref. 8 and unpublished studies.

All mfps contain the modified redox-sensitive amino acid known as Dopa, and mfp-2 and mfp-6 contain redox-sensitive cysteine as well. Surrounding the chamber is bulk seawater, with pH 8.2 and a highly oxidizing E_0' of between +0.4 to +0.7 V SHE.[35] If most of the seawater and O_2 have been extruded from the chamber, the abundance of Dopa[36] and Cys create a strongly reducing environment within the chamber for O_2 reduction, with Cys being the most energetic electron donor:[37]

$$\tfrac{1}{2}\,O_2 + 2e^- + H^+ \rightarrow H_2O \quad E_0' = +0.82\ \text{V} \tag{9.1}$$

$$\text{Dopa} \rightarrow \text{DopaQuinone} + 2e^- + H^+ \quad E_0' = -0.33\ \text{V} \tag{9.2}$$

$$\text{Dopa} + \tfrac{1}{2}\,O_2 \rightarrow \text{DopaQuinone} + H_2O \quad E_0' = +0.49\ \text{V} \tag{9.3}$$

$$\tfrac{1}{2}\,O_2 + 2e^- + H^+ \rightarrow H_2O \quad E_0{}' = +0.82\ V \tag{9.1}$$

$$2Cys \rightarrow Cys\text{-}S\text{-}S\text{-}Cys + 2e^- + H^+ \quad E_0{}' = +0.22\ V \tag{9.4}$$

$$2Cys + \tfrac{1}{2}\,O_2 \rightarrow Cys\text{-}S\text{-}S\text{-}Cys + H_2O \quad E_0{}' = +1.02\ V \tag{9.5}$$

Prima facie, the O_2 would be quickly scavenged by one or both reactions, but at what cost to adhesion? The cost for reactions in eqn (9.3) and (9.5) is still unclear; however, there is experimental evidence that the reaction in eqn (9.3) effectively undermines adhesion *in vitro* by reducing the contribution of Dopa to adhesion in an SFA (Figure 9.7). To prevent this from happening, eqn (9.4) is again recruited to restore Dopa and adhesion with mfp-6 providing reducing cysteines:

$$2Cys \rightarrow Cys\text{-}S\text{-}S\text{-}Cys + 2e^- + H^+ \quad E_0{}' = +0.22\ V \tag{9.4}$$

$$DopaQuinone + 2e^- + H^+ \rightarrow Dopa \quad E_0{}' = +0.33\ V \tag{9.6}$$

$$DopaQuinone + 2Cys \rightarrow Dopa + Cys\text{-}S\text{-}S\text{-}Cys \quad E_0{}' = +0.55\ V \tag{9.7}$$

In this way, the content of mfp-6 represents a reservoir of reducing activity to scavenge oxygen and reduce quinones back to Dopa. Some will object that eqn (9.7) would not be necessary if O_2 had been effectively scavenged at the outset. However, other oxidants besides O_2 are present at the interface, *e.g.* $Fe^{3+} + e^- \rightarrow Fe^{2+}$ ($E_0{}' = +0.77$ V). A picture is emerging: a chemical adhesion based on Dopa carries certain risks, that is, a tendency to become auto-oxidized. Notwithstanding this, mussel adhesion works well enough to justify investment in a rescue system based on mfp-6 to keep interfacial Dopa reduced.

A closer look at the system, however, does pique curiosity. The initial pH within the chamber is not 7 but 2–3, and the reactions are non-equilibrium, often irreversible, and involve other intermediates that can become important end products. The acidic pH represents a problem in that thiolates, not thiols, are the operative reductants, thus cannot reduce at pH values below their typical pK_a, *i.e.* 8–9.[38] This is resolved by proteins, including mfp-6, by sequence specific effects such as flanking cationic groups that reduce the Cys pK_a, *e.g.* redox active Cys-59 in DsB-A has a pK_a of 3.5.[39] Several Cys residues in mfp-6 are acidic,[29] but specific pK_a values have yet to be measured. The non-equilibrium, irreversible nature of the oxidation reactions is a particular problem with Dopa and other catechols. Indeed, the chemical fate of catechols in mussel byssus is highly dependent on their location. In the cuticle, the fate of Dopa appears to be tris catecholato-Fe^{3+} complexes;[40] in the thread and plaque core, Dopa forms covalent cross-links after oxidation to quinones,[41] whereas at the plaque-substratum interface, it is some combination of metal chelates and reduced H-bonded Dopa on metal oxide surfaces. The reducing capacity of mfp-6 plays a role in maximizing the latter and is astonishingly sustained, *i.e.* >21 days.[42]

One intriguing redox side reaction is the tautomerization of DopaQuinone to α, β-dehydroDopa[43] (Figure 9.8D). The reaction is notable because after Dopa oxidation to quinone, a catechol (vinyl-catechol) side-chain is restored without relying on intermolecular reduction. The role of dehydroDopa in adhesion is difficult to fathom at this time. On the one hand, tautomerization from quinones at low pH is not favored, thus it is an unlikely initial interfacial player. Later, however, when the pH increases, dehydroDopa formation is more favorable. Some chemical effects are predicable: vinyl catechol retains the bidentate H-bonding and Fe-coordination of Dopa, ultraviolet absorbance is shifted to higher wavelengths ($\lambda_{max} = 350$ nm), and the oxidation potential is decreased to 300 mV lower than Dopa at pH 3.[8] The migration of the double bond to the side chain imposes planarity on the N-αC-C grouping of Dopa, limiting the dihedral angles to 180°, with associated conformational consequences, *e.g.* 3_{10} helix and β-sheet.[44] The decrease in oxidation potential makes dehydroDopa a better reducing agent and more vulnerable to oxidative loss. Is this relevant to mfps, and if so, is it good or bad for adhesion? A recent cyclic voltammetry analysis has revealed that, although mfp-3f has an oxidation potential of 400 mV at pH 4, consistent with Dopa at pH 7, the oxidation potential of 200 mV matches that of dehydroDopa and is 150 mV below that of Dopa. In other words, dissolving mfp-3f in solution at neutral pH without antioxidants leads to a nearly spontaneous oxidation and tautomerization of Dopa to dehydroDopa. Quartz crystal microbalance analysis of the adsorption of Dopa and dehydroDopa derivatives to hydroxyapatite (HAP, a bone-like mineral), has indicated that the adsorption capacity of dehydroDopa on HAP is 10-fold greater than that of Dopa.[45]

The trouble with redox chemistry is that it gives the false impression that only a few reactions occur and that specific outcomes can be predicted with confidence. In a complex mixture of molecules, this is rarely true, in part because biology is not about equilibria and because the reactions shown are not as simple as written. With respect to the first caveat, redox potentials can be converted into the free energy changes associated with the reactants and products, but activation energies are necessary to investigate the actual kinetic effects.

9.6 Summary

In biomimetics, including mussel adhesion, the good translation of bio-inspired motifs requires a firm scientific understanding of the underlying biology. At present, there is an unwholesome rush to translate intriguing motifs with little regard to their context in nature. The much imitated Dopa functionalities of mussel adhesive proteins require peculiarly acidic and reducing conditions to bind target surfaces, which could never have been surmised. Even the primary sequence flanking Dopa in mussel foot proteins can dramatically affect the adhesive performance. Defining the natural context of bio-adhesion deserves the highest priority. Biomimetic pursuits without such insight from the natural context are highly likely to squander research funding.

Acknowledgements

The author thanks W. Wei, R. Mirshafian, J. Yu, and N. Martinez for discussions related to this review.

References

1. P. Lee, P. B. Messersmith, J. N. Israelachvili and J. H. Waite, *Annu. Rev. Mater. Res.*, 2011, **41**, 99.
2. E. Carrington, J. H. Waite, G. L. Sara and K. P. Sebens, *Ann. Rev. Mar. Sci.*, 2015, **7**, 443.
3. H. G. Silverman and F. F. Roberto, *Mar. Biotechnol.*, 2007, **9**, 661.
4. Z. Qin and M. Buehler, *Nat. Commun.*, 2013, **4**, 2187.
5. J. H. Waite, H. C. Lichtenegger, G. Stucky and P. K. Hansma, *Biochemistry*, 2005, **43**, 7653.
6. A. V. Spuskanyuk, R. M. McMeeking, V. S. Deshpande and E. Arzt, *Acta Biomater.*, 2008, **4**, 1669.
7. K. Desmond, N. Zacchia, J. H. Waite and M. Valentine, *Soft Matter*, 2015, **11**, 6832.
8. J. Yu, W. Wei, J. N. Israelachvili and J. H. Waite, *Adv. Mater.*, 2011, **23**, 2362.
9. P. Roach, D. Eglin, K. Rohde and C. C. Perry, *J. Mater. Sci: Mater. Med.*, 2007, **18**, 1263.
10. G. I. Loeb and R. A. Neihof, in *Applied Chemistry at Protein Interfaces*, ed. R. E. Baier, ACS Advances in Chemistry Series, The American Chemical Society, Washington, 1975, vol. 145, p. 319.
11. I. Sutherland, in *Polysaccharides: Structural Diversity and Functional Versatility*, ed. S. Dumitriu, 2004, CRC Press, Boca Raton, FL, p. 482.
12. N. Siboni, M. Lidor, E. Kramarsky-Winter and A. Kushmaro, *FEMS Microbiol. Lett.*, 2007, **274**, 24.
13. R. M. Donlan, *Emerging Infect. Dis.*, 2002, **8**, 881.
14. N. Hutchinson *et al.*, *Aquat. Microbiol. Ecol.*, 2006, **42**, 187.
15. J. S. Maki, D. Rittschoff, J. D. Costlow and R. Mitchell, *Mar. Biol.*, 1988, **97**, 199.
16. N. Martinez-Rodriguez, PhD thesis, University of California Santa Barbara, 2014.
17. S. H. Marshall and G. Arenas, *Electron. J. Biotechnol.*, 2003, **6**, 271–284.
18. P. Benjamin and C. Weaver, *Proc. Roy. Soc. London, Ser. A*, 1960, **254**, 177.
19. A. Omoike, J. Chorover, K. D. Kwon and J. D. Kubicki, *Langmuir*, 2004, **20**, 11108.
20. Y. P. Tan, S. Hoon, P. A. Guerette, W. Wei, A. Ghadban, C. Hao, A. Miserez and J. H. Waite, *Nat. Chem. Biol.*, 2015, **11**, 488.
21. B. Ratner, *Biomaterials Science: An Introduction to Materials in Medicine*, Elsevier Academic Press, San Diego, 2004, p. 574.

22. Y. Akdogan, W. Wei, K.-Y. Huang, Y. Kageyama, E. W. Danner, D. R. Miller, N. Martinez-Rodriguez, J. H. Waite and S. I. Han, *Angew. Chem., Int Ed.*, 2014, **53**, 11253.
23. W. Wei, L. Petrone, A. Miserez and J. H. Waite, *Adv. Func. Mat.*, 2016, **26**, 3496–3507.
24. W. Stumm and J. J. Morgan, in *Aquatic Chemistry*, John Wiley & Sons, N.Y., 1981, p. 625.
25. G. Maier, M. Rapp, J. H. Waite, J. N. Israelachvili and A. Butler, *Science*, 2015, **349**, 628.
26. M. J. O'Donnell, M. N. George and E. Carrington, *Nat. Clim. Change*, 2012, **3**, 587–590.
27. N. Martinez-Rodriguez, S. Das, Y. Kaufman, J. N. Israelachvili and J. H. Waite, *Biofouling*, 2015, **31**, 221.
28. A. Rising and J. Johansson, *Nat. Chem. Biol.*, 2015, **11**, 309.
29. J. Yu, W. Wei, E. Danner, R. Ashley, J. N. Israelachvili and J. H. Waite, *Nat. Chem. Biol.*, 2011, **7**, 588.
30. Y. Kan, E. Danner, J. N. Israelachvili, Y. Chen and J. H. Waite, *PLoS One*, 2014, **9**, e1088869.
31. J. P. O'Reilly, C. P. Butts, I. A. I'Anson and A. M. Shaw, *J. Am. Chem. Soc.*, 2005, **127**, 1632.
32. P. B. Petersen and R. J. Saykally, *J. Phys. Chem. B*, 2006, **110**, 15037.
33. M. Rooth and A. M. Shaw, *Phys. Chem. Chem. Phys.*, 2006, **8**, 4741.
34. G. S. Longo, M. Olivera de la Cruz and I. Szleifer, *Macromolecules*, 2011, **44**, 147.
35. L. H. N. Cooper, *J. Mar. Biol. Ass. UK*, 1937, **22**, 167.
36. M. Eslami, H. R. Zare and M. Namazian, *J. Phys. Chem. B*, 2012, **116**, 12557.
37. I. H. Segel, *Biochemical Calculations*, John Wiley & Sons, N.Y., 2nd edn, 1976, pp. 414–415.
38. S. C. T. Nicklisch and J. H. Waite, *Biofouling*, 2012, **28**, 865–877.
39. K. S. Jensen, R. E. Hansen and J. R. Winther, *Antioxid. Redox Signaling*, 2009, **11**, 1047.
40. M. J. Harrington, A. Masic, N. Holten-Andersen, J. H. Waite and P. Fratzl, *Science*, 2010, **328**, 216–220.
41. L. M. McDowell, L. A. Burzio, J. H. Waite and J. Schaefer, *J. Biol. Chem.*, 1999, **274**, 20293.
42. D. R. Miller, J. E. Spahn and J. H. Waite, *J. R. Soc., Interface*, 2015, **12**, 20150614.
43. L. M. Rzepecki, T. Nagafuchi and J. H. Waite, *Arch. Biochem. Biophys.*, 1991, **285**, 17.
44. P. Mathur, S. Ramakumar and V. S. Chauhan, *Biopolymers*, 2004, **76**, 150.
45. R. Mirshafian, W. Wei, J. N. Israelachvili and J. H. Waite, unpublished data.

CHAPTER 10

Mussel Adhesive-inspired Polymers

BRUCE P. LEE

Department of Biomedical Engineering, Michigan Technological University, 1400 Townsend Dr., Houghton, MI 49931, USA
Email: bplee@mtu.edu

10.1 Introduction

Adhesives are used ubiquitously in various industries (*i.e.*, automobile, manufacturing, biomedical) to join two dissimilar materials together. However, the performance of most synthetic adhesives is significantly compromised in the presence of moisture. Water is traditionally treated as a surface contaminant and needs to be removed to prevent the formation of a weak boundary layer.[1] In many applications, and particularly those in the medical field, development of moisture-resistant adhesives is in high demand.

Mussels secrete proteinaceous fluids that harden to form an adhesive plaque and byssal thread complex that enables these organisms to anchor themselves to surfaces in wet, saline, and turbulent environments.[2,3] These adhesive proteins contain a large abundance of a unique amino acid, 3,4-dihydroxyphenylalanine (DOPA, Figure 10.1), which is post-translationally modified from tyrosine. DOPA is responsible for both the rapid curing and interfacial binding properties of these adhesive proteins.[4–6] The catechol side chain of DOPA is capable of binding to both inorganic and organic surfaces through either strong reversible or irreversible covalent bonds.[3,7]

RSC Polymer Chemistry Series No. 22
Bio-inspired Polymers
Edited by Nico Bruns and Andreas F. M. Kilbinger

Published by the Royal Society of Chemistry, www.rsc.org

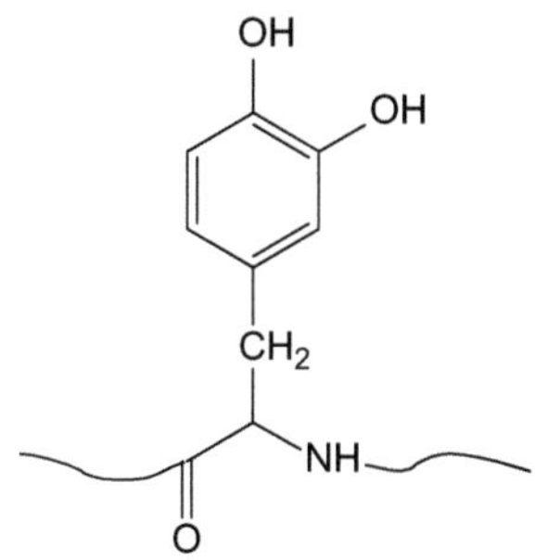

Figure 10.1 The chemical structure of protein-bound DOPA containing the catechol side chain.

Adhesive proteins extracted from molluscs such as common blue mussel (*Mytilus edulis*) have yielded strong adhesion.[8,9] However, several thousand specimens of mussels are needed to produce one gram of protein,[10] which makes it highly impractical for manufacturing adhesives through this technique. Recent advances have focused on the functionalization of polymers with DOPA and various catechol derivatives for creating functional, adhesive materials. Modification of catechol moieties onto inert, synthetic polymers has imparted these materials with strong, water-resistant adhesive properties to various biological substrates, including soft tissue,[11–13] mucin,[14] bone,[15] and various metallic[16–22] and polymeric[23–26] substrates. This chapter summarizes the existing understanding of the catechol side chain chemistries and reviews the methods for the preparation of catechol-functionalized polymers. Finally, applications of these adhesive materials as tissue adhesives, antifouling coatings, delivery vehicles for therapeutic drugs and cells, and environmental cue-responsive smart materials are reviewed.

10.2 Catechol Side Chain Chemistry

Catechol is a unique and versatile adhesive molecule capable of forming reversible physical interactions and irreversible covalent bonds (Figure 10.2). In this section, various catechol chemical interactions are introduced. Additionally, various chemical modifications used to modulate catechol side chain reactivity and in the preparation of catechol-functionalized polymers are reviewed.

10.2.1 Reversible Catechol Interactions

Catechol interacts with other aromatic rings through π–π electron interactions (Figure 10.2A), which may contribute to cohesive properties of catechol-containing adhesives and binding to surfaces rich in aromatic compounds such as polystyrene[27] and gold substrates.[28] The hydroxyl groups of catechol can function both as a hydrogen bond donor and acceptor (Figure 10.2B). The ability to form hydrogen bonds allows catechol

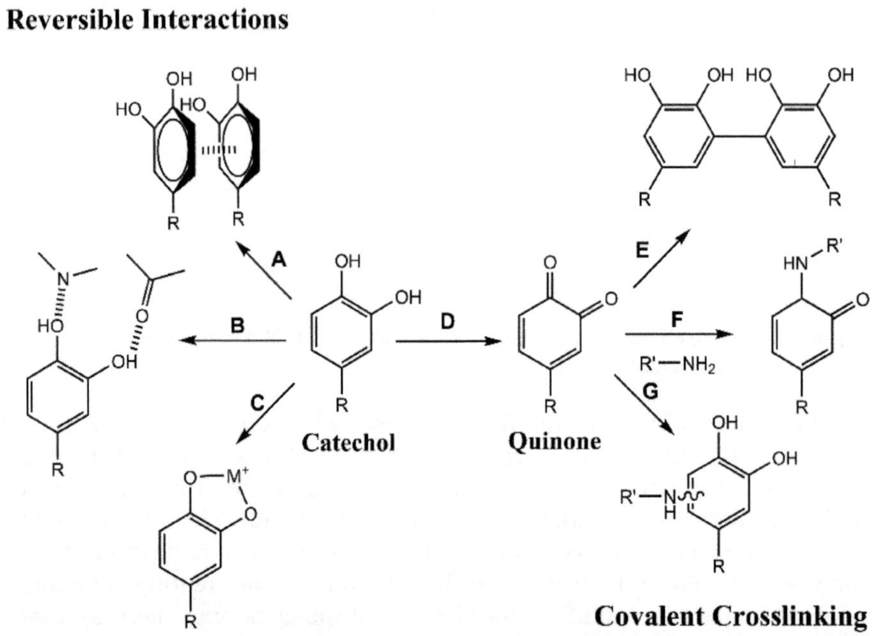

Figure 10.2 The catechol side chain of DOPA is capable of forming reversible interactions and irreversible covalent bonds. The benzene ring of the catechol is capable of π–π interactions (**A**). Catechol -OH groups can function both as a hydrogen bond donor and acceptor (**B**). Catechol forms strong coordination complexes with metal ions (**C**). When catechol is oxidized to form highly reactive quinone (**D**), it can undergo dimer formation (**E**) and subsequently polymerize into oligomers. Quinone can form intermolecular crosslinking with nucleophile such as $-NH_2$ through Schiff base substitution (**F**) and Michael-type addition (**G**).

to compete with water for hydrogen bonding sites and absorb onto mucosal tissues[14,29,30] and hydroxyapatite.[31] Additionally, catechol is capable of forming strong complexes with various metal ions (Figure 10.2C) with log stability constants that are significantly higher than those with polymeric acid- or amino acid-based ligands (Table 10.1). Catechol forms complexes with different stoichiometries (*i.e.*, mono-, bis- and tris-catecholate–metal ion complexes) depending on the molar ratio of catechol to the metal ion, metal ion type and valency, and pH.[32–34] Breaking and reforming strong, reversible catechol–metal ion complexes contribute to the wear resistance properties of the protective coatings on mussel byssus threads, which experience large cyclic strains in the turbulent intertidal zone.[35,36] This interaction has been exploited to form self-healing networks with elastic moduli approaching those of covalently crosslinked hydrogels.[37,38]

The affinity of catechols for metal ions is also extended to metal oxide surfaces.[44–47] A single DOPA moiety forms a strong, reversible bond with the titanium surface with a bond strength averaging around 800 pN.[48] This is the

Table 10.1 Log stability constant of catechol–metal ion complexes.

Ligand	Metal	Mono	Bis	Tris	Ref.
DOPA	Cu^{2+}	15	27	—	34
Catechol	Zn^{2+}	13	27	—	39
Catechol	Mn^{2+}	15	32	—	39
DOPA	Fe^{3+}	18	33	41	34
DOPA	V^{3+}	19	33	41	34
Catechol	Ti^{3+}	—	—	40	40
Catechol	Ti^{4+}	—	25	62	41
Poly(acrylic acid)	Cu^{2+}		6.6		42
EDTA[a]	Fe^{3+}		25		43

[a]Ethylenediaminetetraacetic acid (EDTA).

strongest reversible interaction measured to-date for a biomolecule, which approximates 40% the bond strength of a covalent bond (2000 pN for the carbon–silicon bond). Density functional theory analysis revealed that catechol readily displaces water molecules and adsorbs directly to the metal oxide surface with binding energies reaching 20–30 kcal mol^{-1}.[49–51] The ability to adsorb to various metal oxide surfaces makes catechol a versatile anchoring group for surface modification. Catechol functions as a robust anchor group for various metal substrates (Au_2O_3, Al_2O_3, SiO_2 TiO_2, NiTi, and stainless steel).[16–18] However, the binding strength of catechol to the metal surface is highly dependent on the oxidation state of the catechol side chain. Experiments using contact mechanical testing,[52,53] surface forces apparatus,[54] and atomic force microscopy[48] have demonstrated that the interfacial binding properties of catechol to metal substrates are drastically reduced when the catechol is oxidized either by a chemical oxidant or under basic conditions.

10.2.2 Oxidation-induced Covalent Crosslinking

Catechol can be easily oxidized by either enzymatic (*e.g.* tyrosinase, peroxidase) or chemical (*e.g.*, periodate) oxidants to form highly reactive quinone (Figure 10.2D). Quinone can crosslink with other catechol moieties, resulting in the formation of dimers (Figure 10.2E) and subsequent polymerization of the catechol moieties.[2,55] Oxidation-induced intermolecular covalent crosslink formation is critical for the rapid hardening of catechol-containing adhesives. The crosslinking rate and degree of polymerization can be tailored by the oxidant type and concentration.[56] Additionally, quinone forms covalent crosslinks with various functional groups (*i.e.*, $-NH_2$, –SH, imidazole) found on biological substrates resulting in the formation of interfacial covalent crosslinks through either Schiff base substitution (Figure 10.2F) or Michael-type addition (Figure 10.2G).[57–59] Catechols that contain a free amine group (*i.e.*, dopamine) undergo intramolecular cyclization between the catechol side chain and the amine group to form a leukochrome that can undergo polymerization similar to melanin formation.[56] The use of polydopamine as an adhesive primer and grafting of multifunctional polymers through a simple dip-coating process in an aqueous medium has been reported.[60] The use of polydopamine as a facile and robust approach in designing multifunctional materials are reviewed elsewhere,[61–64] and is not the focus of the current review.

Cencer *et al.*[65] recently reported that the oxidation-mediated crosslinking of catechol is highly dependent on the pH. Under mild acidic conditions (pH = 5.7–6.7), catechol crosslinked more slowly than at neutral or basic pH. The oxidation intermediate of catechol (*i.e.*, quinone methide) is likely stabilized under acidic conditions,[66] which retards the subsequent crosslinking reactions. Similarly, pH also affected the formation of interfacial covalent bonds with soft tissue substrates.[65] Under mildly acidic conditions, the nucleophilic functional groups are protonated (*e.g.*, pK_a of ε-lysine ∼10), which reduces their ability to form interfacial bonds. The reduced ability of catechol to cure and adhere under acidic conditions may limit its application for repairing relatively acidic tissues (pH = 4–6 for skin,[67] pH = 6.7–7.1 for subcutaneous tissue,[68] pH = 7 for dysoxic tissue due to extensive hemorrhage or ischemia[69]) or deliver therapeutic drug particles to cancer tissues (pH ≤ 7).[70] However, the interaction between catechol and inorganic substrates (*i.e.*, metal oxide) is strongest under acidic conditions as the oxidized form of catechol (*i.e.*, quinone) adheres poorly to these surfaces.[48,71]

During the process of catechol oxidation, reactive oxygen species (ROS) such as super oxide anion ($O_2^{\bullet-}$) and hydrogen peroxide (H_2O_2) are generated as byproducts (Figure 10.3).[72,73] Hydrogels containing 1–3 wt% dopamine released 10^2–10^3 μM of H_2O_2 over 48 h as the network-bound catechol underwent auto-oxidation, which was cytotoxic to cells in culture.[72] The cytotoxic effect was counteracted by the introduction of catalase, an enzyme commonly found in many living organisms that can catalyze H_2O_2 into water and oxygen.[74] When the catechol-containing hydrogel was implanted subcutaneously in rats, H_2O_2 generation stimulated the recruitment of macrophages to the tissue–hydrogel interface after 1 week of implantation.[72] ROS has been found to have both beneficiary (*i.e.*, promote wound healing, antimicrobial effects, normal cell signaling) and deleterious (*i.e.*, chronic inflammation, promote tumor initiation) effects.[75–78] Therefore, ROS concentration needs to be

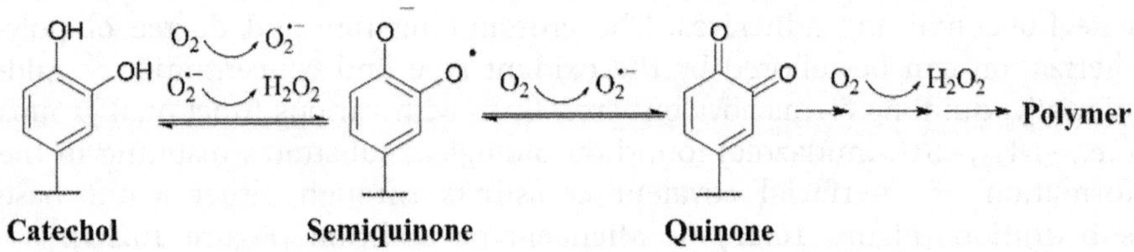

Figure 10.3 Auto-oxidation and polymerization of catechol, while generating reactive oxygen species such as superoxide anion ($O_2^{\bullet-}$) and hydrogen peroxide (H_2O_2).

rigorously regulated for catechol-containing biomaterials depending on the intended application.

10.2.3 Chemical Modification of Catechol

Similar to marine mussels, sandcastle worms also secrete DOPA-containing adhesive proteins that they utilize to cement sand fragments into tubed-shaped dwellings.[79] The catechol side chains in these adhesives are further modified with an electron withdrawing chloro-functional group (*e.g.*, 2-chloro-4,5-dihydroxyphenylalanine, Figure 10.4B), which has been proposed as a natural adaptation to increase interfacial binding strength.[79] Substituting –H with electron withdrawing groups (EWGs) lowers the dissociation constant (pK_a) of the catechol hydroxyl groups, which greatly enhance its reactivity for adhesion.[80] Similarly, nitro-functionalized catechol (Figure 10.4C) exhibits an increased rate of covalent crosslinking,[81] binds to tissue substrates over a wider range of pH,[81] and hastens the rate of adhesive degradation[82] when compared to unmodified catechol. Nitro-group modified catechol also binds more tightly to inorganic surfaces and is more resistant to oxidation and elevated temperature treatments.[22,71,80] Lowering the pK_a also allows catechol to form metal ion complexes at a higher stoichiometry at reduced pH values.[83] Additionally, these modifications have resulted in adhesive biomaterials with unique properties, such as light-induced degradation[84] and antimicrobial properties[85] for the $-NO_2$ and –Cl modified catechols, respectively. Adhesion by –OH modified catechols (Figure 10.4D and E) has also been demonstrated.[86] However, these side chain modifications can increase the steric hindrance and affect the chemical crosslinking. For example, $-NO_2$ modified catechol forms only dimers as opposed to the oligomers obtained from unmodified catechol, which affects the crosslinking density of the cured adhesive.[81] Finally, pyridine modified quinone (Figure 10.4F) has also exhibited the ability to form strong metal ion complexes[83] and interfacial binding to inorganic surfaces.[22,80]

One of the first reported catechol-modified polymers involved coupling the amino acid form of DOPA (Figure 10.4G) onto amine-functionalized poly(ethylene glycol) (PEG) through amide bond formation.[56] However, this synthetic approach required the use of DOPA with its amine group protected with an *N*-*t*-butoxycarbonyl functional group. Subsequent approaches utilized catechols containing either an acid (Figure 10.4H)[87] or an amine (Figure 10.4I),[65,81,88] which simplified the synthesis. Additionally, the amine group can be further modified with an initiator (Figure 10.4J)[89,90] or a polymerizable methacrylate group (Figure 10.4K),[23] both of which can be used in the preparation of polymeric materials through free-radical polymerization. Similarly, catechol functionalized with a vinyl group has also been used (Figure 10.4L).[91] Finally, DOPA amino acid can be modified to form *N*-carboxyanhydride (NCA, Figure 10.4M), which can be polymerized in the presence of a weak base to form DOPA-containing peptides.[92,93] The use

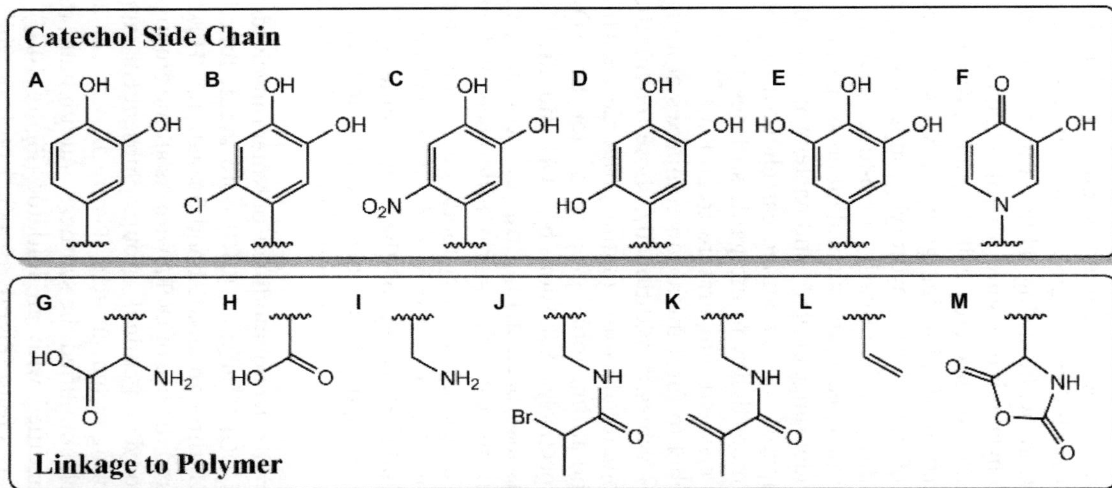

Figure 10.4 Diversity of chemistry structures utilized to create biomimetic adhesive polymers. Catechol side chain (**A**) modification alters its interfacial binding strength and reactivity. Substitution can be achieved by replacing –H at the *para* position with chloro- (**B**), nitro- (**C**) and hydroxyl (**D**) groups or a hydroxyl group at the *meta* position (**E**). The benzene ring can be substituted with a pyridine group (**F**). Linking the catechol with a polymer can be achieved *via* reaction of the amino acid (**G**), acid (**H**), or amine (**I**) groups. Catechol modified with a bromide propionamide initiator to initiate polymerization (**J**) or functionalized with polymerizable methacrylamide (**K**), vinyl (**L**), and *N*-carboxyanhydride (NCA, **M**) groups.

of these different catechol species in the preparation of adhesive polymers is reviewed in the following section.

10.3 Preparation of Catechol Functionalized Polymers

Simple modification of catechol onto various types of polymers can impart these inert polymers with desired moisture-resistant interfacial binding properties and the ability to induce covalent crosslinking on command. Here, methods used in protecting the reactive catechol side chain are introduced. Additionally, three general approaches in creating catechol-modified polymers are reviewed. These approaches include (1) direct functionalization of polymers with catechol, (2) polymerization of catechol-modified monomers, and (3) the use of catechol-functionalized initiators to initiate polymerization. Other approaches in preparing catechol-modified peptide or proteins with controlled sequences through the solid phase peptide synthesis approach[94–96] and recombinant genetic engineering techniques[97,98] are not reviewed here.

10.3.1 Catechol Side Chain Protection

One of the challenges in preparing catechol-functionalized polymers is preserving the reactivity of the catechol side chain, which may be oxidized during chemical reactions. Under basic and oxygenated conditions, catechol transforms into quinone, which undergoes irreversible self-polymerization or nucleophilic addition reactions with amino or sulfhydryl groups.[60] Proper protection of the catechol group is often required during the preparation of catechol-functionalized polymers. Table 10.2 summarizes a list of common protecting groups and their deprotection methods. These protecting groups include acetonide,[99,100] acetyl,[53,101] *t*-butyldimethylsilyl (TBDMS),[102] carboxybenzyl (Cbz),[93] cyclic ethyl orthoformate (Ceof),[103] and methyl ether.[91] It is critical that the protecting group is stable during the desired chemical reaction, while the deprotection method is compatible with the chemical structure of the resulting catechol-functionalized polymer. Most notably, it is important that the deprotection method does not adversely affect the reactivity of the catechol side chain. The use of acid during the deprotection of catechol (*e.g.*, HBr in acetic acid for removing the Cbz group) is ideal in preserving the reduced state and reactivity of catechol. However, deprotection methods that utilize strong bases (*e.g.*, pyridine for removing the acetyl group) require special care to remove molecular oxygen in order to prevent catechol oxidation and self-polymerization.[53]

The aforementioned protecting groups are chemically stable and are well suited for multi-step chemistries. Alternatively, borate forms complexes with the catechol side chain at an elevated pH[104] and is suitable for functioning as a temporary protecting group for chemical reactions carried out in a basic

Table 10.2 Common catechol side chain protecting groups.

Protecting group		Deprotection method	Ref.
Acetonide	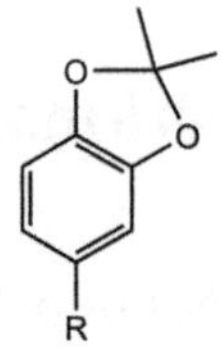	Trifluoroacetic acid in aqueous methanol mixture	99, 100
Acetyl	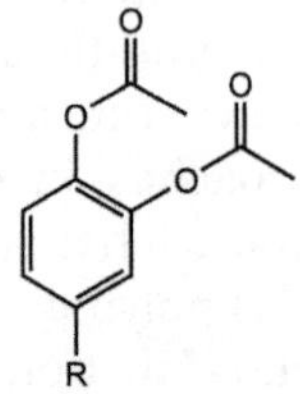	5% pyridine	53, 101
Borate	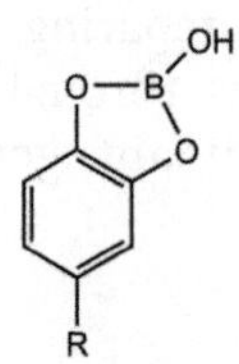	Reduce pH to less than 3	23, 90, 102
t-Butyldimethylsilyl (TBDMS)	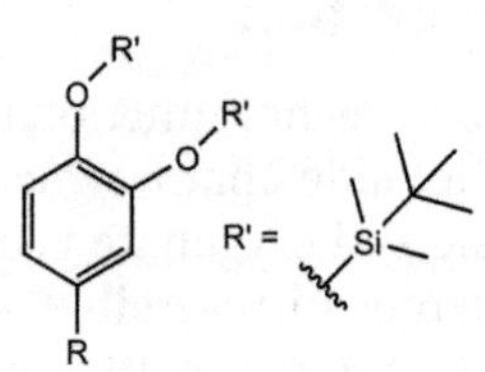	Tetrabutylammonium fluoride	102
Carboxybenzyl (Cbz)	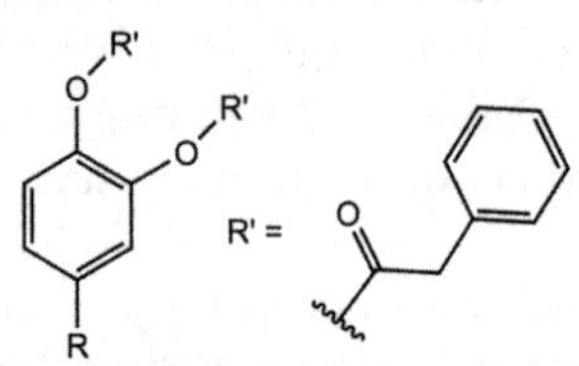	HBr in acetic acid	93
Cyclic ethyl orthoformate (Coef)	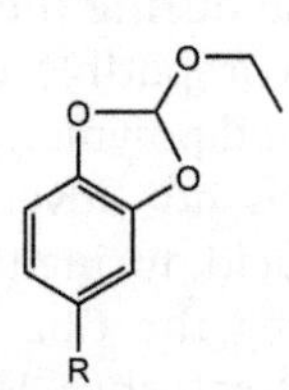	Trimethylsilyl bromide in trifluoroacetic acid	103
Methyl ether	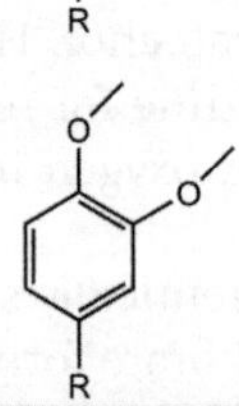	Boron tribromide	91

aqueous solution.[23,90,102] Borate can be easily removed by acidifying the reaction mixture. The pH-responsive nature of catechol–borate complexation has been exploited for designing self-healing hydrogels[105] and drug delivery applications.[106]

10.3.2 Direct Functionalization of Catechol

Catechol can be directly coupled to a polymer with a suitable functional group (*i.e.*, $-NH_2$, $-COOH$, or $-OH$) using carbodiimide chemistry to form an amide linkage[56] or through the formation of urethane[14] linkages (Figure 10.5). The first reported catechol-functionalized polymer through the direct functionalization approach involved end-modification of PEG of both linear and branched structures with DOPA.[56] PEG is a hydrophilic and chemically inert polymer and has been used to create polymeric model systems to study the crosslinking chemistries of catechol.[56,65,81,107] Additionally, PEG is well known for its biocompatibility and serves as an excellent macromolecular support in developing tissue adhesives and sealants.[88,108,109] The bioinert nature and ability to interact strongly with water molecules make PEG-based polymers suitable to function as antifouling coatings.[16,26]

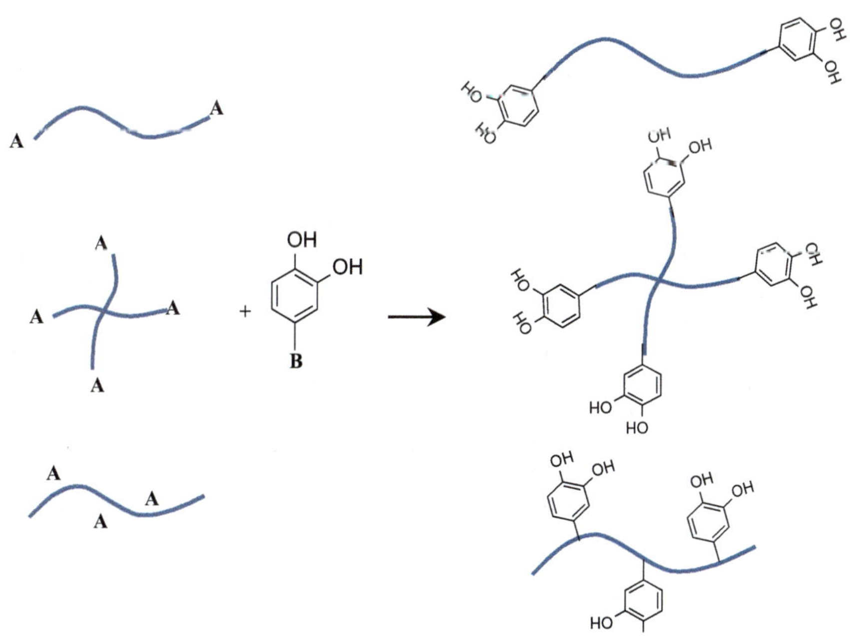

Figure 10.5 Schematic representation of the direct functionalization of polymers of different architectures with catechol, where functional group "**A**" on the polymer reacts with functional group "**B**" of the catechol.

In addition to hydrophilic PEG, adhesives based on amphiphilic block copolymers composed of PEG and polycaprolactone (PCL),[11,12] PEG and polypropylene oxide (PPO),[14,110] polyurethane of PEG and oligo alkyl chains,[111] and polymethacylate and poly(methyl mehacrylate)[52] have also been described. The ability for these block copolymers to self-assemble in water has been exploited for the creation of materials with thermal responsive properties,[14] increased bulk materials properties,[11,12] and controlled swelling.[110] However, these synthetic polymers are non-bioactive, do not support cell attachment, and may have undesirable degradation rates. Polysaccharides-based bioadhesive polymers have been created by direct coupling of either amine or acid-functionalized catechols onto alginate,[112,113] chitosan,[114,115] dextran,[116,117] and hyaluronic acid[118,119] to create bioactive bioadhesives suitable for drug delivery and tissue engineering applications. Similarly, modification of gelatin, which is the denatured form of the extracellular matrix protein collagen, with catechol has also been reported.[120,121]

10.3.3 Polymerization of Catechol-modified Monomers

Methacrylated dopa[102] or dopamine[23] (Figure 10.4K) can be copolymerized into acrylate-based polymers through photo-initiated or heat-activated free-radical polymerization (Figure 10.6A). A variety of co-monomers with neutral,[23] charged,[15,122] environmental cue-responsive,[120] and oligomeric ethylene glycol[25] side chains can be copolymerized with these catechol-functionalized monomers to form polymers with a wide range of physical properties. Similarly, vinyl-containing catechol (*e.g.*, 3,4-dihyroxystyrene, Figure 10.4L) has also been used to create adhesive polystyrene-based polymers.[91,123–125] The catechol side chain can retard and inhibit free-radical polymerization reactions in the presence of atmospheric oxygen.[102] In order to create polymers with elevated molecular weight (MW) and high catechol content, the removal of molecular oxygen[23] or catechol side chain protection[91] are required. Additionally, Lee *et al.*[53] found that the use of amphiphilic, multiblock copolymers to separate the polymerizable acrylate group from catechols is an effective way to reduce the inhibitive effect of the catechol side chain. When bifunctional crosslinkers are incorporated, polymerization results in the formation of three-dimensional hydrogel networks functionalized with catechol moieties (Figure 10.6C).[72,126]

In addition to free-radical polymerization, synthetic peptides have been prepared through ring-opening polymerization of DOPA NCA (Figure 10.4M).[127] Water soluble peptides have been created by copolymerization of lysine NCA.[92,93,101] Peptides with MWs up to 500 000 Da and a very narrow MW distribution (polydispersity <1.2) can be created. Additionally, polymer-bound initiators have been utilized to create block copolymers functionalized with oligomeric poly(DOPA)[53,128] and poly(DOPA-lysine) copolypeptide (Figure 10.6B).[24,26,53]

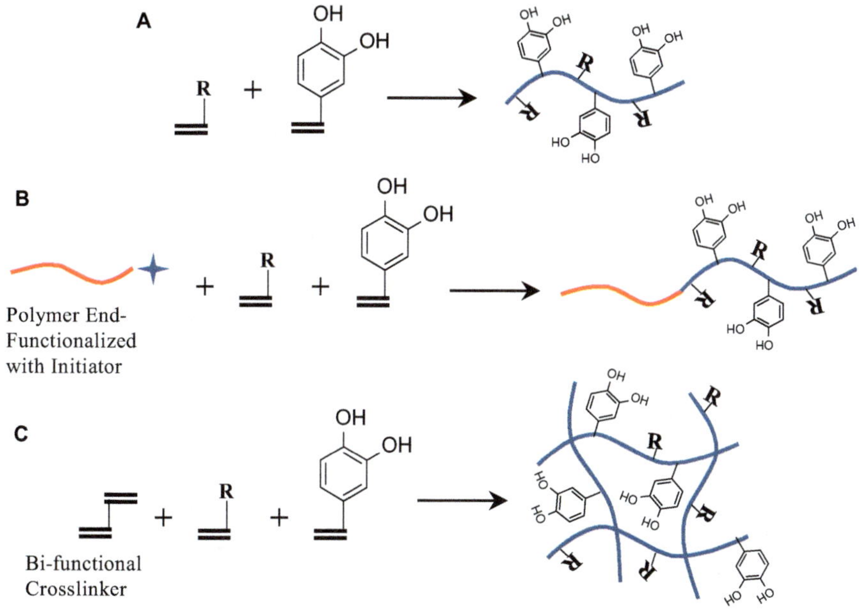

Figure 10.6 Schematic representation of polymer preparation using catechol-modified and polymerizable monomers (**A**) and using a polymer-bound initiator to form a block copolymer (**B**). In the presence of bi-functional crosslinkers, polymerization forms a three-dimensional polymer network containing catechol (**C**). **R** represents the side chain of the monomers used to copolymerize with the catechol-functionalized monomers.

10.3.4 Catechol-functionalized Initiator

Dopamine functionalized with alkyl bromine has been used as an initiator to prepare poly(mathacrylate)- and poly(acrylate)-based polymers using atom transfer radical polymerization (ATRP) (Figure 10.7).[89,90] The use of a catechol-functionalized initiator is ideal for preparing a densely packed polymer coating using the so-called "graft-from" approach, where the catechol group serves as a robust anchor to graft the initiator onto a substrate surface to perform surface-initiated polymerization.[129] This approach is advantageous when compared to the traditional "graft-to" approach of anchoring a polymer on to a surface, where the steric hindrance amongst large polymer chains prevents the formation of dense polymer coatings.

Surface-initiated polymerization is facile and highly tunable. When combined with photolithography techniques, surface patterns of various shapes (*e.g.*, squares, lines) in the micron-scale can be created.[90,130] Additionally, surface-initiated ATRP can be used to graft polymer brushes with various side chain functional groups (*e.g.*, oligomeric ethylene glycol,[89,90] zwitterionic,[131] and thermo-responsive[130]). These coatings have rendered surfaces such as titanium and gold highly resistant to fouling from cells and

Figure 10.7 Schematic representation of initiator-modified catechol to prepare polymers end-functionalized with catechol. **R** represents the side chain of the monomers used during polymerization.

proteins.[89,131] In addition to tethering polymers to planar substrates, surface-initiated polymerization has been used to anchor polymers to nanoparticles,[90,130] which has further expanded its application in designing novel materials. In addition to ATRP chemistry, Ye *et al.*[132] utilized surface-initiated ring-opening metathesis polymerization to graft poly(pentadecafluorooctyl-5-norbornene-2-carboxylate) brushes onto TiO_2 nanowires to create a superhydrophobic surface capable of repelling water and alkane liquids. In addition to initiators modified with dopamine, poly(DOPA-lysine) oligopeptides end-functionalized with an initiator have been used to prepare coatings with improved stability.[133]

10.4 Application of Catechol Functionalized Polymers

The simplicity and the wide range of chemistries that catechol can participate in have permitted scientists to design adhesives with tunable physical, adhesive, and mechanical properties necessary to tailor materials for specific applications. The use of catechol-containing polymers as biomedical adhesives, drug delivery carriers, antifouling coatings, biomaterials for delivering therapeutic cells, and smart hydrogel actuators and adhesives are reviewed here. Although the current review focuses on the application of mussel-mimetic polymers for biomedical applications, catechol-functionalized materials have found utility beyond biomedical fields (*e.g.*, drinking water purification,[134] controlled release of fertilizers,[135] nanocomposites for tire rubbers,[136] adhesives for batteries[137] and plastics[91]).

10.4.1 Biomedical Adhesives

Tissue adhesives are universally applied in surgery. Tissue adhesives can overcome challenges associated with traditional mechanical perforating devices (*e.g.*, sutures, tacks, and staples), which are unable to stop leakage or reconnect tissues with low cohesive properties (*e.g.*, lung, spleen), cause localized stress concentration that lead to failure, and cause persistent pain and nerve damage.[138–140] However, existing tissue adhesives are hampered by weak adhesive strength (*i.e.*, fibrin glue), and poor biocompatibility (*i.e.*, cyanoacrylate).[141–143] Thus, there is a continued need in developing biocompatible and moisture-resistant bioadhesives.

One of the first reported mussel-mimetic tissue adhesives was PEG end-functionalized with DOPA and its derivatives (*e.g.*, dopamine, 3,4-dihydroxy-hydrocinnamic acid).[13,87] PEG is a water soluble polymer widely used in the biomedical field due to its non-toxicity and non-immunogenicity.[144] PEG serves as a biocompatible macromolecular support for designing injectable tissue adhesives and sealants. PEG–catechol adhesives have outperformed commercial fibrin glue in various adhesion tests (*e.g.*, lap shear, burst strength, and peel adhesion tests, *etc.*). By controlling variables, such as PEG architecture (*i.e.* 4- to 8-arm) and MW (10–20 kDa), the number of terminal catechol residues (up to 4) and linking groups (*i.e.* ester, urethane, amide), adhesives with varying properties (*i.e.* rate of curing, extent of swelling, mechanical properties, and degradation rate) can be obtained.[87,144,145] PEG–catechol adhesives have successfully sealed human fetal membrane defects while demonstrating no cytotoxicicity.[146,147]

Although PEG–catechol adhesives are effective in functioning as bioadhesives, due to the affinity of PEG for molecular water, crosslinked PEG-based adhesives can swell extensively. Excessive swelling of medical adhesives can lead to local nerve compression and reduced bulk mechanical properties. To control the swelling behavior of catechol-based adhesives, a thermal responsive Tetronic polymer was functionalized with catechol to develop a zero- or negative swelling adhesive.[110] Tetronic is a 4-armed, amphiphilic block copolymer with a central, hydrophobic PPO block and peripheral, hydrophilic PEG blocks. At body temperature, the hydrophobic PPO self-assembles to induce a volume contraction while increasing the bulk mechanical properties of the crosslinked adhesive. Tetronic–catechol adhesives have demonstrated a 3–5 fold increase in measured adhesive strength[110] and increased resistance to dynamic loading[109] when compared to PEG–catechol adhesives.

Similarly, catechol-modified multiblock copolymers were prepared with repeating 4-arm PEG and linear PCL polymers (Figure 10.8).[12] The hydrophobic PCL allowed the multiblock copolymer to form stable adhesive films that could be coated onto biologic meshes such as bovine pericardium or porcine dermal tissues, commonly used in soft tissue reconstruction. These bioadhesive constructs exhibited adhesive strength an order of magnitude higher than those of fibrin glue, while reaching 60% that of cyanoacrylate-based adhesives.[12] Additionally, these constructs were used to augment the primary suture repair of transected Achilles tendons.[11]

The degradation behavior of a tissue adhesive is critical to its ability to repair a wound while minimizing a prolonged inflammatory response. Degradation of PEG–catechol is typically achieved through the introduction of hydrolysable ester linkages between PEG and the end-functionalized catechol moieties.[82,148] While it is feasible to tailor the hydrolytic degradation rate of synthetic PEG-based adhesives by tuning its composition,[87,149] it is difficult to match the adhesive degradation rate with the rate of tissue ingrowth for satisfactory wound healing. PEG–catechol adhesive inserted with

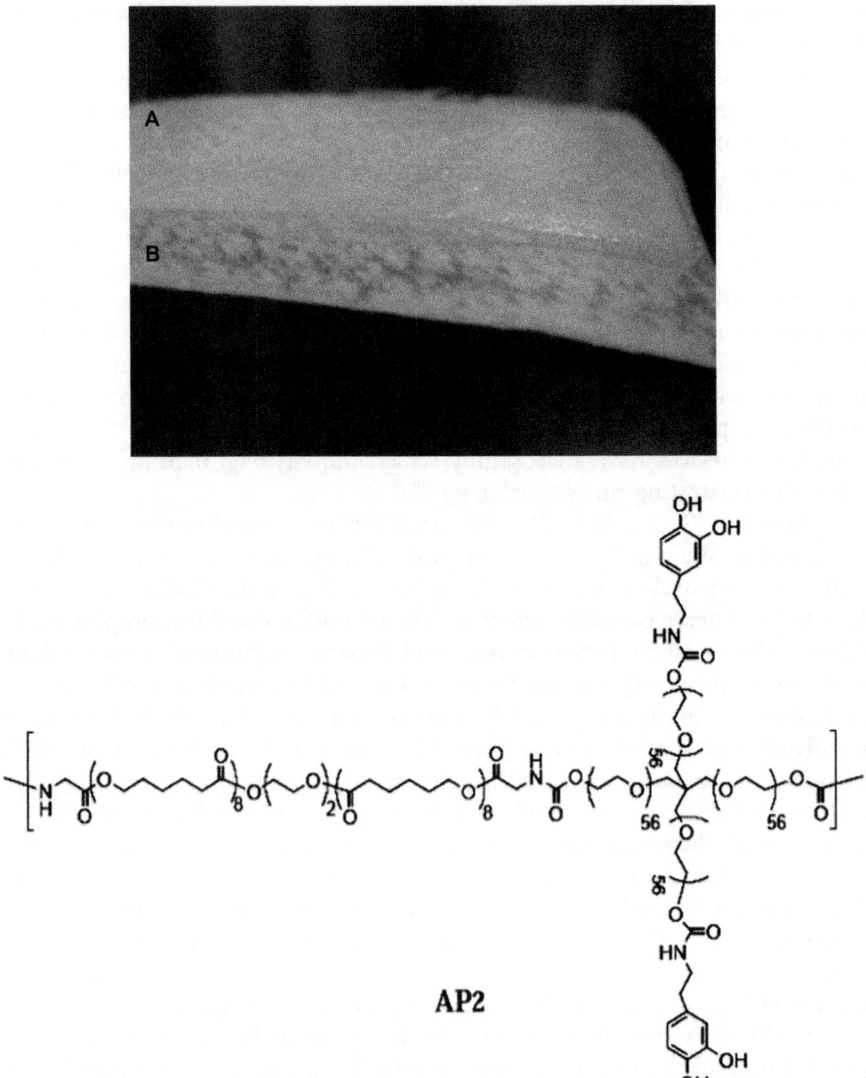

Figure 10.8 Photograph of adhesive film (4 cm×8 cm, **A**) coated onto a 6 cm×8 cm segment of porcine dermal tissue (**B**). Chemical structure of amphiphilic multiblock copolymers modified with catechol (**C**). Reprinted from ref. 11. © IOP Publishing.

an alanine–alanine (Ala–Ala) dipeptide linker exhibited cell-mediated degradation.[150] The Ala–Ala peptide can be selectively cleaved by elastase secreted by activated neutrophils at the site of the injury. From a subcutaneous implantation study in a mouse model, immunohistochemical staining for elastase was found at the adhesive–tissue interface and the adhesive slowly degraded while illiciting minimal inflammatory response.

One approach to simultaneously improve the bioactivity and the mechanical and adhesive properties of PEG–catechol adhesives involved the incorporation of nano-silicate, Laponite ($Na^{0.7+}(Mg_{5.5}Li_{0.3}Si_8)O_{20}(OH)_4)^{0.7-}$) (Figure 10.9).[151] Incorporation of up to 2 wt% of Laponite formed an injectable nanocomposite adhesive that exhibited an increased work of adhesion to wetted tissue, which was 5-times higher than that of the Laponite-free adhesive. This increase in measured adhesive properties was attributed to the formation of strong interfacial bonds between the polymer-bound

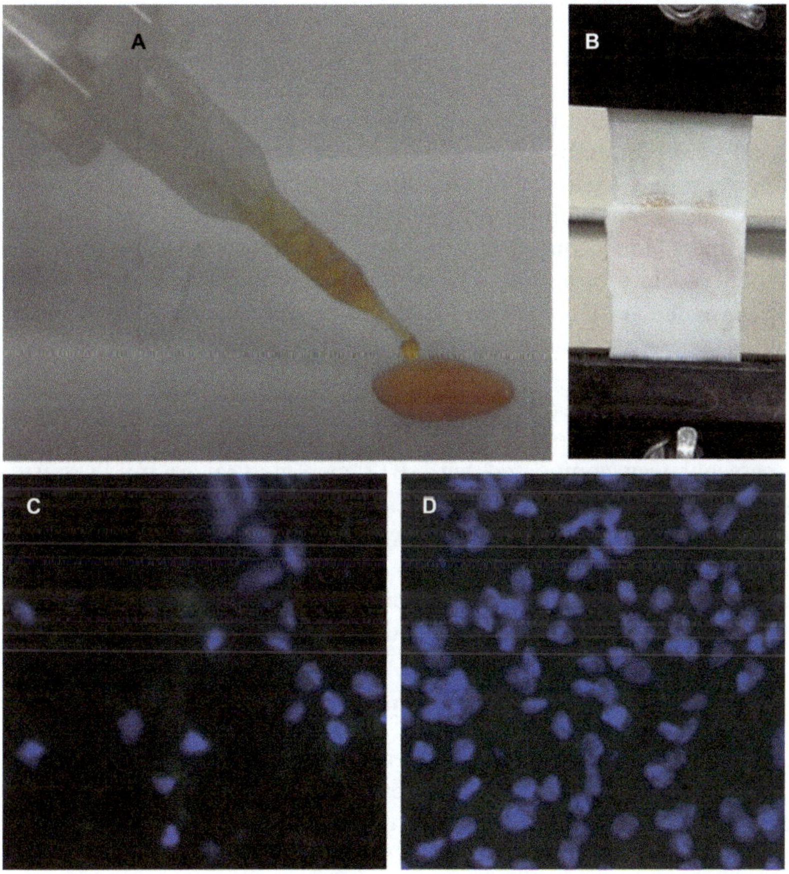

Figure 10.9 Photograph of nanocomposite PEG–catechol adhesive injected through a dual-barrel syringe (**A**) and the adhesive joining two pieces of bovine pericardium tissue (**B**). Immunohistochemical staining images of cells infiltrated into adhesive containing 0 (**C**) and 2 (**D**) wt% Laponite after 8 weeks of subcutaneous implantation in a rat. Cell nuclei were stained by DAPI (blue), fibroblasts were stained by marker S100A4 (green).
Reproduced with permission from Y. Liu, H. Meng, S. Konst, R. Sarmiento, R. Rajachar and B. P. Lee, *ACS Appl. Mater. Interfaces*, 2014, **6**, 16982–16992.[88] Copyright (2014) American Chemical Society.

catechol and the encapsulated nanoparticles.[71,126] The incorporation of Laponite provided binding sites for cellular attachment and the nanocomposite adhesive exhibited increased cellular infiltration into the adhesive *in vivo*.[151]

10.4.2 Drug Delivery

One of the key requirements in the design of drug delivery carriers is the ability to localize the delivery to the targeted tissue to increase the drug efficacy while minimizing its systemic toxicity. One example of exploiting the bioadhesive properties of catechol in drug carrier design involved catechol-functionalized alginates.[152] The adhesive hydrogel demonstrated strong affinity to endothelium surfaces and was successfully coated onto different types of arteries, including the common carotid, the bifurcation of the carotid, and the abdominal aorta. In a mouse model with atherosclerosis, inflamed plaques treated with steroid-eluting adhesive hydrogels demonstrated reduced macrophage content and developed protective fibrous caps covering the plaque core. The drug-eluting adhesive also lowered plasma cytokine levels and biomarkers of inflammation in the plaque.

The pH-sensitive nature of catechol–borate interactions was exploited for the targeted delivery of an anticancer drug, bortezomib (BTZ), to cancer cells (Figure 10.10).[106] BTZ is a dipeptide functionalized with boronic acid and can inhibit the cancer cell proteasome through the direct binding of its boronic acid group and the threonine residues in the active sites of several proteases. PEG–catechol was conjugated with BTZ at a neutral to mildly basic pH through catechol–borate complexation. BTZ dissociated from PEG–catechol in response to the mildly acidic microenvironment of cancer tissues, liberating the free drug, which could be taken up by cancer cells. A cancer-cell-targeting ligand, biotin, was presented on the polymer carriers to facilitate the targeted entry of the drug-loaded polymer carriers into the cancer cells.

10.4.3 Coatings for Reducing Biofouling

Controlling interfacial adsorption events is critical to the success of many implantable medical devices and biomaterials.[153] Uncontrolled and nonspecific adsorption of biomolecules, bacteria, and cells can severely compromise the function of these implants. One strategy in minimizing surface fouling is to functionalize these surfaces with antifouling polymers such as PEG, which can reduce the adsorption of cells,[154] microbes,[155,156] and macromolecules.[157] The hydrophilic nature of these polymers strongly couples water molecules, thus limiting the access of approaching cells and macromolecules to the surface adsorption sites. However, many surface anchoring strategies that involve the formation of robust chemical bonds between the polymer and surface are limited by substrate chemical specificity,[158,159] complicated and expensive protocols,[160,161] and susceptibility to thermal or hydrolytic degradation.[162,163]

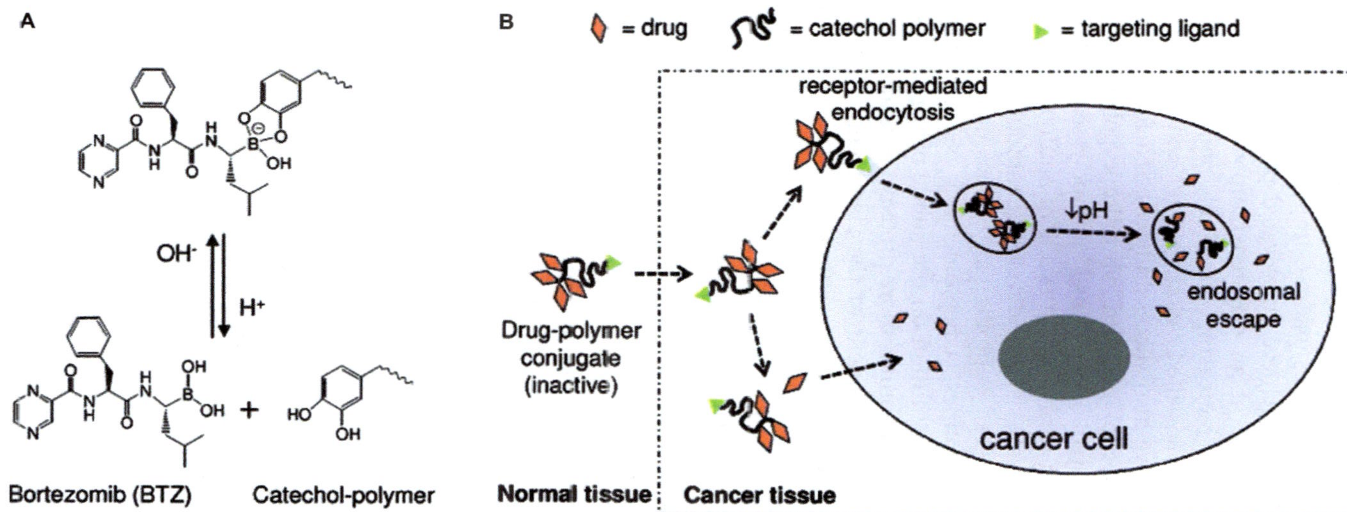

Figure 10.10 pH-sensitive polymer–drug conjugates based on catechol–borate complexation (**A**). The catechol and the boronic acid structure in BTZ form a stable conjugate at neutral and alkaline pH, but this structure dissociates in acidic environments to release the free active drug (**B**). The catechol polymer–BTZ conjugate may dissociate in response to the mildly acidic cancer tissue microenvironment to liberate the free drug, which can be taken up by cancer cells.
Reproduced with permission from J. Su, F. Chen, V. L. Cryns and P. B. Messersmith, *J. Am. Chem. Soc.*, 2011, **133**, 11850–11853.[106] Copyright (2011) American Chemical Society.

The ability of catechol to adhere to both organic and inorganic surfaces makes it a versatile anchoring group for tethering antifouling polymers. PEG-functionalized with catechol and DOPA-containing oligo-peptides can attach to a wide range of surface substrates using a simple dip-coating approach in aqueous buffer.[16,17,24,26,128] To achieve an elevated polymer coating density, dip-coating can be performed at an elevated temperature or salt concentration, where PEG becomes increasingly insoluble and occupies a smaller hydrodynamic volume (the so-called "cloud point" condition).[18] Catechol-functionalized polymers have also successfully resisted fouling of microbes *in vivo*.[25] Additionally, catechol-modified zwitterionic polymers,[164–166] polymers with oligo-ethylene glycol side chains,[89] and polysaccharides (*e.g.*, dextran)[117] have also conferred surfaces with excellent antifouling properties.

Messersmith and collegues[167,168] have developed antifouling coatings based on peptoids (Figure 10.11). Peptoids are biocompatible and non-immunogenic mimics of peptides. A peptoid has a peptide-like polyamide backbone with side chain substitution on the amide nitrogen instead of the α-carbon. This change makes peptoids more protease-resistant than peptides, likely due to the spatial misalignment of the side chains.[169] Peptoids synthesized with ethylene glycol[167,168] and zwitterionic[170] side chains have demonstrated the ability to repels non-specific binding of proteins, cells, and bacteria. Finally, peptoids containing additional antimicrobial sequences exhibited added active bactericidal properties in addition to passive antifouling capability.[171]

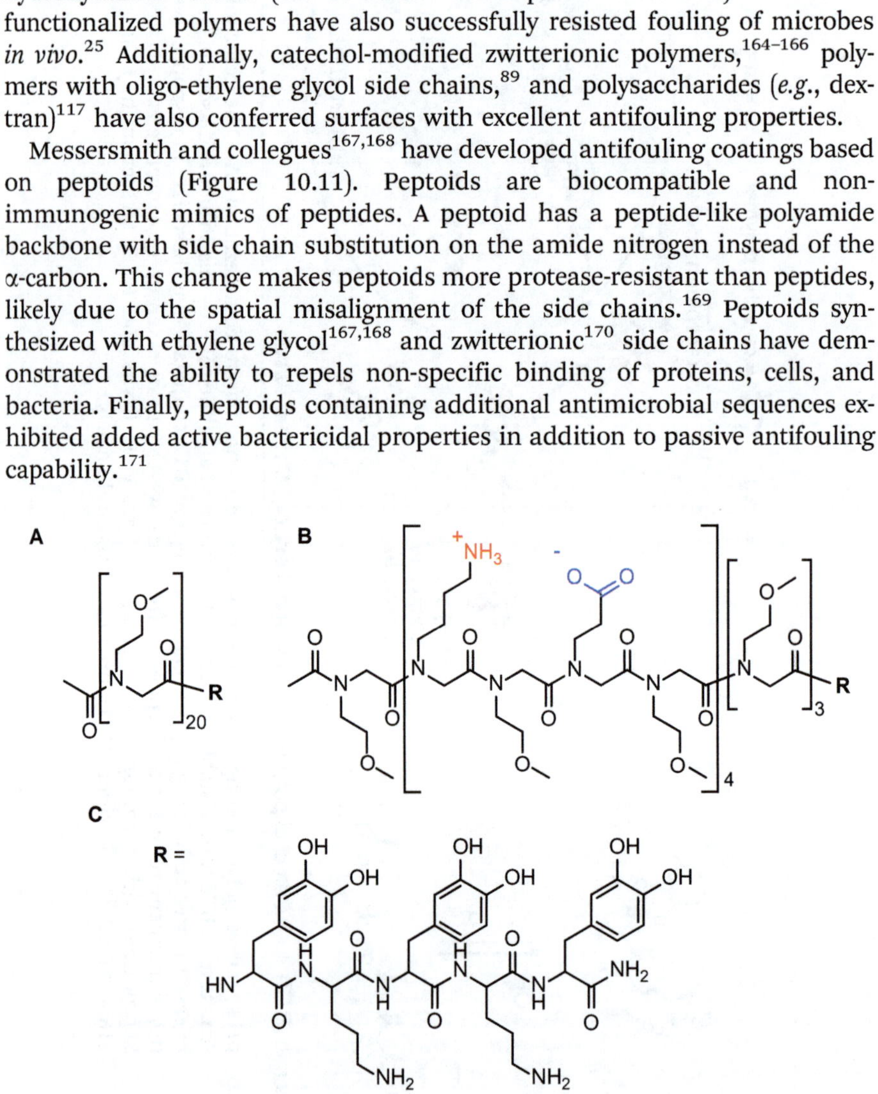

Figure 10.11 Chemical structure of antifouling peptoids with ethylene glycol (**A**) and zwitterionic (**B**) side chains. Petoids are end-functionalized with oligo-peptides containing repeating DOPA and lysine units (**C**). Reproduced from ref. 170 with permission from John Wiley and Sons. Copyright © 2014, John Wiley and Sons.

10.4.4 Delivery of Therapeutic Cells

Tissue and cell engineering combine living cells and a suitable polymeric scaffold to regenerate healthy tissues or to replace the lost function of diseased organs.[172] However, one of the challenges in advancing these therapeutic approaches is associated with the delivery and fixation of tissue engineered biomaterials. The ability of catechol to function as a crosslinking precursor and adhere to various tissue types has been exploited for delivering and immobilizing therapeutic cells. Catechol-modified PEG and polysaccharides (*e.g.*, hyaluronic acid, alginate) have been used to encapsulate various human primary cells, including stem cells, into a three dimensional matrix.[108,112,118] Injectable PEG–catechol adhesives demonstrated the ability to immobilize islet cells onto adipose tissue surfaces in a diabetic mice model.[108] Transplanted islet cells produced insulin and the mice remained normoglycemic for over 16 weeks post-transplantation. Upon removing the transplanted cells, these mice immediately became hyperglycemic, indicating that the islet cells were responsible for blood sugar regulation. Similarly, hyaluronic acid-functionalized catechol (HA-CA) adhesives successfully transplanted hepatocytes onto the liver of athymic mice and improved their hepatic function for over two weeks.[173] Additionally, HA-CA also adhered human adipose-derived stem cells onto the ischemic myocardium of a rat heart, resulting in enhanced angiogenesis of the myocardium.

10.4.5 Hydrogel Actuators

Hydrogels that can change their shape and physical properties in response to environmental stimuli (*e.g.*, temperature, pH, electricity) are currently being explored as soft robotic components, biosensors, artificial muscle tissues, and in controlled drug delivery applications.[174–176] Recently, Lee *et al.*[177] reported for the first time a hydrogel actuator that exploits the catechol-metal ion coordination chemistry (Figure 10.2C). Hydrogels modified with catechol were locally printed with ferric (Fe^{3+}) ions by applying an electric potential (2.5–25 V) using an iron electrode as the anode and aluminum foil as the counter electrode (cathode). Locally deposited metal ions were captured by the network-bound catechol (dark bands in Figure 10.12). When the hydrogels were submerged in a basic solution (pH 9.5), formation of tris-catecholate-Fe^{3+} ion complex increased the local crosslinking density at the site of ionoprinting and resulted in sharp hydrogel bending.[177–179] When the hydrogel was further submerged in an acidic solution (12 mM HCl), the hydrogel returned to its original shape, due to the formation of mono-catecholate-Fe^{3+} ion complexes with reduced crosslinking density. Hydrogels can adopt multiple 3-dimensional shapes depending on the ion patterns. The dynamic response of hydrogel actuation can be controlled by multiple factors, which include the catechol and metal ion content, pH, and hydrogel thickness.[178] Finally, Fe^{3+} ions are reversibly bound to catechol

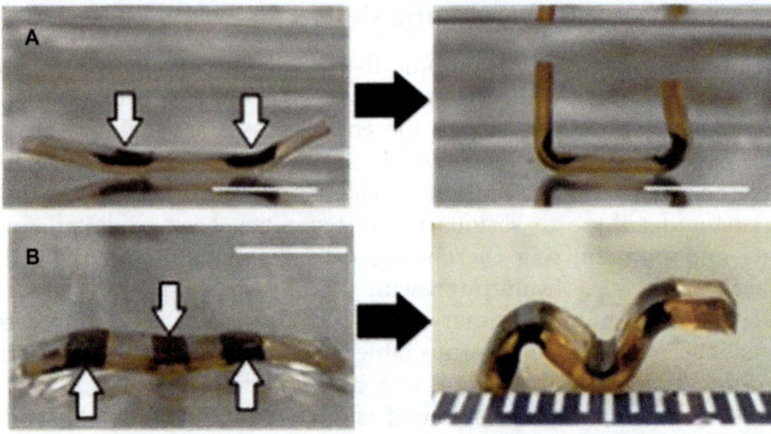

Figure 10.12 Photographs of hydrogel ionoprinted with Fe^{3+} ions and forming a "U" (**A**) and "M" (**B**) shape before (left panels) and after (right panels) incubation in a pH 9.5 solution to induce local tris-catecholate-Fe^{3+} ion complex formation. The white arrows indicate the direction and location of ionoprinting. Scale bars = 5 mm.
Reproduced from ref. 177 with permission from John Wiley and Sons.

and can be removed through ethylenediaminetetraacetic acid (EDTA) chelation. Metal ions can be reintroduced in a different pattern, so that the hydrogel can be reprogrammed to transform into a different 3-dimentional shape.[177]

10.4.6 Smart Adhesives

Smart adhesives that are capable of surface bonding and debonding on command have been met with significant interest in various fields of materials science and engineering,[180–182] including manufacturing (*e.g.*, disassembly of bonded parts in automatized production processes), robotic locomotion (*i.e.*, attachment and detachment of robotic foot), repair of complex structural components, and the development of painless removal of wound dressings. However, existing smart adhesives are limited by the need for extreme conditions to promote debonding (*e.g.*, elevated temperature[181]), adhesion to a specific type of substrate,[183] or weakened adhesive strength under moist conditions.[182] Most smart adhesives reported to-date have demonstrated adhesion predominately to dry substrates, and catechol-based smart adhesives with demonstrated moisture-resistant adhesive properties have only been reported in the last several years. Catechol-modified photodegradable polymer backbones containing di-*o*-nitrobenzaldehyde can debond from the substrate's surface with ultraviolet light irradiation.[184] Similarly, nitrodopamine-modified PEG contains photocleavable *o*-nitrophenyl ethyl linkages between the PEG and the catechol groups (Figure 10.13).[84] These adhesives can cure rapidly though

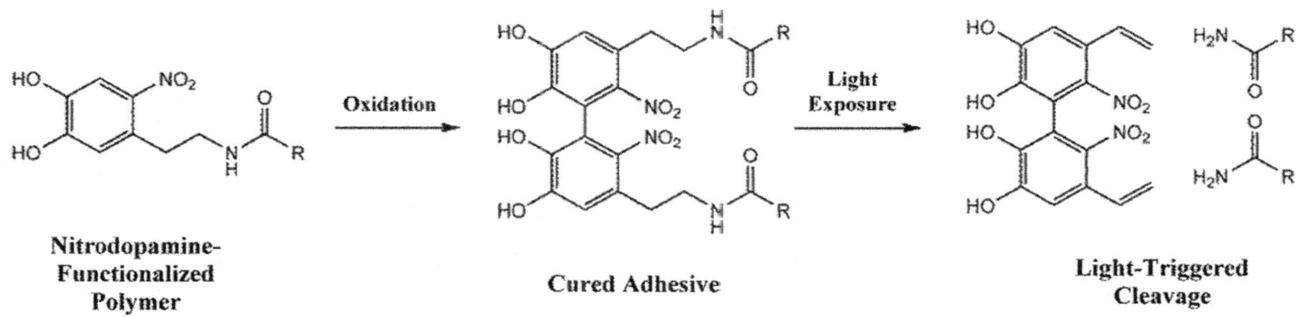

Figure 10.13 Chemical structures of nitrodopamine-functionalized polymer forming a cured adhesive through dimerization of the catechol side chain. The cured adhesive is degraded upon light exposure. R = polymer chain.

the oxidation-mediated crosslinking of catechol and the adhesive degrades when exposed to light. Tyrosine-modified adhesives were demonstrated to be activated to their adhesive catechol form with the addition of tyrosinase, which further oxidized the catechol for crosslinking reactions.[185,186] Although these adhesives exhibit stimulus- or enzyme-responsive adhesive properties under wet conditions, these technologies have only been demonstrated for one cycle of reversibility.

10.5 Future Outlook

Mussel-mimetic polymers reported to-date have predominantly focused on exploiting catechol chemistries for the design of advanced and functional materials. However, the contribution of other amino acids toward adhesion is not well understood and have only been studied recently.[187–189] Specifically, mussel adhesive proteins form complex coacervates due to charge interactions of the ionic side chains, which may be critical for preserving the reactivity of catechol and the deployment of adhesive proteins.[15,190] Additionally, mussel byssus threads are composed of high histidine contents.[36] The imidazole side chain of histidine forms reversible bonds with transitional metal ions (*e.g.*, zinc and copper), which can be broken and reformed while minimizing permanent structural damage. This type of interaction has been recently exploited to create a self-healing hydrogel.[191] Finally, the byssus thread and the adhesive plaque each exhibit gradation in their material properties to prevent structural damage due to contact deformation between two materials with drastically different stiffness.[192,193] These complex and hierarchical structures have not been duplicated in existing synthetic mussel-mimetic materials. Recently, Zhong *et al.*[194] combined protein sequences from *Escherichia coli* fiber-forming amyloids and mussel adhesive proteins to form a chimeric peptide capable of self-assembling into higher-ordered, adhesive nanofibers. This approach may pave the way for future designs of synthetic adhesives with unique hierarchical structures and nano-sized domains that can be assembled using the "bottom-up" approach.

10.6 Summary

The catechol chemistries employed by marine organisms provide a simple and robust strategy for designing adhesive polymers suited for various interfacial applications. Incorporation of a single DOPA amino acid or other catechol derivatives can impart inert polymer materials with strong, water-resistant adhesive properties and the ability to form predictable and controllable chemical crosslinks. This is a unique feature that differs drastically from other biomimetic strategies where a peptide sequence needs to be duplicated and preserved to mimic the function of a naturally derived protein (*e.g.*, arginine-glycine-aspartic acid for cellular adhesion, β-sheet forming sequences in silk proteins, *etc.*). The versatility of catechol chemistries has resulted in the development of various catechol-functionalized polymers for biomedical applications (*e.g.*, biomedical adhesives, drug delivery, antifouling coatings, tissue engineering, and smart biomaterials) as well as applications beyond the medical field (*e.g.*, automobile, agricultural, energy, *etc.*).

Acknowledgements

The authors would like to acknowledge the National Institutes of Health (GM 104846 and GM 112082) for financial support.

References

1. J. Comyn, in *Developments in Adhesives*, ed. A. J. Kinloch, Applied Science Publishers, Barking, UK, 1981, vol. 2, pp. 279–313.
2. J. H. Waite, *Int. J. Adhes. Adhes.*, 1987, **7**, 9–14.
3. B. P. Lee, P. B. Messersmith, J. N. Israelachvili and J. H. Waite, *Annu. Rev. Mater. Res.*, 2011, **41**, 99–132.
4. M. Yu, J. Hwang and T. J. Deming, *J. Am. Chem. Soc.*, 1999, **121**, 5825–5826.
5. T. J. Deming, M. Yu and J. Hwang, *Polym. Mater.: Sci. Eng.*, 1999, **80**, 471–472.
6. J. H. Waite, *Chem. Ind.*, 1991, 607–611.
7. B. P. Lee, J. L. Dalsin and P. B. Messersmith, in *Biological Adhesives*, ed. A. M. Smith and J. A. Callow, Springer, Berlin Heidelberg, Berlin, Germany, 2006, pp. 257–278.
8. J. H. Waite and S. O. Andersen, *Biochim. Biophys. Acta*, 1978, **541**, 107–114.
9. L. Ninan, J. Monahan, R. L. Stroshine, J. J. Wilker and R. Shi, *Biomaterials*, 2003, **24**, 4091–4099.
10. L. O. Burzio, V. A. Burzio, T. Silva, L. A. Burzio and J. Pardo, *Curr. Opin. Biotechnol.*, 1997, **8**, 309–312.
11. M. Brodie, L. Vollenweider, J. L. Murphy, F. Xu, A. Lyman, W. D. Lew and B. P. Lee, *Biomed. Mater.*, 2011, **6**, 015014.
12. J. L. Murphy, L. Vollenweider, F. Xu and B. P. Lee, *Biomacromolecules*, 2010, **11**, 2976–2984.
13. S. A. Burke, M. Ritter-Jones, B. P. Lee and P. B. Messersmith, *Biomed. Mater.*, 2007, **2**, 203–210.
14. K. Huang, B. P. Lee, D. Ingram and P. B. Messersmith, *Biomacromolecules*, 2002, **3**, 397–406.
15. H. Shao, K. N. Bachus and R. J. Stewart, *Macromol. Biosci.*, 2009, **9**, 464–471.
16. J. L. Dalsin, B.-H. Hu, B. P. Lee and P. B. Messersmith, *J. Am. Chem. Soc.*, 2003, **125**, 4253–4258.
17. J. L. Dalsin, L. Lin and P. B. Messersmith, *PMSE Prepr.*, 2004, **90**, 247–248.

18. J. L. Dalsin, L. Lin, S. Tosatti, J. Voeroes, M. Textor and P. B. Messersmith, *Langmuir*, 2005, **21**, 640–646.
19. P. Z. Araujo, P. J. Morando and M. A. Blesa, *Langmuir*, 2005, **21**, 3470–3474.
20. C. Creutz and M. H. Chou, *Inorg. Chem.*, 2008, **47**, 3509–3514.
21. T. Kawabata, V. Schepkin, N. Haramaki, R. S. Phadke and L. Packer, *Biochem. Pharmacol.*, 1996, **51**, 1569–1577.
22. E. Amstad, T. Gillich, I. Bilecka, M. Textor and E. Reimhult, *Nano Lett.*, 2009, **9**, 4042–4048.
23. H. Lee, B. P. Lee and P. B. Messersmith, *Nature*, 2007, **448**, 338–341.
24. K. Chawla, S. Lee, B. P. Lee, J. L. Dalsin, P. B. Messersmith and N. D. Spencer, *J. Biomed. Mater. Res., Part A*, 2009, **90A**, 742–749.
25. A. Pechey, C. N. Elwood, G. R. Wignall, J. L. Dalsin, B. P. Lee, M. Vanjecek, I. Welch, R. Ko, H. Razvi and P. A. Cadieux, *J. Urol.*, 2009, **182**, 1628–1636.
26. J. L. Dalsin, D. L. Sherman, B. P. Lee and P. B. Messersmith, *PMSE Prepr.*, 2006, **94**, 854–855.
27. A. M. Baty, P. K. Leavitt, C. A. Siedlecki, B. J. Tyler, P. A. Suci, R. E. Marchant and G. G. Geesey, *Langmuir*, 1997, **13**, 5702–5710.
28. B. R. Baker, A. N. Laiwalla, J.-Y. Yoon, J. Canavate and R. L. Garrell, *Polym. Mater.: Sci. Eng.*, 2001, **85**, 115–116.
29. M. P. Deacon, S. S. Davis, J. H. Waite and S. E. Harding, *Biochemistry*, 1998, **37**, 14108–14112.
30. J. Schnurrer and C.-M. Lehr, *Int. J. Pharm.*, 1996, **141**, 251–256.
31. W. M. Chirdon, W. J. O'Brien and R. E. Robertson, *J. Biomed. Mater. Res., Part B*, 2003, **66B**, 532–538.
32. S. W. Taylor, G. W. Luther, III and J. H. Waite, *Inorg. Chem.*, 1994, **33**, 5819–5824.
33. S. W. Taylor, D. B. Chase, M. H. Emptage, M. J. Nelson and J. H. Waite, *Inorg. Chem.*, 1996, **35**, 7572–7577.
34. M. J. Sever and J. J. Wilker, *Dalton Trans.*, 2006, **14**, 813–822.
35. M. J. Harrington, A. Masic, N. Holten-Andersen, J. H. Waite and P. Fratzl, *Science*, 2010, **328**, 216–220.
36. M. J. Harrington, H. S. Gupta, P. Fratzl and J. H. Waite, *J. Struct. Biol.*, 2009, **167**, 47–54.
37. N. Holten-Andersen, M. J. Harrington, H. Birkedal, B. P. Lee, P. B. Messersmith, K. Y. C. Lee and J. H. Waite, *Proc. Natl. Acad. Sci. U. S. A.*, 2011, **15**, 2651–2655.
38. N. Holten-Andersen, A. Jaishankar, M. J. Harrington, D. E. Fullenkamp, G. DiMarco, L. He, G. H. McKinley, P. B. Messersmith and K. Y. C. Lee, *J. Mater. Chem. B*, 2014, **2**, 2467–2472.
39. C. A. Tyson and A. E. Martell, *J. Am. Chem. Soc.*, 1968, **90**, 3379–3386.
40. B. A. Borgias, S. R. Cooper, Y. B. Koh and K. N. Raymond, *Inorg. Chem.*, 1984, **23**, 1009–1016.
41. L. Sommer, *Z. Anorg. Allg. Chem.*, 1963, **321**, 191–197.

42. C. Morlay, M. Cromer, Y. Mouginot and O. Vittori, *Talanta*, 1998, **45**, 1177–1188.
43. O. K. Borggaard, *Acta Chem. Scand.*, 1972, **26**, 393–395.
44. R. Kummert and W. Stumm, *J. Colloid Interface Sci.*, 1980, **75**, 373–385.
45. M. P. Soriaga and A. T. Hubbard, *J. Am. Chem. Soc.*, 1982, **104**, 3937–3945.
46. D. C. Hansen, S. C. Dexter and J. H. Waite, *Corros. Sci.*, 1995, **37**, 1423–1441.
47. B. P. Lee, C.-Y. Chao, F. N. Nunalee, E. Motan, K. R. Shull and P. B. Messersmith, *Maclomolecules*, 2006, **39**, 1740–1748.
48. H. Lee, N. F. Scherer and P. B. Messersmith, *Proc. Natl. Acad. Sci. U. S. A.*, 2006, **103**, 12999–13003.
49. S. A. Mian, X. Gao, S. Nagase and J. Jang, *Theor. Chem. Acc.*, 2011, **130**, 333–339.
50. S. A. Mian, L. C. Saha, J. Jang, L. Wang, X. Gao and S. Nagase, *J. Phys. Chem. C*, 2010, **114**, 20793–20800.
51. M. Vega-Arroyo, P. R. LeBreton, T. Rajh, P. Zapol and L. A. Curtiss, *Chem. Phys. Lett.*, 2005, **406**, 306–311.
52. M. Guvendiren, P. B. Messersmith and K. R. Shull, *Biomacromolecules*, 2008, **9**, 122–128.
53. B. P. Lee, C.-Y. Chao, F. N. Nunalee, E. Motan, K. R. Shull and P. B. Messersmith, *Macromolecules*, 2006, **39**, 1740–1748.
54. J. Yu, W. Wei, M. S. Menyo, A. Masic, J. H. Waite and J. N. Israelachvili, *Biomacromolecules*, 2013, **14**, 1072–1077.
55. L. M. McDowell, L. A. Burzio, J. H. Waite and J. Schaefer, *J. Biol. Chem.*, 1999, **274**, 20293–20295.
56. B. P. Lee, J. L. Dalsin and P. B. Messersmith, *Biomacromolecules*, 2002, **3**, 1038–1047.
57. J. H. Waite, *Comp. Biochem. Physiol., Part B: Biochem. Mol. Biol.*, 1990, **97**, 19–29.
58. M. Sugumaran, H. Dali and V. Semensi, *Arch. Insect Biochem. Physiol.*, 1989, **11**, 127–137.
59. H. Lee, N. F. Scherer and P. B. Messersmith, *Proc. Natl. Acad. Sci. U. S. A.*, 2006, **103**, 12999–13003.
60. H. Lee, S. M. Dellatore, W. M. Miller and P. B. Messersmith, *Science*, 2007, **318**, 426–430.
61. Y. Liu, K. Ai and L. Lu, *Chem. Rev.*, 2014, **114**, 5057–5115.
62. M. E. Lynge, R. van der Westen, A. Postma and B. Stadler, *Nanoscale*, 2011, **3**, 4916–4928.
63. H. Meng, J. Gazella and B. P. Lee, in *Bio- and Bioinspired Nanomaterials* ed. D. Ruiz-Molina, F. Novio and C. Roscini, Wiley-VCN, Weinheim, Germany, 2014, pp. 309–334.
64. H. Meng, Y. Liu, M. M. Cencer and B. P. Lee, in *Bioadhesion and Biomimetics: From Nature to Applications*, ed. H. Bianco-Peled and M. Davidovich-Pinhas, Pan Stanford Publishing, 2015, pp. 131–166.

65. M. M. Cencer, Y. Liu, A. Winter, M. Murley, H. Meng and B. P. Lee, *Biomacromolecules*, 2014, **15**, 2861–2869.
66. J. Li and B. M. Christensen, *J. Electroanal. Chem.*, 1994, **375**, 219–231.
67. H. Ohman and A. Vahlquist, *Acta Derm.-Venereol.*, 1994, **74**, 375–379.
68. B. R. Soller, R. H. Micheels, J. Coen, B. Parikh, L. Chu and C. Hsi, *J. Clin. Monit. Comput.*, 1996, **12**, 387–395.
69. B. R. Soller, T. Khan, J. Favreau, C. Hsi, J. C. Puyana and S. O. Heard, *J. Surg. Res.*, 2003, **114**, 195–201.
70. I. F. Tannock and D. Rotin, *Cancer Res.*, 1989, **49**, 4373–4384.
71. X. Ding, G. K. Vegesna, H. Meng, A. Winter and B. P. Lee, *Macromol. Chem. Phys.*, 2015, **216**, 1109–1119.
72. H. Meng, Y. Li, M. Faust, S. Konst and B. P. Lee, *Acta Biomater.*, 2015, **17**, 160–169.
73. M. Mochizuki, S.-i. Yamazaki, K. Kano and T. Ikeda, *Biochim. Biophys. Acta, Gen. Subj.*, 2002, **1569**, 35–44.
74. P. Chelikani, I. Fita and P. C. Loewen, *CMLS, Cell. Mol. Life Sci.*, 2004, **61**, 192–208.
75. N. Brian, H. Ahswin, N. Smart, Y. Bayon, S. Wohlert and J. A. Hunt, *Eur. Cells Mater.*, 2012, **24**, 249–265.
76. A. M. Cooper, B. H. Segal, A. A. Frank, S. M. Holland and I. M. Orme, *Infect. Immun.*, 2000, **68**, 1231–1234.
77. A. E. K. Loo, Y. T. Wong, R. Ho, M. Wasser, T. Du, W. T. Ng and B. Halliwell, *PLoS One*, 2012, **7**, e49215.
78. S. Chigurupati, M. R. Mughal, E. Okun, S. Das, A. Kumar, M. McCaffery, S. Seal and M. P. Mattson, *Biomaterials*, 2013, **34**, 2194–2201.
79. C. J. Sun, A. Srivastava, J. R. Reifert and J. H. Waite, *J. Adhes.*, 2009, **85**, 126–138.
80. E. Amstad, A. U. Gehring, H. Fischer, V. V. Nagaiyanallur, G. Hahner, M. Textor and E. Reimhult, *J. Phys. Chem. C*, 2011, **115**, 683–691.
81. M. Cencer, M. Murley, Y. Liu and B. P. Lee, *Biomacromolecules*, 2015, **16**, 404–410.
82. J. Anderson, M.-H. Lin, C. Privette, M. Flowers, M. Murley, B. P. Lee and K. G. Ong, *ScienceJet*, 2015, **4**, 80.
83. M. S. Menyo, C. J. Hawker and J. H. Waite, *Soft Matter*, 2013, 10314–10323.
84. Z. Shafiq, J. Cui, L. Pastor-Pérez, V. S. Miguel, R. A. Gropeanu, C. Serrano and A. del Campo, *Angew. Chem., Int. Ed.*, 2012, **51**, 4332–4335.
85. L. García-Fernández, J. Cui, C. Serrano, Z. Shafiq, R. A. Gropeanu, V. S. Miguel, J. I. Ramos, M. Wang, G. K. Auernhammer, S. Ritz, A. A. Golriz, R. Berger, M. Wagner and A. del Campo, *Adv. Mater.*, 2013, **25**, 529–533.
86. P. F. Ren, H. C. Yang, H. Q. Liang, X. L. Xu, L. S. Wan and Z. K. Xu, *Langmuir*, 2015, **31**, 5851–5858.
87. J. L. Dalsin, B. P. Lee, L. Vollenweider, S. Silvary, J. L. Murphy, F. Xu, A. Spitz, and A. Lyman, *US Pat.*, 8,119,742, 2012.

88. Y. Liu, H. Meng, S. Konst, R. Sarmiento, R. Rajachar and B. P. Lee, *ACS Appl. Mater. Interfaces*, 2014, **6**, 16982–16992.
89. X. Fan, L. Lin and P. B. Messersmith, *Biomacromolecules*, 2006, **7**, 2443–2448.
90. X. Fan, L. Lin, J. L. Dalsin and P. B. Messersmith, *J. Am. Chem. Soc.*, 2005, **127**, 15843–15847.
91. G. Westwood, T. N. Horton and J. J. Wilker, *Macromolecules*, 2007, **40**, 3960–3964.
92. T. J. Deming, *Nature*, 1997, **390**, 386–389.
93. M. Yu and T. J. Deming, *Macromolecules*, 1998, **31**, 4739–4745.
94. H. Yamamoto, *J. Adhes. Sci. Technol.*, 1987, **1**, 177–183.
95. H. Yamamoto and H. Tatehata, *J. Mar. Biotechnol.*, 1995, **2**, 95–100.
96. H. Tatehata, A. Mochizuki, T. Kawashima, S. Yamashita and H. Yamamoto, *J. Appl. Polym. Sci.*, 2000, **76**, 929–937.
97. B. Yang, D. G. Kang, J. H. Seo, Y. S. Choi and H. J. Cha, *Biofouling*, 2013, **29**, 483–490.
98. B. J. Kim, D. X. Oh, S. Kim, J. H. Seo, D. S. Hwang, A. Masic, D. K. Han and H. J. Cha, *Biomacromolecules*, 2014, **15**, 1579–1585.
99. Z. Liu, B.-H. Hu and P. B. Messersmith, *Tetrahedron Lett.*, 2008, **49**, 5519–5521.
100. K. C. L. Black, Z. Liu and P. B. Messersmith, *Chem. Mater.*, 2011, **23**, 1130–1135.
101. W. D. Fuller, M. S. Verlander and M. Goodman, *Biopolymers*, 1978, **17**, 2939–2943.
102. B. P. Lee, K. Huang, F. N. Nunalee, K. R. Shull and P. B. Messersmith, *J. Biomater. Sci., Polym. Ed.*, 2004, **15**, 449–464.
103. B.-H. Hu and P. B. Messersmith, *Tetrahedron Lett.*, 2000, **41**, 5795–5798.
104. J. H. Waite and C. V. Benedict, *Methods Enzymol.*, 1984, **107**, 397–413.
105. L. He, D. E. Fullenkamp, J. G. Rivera and P. B. Messersmith, *Chem. Commun.*, 2011, **47**, 7497–7499.
106. J. Su, F. Chen, V. L. Cryns and P. B. Messersmith, *J. Am. Chem. Soc.*, 2011, **133**, 11850–11853, DOI: 10.1021/ja203077x.
107. D. G. Barrett, D. E. Fullenkamp, L. He, N. Holten-Andersen, K. Y. C. Lee and P. B. Messersmith, *Adv. Funct. Mater.*, 2013, **23**, 1111–1119.
108. C. E. Brubaker, H. Kissler, L.-J. Wang, D. B. Kaufman and P. B. Messersmith, *Biomaterials*, 2010, **31**, 420–427.
109. M. Perrini, D. Barrett, N. Ochsenbein-Koelble, R. Zimmermann, P. Messersmith and M. Ehrbar, *J. Mech. Behav. Biomed. Mater.*, 2016, **58**, 57–64.
110. D. G. Barrett, G. G. Bushnell and P. B. Messersmith, *Adv. Healthcare Mater.*, 2013, **2**, 745–755.
111. P. Sun, J. Wang, X. Yao, Y. Peng, X. Tu, P. Du, Z. Zheng and X. Wang, *ACS Appl. Mater. Interfaces*, 2014, **6**, 12495–12504.
112. C. Lee, J. Shin, J. S. Lee, E. Byun, J. H. Ryu, S. H. Um, D. I. Kim, H. Lee and S. W. Cho, *Biomacromolecules*, 2013, **14**, 2004–2013.

113. C. J. Kastrup, M. Nahrendorf, J. L. Figueiredo, H. Lee, S. Kambhampati, T. Lee, S.-W. Cho, R. Gorbatov, Y. Iwamoto, T. T. Dang, P. Dutta, J. H. Yeon, H. Cheng, C. D. Pritchard, A. J. Vegas, C. D. Siegel, S. MacDougall, M. Okonkwo, A. Thai, J. R. Stone, A. J. Coury, R. Weissleder, R. Langer and D. G. Anderson, *Proc. Natl. Acad. Sci. U. S. A.*, 2012, **109**, 21444–21449.
114. E. Kim, Y. Liu, X. W. Shi, X. H. Yang, W. E. Bentley and G. F. Payne, *Adv. Funct. Mater.*, 2010, **20**, 2683–2694.
115. K. Kim, J. H. Ryu, D. Y. Lee and H. Lee, *Biomater. Sci.*, 2013, **1**, 783–790.
116. T. M. Shazly, A. B. Baker, J. R. Naber, A. Bon, K. J. Van Vliet and E. R. Edelman, *J. Biomed. Mater. Res., Part A*, 2010, **95A**, 1159–1169.
117. J. Y. Park, J. Yeom, J. S. Kim, M. Lee, H. Lee and Y. S. Nam, *Macromol. Biosci.*, 2013, **13**, 1511–1519.
118. S. Hong, K. Yang, B. Kang, C. Lee, I. T. Song, E. Byun, K. I. Park, S.-W. Cho and H. Lee, *Adv. Funct. Mater.*, 2013, **23**, 1774–1780.
119. A. I. Neto, A. C. Cibrao, C. R. Correia, R. R. Carvalho, G. M. Luz, G. G. Ferrer, G. Botelho, C. Picart, N. M. Alves and J. F. Mano, *Small*, 2014, **10**, 2459–2469.
120. B. P. Lee, *US Pat.*, 7,622,533, 2009.
121. B. P. Lee, *US Pat.*, 8,030,413, 2011.
122. H. Shao and R. J. Stewart, *Adv. Mater.*, 2010, **22**, 729–733.
123. C. R. Matos-Perez, J. D. White and J. J. Wilker, *J. Am. Chem. Soc.*, 2012, **134**, 9498–9505.
124. C. L. Jenkins, H. J. Meredith and J. J. Wilker, *ACS Appl. Mater. Interfaces*, 2013, **5**, 5091–5096.
125. C. R. Matos-Pérez and J. J. Wilker, *Macromolecules*, 2012, **45**, 6634–6639.
126. S. Skelton, M. Bostwick, K. O'Connor, S. Konst, S. Casey and B. P. Lee, *Soft Matter*, 2013, **9**, 3825–3833.
127. H. Yamamoto and T. Hayakawa, *Biopolymers*, 1977, **16**, 1593–1607.
128. R. Ko, P. A. Cadieux, J. L. Dalsin, B. P. Lee, C. N. Elwood and H. Razvi, *J. Endourol.*, 2008, **22**, 1153–1160.
129. J. L. Dalsin and P. B. Messersmith, *Mater. Today*, 2005, **8**, 38–46.
130. X. Wang, Q. Ye, T. Gao, J. Liu and F. Zhou, *Langmuir*, 2011, **28**, 2574–2581.
131. C. Gao, G. Li, H. Xue, W. Yang, F. Zhang and S. Jiang, *Biomaterials*, 2010, **31**, 1486–1492.
132. Q. Ye, X. Wang, S. Li and F. Zhou, *Macromolecules*, 2010, **43**, 5554–5560.
133. J. Kuang and P. B. Messersmith, *Langmuir*, 2012, **28**, 7258–7266.
134. M. Lee, J. Rho, D.-E. Lee, S. Hong, S.-J. Choi, P. B. Messersmith and H. Lee, *ChemPlusChem*, 2012, **77**, 987–990.
135. X. Jia, Z. Y. Ma, G. X. Zhang, J. M. Hu, Z. Y. Liu, H. Y. Wang and F. Zhou, *J. Agric. Food Chem.*, 2013, **61**, 2919–2924.
136. X.-D. Pan, Z. Qina, Y.-Y. Yana and P. Sadhukhana, *Polymer*, 2010, **51**, 3453–3461.
137. M.-H. Ryou, J. Kim, I. Lee, S. Kim, Y. K. Jeong, S. Hong, J. H. Ryu, T.-S. Kim, J.-K. Park, H. Lee and J. W. Choi, *Adv. Mater.*, 2013, **25**, 1570.

138. R. Schwab, A. Willms, A. Kroger and H. P. Becker, *Hernia*, 2006, **10**, 272–277.
139. F. H. Berndsen, U. Petersson, D. Arvidsson, C. E. Leijonmarck, C. Rudberg, S. Smedberg and A. Montgomery, *Hernia*, 2008, **12**, 445–446.
140. J. Koniger, J. Redecke and M. Butters, *Langenbeck's Arch. Surg.*, 2004, **389**, 361–365.
141. Y. Ikada, in *Wound Closure Biomaterials and Devices*, ed. C. C. Chu, J. A. von Fraunhofer and H. P. Greisler, CRC Press, Inc., Boca Raton, Florida, 1997, pp. 317–346.
142. R. H. Fortelny, A. H. Petter-Puchner, N. Walder, R. Mittermayr, W. Öhlinger, A. Heinze and H. Redl, *Surg. Endosc.*, 2007, **21**, 1781–1785.
143. C. E. Evans, G. C. Lees and I. A. Trail, *J. Hand Surg.*, 1999, **24B**, 658–661.
144. J. M. Harris, J. M. Dust, R. A. McGill, P. A. Harris, M. J. Edgell, R. M. Sedaghat-Herati, L. J. Karr and D. L. Donnelly, *Water-Soluble Polymer*, 1991, vol. 467, pp. 418–429.
145. P. B. Messersmith, J. L. Dalsin, B. P. Lee and S. A. Burke, *US Pat.*, 8,673,286, 2014.
146. G. Bilic, C. Brubaker, P. B. Messersmith, A. S. Mallik, T. M. Quinn, C. Haller, E. Done, L. Gucciardo, S. M. Zeisberger, R. Zimmermann, J. Deprest and A. H. Zisch, *Am. J. Obstet. Gynecol.*, 2010, **202**, 85.e81–85.e89.
147. C. M. Haller, W. Buerzle, A. Kivelio, M. Perrini, C. E. Brubaker, R. J. Gubeli, A. S. Mallik, W. Weber, P. B. Messersmith, E. Mazza, N. Ochsenbein-Koelble, R. Zimmermann and M. Ehrbar, *Acta Biomater.*, 2012, **8**, 4365–4370.
148. M.-H. Lin, J. Anderson, R. Pinnaratip, H. Meng, S. Konst, A. J. DeRouin, R. Rajachar, K. G. Ong and B. P. Lee, *IEEE Trans. Biomed. Eng.*, 2015, **62**, 1838–1842.
149. P. B. Messersmith, J. L. Dalsin, B. P. Lee and S. A. Burke, *US Pat.* Application, 20080247984, 2008.
150. C. E. Brubaker and P. B. Messersmith, *Biomacromolecules*, 2011, **12**, 4326–4334.
151. Y. Liu, H. Zhan, S. Skelton and B. P. Lee, *MRS Proc.*, 2013, **1569**, 33–38.
152. C. J. Kastrup, M. Nahrendorf, J. L. Figueiredo, H. Lee, S. Kambhampati, T. Lee, S. W. Cho, R. Gorbatov, Y. Iwamoto, T. T. Dang, P. Dutta, J. H. Yeon, H. Cheng, C. D. Pritchard, A. J. Vegas, C. D. Siegel, S. MacDougall, M. Okonkwo, A. Thai, J. R. Stone, A. J. Coury, R. Weissleder, R. Langer and D. G. Anderson, *Proc. Natl. Acad. Sci. U. S. A.*, 2012, **109**, 21444–21449.
153. G. D. Bixler and B. Bhushan, *Philos. Trans. R. Soc., A*, 2012, **370**, 2381–2417.
154. M. Q. Zhang, T. Desai and M. Ferrari, *Biomaterials*, 1998, **19**, 953–960.
155. N. P. Desai, S. F. A. Hossainy and J. A. Hubbell, *Biomaterials*, 1992, **13**, 417–420.
156. L. G. Harris, S. Tosatti, M. Wieland, M. Textor and R. G. Richards, *Biomaterials*, 2004, **25**, 4135–4148.

157. S. Pasche, S. M. De Paul, J. Voros, N. D. Spencer and M. Textor, *Langmuir*, 2003, **19**, 9216–9225.
158. C. D. Bain and G. M. Whitesides, *Science*, 1988, **240**, 62–63.
159. Y. Zhang, N. Kohler and M. Zhang, *Biomaterials*, 2002, **22**, 1553–1561.
160. F. Zhang, E. T. Kang, K. G. Neoh, P. Wang and K. L. Tan, *Biomaterials*, 2001, **22**, 1541–1548.
161. M. C. Shen, L. Martinson, M. S. Wagner, D. G. Castner, B. D. Ratner and T. A. Horbett, *J. Biomater. Sci., Polym. Ed.*, 2002, **13**, 367–390.
162. M. H. Schoenfisch and J. E. Pemberton, *J. Am. Chem. Soc.*, 1998, **120**, 4502–4513.
163. M. J. Tarlov and J. G. Newman, *Langmuir*, 1992, **8**, 1398–1405.
164. G. Li, H. Xue, C. Gao, F. Zhang and A. S. Jiang, *Macromolecules*, 2010, **43**, 14–16.
165. G. Li, G. Cheng, H. Xue, S. Chen, F. Zhang and S. Jiang, *Biomaterials*, 2008, **29**, 4592–4597.
166. N. D. Braulta, C. Gaoa, H. Xuea, M. Piliarik, J. R. Homolab, S. Jianga and Q. Yua, *Biosens. Bioelectron.*, 2010, **25**, 2276–2282.
167. A. R. Statz, A. E. Barron and P. B. Messersmith, *Soft Matter*, 2008, **4**, 131–139.
168. A. R. Statz, R. J. Meagher, A. E. Barron and P. B. Messersmith, *J. Am. Chem. Soc.*, 2005, **127**, 7972–7973.
169. S. M. Miller, R. J. Simon, S. Ng, R. N. Zuckermann, J. M. Kerr and W. H. Moos, *Drug Dev. Res.*, 1995, **35**, 20–32.
170. K. H. A. Lau, T. S. Sileika, S. H. Park, A. M. L. Sousa, P. Burch, I. Szleifer and P. B. Messersmith, *Adv. Mater. Interfaces*, 2015, **2**, n/a–n/a.
171. A. R. Statz, J. P. Park, N. P. Chongsiriwatana, A. E. Barron and P. B. Messersmith, *Biofouling*, 2008, **24**, 439–448.
172. Y. Li, H. Meng, Y. Liu and B. P. Lee, *Sci. World J.*, 2015, **2015**, 685690.
173. J. Shin, J. S. Lee, C. Lee, H.-J. Park, K. Yang, Y. Jin, J. H. Ryu, K. S. Hong, S.-H. Moon, H.-M. Chung, H. S. Yang, S. H. Um, J.-W. Oh, D.-I. Kim, H. Lee and S.-W. Cho, *Adv. Funct. Mater.*, 2015, **25**, 3814–3824.
174. D. H. Gracias, *Curr. Opin. Chem. Eng.*, 2013, **2**, 112–119.
175. H. L. Lim, Y. Hwang, M. Kar and S. Varghese, *Biomater. Sci.*, 2014, **2**, 603–618.
176. L. Ionov, *Adv. Funct. Mater.*, 2013, **23**, 4555–4570.
177. B. P. Lee and S. Konst, *Adv. Mater.*, 2014, **26**, 3415–3419.
178. B. P. Lee, M.-H. Lin, A. Narkar, S. Konst and R. Wilharm, *Sens. Actuators, B*, 2015, **206**, 456–462.
179. B. P. Lee, Y. Liu and S. Konst, *MRS Proc.*, 2014, **1710**, DOI: 10.1557/opl.2014.511.
180. C. Heinzmann, S. Coulibaly, A. Roulin, G. L. Fiore and C. Weder, *ACS Appl. Mater. Interfaces*, 2014, **6**, 4713–4719.
181. M. T. Northen, C. Greiner, E. Arzt and K. L. Turner, *Adv. Mater.*, 2008, **20**, 3905–3909.
182. X. Luo, K. E. Lauber and P. T. Mather, *Polymer*, 2010, **51**, 1169–1175.

183. G. Sudre, L. Olanier, Y. Tran, D. Hourdet and C. Creton, *Soft Matter*, 2012, **8**, 8184–8193.
184. Y.-Z. Wang, L. Li, F.-S. Du and Z.-C. Li, *Polymer*, 2015, **68**, 270–278.
185. P. Wilke, N. Helfricht, A. Mark, G. Papastavrou, D. Faivre and H. G. Börner, *J. Am. Chem. Soc.*, 2014, **136**, 12667–12674.
186. P. Wilke and H. G. Börner, *ACS Macro Lett.*, 2012, **1**, 871–875.
187. W. Wei, J. Yu, M. A. Gebbie, Y. Tan, N. R. Martinez Rodriguez, J. N. Israelachvili and J. H. Waite, *Langmuir*, 2015, **31**, 1105–1112.
188. G. P. Maier, M. V. Rapp, J. H. Waite, J. N. Israelachvili and A. Butler, *Science*, 2015, **349**, 628–632.
189. J. D. White and J. J. Wilker, *Macromolecules*, 2011, **44**, 5085–5088.
190. S. Seo, S. Das, P. J. Zalicki, R. Mirshafian, C. D. Eisenbach, J. N. Israelachvili, J. H. Waite and B. K. Ahn, *J. Am. Chem. Soc.*, 2015, **137**, 9214–9217.
191. D. E. Fullenkamp, L. He, D. G. Barrett, W. R. Burghardt and P. B. Messersmith, *Macromolecules*, 2013, **46**, 1167–1174.
192. E. Bell and J. Gosline, *J. Exp. Biol.*, 1996, **199**, 1005–1017.
193. A. Tamarin, P. Lewis and J. Askey, *J. Morphol.*, 1976, **149**, 199–221.
194. C. Zhong, T. Gurry, A. A. Cheng, J. Downey, Z. Deng, C. M. Stultz and T. K. Lu, *Nat. Nano*, 2014, **9**, 858–866.

CHAPTER 11

Self-reporting Polymeric Materials with Mechanochromic Properties

JOSE V. ARAUJO,† OMAR RIFAIE-GRAHAM,† EDWARD A. APEBENDE† AND NICO BRUNS*

Adolphe Merkle Institute, University of Fribourg, Chemin des Verdiers 4, 1700 Fribourg, Switzerland
*Email: nico.bruns@unifr.ch

11.1 Introduction

Mechanical forces are ubiquitous in the living world, as the physical environment interacts with living beings. Moreover, animals, plants, bacteria, and cells exert forces on their environment *e.g.* to move, crawl or communicate. Thus, it is not surprising that mechanisms to sense stress, strain, tension, shear forces, and displacement have evolved. Mechanobiological sensing events include, amongst others, the sense of touch, the sense of hearing, pressure sensing in the cardiovascular system, strain sensing in muscles and in the cytoskeleton, as well as the stimulation of certain marine plankton to emit chemoluminescence upon mechanical perturbation. Most of these mechanisms are based on efficient mechanotransduction on the molecular scale, which leads to biochemical signalling cascades. However, some mechanical events also result in a colour change at the macroscopic

†Authors contributed equally.

RSC Polymer Chemistry Series No. 22
Bio-inspired Polymers
Edited by Nico Bruns and Andreas F. M. Kilbinger

Published by the Royal Society of Chemistry, www.rsc.org

level, *e.g.* to allow for camouflage, or for inter- and intraspecies communication (as in the case of marine glow). Moreover, mechanical stimuli can induce non-specific colour changes in living matter, such as the formation of bruises on skin, or the bleeding of a wound. While these events are a coincidence of tissue damage, humans and animals have learned to react to these warning signals, therefore allowing to easily assess whether a tissue has suffered damage or not. Thus, nature is a rich source of inspiration for the development of polymeric materials that are able to autonomously detect and report deformation as well as the occurrence of damage, such as cracks, fractures, and delamination. Such self-reporting materials (also termed self-assessing materials) provide intriguing opportunities to create safety features that could prevent catastrophic materials failure, to develop simple and cheap force sensors, and as analytical tools to measure stress distributions within materials.

Although the responsiveness in mechano-responsive materials is triggered by a mechanical input (*e.g.* impact, tension, ultrasound), the output can be of various kinds. For instance, force-induced perturbations in the structure of polymeric materials can provide the activation energy for specific chemical reactions such as the scission of labile covalent bonds or ring-opening reactions. Moreover, mechanical forces can cause changes in molecular arrangements on the supramolecular level and influence the conformation or structure of proteins. Forces can also be transferred to microscopic objects that are embedded in the bulk polymer, such as reagent-filled microcapsules, hollow fibres, and microvascular networks. If moieties that undergo force-induced changes in their photophysical properties were chosen, externally applied forces would result in spectral changes in the UV-vis range. This allows for the development of mechanochromic polymers, *i.e.* materials in which the response to mechanical perturbations is observed as a change in colour, fluorescence, photoluminescence, or chemoluminescence.

In this chapter, we will discuss the design, synthesis, and the properties of mechano-responsive polymeric materials with focus on mechanochromic systems. Initially, natural mechanochromic events will be described in order to show how these mechanisms can serve as inspiration in the design of artificial devices. Following, mechanical responsiveness in the context of mechanochromic materials will be reviewed. The effect of mechanical forces on the molecular, supramolecular, microscopic and macroscopic levels will be taken into consideration, as well as the use of mechanically responsive biological macromolecules for the preparation of self-reporting polymer–protein hybrid materials.

11.2 Learning from Nature

The outstanding self-reporting and mechano-responsive properties of biological systems are the result of millions of years of evolution and adaptation. This involves complex molecular mechanisms, which are not completely understood or difficult to replicate given the current technology available. Nevertheless, one of the challenges of artificial self-reporting and mechano-responsive materials is to pursue systems which can mimic the behaviour of natural systems, albeit with simpler molecular mechanisms. Figure 11.1 shows some prominent mechanochromic events in nature. Moreover, it highlights examples of polymeric materials and composites that imitate such self-reporting properties.

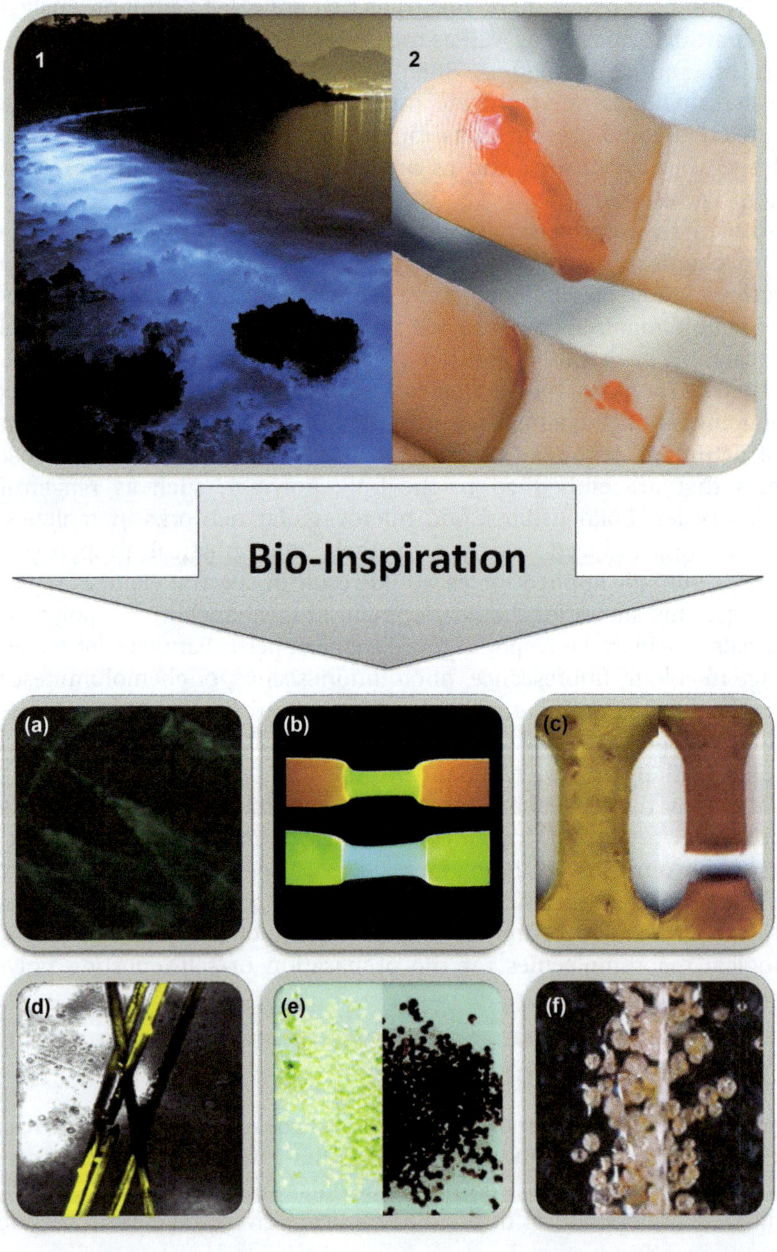

Cells are the main units of living organisms.[1] It is well known that external signals generate changes in the internal chemical composition of cells, making them adapt to the surrounding environment. The routes for cell adaptation involve different macromolecular complexes called receptors and channels.[2] These complexes can be composed of one or more proteins, and can also be combined with sugar chains forming glycoproteins.[3] Receptors have been classified according to the types of stimuli they respond to, such as temperature, pH variations, and light.[4] Of particular interest for this review are receptors that respond to mechanical perturbations. In general, natural mechano-responsive systems are activated through the transmission of force to the cell membrane and the tension is sensed by these receptors.[5] The response of the receptors can be diverse. The stimulus may initiate a signal cascade in which biomolecules are generated or transformed. However, the most general response occurs when a receptor activates directly an ion channel.[6] In response to a mechanical stimulus, the channel is opened, altering the normal concentration of different ions at both sides of the cell membrane,[7] which results in the sense of touch,[8] sound,[9] proprioception,[10] or the control of blood pressure.[11] Membrane-bound force receptors are some of the most common biological mechanosensors. However, their property to induce a chemical or electrophysiological signal in response to a mechanical stimulation is difficult to translate into responsive polymeric systems.

Microorganisms also respond to mechanical stimuli. A fascinating event is the marine glow (Figure 11.1). It occurs when dinoflagellates are mechanically disturbed upon a velocity gradient generated by external factors, such

Figure 11.1 Force-induced mechanochromic events in nature serve as inspiration for self-reporting materials. (1) Marine glow caused by mechanical perturbation of dinoflagellates. (2) Bleeding wound caused by a sharp object. (a) Fluorescence of cracks formed in a film of 1,1,1-tris(cinnamoyloxymethyl)ethane. (b) Polymeric blends containing cyano-OPVs. (c) Mechanical response in PMA films containing spiropyranes. (d) Mechanical unfolding of a fluorescent protein in carbon-fibre-reinforced composites. (e) 1,3,5,7-Cyclooctatetraene ring opening metathesis polymerisation (ROMP) after mechanical damage of capsules in contact with Grubbs–Love catalyst. (f) Epoxy coating with microcapsules that contain 2,7-dichlorofluorescein (DCF). (1) and (2) © Shutterstock. (a) Reprinted from S.-Y. Cho, J.-G. Kim and C.-M. Chung, A fluorescent crack sensor based on cyclobutane-containing crosslinked polymers of tricinnamates, *Sens. Actuators*, **134**, 822–825, Copyright 2008 with permission from Elsevier.[27] (b) Reproduced with permission from B. R. Crenshaw and C. Weder, *Chem. Mater.*, 2003, **15**, 4717–4724. Copyright (2008) American Chemical Society.[28] (c) Reprinted by permission of Macmillan Publishers Ltd: *Nature*,[29] Copyright 2009. (d) Reproduced from ref. 30 with permission from the Royal Society of Chemistry. (e) Reproduced with permission from S. A. Odom, A. C. Jackson, A. M. Prokup, S. Chayanupatkul, N. R. Sottos, S. R. White and J. S. Moore, *ACS Appl. Mater. Interfaces*, 2011, **3**, 4547–4551. Copyright (2011) American Chemical Society.[31] (f) Reproduced from ref. 32 with permission from John Wiley and Sons. Copyright © 2016 WILEY-VCH Verlag GmbH & Co. KGaA, Weinheim.

as swimming animals, waves, or ships.[12,13] These events cause the cell membrane of the protists to deform[14] and activate a GTP-depending protein[15] which induces a signal cascade, mediated by the change of Ca^{2+} concentration.[16] This results in a change of cytoplasmic pH, favouring a bioluminescence reaction. Dinoflagellates have been used as markers for local velocity gradients of water in the field[13] and in the laboratory.[12,17] Microplankton would, however, not be of much use as force-sensors in solid polymeric materials. However, various chemical reactions are known to result in chemoluminescence. Moreover, it has been shown that such reactions can be triggered by the application of forces, so that polymeric materials that react to mechanical forces with the emission of light have been realised.[18]

A simple natural response to mechanical stimuli that is not based on sophisticated force-induced signalling cascades can be observed in the human skin. When skin is cut, blood is released from the capillary network in the dermal layer. Its intense red colour indicates damage (Figure 11.1). At a later stage, a series of biochemical pathways are activated to induce healing of the wound. First, the blood platelets adhere to the damaged surface and change their morphology to promote blood clotting. The healing of the wound is enhanced by the activation of fibrinogen that turns into fibrin, forming a matrix for the settling of the platelets. Different biochemical signals are then emitted for the remodelling of the new tissue ending in healing.[19] The stages of healing are accompanied by typical morphologies, textures, and colour of the blood clot. Any trained human can assess the stage of its healing by looking at the healing wound. Direct visual indication of damage in a synthetic material is the main strategy to create self-reporting polymers and many examples will be discussed throughout this chapter.[20–22] Even the bleeding of a wound has been mimicked by the release of a liquid dye from reservoirs in "bleeding composites".[23,24]

When skin is hit with force, a bruise forms (Figure 11.1). They are caused by the extravasation of haemoglobin from capillaries after a blunt mechanical stimulus that causes damage. Upon breakdown of the macromolecule due to phagocytosis, the bruise changes colour, coinciding with the presence of the different degraded species biliverdin, bilirubin, and hemosiderin, going from purple (haemoglobin) to green, yellow, and brown, respectively.[25] During the healing process, the pigments are slowly transported away, and the absence of colour indicates that the tissue is healed. Bruising coatings and materials have been realised with dye-filled microcapsules that release their content into the damaged area of the material.[26,27] Although these colour-indicators have been combined with healing agents, the vanishing of a hit mark within a material upon successful self-healing remains a challenge yet to be addressed.

11.3 Mechano-responsiveness

In order to better understand how polymeric materials respond to mechanical perturbations, the next section will be dedicated to discuss mechano-responsiveness in detail. It is the ability of a material to change its properties in response to a mechanical input or stimulus. In the last decade, several groups have begun to use known mechanical transformations to impart mechano-responsiveness on polymeric systems.[20,33–36] In this field, polymer chains are used to direct macroscopic forces to mechanically labile functional groups within the material, whose activation leads to transformations in chemical composition or structure. These responses to mechanical stimuli can be a change in colour,[37] mechanical properties,[38] or shape.[39] Mechano-responsiveness can also be the ability to self-repair or "heal" when an applied mechanical stimulus causes a material to craze or crack.[40]

Mechanical transduction in these systems can be at the molecular, supramolecular or the micro/macro levels. For instance, chemical and structural changes caused by mechanical stimuli, such as flow- and ultrasound-induced chain scission, are interesting phenomena that can be taken into account in the development of mechano-responsive materials. In the following section, we will look closely at the types of response to mechanical stimuli and the underlying mechanisms involved either in mechanochemical or physical transformations.

11.3.1 Mechanical Input and Methods to Measure Mechanically-induced Changes in Polymers

As a general rule, polymers can be characterised in solution, as solid materials, or in the melt. In the following section, the commonly applied methods to study the response of polymers to mechanical forces are introduced, starting with solution measurements and then moving on to bulk measurements. The effect of shear forces on polymer chains in melts is fascinating, but has not been studied with the help of mechanochromic systems and is therefore not covered in this chapter.

In order to gain a fundamental understanding of force-induced changes in polymer chains, solution studies are best suited because the theoretical framework to describe the general behaviour of polymer chains in solution is well developed and because the experimental system is a homogeneous solution. Flow induced chain scission and ultrasound mechano-chemistry are the common methods used to study the response of polymers to mechanical forces in solution. Although this chapter focuses on mechano-responsive polymers, *i.e.* polymers that have been designed to undergo specific changes in response to mechanical forces, it is important to understand that strong enough forces will cause breaking of bonds in the polymer backbone of any polymer. In fact, the first evidences of polymer chain scission were observed in solution studies.[41–43] Therefore, knowing the fundamental principles of force-induced chain scission allows designing effective mechano-responsive polymer systems.

The mechanical input in the flow-induced scission method can be an elongation (uniform) or a turbulent (non-uniform) field. For instance, the passage of a coiled polymer through an elongation flow field results in a coil-stretch transition that is influenced by strong hydrodynamic forces. Polymer elongation starts when the rate of chain stretching exceeds the entropic elasticity, which causes the molecules to orient in the direction of the flow.[44] If elongation forces are strong enough, chain scission takes place. The first evidence of cleavage of covalent bonds in a turbulent flow was observed by Toms.[45] The author reported that diluted solutions of poly(methyl methacrylate) (PMMA) in chlorobenzene presented a considerable reduction in turbulent flow pressure when compared to that of the pure solvent. It is of common agreement that such behaviour is caused by shearing forces, which promote a partial chain extension and, eventually, scission.[20,46] There are different parameters which can influence the rate of chain scission, namely the type of solvent,[47] polymer concentration,[48,49] and temperature.[50] Chain scission can also be promoted by ultrasound, which can be applied to polymer solutions using either an ultrasonic bath or ultrasonic probe. In the first case, a vessel with the solution is immersed into an ultrasonic bath without direct contact between the samples and equipment. In the latter case, the probe is directly introduced into the polymer solution.[51] It is assumed that polymer chain scission occurs by cavitation according to the following mechanism[52,53] (Figure 11.2). The acoustic field creates microbubbles in solution. The collapse (or implosion) of such bubbles results in solvodynamic shear, which causes the polymer chain to elongate, undergoing a partial coil-stretch transition. It is generally agreed that scission occurs next to the midpoint of the chain.[54,55] Like flow induced chain scission, different parameters affect the cleavage of covalent bonds upon application of acoustic fields, namely the nature of the solvent (*i.e.* vapour pressure and surface tension), temperature, polymer concentration and viscosity, polymer molecular weight and, as expected, the intensity of the ultrasound power.

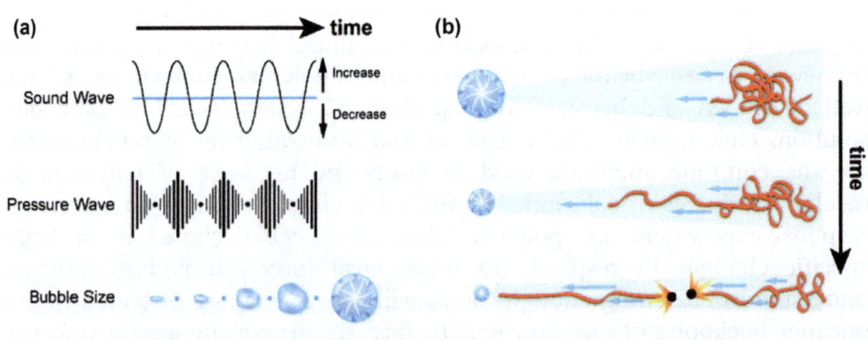

Figure 11.2 Polymer chain scission induced by ultrasound. (a) Pressure variation results in the formation of bubbles. (b) Solvodynamic shear is created by the rapid collapse of bubbles.
Adapted with permission from M. M. Caruso, D. A. Davis, Q. Shen, S. A. Odom, N. R. Sottos, S. R. White and J. S. Moore, *Chem. Rev.*, 2009, **109**, 5755–5798.[20] Copyright 2009, American Chemical Society.

Although these methods help us understand the changes that occur at the molecular level, most polymers find applications in the solid state. It is therefore important that mechano-responsive polymeric systems are tested with the kind of forces that they will experience when used. However, studies of mechanochromic behaviour in bulk polymers are complicated by inhomogeneities in the materials (*e.g.* crystalline *vs.* amorphous regions) and by the fact that it is nearly impossible to address all polymer chains in a sample with the same force, since they are differently oriented to the force vector, show entanglements, *etc.* Mechano-responsive bulk materials can be studied by subjecting them to elongation and compressional forces. Elongation forces are usually provided by tensile testing equipment and rheometers, while compressional forces are achieved in hydraulic presses and pressure cell, among others. Moreover, polymeric materials can be subjected to mechanical forces in impact tests and in bending tests. Quite often, the instruments can be coupled to spectrometers to allow for *in situ* measurements of the optical changes that occur as the material is subjected to a force. Alternatively, deformation or damage can be visualized post-loading, *e.g.*, by microscopic methods. The application of force to solid samples has been utilised to observe several kinds of mechanochromic phenomena, such as the formation or disruption of excimers, bond scission, chain alignment, distortion of embedded biomolecules, and the disruption of aggregates that are formed by non-covalent interactions (*e.g.* hydrogen bonding, ionic interactions).

11.3.2 Mechano-responsiveness at the Molecular Level

The ability of mechanical forces to cleave covalent bonds can be utilised to design force-responsive polymers.[32] This can be achieved by covalently linking mechanically labile moieties, also known as mechanophores, into polymer chains. The polymer facilitates the transduction of force to the mechanophores. The mechanophore can be designed to act as a specific cleavage point in the polymer backbone, to liberate small molecules, to initiate a subsequent chemical reaction or to change its photophysical properties, such as colour or fluorescence.[56] The latter results in self-reporting polymers. The chemical mechanisms by which these forced-induced transformations occur can be categorised into three groups, namely cycloreversions, electrocyclic ring opening, and stable radical formation.

Examples of mechanically triggered cycloreversion reactions reported in the literature include [2 + 2] cycloreversions, [4 + 2] cycloreversions (*e.g.* retro Diels–Alder reactions), and [4 + 4] cyloreversions.[20] Here, we will discuss selected mechanisms that are important in the context of self-reporting materials. Kryger *et al.* demonstrated that mechanical forces could trigger formal [2 + 2] cycloreversion when they sought to produce reactive species within polymers which could be utilised for self-healing applications.[57] Dicyano-substituted cyclobutane was incorporated as a mechanophore at the chain centre of poly(methyl acrylate) (Scheme 11.1). The strained ring

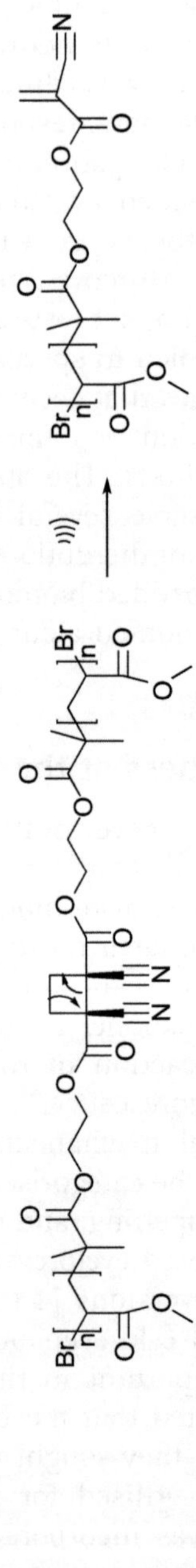

Scheme 11.1 Poly(methyl acrylate) with a centrally positioned dicyano-substituted cyclobutane mechanophore cleaved under the influence of ultrasound.[57]

could revert under the influence of mechanical force to give cyanoacrylates, which readily undergo addition reactions and could therefore be used to self-repair polymers. A greater rate of ultrasound-induced polymer chain scission was observed for the mechanophore-containing polymer than for the unmodified control polymer, which showed that the cyclobutane moiety acted as a mechanophore. Furthermore, trapping tests with a Michael donor confirmed that formal [2 + 2] cleavage of the mechanophore-containing polymer had occurred.

In another case, cyclobutane-containing polymers of tricinnamates with the ability to sense and report cracks were reported by Chung and coworkers.[28] The [2 + 2] cycloaddition of cinnamoyl groups forms dimers that contain a cyclobutane group.[58] This chemistry was used to create crosslinked polymeric samples from the monomer 1,1,1-tris(cinnamoyloxymethyl)ethane. Cyclobutanes within this material reverted under mechanical activation upon grinding or within the cracks of a polymer film. As cinnamates are fluorescent, while their cycloadducts are not, the retrocycloaddition caused the cracks to become fluorescent, which allowed visualising the damage in these films by fluorescence microscopy (Figure 11.3). The same concept was later applied by Dai and coworkers to prepare damage-detecting epoxy resins[59] and composites.[60]

Anthracene and its derivatives undergo a [4 + 4] photocycloaddition to cyclooctane-containing dimers. This reaction can be reversed with mechanical forces, yielding fluorescent anthracenes. Poly(vinyl alcohol) was crosslinked with anthracene dimers to give polymers that could report microcracks in a similar way as the tricinnamate-containing polymers described above[61] (Figure 11.3).

The ring opening of Diels–Alder adducts are other examples of mechanically triggered cycloreversions in the literature.[62–64] Although thermoresponsive materials based on the cycloreversion of Diels–Alder adducts had been reported,[65,66] Lyu *et al.* showed that the cycloreversion of Diels–Alder adducts of furan and maleic anhydride derivatives could be achieved more selectively and efficiently with force.[64] In another report, Larsen and Boydston elegantly demonstrated the cycloreversion of Diels–Alder adducts of furan and methacryloyl anhydride, which served as both mechanophores and crosslinkers within poly(methyl acrylate) chains.[63] Their system utilises the polymer backbone to transfer the compressional force to the mechanophore, causing bond angle distortions that eventually lead to scission. This unique method of activating the mechanophore by bending its bonds is a more efficient means of mechanochemical transduction than conventional approaches involving bond stretching. Boydston and coworkers,[67] as well as Moore and coworkers,[68] showed that Diels–Alder adducts of anthracenes and maleimides can be mechanically cleaved. The former group used such mechanophore to compare chain scission rates of linear polymers with those of three-arm branched polymers, as the resulting anthracene-terminated polymer chains can be easily observed by their UV absorbance and fluorescence. The latter group took advantage of these self-reporting properties

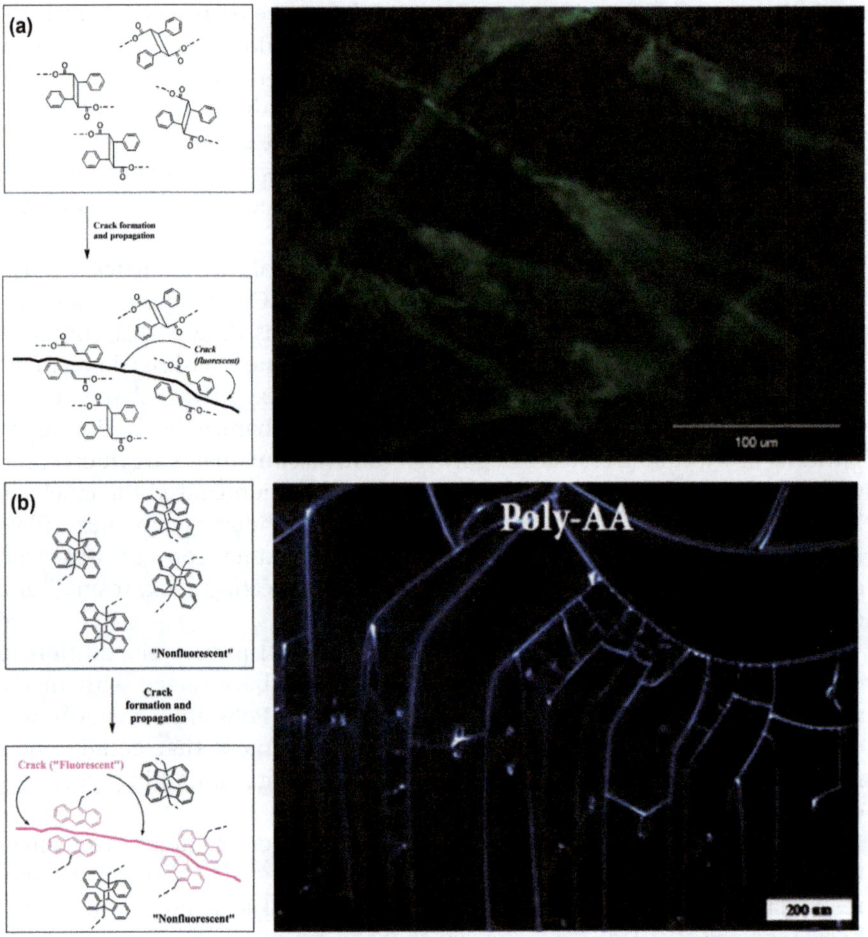

Figure 11.3 Detection of cracks in polymers by cycloreversion reactions. (a) [2 + 2] retrocycloaddition of a cyclobutane. (b) [4 + 4] retrocycloaddition of a cyclooctane. Fluorescent products form in cracks and are imaged by fluorescence microscopy.
Reproduced from S.-Y. Cho *et al.*, A fluorescent crack sensor based on cyclobutane-containing crosslinked polymers of tricinnamates, *Sens. Actuators, A*, **134**, 822–825, Copyright 2008 with permission from Elsevier[28] and from ref. 61 with permission from the Royal Society of Chemistry.

of the mechanophore to measure the detachment of the polymer chains from the surface of silica nanoparticles, to which the polymer had been grafted *via* maleimide-anthracene linkers.

The prime example of self-reporting mechanophores that undergo force-activated electrocyclic ring opening are spirospyrans. These molecules are well known for their photochromism[69,70] and have more recently been utilised to develop thermoplastic, elastomeric, and glassy materials in which mechanical activation leads to a visual colour and fluorescence change[32,71–83] (Scheme 11.2). In their pioneering work, White, Moore, Sottos, and coworkers synthesised poly(methyl acrylate) (PMA) from a bifunctional spiropyran ATRP initiator in order to position the mechanophore centrally within the polymer chain.[32] Dog-bone samples of this material were observed to change colour from light brown to red under mechanical loading, which was caused by the electrocyclic ring opening of the colourless spiropyran to its coloured merocyanine derivative (Figure 11.4). The colour change could be reversed after exposure to fluorescent light for about six hours, which is in line with the photochromism of the compound. Spiropyran mechanophores have since been used to develop a multitude of self-reporting materials, such as rubber toughened PMMA with the ability to indicate plastic deformation ahead of crack tips,[77] stress-sensing and self-healing metallosupramolecular polymers,[78] 3D-printed mechanochromic materials,[79] and polystyrene coatings that evolve colour and fluorescence in areas subjected to laser-generated shock waves.[80] These reports show that spiropyranes can be used to visualise stress and strain distributions within a material. A fascinating application for this property was reported by Craig and coworkers. Soft robotic grippers were prepared from spiropyran-containing poly(dimethylsiloxane) (PDMS). The material developed colour and fluorescence in areas with the highest deformation, thus allowing to identify regions that could be at risk to fail if the gripper is overloaded (Figure 11.5).[81] A similar approach was followed to prepare silicon rubbers that indicate localised compression[82] (Figure 11.6). Coatings of these materials on white surfaces resulted in paper-sized boards for drawing and writing with a stylus. As the colour can be erased by irradiation with strong white light, the drawing board should be reusable. The mechanochromic properties of spiropyrans were also exploited to create colour-changing materials that were inspired by cephalopods.[83] Electric fields induced deformation of electro-mechano-chemically responsive elastomers, which caused the mechanophore to undergo electrocyclic ring-opening. As a result, areas of high deformation became fluorescent (Figure 11.7). The appearance of a fluorescent pattern on demand could be used for camouflage, in analogy to the colour-changing properties of squids. Alternatively, such electronically switchable materials could be developed into flexible displays for use in wet environments.

All the previously described mechanisms occur by heterolytic cleavage of the labile bonds of the mechanophores. However, Imato *et al.* have recently demonstrated a mechanochromic system in which force induced cleavage of the mechanophore's labile bonds is a homolytic process.[84] They synthesised segmented polyurethane elastomers with diaryldibenzofuranone covalently positioned in the soft segments. Dog-bone samples of the polymer were subjected to uniaxial tensile loading. The cleavage of the bond led to the formation of stable radicals with the corresponding change from yellow to blue (Figure 11.8). The colour change was observed instantly and the colour reverted to its initial yellow when left at room temperature for 300 minutes. Electron paramagnetic resonance (EPR) spectroscopy before, during, and

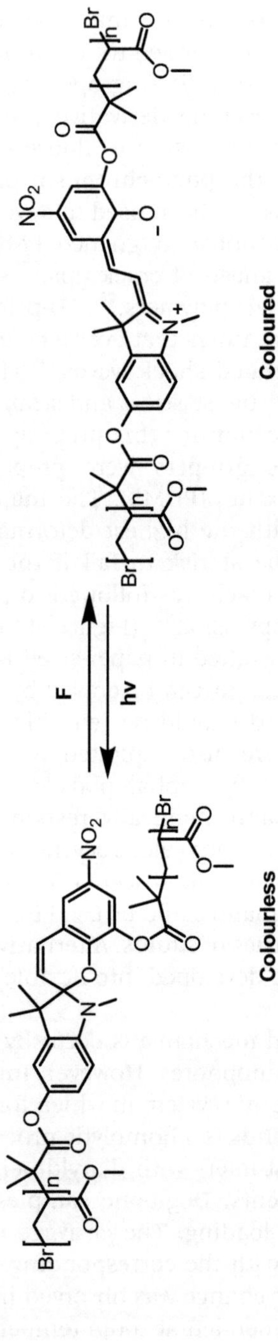

Scheme 11.2 Electrolytic ring opening of spiropyran mechanophore facilitated by force and the reverse ring closing facilitated by UV light.[32]

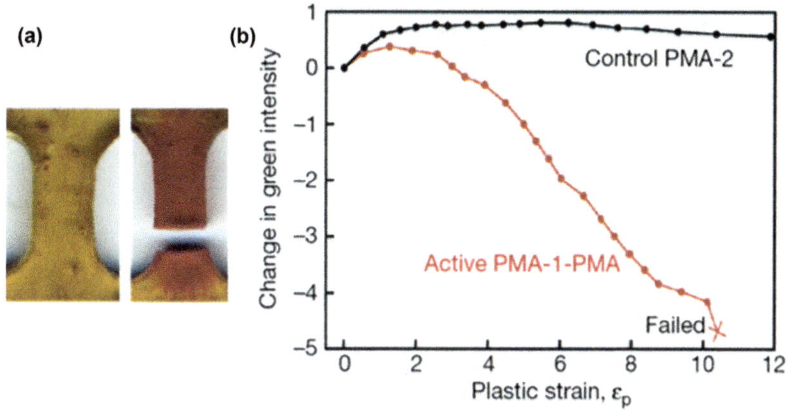

Figure 11.4 Mechanochromic spiropyran-PMA. (a) Dog-bone samples before and after uniaxial elongation. (b) Change in intensity of the green channel (of a RGB image) with the increase in elongation strain for a mechanochromic sample (active sample; red) and a control sample (black).
Adapted by permission of Macmillan Publishers Ltd: *Nature*,[32] Copyright 2009.

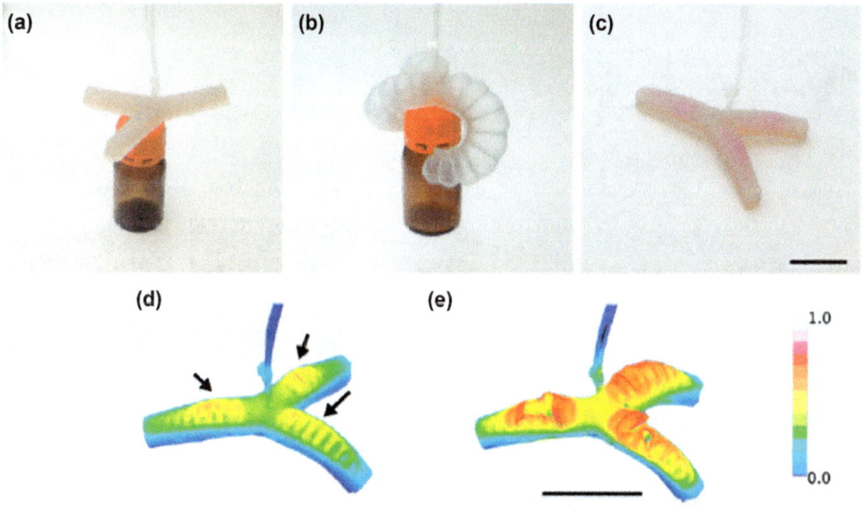

Figure 11.5 Self-reporting soft robotic gripper that reports areas of large deformation by spiropyran-based colour and fluorescence change. (a,b) A spiropyran-PDMS gripper can hold cargo when inflated. (c,d) After deflation, the colour and fluorescence change of the mechanophore persists, marking those areas of the gripper that have experienced the largest deformation. (e) Fluorescence image of a gripper that was overinflated until it ruptured. The damage occurred in those areas that were most coloured and fluorescent.
Reproduced with permission from G. R. Gossweiler, C. L. Brown, G. B. Hewage, E. Sapiro-Gheiler, W. J. Trautman, G. W. Welshofer and S. L. Craig, *ACS Appl. Mater. Interfaces*, 2015, 7, 22431–22435.[81] Copyright 2015, American Chemical Society.

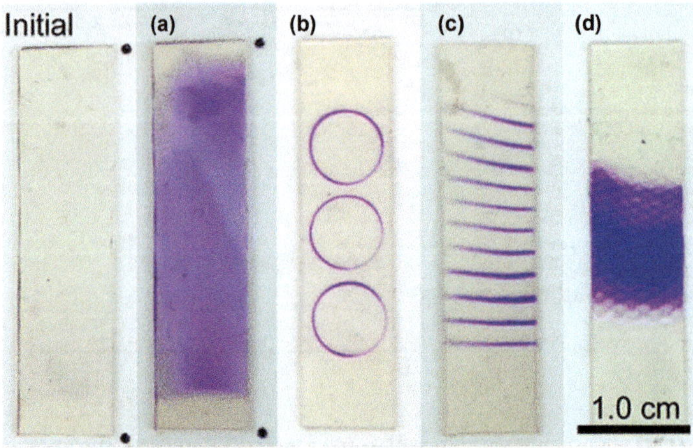

Figure 11.6 PDMS elastomer that reports localized compression by force-induced ring opening of spiropyran. The elastomer was subjected to (a) tension, and (b–d) rolling of patterned objects across its surface.
Reproduced with permission from G. R. Gossweiler, G. B. Hewage, G. Soriano, Q. Wang, G. W. Welshofer, X. Zhao and S. L. Craig, *ACS Macro Lett.*, 2014, **3**, 216–219.[82] Copyright 2014, American Chemical Society.

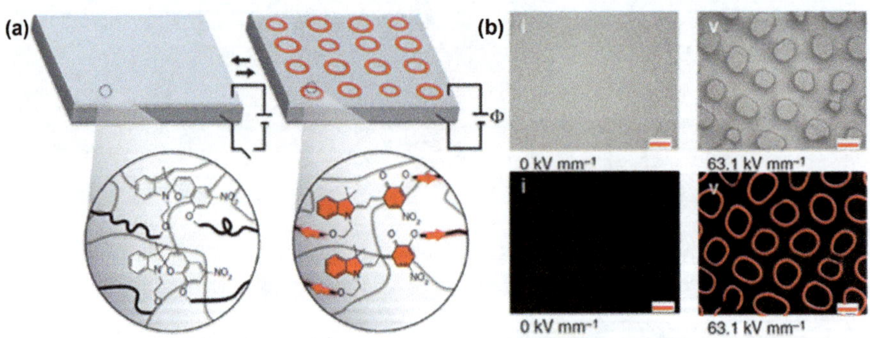

Figure 11.7 On-demand fluorescence patterning in an electronically addressable self-reporting elastomer. (a) Schematic illustration of a cephalopod-inspired device that contains a layer of electro-mechano-chemically responsive elastomer. (b) Optical microscopy and fluorescence microscopy images of such device before and during application of voltage. The responsive elastomer undergoes wrinkling, followed by the formation of craters. At their edge, the deformation of the elastomer is the highest, resulting in a mechanochromic response.
Adapted by permission of Macmillan Publishers Ltd: *Nat. Commun.*,[83] Copyright 2014.

after mechanical loading confirmed the generation of radicals and their subsequent recombination.

Dinoflagellates transduce mechanical stimulation into the emission of light. This behaviour was mimicked with chemiluminescent polymeric

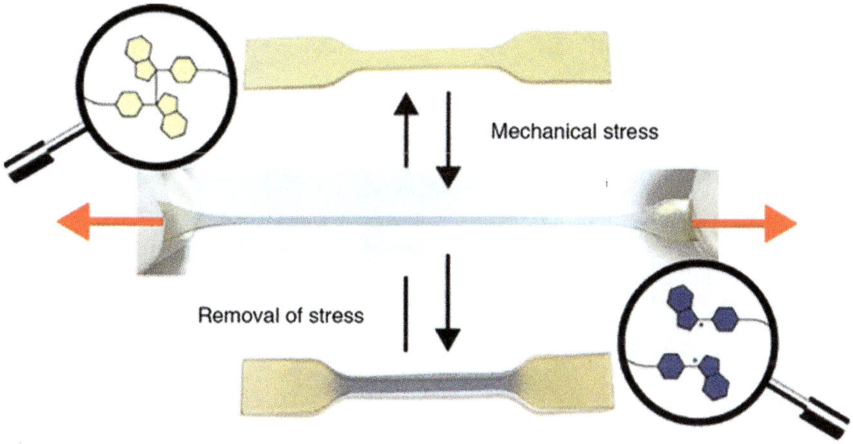

Figure 11.8 Self-reporting polyurethane that indicates deformation by the formation of stable radicals, which appear blue.
Reproduced with permission from K. Imato, T. Kanehara, T. Ohishi, M. Nishihara, H. Yajima, M. Ito, A. Takahara and H. Otsuka, *ACS Macro Lett.*, 2015, **4**, 1307–1311.[84] Copyright 2015, American Chemical Society.

materials based on 1,2-dioxetane by Sijbesma and coworkers.[18] They prepared rubber-like PMA crosslinked with bis(adamantyl)-1,2-dioxetane bisacrylate, and linear PMA with the mechanophore incorporated into the main chain. Mechanical deformation or sonication resulted in the cleavage of the four-membered 1,2-dioxetane ring into two adamantanone groups. One of these ketones was in the excited state and subsequently emitted bright-blue light (Figure 11.9). The intensity and colour of the emitted light could be further tuned by mixing in suitable acceptors. During uniaxial deformation of the materials, chemiluminescence occurred throughout the material. Emission of light increased markedly at the location of a fracture upon material failure. Thus, such self-reporting chemiluminescent materials can be used to map stress distributions and breaking of sacrificial bonds in real time without the need for an external light source, as demonstrated by Sijbesma, Creton, and coworkers.[85] They produced a series of elastomers that contained bis(adamantyl)-1,2-dioxetane bisacrylate crosslinkers and observed how bond scission happened at the crack tip of notched samples. Moreover, this technique allowed the identification of differences in the area of force dissipation between elastomers of various molecular architectures, which can guide the design of elastomers that are toughened by sacrificial bonds.

11.3.3 Mechano-responsiveness at the Supramolecular Level

The previous section demonstrated how polymers respond to mechanical forces at the molecular level through the cleavage of covalent bonds. Changes in the polymer structure upon mechanical stimuli, such as impact

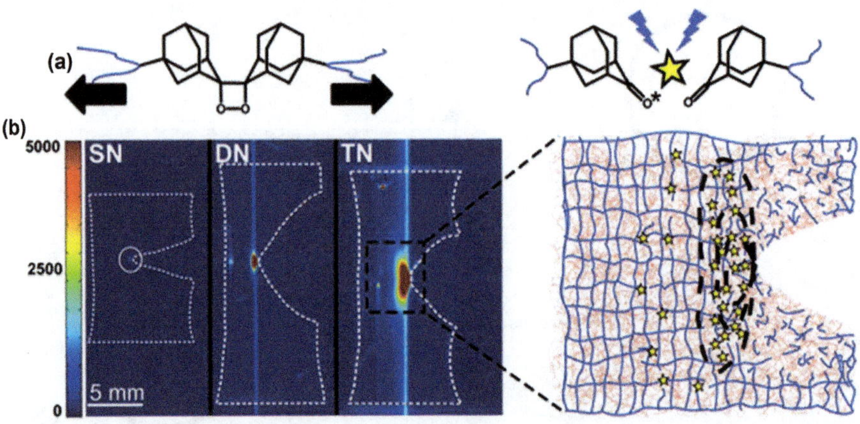

Figure 11.9 Mapping of bond cleavage in self-reporting chemiluminescent elastomers that are toughened by sacrificial bonds. (a) Bis(adamantyl)-1,2-dioxetane breaks under a mechanical force, resulting in chemiluminescence. (b) Intensity-coloured images of polymer networks during crack propagation of notched samples and schematic depiction of bond breaking around the crack tip. SN, DN, TN label elastomers of different molecular architecture ("single network", "double network" and "triple network"). The dashed line indicates the perimeter of the sample. Vertical lines are artefacts of the detector. (Adapted with permission from ref. 85. Copyright © 2014, The American Association for the Advancement of Science).

and deformation, may also take place in the absence of covalent bond scission. This occurs on a larger length scale and generally involves rearrangements of supramolecular motifs. In order to show how such changes can result in mechanochromic materials, the role of mechanical perturbation on mechanochromic blends will be discussed, followed by a summary of mechanophores based on the cleavage of hydrogen bonds and ionic interactions.

11.3.3.1 Mechanochromic Blends

Some dyes have different photoluminescent properties when the dye molecules are aggregated in the solid state or when they are molecularly dispersed, *e.g.* in solution.[35] For example, cyano oligo(*p*-phenylene vinylenes) (cyano-OPVs) form excimers (excited dimers) in the crystalline state due to π–π stacking, which leads to a red-shift of their fluorescence emission (Scheme 11.3). Mechanochromic polymer blends[22,35] can be produced from such excimer-forming dyes by mixing them into a polymer solution or melt, as pioneered by Weder and coworkers.[29,37,86–91] After solidification of the polymer by solvent evaporation or by cooling below the melting point, polymeric materials are obtained in which the dye forms nanoscale aggregates within the polymer matrix. These aggregates convert to molecular mixtures upon deformation of the polymer matrix, exhibiting a shift from excimer- to

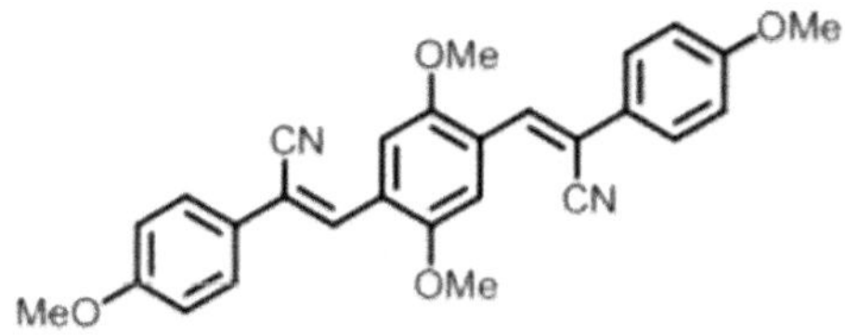

BCMDB

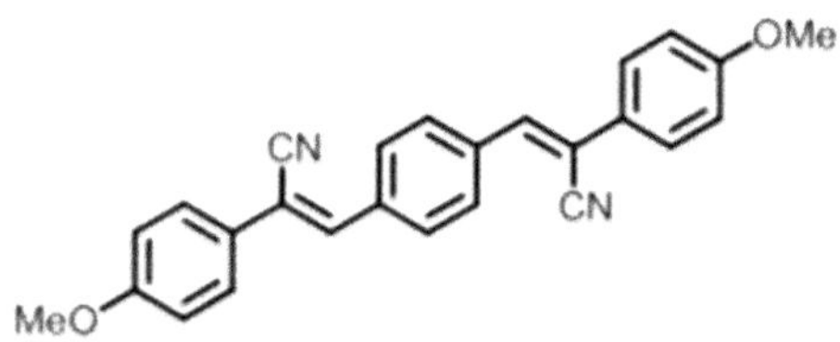

BCMB

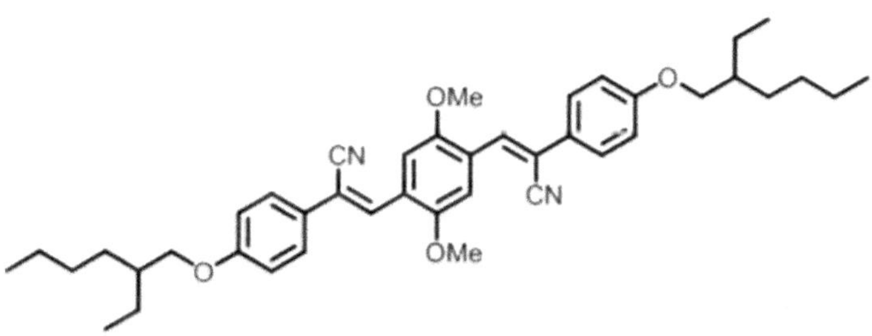

BCEDB

Scheme 11.3 Chemical structures of various cyano-OPVs: 1,4-bis(α-cyano-4-methoxystyryl)-2,5-dimethoxybenzene (BCMDB), 1,4-bis(α-cyano-4-methoxystyryl)-benzene (BCMB), and 1,4-bis(α-cyano-4-(2-ethylhexyloxystyryl))-2,5-dimethoxybenzene (BCEDB).[29]

monomer-dominated emission. For example, blends of semi-crystalline linear low-density polyethylene (LLDPE) containing varied concentrations of different cyano-OPVs have been prepared (Figure 11.10).[29] Excimers formed upon rapid quenching and plasticisation after melt processing. Increases in dye concentration resulted in an increase of the photoluminescence intensity of the excimer band. Mechanical deformation of cyano-OPV/LLDPE films led to a considerable change in their photoluminescence due to the transformation of aggregates of the photoluminescent guest into molecular mixtures within the polymer, thus causing changes in the fluorescence colour from yellow (intact region) to blue (stretched region). Indeed, monomer to excimer emission ratios increased by a factor of up to 10.

Blends of cyano-OPVs and amorphous polymers (PMMA and poly(bisphenol A carbonate)) were also produced by melt-processing and rapid quenching.[87]

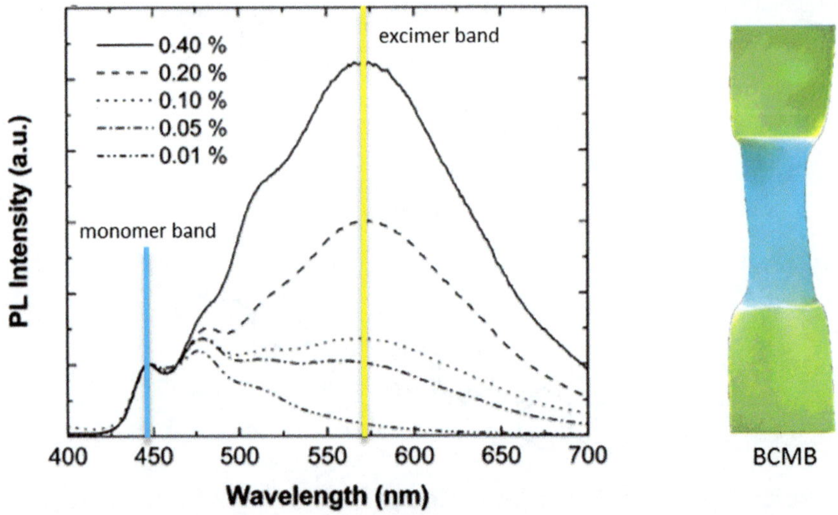

Figure 11.10 Deformation induced colour changes in mechanochromic blends. Photoluminescence emission spectra (left) of freshly prepared blend films of LLDPE and BCMB as a function of dye concentration (given in % w/w), and image (right) of a blend film of LLDPE and 0.20% w/w BCMB stretched at room temperature to a draw ratio of 500% (irradiated with UV light of $\lambda = 365$ nm). All spectra were normalised to the intensity of the monomer peak.
Adapted with permission from B. R. Crenshaw and C. Weder, *Chem. Mater.*, 2003, **15**, 4717–4724.[29] Copyright 2003, American Chemical Society.

Contrary to the previous case, where the semi-crystalline host allowed the formation of excimers by diffusion of the OPV molecules, the formation of excimers in the amorphous polymer was not observed. Thus, blends of cyano-OPV and amorphous polymers display monomer emission. Formation of excimers, and the consequent change in colour of the blend from blue (monomer) to green (excimer), was achieved as a result of phase separation upon annealing the blends at temperatures above their glass-transition temperature (T_g). A similar behaviour was observed in binary blends of cyano-OPVs and polyesters, such as polyethylene terephthalate (PET) and amorphous polyethylene terephthalate glycol (PETG), when polymers were melt-processed with small amounts of the dyes (0.2–2% w/w).[88] In this particular case, the length of the dye's alkyl chains was increased in order to reduce the solubility of the dyes in the polyesters, which resulted in the formation of excimers when blends were annealed. Similarly to previous examples, mechanical deformation of the annealed materials led to changes in the photoluminescence spectra from excimer-rich to monomer-dominated emission. However, the increase in length of the cyano-OPV alkyl chains does not always favour the formation of small aggregates within the polymer matrix. For example, cyano-OPVs with terminal alkyl tails of varied lengths were blended with polyfluorinated polymers.[89] Contrary to the previous case, cyano-OPVs with longer alkyl tails exhibited large aggregates upon blending with fluoropolymers and featured no mechanochromic response. On the other hand, cyano-OPVs with shorter alkyl tails led to the formation of smaller aggregates with pronounced mechanochromic response. This suggests that cyano-OPV nucleation is greatly dependent on the polymer matrix into which they are dispersed, rather than on the dye structure.

The built-in deformation sensitivity of cyano-OPVs upon dispersion in thermoplastic elastomers (polyurethanes) was also investigated.[90] Dyes were dispersed within the polymer by either physical or covalent incorporation. In the first case, only a very modest and irreversible change in the photoluminescence emission was observed upon deformation. Although no molecular mechanism was proposed, the authors attributed such a small change in photoluminescence emission to the presence of large aggregates of guest molecules, which did not disperse properly when deformed. In fact, while excimers are shattered upon polymer stretching (which promotes orientation of single molecules),[92] microsized aggregates are not easily dispersed during mechanical deformation, thus limiting the anisotropic arrangement of the dyes.[93,94] Such behaviour does not seem to occur when the dye is covalently linked to the polymer. The covalent integration of the dyes into the elastomer chains seems to originate in very small aggregates, which results in blends with predominant excimer emission when unstretched. The significant fluorescence change achieved upon deformation was largely reversible.

Cyano-OPVs are not the only fluorescent and photoluminescent dyes that can be used to prepare mechanochromic blends.[22] Alternative dyes are, *e.g.*, bis(benzoxazolyl) stilbene (BBS)[93,95] and perylene derivatives.[96] For instance, Pucci *et al.* produced blends of BBS and polypropylene (BBS/PP) containing different concentrations of the dye.[97] Excimer formation was observed when the concentration of BBS was higher than 0.2 wt%. Similarly to OPV-based blends, BBS/PP films presented changes in their photoluminescence emission upon stretching, with luminescence changing from green to blue in the deformed region, which is the characteristic luminescence of the isolated BBS molecules. Studies were extended to BBS-poly(1,4-butylene succinate) (BBS/PBS) blends.[95] Similar results were obtained after deformation of films containing BBS at concentrations over 0.05 wt%. More recently, BBS was dispersed into thermoplastic polyurethane.[98] Different concentrations of the dye were used during blend preparation by either solution casting or melt compounding approaches. Mechanochromic changes from cyan to dark blue were observed when blend films containing 0.5% of BBS were stretched from 0 to above 300%. The initial emission was fully recovered upon film relaxation, followed by a short annealing at 120 °C.

As many authors suggest, mechanochromic blends may find numerous applications, *e.g.*, as strain sensors or as built-in threshold temperature sensors in polymer-based materials, therefore representing a promising technology for the production of devices with security purposes.

11.3.3.2 Hydrogen Bonds

Hydrogen bridges are weak non-covalent bonds in which a heteroatom of a functional group has an electron density which permits it to attract a proton from another functional group. Tuning the quantity of hydrogen bonds and the nature of the proton donors and acceptors in a polymeric unit is a valuable means of controlling the threshold needed for the material to react to the application of force.[99,100] Although self-reporting materials could be based on the cleavage of single or multiple hydrogen bonds in mechanochromic H-bonding motifs, hydrogen bonds have been more prominently used to tune the threshold of forces needed to activate mechanochromic systems. Kim *et al.* proposed a concept for a self-reporting carbon-fibre reinforced composite by computational modelling.[101] A linker, which contained a chromophore based on a *cis* azo compound, could be attached to the surface of a fibre. This isomer would be stabilised by the formation of a loop *via* hydrogen bonding units in the polymer backbone. The polymer would then continue into the bulk. Upon application of a mechanical load, the system would elongate the loop and the azo chromophore would rearrange into its most stable conformation, the *trans* form. This would result in the desired colour change. Unfortunately, this concept was never translated into experimental work and therefore it remains to be proven that it works in practice. Further work was instead carried out synthesising a thermoplastic elastomer based on a polyurethane-*block*-copolyamide copolymer which presented azobenzenes within its backbone.[102] Upon mechanical damage, the UV absorption changed due to the *cis*-to-*trans* isomerisation of azobenzenes. The copolyamide blocks were introduced as the soft segments of a hard polyurethane network. The system relied on the transduction of forces from the polyurethane network to the copolyamide block, where all the functional groups of the polymer are found. The design of the copolyamides included three types of functionalities: azobenzenes, which gave the self-reporting properties; hydrogen bonding units, which acted as intramolecular and intermolecular latches; and reactive modules, such as amine and carboxyl groups at the chain ends, which would permit the formation of a fibre-reinforced composite. The chromophore used was *p*,*p*′-diaminoazobenzene, which undergoes a photochemical isomerisation from its most stable isomer, *trans*, to the less stable one, *cis*. The *cis* isomer reverts to the *trans* form in the presence of visible light or at room temperature. However, the *cis* form was stabilised *via* the formation of intramolecular hydrogen bonds within the polymer backbone. When the strain was large enough to break these hydrogen bonds, the chromophore was not stabilised anymore, passing irreversibly to the *trans* form. The main drawback of this system is that the half-life of the photo-produced *cis*-form is between 5 and 6 hours, limiting its use in real applications.

A particularly interesting system in which spiropyran containing pre-polymers were end-capped with ureidopyrimidone moieties (UPy) has been recently published by Weng and coworkers.[71] Their work shows that the force-induced activation of the spiopyran is enhanced by the supramolecular H-bonding interactions between UPy moieties, meaning a higher sensitivity towards mechanical stimuli of a certain magnitude. The fraction of mechanically activated mechanophores was around 4 to 5 times higher than in the control polymer without the UPy moieties. The inclusion of UPys permitted the formation of stress-hardening regions in the polymeric material, where the major colour change was observed. This phenomenon can occur in highly oriented polymers or in strain-induced crystallisation. Therefore, the orientation of the polymer chains was analysed by small angle X-ray scattering (SAXS). Before deformation, randomly oriented UPy stacks were observed. As strain was increased, shear caused the UPy stacks to rotate and fragment into smaller pieces. The SAXS patterns further revealed that the fragmented stacks were oriented perpendicularly to the applied force axis with highly orientated soft segments lying parallel to the tensile axis. These regions further nucleated to crystalize in the soft phases. The force needed to break a single UPy dimer is around 100–200 pN, while the strength needed to activate the spiropyran is between 2–3 nN. The resulting crystallites ended up acting as cross-linkers that held a high quantity of the load, and therefore the force was effectively transferred to the spiropyrans, leading to their higher activation.

As a general remark, hydrogen bond moieties have been extensively used in self-healing materials,[100] but not in self-reporting materials. Due to the lability in comparison to covalent bonds (spiropyrans, cyclobutene, cyclooctane), the threshold for the disruption of one hydrogen bond in a mechanochromic motif would be too low for many applications. However, two points should be considered. First, as mentioned above, hydrogen bonding units are interesting for tuning the effectivity of other mechanochromic units. Second, tuning the quantity of hydrogen bonds per potential mechanophore can increase this threshold, thus allowing for the development of new mechanochromic groups.

11.3.3.3 Ionic Interactions

Attractive electrostatic interactions form strong non-covalent bonds between atoms or molecules. These interactions are weaker than covalent bonds and can be reversible. This feature can be exploited for the preparation of self-reporting materials. The Huck group pioneered the study of pressure-sensing surfaces based on electrostatic interactions. The system was based on polyelectrolyte brushes of poly([2-(methacroyl-oxy)ethyl]trimethylammonium chloride) (PMETAC) which were covalently attached to a PDMS surface.[103] The polycationic chains adopted a completely stretched conformation in water due to their high local concentration and due to the repulsion between their charges. Addition of the electrolyte bromothymol blue (BTB), a pH-sensitive dye, led to the screening of these charges. Applying pressure to this kind of surface layer reduced the local volume, collapsed the polymer brushes and, therefore, increased the differential concentration of the dye and

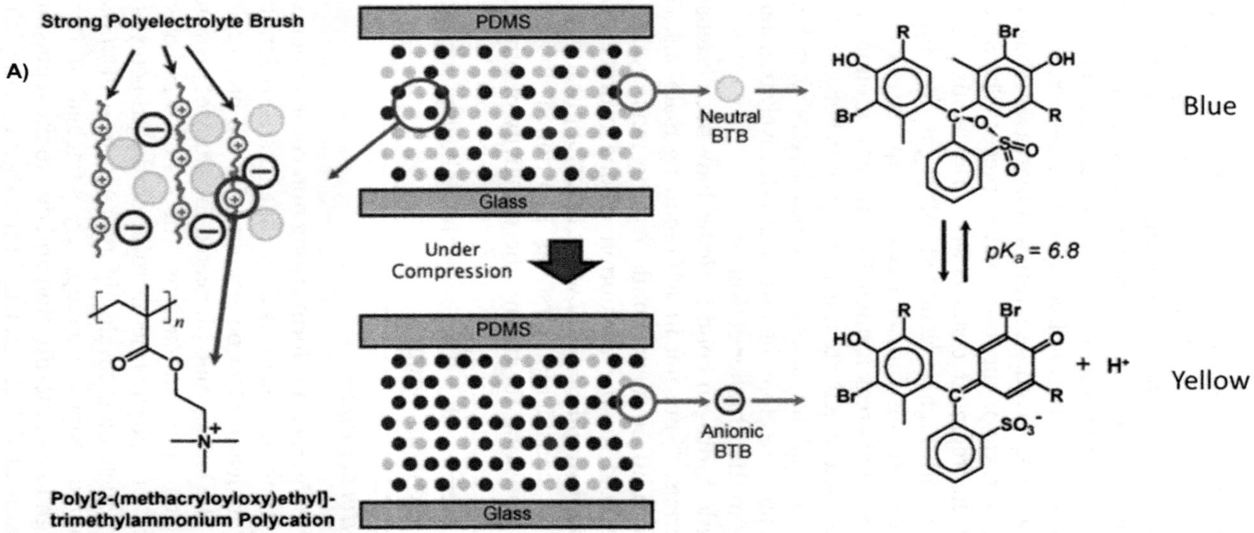

Figure 11.11 Self-reporting pressure sensing surface based on polyelectrolyte brushes with the embedded dye probe BTB. Neutral dye molecules (light circles) convert to anionic species (dark circles) within PMETAC brushes upon compression. Bottom left: chemical structure of cationic polyelectrolyte brush. Right: neutral (top) and anionic (bottom) forms of BTB. Adapted from ref. 103 with permission from John Wiley and Sons. Copyright 2006, Wiley-VCH Verlag GmbH & Co. KGaA.

the polyelectrolyte (Figure 11.11). As a result, dye molecules passed from a non-ionised state, which is yellow, to an ionised state, which is blue, to compensate the new concentration of cations. Once collapsed, the system did not revert to the state prior to the applied pressure because the polymer brushes had collapsed and strong ion pairs had formed. Apart from being an elegant proof-of-concept that demonstrates the feasibility of force transduction onto polymer brushes, the authors suggested that this approach could also be applied to bulk systems if nanoconfinement of the polyelectrolytes can be achieved.

In a follow up study, PMETAC brushes were combined with the fluorescent dye carboxyfluorescein (CF).[104] In this case, the dye was covalently conjugated to the polyelectrolyte. The system was compressed using soft colloidal probe atomic force microscopy (AFM) with an elastomeric particle. This particle was pushed against the composite surface and the area of contact was imaged by confocal laser scanning microscopy (CLSM) (Figure 11.12). The contact area between the polyelectrolyte-covered surface and the probe lost its fluorescence. When the particle was retracted, a fluorescent "rim" was observed around the area that was previously compressed. The authors explain these observations by proposing that CF generates electrostatic non-fluorescent complexes with the quaternary ammonium groups of the polymer when under pressure. Upon compression of the system, the partial concentration of ammonium groups increased, leading to more non-fluorescent complexes. During retraction of the colloidal probe, the polymer adhered to the beads, which induced the stretching of the chains. This resulted in a lower concentration of the ammonium groups and increased the concentration of fluorescent CF, giving a brighter fluorescence at the rim. The simplicity of these polymer-brush systems could be a key advantage for their utilisation, *e.g.*, in low-cost mechanochromic pressure-sensing devices.

11.3.4 Mechanobiochemistry

Mechanobiochemistry comprehends the study of mechanically responsive macromolecules of biological origin, such as proteins, nucleic acids and polysaccharides. The role of those biomacromolecules in translating mechanical signals into chemical responses is paramount in many natural processes, as highlighted in the introduction of this chapter and as summarised in excellent reviews.[6,105] Using biomacromolecules as force-responsive entities in synthetic polymeric materials has attracted particular interest in recent years.[106] The unsurpassed degree of control over biomacromolecular structures allows to precisely tune the force-response of biological mechanophores and to tune their force thresholds.[107] For example, proteins can be selectively stabilised by protein engineering methods, *e.g.* by introducing disulphide bridges[108] or by tuning non-covalent interactions (*e.g.* hydrophobic and electrostatic) between protein subunits or protein domains, while DNA behaves as a molecular spring, which can be used as a tension gauge tether with tuneable tension threshold.[109] Moreover, a growing number of biomacromolecules has been characterised by single

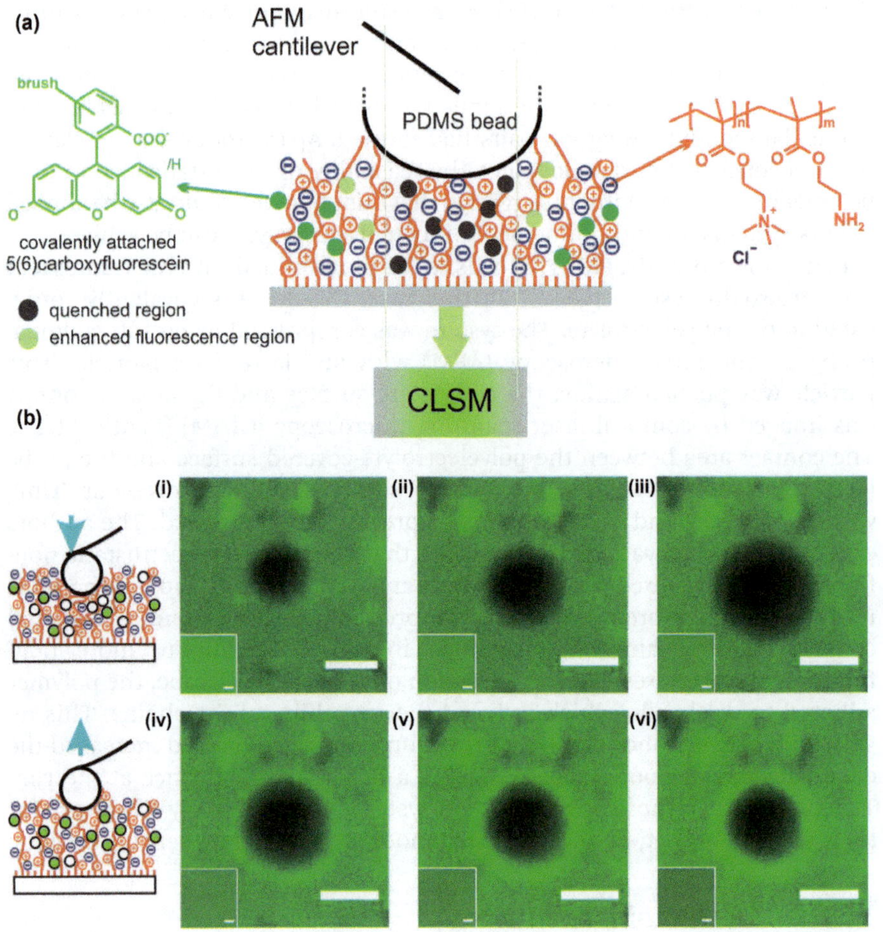

Figure 11.12 Self-reporting pressure sensing surface based on PMETAC polyelectrolyte brushes and carboxyfluorescein. (a) Experimental design to determine the fluorescence-based output upon mechanical compression of functionalised polyelectrolyte brushes using a soft colloidal probe. (b) Fluorescence images obtained from a compression experiment on fluorescently labelled PMETAC in water: i–iii, during approach, with loads of 0.7, 2.8, and 5.6 μN; iv–vi, during retraction, with loads of 2.8, 0.7, and −1.4 μN. Scale bar: 5 μm.
Adapted from ref. 104 with permission from John Wiley and Sons. Copyright 2011, Wiley-VCH Verlag GmbH & Co. KGaA.

force spectroscopy, so that detailed force maps are available that describe the threshold forces to unfold a protein along various pulling axes.[110] Many methods have been developed over the years to synthesise hybrid materials of synthetic polymers and proteins, peptides, nucleic acids, *etc.*[111–113] Therefore, it has become feasible to prepare mechano-responsive materials that rely on mechanobiochemical effects.

Bruns *et al.* pioneered self-reporting materials based on the mechanical perturbation of proteins within polymeric matrices.[114,115] The protein cage thermosome (THS), a chaperonin from the archaea *Thermoplasma acidophilum*, is composed of sixteen protein subunits that form two hemispheres with the ability to enclose macromolecular guests (Figure 11.13).[116] THS is approx. 16 nm in diameter. The hemispheres of chaperonins are connected through a mechanically weak plane, so that the two halves of these protein cages can be separated from each other by mechanical forces, as demonstrated in AFM experiments.[117] THS was turned into a sensor for mechanical deformation by covalently linking two proteins, enhanced cyan fluorescent protein (eCFP) and enhanced yellow fluorescent protein (eYFP), into the

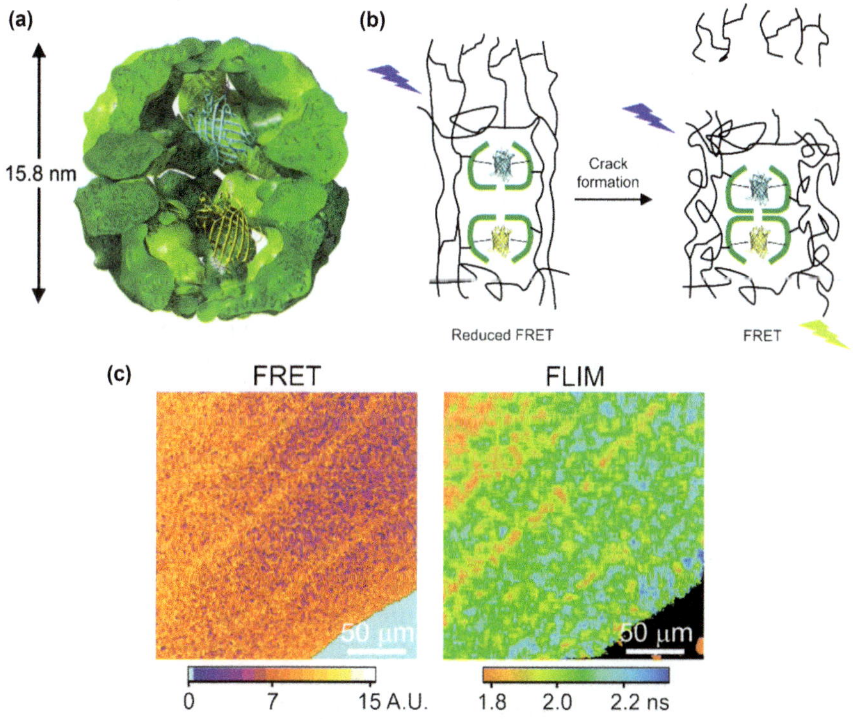

Figure 11.13 The protein cage THS, in combination with a pair of encapsulated fluorescent proteins, acts as a reporting system for structural deformation in a polymeric material. (a) Cut away model of THS in its closed conformation with eCFP and eYFP covalently bound into the protein cage. (b) Schematic depiction of the proposed sensing mechanism (green thick lines: THS; black lines: polymer network). The formation of cracks releases internal stresses of the polymer and the embedded chaperonin in the surroundings of the cracks, which increases FRET. (c) FRET intensity and FLIM images of microcracks surrounding a fracture site in a polyacrylamide/THS-eYFP-eCFP hybrid material. Adapted from ref. 114 with permission from John Wiley and Sons. Copyright 2009, Wiley-VCH Verlag GmbH & Co. KGaA and ref. 115. Copyright 2011, Swiss Chemical Society.

cavities of THS. The proximity between the two fluorescent proteins allows for fluorescence resonance energy transfer (FRET). FRET occurs when an excited fluorescence donor transfers energy to a fluorescence acceptor without emitting radiation, causing a reduction in donor emission and an increase in acceptor emission. FRET efficiency is highly distance-dependent and does not occur if the donor–acceptor distance exceeds 10 nm. Thanks to these properties, FRET has found various applications as a "molecular ruler" in areas such as molecular imaging[118] and biosensors.[119–121] The THS-eCFP-eYFP conjugate was modified with acrylamide groups and then copolymerised into polyacrylamide gels. Upon drying, transparent and brittle polymer samples were obtained. The samples were uniaxially strained until they broke and analysed *post mortem* with fluorescence microscopy and fluorescence lifetime-imaging (FLIM). FRET efficiency of the protein cage with its fluorescent guests increased 1.5-fold in the vicinity of microcracks, while FRET efficiency in undeformed regions remained low (Figure 11.13).[114,115] The observation can be explained by the protein cage adopting a deformed conformation in the polymer specimen due to local stress build-up during drying. In the case where a force strong enough to create cracks was applied, the polymer and the chaperonin in the vicinity of the crack relaxed, causing the protein cage to recover its initial conformation, thus increasing FRET.

In this system, THS acted as a mechano-responsive scaffold, while the fluorescent proteins were needed to detect the deformation of the protein cage. Their structure remained unperturbed. However, fluorescent proteins are mechanophores themselves, as proven by single force spectroscopy and by computational methods.[108,122–126] They may have their fluorescence tuned upon external stimuli due to conformational changes within the protein structure. Changes in the surroundings of the chromophore greatly affect the protein fluorescence, making fluorescent proteins the ultimate biological mechanosensor. Green fluorescent protein (GFP) is the archetype fluorescent protein from which most other coloured fluorescent proteins were derived by genetic engineering.[127,128] The GFP structure contains eleven β-strands displaced in order to form a β-barrel (Figure 11.14). In the centre of the barrel, a helix contains the chromophore 4-(*p*-hydroxybenzylidene)imidazolidin-5-one.[129,130] The chromophore is composed of Ser65, Tyr66, and Gly67 residues[131] and stabilised through interactions (hydrogen bonding) between Thr203 and water molecules.[132] Mutations of those residues lead to genetic derivatives. For example, mutation in Tyr66 to Trp originates the blue fluorescent protein (BFP), whereas the conversion of Thr203 into Tyr results in the yellow fluorescent protein (YFP).[21,132,133]

Dietz and Rief studied the unfolding mechanism of GFP using single-molecule force spectroscopy.[122] These authors showed that mechanical unfolding of the fluorescent protein occurs through at least two different metastable intermediate states. The first unfolding step (detachment of the small helix at the top of the β-barrel) occurs at a force of approx. 35 pN. The protein can withstand a maximum load of 104 ± 40 pN at a pulling speed of 300 nm s^{-1}, above which the protein unfolds into the denatured state.

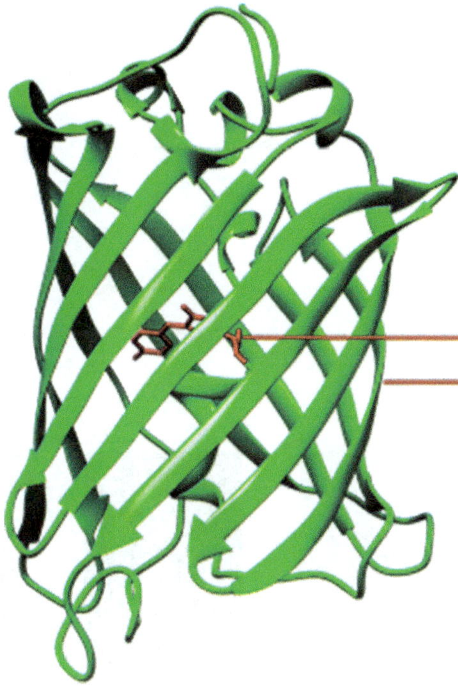

Figure 11.14 Green fluorescent protein (GFP) structure. β-Barrel formed by 11 β-strands that contains a 4-(*p*-hydroxybenzylidene)imidazolidin-5-one chromophore in its centre.

For eYFP, Fernandez and coworkers determined an unfolding force of approx. 60 pN.[123] As the fluorescence is closely linked to the native state of the protein, mechanical unfolding switches its fluorescence off.[126]

Based on those findings, Bruns and coworkers implemented fluorescent proteins as mechanophores in polymeric materials.[31,134] In order to achieve efficient force transfer from the material onto the proteins, they were placed at the interface between fibres and the polymer in fibre-reinforced composites. More specifically, eYFP was covalently linked to glass fibres[134] and carbon fibres,[31] which were then embedded in an epoxy resin. During curing, the resin reacted with functional groups on the surface of the protein, thus creating covalent bonds between the polymer and the protein. Low velocity impacts perpendicularly to the fibre orientation induced the formation of microdamages within the composites, while the composite surface remained intact. Fluorescence vanished from the fractured regions of the fibres and also from the surrounding areas. This observation can be attributed to the debonding of the epoxy resin from the fibre surfaces, which induced unfolding of the eYFP as depicted in Figure 11.15. A difference between carbon fibres and glass fibres is that carbon surfaces tend to quench the fluorescence of fluorophores. Fortunately, no considerable quenching was observed in eYFP-coated carbon fibres, as the three-dimensional

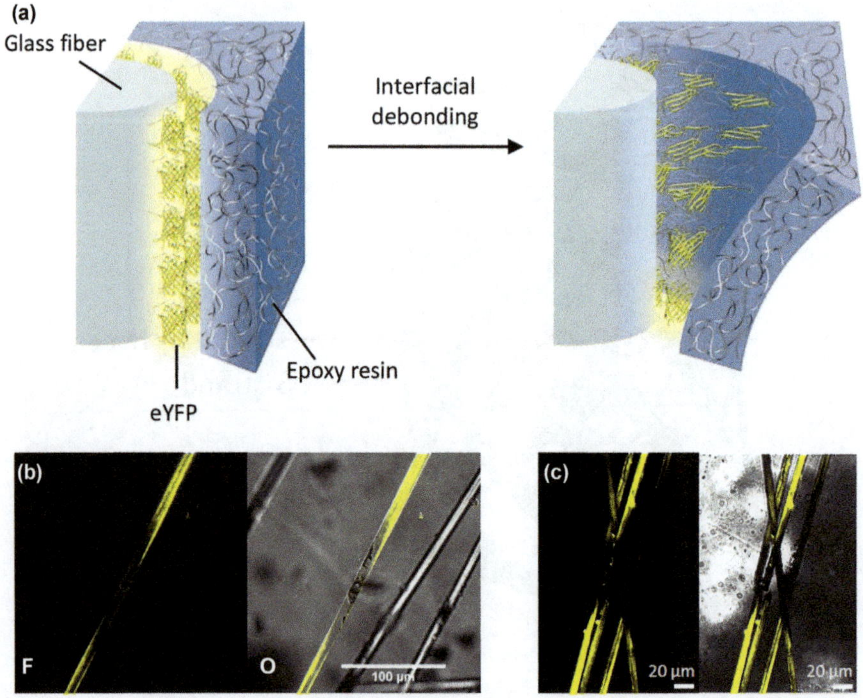

Figure 11.15 Fluorescent protein as a mechanophore at the fibre-epoxy resin interface in self-reporting fibre-reinforced composites. (a) The formation of microdamages promotes interfacial debonding between resin and fibre, therefore causing the protein to unfold and to lose its fluorescence. (b) Confocal fluorescence microscopy image of a damaged glass fibre-eYFP/epoxy composite. (c) Z-stack projection of confocal fluorescence microscopy images of a damaged carbon fibre-eYFP/epoxy composite. (F: yellow fluorescence channel, O: overlay of fluorescence and transmission images).
Adapted from ref. 134 with permission from John Wiley and Sons. Copyright 2013, Wiley-VCH Verlag GmbH & Co. KGaA and ref. 31 with permission from the Royal Society of Chemistry.

structure of the protein kept the fluorophore sufficiently distant from the carbon surfaces.

Longo *et al.* developed a GFP-based material with a mechano-responsive surface.[106] In this approach, modified GFP was covalently linked to PEG brushes grafted onto functionalised PDMS. Unidirectional stretching of the GFP-modified PDMS surfaces by 10 to 30% resulted in a decrease in fluorescence of 23 to 60%, respectively. Interestingly, the materials were able to recover their fluorescence when relaxed to 0% strain. These results suggest that changes in fluorescence upon stretching are due to small and reversible conformation changes of the GFP molecules.

Not only stretching, but also pressure has an effect on the structure of YFP. A shift in the fluorescence peak of YFP from 527 to 510 nm was observed as the pressure on protein solutions was increased from ambient to values above 350 MPa.[135] In YFP, the chromophore undergoes π–π interactions with the phenol group of Tyr203, which results in a shift of the fluorescence to longer wavelengths (compared to GFP).[128] The pressure resulted in conformational changes, which reduced the influence of the phenol group over the chromophore. In order to better understand that phenomenon, Gruner *et al.* evaluated the motion of the chromophore residues upon application of high pressures.[136] As such, protein crystals were obtained at different pressures, cooled down to 77 K to retain their structures, and characterised using a custom microspectrophotometer. The YFP β-barrel is composed of two atomic clusters (Figure 11.16), whose positions were observed to vary with pressure. Tyr203 and the main chromophore attach to different clusters and, therefore, motion and rotation under pressure cause perturbations within the protein structure, thus resulting in a fluorescence shift. The pressure

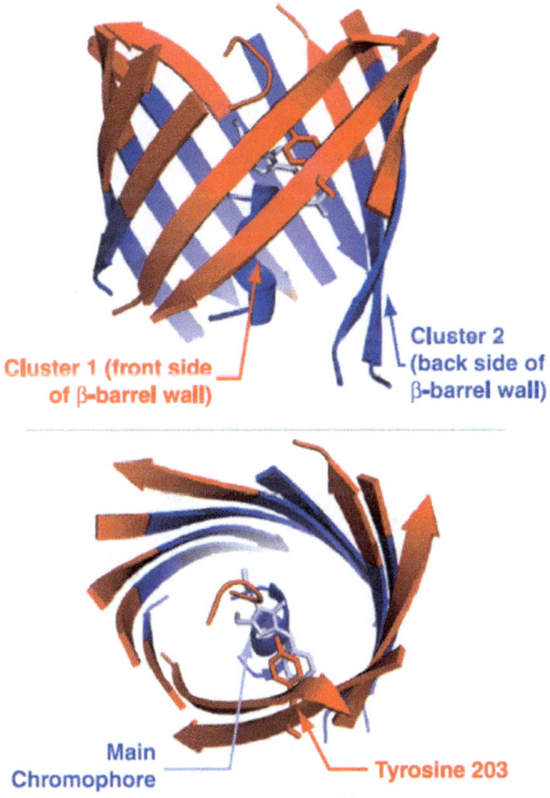

Figure 11.16 Structural clusters in the YFP β-barrel are responsible for the protein's response to pressure. Tyr203 is attached to cluster 1, whereas the main chromophore is attached to cluster 2. Pressure deforms the protein so that the relative position of the two clusters changes, which results in a hypsochromic shift of the fluorescence.
Reprinted from B. Barstow *et al.*, Coupling of pressure-induced structural shifts to spectral changes in a yellow fluorescent protein, *Biophys. J.*, **97**, Copyright 2009 with permission from Elsevier.[136]

response of YFP was exploited in polymeric materials by Brantley *et al.* YFP was embedded into PMMA to obtain mechano-responsive biocomposites.[137] Compression of the biocomposites resulted in a gradual shift in the fluorescence from 539 nm (at 0 MPa) to 534 nm (at 360 MPa). Again, distortion of the interaction between the chromophore and Tyr203 were most likely the cause of the hypsochromic shift.

Recently, Watanabe *et al.* developed a circular permuted variant of YFP (cpYFP) capable to sense and report intramolecular strain in solution.[133] YFP was split into two parts, one containing the chromophore and the other containing Tyr203. These fragments were linked with a flexible peptide linker (at the top of the β-barrel), which resulted in changes of the emission spectrum. Inclusion of a peptide bridge pair (β-hairpin) between the N- and C-termini of the protein, which were initially apart, recovered the yellow fluorescence that resembled that of the wild-type YFP (Figure 11.17). Application of force to the protein increased the distance between the

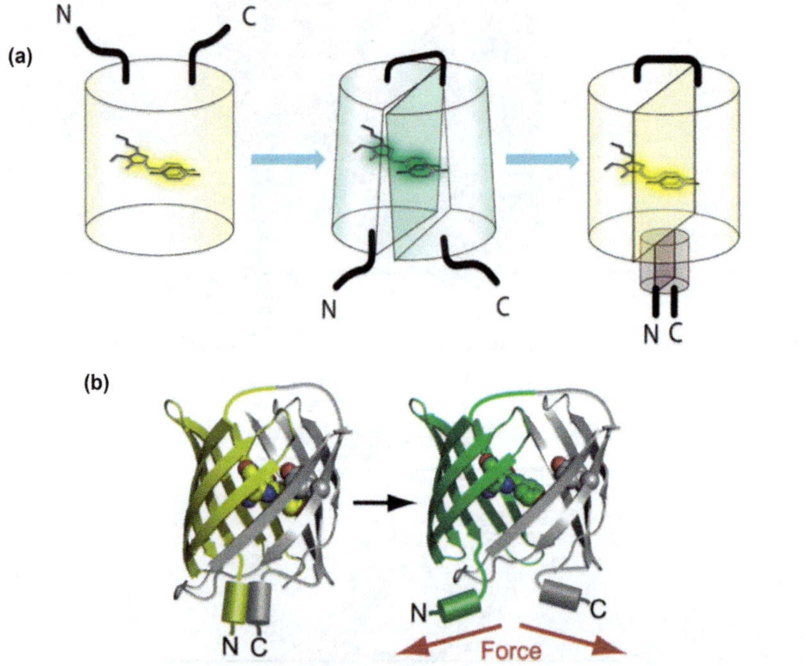

Figure 11.17 Strategy to report intramolecular strain with a fragmented YFP. (a) Construction of the YFP strain sensor: YFP is split into two parts by circular permutation, one containing the chromophore and the other one containing Tyr203. The fragments are linked together with a flexible linker (emission spectrum shifts) and the N- and C-termini are fused (yellow fluorescence is recovered). (b) Strain-sensing: Tyr203 is displaced relative to the fluorophore under force causing a hypsochromic shift in the fluorescence.
Adapted from ref. 133 and 21 with permission from the Royal Society of Chemistry.

two termini. Therefore, it also increased the space between the chromophore and Tyr203, promoting changes in fluorescence. The split-YFP provided an interesting tool to measure intramolecular strain. Fusion of the modified cpYFP with kinesin, a microtubule-associated motor protein, resulted in a chimera that showed a shift in the excitation spectrum when exposed to adenosine triphospate (ATP). This was credited to variations in the YFP conformation due to the external forces that were exerted by the motor protein during hydrolysis of ATP. This system elegantly demonstrates how advanced protein engineering can be used to create novel biomechanophores. It has, however, not yet been employed as a force or deformation sensor in solid materials.

Proteins that change their fluorescence wavelength, intensity, or FRET efficiency upon mechanical deformation are versatile entities for the creation of self-reporting materials, and the examples that are discussed herein highlight their potential. For instance, the utilisation of fluorescent proteins as mechanophores in fibre-reinforced composites can be implemented as a safety feature capable to report and monitor damage propagation and stress distribution, thereby preventing materials failure. However, processing should be carefully conducted as proteins can denatured upon contact with solvents and at high temperatures. Nevertheless, fluorescent proteins are robust proteins and stable up to 70 °C in aqueous solution. For instance, the midpoint for the unfolding and denaturation transition of GFP was measured to be 78 °C.[138] When embedded into a solid polymer, the stability of proteins increases substantially.[139] Thus, biohybrid materials that are stable to at least 100 °C should be possible.

Biological mechanophores should not only be considered as functional additives in polymeric materials, but also as concept generators for the design of novel chemical mechanophores. For example, it appears to be feasible to synthesise a supramolecular mechanophore which has a fluorescence emission shift upon displacement of a neighbouring functional group, similar to that of YFP and circular permutated YFP.

11.3.5 Mechano-responsiveness at the Microscopic Level

A bruise forms when a blunt object hits the skin. Similarly, tissue starts to bleed when it is cut. Both processes are caused by liquid blood flowing out of damaged blood vessels into or onto the surrounding tissue, starting a healing process and inducing a clearly visible colour change. These mechanisms can be mimicked in synthetic materials by embedding reservoirs into the solid material that are filled with liquid healing and colour-forming reagents. The most common approaches are liquid-filled microcapsules, microchannels, and hollow fibres, which will be reviewed in detail in the following sections. When the composite material is damaged, *e.g.* by an impact or by a cut, the liquid is released from these reservoirs into the site of damage, thus causing self-healing and a mechanochromic response.

11.3.5.1 Composites Containing Dye-filled Microcapsules

Microcapsules have become tremendously popular ever since White, Sottos, Moore, and coworkers published their ground-breaking report on self-healing composites based on microcapsules filled with healing agents.[140] The field is summarised in good reviews,[26,141] so we will only highlight some important work that demonstrates the underlying principles and current trends, with a focus on self-reporting materials. Before going into details, it should be noted, however, that the concept of force-triggered dye-release from microcapsules is not new. For instance, water-soluble microcapsules consisting of gel-like polymers have been used for carbonless copy paper since 1957.[142] These microcapsules are filled with a leuco dye, which is liberated and gets in contact with a developer when pressure is applied. As paper consists of biopolymers, writing on carbonless copy paper can be considered as one of the oldest applications of self-reporting polymeric materials.

In their initial report, White *et al.* encapsulated a monomer for ROMP into urea-formaldehyde microcapsules.[140] The capsules were dispersed into an epoxy resin containing ruthenium-based Grubbs catalysts. Upon infliction of damage, the microcapsules broke and released their content into the cracks, where the monomers got into contact with the catalyst. Polymerisation formed a polymer within the crack. The self-healed specimen recovered up to 75% of their original toughness. Although the focus of this study was on the recovery of the mechanical properties after damage, the authors also added a dye to visualise the release of the capsule content into the micro-cracks, thus also creating a self-reporting material. However, the dye had the same colour within the capsules and when released into the material. Therefore, the colour contrast between damaged and non-damaged areas was low. In order to increase the contrast, recent developments aim at inducing a colour change when the capsule content is released.

This can be achieved by encapsulating a dye that reacts with some component of the polymer matrix, in analogy to the principle of carbonless copy paper. For example, pH responsive dyes that are nearly colourless within the capsules can develop colour when getting in contact with acidic or basic groups in the polymer matrix. Such "bruising coating" has been described in the patent literature,[143] and was considered as impact detection coating for aircrafts.[144,145] Moreover, Vidinejev *et al.* reported self-reporting acrylic coatings in which impact damage caused microencapsulated crystal violet lactone to get into contact with methyl 4-hydroxybenzoate or silica gel in the bulk of the polymer.[146] The developers provided an acidic environment, so that the leuco dye transformed from the colourless form to its blue isomer. The same concept was applied to render composites of nylon fabric and epoxy resin self-reporting for impact damage.[147]

White, Sottos, and coworkers managed to greatly enhance the colour contrast and the colour stability in such pH-switchable systems.[27] 2′,7′-Dichlorofluorescein (DCF) reacts with amines to form a colourful derivative,

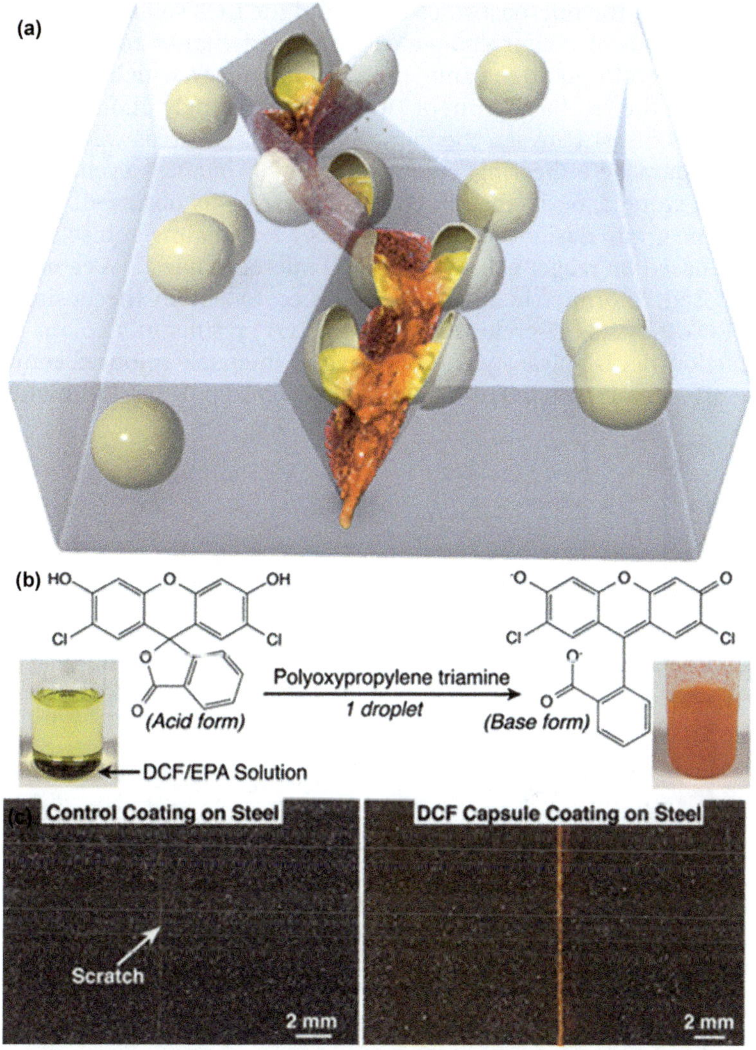

Figure 11.18 Microcapsule-based self-reporting material: amine cured epoxy coating mixed with microcapsules that contained 2′,7′-dichlorofluorescein (DCF). (a) Schematic of the release of DCF solution from broken microcapsules into a crack or scratch where the dye encounters the amine groups of the resin, resulting in colour change. (b) Colour change of a DCF solution in the presence of an amine. (c) Self-reporting properties of the described material.
Adapted from ref. 27 with permission from John Wiley and Sons. Copyright © 2016, John Wiley and Sons.

passing from yellow to bright red (Figure 11.18). DCF was dissolved in ethyl phenyl acetate and then encapsulated into double-walled polyurethane/poly(urea-formaldehyde) microcapsules. They were dispersed in an epoxy resin containing an amine-based curing agent. Upon mechanical damage of the cured films, the microcapsules broke and the DCF solution spread along the crack or cut. A colour change was observed thanks to the proton exchange between DCF and the amine containing matrix, which resulted in the formation of the red basic form of DCF. As it is strongly photoluminescent, the colour contrast between the damaged and intact areas was high. Moreover, the damage-indicating colour was highly stable. Even after eight months of storage, no change in colour intensity was observed.

The same group designed a simultaneously self-healing and self-reporting material based on reagent-filled core–shell microcapsules.[30] A catalyst and a ROMP-based colorimetric assay were selected to obtain the desired properties. ROMP of 1,3,5,7-cyclooctratetraene (COT) results in a colour change. The reaction yields polyacetylene, which is an intensely coloured conjugated polymer (Figure 11.19). COT was incorporated into poly(urea-formaldehyde)

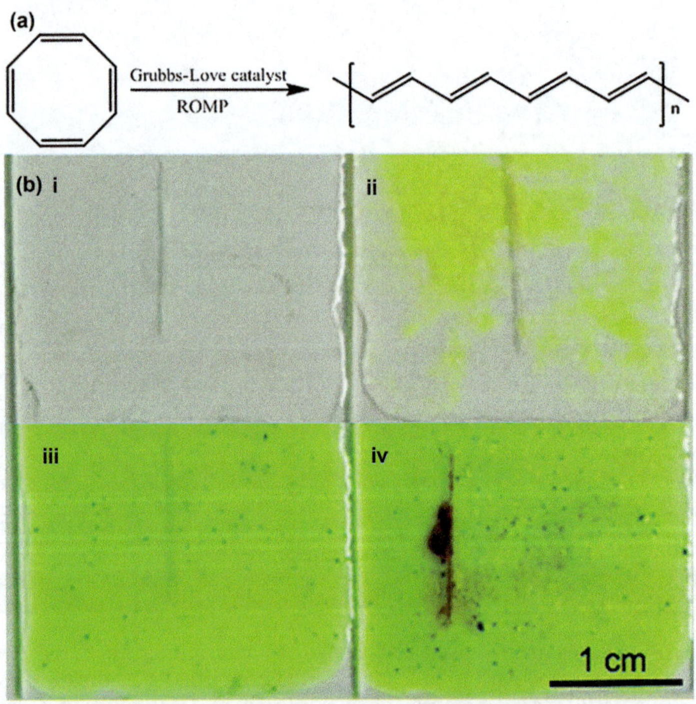

Figure 11.19 Formation of polyacetylene as a means to visually indicate damage in microcapsule-based self-reporting and self-healing films. (a) The colourless COT monomer undergoes ROMP in the presence of the Grubbs–Love catalyst to yield purple polyacetylene. (b) Photographs of damaged poly(acrylic acid) films containing (i) no catalyst or microcapsules, (ii) only COT-filled microcapsules, (iii) only Grubbs–Love catalyst, and (iv) both COT-filled microcapsules and Grubbs–Love catalyst. All the films were cut in the centre and only (iv) showed the expected colour change.
Adapted with permission from S. A. Odom, A. C. Jackson, A. M. Prokup, S. Chayanupatkul, N. R. Sottos, S. R. White and J. S. Moore, *ACS Appl. Mater. Interfaces*, 2011, **3**, 4547–4551.[30] Copyright (2011) American Chemical Society.

via emulsification condensation polymerisation to form substrate-filled core–shell microcapsules. These capsules were then dispersed, together with Grubbs–Love catalyst, into poly(acrylic acid) films. The reaction started when the film was scratched with a razor blade. A red-orange colour was observed when the film was damaged, indicating the progress of the chemical reaction. After more time, the damaged region became purple-black. In addition to this visual indication of damage, the formation of the polymer healed the damaged area.

The Turri group showed another approach towards mechanical sensing based on dye-filled microcapsules.[148] In this case, the mechanical damage became visible when UV-A light was applied. For this purpose, polyurea microcapsules with the ability to screen UV light were employed. These microcapsules were filled with a solution of a spiropyran-based dye in sunflower oil, and then embedded within a poly(pentaethyleneglycol monomethacrylate) matrix. Upon local mechanical damage, the microcapsules released their content. Because of the ability of sunflower oil to diffuse through the cracks of the polymer matrix, the entire damaged surface was covered with the spiropyrane. No colour change was observed until the material was irradiated with UV-A light inducing the electrocyclic ring opening to the merocyanine form (Figure 11.20). The photoreaction was concluded within seconds, passing from yellow to purple, and only occurred in the damaged areas where the dye was not screened from UV light within intact capsules.

Another example for microcapsule-based self-reporting materials involved the use of charge-transfer complexes (CTCs).[149] CTCs form between an electron donor and an electron acceptor. The transfer of charge results in the attraction between the two species and a change in colour. Poly(ureaformaldehyde) microcapsules were filled separately with a donor, hexamethylbenzene (HMB), and with an acceptor, chloranil (CA) (Figure 11.21). Both dyes were dissolved in toluene. The capsules were embedded into a PDMS matrix. They ruptured and released the dyes upon mechanical stimulus. This resulted in the formation of CTCs which changed the colour of scratches and of deformed areas from yellow to red.

As these examples show, dye-filled microcapsules offer the advantage of the self-reporting signals being permanent and highly localised. A key factor for the translation of such technology into practical applications will be a high stability of the systems also under outdoor conditions (*e.g.* long exposure to sunlight, changes in humidity, temperature range between −30 °C and 60 °C or more) in order to prevent a premature release of the dye/healing agent and to avoid that the encapsulated reagents lose their reactivity and colour over time.

11.3.5.2 Microvascular and Hollow Fibre Composites

A drawback of dye-filled microcapsules is that, once released, their content cannot be refilled. Therefore, it is not possible to indicate or heal damage a second time in the same area, *e.g.* after repetitive impacts or scratches. An

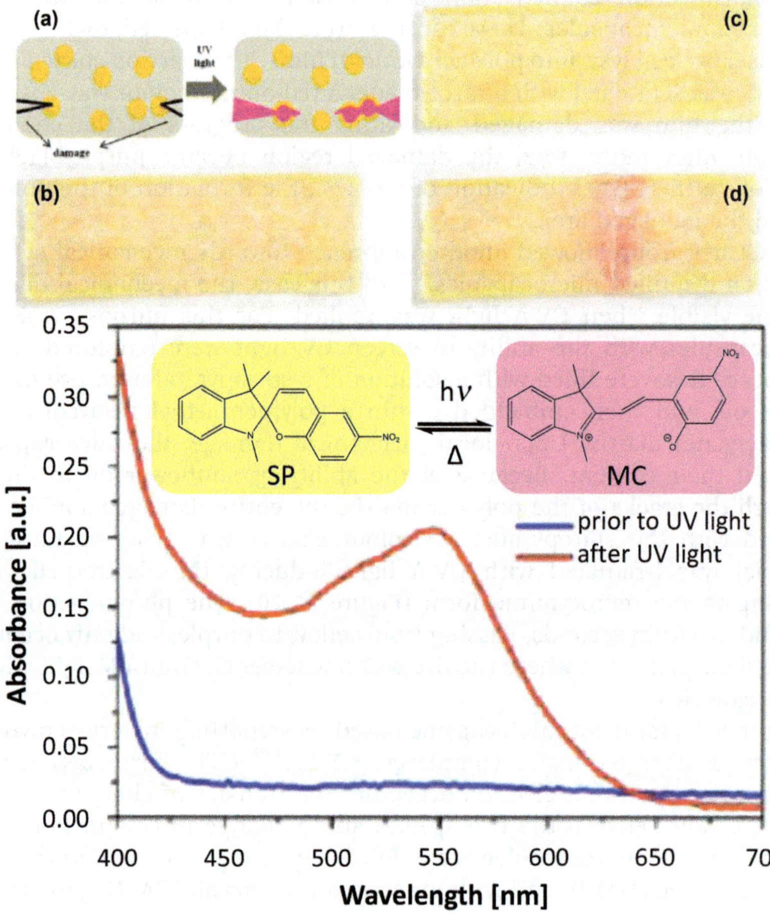

Figure 11.20 UV-responsive self-reporting polymer based on spiropyrane-filled microcapsules. (a) Scheme showing the release of dye/sunflower oil solution from damaged microcapsules into cracks and subsequent colour development induced by UV light. (b) Image of intact poly(pentaethyleneglycol monomethacrylate) with spiropyrane-filled microcapsules. (c) Scratched film. (d) Colour change in damaged area after irradiation with UV-A light. (e) UV-vis spectra showing the increase in absorbance around 550 nm, which results from the electrocyclic ring opening of the spiropyran to its merocyanine form. Adapted with permission from B. Di Credico, G. Griffini, M. Levi and S. Turri, *ACS Appl. Mater. Interfaces*, 2013, **5**, 6628–6634.[148] Copyright 2013, American Chemical Society.

already bruised or cut skin, however, can take another blow, simply because blood will again flow to the damaged site through a network of blood vessels. This observation served as inspiration to create composites with a network of microchannels through which healing agents or dye solutions can be

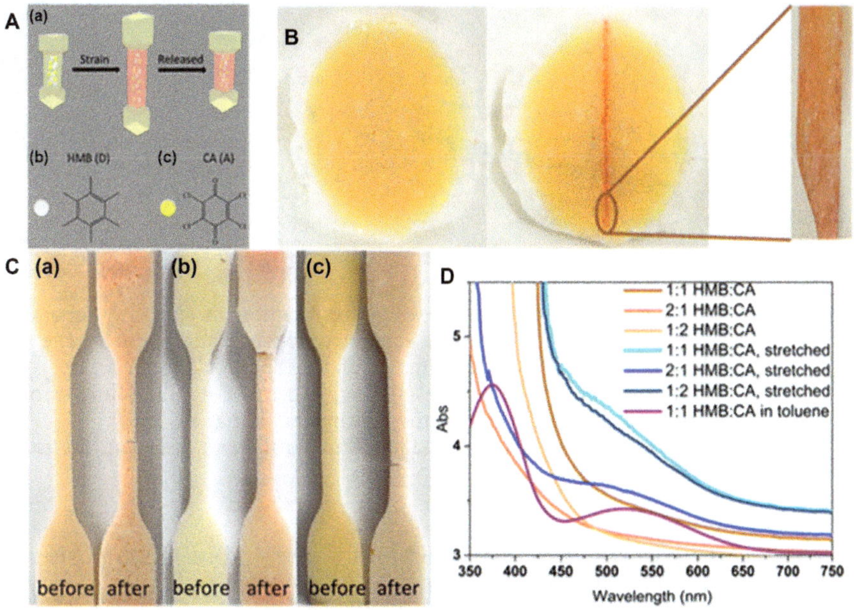

Figure 11.21 Charge-transfer complexes (CTCs) as indicators for damage and deformation in microcapsule-filled PDMS. (A) A mixture of microcapsules containing HMB (donor) or CA (acceptor), respectively, is embedded in PDMS films. Mechanical perturbation of the capsules releases the two species, which then form CTCs in the material. (B) Local colour change after cutting a PDMS film with a razor blade. (C) PDMS films before and after uniaxial tensile tests with different proportions of donor capsules and acceptor capsules. (D) UV-vis spectra showing changes in the absorbance corresponding to the formation of CTCs after stretching of the films.
Adapted with permission from A. Lavrenova, J. Farkas, C. Weder and Y. C. Simon, *ACS Appl. Mater. Interfaces*, 2015, 7, 21828–21834.[149] Copyright 2015 American Chemical Society.

transported to the site of damage. White, Sottos, Moore, and collaborators created such microvascular architectures for repairing microfractures in epoxy based composites (Figure 11.22).[150,151] As such, an epoxy coating mixed with solid catalyst particles was deposited on a three-dimensional microvascular substrate filled with a liquid healing agent. The formation of cracks induced by mechanical forces caused the microchannels to rupture. The healing agent that was released from the channels polymerised upon interaction with the catalyst particles in the coating, thus rebonding the crack faces.

Since these early reports, the design of vascularised structural composites has been greatly refined and the reader is referred to an excellent review on this topic for an overview of the field.[152] It should be mentioned that dyes are often added to healing agents as a means to investigate the efflux of healing agents from the microchannels.[153] For example, three-dimensional

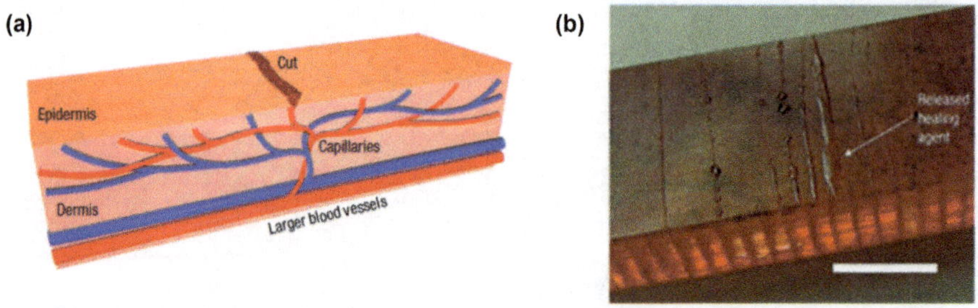

Figure 11.22 Self-healing material mimicking the function of skin. (a) Schematic diagram of skin, showing a capillary network and a cut in the epidermis layer. (b) Optical image of the self-healing structure upon formation of cracks in the coating; excess of healing fluid can be observed on the surface of the coating (scale bar = 5 mm).
Adapted by permission of Macmillan Publishers Ltd: *Nat. Mater.*,[151] Copyright 2007.

microvascular networks were integrated into fibre reinforced composites by stitching sacrificial poly(lactic acid) filaments into woven glass fibre fabric.[154] Embedding of these fabrics into epoxy resin, followed by vaporization of the sacrificial filaments resulted in a vascularised composite. When the two interwoven channels were filled with two components of a healing chemistry, the composites could be healed several times after interlaminar delamination. The mixing of the two healing agents at damaged sites was visualised by adding two fluorescent dyes, nile red and fluorescein, to the healing agents. With this method, it could be shown that a herringbone pattern of the channels was more effective to deliver and mix the two components at a damaged site than a parallel layout of the channels.

Also in the context of bleeding materials, efforts have been made to detect and repair barely visible impact damage (BVID) in fibre-reinforced composites. Such damage is caused by low velocity impacts, *e.g.* by a dropped tool during service and maintenance of a composite structure. It often results in delamination and matrix cracking, leading to a reduction in the local strength and eventually catastrophic failure.[24] It is therefore of ultimate importance to detect and repair these damages at their early stages.[155] In light of this, bleeding fibrous composites that mimic the self-reporting and self-repairing properties of skin were developed (Figure 11.23).[23,24,156] In the original approach by Bond and Pang, epoxy-based plies containing glass hollow fibres filled with either uncured resin or a mixture of hardener/UV fluorescent dye were alternately stacked together through a hand lay-up process to form laminates.[23,24] The authors theorised that, upon damage, BVIDs would be easily visualised by bleeding of the fluorescent dye, whereas mechanical properties would be restored by a self-activated healing process. Impact tests and flexural tests revealed that the UV fluorescent dye was released from the fractured hollow fibres and bled into the damaged area. Simultaneously, cracks were healed by interaction of uncured resin with a

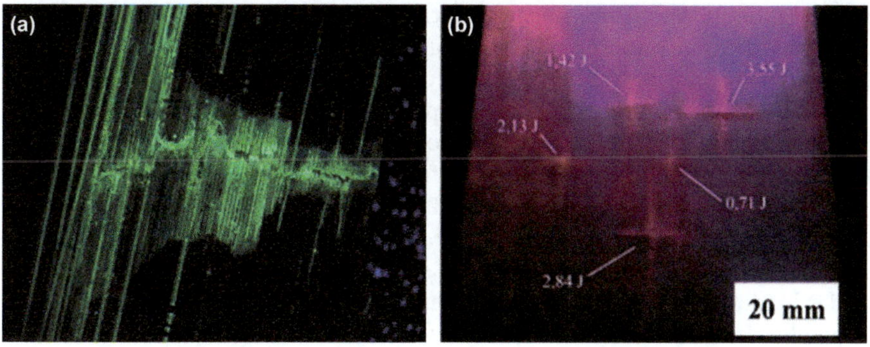

Figure 11.23 Bleeding of fluorescent dye from hollow glass fibres into a damaged fibre-reinforced composite allows for visualization of damage. (a) Fluorescent microscopy image of a bleeding composite after flexural bending that led to fracture of the specimen (45× magnification). (b) Photograph of a self-reporting composite after impact events. Photo taken from the opposite side of impacts under UV-C light illumination.
Reprinted from J. W. C. Pang and I. P. Bond, 'Bleeding composites' – damage detection and self-repair using a biomimetic approach, *Composites, Part A*, **36**, Copyright 2004 with permission from Elsevier[24] and from S. King and T. Czigany, Damage detection and self-repair in hollow glass fiber fabric-reinforced epoxy composites *via* fiber filling, *Compos. Sci. Technol.*, **99**, Copyright 2014 with permission from Elsevier.[156]

hardener that was released from damaged hollow fibres. While it is clear that the bleeding agent allows for the rapid visualisation of the damage at the composite surfaces upon irradiation with UV light, the healing mechanism may not eradicate the effects of damage permanently but rather inhibit further crack propagation.

11.4 Conclusions and Future Perspectives

In nature, self-reporting events can indicate that a natural structure (*e.g.* tissue, skin, organs) has experienced mechanical stress or damage. Frequently, the response of natural systems to mechanical perturbation is accompanied by visual indicators, such as changes in colour, fluorescence, or chemoluminescence. This chapter shows how these natural processes can serve as inspiration for the development of a wide range of polymeric self-reporting materials with mechanochromic properties. Such artificial materials respond to mechanical stimuli by either chemical or physical means. In these systems, mechanical forces cause bond scission of mechanophores, the disaggregation of molecules in supramolecular systems, perturbation of the structure of proteins, or rupture of microcapsules, hollow fibres and microchannels. These events result in new chemical entities, supramolecular conformations, or the release of a dye into damaged areas, which cause optically detectable changes in colour, fluorescence, or chemoluminescence.

Such signals allow to map and quantify damage, stress, and strain distributions within the materials, and some of the examples reviewed herein show how useful this can be, *e.g.* to guide the design of soft robotics, for non-destructive damage evaluation in composite materials, or the preparation of impact-detection coatings. As self-reporting materials autonomously report mechanical stress without the need for sophisticated equipment such as ultrasound probes, X-rays, or micro CT scanners, they represent a "first-line defence" in the detection of critical microdamage in materials and can serve as a tool to monitor the evolution of stress and crack growth in real time under load. The future use of self-reporting materials will have deep impact at two levels. The most obvious effect will be as a safety measure. Materials with mechanochromic properties are promising for the manufacture of devices in which, due to extensive use, a certain part or area will be continuously subjected to mechanical stress. When the mechanical stimulus achieves a threshold for the material to show its self-reporting properties, we will know it is the moment to replace it. Changing a material at the right moment, before catastrophic failure, can save lives in critical situations, such as in airplanes or in the automotive industry. The second implication is economical. The introduction of a molecular mechanism that visually indicates damage or wear and tear permits to use a material until it has reached its actual end of life, *i.e.* until it has accumulated a certain threshold of microdamages or until its mechanical properties fall below a threshold value. Compared to the current praxis, where most materials are assigned a theoretical lifetime, the self-reporting approach would avoid the replacement of components based on regular service intervals and lifetime assumptions, thus allowing to use them for longer periods of time.

To achieve the practical implementation of self-reporting materials in real-world products, several key factors have yet to be addressed. For example, it has to be made sure that the self-reporting mechanisms are stable over long periods of time and that they do not produce false positive or false negative results. Moreover, in many cases the threshold forces to trigger mechanochromic events in materials have not yet been determined in detail, as they depend on a complex interplay of the properties of the mechanochromic reporter, the mechanical properties of the material, and the efficiency of force transduction from the material onto the mechanophores, proteins, microcapsules, *etc.* Given the current vibrant dynamic of the research field, it can be expected that such questions will be addressed in future research. This will also require the refinement of known mechanophores and the development of new mechanochromic systems.

Another interesting field of application of the self-reporting approach would be biomedical materials. However, to the best of our knowledge, mechanochromic polymer-based materials have not yet been designed for such applications. For example, the synthesis of mechano-responsive nanoparticles, which would exert their biomedical function in response to a mechanical stimulus and simultaneously generate a change in colour, would permit to develop new theranostics tools. Such systems could lead to a better understanding of mechanical processes within tissues and cells. Another useful application of self-reporting materials could be their implementation in artificial heart valves or, if some day realised, artificial hearts. These manufactured organs are continuously submitted to mechanical stress which could finally end in failure of the material. If a self-reporting mechanism could be included in these components, they could indicate that these organs are damaged, *e.g.* by releasing a colourful compound to physiological fluids, or by increasing their X-ray or ultrasound contrast. This would be the indicator for replacement surgery.

In conclusion, we believe that specificity to mechanical stimuli, long term stability, and control of the threshold at which they start to show their self-reporting properties are the key features that will bring self-reporting materials into real life applications. In the coming years, we expect to see more research efforts geared towards filling this knowledge gap. Furthermore, the possibility of manufacturing these materials based on polymers will permit the scientific community to develop low-cost materials for the sensing of critical situations.

Acknowledgements

This work was supported by the Swiss National Science Foundation through the National Centre of Competence in Research (NCCR) Bio-Inspired Materials and through project PP00P2_144697.

References

1. J. P. Schrum, T. F. Zhu and J. W. Szostak, *Cold Spring Harbor Perspect. Biol.*, 2010, **2**, a002212.
2. I. J. Uings and S. N. Farrow, *Mol. Pathol.*, 2000, **53**, 295–299.
3. R. D. Marshall, *Annu. Rev. Biochem.*, 1972, **41**, 673–702.
4. D. Wicher, *Front. Cell. Neurosci.*, 2010, **4**, 25.
5. Y. Tang, G. Cao, X. Chen, J. Yoo, A. Yethiraj and Q. Cui, *Biophys. J.*, 2006, **91**, 1248–1263.
6. S. Sukharev and F. Sachs, *J. Cell Sci.*, 2012, **125**, 3075–3083.
7. P. A. Gottlieb and F. Sachs, *Nature*, 2012, **483**, 163–164.
8. L. Bianchi, *Mol. Neurobiol.*, 2007, **36**, 254–271.
9. P. Kazmierczak and U. Müller, *Trends Neurosci.*, 2012, **35**, 220–229.
10. J. C. de Nooij, C. M. Simon, A. Simon, S. Doobar, K. P. Steel, R. W. Banks, G. Z. Mentis, G. S. Bewick and T. M. Jessell, *J. Neurosci.*, 2015, **35**, 3073–3084.
11. C. Hahn and M. A. Schwartz, *Nat. Rev. Mol. Cell Biol.*, 2009, **10**, 53–62.
12. M. D. Stokes, G. B. Deane, M. I. Latz and J. Rohr, *J. Geophys. Res.: Oceans*, 2004, **109**, C01004.
13. J. Rohr, M. Hyman, S. Fallon and M. I. Latz, *Deep Sea Res., Part I*, 2002, **49**, 2009–2033.

14. S. Mallipattu, M. Haidekker, P. Von Dassow, M. Latz and J. Frangos, *J. Comp. Physiol., A*, 2002, **188**, 409–416.
15. A. K. Chen, M. I. Latz, P. Sobolewski and J. A. Frangos, *Am. J. Physiol. Regul., Integr. Comp. Physiol.*, 2007, **292**, R2020–R2027.
16. P. von Dassow and M. I. Latz, *J. Exp. Biol.*, 2002, **205**, 2971–2986.
17. A. K. Chen, M. I. Latz and J. A. Frangos, *Biotechnol. Bioeng.*, 2003, **83**, 93–103.
18. Y. Chen, A. J. H. Spiering, S. Karthikeyan, G. W. M. Peters, E. W. Meijer and R. P. Sijbesma, *Nat. Chem.*, 2012, **4**, 559–562.
19. W. K. Stadelmann, A. G. Digenis and G. R. Tobin, *Am. Chem. J. Surg.*, 1998, **176**, 26S–38S.
20. M. M. Caruso, D. A. Davis, Q. Shen, S. A. Odom, N. R. Sottos, S. R. White and J. S. Moore, *Chem. Rev.*, 2009, **109**, 5755–5798.
21. J. N. Brantley, C. B. Bailey, K. M. Wiggins, A. T. Keatinge-Clay and C. W. Bielawski, *Polym. Chem.*, 2013, **4**, 3916–3928.
22. Y. Sagara, S. Yamane, M. Mitani, C. Weder and T. Kato, *Adv. Mater.*, 2016, **28**, 1073–1095.
23. J. W. C. Pang and I. P. Bond, *Compos. Sci. Technol.*, 2005, **65**, 1791–1799.
24. J. W. C. Pang and I. P. Bond, *Composites, Part A*, 2005, **36**, 183–188.
25. L. L. Randeberg, O. A. Haugen, R. Haaverstad and L. O. Svaasand, *Lasers Surg. Med.*, 2006, **38**, 277–289.
26. D. Y. Zhu, M. Z. Rong and M. Q. Zhang, *Prog. Polym. Sci.*, 2015, **49–50**, 175–220.
27. W. Li, C. C. Matthews, K. Yang, M. T. Odarczenko, S. R. White and N. R. Sottos, *Adv. Mater.*, 2016, **28**, 2189–2194.
28. S.-Y. Cho, J.-G. Kim and C.-M. Chung, *Sens. Actuators, B*, 2008, **134**, 822–825.
29. B. R. Crenshaw and C. Weder, *Chem. Mater.*, 2003, **15**, 4717–4724.
30. S. A. Odom, A. C. Jackson, A. M. Prokup, S. Chayanupatkul, N. R. Sottos, S. R. White and J. S. Moore, *ACS Appl. Mater. Interfaces*, 2011, **3**, 4547–4551.
31. S. Lörcher, T. Winkler, K. Makyła, C. Ouellet-Plamondon, I. Burgert and N. Bruns, *J. Mater. Chem. A*, 2014, **2**, 6231–6237.
32. D. A. Davis, A. Hamilton, J. Yang, L. D. Cremar, D. Van Gough, S. L. Potisek, M. T. Ong, P. V. Braun, T. J. Martínez and S. R. White, *Nature*, 2009, **459**, 68–72.
33. T. J. Kucharski and R. Boulatov, *J. Mater. Chem.*, 2011, **21**, 19413–19415.
34. A. L. Black, J. M. Lenhardt and S. L. Craig, *J. Mater. Chem.*, 2011, **21**, 1655–1663.
35. F. Ciardelli, G. Ruggeri and A. Pucci, *Chem. Soc. Rev.*, 2013, **42**, 857–870.
36. X. Jin, M. Götz, S. Wille, Y. K. Mishra, R. Adelung and C. Zollfrank, *Adv. Mater.*, 2013, **25**, 1342–1347.
37. C. Löwe and C. Weder, *Adv. Mater.*, 2002, **14**, 1625–1629.
38. A. L. B. Ramirez, Z. S. Kean, J. A. Orlicki, M. Champhekar, S. M. Elsakr, W. E. Krause and S. L. Craig, *Nat. Chem.*, 2013, **5**, 757–761.
39. L. Sun, W. Huang, C. Wang, Z. Ding, Y. Zhao, C. Tang and X. Gao, *Liq. Cryst.*, 2014, **41**, 277–289.

40. R. P. Wool, *Soft Matter*, 2008, **4**, 400–418.
41. H. Staudinger, *Ber. Dtsch. Chem. Ges. A/B*, 1930, **63**, 921–934.
42. H. Staudinger and H. F. Bondy, *Ber. Dtsch. Chem. Ges. A/B*, 1930, **63**, 734–736.
43. H. Staudinger and E. O. Leupold, *Ber. Dtsch. Chem. Ges. A/B*, 1930, **63**, 730–733.
44. B. A. Buchholz, J. M. Zahn, M. Kenward, G. W. Slater and A. E. Barron, *Polymer*, 2004, **45**, 1223–1234.
45. B. A. Toms, *Proceedings of the 1st International Congress on Rheology*, 1948, vol. 2, pp. 135–141.
46. A. F. Horn, *Nature*, 1984, **312**, 140–141.
47. T. Moussa, C. Tiu and T. Sridhar, *J. Non-Newtonian Fluid Mech.*, 1993, **48**, 261–284.
48. T. Moussa and C. Tiu, *Chem. Eng. Sci.*, 1994, **49**, 1681–1692.
49. W. Brostow, *J. Ind. Eng. Chem.*, 2008, **14**, 409–416.
50. F. Hadri, A. Besq, S. Guillou and R. Makhloufi, *J. Non-Newtonian Fluid Mech.*, 2011, **166**, 326–331.
51. M. Tanigawa, M. Suzuto, K. Fukudome and K. Yamaoka, *Macromolecules*, 1996, **29**, 7418–7425.
52. K. S. Suslick and G. J. Price, *Annu. Rev. Mater. Sci.*, 1999, **29**, 295–326.
53. K. S. Suslick, *Science*, 1990, **247**, 1439–1445.
54. P. A. R. Glynn and B. M. E. van der Hoff, *J. Macromol. Sci., Part A*, 1973, 7, 1695–1719.
55. P. A. R. Glynn, B. M. E. Van Der Hoff and P. M. Reilly, *J. Macromol. Sci., Part A*, 1972, **6**, 1653–1664.
56. J. N. Brantley, K. M. Wiggins and C. W. Bielawski, *Polym. Int.*, 2013, **62**, 2–12.
57. M. J. Kryger, M. T. Ong, S. A. Odom, N. R. Sottos, S. R. White, T. J. Martinez and J. S. Moore, *J. Am. Chem. Soc.*, 2010, **132**, 4558–4559.
58. P. Egerton, E. Hyde, J. Trigg, A. Payne, P. Beynon, M. Mijovic and A. Reiser, *J. Am. Chem. Soc.*, 1981, **103**, 3859–3863.
59. J. Zou, Y. Liu, B. Shan, A. Chattopadhyay and L. Dai, *Smart Mater. Struct.*, 2014, **23**, 095038.
60. J. Zou, Y. Liu, A. Chattopadhyay and L. Dai, Proceedings of SPIE 9432, *Behavior and Mechanics of Multifunctional Materials and Composites*, 2015, 943204.
61. Y.-K. Song, K.-H. Lee, W.-S. Hong, S.-Y. Cho, H.-C. Yu and C.-M. Chung, *J. Mater. Chem.*, 2012, **22**, 1380–1386.
62. M. B. Larsen and A. J. Boydston, *J. Am. Chem. Soc.*, 2014, **136**, 1276–1279.
63. M. B. Larsen and A. J. Boydston, *J. Am. Chem. Soc.*, 2013, **135**, 8189–8192.
64. B. Lyu, W. Cha, T. Mao, Y. Wu, H. Qian, Y. Zhou, X. Chen, S. Zhang, L. Liu, G. Yang, Z. Lu, Q. Zhu and H. Ma, *ACS Appl. Mater. Interfaces*, 2015, **7**, 6254–6259.
65. X. Chen, M. A. Dam, K. Ono, A. Mal, H. Shen, S. R. Nutt, K. Sheran and F. Wudl, *Science*, 2002, **295**, 1698–1702.

66. A. M. Schenzel, C. Klein, K. Rist, N. Moszner and C. Barner-Kowollik, *Adv. Sci.*, 2016, **3**, 1500361.
67. D. C. Church, G. I. Peterson and A. J. Boydston, *ACS Macro Lett.*, 2014, **3**, 648–651.
68. J. Li, T. Shiraki, B. Hu, R. A. E. Wright, B. Zhao and J. S. Moore, *J. Am. Chem. Soc.*, 2014, **136**, 15925–15928.
69. B. Lukyanov and M. Lukyanova, *Chem. Heterocycl. Compd.*, 2005, **41**, 281–311.
70. K. Schöller, S. Küpfer, L. Baumann, P. M. Hoyer, D. de Courten, R. M. Rossi, A. Vetushka, M. Wolf, N. Bruns and L. J. Scherer, *Adv. Funct. Mater.*, 2014, **24**, 5194–5201.
71. Y. Chen, H. Zhang, X. Fang, Y. Lin, Y. Xu and W. Weng, *ACS Macro Lett.*, 2014, **3**, 141–145.
72. R. Klajn, *Chem. Soc. Rev.*, 2014, **43**, 148–184.
73. C. K. Lee, D. A. Davis, S. R. White, J. S. Moore, N. R. Sottos and P. V. Braun, *J. Am. Chem. Soc.*, 2010, **132**, 16107–16111.
74. C. M. Kingsbury, P. A. May, D. A. Davis, S. R. White, J. S. Moore and N. R. Sottos, *J. Mater. Chem.*, 2011, **21**, 8381–8388.
75. L. Florea, D. Diamond and F. Benito-Lopez, *Macromol. Mater. Eng.*, 2012, **297**, 1148–1159.
76. C. K. Lee, C. E. Diesendruck, E. Lu, A. N. Pickett, P. A. May, J. S. Moore and P. V. Braun, *Macromolecules*, 2014, **47**, 2690–2694.
77. A.-D. N. Celestine, B. A. Beiermann, P. A. May, J. S. Moore, N. R. Sottos and S. R. White, *Polymer*, 2014, **55**, 4164–4171.
78. G. Hong, H. Zhang, Y. Lin, Y. Chen, Y. Xu, W. Weng and H. Xia, *Macromolecules*, 2013, **46**, 8649–8656.
79. G. I. Peterson, M. B. Larsen, M. A. Ganter, D. W. Storti and A. J. Boydston, *ACS Appl. Mater. Interfaces*, 2015, **7**, 577–583.
80. M. E. Grady, B. A. Beiermann, J. S. Moore and N. R. Sottos, *ACS Appl. Mater. Interfaces*, 2014, **6**, 5350–5355.
81. G. R. Gossweiler, C. L. Brown, G. B. Hewage, E. Sapiro-Gheiler, W. J. Trautman, G. W. Welshofer and S. L. Craig, *ACS Appl. Mater. Interfaces*, 2015, **7**, 22431–22435.
82. G. R. Gossweiler, G. B. Hewage, G. Soriano, Q. Wang, G. W. Welshofer, X. Zhao and S. L. Craig, *ACS Macro Lett.*, 2014, **3**, 216–219.
83. Q. Wang, G. R. Gossweiler, S. L. Craig and X. Zhao, *Nat. Commun.*, 2014, **5**, 4899.
84. K. Imato, T. Kanehara, T. Ohishi, M. Nishihara, H. Yajima, M. Ito, A. Takahara and H. Otsuka, *ACS Macro Lett.*, 2015, **4**, 1307–1311.
85. E. Ducrot, Y. Chen, M. Bulters, R. P. Sijbesma and C. Creton, *Science*, 2014, **344**, 186–189.
86. C. Löwe and C. Weder, *Synthesis*, 2002, **2002**, 1185–1190.
87. B. R. Crenshaw and C. Weder, *Adv. Mater.*, 2005, **17**, 1471–1476.
88. M. Kinami, B. R. Crenshaw and C. Weder, *Chem. Mater.*, 2006, **18**, 946–955.
89. J. Lott and C. Weder, *Macromol. Chem. Phys.*, 2010, **211**, 28–34.

90. B. R. Crenshaw and C. Weder, *Macromolecules*, 2006, **39**, 9581–9589.
91. A. Lavrenova, A. Holtz, Y. C. Simon and C. Weder, *Macromol. Mater. Eng.*, 2016, **301**, 549–554.
92. N. Tirelli, S. Amabile, C. Cellai, A. Pucci, L. Regoli, G. Ruggeri and F. Ciardelli, *Macromolecules*, 2001, **34**, 2129–2137.
93. F. Donati, A. Pucci, C. Cappelli, B. Mennucci and G. Ruggeri, *J. Phys. Chem. B*, 2008, **112**, 3668–3679.
94. F. Donati, A. Pucci, L. Boggioni, I. Tritto and G. Ruggeri, *Macromol. Chem. Phys.*, 2009, **210**, 728–735.
95. A. Pucci, F. Di Cuia, F. Signori and G. Ruggeri, *J. Mater. Chem.*, 2007, **17**, 783–790.
96. F. Donati, A. Pucci and G. Ruggeri, *Phys. Chem. Chem. Phys.*, 2009, **11**, 6276–6282.
97. A. Pucci, M. Bertoldo and S. Bronco, *Macromol. Rapid Commun.*, 2005, **26**, 1043–1048.
98. S. Bao, J. Li, K. I. Lee, S. Shao, J. Hao, B. Fei and J. H. Xin, *ACS Appl. Mater. Interfaces*, 2013, **5**, 4625–4631.
99. D. Montarnal, P. Cordier, C. Soulié-Ziakovic, F. Tournilhac and L. Leibler, *J. Polym. Sci., Part A: Polym. Chem.*, 2008, **46**, 7925–7936.
100. F. Herbst, D. Döhler, P. Michael and W. H. Binder, *Macromol. Rapid Commun.*, 2013, **34**, 203–220.
101. D. H. Reneker, W. L. Mattice, R. P. Quirk and S. Kim, *Smart Mater. Struct.*, 1992, **1**, 84.
102. S.-J. Kim and D. Reneker, *Polym. Bull.*, 1993, **31**, 367–374.
103. O. Azzaroni, B. Trappmann, P. van Rijn, F. Zhou, B. Kong and W. T. S. Huck, *Angew. Chem., Int. Ed.*, 2006, **45**, 7440–7443.
104. J. Bünsow, J. Erath, P. M. Biesheuvel, A. Fery and W. T. S. Huck, *Angew. Chem., Int. Ed.*, 2011, **50**, 9629–9632.
105. B. D. Hoffman, C. Grashoff and M. A. Schwartz, *Nature*, 2011, **475**, 316–323.
106. J. Longo, C. Yao, C. Rios, N. T. T. Chau, F. Boulmedais, J. Hemmerlé, P. Lavalle, S. M. Schiller, P. Schaaf and L. Jierry, *Chem. Commun.*, 2015, **51**, 232–235.
107. A.-L. Cost, P. Ringer, A. Chrostek-Grashoff and C. Grashoff, *Cell. Mol. Bioeng.*, 2014, **8**, 96–105.
108. M. Mickler, I. Dima Ruxandra, H. Dietz, C. Hyeon, D. Thirumalai and M. Rief, *Proc. Natl. Acad. Sci. U. S. A.*, 2007, **104**, 20268–20273.
109. X. Wang and T. Ha, *Science*, 2013, **340**, 991–994.
110. G. Žoldák and M. Rief, *Curr. Opin. Struct. Biol.*, 2013, **23**, 48–57.
111. K. L. Heredia and H. D. Maynard, *Org. Biomol. Chem.*, 2007, **5**, 45–53.
112. I. Cobo, M. Li, B. S. Sumerlin and S. Perrier, *Nat. Mater.*, 2015, **14**, 143–159.
113. T. Schnitzler and A. Herrmann, *Acc. Chem. Res.*, 2012, **45**, 1419–1430.
114. N. Bruns, K. Pustelny, L. M. Bergeron, T. A. Whitehead and D. S. Clark, *Angew. Chem., Int. Ed.*, 2009, **48**, 5666–5669.
115. N. Bruns and D. S. Clark, *Chimia*, 2011, **65**, 245–249.

116. L. Ditzel, J. Löwe, D. Stock, K.-O. Stetter, H. Huber, R. Huber and S. Steinbacher, *Cell*, 1998, **93**, 125–138.
117. F. Valle, J. A. DeRose, G. Dietler, M. Kawe, A. Plückthun and G. Semenza, *Ultramicroscopy*, 2002, **93**, 83–89.
118. H. Wallrabe and A. Periasamy, *Curr. Opin. Biotechnol.*, 2005, **16**, 19–27.
119. C. Grashoff, B. D. Hoffman, M. D. Brenner, R. Zhou, M. Parsons, M. T. Yang, M. A. McLean, S. G. Sligar, C. S. Chen and T. Ha, *Nature*, 2010, **466**, 263–266.
120. M. Morimatsu, A. H. Mekhdjian, A. S. Adhikari and A. R. Dunn, *Nano Lett.*, 2013, **13**, 3985–3989.
121. M. Massey, W. R. Algar and U. J. Krull, *Anal. Chim. Acta*, 2006, **568**, 181–189.
122. H. Dietz and M. Rief, *Proc. Natl. Acad. Sci. U. S. A.*, 2004, **101**, 16192–16197.
123. R. Perez-Jimenez, S. Garcia-Manyes, S. R. K. Ainavarapu and J. M. Fernandez, *J. Biol. Chem.*, 2006, **281**, 40010–40014.
124. M. Bertz, A. Kunfermann and M. Rief, *Angew. Chem., Int. Ed.*, 2008, **47**, 8192–8195.
125. M. Caraglio, A. Imparato and A. Pelizzola, *Phys. Rev. E: Stat., Nonlinear, Soft Matter Phys.*, 2011, **84**, 021918.
126. J. Saeger, V. P. Hytönen, E. Klotzsch and V. Vogel, *PloS One*, 2012, **7**, 1.
127. M. Zimmer, *Chem. Rev.*, 2002, **102**, 759–781.
128. R. N. Day and M. W. Davidson, *Chem. Soc. Rev.*, 2009, **38**, 2887–2921.
129. V. Helms, T. Straatsma and J. A. McCammon, *J. Phys. Chem. B*, 1999, **103**, 3263–3269.
130. W. W. Ward, C. W. Cody, R. C. Hart and M. J. Cormier, *Photochem. Photobiol.*, 1980, **31**, 611–615.
131. C. W. Cody, D. C. Prasher, W. M. Westler, F. G. Prendergast and W. W. Ward, *Biochemistry*, 1993, **32**, 1212–1218.
132. G. Jung, J. Wiehler and A. Zumbusch, *Biophys. J.*, 2005, **88**, 1932–1947.
133. T. Ichimura, H. Fujita, K. Yoshizawa and T. M. Watanabe, *Chem. Commun.*, 2012, **48**, 7871–7873.
134. K. Makyła, C. Müller, S. Lörcher, T. Winkler, M. G. Nussbaumer, M. Eder and N. Bruns, *Adv. Mater.*, 2013, **25**, 2701–2706.
135. B. Barstow, N. Ando, C. U. Kim and S. M. Gruner, *Proc. Natl. Acad. Sci. U. S. A.*, 2008, **105**, 13362–13366.
136. B. Barstow, N. Ando, C. U. Kim and S. M. Gruner, *Biophys. J.*, 2009, **97**, 1719–1727.
137. J. N. Brantley, C. B. Bailey, J. R. Cannon, K. A. Clark, D. A. Vanden Bout, J. S. Brodbelt, A. T. Keatinge-Clay and C. W. Bielawski, *Angew. Chem., Int. Ed.*, 2014, **53**, 5088–5092.
138. H. Herberhold, S. Marchal, R. Lange, C. H. Scheyhing, R. F. Vogel and R. Winter, *J. Mol. Biol.*, 2003, **330**, 1153–1164.
139. Y. D. Kim, J. S. Dordick and D. S. Clark, *Biotechnol. Bioeng.*, 2001, **72**, 475–482.

140. S. R. White, N. R. Sottos, P. H. Geubelle, J. S. Moore, M. R. Kessler, S. R. Sriram, E. N. Brown and S. Viswanathan, *Nature*, 2001, **409**, 794–797.
141. B. J. Blaiszik, S. L. B. Kramer, S. C. Olugebefola, J. S. Moore, N. R. Sottos and S. R. White, *Annu. Rev. Mater. Res.*, 2010, **40**, 179–211.
142. M. A. White, *J. Chem. Educ.*, 1998, **75**, 1119.
143. Bryan E. Koene, Luna Innovations Inc., Martin E. Rogers, WO 2006105290, 2006.
144. Air Force SBIR Program: Impact Indicator Paint for Composites, http://www.afsbirsttr.com/Publications/Documents/Innovation-091908-Luna-AF04-133.pdf (accessed 02/2016).
145. L. Monier, K. Jeune, I. Kondolff and G. Vilaça, in *Smart Intelligent Aircraft Structures (SARISTU): Proceedings of the Final Project Conference*, ed. C. P. Wölcken and M. Papadopoulos, Springer International Publishing, Cham, 2016, DOI: 10.1007/978-3-319-22413-8_36, pp. 735–743.
146. S. Vidinejevs, A. N. Aniskevich, A. Gregor, M. Sjöberg and G. Alvarez, *J. Intell. Mater. Syst. Struct.*, 2012, **23**, 1371–1377.
147. S. Vidinejevs, O. Strekalova, A. Aniskevich and S. Gaidukov, *Mech. Compos. Mater.*, 2013, **49**, 77–84.
148. B. Di Credico, G. Griffini, M. Levi and S. Turri, *ACS Appl. Mater. Interfaces*, 2013, **5**, 6628–6634.
149. A. Lavrenova, J. Farkas, C. Weder and Y. C. Simon, *ACS Appl. Mater. Interfaces*, 2015, **7**, 21828–21834.
150. C. J. Hansen, W. Wu, K. S. Toohey, N. R. Sottos, S. R. White and J. A. Lewis, *Adv. Mater.*, 2009, **21**, 4143–4147.
151. K. S. Toohey, N. R. Sottos, J. A. Lewis, J. S. Moore and S. R. White, *Nat. Mater.*, 2007, **6**, 581–585.
152. M. Scheiner, T. J. Dickens and O. Okoli, *Polymer*, 2016, **83**, 260–282.
153. C. J. Norris, G. J. Meadway, M. J. O'Sullivan, I. P. Bond and R. S. Trask, *Adv. Funct. Mater.*, 2011, **21**, 3624–3633.
154. J. F. Patrick, K. R. Hart, B. P. Krull, C. E. Diesendruck, J. S. Moore, S. R. White and N. R. Sottos, *Adv. Mater.*, 2014, **26**, 4302–4308.
155. U. Polimeno and M. Meo, *Compos. Struct.*, 2009, **91**, 398–402.
156. S. Kling and T. Czigány, *Compos. Sci. Technol.*, 2014, **99**, 82–88.

CHAPTER 12

Mechanically Adaptive Nanocomposites Inspired by Sea Cucumbers

DAFNI MOATSOU* AND CHRISTOPH WEDER*

Adolphe Merkle Institute, University of Fribourg, Chemin des Verdiers 4, 1700 Fribourg, Switzerland
*Email: dafni.moatsou@unifr.ch; christoph.weder@unifr.ch

12.1 Introduction

The last two decades have witnessed significant activities focused on the development of materials that exhibit a predefined change of one or multiple physicochemical properties when exposed to an external stimulus, and which are thus referred to as "stimulus-responsive", "adaptive", "smart", or "intelligent".[1,2] Mechanically adaptive materials, which change their mechanical characteristics "on command", represent an important subset of this materials class.[3] Such systems are potentially useful in a plethora of technical applications ranging from portable knee braces to actuators to intracortical microelectrodes.[4] Various different design approaches to create mechanically morphing materials have been pursued, ranging from simple plasticization of polymers to sophisticated shape-memory mechanisms, and actuators. Many of the pertinent studies have been compiled in a handful of excellent recent reviews.[5–9]

RSC Polymer Chemistry Series No. 22
Bio-inspired Polymers
Edited by Nico Bruns and Andreas F. M. Kilbinger

Published by the Royal Society of Chemistry, www.rsc.org

Mechanical morphing is also an essential feature in many biological materials.[10] Indeed, nature has developed a range of different pathways to realize this important effect.[11] Perhaps the most simple and widespread process is plasticization, *i.e.* a stiffness reduction that is caused by infusing a material with a softening component. This effect, which is well known in the domain of polymeric materials,[12] can be observed in many natural materials, most prominently in hemicelluloses, whose Young's modulus can decrease by up to nearly three orders of magnitude when exposed to humid conditions.[13] Similarly, spider silk, a semicrystalline polymer, has been shown to alter its mechanical properties as a result of changes in relative humidity, which causes a reduction of the semicrystalline polymer's glass transition temperature (T_g).[14] Another element that governs the macroscopic mechanical properties of polymeric materials is the introduction of crosslinks, which effectively connect macromolecules into network structures, which give rise to materials characteristics that are very different than those of their linear analogues. This general principle can for example be found in plant cell walls, where crosslinking *via* borate chemistry moderates the mechanical properties of the tissue.[15] Indeed, it has been shown that in the absence of the borate crosslinker, the growth capacity of a plant is limited, resulting in a larger pore size.[16] Another general design feature used by nature to modulate mechanical properties over wide ranges is the formation of composite materials, which often comprise components in a hierarchically organized manner. While not a morphing material, one example where hierarchical organization imparts unique mechanical properties is the abalone shell nacre that was found to consist of six structural levels of hierarchical organization in different scales (from *ca.* 30 nm up to 0.5 μm, Figure 12.1).[17] The transition from one scale to the next represents a stiff ness variation, a characteristic that can only be modelled by simulations, which predict the stiffness dependence on the levels of hierarchical organization.[18] These examples (although not exclusively) show strikingly that biological species are masters in achieving complexity while only relying on abundant and non-hazardous materials.[4,19] Indeed, as living organisms synthesize their materials from only a few chemical elements, hierarchical structuring has evolved as an important alternate design element to achieve complex functions.[20]

12.2 Mechanical Morphing of the Sea Cucumber Dermis

A prime example of how nature uses hierarchical structuring to achieve mechanical morphing is found in sea cucumbers (echinoderms from the class *holothuroidea*), marine animals which have the ability to rapidly and reversibly alter the mechanical properties of their dermis (Figure 12.2).[21] This tissue is composed of rigid, high-aspect-ratio collagen fibrils which are arranged into interwoven bundles and embedded within a viscoelastic

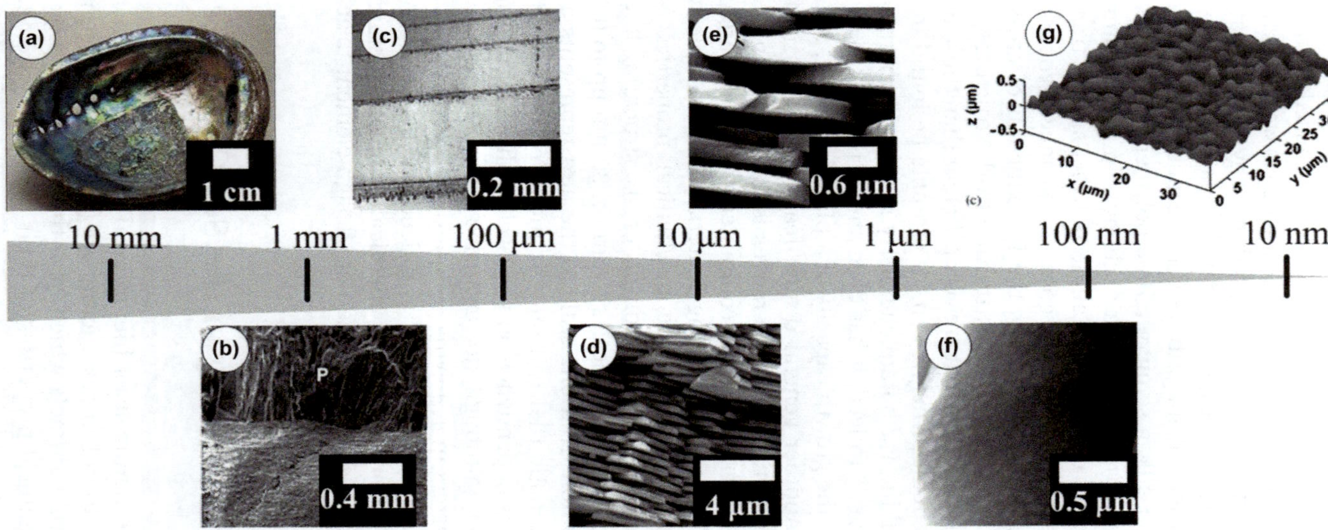

Figure 12.1 Photograph (a), scanning electron microscopy images (b–f), and laser profilometry topology (g) illustrating the six structural levels of the hierarchical organization in nacre.
Images reprinted in part with permission from M. A. Meyers *et al.*, Biological Materials: A materials science approach, *J. Mech. Behav. Biomed. Mater.*, **4**, 626–657, Copyright 2011 with permission from Elsevier;[137] X. Li, W.-C. Chang, Y. J. Chao, R. Wang and M. Chang, *Nano Lett.*, 2004, **4**, 613–617.[138] Copyright 2004 American Chemical Society; R. Menig *et al.*, Quasi-static and dynamic mechanical response of Haliotis rufescens (abalone) shells, *Acta Mater.*, **48**, 2383–2398, Copyright 2000 with permission from Elsevier;[139] and F. Barthelat *et al.*, On the mechanics of mother-of-pearl: A key feature in the material hierarchical structure, *J. Mech. Phys. Solids*, **55**, 306–337, Copyright 2007 with permission from Elsevier.[140]

Figure 12.2 Pictures of a sea cucumber in its relaxed, soft state (left) and, feeling threatened, in its stiff form (right).
From J. R. Capadona, K. Shanmuganathan, D. J. Tyler, S. J. Rowan and C. Weder, *Science*, 2008, **319**, 1370–1374. Reprinted with permission from AAAS.

matrix comprising water and fibrillin-like microfibrils.[22] The stiffness of the tissue depends on the ability of adjacent collagen fibrils to transfer stress *via* transiently established interactions, which are controlled by regulatory molecules discharged by neurosecretory cells.[23–26] The glycoprotein stiparin has been found to affect the aggregation of collagen fibers by forming weak bonds with proteoglycans on the surface of adjacent fibrils.[27,28] As such, a percolating network of collagen is formed, causing a significant increase of the elastic modulus of the tissue. The reversibility of the stiffening is regulated by a stiparin inhibitor, which disrupts the weak crosslinking bonds and restores the original soft state. In an *ex vivo* study, the inner and outer regions of the dermis were separated and extracts from these specimen were applied to intact pieces of dermis, one invoking a stiff mechanical state and the other invoking a compliant state. Dynamic oscillatory shear tests were used to probe the dynamic shear stiffness and the relative damping of the tissue, and these parameters were found to exhibit an inverse relationship. A theoretical viscoelastic model served to interpret the findings and support the conclusion that the mechanical changes of the tissue modulus of an order of magnitude from *ca.* 5 to *ca.* 50 MPa involved interactions between elastic elements, rather than a change of the viscous components.[26] *In vivo* experiments demonstrated that stiparin alone cannot induce stiffening of live specimens, an effect that another protein, tensilin, was found to exhibit.[29,30] While many details regarding the mechanical switching mechanism of sea cucumbers remain unclear, the architecture and switching concept have been recognized as an intriguing model for artificial materials with mechanical morphing capabilities.[31] In the following sections, the various attempts to mimic the structure and function of such sea cucumber mimics are summarized. Early work on such materials and details of models utilized to describe the mechanical characteristics have been the subject of previous reviews,[32,33] and therefore this chapter emphasizes the more recent work on advanced versions of such materials.

12.3 Water-responsive Sea Cucumber-mimicking Nanocomposites

Virtually all sea cucumber-mimicking materials comprise cellulose nanocrystals (CNCs) as the rigid filler that emulates the function of the collagen fibrils.[32,34] Cellulose is a linear syndiotactic homopolymer composed of D-anhydroglucopyranose monomers connected by glycosidic bonds (Figure 12.3a). Naturally occurring cellulose is a structurally complex, hierarchically organized material, which at the lowest hierarchical level contains highly oriented, hydrogen-bonded crystalline domains that are separated by amorphous regions. The latter are more readily hydrolyzed upon exposure to mineral acids than the crystalline portions, thus permitting the isolation of the rod-like CNCs in a fairly straightforward manner (Figure 12.3b and c). Their dimensions and mechanical characteristics depend on the cellulose source (*e.g.*, wood, cotton, straw, tunicate, algae) and the isolation method, in particular the nature of the mineral acid and the hydrolysis conditions employed.[35–44] CNCs exhibit a high on-axis stiffness, with a tensile modulus of *ca.* 80 to 150 GPa,[45] and dimensions in the nanometer scale (5–20 × 130–1200 nm) with typical aspect ratios of *ca.* 10–100 (Figure 12.3d).[39,43] Their surface is rich in hydrogen-bonding hydroxyl groups, and (in the prominent cases where sulfuric or phosphoric acid are used for the hydrolysis) may contain sulfate or phosphate groups, which moderate the interactions among the CNCs, between CNCs and solvents, as well as between CNCs and

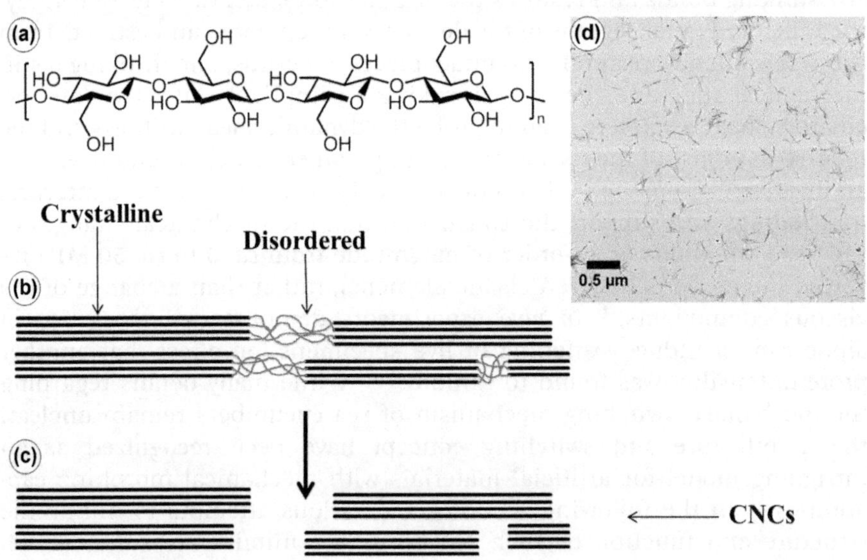

Figure 12.3 Chemical structure of cellulose (a), schematic representation of the crystalline and disordered mixed morphology of cellulose before (b) and after (c) isolation of the CNCs, and a typical transmission electron microscopy image of cotton-derived CNCs (d).

the polymeric matrices in which they are embedded.[46–50] There is an increasing body of work that suggests that nanocelluloses offer biocompatibility, biodegradability, and low cytotoxicity.[43,51,52]

CNCs (isolated from tunicates) were first used two decades ago by Favier *et al.* to reinforce a rubbery polymer latex.[53,54] The immense stiffness increase, which has since been exploited in a plethora of nanocomposites formed by a broad range of polymers and CNC types,[41,53,55–58] was interpreted to result from the formation of a continuous CNC network held together by hydrogen bonds. The stiffness increase was well-described by a mechanical percolation model,[53,54] which considered the soft matrix polymer and the percolating network formed by CNCs as separate components.[53,59,60] As the mechanical properties of short-fiber polymer composites, in which the filler particles are integrated in a discrete, non-percolating fashion, are much better described by the Halpin–Kardos model,[61] the agreement with the percolation model is usually taken as indication of the formation of a percolating filler network, which is often difficult to confirm by microscopic techniques.[62]

Based on the mechanism at play in the case of the stiffening sea cucumber skin, and anticipating that stress transfer among CNCs within a hydrogen-bonded CNC network could be moderated through a competitive hydrogen bonding agent, Capadona *et al.* developed the first bio-inspired sea cucumber mimics, which were based on an ethylene oxide/epichlorohydrin copolymer (EO–EPI) matrix into which tunicate-sourced CNCs were dispersed.[63] In the dry state, the materials displayed the reinforcement predicted by the percolation model, with an increase of the tensile storage modulus (E') from *ca.* 4 MPa (neat matrix polymer) to 800 MPa (nanocomposite with 19% v/v CNCs). However, upon immersion in water, E' significantly decreased to *ca.* 20 MPa, a process that was found to be fully reversible upon drying. Importantly, the contrast in the mechanical properties of the dry and the water-swollen materials was not primarily the result of plasticization by water, although substantial aqueous swelling (up to *ca.* 30% v/v) was observed. This was demonstrated by modelling the mechanical data of the composites as a function of the filler content with the above-discussed mechanical models. In the dry state, the data perfectly agreed with the percolation model[64] whereas, in the hydrated state, a much closer match with predictions made on the basis of the Halpin–Kardos model was found, suggesting the absence of pronounced filler–filler interactions.[61] Thus, the incorporation of water into the composites indeed impacted the interactions among the filler particles (and possibly between the filler particles and the matrix polymer), as intended (Figure 12.4). In the same study, it was shown that the mechanical contrast between a dry rigid and a water-swollen soft state could be further increased by using a glassy polymer matrix, which was plasticized upon exposure to water so that the T_g was reduced.[63] As such, when CNCs isolated from tunicates were integrated into poly(vinyl acetate) (PVAc), glassy nanocomposites were obtained which, in the dry state, exhibited a modulus of up to *ca.* 4.2 GPa (12.2% v/v CNCs) and, upon exposure to an aqueous solution, featured an E' of only a few MPa (*ca.* 1–60 MPa).[32,63]

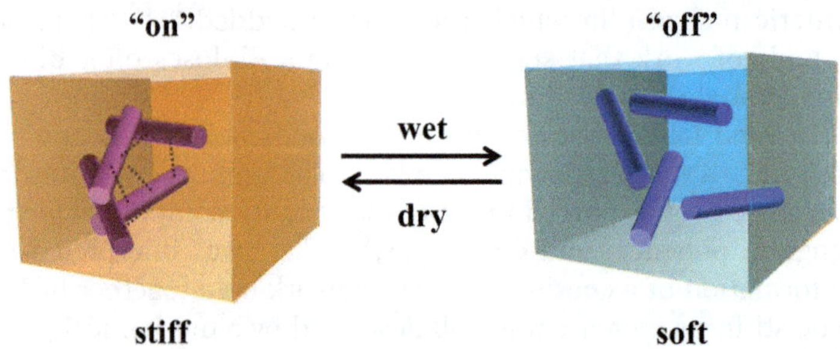

Figure 12.4 Schematic representation of the mechanism at play in CNC-based mechanically adaptive nanocomposites. The reversible switching of the interactions among CNCs is achieved by wetting and drying, resulting in the dissociation and formation, respectively, of a percolating CNC network, concomitant with softening and stiffening of the composite.

The ability of polymer composites with CNCs to adapt their mechanical properties upon exposure to external stimuli has since been demonstrated in a broad range of systems that involve different polymers and CNC types, of which some, after appropriate surface modification, respond to stimuli other than water. It should be noted that although the original study on polymer/CNC composites exhibiting mechanically adaptive properties exploited the highly anisotropic CNCs obtained from tunicates, it was also shown that conventional cellulose sources, such as cotton, can also be used to obtain CNCs with comparable reinforcing and responsive properties, albeit less pronounced due to their lower aspect ratio.[65]

Based on the intriguing mechanically adaptive properties displayed by EO–EPI/CNC and PVAc/CNC nanocomposites, a further evaluation of the effect of the T_g of the polymer matrix on the mechanical switching of the composite material upon wetting was carried out by Shanmuganathan *et al.*[66] With the objective of also reducing the extent of aqueous swelling (which in the aforementioned nanocomposites containing *ca.* 16% v/v CNCs isolated from tunicates was *ca.* 30 wt% for EO–EPI and *ca.* 80 wt% for PVAc) poly(butyl methacrylate) (PBMA) was used as a hydrophobic matrix. PBMA displays a T_g of 20 °C, which is only slightly lower than that of PVAc (30 °C). Thus, the two dry nanocomposite series display comparable mechanical reinforcement at room temperature, where both materials are in their glassy state, although the absolute stiffness of the PBMA-based nanocomposites is only about half that of the PVAc-based materials (*ca.* 2 *vs.* *ca.* 4 GPa for composites containing *ca.* 12 wt% CNCs). However, upon immersion in water, the T_g of the PBMA-based composites remained well-above room or the body temperature, unlike in the case of PVAc, and on account of reduced plasticization, the mechanical contrast was very moderate. This was attributed to the greater hydrophobic character of the PBMA matrix, which does not allow an effective uptake of water. As a consequence, the mechanical contrast between the dry and the hydrated states at ambient or physiological temperature was minimal. Nonetheless, heating the PBMA-based composites above their T_g resulted in a significant modulus reduction, thus indicating the decoupling of the CNC network (Figure 12.5). These results indicate that careful selection of the polymer can fine tune the temperature

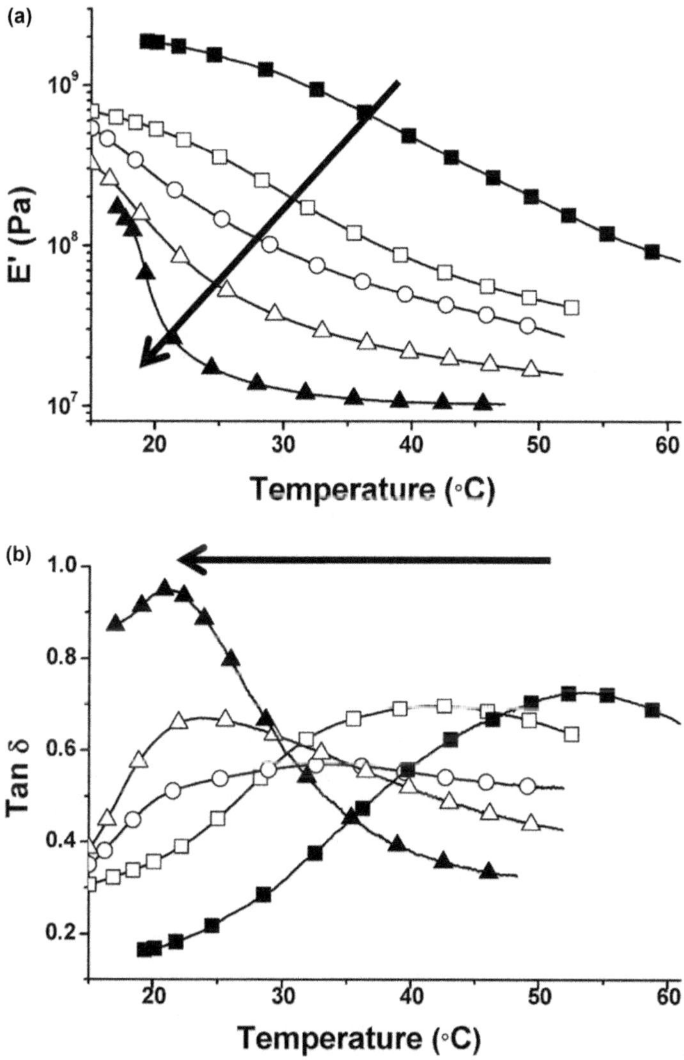

Figure 12.5 Dynamic mechanical analysis (DMA) data of water-swollen PVAc–PBMA/CNC nanocomposites as a function of the PVAc/PBMA ratio. Upon increasing the PVAc content, the tensile modulus (a) decreases, while the loss tangent (tan δ) shifts to lower temperatures (b). The PVAc/PBMA ratio was 100/0 (▲), 60/40 (△), 40/60 (○), 20/80 (□), 0/100 (■). Adapted with permission from K. Shanmuganathan, J. R. Capadona, S. J. Rowan and C. Weder, *ACS Appl. Mater. Interfaces*, 2010, **2**, 165–174,[66] Copyright 2009 American Chemical Society.

at which the mechanical switching occurs. To further demonstrate this, PBMA–PVAc blended matrices with different polymer ratios were shown to directly affect the moduli of the materials at given temperatures.[66]

In an attempt to clarify the effect of the polymer matrix polarity on the ability of the CNC percolating network to uptake water, thus inducing switching of the mechanical properties, low T_g hydrophobic polymers, namely poly(styrene-*co*-butadiene) and polybutadiene, were explored as the polymer matrix.[67] These composites were fabricated using a template approach[64,68] in order to achieve the formation of a percolating network of CNCs within the polymer matrix. Similarly to the composites with the more hydrophilic soft EO–EPI matrix, the mechanical reinforcement effect of the CNCs was significant (an increase of E' of two orders of magnitude was reported) as a result of the formation of a three-dimensional cellulose network. Gratifyingly, the composites were found to exhibit significant water-induced mechanical switching. It is noteworthy that the softening of the composites was achieved only after days of immersion in water, as opposed to minutes in the case of the composites with more polar matrices, and that the process was found to be reversible upon drying. Through comparative swelling experiments, the study further confirmed that the highly polar CNCs serve as hydrophilic "channels" that promote water transfer through the hydrophobic matrix and that the (moderate) water uptake is related to the content of CNCs. Similar results were obtained when chitin nanocrystals were employed as rigid nanofillers for a styrene/butadiene copolymer matrix containing carboxylic acid groups. Once again, no significant change in T_g was observed; however, upon immersion in water, a ten-fold reduction of the tensile modulus was observed, consistent with a change in the interactions between the constituting elements of the percolating chitin network within the polymer matrix.[69]

Jorfi *et al.* explored mechanically adaptive stimulus–responsive nanocomposites based on poly(vinyl alcohol) (PVOH) as the matrix and CNCs derived from tunicates and cotton as the filler.[49] This design was based on the hypothesis that the use of a highly polar glassy polymer that promotes significant hydrogen-bonding between the matrix and the CNCs would result in stiffer nanocomposites than with previously employed matrices. PVOH is known to be highly hydrophilic (and in some circumstances even water soluble)[70] and its numerous hydroxyl groups can interact with the surface hydroxyl groups of CNCs. Moreover, PVOH is biocompatible, non-toxic, and already used in a wide variety of biomedical applications.[71–74] Several other studies have reported PVOH-based nanocomposites with tunicate CNCs[75] and cotton CNCs,[76–78] and demonstrated changes of the mechanical properties of such nanocomposites when these materials were exposed to humid atmosphere. By contrast, Jorfi *et al.* investigated the mechanical properties of such materials after submersion into a simulated physiological fluid to explore the properties under conditions relevant for use in biomedical applications. Indeed, the mechanical contrast between the dry and the wet state of these materials is significantly greater (14 GPa and 10 MPa, respectively) than for any CNC-containing polymer studied before. It was also demonstrated that by modifying the processing conditions, it is possible to control the aqueous swelling of the material and the degree of softening it exhibits, on account of the formation of crosslinks.

Amorphous cellulose has also been used to formulate mechanically adaptive materials. Luo *et al.* synthesized interpenetrating networks of cellulose and poly(acrylic acid) that were permeable by water, thus disrupting the hydrogen bonds of the components that were initially responsible for the high storage modulus of the material. The process was shown to be fully reversible upon consecutive wetting-drying cycles.[79] Poly(acrylic acid) was also used by Spoljaric *et al.*, who prepared covalently crosslinked nanocomposites with different concentrations of nanofibrillated cellulose and extensively studied the effect of water on the swelling and the mechanical properties of the materials. When compared to non-crosslinked composites, the materials were found to better preserve their mechanical properties upon immersion in water, which was attributed to the retention of their crosslinked network morphology, thus indicating the importance of water mobility in the contrasting effect of such mechanically adaptive materials.[33]

Slightly diverting from the sea cucumber mimics, but relying on the same mechanical switching principle discussed so far, another biologically inspired model exploiting the remarkable properties of CNCs aspires to imitate the unique mechanical gradient of the squid beak, which serves as a compatibilizer between stiff and soft tissues. In this example, a PVAc/CNC nanocomposite was covalently crosslinked *via* incorporation of allyl groups and a thiol crosslinker that was activated by UV irradiation. By controlling the exposure times, a gradient in crosslinking density across the material was achieved. The resulting material exhibited properties similar to the PVAc/CNC composites but the extent of softening upon exposure to humidity was also found to follow a gradient pattern.[80]

12.3.1 Stress Transfer in Mechanically Adaptive Materials

To gain quantitative insight into the stress-transfer mechanisms that determine the mechanical morphing of PVAc/CNC nanocomposites, Raman spectroscopy has been employed.[81] It was shown that the extent of stress transfer is influenced by the local orientation of the CNCs, which in turn is governed by the processing conditions used to fabricate the nanocomposites. While solution-cast materials displayed no microscopic anisotropy, samples that were cast and subsequently compression-molded featured both isotropic regions as well as domains in which the CNCs had some preferential (local) orientation. Polarized Raman spectroscopy of the 1095 cm^{-1} band, which is associated with C–O ring stretching of the cellulose backbone, was used to quantify the local orientation of the CNCs. Clear and discernable shifts of this Raman band upon uniaxial deformation of nanocomposite samples were used to monitor the level of stress experienced by the CNCs, which ultimately reflected the stress transfer between the PVAc matrix and the CNCs, and between the CNCs within the network. In the isotropic regions, where the CNCs are assumed to percolate, the Raman shift rate with respect to macroscopic strain was smaller than in the regions in which the CNCs adopted a local uniaxial orientation, and in which no such networks are thought to exist. The Raman shift rate was strongly affected by exposure of the materials to water, and no appreciable stress transfer could be detected by the Raman technique in samples that were fully hydrated. Heating the nanocomposites above the T_g of the PVAc matrix also reduced the stress experienced by the individual CNCs.

In a subsequent study, the aspect ratio and surface charge effects on the stress transfer properties of CNCs on the polymeric matrix were evaluated, again using Raman spectroscopy as the main tool. The comparison of epoxy-based nanocomposites comprising tunicate- or cotton-derived CNCs demonstrated greater stress transfer levels in the former, attributed to the higher aspect ratio of tunicate-derived CNCs. Furthermore, a comparison of the epoxy composites with CNCs isolated by either sulfuric acid or hydrochloric acid showed that uncharged CNCs (obtained by HCl hydrolysis) lead to minimal stress transfer from the matrix, presumably due to their tendency to aggregate, thus decreasing their effective surface area. By contrast, composites with CNCs isolated by sulfuric acid hydrolysis (which feature deprotonated sulfate ester surface groups at a concentration of 38 and 85 mmol SO_4^- kg^{-1} cellulose for cotton- and tunicate-derived CNCs, respectively) were found to exhibit limited aggregation, potentially as a result of the balanced repulsive and attractive forces of the sulfate and hydroxyl surface groups, respectively. Consequently, the effective interphase area of the composite between the filler and the polymer matrix remains high and greater stress transfer levels are achieved. Thus, although charges reduce the CNC–CNC interactions, they are an important factor in the reinforcement of the polymer matrix. More recently, it was proposed that water at the interface between hydrophilic CNC surfaces has a negligible effect on the tensile separation adhesion energy; however, under shear loading, it acts as a lubricant by reducing the atomic friction.[82]

It is apparent that the presence of a molecule that disrupts the interactions between CNCs is the key factor to trigger the switching of the mechanical properties of the composites. Nonetheless, the extent to which the mechanical characteristics of the materials are altered, *i.e.* the mechanical contrast, is influenced by several parameters. In polymer/CNC composites, the polymer used for the formulation was found to significantly affect the extent of reinforcement and thus the mechanical contrast, as well as the swelling behavior and kinetics of the switching, with the main contributing factors being the polarity and the T_g of the polymeric matrix, as well as the surface area and charge density of the filler.[83] Furthermore, important to the mechanical properties of the composites are the filler architecture, as well as the interactions between the filler and the matrix.[84,85] As such, other stimuli that can impact these parameters can also trigger, to an extent, the mechanical response in these nanocomposites.

12.4 Mechanically Adaptive Sea Cucumber Mimics with Specific Responsiveness

12.4.1 pH-responsive Composites

In all the aforementioned examples, the interactions among the CNCs, between the CNCs and the polymer matrix, and the CNCs and the regulating agent (*i.e.* water, in virtually all cases) were dominated by hydrogen bonds that involved the hydroxyl groups present on the CNC surfaces. As such, both the nature of the interactions as well as the effect of the stimulus were relatively unspecific. Way *et al.* speculated that it should be possible to modify the surface chemistry of the CNCs so that a more specific stimulus could be used to influence the various interactions and, therewith, the mechanical properties of their corresponding nanocomposites. Indeed, the authors showed that the functionalization of the surface of CNCs with either carboxylic acid or amine groups rendered them pH-responsive. At low pH values, at which the amine groups are protonated, the aminated CNCs form aqueous dispersions in water due to the electrostatic repulsions of the ammonium moieties that inhibit aggregation. However, gelation was observed at higher pH values, where the amine groups are neutral and attractive forces based on hydrogen bonding dominate. The opposite behavior was observed for carboxylated CNCs, which were found to disperse in water at high pH values and form gels in an acidic environment (Figure 12.6). These pH-responsive CNCs were incorporated into a PVAc matrix and the mechanical morphing of these materials was also found to be pH-dependent. More specifically, amine-functionalized CNC composites exhibited a decrease in modulus at low pH values due to the electrostatic repulsion of the charged amines, while the modulus of the corresponding carboxylic acid-functionalized CNC composites decreased at high pH values.[86]

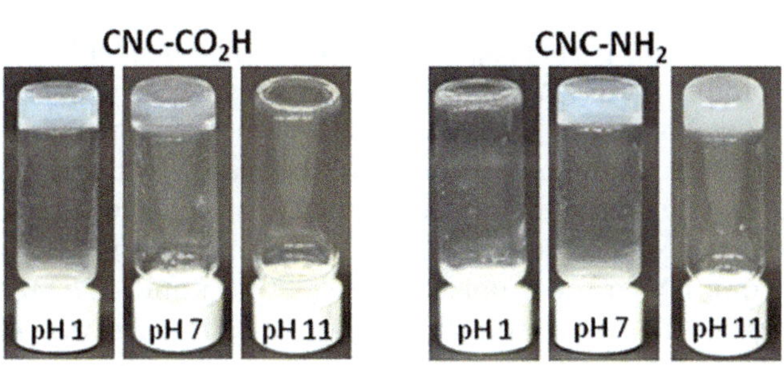

Figure 12.6 Pictures of aqueous dispersions of carboxylic acid- and amine-functionalized CNCs at different pH values, indicating the gelation of the material upon neutralizing the respective functional groups. Adapted with permission from A. E. Way, L. Hsu, K. Shanmuganathan, C. Weder and S. J. Rowan, *ACS Macro Lett.*, 2012, **1**, 1001–1006,[86] Copyright 2012 American Chemical Society.

Similarly, in a different example, positively charged chitosan was adsorbed onto the negatively charged surface of cotton-derived CNCs, and the resulting chitosan-modified CNCs were shown to exhibit pH-responsive behavior. Furthermore, the mechanical contrast between the dry and the wet material was significantly more prominent at either acidic or basic pH values, as opposed to the moderate stiffness contrast obtained at neutral pH values.[87] This was attributed to the presence of both positively and negatively charged groups in the composite owing to the chitosan and CNC surface charges, respectively.

12.4.2 Light-responsive Composites

A different stimulus that has been used to trigger the mechanical morphing of CNC-based nanocomposites is the irradiation with ultraviolet (UV) light, which is a "precise stimulus" as it can be applied in a spatially resolved manner with controlled intensity and can be delivered to discrete chemical entities using matched wavelength and polarization direction. In addition, optical stimulation is an elegant approach that permits non-contact control, thus can be triggered by a simple action of switching an external irradiation source on/off, while parameters such as intensity and wavelength of irradiation can be easily adjusted. Furthermore, light-triggered reactions generally generate limited amounts of byproducts since no additional reagents are needed.[88,89] The first system reported to combine the "sea cucumber approach" with optical switches involved the functionalization of CNCs with coumarin moieties before these were integrated into a rubbery EO–EPI matrix. Owing to the photodimerization of the coumarin motifs (and thereby establishing covalent crosslinks between CNCs) by way of [2 + 2] cycloaddition reactions, the E' of the composites increased both in the dry (199 to 291 MPa) and the water swollen (38 to 80 MPa) states upon exposure to UV irradiation ($\lambda > 300$ nm).[90]

Using a similar approach, light-responsive, mechanically switchable, photopatternable nanocomposites based on benzophenone-functionalized CNCs were designed, prepared, and studied.[91] The benzophenone-carrying CNCs were found to be highly photoreactive and, very much like other benzophenone-containing species, undergo radical-mediated reactions upon UV exposure. Nanocomposites were fabricated by incorporating 10 or 20% w/w of the benzophenone-carrying CNCs into an EO–EPI matrix. This caused a pronounced stiffness increase – for example, in the dry state, the E' increased from 4 MPa (neat polymer) to 222 MPa and 407 MPa for nanocomposites with 10% w/w and 20% w/w CNCs, respectively. E' increased further to 293 MPa and 508 MPa, respectively, upon irradiation with 365 nm UV light, on account of the formation of covalent bonds between the benzophenone-modified CNCs and between the benzophenone-modified CNCs and the EO–EPI matrix. The photoreaction reduced the level of aqueous swelling of the nanocomposites and therewith the extent of water-induced softening. The use of UV light to trigger these reactions permitted to

Figure 12.7 Pictures of EO–EPI/CNC composites containing benzophenone-functionalized CNCs, demonstrating the temporary programmed coil shape (a), which was partially irradiated to induce permanent cross-linking. Upon wetting (b), the non-irradiated part of the material returned to its permanent shape.
Reproduced from ref. 91 with permission from the Royal Society of Chemistry.

change the mechanical properties in a spatially-resolved manner. It is noteworthy that the benzophenone-EO–EPI/CNC composites also exhibited moderate shape-memory properties. These were a result of "fixing" the material to a temporary shape *via* wetting (thus softening) before shaping and drying and, upon immersion into water, the material returned to its original shape (Figure 12.7). This shape shifting was not observed for UV irradiated parts of the material, as covalent crosslinking imparted a permanent shape.

12.5 Application of Mechanically Adaptive Nanocomposites in Cortical Implants

While several potential applications for mechanically adaptive materials have been considered, one that has been further explored is the use of such composites in softening medical implants and, in particular, in the context of intracortical electrodes. A variety of such implants have been designed to provide an interface with the brain that is useful for neural recording and/or stimulating purposes,[92,93] but a common problem revolves around the limited long-term reliability of such probes. This is a result of the inflammatory response of the surrounding tissue upon insertion and post-implantation, not only due to chemical cues, but also micromotions and mechanical mismatch between the rigid implants and the soft cortical tissue.[94] To circumvent this issue, it was proposed that mechanically adaptive materials, which are sufficiently rigid to permit insertion but rapidly soften upon exposure to physiological conditions and ideally match the modulus of the tissue, are used.[34,95,96] Further considerations when designing such implants involve the compatibility of the material with the inflammation combating mechanisms of the immune-privileged cortical tissue, as often long term use of implants is limited by the decrease of

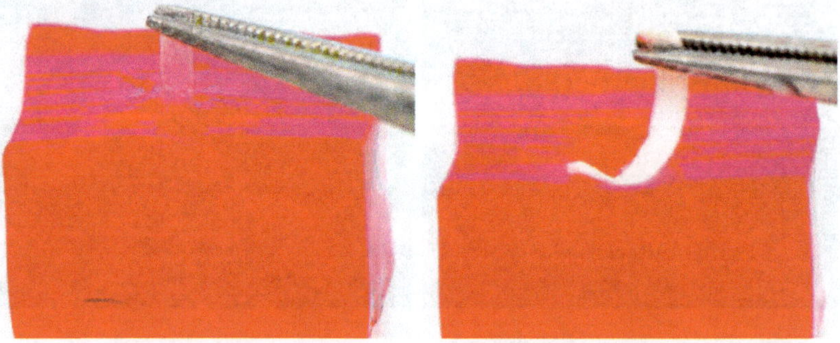

Figure 12.8 Pictures demonstrating the successful insertion (left) and the buckling (right) of a dry and a wet, respectively, PVAc/CNC nanocomposite intended for use in intracortical implants. The red gel is an artificial mechanical mimic of the brain.
Reprinted from K. Shanmuganathan, J. R. Capadona, S. J. Rowan and C. Weder, *ACS Appl. Mater. Interfaces*, 2010, **2**, 165–174,[66] Copyright 2009 American Chemical Society.

valuable proximal neuronal density.[97] This concept and other considerations for implantable intracortical devices are extensively discussed in recent reviews offering different strategies to improve implant performance.[98,99]

The first examples of cortical implants based on CNC-based nanocomposites relied on PVAc/CNC composites (see Section 12.3), which were indeed sufficiently rigid to allow for the insertion of shanks with a cross-section of only 100 μm×200 μm (Figure 12.8). Rapid softening of the initially rigid implants was not only observed *in vitro*, but also upon insertion into a rat cortex. Upon contact with the tissue, the samples uptake the brain fluid which, as previously discussed, plasticizes the PVAc and disengages the reinforcing CNC network, thereby causing a significant stiffness reduction.[100] It is noteworthy that cortical implants based on mechanically adaptive materials composed of PVAc/CNCs have indeed been found to reduce inflammation when compared to rigid silicon-based reference implants.[101,102] Nguyen *et al.* recently reported an in-depth histological evaluation of the neuroinflammatory response to PVAc/CNC implants in comparison to rigid PVAc-coated reference implants and found a significant attenuation of microglia activation accompanied by retention of a high neuron density around the mechanically compliant implants.[103]

In parallel to these encouraging findings, nanocomposites based on PVAc and either tunicate- or cotton-derived CNCs were utilized to fabricate variable-modulus microstructures. Despite the miniaturization of the probes, compared to previously studied bulk materials, the microstructures exhibited a similarly reversible change in modulus upon exposure to water, albeit with the switching occurring significantly faster. The probes were also successfully decorated with metal features using conventional photolithography methods, thus demonstrating the usefulness of CNC composites in microelectromechanical systems with mechanically adaptive properties.[104,105]

To further reduce the inflammatory response resulting from the acute trauma associated with the insertion of the implant into the brain, anti-inflammatory drugs such as resveratrol and curcumin were incorporated into the mechanically adaptive composites. This approach targets both mechanical and chemical factors that are believed to contribute to the chronic instability of the blood–brain barrier and the neuroinflammatory response associated with implanted intracortical microelectrodes. In a first study, physiologically responsive, mechanically adaptive polymer implants based on PVOH were modified to also release the anti-oxidant curcumin.[106] An *in vivo* study based on rats explored whether the combination of two independently effective mechanisms – softening of the implant and anti-oxidant release – led to synergistic effects. Over the first four weeks of the implantation, curcumin-releasing, mechanically adaptive implants led to higher neuron survival and a more stable blood–brain barrier than the controls made from neat PVOH. Twelve weeks post implantation, the benefits of the curcumin release were lost, and both sets of compliant materials (with and without curcumin) had no statistically significant differences in neuronal density distribution profiles. The second series of implants investigated were based on a mechanically adaptive, compliant nanocomposite based on PVAc/CNC, which reduced the tissue response and tissue strain.[107] This material was doped with various concentrations of the antioxidant resveratrol, with the objective of local and rapid delivery. Implants made from the neat nanocomposite reference and probes containing resveratrol were implanted into the cortical tissue of rats for up to sixteen weeks. Histochemical analysis suggested that at three days post-implantation, neither materials nor therapeutic approaches (independently or in combination) could alter the initial wound healing response. However, two weeks after implantation, the resveratrol-releasing nanocomposite showed a reduction in activated microglia/macrophages and an improvement in neuron density at the tissue–implant interface when compared to the neat nanocomposite reference. However, sixteen weeks post-implantation, when the antioxidant was exhausted, both implants exhibited similar tissue responses. Thus, the data show that antioxidant-releasing mechanically adaptive nanocomposites provide short-term, short-lived benefits due to drug release, and a long-term reduction in neuroinflammation on account of their mechanical adaptive, compliant nature.

12.6 Mechanically Adaptive Nanocomposites with Other Functions

12.6.1 Healable Materials

Enabled by the plasticizing effect of water within a cellulose network, a poly(vinyl amine)/nanofibrillated cellulose composite, formed *via* layer-by-layer assembly, was reported to exhibit mechanical switching upon exposure to water.[108] The mechanical reinforcement was dependent on the pH of the solution used to deposit the poly(vinyl amine) layers, as higher pH values enabled higher incorporation of the polymer within the composite. Furthermore, although the prepared films were brittle, they showed the propensity for self-healing upon wetting, which was attributed to the softening of the material and the subsequent mobility of the cationic polymer chains.

Another elegant self-healing system was developed utilizing ureidopyrimidone-functionalized CNCs and a telechelic polyethylene-polybutylene copolymer, also containing ureidopyrimidone (UPy) moieties.[109] The UPy moieties form strong quadruple hydrogen bonds with one another, which leads to the formation of supramolecular nanocomposites through chain-extension of the UPy-terminated telechelic polymer chains, and network formation that involves the UPy-decorated CNCs as network points. These nanocomposites showed significantly improved mechanical properties in comparison to the supramolecular polymer formed by assembly of the UPy-terminated telechelic chains alone. For example, at temperatures below the T_g of the polymer matrix, the E' of the composites was more than twice the E' value of the neat polymer (3.0 GPa and 1.2 GPa, respectively). At temperatures above the T_g of the polymer, the mechanical reinforcement due to the CNCs was more pronounced with the modulus of the composites being 247 MPa, whereas that of the neat polymer was merely 10 MPa. Upon exposure to UV radiation, the UPy motifs were excited and the absorbed energy was converted into heat. This caused temporary disengagement of the hydrogen-bonded UPy dimers, and led to a reversible decrease of the molecular weight and viscosity of the supramolecular polymers. As a result, deliberately introduced defects could be healed quickly and efficiently, even at a filler content of 20% w/w (Figure 12.9).

12.6.2 Shape Memory

Shape-memory materials have gained traction over the years as they have the capability of memorizing a permanent shape, while being programmable to adopt different temporary shapes. Upon exposure to an external stimulus, shape-memory materials will spontaneously recover their permanent shape as a direct result of their architecture, which consists of netpoints and switch units (Figure 12.10). The netpoints determine the permanent shape and can be of chemical or physical nature.[110–112] Unfortunately, many shape-memory materials display rather low strength and stiffness; combined with the fact that most shape-memory materials utilize heat as the trigger, this limit the scope of possible applications.[112] Incorporating stiff fillers or fibers is an effective way to increase the mechanical properties of SMPs and, in several recent examples, CNCs were chosen to fulfill that purpose.

Water-activated shape-memory materials are a particularly attractive subset of shape-memory polymers, as they can be considered for applications where temperature activation is not desirable. Combining the mechanically adaptive properties of CNC-containing composites and the shape-shifting abilities of shape-memory materials, Mendez *et al.*[113]

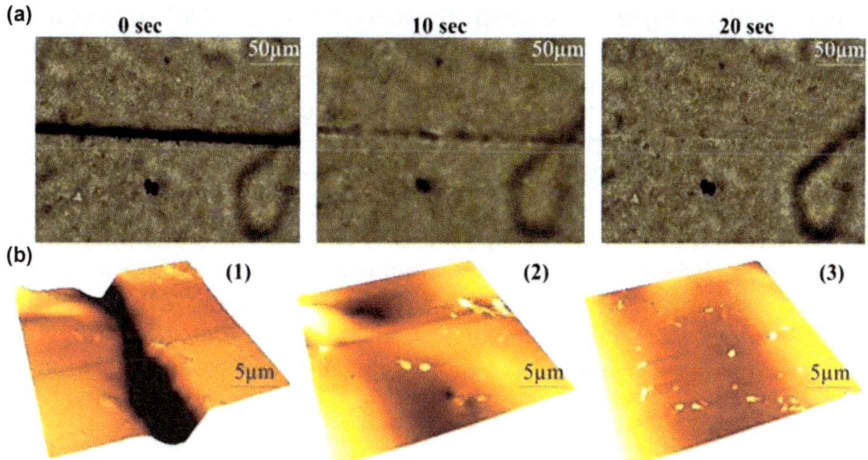

Figure 12.9 Optical microscopy (top) and atomic force microscopy (bottom) images of a scratched polymer/CNC nanocomposite film containing UPy-functionalized CNCs before (left), during (middle), and after (right) exposure to UV irradiation that results in the healing of the material. Reprinted with permission from M. V. Biyani, E. J. Foster and C. Weder, *ACS Macro Lett.*, 2013, **2**, 236–240,[109] Copyright 2013 American Chemical Society.

Figure 12.10 Pictures of the water-induced shape-memory effect displayed by a polyurethane/CNC-based composite, suggesting its potential use as self-propelled fishing lure. Reprinted with permission from J. Mendez, P. K. Annamalai, S. J. Eichhorn, R. Rusli, S. J. Rowan, E. J. Foster and C. Weder, *Macromolecules*, 2011, **44**, 6827–6835,[113] Copyright 2011 American Chemical Society.

prepared polyurethane/CNC nanocomposites whose temporary shape could be programmed and released upon exposure to water. A series of materials with CNC concentrations of 2–20% v/v were produced by solution blending the CNCs with the PU and films were prepared by compression molding. The introduction of CNCs led to a significant increase of the E' in the dry nanocomposites, from 14 MPa to 1.08 GPa for the 20% v/v composite. The level of reinforcement scaled with the CNC content; the stiffness was well described by the Halpin–Kardos model below the percolation limit, whereas the percolation model described correctly the characteristics above this limit. Upon exposure to water, the materials with a CNC content above the percolation limit swelled slightly and showed a significant decrease of E', for example from 1.08 GPa to 144 MPa in the case of the nanocomposite containing 20% v/v CNCs. As in many materials discussed herein, this effect is the result of competitive hydrogen bonding between water and the CNCs, which weakens the CNC network that is responsible for the reinforcement in the dry state. The mechanically adaptive behavior and high elasticity of wet materials are the basis for a shape-memory effect: in the water-swollen soft state, tensile deformation of the nanocomposites causes a significant level of uniaxial orientation of the CNCs, which is largely retained upon drying. Indeed, drying causes the formation of an elongated CNC network, which fixates a temporary shape and, upon wetting, the CNC network is again disengaged and the temporary shape relaxes to the original (permanent) shape, due to the elastic polymer matrix employed. Despite the slow shape shifting and significant hysteresis (which is common for polyurethane-based shape-memory polymers), this example showcases the potential evolution of shape-memory materials to also vary the mechanical properties, thus opening up the field of mechanically adaptive water-activated shape-memory composites.[114–121] Another recent example involves the preparation of nanocomposites consisting of a hydrophobic poly(glycerol sebacate urethane) polymer matrix and CNCs. As expected, the presence of the CNCs improved the tensile strength and the modulus of the polymer as a result of the CNC percolating network while, upon exposure to water, the mechanical properties of the composite were significantly inferior. Furthermore, the material exhibited shape-memory properties with a high stable shape fixity (up to 98%) and high shape recovery ratio (92–99%) being reported.[122] CNCs were covalently crosslinked *via* the immobilization of the end groups of telechelic polyethylene glycol and polycaprolactone polymers to their surface. Upon hydration, softening of the crosslinked composite was observed while water was also used to induce a shape-memory response whereby the CNCs imparted the rigid phase. Due to the biocompatible nature of all the components of the material, the cytotoxicity of the samples was found to be minimal, thus suggesting the potential use of such mechanically adaptive, water activated shape-memory materials in biomedical applications.[123,124] In another example, the surface of the CNCs was functionalized with carboxylic and pyridine moieties to allow the pH-responsive regulation of the CNC network rigidity. When combined with a polycaprolactone–polyethylene glycol–polycaprolactone triblock polymer matrix, the composite material obtained shape-memory properties which were dependent on the pH of the solution.[125]

12.6.3 Actuators

Owing to the effect of water on the hydrogen bonding abilities of cellulose, a few systems have emerged whereby humidity is used as a trigger for actuators. As such, it was recently shown that thin films composed of cellulose fibers rapidly and directionally bend upon exposure to water vapors due to the different water content on opposite film sides.[126] Such films have also been reinforced with carbon nanotubes to enhance their mechanical properties and response to humidity.[127] In a similar concept, films composed of microcrystalline cellulose that had been modified with stearoyl moieties to endow partial hydrophobicity were used. Depending on the extent of stearoyl surface functionalization, the films were found to deform in the presence of humidity, thus demonstrating their potential use in soft water-activated actuators.[128]

12.7 Related Examples of Mechanically Adaptive Materials

Expanding on the aforementioned concept of CNC-composite intracortical implants, fibers consisting solely of PVOH were fabricated and were also found to exhibit mechanically adaptive properties in response to exposure to water. Upon slight swelling, the stiffness of the small fibers was found reduced *ca.* 200 times, nonetheless retaining their optical properties. Such materials may be the basis for a new generation of adaptable optical fibers for optogenetics.[129] With regards to the biocompatibility of the fibers, a known anti-inflammatory, curcumin, was loaded into polyvinyl alcohol implants and subsequently released upon implantation in rat brains. As a result, decreased implant-related neuroinflammation was reported, while the implants maintained their mechanically adaptive properties.[106]

Poly(vinyl alcohol) has also been shown to provide mechanically adaptive properties to composite films comprised of a PVOH electrospun mat and a PVAc or EO–EPI matrix. Films made from such composites exhibited controlled response to hydration that resulted in a two-fold decrease of the storage modulus which, in the case of the PVAc composite, was fully reversible.[130,131] Expanding on this concept and in a similar motif to the aforementioned CNC composites, Korley and coworkers have recently showcased the potential use of montmorillonite in mechanically adaptive composite materials. Montmorillonite is a layered silicate with high aspect ratio and surface area and is often used in polymer composites to mechanically reinforce the resulting material, while its dispersibility and tunable surface chemistry make its use particularly attractive.[132] In this case, the incorporation of montmorillonite in the PVOH fibrous filler resulted in enhanced moduli when the polymer matrix consisted of EO–EPI, as opposed to the limited enhancement observed in the case of a PVAc matrix. Nonetheless, both composites showed mechanically switching properties upon hydration, with the weight fraction of montmorillonite affecting the mechanical contrast as well as the response time.[133]

Another example of PVOH being used in a mechanically adaptive system involves the formation of a PVOH/ferritin microfiber hydrogel *via* electrospinning. Ferritin is an iron-containing protein with both amine and carboxylic acid groups on its surface available for hydrogen bonding. When combined with the PVOH matrix, ferritin was found to reinforce the material as a result of these non-covalent interactions, without however affecting its swelling in water. Furthermore, upon exposure to basic and acidic aqueous solutions, the hydrogel contracted and expanded, respectively, thus demonstrating its pH-regulated actuating properties.[134]

Water-activated shape-memory properties were realized for nanocomposites composed of poly(vinyl alcohol) particles and polyurethane. PVOH acts as the activating phase which is plasticized upon exposure to water leading to a reversible and dramatic decrease in the material modulus. It was noted that the moduli of the composites containing different amounts of PVOH could be predicted by the percolation model, suggesting the existence of a percolating network within the material. Furthermore, due to the resilient nature of the urethane content, the composites exhibited water-activated shape-memory properties.[135] Recent work carried out in our group has also shown thermally induced shape-memory behavior of composites made from electrospun PVOH mats that were compacted with a thermoplastic polyether-amide elastomer. Indeed, at a composition of 20 wt% PVOH the material exhibited a fixity of *ca.* 66% at low strain and a recovery of *ca.* 98%, while the storage modulus was 4–5 times greater than that of the neat elastomer, thus showing the aforementioned mechanical reinforcement properties of PVOH.[136]

12.8 Summary and Outlook

Nature has inspired science to generate a wealth of "smart" materials that, upon benign cues, drastically alter their properties. Mechanically adaptive polymers and composites are a class of such materials that take advantage of energetically insignificant stimuli to alter their stiffness. The uniqueness of such adaptable materials is the fine line that separates a remarkably stiff solid from a compliant elastomer, which can be as fine as the pia mater of the brain. Even though to this day the amount and variety of systems that bear mechanically adaptive properties is limited, it is not inconceivable that more will emerge, while existing systems are constantly being improved to adapt to their corresponding applications. To that extent, it is apparent that polymer/CNC polymer composites have gained ground on their use in biomedical applications, being intracortical implants the most prominent one. Nonetheless, with actuators and self-healing materials being in the spotlight of advance materials science, the applicability of mechanically adaptive composites will potentially expand.

Acknowledgements

This work was supported by the Swiss National Science Foundation through the National Centre of Competence in Research (NCCR) Bio-Inspired Materials and the Adolphe Merkle Foundation.

References

1. M. A. C. Stuart, W. T. S. Huck, J. Genzer, M. Muller, C. Ober, M. Stamm, G. B. Sukhorukov, I. Szleifer, V. V. Tsukruk, M. Urban, F. Winnik, S. Zauscher, I. Luzinov and S. Minko, *Nat. Mater.*, 2010, **9**, 101–113.
2. D. Roy, J. N. Cambre and B. S. Sumerlin, *Prog. Polym. Sci.*, 2010, **35**, 278–301.
3. L. Hsu, C. Weder and S. J. Rowan, *J. Mater. Chem.*, 2011, **21**, 2812–2822.
4. N. D. Wanasekara and L. T. J. Korley, *J. Polym. Sci., Part B: Polym. Phys.*, 2013, **51**, 463–467.
5. E. A. Peraza-Hernandez, D. J. Hartl, J. Richard, J. Malak and D. C. Lagoudas, *Smart Mater. Struct.*, 2014, **23**, 094001.
6. H. Meng and G. Li, *J. Mater. Chem. A*, 2013, **1**, 7838–7865.
7. M. D. Hager, S. Bode, C. Weber and U. S. Schubert, *Prog. Polym. Sci.*, 2015, **49–50**, 3–33.
8. Q. Zhao, H. J. Qi and T. Xie, *Prog. Polym. Sci.*, 2015, **49–50**, 79–120.
9. A. R. Studart and R. M. Erb, *Soft Matter*, 2014, **10**, 1284–1294.
10. A. R. Studart, *Angew. Chem., Int. Ed.*, 2015, **54**, 3400–3416.
11. E. I. Saavedra Flores, M. I. Friswell and Y. Xia, *J. Intell. Mater. Syst. Struct.*, 2013, **24**, 529–540.
12. A. J. P. Norbert, *Plasticization and Plasticizer Processes*, American Chemical Society, 1965.
13. W. J. Cousins, *Wood Sci. Technol.*, 1978, **12**, 161–167.
14. G. R. Plaza, G. V. Guinea, J. Pérez-Rigueiro and M. Elices, *J. Polym. Sci., Part B: Polym. Phys.*, 2006, **44**, 994–999.
15. M. A. O'Neill, S. Eberhard, P. Albershcim and A. G. Darvill, *Science*, 2001, **294**, 846 849.
16. A. Fleischer, M. A. O'Neill and R. Ehwald, *Plant Physiol.*, 1999, **121**, 829–838.
17. G. M. Luz and J. F. Mano, *Philos. Trans. R. Soc., A*, 2009, **367**, 1587 1605.
18. Z. Zhang, Y.-W. Zhang and H. Gao, *Proc. R. Soc. B*, 2011, **278**, 519–525.
19. A. R. Studart, *Adv. Mater.*, 2012, **24**, 5024–5044.
20. D. A. Stone and L. T. J. Korley, *Macromolecules*, 2010, **43**, 9217–9226.
21. T. Motokawa, *Comp. Biochem. Physiol., C: Comp. Pharmacol.*, 1981, **70**, 41–48.
22. F. Thurmond and J. Trotter, *J. Exp. Biol.*, 1996, **199**, 1817–1828.
23. J. Trotter and T. Koob, *J. Exp. Biol.*, 1995, **198**, 1951–1961.
24. T. Motokawa and A. Tsuchi, *Biol. Bull.*, 2003, **205**, 261–275.
25. T. Motokawa, *Comp. Biochem. Physiol., B: Comp. Biochem.*, 1994, **109**, 613–622.
26. G. K. Szulgit and R. E. Shadwick, *J. Exp. Biol.*, 2000, **203**, 1539–1550.
27. J. A. Trotter, G. Lyons-Levy, K. Chino, T. J. Koob, D. R. Keene and M. A. L. Atkinson, *Matrix Biol.*, 1999, **18**, 569–578.
28. J. A. Trotter, G. Lyons-Levy, D. Luna, T. J. Koob, D. R. Keene and M. A. L. Atkinson, *Matrix Biol.*, 1996, **15**, 99–110.

29. J. P. Tipper, G. Lyons-Levy, M. A. Atkinson and J. A. Trotter, *Matrix Biol.*, 2002, **21**, 625–635.
30. T. J. Koob, M. M. Koob-Emunds and J. A. Trotter, *J. Exp. Biol.*, 1999, **202**, 2291–2301.
31. J. A. Trotter, J. Tipper, G. Lyons-Levy, K. Chino, A. H. Heuer, Z. Liu, M. Mrksich, C. Hodneland, W. S. Dillmore, T. J. Koob, M. M. Koob-Emunds, K. Kadler and D. Holmes, *Biochem. Soc. Trans.*, 2000, **28**, 357–362.
32. K. Shanmuganathan, J. R. Capadona, S. J. Rowan and C. Weder, *Prog. Polym. Sci.*, 2010, **35**, 212–222.
33. S. Spoljaric, A. Salminen, N. Luong and J. Seppälä, *Cellulose*, 2013, **20**, 2991–3005.
34. J. R. Capadona, D. J. Tyler, C. A. Zorman, S. J. Rowan and C. Weder, *MRS Bull.*, 2012, **37**, 581–589.
35. I. A. Sacui, R. C. Nieuwendaal, D. J. Burnett, S. J. Stranick, M. Jorfi, C. Weder, E. J. Foster, R. T. Olsson and J. W. Gilman, *ACS Appl. Mater. Interfaces*, 2014, **6**, 6127–6138.
36. L. P. Novo, J. Bras, A. García, N. Belgacem and A. A. S. Curvelo, *ACS Sustainable Chem. Eng.*, 2015, **3**, 2839–2846.
37. J. Sugiyama, J. Persson and H. Chanzy, *Macromolecules*, 1991, **24**, 2461–2466.
38. M. A. Hubbe, O. J. Rojas, L. A. Lucia and M. Sain, *BioResources*, 2008, **3**, 929–980.
39. R. J. Moon, A. Martini, J. Nairn, J. Simonsen and J. Youngblood, *Chem. Soc. Rev.*, 2011, **40**, 3941–3994.
40. I. Siró and D. Plackett, *Cellulose*, 2010, **17**, 459–494.
41. S. J. Eichhorn, A. Dufresne, M. Aranguren, N. E. Marcovich, J. R. Capadona, S. J. Rowan, C. Weder, W. Thielemans, M. Roman, S. Renneckar, W. Gindl, S. Veigel, J. Keckes, H. Yano, K. Abe, M. Nogi, A. N. Nakagaito, A. Mangalam, J. Simonsen, A. S. Benight, A. Bismarck, L. A. Berglund and T. Peijs, *J. Mater. Sci.*, 2010, **45**, 1–33.
42. Y. Habibi, L. A. Lucia and O. J. Rojas, *Chem. Rev.*, 2010, **110**, 3479–3500.
43. M. M. de Souza Lima and R. Borsali, *Macromol. Rapid Commun.*, 2004, **25**, 771–787.
44. J. R. Capadona, K. Shanmuganathan, S. Trittschuh, S. Seidel, S. J. Rowan and C. Weder, *Biomacromolecules*, 2009, **10**, 712–716.
45. A. Šturcová, G. R. Davies and S. J. Eichhorn, *Biomacromolecules*, 2005, **6**, 1055–1061.
46. S. Camarero Espinosa, T. Kuhnt, E. J. Foster and C. Weder, *Biomacromolecules*, 2013, **14**, 1223–1230.
47. W. T. Y. Tze, D. J. Gardner, C. P. Tripp and S. C. O'Neill, *J. Adhes. Sci. Technol.*, 2006, **20**, 1649–1668.
48. M.-C. Li, Q. Wu, K. Song, S. Lee, Y. Qing and Y. Wu, *ACS Sustainable Chem. Eng.*, 2015, **3**, 821–832.
49. M. Jorfi, M. N. Roberts, E. J. Foster and C. Weder, *ACS Appl. Mater. Interfaces*, 2013, **5**, 1517–1526.

50. K. L. Dagnon, A. E. Way, S. O. Carson, J. Silva, J. Maia and S. J. Rowan, *Macromolecules*, 2013, **46**, 8203–8212.
51. M. Jorfi and E. J. Foster, *J. Appl. Polym. Sci.*, 2015, **132**, 41719.
52. D. Klemm, F. Kramer, S. Moritz, T. Lindström, M. Ankerfors, D. Gray and A. Dorris, *Angew. Chem., Int. Ed.*, 2011, **50**, 5438–5466.
53. V. Favier, H. Chanzy and J. Y. Cavaille, *Macromolecules*, 1995, **28**, 6365–6367.
54. V. Favier, J. Y. Cavaille, G. R. Canova and S. C. Shrivastava, *Polym. Eng. Sci.*, 1997, **37**, 1732–1739.
55. M. Mariano, N. El Kissi and A. Dufresne, *J. Polym. Sci., Part B: Polym. Phys.*, 2014, **52**, 791–806.
56. R. Sinko, X. Qin and S. Keten, *MRS Bull.*, 2015, **40**, 340–348.
57. R. T. Woodhams, G. Thomas and D. K. Rodgers, *Polym. Eng. Sci.*, 1984, **24**, 1166–1171.
58. M. A. S. Azizi Samir, F. Alloin and A. Dufresne, *Biomacromolecules*, 2005, **6**, 612–626.
59. N. Ouali, J. Y. Cavaille and J. Perez, *Plast., Rubber Compos. Process. Appl.*, 1991, **16**, 55–60.
60. M. Takayanagi, S. Minami and S. Uemura, *J. Polym. Sci., Part C: Polym. Symp.*, 1964, 113–122.
61. J. C. Halpin and J. L. Kardos, *J. Appl. Phys.*, 1972, **43**, 2235–2241.
62. M. R. Kamal and V. Khoshkava, *Carbohydr. Polym.*, 2015, **123**, 105–114.
63. J. R. Capadona, K. Shanmuganathan, D. J. Tyler, S. J. Rowan and C. Weder, *Science*, 2008, **319**, 1370–1374.
64. J. R. Capadona, O. Van Den Berg, L. A. Capadona, M. Schroeter, S. J. Rowan, D. J. Tyler and C. Weder, *Nat. Nanotechnol.*, 2007, **2**, 765–769.
65. K. Shanmuganathan, J. R. Capadona, S. J. Rowan and C. Weder, *J. Mater. Chem.*, 2010, **20**, 180–186.
66. K. Shanmuganathan, J. R. Capadona, S. J. Rowan and C. Weder, *ACS Appl. Mater. Interfaces*, 2010, **2**, 165–174.
67. K. L. Dagnon, K. Shanmuganathan, C. Weder and S. J. Rowan, *Macromolecules*, 2012, **45**, 4707–4715.
68. J. Sapkota, M. Jorfi, C. Weder and E. J. Foster, *Macromol. Rapid Commun.*, 2014, **35**, 1747–1753.
69. M. Liu, Q. Peng, B. Luo and C. Zhou, *Eur. Polym. J.*, 2015, **68**, 190–206.
70. E. Chiellini, P. Cinelli, V. I. Hieva and M. Martera, *Biomacromolecules*, 2008, **9**, 1007–1013.
71. S.-H. Hyon and Y. Ikada, *US Pat.*, 4874562, 1989.
72. P. I. Lee, *US Pat.*, 4559186, 1985.
73. R. A. Janssen, P. I. Lee and E. M. Ajello, *US Pat.*, 5174929, 1992.
74. A. R. Nectow, K. G. Marra and D. L. Kaplan, *Tissue Eng., Part B*, 2012, **18**, 40–50.
75. A. J. Uddin, J. Araki and Y. Gotoh, *Polym. Int.*, 2011, **60**, 1230–1239.
76. A. J. Uddin, J. Araki and Y. Gotoh, *Biomacromolecules*, 2011, **12**, 617–624.

77. S. A. Paralikar, J. Simonsen and J. Lombardi, *J. Membr. Sci.*, 2008, **320**, 248–258.
78. M. Roohani, Y. Habibi, N. M. Belgacem, G. Ebrahim, A. N. Karimi and A. Dufresne, *Eur. Polym. J.*, 2008, **44**, 2489–2498.
79. H. Luo, J. Hu, Y. Zhu, J.-Y. Wu, S. Zhang, Y. Fan and G. Ye, *J. Appl. Polym. Sci.*, 2013, **127**, 675–681.
80. J. D. Fox, J. R. Capadona, P. D. Marasco and S. J. Rowan, *J. Am. Chem. Soc.*, 2013, **135**, 5167–5174.
81. R. Rusli, K. Shanmuganathan, S. J. Rowan, C. Weder and S. J. Eichhorn, *Biomacromolecules*, 2010, **11**, 762–768.
82. R. Sinko and S. Keten, *Appl. Phys. Lett.*, 2014, **105**, 243702.
83. P. K. Annamalai, K. L. Dagnon, S. Monemian, E. J. Foster, S. J. Rowan and C. Weder, *ACS Appl. Mater. Interfaces*, 2014, **6**, 967–976.
84. Y. Bréchet, J. Y. Cavaillé, E. Chabert, L. Chazeau, R. Dendievel, L. Flandin and C. Gauthier, *Adv. Eng. Mater.*, 2001, **3**, 571–577.
85. I. S. Gunes, C. Perez-Bolivar, F. Cao, G. A. Jimenez, P. Anzenbacher and S. C. Jana, *J. Mater. Chem.*, 2010, **20**, 3467–3474.
86. A. E. Way, L. Hsu, K. Shanmuganathan, C. Weder and S. J. Rowan, *ACS Macro Lett.*, 2012, **1**, 1001–1006.
87. T. Wu, Y. Su and B. Chen, *ChemPhysChem*, 2014, **15**, 2794–2800.
88. H. Y. Jiang, S. Kelch and A. Lendlein, *Adv. Mater.*, 2006, **18**, 1471–1475.
89. J.-M. Schumers, C.-A. Fustin and J.-F. Gohy, *Macromol. Rapid Commun.*, 2010, **31**, 1588–1607.
90. M. V. Biyani, C. Weder and E. J. Foster, *Polym. Chem.*, 2014, **5**, 5501–5508.
91. M. V. Biyani, M. Jorfi, C. Weder and E. J. Foster, *Polym. Chem.*, 2014, **5**, 5716–5724.
92. D. A. Robinson, *Proc. IEEE*, 1968, **56**, 1065–1071.
93. M. J. Nelson, P. Pouget, E. A. Nilsen, C. D. Patten and J. D. Schall, *J. Neurosci. Methods*, 2008, **169**, 141–157.
94. D. E. Arreaga-Salas, A. Avendaño-Bolívar, D. Simon, R. Reit, A. Garcia-Sandoval, R. L. Rennaker and W. Voit, *ACS Appl. Mater. Interfaces*, 2015, **7**, 26614–26623.
95. A. Mercanzini, K. Cheung, D. Buhl, M. Boers, A. Maillard, P. Colin, J. C. Bensadoun, A. Bertsch, A. Carleton and P. Renaud, Proc. IEEE 20th Int. MEMS Conf., 2007.
96. S. Jeyakumar, C. M. David and R. K. Daryl, *J. Neural. Eng.*, 2005, **2**, 103.
97. G. C. McConnell, H. D. Rees, A. I. Levey, C.-A. Gutekunst, R. E. Gross and R. V. Bellamkonda, *J. Neural. Eng.*, 2009, **6**, 056003.
98. M. Jorfi, J. L. Skousen, C. Weder and J. R. Capadona, *J. Neural. Eng.*, 2015, **12**, 011001.
99. D. R. Merrill, *Curr. Opin. Solid State Mater. Sci.*, 2014, **18**, 329–336.
100. J. P. Harris, A. E. Hess, S. J. Rowan, C. Weder, C. A. Zorman, D. J. Tyler and J. R. Capadona, *J. Neural. Eng.*, 2011, **8**, 046010.
101. A. Sridharan, J. K. Nguyen, J. R. Capadona and J. Muthuswamy, *J. Neural. Eng.*, 2015, **12**, 036002.

102. J. P. Harris, J. R. Capadona, R. H. Miller, B. C. Healy, K. Shanmuganathan, S. J. Rowan, C. Weder and D. J. Tyler, *J. Neural. Eng.*, 2011, **8**, 066011.
103. J. K. Nguyen, D. J. Park, J. L. Skousen, A. E. Hess-Dunning, D. J. Tyler, S. J. Rowan, C. Weder and J. R. Capadona, *J. Neural. Eng.*, 2014, **11**, 056014.
104. A. E. Hess, J. R. Capadona, K. Shanmuganathan, L. Hsu, S. J. Rowan, C. Weder, D. J. Tyler and C. A. Zorman, *J. Micromech. Microeng.*, 2011, **21**, 054009.
105. A. E. Hess, K. Shanmuganathan, J. R. Capadona, L. Hsu, S. Rowan, C. Weder, D. J. Tyler and C. A. Zorman, Proc. IEEE 24th Int. MEMS Conf., 2011.
106. K. A. Potter, M. Jorfi, K. T. Householder, E. J. Foster, C. Weder and J. R. Capadona, *Acta Biomater.*, 2014, **10**, 2209–2222.
107. J. K. Nguyen, M. Jorfi, K. L. Buchanan, D. J. Park, E. J. Foster, D. J. Tyler, S. J. Rowan, C. Weder and J. R. Capadona, *Acta Biomater.*, 2016, **29**, 81–93.
108. R. Merindol, S. Diabang, O. Felix, T. Roland, C. Gauthier and G. Decher, *ACS Nano*, 2015, **9**, 1127–1136.
109. M. V. Biyani, E. J. Foster and C. Weder, *ACS Macro Lett.*, 2013, **2**, 236–240.
110. A. Lendlein and S. Kelch, *Angew. Chem., Int. Ed.*, 2002, **41**, 2034–2057.
111. J. Leng, X. Lan, Y. Liu and S. Du, *Prog. Mater. Sci.*, 2011, **56**, 1077–1135.
112. C. Liu, H. Qin and P. T. Mather, *J. Mater. Chem.*, 2007, **17**, 1543–1558.
113. J. Mendez, P. K. Annamalai, S. J. Eichhorn, R. Rusli, S. J. Rowan, E. J. Foster and C. Weder, *Macromolecules*, 2011, **44**, 6827–6835.
114. Q. Bai, G. Zhang, B. Xu, X. Feng, H. Jiang and H. Li, *RSC Adv.*, 2015, **5**, 91213–91217.
115. S. Chen, F. Mo, F. J. Stadler, S. Chen, Z. Ge and H. Zhuo, *J. Mater. Chem. B*, 2015, **3**, 6645–6655.
116. W. Wang, H. Lu, Y. Liu and J. Leng, *J. Mater. Chem. A*, 2014, **2**, 5441–5449.
117. X. Qi, X. Yao, S. Deng, T. Zhou and Q. Fu, *J. Mater. Chem. A*, 2014, **2**, 2240–2249.
118. Y. Zhu, J. Hu, H. Luo, R. J. Young, L. Deng, S. Zhang, Y. Fan and G. Ye, *Soft Matter*, 2012, **8**, 2509–2517.
119. I. Dallmeyer, S. Chowdhury and J. F. Kadla, *Biomacromolecules*, 2013, **14**, 2354–2363.
120. Y. Liu, Y. Li, H. Chen, G. Yang, X. Zheng and S. Zhou, *Carbohydr. Polym.*, 2014, **104**, 101–108.
121. H. Luo, J. Hu, Y. Zhu, S. Zhang, Y. Fan and G. Ye, *J. Appl. Polym. Sci.*, 2012, **125**, 657–665.
122. T. Wu, M. Frydrych, K. O'Kelly and B. Chen, *Biomacromolecules*, 2014, **15**, 2663–2671.
123. Y. Liu, Y. Li, G. Yang, X. Zheng and S. Zhou, *ACS Appl. Mater. Interfaces*, 2015, **7**, 4118–4126.

124. W. M. Huang, B. Yang, L. An, C. Li and Y. S. Chan, *Appl. Phys. Lett.*, 2005, **86**, 114105.
125. Y. Li, H. Chen, D. Liu, W. Wang, Y. Liu and S. Zhou, *ACS Appl. Mater. Interfaces*, 2015, **7**, 12988–12999.
126. M. Wang, X. Tian, R. H. A. Ras and O. Ikkala, *Adv. Mater. Interfaces*, 2015, **2**, DOI: 10.1002/admi.201500080.
127. C. Echeverria, L. E. Aguirre, E. G. Merino, P. L. Almeida and M. H. Godinho, *ACS Appl. Mater. Interfaces*, 2015, **7**, 21005–21009.
128. K. Zhang, A. Geissler, M. Standhardt, S. Mehlhase, M. Gallei, L. Chen and C. Marie Thiele, *Sci. Rep.*, 2015, **5**, 11011.
129. M. Jorfi, G. Voirin, E. J. Foster and C. Weder, *Opt. Lett.*, 2014, **39**, 2872–2875.
130. N. D. Wanasekara, D. A. Stone, G. E. Wnek and L. T. J. Korley, *Macromolecules*, 2012, **45**, 9092–9099.
131. D. A. Stone, N. D. Wanasekara, D. H. Jones, N. R. Wheeler, E. Wilusz, W. Zukas, G. E. Wnek and L. T. J. Korley, *ACS Macro Lett.*, 2012, **1**, 80–83.
132. M. Alexandre and P. Dubois, *Mater. Sci. Eng., R*, 2000, **28**, 1–63.
133. N. D. Wanasekara, L. E. Matolyak and L. T. J. Korley, *ACS Appl. Mater. Interfaces*, 2015, **7**, 22970–22979.
134. M. K. Shin, G. M. Spinks, S. R. Shin, S. I. Kim and S. J. Kim, *Adv. Mater.*, 2009, **21**, 1712–1715.
135. T. Wu, K. O'Kelly and B. Chen, *Eur. Polym. J.*, 2014, **53**, 230–237.
136. A. Shirole, J. Sapkota, E. J. Foster and C. Weder, *ACS Appl. Mater. Interfaces*, 2016, **8**, 6701–6708.
137. M. A. Meyers, P.-Y. Chen, M. I. Lopez, Y. Seki and A. Y. M. Lin, *J. Mech. Behav. Biomed. Mater.*, 2011, **4**, 626–657.
138. X. Li, W.-C. Chang, Y. J. Chao, R. Wang and M. Chang, *Nano Lett.*, 2004, **4**, 613–617.
139. R. Menig, M. H. Meyers, M. A. Meyers and K. S. Vecchio, *Acta Mater.*, 2000, **48**, 2383–2398.
140. F. Barthelat, H. Tang, P. D. Zavattieri, C. M. Li and H. D. Espinosa, *J. Mech. Phys. Solids*, 2007, **55**, 306–337.

CHAPTER 13

Bio-inspired Polymer Artificial Muscles

SINA NAFICY,*[a] GEOFFREY M. SPINKS[a] AND RAY H. BAUGHMAN[b]

[a] Intelligent Polymer Research Institute and ARC Centre of Excellence for Electromaterials Science, University of Wollongong, Wollongong NSW 2500, Australia; [b] Alan G. MacDiarmid NanoTech Institute, University of Texas at Dallas, Richardson TX 75083, USA
*Email: snaficy@uow.edu.au

13.1 Introduction

The application of muscle to voluntarily control movement and allow locomotion in its various forms is fundamental to animal life and an essential distinguishing feature between animals and plants. One type of animal – humans – seems inherently fascinated by the challenge to emulate nature, and various different types of 'artificial muscles' have been introduced over the past several decades. Some of these different artificial muscle materials can produce muscle-like performances that match, and even exceed in some cases, the performance of natural muscle. As a result of this work, we are getting closer to simultaneously reproducing muscle performance in all of its key attributes: force, movement, speed, efficiency, and scalability.

Artificial muscles are driven by materials that are able to change their shape or volume when stimulated. Many different types of artificial muscles are available and may be distinguished by the nature of the input

RSC Polymer Chemistry Series No. 22
Bio-inspired Polymers
Edited by Nico Bruns and Andreas F. M. Kilbinger

Published by the Royal Society of Chemistry, www.rsc.org

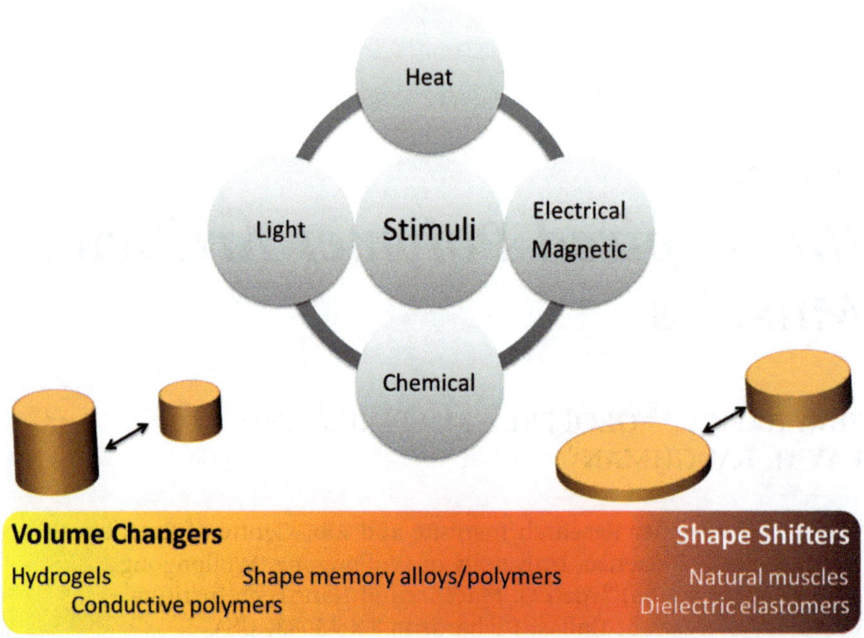

Figure 13.1 Different types of actuator materials and their stimuli.

stimulation and type of response. As illustrated in Figure 13.1, the stimulation can be electrical, thermal, chemical, photonic, or magnetic, and the response is a volume change and/or change in shape. The changing dimensions of the actuator material can be used to generate movement and force and perform useful mechanical work. Interestingly, natural muscle undergoes little volume change during its 'contraction' and can be considered as a shape-shifting type of actuator material.

In common with muscle, actuator materials have particular advantages over conventional motors and engines, particularly in applications where space is limited. The motor's power to weight ratio is a convenient comparison, since power output is defined by the motor's force, displacement, and speed, and a clear correlation exists between the performance of all classes of motors and engines and their respective size (Figure 13.2).[1] For engines where expanding gases or pressurized fluids push pistons or turn turbines, down-scaling is unfavourable since the maximum force exerted decreases with the cross-sectional area and the velocity decreases with the linear dimensions. Similarly, the torque from electric motors decreases with the motor size. Consequently, the power-weight ratio tends to scale with the motor mass so that tiny powerful motors cannot be built using conventional systems. In contrast, both muscle and some actuator materials show power-weight ratios of over 100 $W\,kg^{-1}$, which is significantly higher than for similarly sized motors of 1 kg or less. Simple construction and noiseless operation can be additional advantages of artificial muscles.

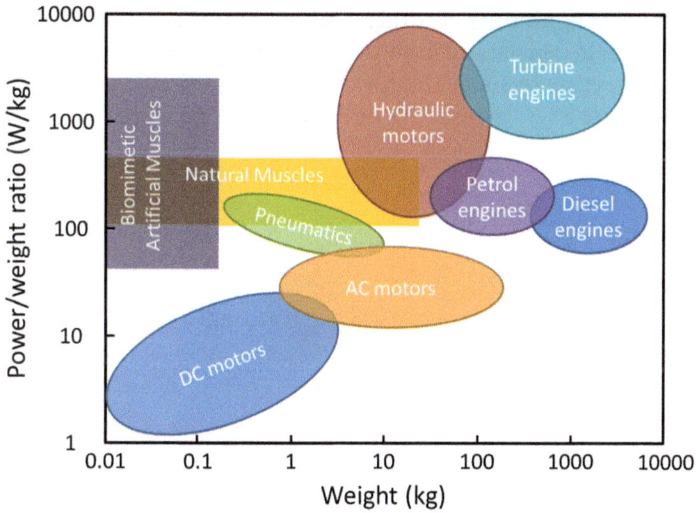

Figure 13.2 Comparison of power-to-weight ratios of various engines and motors, natural skeletal muscle and high-strain artificial muscles.

Artificial muscles are truly bio-inspired materials. This chapter compares and contrasts various different polymer-based artificial muscles (PAMs) with each other and with natural skeletal muscle. The different mechanisms of shape and/or volume change are described, and the performances in terms of dimensional changes, forces generated, and speeds of response are presented. The progress made in recent years with regards to the performance of artificial muscles is summarized along with suggestions for further work. Since many different types of polymer based artificial muscles have been developed it is useful to compare their performances and attributes. First, however, the structure and function of natural skeletal muscle is described to provide a benchmark for the evaluation of artificial muscle systems. Finally, bending type PAMs are briefly reviewed as they have been used to develop some simple, biologically-inspired swimming and crawling robots.

13.2 Natural Muscle

The enormous scale of the challenge to build a truly biomimetic muscle can be appreciated by considering what is known about natural muscle structure. The contractions of these motors involve a highly complex and co-ordinated sequence of electrical, chemical, and physical phenomena within a composite, gel-like polymeric material that is known to possess a detailed hierarchical structure, stretching from the nano-scale assembly of proteins through the cellular fibrillar textures to the macroscopic tissue. The exact function of many of these features is still the subject of on-going research. In general, the skeletal muscles consist of *tendons* (non-active) and *muscle belly* (active). While tendons mainly provide muscle connectivity to hard bones, the muscle belly is able to contract and generate a contractile force. The muscle belly itself consists of bundles of muscle cells, called *muscle fibres*, and other connecting tissues. The hierarchical organization of a muscle belly is shown in Figure 13.3,[2] where the individual muscle fibres surrounded by a

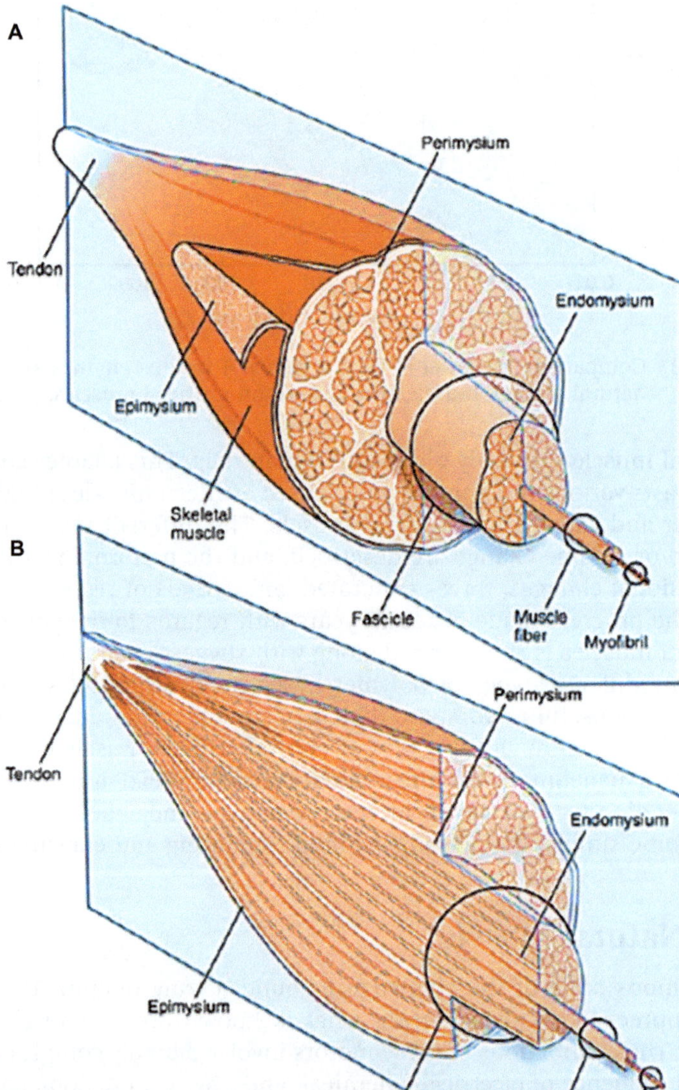

Figure 13.3 The hierarchical structure of skeletal muscle. (Reproduced from ref. 2 with permission from John Wiley and Sons. Copyright © 2011 Wiley Periodicals, Inc.)

connective tissue (endomysium) are grouped together to form bundles known as *fascicles*. These fascicles are held together within perimysium sheaths. Finally, all fascicle bundles are bound together within the outer-most layer of another connective tissue (epimysium) that encloses the entire muscle.

As the main element of muscle contraction, a skeletal muscle fibre is a multinucleated cell consisting of smaller filaments called *myofibril*. Roughly aligned parallel to the muscle fibre, myofibrils are made of subunits known as *sarcomeres*. The sarcomere units are connected end-to-end through *Z-lines* and each one is composed of two types of filaments (myofilaments): thicker *myosin* proteins and thinner *actin* proteins. Bonding of actin to myosin triggered by a chain of stimulations results in sliding of myosin against actin proteins and, hence, shortening of sarcomeres. This phenomenon is the basic mechanism responsible for muscle contraction.

Since sarcomeres are arranged end-to-end along the muscle fibres, the contraction of a muscle fibre is considered the summation of the contraction of individual sarcomere units. In general, human sarcomeres have a length of a few micrometres (2.4–5 μm)[3–5] and can shorten to the size of their comprising myofilaments (1–2 μm). Thus, the total contraction strain of a single sarcomere is approximately 50 to 60%, meaning that the overall actuation of a single muscle fibre should be around the same value. The speed of sarcomere length change is in the scale of a few $\mu m\ s^{-1}$, and varies with the applied force.[6–8] Muscle fibre strain rates can then be in excess of 50% per second.

Similar to the shortening of individual sarcomeres, which corresponds to the shortening of the muscle itself, the amount of force generated by a muscle can also be traced down to the actual number of connections (crossbridges) formed between the actin and myosin proteins in sarcomere units. Since sarcomeres of a myofibril work in series (*i.e.* the force exerted on each sarcomere remains the same along the length of myofibril), the total amount of force produced in a muscle is roughly proportional to the length of actin and myosin chains available in a sarcomere to form the crossbridges (the sarcomere length). It is believed that, for human muscle fibres, this stress peaks at ~200–250 kPa when the sarcomere length is ~2.8 μm.[9] In addition to the sarcomere length, the generated force is directly related to the number of myofibrils and muscle fibres available in the muscle (that is, the muscle diameter), and also the orientation of the muscle fibres (the muscle architecture). The overall force generated by a muscle is the vector summation of all forces generated by each individual muscle fibre. The work performed by the muscles is a function of the force, the displacement, and the frequency at which the muscles are functioning. The maximum work density achievable from mammalian skeletal muscle is around 40 $kJ\ m^{-3}$ (~39 $J\ kg^{-1}$) and the specific power output is around 320 $W\ kg^{-1}$.[10]

Muscle is regarded as highly efficient in converting chemical energy into mechanical work. Determining the exact efficiency is difficult, however, and the subject of on-going research.[11] The simplest technique involves whole animal experiments where the mechanical work output of a specific muscle during contraction is measured and the energy input is estimated by direct heat or oxygen consumption measurements. In the latter case, the chemical energy expended during exercise is calculated from the mitochondrial efficiency (to determine the consumption of adenosine triphosphate, ATP) and the known molar free energy released during conversion of ATP to adenosine diphosphate (ADP), which is the chemical process involved in sarcomere contraction. Such measurements on humans return a muscle energy conversion efficiency of ~20%.[12] However, these measurements are complicated by negative and positive contributions, including the baseline energy cost of maintaining body functions, the work expended against other muscles and joints, and the elastic energy stored in tendons that can be released to aid muscle contraction. More detailed studies of individual muscle efficiency in which changes in ATP concentration have been measured from muscle biopsies have returned contraction coupling efficiencies as high as 68%.[11]

In summary, the human skeletal muscle is capable of generating up to 40–60% strain and sustaining maximum stresses of ~300 kPa. It is important to note that the peak stress generated by mammalian skeletal muscle is fairly constant across animals and organisms, while the specific power varies widely amongst species. The strain rate observed for mammalian skeletal muscle can exceed 50% per second, while the typical work density is around 8 kJ m^{-3} and can reach as high as 40 kJ m^{-3}. The specific power of muscles ranges between 50 to 320 W kg^{-1}.[13,14] Skeletal muscle is inherently scalable since the basic functional unit (the sarcomere) is formed in series in muscle fibres to increase the stroke and the muscle fibres are arranged in parallel to increase the force. The same basic muscle architecture is found in all animals, from ants to elephants.

13.3 Types of Polymer Artificial Muscles

When considering electroactive PAMs, one popular distinction is to differentiate these materials into 'ionic' or 'electronic' based on the nature of the dominant charge carrier. The latter essentially operate like capacitors where the active polymer is the dielectric between two electrodes. Electronic charging of the electrodes generates a volume or shape change and this category includes the electrostrictive polymers and the dielectric elastomers. Ionic PAMs incorporate an electrolyte that is in contact with the polymer. Charging the polymer causes the movement of ions and solvent within the polymer and, in some cases, between the polymer and an external electrolyte reservoir. These types of actuators include the ionic-polymer-metal composites (IPMCs) and conducting polymers. Carbon nanotube assemblies in the form of sheets and fibres also operate similarly to conducting polymers.

Other forms of stimuli are also used with PAMs. Volume changes caused by heating/cooling, irradiation, solvent absorption, or exposure to other chemical environments (*e.g.* pH changes) can initiate actuation in oriented polymer fibres, hydrogels, and conducting polymers. Electrothermal heating is a useful way to stimulate thermally-responsive actuators, such as twisted oriented polymer fibres and their related polymer coil muscles. Pneumatically and hydraulically operated braided elastomer cylinders are also sometimes referred to as artificial muscles (or 'McKibben muscles'), although they require a means for injecting and removing pressurised fluid – most typically using a compressor or pump.

Finally, PAMs are also sometimes differentiated in terms of the type of movement that they generate. Three different actuation modes have been reported: 'benders', linear or 'tensile' actuators, and 'twisters' or torsional actuators. Benders operate like bi-metallic strips, although bending PAMs normally involve three layers. Small strains of opposite direction in the outer layers result in bending displacement, which can be large if the bending stiffness is low but at the cost of small tip forces. IPMCs, electrostrictive polymers, conducting polymers, and certain carbon nanotube assemblies are typically designed to operate in the bending mode. Tensile actuators can produce appreciable strains in at least one direction and include polymer coil muscles and dielectric elastomers. These systems mimic natural skeletal muscle most closely. Torsional muscles form the final category and have only recently been described. First observed in twisted carbon nanotube yarns, the torsional mode has now been observed in twisted polymer fibres and forms the basis of large tensile actuation when the twisted fibres are formed into polymer coils (as described in detail below).

Table 13.1 compares the main performance metrics for the various different PAMs. Direct comparisons are difficult, because the data is incomplete in some cases and different test methods have been used in the various studies. However, the very large differences in performance between

Table 13.1 Typical performance parameters for natural skeletal muscle and several different polymer artificial muscles.

	Natural muscle	Dielectric elastomer	Polymer coil muscle	Conducting polymer	IPMC	Electro-strictive polymer
Stress (MPa)	0.1–0.35	0.3–7.7	20–30	5–35	0.20–5	3
Strain (%)	20–60	20–350	20–60	1–10	0.5–3.3	7
Strain rate (% s^{-1})	50–500	450–3.4×10^4	200	0.1–10	<3.5	—
Work density (kJ m^{-3})	8–40	Up to 3.5×10^3	Up to 2.9×10^3	100	<5.5	1.5
Specific power (W kg^{-1})	50–320	Up to 5×10^3	Up to 5×10^4	150	<2.5	—
Efficiency (%)	20–68	Up to 90	1–2	1	1.5–3	—
Life cycle	$>10^9$	10^6	$>10^6$	$<10^5$	—	—
Operational condition	—	>1 kV	<100 V	1–10 V	1–10 V	>1 kV
Actuation type	Tensile	Tensile	Tensile	Tensile bending	Bending	Bending

the various PAMs mean that indicative values are sufficient for providing a comparative analysis. A key distinguishing feature of the PAMs listed in Table 13.1 is the magnitude of the generated strain with some PAMs categorised as high strain (polymer coils and dielectric elastomers) and others as small strain (conducting polymers, IPMC, and electrostrictive polymers). The latter types are usually operated in bending mode so that useful displacements can be achieved. The PAMs that are most applicable in biomimetic applications are the polymer coil muscles and dielectric elastomers, as they both generate large tensile strains and biologically relevant forces and speeds. In the following sections, the current state of understanding of both of these systems is considered in more detail.

13.3.1 Polymer Coil Muscles

Artificial muscles discovered by Baughman and co-workers[15] and made by twisting and coiling ordinary polymer fibres represent a new paradigm in actuator technology. The precursor materials are cheap and readily accessible fibres that are used for textiles, ropes, fishing lines, carpets, and a myriad of other applications. Twisting a fibre while maintaining sufficient tension to prevent snarling causes the fibre to form tight spring-like coils (Figure 13.4a). The coiled geometry can be fixed by thermal annealing and tension is usually used during annealing to separate adjacent turns in the coil (Figure 13.4b, left image). Subsequent heating causes the coil to contract in length (Figure 13.4b, right image) with reversible strokes of 35% or more, which is an order of magnitude larger than the contractions measured in untwisted/uncoiled fibres when heated to the same temperature (Figure 13.4c). Even higher strokes of more than 60% are possible in large diameter coils formed by wrapping the twisted fibre around a mandrel (Figure 13.4a, right image). The contractile forces generated by the polymer coil muscles are 100 times higher than those for the same length and weight human muscle, and fast electrothermal heating using an integrated conducting wire (Figure 13.5a) can deliver a mechanical power of 5.3 $kW\,kg^{-1}$ based on muscle mass. This power to weight ratio is similar to a jet engine and more than 5 times higher than a car's petrol engine. The process to convert precursor fibres into polymer coil muscles is simple, scalable (Figure 13.5b), and considerably lower in cost than competing shape-memory alloy (SMA) wires. In addition, polymer coil muscles show negligible hysteresis when compared to SMAs (Figure 13.6) which is important for the fine control of muscle length between the fully extended and fully contracted states. Continuous operation for more than 1 million heat/cool cycles and the delivery of more than 10% contraction strain (Figure 13.6b) demonstrate the robustness of polymer coil muscles.

The tensile actuation in polymer coil muscles is directly linked to torsional actuation derived from the twisted polymer fibre. To illustrate the link between fibre torsional actuation and coil tensile actuation, coils were fabricated as either 'homochiral' or 'heterochiral', depending on whether the

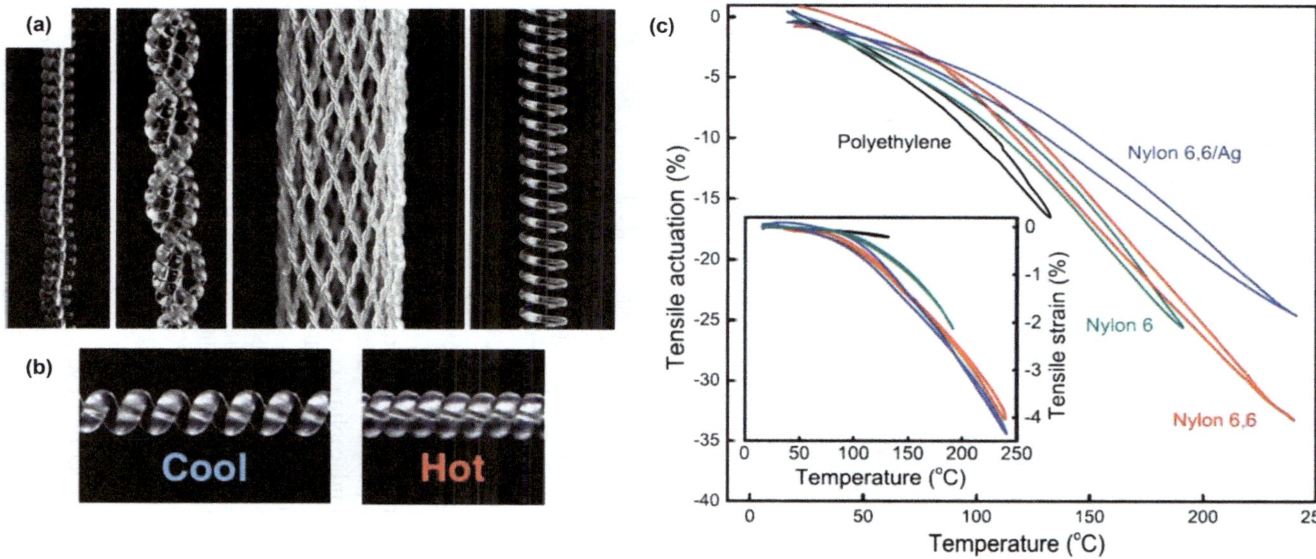

Figure 13.4 (a) Optical images of a coiled fibre made by twist insertion into a 300 μm diameter nylon-6,6 fibre, a two-ply muscle made of the coiled fibres of (a), the braided structure made of 32 two-ply coiled nylon fibres made from a ~100 μm diameter precuror fibre, and a 1.55 mm coil made from a 300 μm diameter precursor nylon fibre by twist insertion followed by mandrel coiling. (b) Optical images of coiled nylon-6,6 (300 μm) before and after heating. (c) Tensile actuation of coiled fibres as a function of the temperature. Inset shows the same data for untwisted fibres.
(From ref. 15: Haines *et al.* "Artificial Muscles from Fishing Line and Sewing Thread", *Science*, 2014, **343**, 868–872. Reprinted with permission from AAAS.)

Figure 13.5 (a) Polymer coil muscle made from nylon-6,6 precursor fibre and wrapped with a copper wire for electrothermal heating. (b) Spools of continuously produced polymer coil muscles. The right hand spool incorporates the fine copper wire.

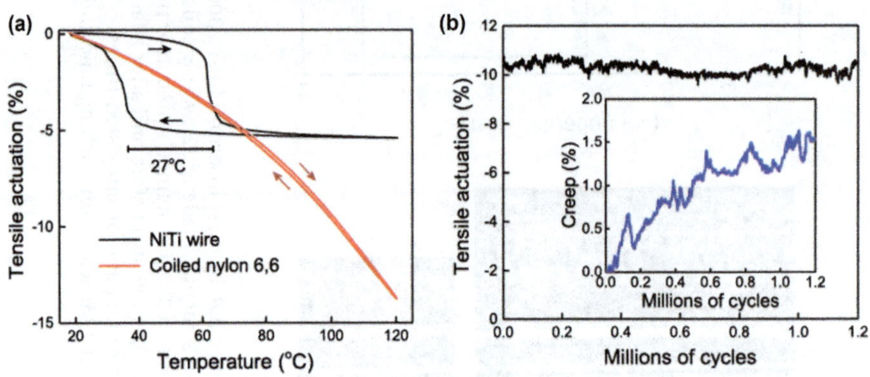

Figure 13.6 (a) Comparison of thermal actuation in a shape-memory alloy (NiTi) wire and a polymer coil muscle made from nylon-6,6 precursor fibre. Both heating and cooling cycles are shown with the typically large hysteresis evident for the SMA wire and negligibly small inherent hysteresis in the polymer coil muscle. (b) Reversible tensile contractions of 10% or more were measured during continuous heating and cooling of coiled nylon-6,6 fibre for 1.2 million cycles without failure and only small creep for the high applied load (inset).
(From ref. 15: Haines *et al.* "Artificial Muscles from Fishing Line and Sewing Thread", *Science*, 2014, **343**, 868–872. Reprinted with permission from AAAS.)

fibre twist direction was the same or was opposite to the coil wrap direction. Heterochiral coils were observed to expand when heated and contract when cooled, which was the opposite to that observed with homochiral coils. The directionality of coil tensile actuation is illustrated in Figure 13.7. In both heterochiral and homochiral coils, the fibre partially untwists when heated

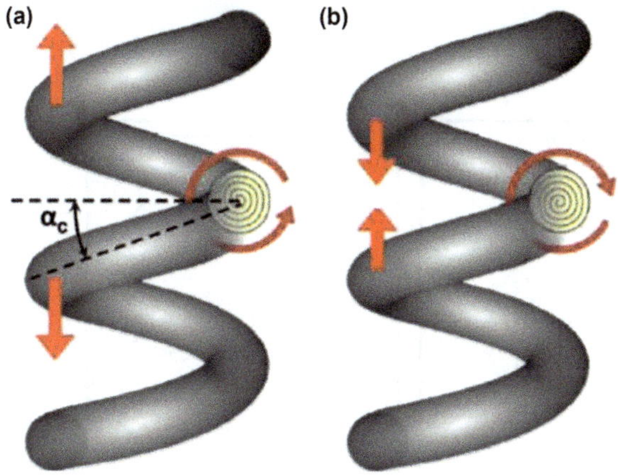

Figure 13.7 Schematic illustration of a twisted then coiled fibre in which the change in fibre twist induces a length change in the coil for (a) heterochiral and (b) homochiral structures. α_c is the bias angle of the coil.
(From ref. 15: Haines *et al.* "Artificial Muscles from Fishing Line and Sewing Thread", *Science*, 2014, **343**, 868–872. Reprinted with permission from AAAS.)

so as to push apart the adjacent turns in a heterochiral coil (Figure 13.7a) or draw closer together the adjacent turns in a homochiral coil (Figure 13.7b). Further evidence to support coupling between fibre torsion and coil tensile actuation is the near equivalence of work density in both modes: 2.1 kJ kg^{-1} torsional work capacity was measured in twisted nylon-6,6 fibre and 2.5 kJ kg^{-1} was measured when the same fibre was formed into a coil and operated in the tensile contraction mode.

A quantitative relationship between fibre twist and coil length has been obtained from spring mechanics,[15] since the stretching or compression of a helical spring is accommodated by torsion within the spring wire. For a homochiral polymer coil muscle containing N turns and made of a twisted fibre containing n_o turns of twist, the total actuator length change ΔL (when the muscle is tethered to prevent end rotation) is:

$$\Delta L = \left(\frac{l}{N}\right)\Delta n, \tag{13.1}$$

where Δn is the change in the number of turns in the twisted fibre that occurs upon stimulation, and l is the initial length of the fibre before twisting. Eqn (13.1) has been shown to accurately model polymer coil contraction and expansion in both homochiral and heterochiral systems.[15]

The discovery of polymer coil muscles was the result of earlier work on the torsional actuation of twisted fibres and yarns. Large stroke torsional actuation was first reported for twisted carbon nanotubes working in an electrolyte[16] and subsequently in air.[17] In the former case, the fibres were immersed in an electrolyte (Figure 13.8a), while in the latter case the twisted

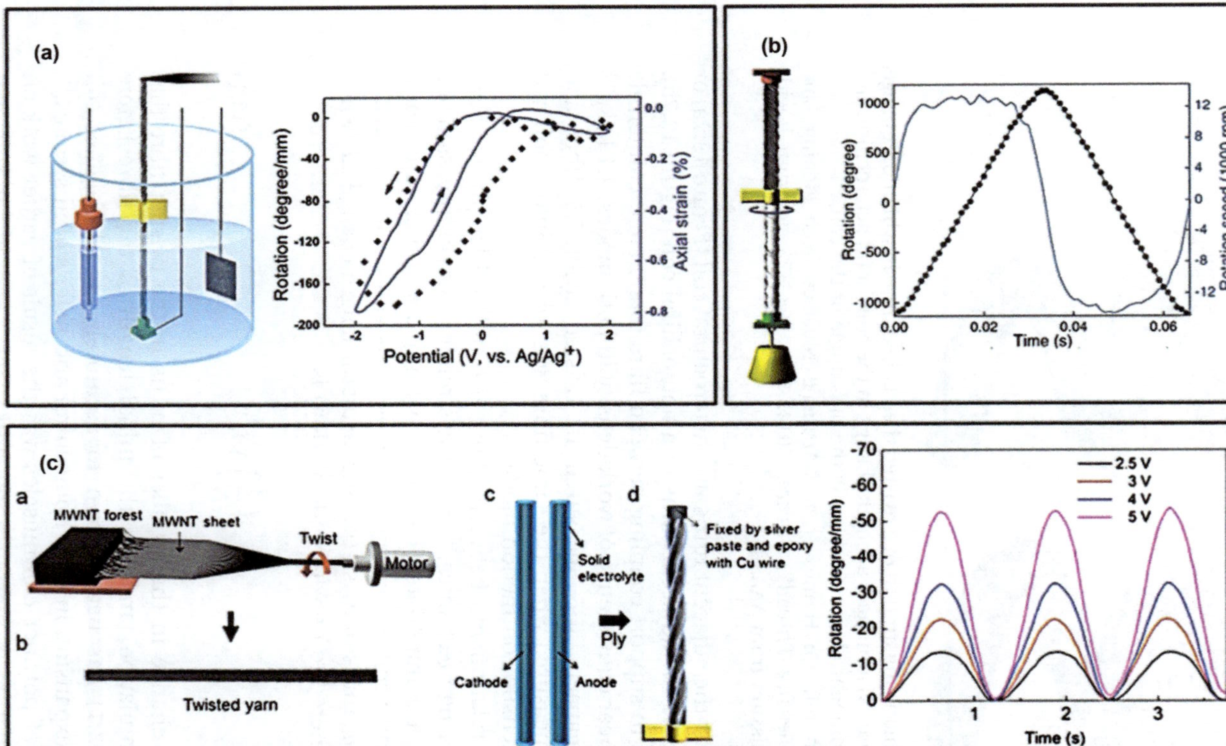

Figure 13.8 Torsional actuation of twisted carbon nanotube yarns. (a) Twisted carbon nanotube yarn operating in 0.2 M tetrabutylammonium hexafluorophosphate/acetonitrile electrolyte. (b) Paraffin impregnated carbon nanotube yarn operating in air. (c) Two-ply carbon nanotube yarn coated with a poly(vinylidene fluoride-*co*-hexafluoropropylene) based tetraethylammonium tetrafluoroborate (TEA.BF_4) solid polyelectrolyte. (Panel (a) is from ref. 16: Foroughi *et al.* "Torsional Carbon Nanotube Artificial Muscles", *Science*, 2011, **334**, 494–497. Reprinted with permission from AAAS. Panel (b) is from ref. 17: Lima *et al.* "Electrically, Chemically, and Photonically Powered Torsional and Tensile Actuation of Hybrid Carbon Nanotube Yarn Muscles", *Science*, 2012, **338**, 928–932. Reprinted with permission from AAAS. Panel (c) is reprinted with permission from ref. 18: Lee *et al.* "All-Solid-State Carbon Nanotube Torsional and Tensile Artificial Muscles", *Nano Lett.*, 2014, **14**, 2664–2669. Copyright 2014 American Chemical Society.)

carbon nanotube fibres were impregnated with paraffin wax (Figure 13.8b). In both examples, the torsional actuation of the multifilament nanotube yarns was related to the yarn volume change. In the paraffin-filled carbon nanotube yarns, the applied electrical field generated heat which resulted in melting of the paraffin wax and its volume expansion. On the other hand, in the absence of paraffin, when the carbon nanotube yarn was immersed in an electrolyte, the applied voltage initiated a double-layer charge at the interface of the nanotubes and the electrolyte, causing a volume change. The carbon nanotube torsional actuator immersed in 0.2 M tetrabutyl ammonium hexafluorophosphate ($TBA \cdot PF_6$) electrolyte provided 340° of reversible rotation per millimetre length of the actuating fibre (when switched between 0–5 V against the Ag/Ag^+ reference). The paraffin-filled twisted carbon nanotube yarn rotated heavy rotors at an average speed of 11 500 revolutions per minute for at least 2 million cycles (using a 15 Hz applied voltage of 40 $V\,cm^{-1}$ of yarn length). The corresponding torsional stoke (32° mm^{-1} of active muscle length) was smaller than for the electrolyte-based muscle. A solid-state electrochemical actuator based on a two-ply twisted carbon nanotube yarn coated with a polymer gel electrolyte has also been demonstrated.[18] Torsional strokes of similar magnitude to those observed in paraffin-filled carbon nanotube yarns were obtained, but with a significantly lower voltage stimulus (Figure 13.8c).

Torsional actuation occurs as a result of a volume change in the twisted fibre, such as by thermal expansion, absorption of small molecules, or electrochemical charging. Figure 13.9 is a helical model that provides some insight into the volume change occurring in a twisted fibre along with changes in the fibre length, radius, and twist. Here, α_o, r_o, and L_o are, respectively, the initial twist bias angle, initial fibre radius, and initial overall length. After a volume change, the new bias angle, radius, and overall length are, respectively, α, r and L. The initial length of untwisted fibre (L_s) is assumed to remain constant during the fibre volume change. The right triangle relationships for the twisted geometries given in Figure 13.9 relate the bias angle to the radius and the number of twists:

$$\cos(\alpha) = L/L_s \tag{13.2a}$$

$$\sin(\alpha) = 2n\pi r/L_s \tag{13.2b}$$

Based on the above relationships, Figure 13.10 illustrates how the number of twists varies when the volume changes. Starting from point A for a stimulation that causes an increase in volume, the number of twists must decrease following line AB if the overall length of the twisted fibre (L) is kept constant. On the other hand, when the overall length of the fibre is allowed to change during stimulation, the number of turns follows path AC (for decreasing length) or AD (for increasing length). The shape of the curves is also highly dependent on the starting bias angle. Thereby, various final states can be achieved using the helical model, when fibre dimension and volume start to vary. Observations to date for carbon nanotube yarns, niobium yarns,

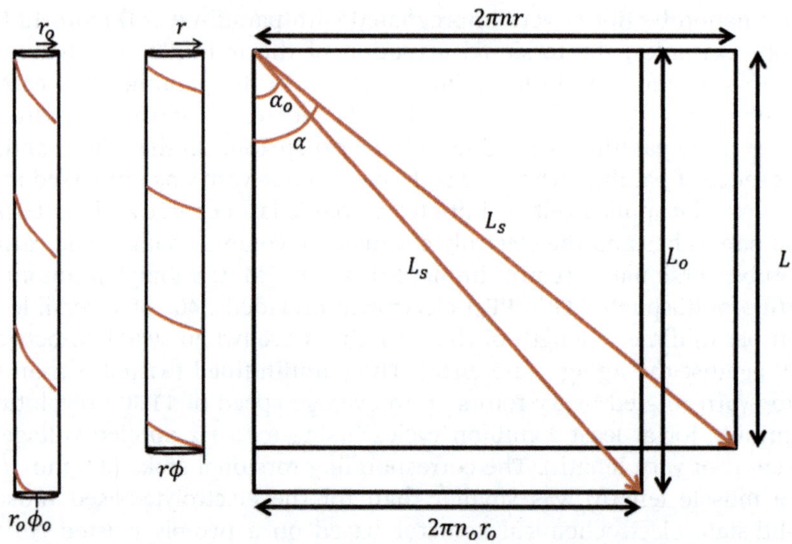

Figure 13.9 Helical model illustrating the correlation between volume change and twisted fibre length, radius, and number of turns of twist.

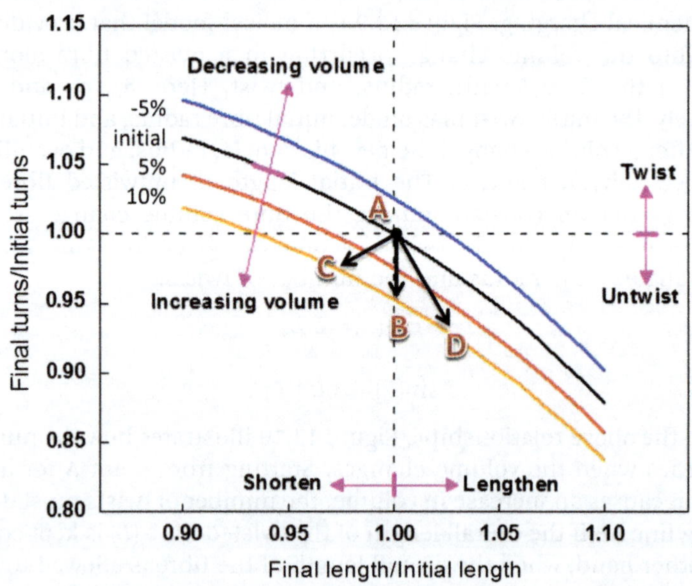

Figure 13.10 The dependence of n/n_o on L/L_o when α_o is 40°. The corresponding volume change causing the actuation is shown on each curve. (The figure is from ref. 16: Foroughi *et al.* "Torsional Carbon Nanotube Artificial Muscles", *Science*, 2011, 334, 494–497. Reprinted with permission from AAAS.)

polymer fibres, and twisted graphene oxide fibres, all indicate that a volume increase causes untwist and length shortening (path AC). The reasons for this particular favoured state upon a volume increase are still to be resolved. However, the effects of external conditions on torsional actuation, such as applied torques and whether the fibre is clamped at one or both ends, are now well understood from torsional mechanics.[19]

The volume expansion needed to stimulate polymer coil muscles is most easily carried out electrothermally using conductive coatings or integrated wires (Figure 13.5a). Although this type of Joule heating[20] can be a fast way of heating fine coiled actuators, it is not the most efficient way due to heat losses to the environment. To date, the best efficiency obtained during the heating/contraction cycle has been 1.32% when fast electrothermal pulses were used on small diameter nylon-6,6 fibres.[15] Cooling the fibres to complete the actuation cycle is also needed. The passive cooling *via* convection is energy efficient but can be very slow, especially for large diameter coiled fibres. Refrigeration can be a much faster way of cooling but requires more energy and complicates the actuator design. Using environmental heat offers a solution in some applications, such as opening or closing louvres or even in smart clothing to adjust fabric porosity.

Recently, Lima *et al.* demonstrated a higher theoretical energy conversion efficiency in silicon rubber-infused carbon nanotube coil actuators operating by swelling/deswelling with various solvents.[21] Large stroke actuations of 30% and 45% were measured, respectively, when 500 μm diameter coiled yarns were stimulated electrothermally or when exposed to hexane vapour for 8 s (Figure 13.11). The contractions were fully reversed during passive cooling or solvent evaporation in air. Much faster reverse expansions were obtained using a solvent exchange system between a high-swelling and a low-swelling solvent. This method also allowed a 'catch state' where the actuator length could be maintained indefinitely without any external stimulus. An energy conversion efficiency of 16% was estimated for the solvent-switch system.

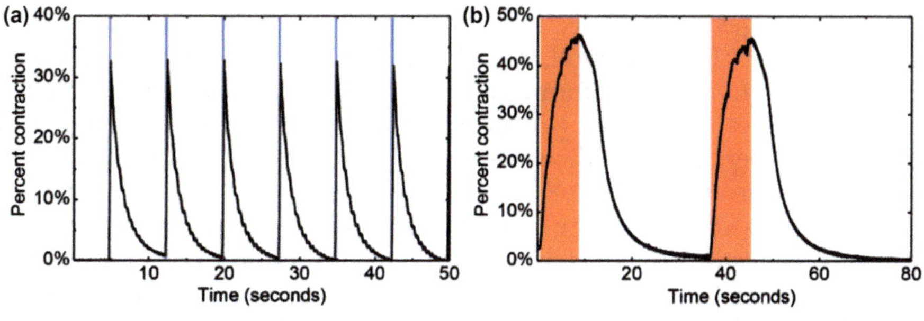

Figure 13.11 Electrothermal (a) and chemical (b) actuation in coiled carbon nanotube yarns infused with silicone rubber.
(Reproduced from ref. 21 with permission from John Wiley and Sons. Copyright © 2015 Wiley Periodicals, Inc.)

Polymer coil muscle actuators provide performances that compare favourably with natural muscle. The maximum achievable actuation strain for these coiled muscles is up to 60%, with a power density of up to 5.3 $kW\,kg^{-1}$. The systems show very small hysteresis and long cycle life. The precursor fibres are cheap, strong, and readily available. Polymer coil muscles can be produced in kilometre lengths using a continuous twisting process. The mechanism of actuation is based on a volume-expansion-induced untwist of the twisted fibre that forms the coil. Volume expansion can also be generated by absorbing small molecules or by electrochemical charging, with the former also providing high energy conversion efficiencies.

13.3.2 Dielectric Elastomer Actuators (DEAs)

Dielectric elastomers are capable of generating large strains (20%–350%), with stresses in the order of several MPa or an order of magnitude larger than skeletal muscle. The strain rate for dielectric elastomers can be very high, ranging from 450% per second (achieved at 100 Hz) to 34 000% per second (achieved at 1400 Hz). The overall work density produced by dielectric elastomers is large and can reach up to several $MJ\,m^{-3}$. The continuous power density that a dielectric elastomer generates is in the range of 400–500 $W\,kg^{-1}$ but the peak power can be as high as 3600 to 5000 $W\,kg^{-1}$. The modulus of dielectric elastomers resembles that of stiffer skeletal muscle, ranging from 100 kPa to 3 MPa.[22–26] Moreover, the materials required for making dielectric elastomers are not necessarily expensive, which reduces costs. The actuation takes place at small currents; however, the required voltage can be as high as several kV which can limit some applications. Also, prestraining is necessary to achieve the large strains mentioned here. This means that, although the actuator is simple and can be lightweight, the additional mass of frame substantially increases the overall volume and mass of the actuator system, reducing the actual work density.

Dielectric elastomers are essentially variable capacitors, where highly stretchable electrodes are separated by a compliant elastomer. When the electrodes are oppositely charged, the electrostatic attraction between opposite charges on each electrode pulls the electrodes towards each other. As a result of this attraction, the elastomeric material between the two electrodes is forced to deform, generating a dimensional change in the actuator. Typically, the voltage applied on the electrodes is in the order of 10–500 MV per unit length of rubber thickness. With rubber thickness typically being around 10–200 μm, the drive voltages can be as high as 100 kV. In the absence of the compliant elastomer, the oppositely charged electrodes are still attracted towards each other. This electrostatic attraction can be balanced by applying an external stress (Maxwell stress). Now, by filling the gap between the two electrodes with an elastomeric material, the induced stress resulting from the electric field squeezes the elastomer, causing it to extend in the direction orthogonal to the applied field, as shown in Figure 13.12.

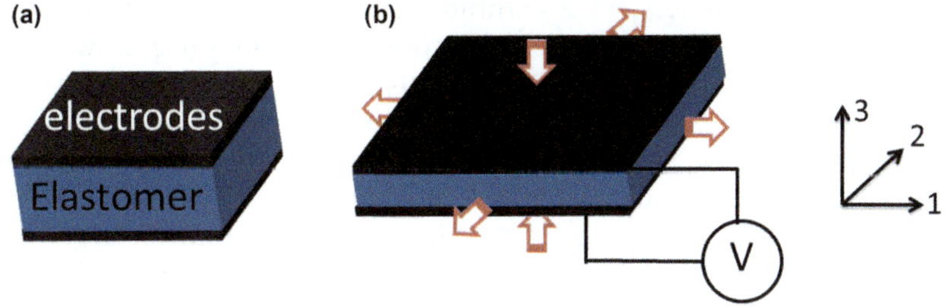

Figure 13.12 Schematic illustration of a dielectric elastomer: (a) before voltage is applied and (b) after voltage is applied, where electrostatic attraction between the electrodes compresses the elastomer in direction 3 (thickness) and simultaneously causing expansion in the 1 and 2 directions.

The induced Maxwell stress (σ_M) can be related to the applied electric field (E) and the dielectric constant of the elastomer (ϵ):[27,28]

$$\sigma_M = \frac{1}{2}\epsilon E^2 \tag{13.3}$$

For an incompressible elastomer undergoing uniaxial deformation, the compressive stress is double the Maxwell stress.[29] Assuming the elastomer is Hookean (*i.e.* $\sigma = Y\varepsilon$, where Y is the elastomer's Young's modulus and ε is the strain), the strain perpendicular to the electrodes (ε_3) is:

$$\varepsilon_3 = \frac{2\sigma_M}{Y} = \frac{\epsilon E^2}{Y} \tag{13.4}$$

If the voltage applied to the electrodes remains constant during actuation, the electric field can be replaced by $E = V/l_3$, where V is the voltage and l_3 is the elastomer thickness ($l_3 < l_{3,o}$). Since the rubber is assumed incompressible, the strains in the other two directions (ε_1 and ε_2) follow the same trend as ε_3 in relation to ϵ, Y, and E. Eqn (13.4) gives a very simplistic picture of a dielectric elastomer and the important parameters that influence the actuation. Stiffer elastomers with a higher modulus (Y) are expected to generate smaller actuation strains when operated at the same voltage and thickness. Meanwhile, a higher voltage can result in larger actuation strains, as long as the system remains stable. Also, elastomers with larger electrical dielectric constants are preferred over the ones with lower dielectric constants. While the simple Hookean model was used to derive eqn (13.4), other models that give a better estimation for actuation performance of dielectric elastomers have been developed[30–34] that are based on more accurate descriptions of non-linear rubber elasticity.[35–39]

Since the early stages of their development, a wide range of rubbers has been used as dielectric elastomers. The most commonly used elastomers belong to the silicone,[22,33] polyurethane,[40,41] and acrylic[23,42,43] families. Other elastomers, such as polyisoprenes, fluorinated elastomers, thermoplastic elastomers, or a combination of elastomeric systems, have also been used.[44–49] Amongst these elastomers, silicones (*e.g.* Dow Corning HS3, Dow Corning Sylgard 186, Nusil CF19-2186) generate actuation strains of ~30–40% in the thickness direction, which is equivalent to that generated by a compressive stress of ~0.5 MPa. These elastomers are capable of operating at higher frequencies due to their high elasticity. Polyurethanes (*e.g.* Deerfield PT6100S, Estane TPU588) normally produce larger stress outputs (~1.5 MPa equivalent compressive stress) but have smaller generated strains (~10% thickness strain). On the other hand, acrylic elastomers (*e.g.* 3M VHB, 3M VHB 4910, 3M VHB 4905) are able to generate large stress (up to ~7 MPa compressive stress) and large actuation strain (up to ~40% thickness strain). However, they have a strong viscous component in their viscoelastic behaviour, which makes operation difficult at high frequencies.

One of the breakthroughs in the field of dielectric elastomers was the observation of considerable positive effects in the actuation performance of dielectric elastomers caused by prestraining (Figure 13.13).[50,51] It was found that by prestraining a 3M VHB 4910 acrylic elastomer (prestrain: 540% in length and 75% in width), area actuation strains as high as ~240% were achievable. Although similar positive effects were observed when silicone and polyurethane-based elastomers were prestrained, the effect of prestraining is not always the same for all elastomeric materials. Generally, acrylic-based elastomers, such as 3M VHBs, perform better than silicone elastomers when prestrained.

Many fundamental studies have been carried out to fully understand the nature of the positive effect of prestraining on the actuation performance of dielectric elastomers. The observed improvements have been mainly attributed to the enhancement of the actuator's stability. The main electromechanical instability in the dielectric elastomers is called the *pull-in*

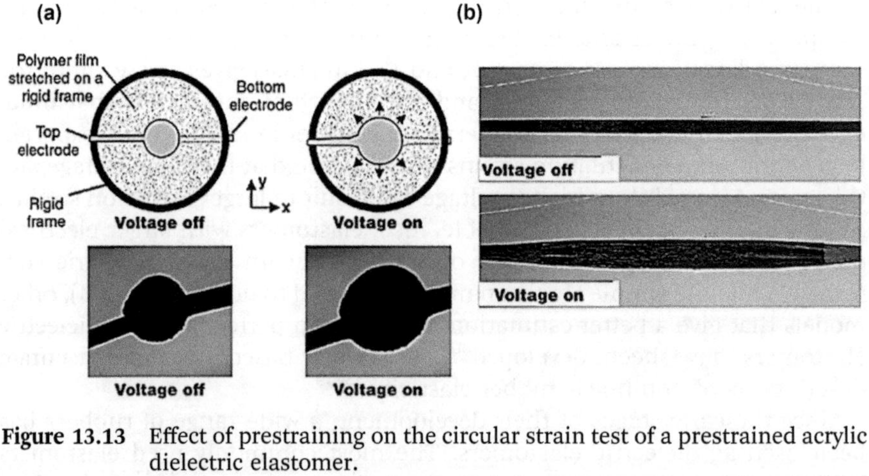

Figure 13.13 Effect of prestraining on the circular strain test of a prestrained acrylic dielectric elastomer.
(Adapted with permission from ref. 50. © 2000 WILEY-VCH Verlag GmbH, Weinheim, Fed. Rep. of Germany.)

phenomenon. The pull-in failure is the result of the positive feedback between the electric field and the continuous thinning of the elastomer. When a high voltage is applied to both sides of the elastomer, the elastomer becomes thinner resulting in a higher electric field strength at the same voltage. This process in turn results in more thinning of the elastomer, destabilising the system and eventually leading to electrical break down. This phenomenon is a well-known mode of failure in insulators. Theoretically, it has been shown that prestraining increases the actuation stretch at which the electromechanical instability occurs.[52] For example, Figure 13.14 shows the stress displacement curve for an ideal elastomer when subjected to an increasing biaxial stretch. In the absence of an applied electric field, the elastomer can be stretched to the mechanical failure point, which is typically a stretch ratio (λ_R) of around 5 or 6. Charge applied across the thickness direction of the elastomer reduces the stress needed for any given extension, or increases the extension for any given stress. The latter effect generates the actuation stroke when a voltage is applied. The actuation stroke increases with the increasing applied voltage until the breakdown limit is reached. Figure 13.14 plots the stress–displacement curves corresponding to the maximum charge that can be accommodated to the point of electrical breakdown (EB) or electromechanical instability (EMI, or 'pull-in' failure). The DEA can therefore operate stably at any stress/displacement

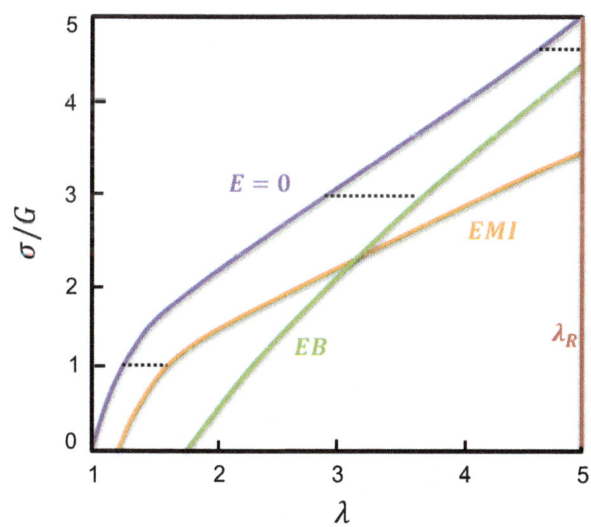

Figure 13.14 Stress normalized to the shear modulus as a function of the extension for a dielectric elastomer. $E = 0$ indicates the rubber elasticity theory when no external electric field is applied. *EMI*, *EB* and λ_R indicate, respectively, the *pull-in*, the electric breakdown and the overstretching failure modes that can occur in a dielectric elastomer actuator. The dashed lines represent the maximum isotonic (constant stress) actuation strokes at three different starting prestrains: longer lines indicate higher strokes.

state between the $E=0$ line and the *EB* or *EMI* lines shown in Figure 13.14. Isotonic actuation strokes are then shown as dashed horizontal lines in Figure 13.14, and it can be seen that the stroke increases with the prestrain until the cross-over point between EB and EMI failure.

While dielectric elastomers have been developed and used in many applications, on-going research work aims to further improve their performance. For example, prestrain-locked networks have been successfully developed to provide the prestrain necessary to achieve higher actuation strokes but without the need for an external frame.[45,47] Prestrain-locked networks are formed when one elastomeric network, usually an acrylic-based elastomer, is strained, followed by introducing a second network within the prestrained elastomer to obtain an interpenetrating polymer network (IPN) in which the first network is highly stretched. A second area of on-going research is the development of compliant electrodes. Response times are limited by the speed at which the stimulus can be applied. Electrical systems are fundamentally limited to the *RC* (resistance$\times$capacitance) time constant. The limited conductivity of the conducting greases and other compliant electrodes used on dielectric elastomers, for example, can limit the response frequency to the kHz region, since resistances can be MΩ and capacitances are only nF. Finally, the operating voltages can be reduced by decreasing the thickness of the elastomeric layers.

13.3.3 Bending Type Polymer Artificial Muscles

Several different PAMs have been developed to operate in the bending mode and some have been used to fabricate simple bio-inspired robots, as described in the next section. Here, the mechanisms of operation for conducting polymers, IPMCs, and electrostrictive polymers are briefly reviewed.

13.3.3.1 Conducting Polymers

Conducting polymer actuators operate electrochemically at low voltages and have the advantage of being configured as either bending type or tensile type artificial muscles. Essentially, conductive polymers are conjugated polymer chains that can be reversibly oxidised and reduced while maintaining mechanical stability. Oxidation and reduction (or 'doping') involves the exchange of small ions (or 'dopants') between the polymer and an external electrolyte. The doping process is reversible: when the oxidation state of the polymer chains is altered, the dopants are forced to leave or re-enter the structure. This migration of charged species to and from the polymer structure is the main mechanism of actuation in conductive polymers.[53–55] The migration of dopants within the polymer structure is also accompanied by the movement of large amounts of solvent molecules from the electrolyte, which consequently results in dimension changes in the conjugated polymer structure.

Bi-layer benders[56,57] can be made by synthesizing a conductive polymer film on a flexible electrode and then operating it by immersing in an electrolyte that also contains a second electrode. Alternatively, tri-layer benders[58–63] are made by synthesizing conductive polymer films on both sides of a porous film (such as poly(vinylidene fluoride)) containing the electrolyte or using solid interpenetrating network electrolyte layers.[64–66] When the required voltage is applied between both sides of the actuator, one side is reduced while the other side is oxidized, initiating the ion and solvent movement from one side to the other, which eventually results in actuator bending towards the less swollen side. Reversing the polarity of the applied voltage causes bending in the opposite direction. The loss of electrolyte through evaporation will reduce the actuation performance over time and encapsulation can be used to increase the cycle life,[67–70] but the added encapsulating layers increase the stiffness and reduce the bending displacement. Alternatively, ionic liquids and solid-state electrolytes have been proposed as non-volatile solvents to increase the life time of the tri-layer actuators working in air, although actuator speed is compromised by the ionic conductivity decrease due to the high viscosity of ionic liquids.[64,71]

13.3.3.2 IPMCs: Ionic Polymer Metal Composites

IPMCs provided one of the first demonstrations of low-voltage activated polymer actuators. These bending-only actuators are in fact solid polyelectrolytes sandwiched between two thin and flexible metal plates. When an electric field is applied between the metal electrodes, a concentration gradient is established within the structure of the layered actuator, causing the solvent to move to a high concentrated area to equilibrate the generated concentration gradient. This mass movement results in actuator bending in the direction of the less swollen side.[72–75]

Typically, the polyelectrolyte is made of a negatively charged polymer, where charges are balanced with counter cations. Nafion®, with a hydrophobic backbone and hydrophilic side groups, is a commonly used example of such a polymer. The polymer-bound anions are far less mobile than the cations so when an electric field is applied, the latter can move more freely than the former and set up the concentration gradient. Both sides of the polyelectrolyte membrane can be loaded with metal nanoparticles[76] (*e.g.* platinum) and then metal plated. The membrane thickness ranges from 30 μm to 2 mm,[77] while the electrode thickness is around 1–5 μm. The metal nanoparticles are 3–10 nm in diameter and are distributed within the membrane up to 10–20 μm from the surface. These nanoparticles can be incorporated physically or by soaking the membrane in a metal salt solution followed by reduction of the salts. An example of tip displacement as a function of the applied voltage for a Nafion®-based IPMC is shown in Figure 13.15, along with modelling results that include the actuator dimensions, bending stiffness, and estimates of the electrostatic stress

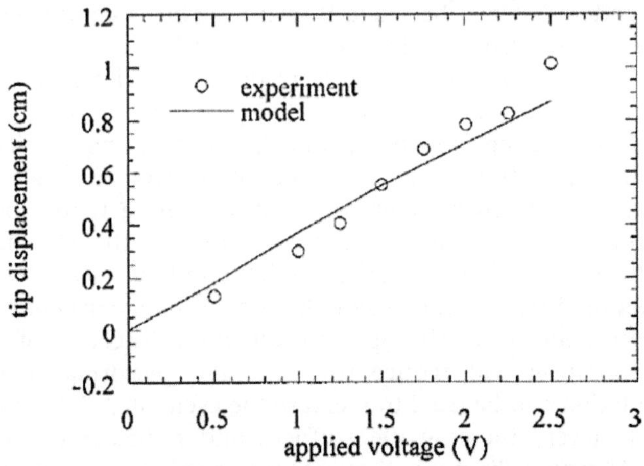

Figure 13.15 Tip displacement as a function of applied voltage for a Nafion®-based IPMC. The effective Young's modulus is 0.5 GPa. The actuator size is 5 mm × 0.2 mm × 20 mm.
(Reprinted with permission from ref. 78: Nemat-Nasser and Li "Electromechanical response of ionic polymer-metal composites", *J. Appl. Phys.*, 2000, **87**, 3321–3331. Copyright 2000, AIP Publishing LLC.)

generated by the charge separation.[78] As shown in Figure 13.15, the tip displacements from IPMCs can be large, but tip forces are limited to the milli-Newton range.

13.3.3.3 Electrostrictive Polymers

One of the oldest types of polymer artificial muscles are based on piezoelectric polymer films. When subject to an external force, an electrical charge or voltage is generated (direct effect). They can also function as actuators where mechanical deformation occurs when the material is subject to an external electric field (inverse effect). The latter effect is known as electrostriction. Piezoelectric materials contain internal electric dipoles, such as what occurs in crystalline materials without crystallographic centrosymmetry regarding the position of positive and negative charges in their crystal lattice. Poling of a piezoelectric material involves applying an increasing electric field in the desired direction generating a strain, as shown by line A–B in Figure 13.16. After the electrical field returns to zero, some polar axes will keep their alignment, leaving some residual strain in the system (line B–C). This residual strain can be cancelled when some electrical field is applied in the opposite direction (line C–D). On the other hand, if the electric field cycles from zero to a set value, the electrostrictive material will respond accordingly with a hysteresis (line C–B). This area is called the working loop of an electrostrictive actuator.

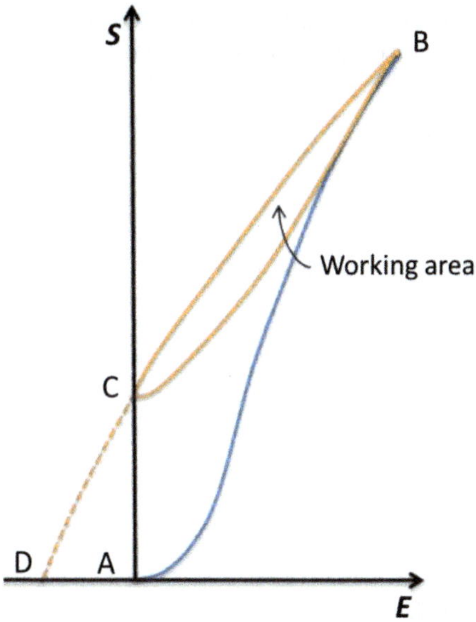

Figure 13.16 Schematic illustration of the strain response (S) of an electrostrictive material to an external field (E).

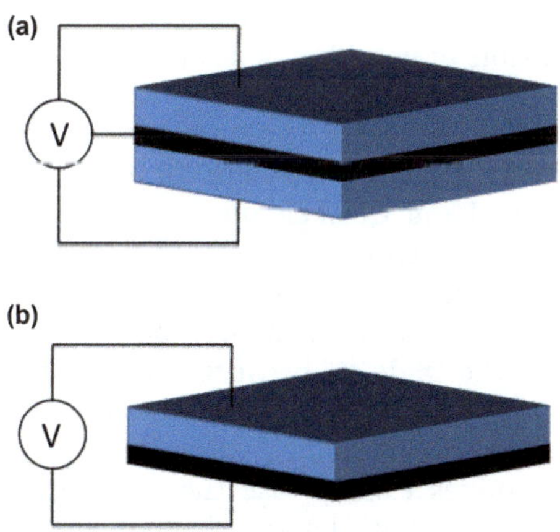

Figure 13.17 Schematic illustration of bending electrostrictive actuators: biomorph (a) and unimorph (b) actuators.

Linear actuators made of electrostrictive materials will generate small actuation strains even when the external voltage is high. On the other hand, some bending electrostrictive actuators can generate large bending deflections even at moderate electric field. In general, the electrostrictive actuators can be divided into two main categories of bimorph and unimorph (Figure 13.17).[79,80] The important parameters that govern the performance of electrostrictive actuators are the geometry of the actuator, the thickness ratio of layers, the Young's modulus ratio of the layers, the piezoelectric constant, and the applied electrical field. The best performing polymer electrostrictive materials are based on poly(vinylidene fluoride) (PVDF). Copolymers of vinylidene fluoride with trifluoroethylene and chlorotrifluoroethylene can generate electrostrictive strains of up to 7% as a result of the very high dielectric constant, electric breakdown strength, and high induced polarization.[81]

13.4 Bio-inspired Applications for Polymer Artificial Muscles

Biomimetic robot research is an active area with the aim of developing robot systems that can swim, fly, climb, run, walk, or crawl.[82] Artificial muscles offer the prospect of lighter, smaller, and noiseless motor systems that can be used in mobile robots and for dexterous robot manipulators. A key advantage of artificial muscles compared to conventional motors and engines is their high energy and power densities for small size/mass systems. Such small motor units are needed when space is limited or mass needs to be minimised. The development of improved artificial muscle materials provides impetus to these research efforts. Several recent examples are described below to highlight the progress made in exploiting the ever-improving artificial muscles in bio-inspired robotic systems.

An enduring interest has been the development of 'soft' robotic hands or grippers that are capable of handling delicate objects. The major obstacle in developing dexterous hands[83] is the need to accommodate 20 or more actuation systems in a relatively small volume within the hand itself and the forearm, where human hand muscles are located. Various robotic hands, prosthetic hands, and hand exoskeletons have been constructed using a wide range of actuation systems, including pneumatic McKibben muscles, shape-memory alloys, and electric motors. All examples demonstrated to date consider the trade-off between the number of actuation units and the size/weight, and also the trade-off between the number of degrees of freedom and the complexity of the control system. As actuator materials improve in performance, more compact hand designs become feasible.

Significant opportunities exist to use the newly described polymer coil muscles in prosthetic and robotic hands. The high energy and power density, long operating life, and low cost of polymer coil actuators are ideally suited to drive tendons in hands. Although these new muscles were only first described in 2014, already research groups are adapting the polymer coil muscles to a range of bio-inspired applications. A single finger device has been evaluated where hot and cold water was used to drive the actuation in the nylon coil.[84] Yip and Niemeyer[85] have built a five-fingered robotic hand driven by conductive nylon coil muscles (Figure 13.18a,b) and developed control models that allow fast grasping manoeuvres with open loop control (Figure 13.18c).

Bending type actuators have also been used in bio-inspired applications. As reviewed by Chu and co-workers,[86] IPMC actuators have been configured into various fish, eel, and ray-like swimming robots. CP bending type

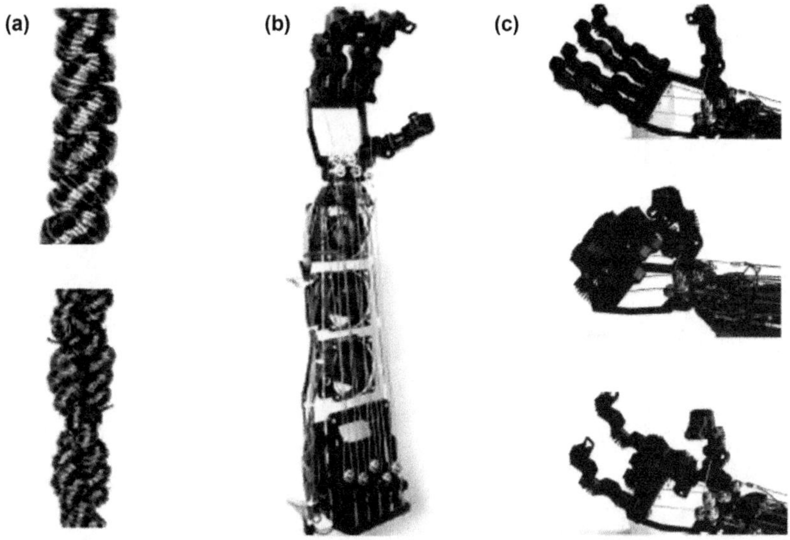

Figure 13.18 Photographs of polymer coil muscles (a) and dexterous prosthetic hand driven by the muscles (b) demonstrating various grasps (c). (© 2015 IEEE. Reprinted with permission from ref. 85: Yip and Niemeyer "High-performance robotic muscles from conductive nylon sewing thread", *IEEE International Conference on Robotics and Automation (ICRA),* 2015, 2313–2318.)

actuators have also been used for robot fish.[87] The most useful of these designs are untethered and carry the battery on-board along with the control module and additional payload. Typically swimming speeds for artificial muscle-powered swimming robots are less than 1 body length per second, or at least an order of magnitude slower than fish. Most designs use a simple caudal (tail) fin oscillation giving an ostraciiform type of swimming motion. Real fish swimming usually involves body undulations as well as caudal oscillations, and these advanced swimming modes are yet to be replicated in artificial muscle propelled fish robots due to the complexity in fabrication and control. In one recent study, however, the modification of the IPMC shape enabled a more complex deformation to create active control surfaces[88] for more efficient swimming action. Significant opportunities exist to further improve the performance of swimming robots, in particular by increasing the muscle mass which is currently as low as 1% of the robot fish.

Terrestrial and flying robots have more demanding operating requirements than aquatic systems, as buoyancy in water greatly reduces the energy cost of propulsion for fish. For artificial muscles to operate in air they additionally require non-volatile solvents and electrolytes, so IPMC actuators are unsuitable unless encapsulated. University of Tartu researchers have recently described inchworm type crawling robots (Figure 13.19) that use non-volatile ionic liquids as the electrolyte layer.[89] High surface-area carbon-based electrodes provide a charged surface to effect a bending

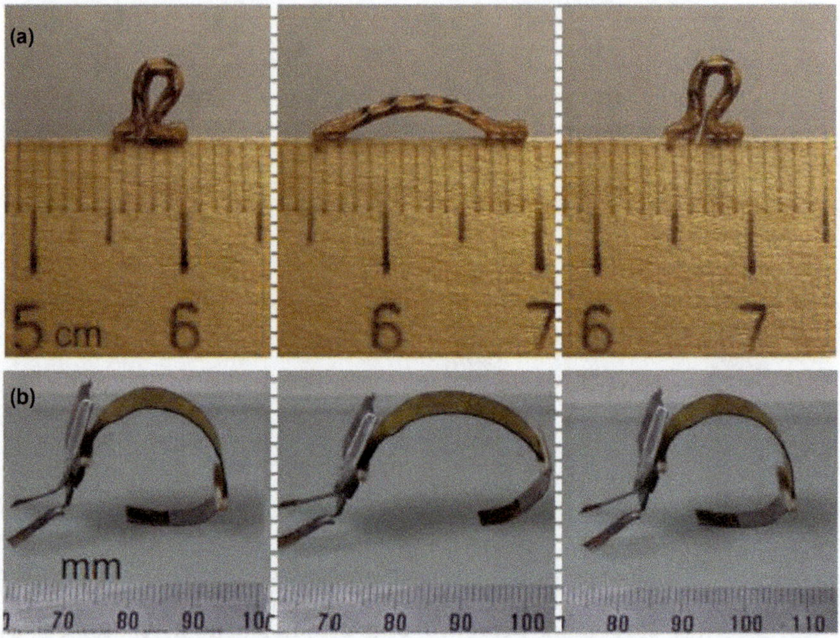

Figure 13.19 "Inchworm" like motion of a bending actuator utilising a similar architecture to IPMC artificial muscles.
(Reproduced from ref. 89 with permission from John Wiley and Sons. Copyright © 2015 Wiley Periodicals, Inc.)

actuation. The robot systems were untethered and carried a battery and micro-processor on board. Flapping wing flying robots are yet to be demonstrated using artificial muscles. A number of prototypes have been evaluated using dielectric elastomer actuators[26] that can generate large wing beat amplitudes at high frequencies. However, the bulk and mass of the power unit and mechanical assembly needed to prestrain the DEA muscle[90] have so far prevented the demonstration of flight. Hovering flight has been demonstrated using piezoelectric ceramic bimorphs to power a small 80 mg flapping wing robot, however, tethering to the power supply was still required.[91]

The application of IPMC bending actuators to hand exoskeletons has been considered in detail and the blocked forces from individual IPMC actuators were found to be too small to match human hand grasp.[92] Stacked actuators have been proposed as a solution to increase force, but not yet demonstrated.

13.5 Conclusions

The pursuit of biomimetic artificial muscles continues to yield significant advances. The general aim is to produce a synthetic material that can be stimulated in a safe manner and match skeletal muscle in all areas including stroke, force, speed, efficiency, and noiseless operation. Many of the artificial muscles described in this chapter can exceed the stress generating ability of muscle. However, only polymer coil muscles and dielectric elastomers can match the large stroke and speed of skeletal muscle. Dielectric elastomer actuators require very high operating voltages and are not suited in applications that involve direct contact with people. Polymer coil muscles require heat activation that may not be suited to all applications, although environmental heat can be used as a 'free' energy source for important needs.

With biomimetic muscle materials available, so too are biomimetic robot systems. Muscle-like propulsion enables sophisticated swimming, slithering, crawling, climbing and flying robots to be designed and evaluated. Parallel developments in sensors (for feedback control), vision systems, and artificial intelligence are driving the rapid development of robotic systems designed to work with and alongside humans.[93] More futuristic applications may include robots for home-help, especially for the elderly. With aging populations in the western societies looming as a huge economic and political challenge, the home-help robots may offer a technological solution. The drive to produce smart materials/smart structures incorporating sensors and adaptive actuators will also draw on improvements in artificial muscles. Mechanical actuators will be used for vibration dampening, compliance shifting, and 'morphing' structures. Already prototype vibration dampening systems are available using ceramic electrostrictive materials, where small strains, high stresses, and rapid responses are provided. Compliance shifting structures may be required when a change in stiffness is desired under changing load conditions. Here, polymer actuators may be useful since they can be readily fabricated into fibres. When used as reinforcing fibres, the fibre modulus (and hence stiffness of the composite) can be changed through the polymer actuation process. Similarly, morphing structures are likely to be built from composites where the dispersed reinforcing phases can be actuated producing stresses that alter the overall shape of the composite.

The global R&D effort devoted to artificial muscle materials regularly exposes new actuator materials, and improvements to existing materials. Continuous research will no doubt produce further performance materials and the goal of matching natural skeletal muscle in all its performance attributes is well within reach.

References

1. S. Hirose, K. Ikuta and Y. Umetani, *Adv. Lab. Autom. Rob.*, 1989, **3**, 3–16.
2. A. R. Gillies and R. L. Lieber, *Muscle Nerve*, 2011, **44**, 318–331.
3. R. L. Lieber, B. O. Ljung and J. Fridén, *J. Exp. Biol.*, 1997, **200**, 19–25.
4. R. L. Lieber, B.-O. Ljung and J. Fridén, *Acta Orthop. Scand.*, 1997, **68**, 249–254.

5. R. L. Lieber, G. J. Loren and J. Friden, *J. Neurophysiol.*, 1994, **71**, 874–881.
6. A. V. Hill, *Proc. R. Soc. B*, 1938, **126**, 136–195.
7. B. Katz, *J. Physiol.*, 1939, **96**, 45–64.
8. M. J. Previs, S. B. Previs, J. Gulick, J. Robbins and D. M. Warshaw, *Science*, 2012, **337**, 1215–1218.
9. D. E. Rassier, B. R. MacIntosh and W. Herzog, *J. Appl. Physiol.*, 1999, **86**, 1445–1457.
10. R. K. Josephson, *Annu. Rev. Physiol.*, 1993, **55**, 527–546.
11. F. E. Nelson, J. D. Ortega, S. A. Jubrias, K. E. Conley and M. J. Kushmerick, *J. Exp. Biol.*, 2011, **214**, 2649–2653.
12. N. P. Smith, C. J. Barclay and D. S. Loiselle, *Prog. Biophys. Mol. Biol.*, 2005, **88**, 1–58.
13. I. W. Hunter and S. Lafontaine, Solid-State Sensor and Actuator Workshop, 1992, 5th Technical Digest., IEEE, 1992.
14. Y. Bar-Cohen, *Electroactive Polymer (EAP) Actuators as Artificial Muscles: Reality, Potential, and Challenges, Society of Photo Optical*, 2004.
15. C. S. Haines, M. D. Lima, N. Li, G. M. Spinks, J. Foroughi, J. D. W. Madden, S. H. Kim, S. Fang, M. Jung de Andrade, F. Göktepe, Ö. Göktepe, S. M. Mirvakili, S. Naficy, X. Lepró, J. Oh, M. E. Kozlov, S. J. Kim, X. Xu, B. J. Swedlove, G. G. Wallace and R. H. Baughman, *Science*, 2014, **343**, 868–872.
16. J. Foroughi, G. M. Spinks, G. G. Wallace, J. Oh, M. E. Kozlov, S. Fang, T. Mirfakhrai, J. D. W. Madden, M. K. Shin, S. J. Kim and R. H. Baughman, *Science*, 2011, **334**, 494–497.
17. M. D. Lima, N. Li, M. Jung de Andrade, S. Fang, J. Oh, G. M. Spinks, M. E. Kozlov, C. S. Haines, D. Suh, J. Foroughi, S. J. Kim, Y. Chen, T. Ware, M. K. Shin, L. D. Machado, A. F. Fonseca, J. D. W. Madden, W. E. Voit, D. S. Galvão and R. H. Baughman, *Science*, 2012, **338**, 928–932.
18. J. A. Lee, Y. T. Kim, G. M. Spinks, D. Suh, X. Lepró, M. D. Lima, R. H. Baughman and S. J. Kim, *Nano Lett.*, 2014, **14**, 2664–2669.
19. S. Aziz, S. Naficy, J. Foroughi, H. R. Brown and G. M. Spinks, *Polym. Test.*, 2015, **46**, 88–97.
20. S. M. Mirvakili, A. Rafie Ravandi, I. W. Hunter, C. S. Haines, N. Li, J. Foroughi, S. Naficy, G. M. Spinks, R. H. Baughman and J. D. W. Madden, *Proceeding SPIE: Electroactive Polymer Actuators and Devices*, 2014.
21. M. D. Lima, M. W. Hussain, G. M. Spinks, S. Naficy, D. Hagenasr, J. S. Bykova, D. Tolly and R. H. Baughman, *Small*, 2015, **11**, 3113–3118.
22. R. D. Kornbluh, R. Pelrine, Q. Pei, S. Oh and J. Joseph, *Proceeding SPIE: Smart Structures and Materials*, 2000.
23. A. M. Tews, K. L. Pope and A. J. Snyder, *Proceeding SPIE: Smart Structures and Materials*, 2003.
24. G. Kofod, The Technical University of Denmark, 2001.
25. P. Sommer-Larsen, G. Kofod, M. H. Shridhar, M. Benslimane and P. Gravesen, *Proceeding SPIE: Smart Structures and Materials*, 2002.

26. R. Pelrine, R. D. Kornbluh, Q. Pei, S. Stanford, S. Oh, J. Eckerle, R. J. Full, M. A. Rosenthal and K. Meijer, *Proceeding SPIE: Smart Structures and Materials*, 2002.
27. J. A. Stratton, *Electromagnetic Theory*, McGraw–Hill, New York, 1941.
28. H. J. Juretschke, *Am. J. Phys.*, 1977, **45**, 277–280.
29. I. Krakovský, T. Romijn and A. Posthuma de Boer, *J. Appl. Phys.*, 1999, **85**, 628–629.
30. X. Zhao and Z. Suo, *J. Appl. Phys.*, 2008, **104**, 123530.
31. M. Wissler and E. Mazza, *Sens. Actuators, A*, 2005, **120**, 184–192.
32. Z. Suo, *Acta Mech. Solida Sin.*, 2010, **23**, 549–578.
33. R. Pelrine, R. Kornbluh, J. Joseph, R. Heydt, Q. Pei and S. Chiba, *Mater. Sci. Eng., C*, 2000, **11**, 89–100.
34. N. Goulbourne, E. Mockensturm and M. Frecker, *J. Appl. Mech.*, 2005, **72**, 899–906.
35. O. H. Yeoh, *Rubber Chem. Technol.*, 1990, **63**, 792–805.
36. R. W. Ogden, *Proc. R. Soc. A*, 1972, **326**, 565–584.
37. M. Mooney, *J. Appl. Phys.*, 1940, **11**, 582–592.
38. A. N. Gent, *Rubber Chem. Technol.*, 1996, **69**, 59–61.
39. E. M. Arruda and M. C. Boyce, *J. Mech. Phys. Solids*, 1993, **41**, 389–412.
40. Q. M. Zhang, J. Su, C. H. Kim, R. Ting and R. Capps, *J. Appl. Phys.*, 1997, **81**, 2770–2776.
41. L. Petit, B. Guiffard, L. Seveyrat and D. Guyomar, *Sens. Actuators, A*, 2008, **148**, 105–110.
42. R. D. Kornbluh, R. Pelrine, Q. Pei, R. Heydt, S. Stanford, S. Oh and J. Eckerle, *Proceeding SPIE: Smart Structures and Materials*, 2002.
43. R. Shankar, T. K. Ghosh and R. J. Spontak, *Macromol. Rapid Commun.*, 2007, **28**, 1142–1147.
44. R. E. Pelrine, R. D. Kornbluh and J. P. Joseph, *Sens. Actuators, A*, 1998, **64**, 77–85.
45. S. M. Ha, W. Yuan, Q. Pei, R. Pelrine and S. Stanford, *Adv. Mater.*, 2006, **18**, 887–891.
46. K. Jung, J. Lee, M. Cho, J. C. Koo, J.-d. Nam, Y. Lee and H. R. Choi, *Smart Mater. Struct.*, 2007, **16**, S288.
47. S. M. Ha, W. Yuan, Q. Pei, R. Pelrine and S. Stanford, *Smart Mater. Struct.*, 2007, **16**, S280.
48. R. Shankar, T. K. Ghosh and R. J. Spontak, *Adv. Mater.*, 2007, **19**, 2218–2223.
49. P. Brochu and Q. Pei, *Macromol. Rapid Commun.*, 2010, **31**, 10–36.
50. R. Pelrine, R. Kornbluh and G. Kofod, *Adv. Mater.*, 2000, **12**, 1223–1225.
51. R. Pelrine, R. Kornbluh, Q. Pei and J. Joseph, *Science*, 2000, **287**, 836–839.
52. X. Zhao and Z. Suo, *Appl. Phys. Lett.*, 2007, **91**, 061921.
53. B. François, N. Mermilliod and L. Zuppiroli, *Synth. Methods*, 1981, **4**, 131–138.
54. M. R. Gandhi, P. Murray, G. M. Spinks and G. G. Wallace, *Synth. Methods*, 1995, **73**, 247–256.

55. A. J. Dominis, G. M. Spinks, L. A. P. Kane-Maguire and G. G. Wallace, *Synth. Methods*, 2002, **129**, 165–172.
56. Q. Pei and O. Inganläs, *Adv. Mater.*, 1992, **4**, 277–278.
57. E. Smela, O. Inganäs, Q. Pei and I. Lundström, *Adv. Mater.*, 1993, **5**, 630–632.
58. K. Kaneto, M. Kaneko, Y. Min and A. G. MacDiarmid, *Synth. Methods*, 1995, **71**, 2211–2212.
59. R. H. Baughman, *Synth. Methods*, 1996, **78**, 339–353.
60. J. M. Sansinena, V. Olazabal, T. F. Otero, J. M. Sansinena, C. N. Polo da Fonseca and M.-A. De Paoli, *Chem. Commun.*, 1997, 2217–2218.
61. T. W. Lewis, L. A. P. Kane-Maguire, A. S. Hutchison, G. M. Spinks and G. G. Wallace, *Synth. Methods*, 1999, **102**, 1317–1318.
62. A. S. Hutchison, T. W. Lewis, S. E. Moulton, G. M. Spinks and G. G. Wallace, *Synth. Methods*, 2000, **113**, 121–127.
63. J. D. Madden, R. A. Cush, T. S. Kanigan and I. W. Hunter, *Synth. Methods*, 2000, **113**, 185–192.
64. F. Vidal, C. Plesse, D. Teyssié and C. Chevrot, *Synth. Methods*, 2004, **142**, 287–291.
65. C. Plesse, F. Vidal, H. Randriamahazaka, D. Teyssié and C. Chevrot, *Polymer*, 2005, **46**, 7771–7778.
66. F. Vidal, C. Plesse, G. Palaprat, A. Kheddar, J. Citerin, D. Teyssié and C. Chevrot, *Synth. Methods*, 2006, **156**, 1299–1304.
67. J. D. Madden, R. A. Cush, T. S. Kanigan, C. J. Brenan and I. W. Hunter, *Synth. Methods*, 1999, **105**, 61–64.
68. S. McGovern, G. Alici, V.-T. Truong and G. Spinks, *Smart Mater. Struct.*, 2009, **18**, 095009.
69. S. T. McGovern, M. Abbot, R. Emery, G. Alici, V.-T. Truong, G. M. Spinks and G. G. Wallace, *Polym. Int.*, 2010, **59**, 357–364.
70. S. Naficy, N. Stoboi, P. G. Whitten, G. M. Spinks and G. G. Wallace, *Smart Mater. Struct.*, 2013, **22**, 075005.
71. W. Lu, A. G. Fadeev, B. Qi, E. Smela, B. R. Mattes, J. Ding, G. M. Spinks, J. Mazurkiewicz, D. Zhou, G. G. Wallace, D. R. MacFarlane, S. A. Forsyth and M. Forsyth, *Science*, 2002, **297**, 983–987.
72. M. Shahinpoor, Y. Bar-Cohen, J. O. Simpson and J. Smith, *Smart Mater. Struct.*, 1998, **7**, R15.
73. M. Shahinpoor and K. J. Kim, *Smart Mater. Struct.*, 2001, **10**, 819.
74. S. Nemat-Nasser and Y. Wu, *J. Appl. Phys.*, 2003, **93**, 5255–5267.
75. C. Jo, D. Pugal, I.-K. Oh, K. J. Kim and K. Asaka, *Prog. Polym. Sci.*, 2013, **38**, 1037–1066.
76. M. Shahinpoor, *Electrochim. Acta*, 2003, **48**, 2343–2353.
77. K. J. Kim and M. Shahinpoor, *Polymer*, 2002, **43**, 797–802.
78. S. Nemat-Nasser and J. Y. Li, *J. Appl. Phys.*, 2000, **87**, 3321–3331.
79. K. Uchino, *Piezoelectric Actuators and Ultrasonic Motors*, Springer US, 1996.
80. S. Takahashi, *Jpn. J. Appl. Phys.*, 1985, **24**, 41.
81. B. Chu, X. Zhou, K. Ren, B. Neese, M. Lin, Q. Wang, F. Bauer and Q. M. Zhang, *Science*, 2006, **313**, 334–336.

82. A. J. Ijspeert, *Science*, 2014, **346**, 196–203.
83. G. Palli, C. Melchiorri, G. Vassura, U. Scarcia, L. Moriello, G. Berselli, A. Cavallo, G. De Maria, C. Natale, S. Pirozzi, C. May, F. Ficuciello and B. Siciliano, *Int. J. Robot. Res.*, 2014, **33**, 799–824.
84. L. Wu, M. Jung de Andrade, R. S. Rome, C. Haines, M. D. Lima, R. H. Baughman and Y. Tadesse, *Proceeding SPIE: Active and Passive Smart Structures and Integrated Systems*, 2015.
85. M. C. Yip and G. Niemeyer, Robotics and Automation (ICRA), 2015 IEEE International Conference on, 2015.
86. W.-S. Chu, K.-T. Lee, S.-H. Song, M.-W. Han, J.-Y. Lee, H.-S. Kim, M.-S. Kim, Y.-J. Park, K.-J. Cho and S.-H. Ahn, *Int. J. Precis. Eng. Manuf.*, 2012, **13**, 1281–1292.
87. C. Fay, K.-T. Lau, S. Beirne, C. Ó. Conaire, K. McGuinness, B. Corcoran, N. E. O'Connor, D. Diamond, S. McGovern, G. Coleman, R. Shepherd, G. Alici, G. Spinks and G. Wallace, *Sens. Actuators, B*, 2010, **150**, 425–435.
88. V. Palmre, J. J. Hubbard, M. Fleming, D. Pugal, S. Kim, K. J. Kim and K. K. Leang, *Smart Mater. Struct.*, 2013, **22**, 014003.
89. I. Must, F. Kaasik, I. Põldsalu, L. Mihkels, U. Johanson, A. Punning and A. Aabloo, *Adv. Eng. Mater.*, 2015, **17**, 84–94.
90. G.-K. Lau, H.-T. Lim, J.-Y. Teo and Y.-W. Chin, *Smart Mater. Struct.*, 2014, **23**, 025021.
91. K. Y. Ma, P. Chirarattananon, S. B. Fuller and R. J. Wood, *Science*, 2013, **340**, 603–607.
92. S. Lee, Y. Kim, G. Jeong, B. Yoon, J. Jho, D. Kim and K. Rhee, *Int. J. Precis. Eng. Manuf.*, 2012, **13**, 2135–2141.
93. J. Bohannon, *Science*, 2014, **346**, 180–181.

CHAPTER 14

Materials for Tissue Engineering and 3D Cell Culture

CAITLIN R. LANGFORD[a] AND NEIL R. CAMERON*[a,b]

[a] Monash University, Department of Materials Science and Engineering, 22 Alliance Lane, Clayton 3800, Victoria, Australia; [b] School of Engineering, University of Warwick, Coventry CV4 7AL, UK
*Email: neil.cameron@monash.edu

14.1 Introduction

At present, there is tremendous interest in the fields of tissue engineering and 3D cell culture.[1–3] In the event of severe injury or failure of a tissue, an effective treatment method must be employed in order to repair or replace the tissue. Tissue grafting and organ transplantation are, at present, the most commonly used treatment methods;[3,4] however, there are several drawbacks.[4,5] The number of donors restricts allografting (the transplantation tissue from a donor to a recipient of the same species), with demand often greatly outstripping the supply of donated organs and tissues, and the risk of rejection results in patients requiring immunosuppressant drugs for life.[5,6] Autografting (the transplantation of tissue from one part of the body to another) is limited by the size and location of the injury, as well as the affected tissue type.[6] The procedure is also painful and can result in severe scarring. Xenografting (the cross-species transplantation of tissues) has been considered as an alternative treatment method; however, the other species used generally have shorter lifespans than humans, and hence their tissues

RSC Polymer Chemistry Series No. 22
Bio-inspired Polymers
Edited by Nico Bruns and Andreas F. M. Kilbinger

Published by the Royal Society of Chemistry, www.rsc.org

age quicker. Xenografted tissues can also be a source of disease transmission[7] or rejection,[8] and introduce a wide range of ethical issues surrounding the permanent alteration of an animal's genetic code.[9] As a result of these drawbacks, the ability to regrow tissues or organs from a patient's own cells in order to transplant them back into the body is becoming increasingly attractive.

In order to culture cells in a 3D environment, a scaffold is required to provide both mechanical support as well as the biological cues found in the cells' native environment. This role is played *in vivo* by the extracellular matrix (ECM), which is a mixture of structural proteins, glycosaminoglycans, and proteoglycans that surrounds the cells. The development of scaffolds for tissue engineering and 3D cell culture is therefore very heavily inspired by biology, since all scaffolds are essentially attempting to mimic the natural ECM, at least to some extent.

There are several features a material must possess before it can be considered for use as a biomaterial,[4] including:

- Biocompatibility
- A surface that allows for sufficient cell attachment
- A morphology that allows for cell infiltration
- A morphology that allows for the transport of nutrients and waste products
- Mechanical properties similar to that of the native tissue
- Biodegradability

There is a wide range of biomaterials that feature the attributes required for tissue engineering and 3D cell culture, including hydrogels,[10–15] porous polymers,[16,17] and electrospun fibres,[18] and such scaffolds have been fabricated from a wide range of materials, including both natural[19] and synthetic polymers.[20] This chapter will focus on the synthesis of fibrous and highly porous polymer scaffolds through the use of electrospinning, thermally induced phase separation, emulsion templating, and the breath figure approach.

14.2 Electrospinning

Electrospinning is a low cost, versatile route for the preparation of fibrous polymer mats, where fibres have diameters in the nanometre to micrometre range.[21] Polymer fibres with a variety of physical and chemical properties can be produced by this method from both natural[22–26] and synthetic polymers.[27–29]

The electrospinning process relies on electrostatics in order to produce polymer fibres from either a polymer solution or melt. The setup requires three basic components: a syringe pump, fitted with a syringe containing the polymer solution or melt; a high voltage DC power supply; and a grounded collector, usually a metal plate, as demonstrated in Figure 14.1.

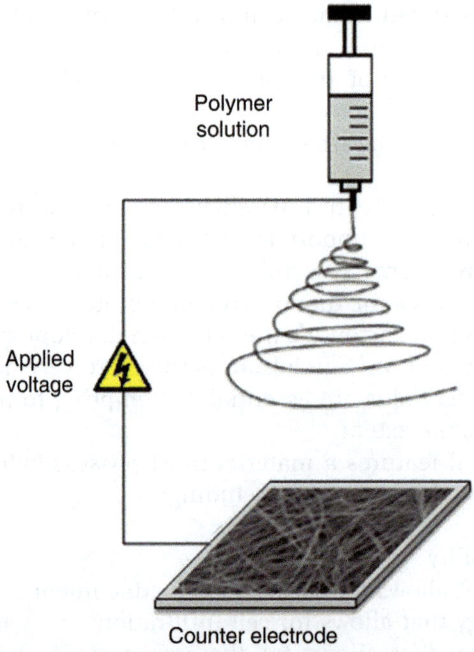

Figure 14.1 Example of an electrospinning set-up.
Reprinted from D. I. Braghirolli, D. Steffens and P. Pranke, Electrospinning for regenerative medicine: a review of the main topics, *Drug Discovery Today*, **19**, 743–753, Copyright 2014 with permission from Elsevier (ref. 30).

A high voltage is applied to the polymer solution or melt, inducing a charge which generates repulsive interactions within the liquid. When these repulsive interactions are balanced by the surface tension of the polymer solution/melt, a Taylor cone is formed.[31] Once the electrical forces become greater than the surface tension, a jet of polymer liquid is ejected from the tip of the Taylor cone. The acceleration of the jet can be up to 600 m s^{-1} and so gravitational forces do not affect the forming polymer fibre.[32] The success of fibre formation depends on a number of parameters[18,33] which can be categorised as follows:

- Processing parameters
- Solution parameters
- Environmental parameters

Processing parameters include the distance between the collector and the tip,[34] the geometry of the collector,[35–37] the electrical potential,[38] and the flow rate.[39,40] Solution parameters include the viscosity, conductivity, and surface tension of the polymer solution or melt,[41] and the environmental parameters include the temperature[42] and humidity[43] in the electrospinning chamber. The concentration of polymer solution determines if a continuous polymer fibre is formed, as a high degree of polymer chain entanglement is required to form continuous polymer fibres. Solutions of low concentration result in discontinuous polymer beads due to the formation of polymer aggregates. As the concentration increases, the degree of chain entanglement increases and hence continuous polymer fibres are produced. As the concentration increases, the fibres become thicker until the increase in viscosity prevents the polymer jet from escaping the Taylor cone.[44–46] As a result of the same considerations, low molecular weight polymers are often more difficult to electrospin than their higher molecular weight counterparts.[47,48]

In order to obtain continuous polymer fibres, highly volatile solvents must be used in order to ensure complete evaporation of the solvent before the jet reaches the collector.[49,50] The temperature and humidity in the electrospinning chamber also impact the rate of solvent evaporation. These, combined with the distance between the tip and the collector, determine whether the obtained polymer fibres are wet or dry, as well as influence their morphology. Fibres obtained at higher temperatures have been found to have smaller diameters due to reductions in the solution viscosity, while increases in humidity result in the formation of pores in the fibres.[42,51–58] Decreases in fibre diameter are also observed when higher voltages and solutions with higher conductivities are used.[59–63]

Electrospun fibres have been widely investigated as scaffolds for tissue engineering due to the morphological similarities of the fibrous meshes and the native ECM. The wide range of polymers that can be processed using this technique have resulted in its use in the fabrication of scaffolds for a variety of tissue types, including nerve,[64,65] skin,[66,67] blood vessels,[68,69] and bone.[70,71] Example scanning electron microscopy (SEM) images of electrospun fibres fabricated from polymers commonly used for tissue engineering applications can be seen in Figure 14.2.[72]

Tubular scaffolds fabricated by electrospinning have been extensively studied as scaffolds for small diameter (<6 mm) vascular grafts. Current artificial grafts, mainly Dacron® and expanded polytetrafluoroethylene (ePTFE), have had some success when used in larger diameter grafts but are limited by issues such as thrombosis when considered for use in smaller diameter grafts.[73–75] Electrospun tubular scaffolds can be fabricated with well-controlled diameters and uniform thickness across the material. The variety of polymers from which these scaffolds can be synthesised has resulted in scaffolds with mechanical properties similar to those of native vascular tissues, and hence these better support blood flow and capillary in-growth.

Synthetic polymers, such as polycaprolactone (PCL), have been popular choices when fabricating electrospun scaffolds for vascular tissue engineering due to their favourable mechanical properties. These polymers are often combined with natural polymers,[76,77] or surface functionalised/coated with biologically relevant molecules, such as peptides containing the integrin binding sequence arginine-glycine-aspartic acid (RGD),[78] the

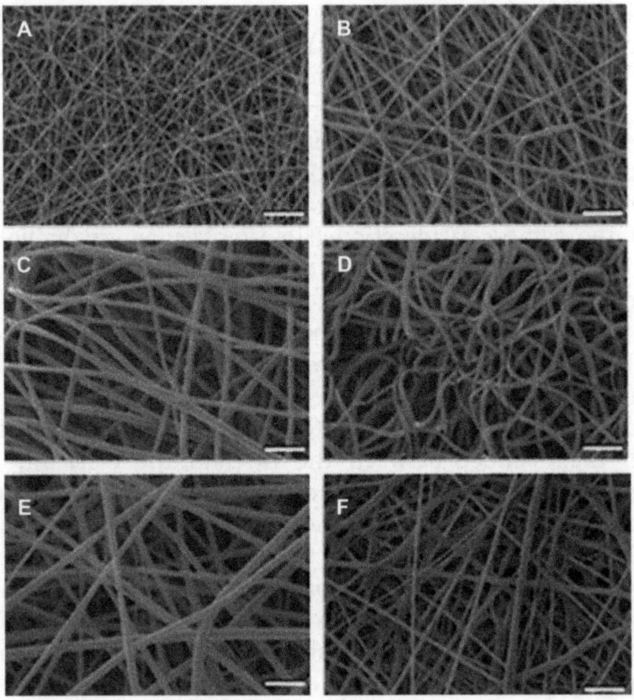

Figure 14.2 Scanning electron micrographs of typical electrospun fibres fabricated from (A) polyglycolic acid, (B) PLGA, (C) PLA, (D) PLGA, (E) PLLA, and (F) PCL. Scale bars = 10 μm.
Reprinted from W.-J. Li *et al.*, Fabrication and characterization of six electrospun poly(α-hydroxy ester)-based fibrous scaffolds for tissue engineering applications, *Acta Biomater.*, **2**, 377–385, Copyright 2006 with permission from Elsevier.[72]

angiogenic vascular endothelial growth factor (VEGF), or fibroblast growth factor 2 (FGF2).[79]

Culture of primary myofibroblasts, harvested from a human vena saphena magna, on electrospun PCL fibres with diameters ranging from 3.4 μm to 12.1 μm showed that cell penetration increased with the increasing fibre diameter over a three day culture period. The increase in penetration is believed to be due to the increase in pore size associated with the increasing fibre diameter.[80,81] Over a 4 week culture period, the PCL scaffolds with fibre diameter 12.1 μm were found to support the formation of homogenous vascular tissues and the deposition of ECM components.[80]

Tubular scaffolds comprised of PCL fibres with diameters ranging from 0.5 μm to 3 μm were coated with the RGD containing peptide Nap-FFGRGD. The hydrophobic naphthalene groups have been shown to self-assemble on the surface of the electrospun PCL fibres, forming an RGD-coating layer on the surface of the material. This RGD coating increases the hydrophilicity of the material, facilitating cell attachment and spreading.[82] Implantation of both PCL and RGD coated PCL grafts into rabbit carotid arteries for both 2 and 4 weeks showed that surface functionalisation with RGD resulted in an increase in the haemocompatibility of the grafts. RGD-coated PCL scaffolds were found to have greatly improved cell infiltration compared to their uncoated counterparts, and endothelialisation occurred at an improved rate.[78]

Tubular nanofibrous PCL/collagen composite scaffolds with mechanical strengths similar to that of native vascular tissues have also been fabricated by electrospinning.[83] The scaffolds were shown to withstand pressurised flow, and the resistance to flow was stable for a period of 4 weeks. The incorporation of collagen into the PCL network resulted in an increase in elasticity, and the resulting materials were found to have a burst pressure strength greater than physiological blood pressure. The materials were shown to support the adhesion and proliferation of both endothelial and smooth muscle cells.[83]

Scaffolds for bone tissue engineering can also be fabricated by electrospinning. As with electrospun materials targeted to other tissues, much of the work has focused on the use of synthetic polymers, such as PCL,[84] polylactic acid (PLA),[85] and polylactic-*co*-glycolic acid (PLGA),[86] and natural polymers, such as collagen.[87] The incorporation of collagen into PCL electrospun fibres was shown to greatly enhance several properties of the nanofibrous material, including the water uptake, mechanical, and degradation properties.[87] The adhesion of MC3T3-E1 osteoblasts was also seen to be enhanced upon the inclusion of collagen into the fibre network, along with the upregulation of collagen type 1, alkaline phosphatase (ALP), and osteopontin (OPN), suggesting that these materials effectively support bone regeneration.[87]

Inorganic components, such as hydroxyapatite (HA)[85,88] and beta-tricalcium phosphate (β-TCP),[89,90] can be incorporated into electrospun polymer networks with relative ease. Both HA and β-TCP have chemical structures similar to that of native bone mineral, and their use in composite scaffolds has been shown to enhance the osteoconductive and mechanical properties of the materials.[91] The inclusion of HA and β-TCP has also been shown to increase the surface roughness of scaffolds,[92] and to improve attachment, proliferation, and differentiation of both osteoblastic cells and mesenchymal stem cells (MSCs).[85,88,90,93,94] In addition, implants, such as hip replacements, are coated in HA to promote integration between living bone and the implant.[95] PLGA/HA composite nanofibres have been shown to increase the ALP activity of human mesenchymal stem cells (hMSCs), and upregulate the expression of osteogenic differentiation genes, including bone sialoprotein (BSP) and osteocalcin (OCN).[96] In addition, a significant increase in calcium deposition was observed, suggesting that PLGA/HA composite materials can support the regeneration of bone.[96]

In addition to the incorporation of bioceramics, such as HA and β-TCP, signalling molecules, such as bone morphogenetic proteins (BMPs), have been incorporated into PLGA/HA[97,98] and poly-L-lactic acid (PLLA)/HA[99] composite electrospun scaffolds. BMPs have been shown to induce both differentiation of MSCs to osteoblasts, and the regeneration of bone *in vivo*. PLLA/HA/BMP-2 composite scaffolds have been shown to support enhanced proliferation of the osteoblastic cell line MG63 when compared to pure PLLA electrospun scaffolds.[99] In addition, implantation of PLLA/HA/BMP-2 into the tibia of rabbits indicated that the combination of both BMP-2 and HA results in the increased formation and mineralisation of new bone, when compared to PLLA, PLLA/HA, and PLLA/BMP-2 scaffolds.[99]

14.3 Thermally Induced Phase Separation

Scaffolding materials with highly porous, interconnected morphologies suitable for cell culture and tissue engineering applications can be fabricated *via* thermally induced phase separation (TIPS). The process, which takes advantage of changes in thermal energy to induce the de-mixing of a polymer solution, occurs in five steps:[100]

- Polymer dissolution
- Phase separation and gelation
- Solvent extraction
- Freezing
- Freeze-drying

Under certain thermal conditions, homogenous polymer solutions will phase separate into polymer-rich and polymer-lean phases by either liquid–liquid phase separation or solid–liquid de-mixing, resulting in materials with either open or closed pore morphologies. Phase separation generally occurs *via* liquid–liquid de-mixing[101] induced by cooling the polymer solution to a temperature below a binodal solubility curve or by exposing the polymer solution to another immiscible solvent. Once phase separation has occurred, the solvent is then extracted by freeze-drying. Controlling the preparation and phase separation conditions allows for control over the final pore diameter and morphology.[102]

Plotting the temperature–composition phase diagram for a binary polymer–solvent system with an upper critical solution temperature results in two curves: a spinodal curve and a binodal curve, as demonstrated in Figure 14.3. The point where these two curves meet represents the critical point of the system, between the spinodal and binodal curves the system is metastable, and under the spinodal curve it is unstable.[103,104] Controlling the region of the curve at which early stage phase separation occurs, controls the pore morphology and degree of interconnectivity obtained. If phase separation occurs in the unstable region under the spinodal curve, liquid–liquid de-mixing occurs rapidly *via* a spinodal decomposition mechanism, resulting in a porous polymer with a fully interconnected morphology. On the other hand, if liquid–liquid de-mixing occurs in the metastable region between the spinodal and binodal curves, de-mixing is controlled by a

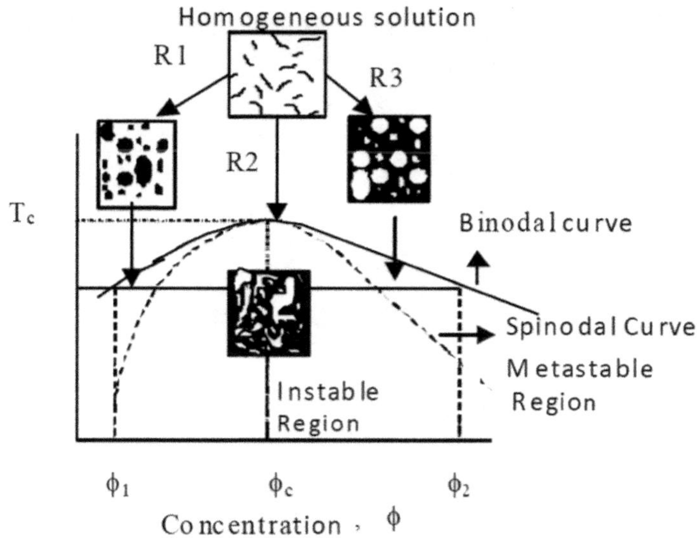

Figure 14.3 Schematic representation of a temperature–concentration phase diagram of a polymer solution.
Reproduced from ref. 106.

nucleation and growth process, and the resulting polymer will have a poorly connected, bead-like pore morphology.[105]

Control over further variables, including the concentration and molecular weight of the polymer used, the rate at which the solution is cooled, the final temperature reached, and the choice of solvent, also impact the morphology of the resulting polymer by varying the phase separation mechanism and the time required for complete phase separation.[107–110] By fine-tuning the phase separation conditions, polymers with hierarchical pore structures suitable for tissue engineering and 3D cell culture applications can be obtained. Typically, macropores in the range 10–80 μm, interconnecting windows in the range of 10 μm and, when phase separation occurs predominantly in the metastable region, a surface microporosity are obtained.[105] Pores with larger diameters can be obtained by increasing the de-mixing time, due to the effects of Ostwald ripening and droplet coalescence during the later stages of phase separation.[111] In addition to coarsening effects, pore size is influenced by the temperature at which phase separation occurs. At lower temperatures, rapid nucleation occurs, resulting in a lower average pore diameter. At higher temperatures, the nucleation process is slower, allowing for growth of the phase separated regions, up to a saturation level.[105,112]

Biodegradable porous polymers have been fabricated by the TIPS method from a range of synthetic and natural polymers including gelatin,[113] chitosan,[114] PLGA,[115] polyurethanes (PU),[116,117] PCL,[118] and PLLA.[112,119] Representative SEM images of the structure of such a phase separated porous

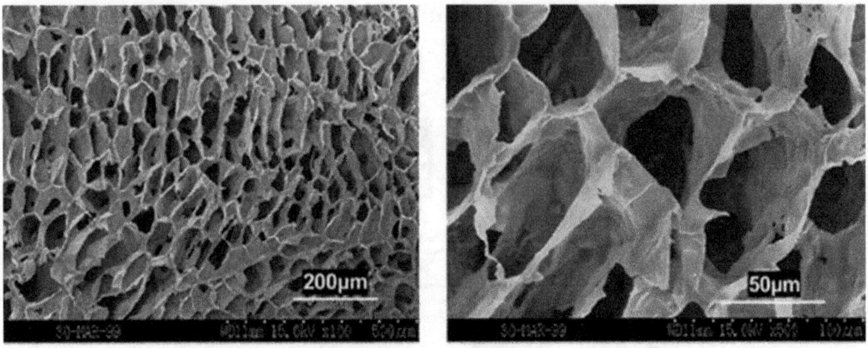

Figure 14.4 SEM images of highly porous PLLA fabricated by thermally induced phase separation at (A) low magnification, and (B) high magnification. Reproduced from ref. 120 with permission from John Wiley and Sons. Copyright © 2000 John Wiley and Sons.

polymer can be seen in Figure 14.4. A range of biologically relevant molecules, including growth factors[116] and HA, have also been incorporated into the scaffolds.

Elastomeric scaffolds with tensile strengths similar to that of the canine thoracic artery have been fabricated from poly(ester urethane) urea (PEUU) and poly(ether ester urethane) urea (PEEUU), synthesised from PCL (PEUU) or polycaprolactone-*b*-poly(ethylene glycol)-*b*-polycaprolactone (PEEUU) plus 1,4-diisocyanate and putrescine, using the FDA approved food stabilizer stannous octoate as the catalyst.[116,121] The resulting porous polymers, phase separated from dimethyl sulfoxide (DMSO), had high porosities, in the range of 80–97%, and an open interconnected pore morphology. Varying the polymer concentration allowed for control over the pore morphology, with 5 and 10 wt% PEUU solutions resulting in irregular pore structures, whereas solutions with a polymer concentration of 8 wt% resulted in a foam with tubular pores with parallel orientation. Both scaffolds were shown to be capable of supporting the growth of smooth muscle cells for periods of seven days, with cells cultured on the PEEUU scaffolds exhibiting higher cell numbers.[121] Basic fibroblast growth factor (bFGF), stabilised with bovine serum albumin (BSA) or heparin, was incorporated into the PEUU scaffold in order to aid angiogenesis. After seven days of culture, the number of rat smooth muscle cells on the bFGF-PEUU scaffold was significantly higher than on the PEUU scaffold with no included growth factors.[116]

The compressive strength of PU foams can be significantly increased through the incorporation of multi-walled carbon nanotubes (CNTs) into the polymer solution prior to phase separation.[117] The CNTs can be observed at the polymer surface, giving the surface a rough "hairy" texture, and appear to be uniformly distributed throughout the bulk of the material. Culture of SaOS-2 osteosarcoma cells on PU/CNT nanocomposite foams indicated that the scaffolds were capable of supporting both the growth and mineralisation of osteoblast-like cells. The presence of CNTs at the polymer surface was also shown to result in an up-regulation of vascular endothelial growth factor (VEGF) in a concentration dependent manner,[117] which has been shown to play an important role in bone repair and remodelling.[122,123]

Highly porous PLLA scaffolds can be fabricated by phase separation of a 5 wt% solution of PLLA in dioxane.[120] The incorporation of HA into the polymer solution prior to phase separation resulted in a composite scaffold with porosity of approximately 90%, and both a higher compressive modulus and higher compressive yield strength than that of the pure PLLA scaffold. The number of MC3T3-E1 osteoblasts on the scaffold was found to be significantly higher on the PLLA/HA composite scaffold than on the PLLA scaffold, and higher expression levels of the osteogenic markers BSP and OCN were also observed.[120]

In addition to the fabrication of porous polymers, TIPS can be used in order to produce scaffolds with a nanofibrous morphology, with fibre diameters generally ranging from 50 nm to 500 nm with porosities up to 98%.[100,124,125] Such nanofibrous architectures are believed to more closely mimic the native ECM environment. Combining phase separation with the addition of a porogen, such as paraffin spheres, yields macroporous scaffolds with a high degree of interconnectivity,[126,127] as highlighted in Figure 14.5. Phase separating PLLA from a mixture of dioxane and pyridine with the addition of paraffin spheres with diameter 250–420 μm resulted in highly porous scaffolds with highly interconnected nanofibrous pore walls.[126] When compared to solid pore walled PLLA scaffolds, the nanofibrous scaffolds were found to adsorb 4.2-fold more proteins from human serum, including fibronectin and vitronectin, which are associated with cell adhesion to the ECM. In addition, MC3T3-E1 osteoblasts were found to

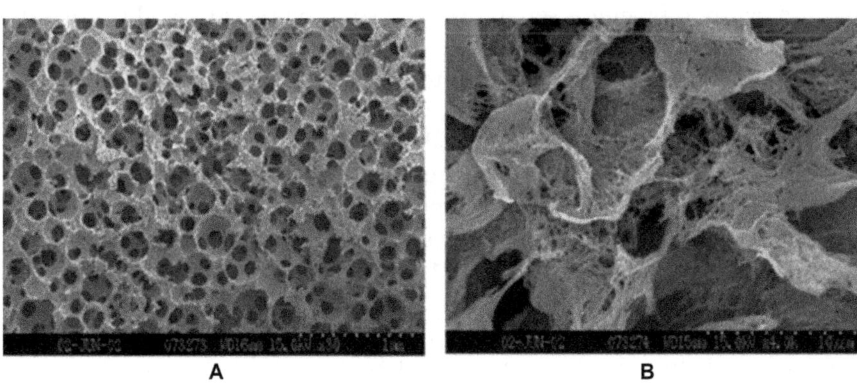

Figure 14.5 SEM images of a nanofibrous PLLA scaffold fabricated by thermally induced phase separation with the addition of paraffin spheres as a porogen at (A) low magnification, and (B) high magnification.
Adapted from ref. 126 with permission from John Wiley and Sons. Copyright © 2003 Wiley Periodicals, Inc.

attach to fetal bovine serum (FBS) pre-treated nanofibrous scaffolds in higher numbers than the solid pore walled scaffolds as a result of the increased level of cell attachment proteins.[126] Neonatal mouse osteoblasts, obtained from frontal and parietal bones, were cultured on the nanofibrous PLLA scaffolds for up to 14 days. Von Kossa staining after 14 days indicated that the fibrous architecture encouraged a much greater degree of mineralisation than the solid walled PLLA scaffolds.[127] In addition, real-time polymerase chain reaction (RT-PCR) showed higher levels of BSP and OCN after both 7 and 10 days of culture on the fibrous scaffold, and Western blot analysis indicated an upregulation of the transcription factor protein runx2, which is associated with the differentiation of osteoblasts. These results suggest that such nanofibrous PLLA scaffolds may be suitable for applications in bone tissue engineering.[127]

HA can be added to the surface of nanofibrous PLLA scaffolds through its inclusion into the PLLA solution prior to phase separation.[128] The presence of HA in the material resulted in an increase in the compressive modulus, giving a value of 420 ± 110 kPa for materials produced from an 80 wt% HA solution. As previously discussed, PLLA scaffolds with nanofibrous pore walls were able to adsorb higher levels of serum proteins than their solid wall counterparts.[126] The addition of HA to the surface of the material was found to further enhance the protein adsorption from BSA when compared to a nanofibrous scaffold.[128] The viability of bone mesenchymal stem cells (BMSCs) cultured on the nanofibrous PLLA scaffolds was determined by MTT assay after seven days. The viability of BMSCs was found to be greatest on the scaffold containing HA.[128]

The conjugate base of citric acid (CA) is an important intermediate in the Krebs cycle, the mechanism by which aerobic organisms, including eukaryotic cells, generate energy. Nanofibrous PLLA scaffolds with CA both inside the polymer network and on the surface of PLLA can be produced *via* the TIPS method by phase separation of a solution of PLLA and CA in THF, followed by solvent exchange with a solution of 0.5 wt% CA in deionised water.[119] The incorporation of CA into the PLLA fibres resulted in the formation of shorter polymer fibres; however, the nanofibre diameter remained unchanged. The presence of CA was also seen to reduce the hydrophobicity of the PLLA scaffold, as measured by a reduction in the water contact angle due to the introduction of polar groups at the polymer surface. The viability of mouse BMSCs cultured on the CA containing PLLA nanofibrous scaffolds was assessed by live/dead staining at 1, 4, and 7 days of culture. After both 4 and 7 days of culture, a greater number of live cells was observed on the PLLA/CA scaffold than on the PLLA only scaffold, as well as a faster rate of proliferation being observed on the PLLA/CA scaffold at 5 and 8 days of culture. Osteogenic differentiation of the BMSCs cultured on the PLLA/CA scaffold was monitored by observing the expression of ALP, OCN, collagen type I, and runx2 between 7 and 28 days of culture. At 14 days of culture, ALP levels were significantly higher on the PLLA/CA scaffolds than on the PLLA scaffolds, indicating that the presence of CA at the polymer surface promotes osteogenic differentiation. At 21 and 28 days, the levels of collagen type I and ALP were found to decrease on the PLLA/CA scaffolds, but remain unchanged on the PLLA scaffold. In contrast, the levels of OCN and runx2 were observed to increase between 21 and 28 days of culture on the PLLA/CA scaffolds. Since collagen type 1 and ALP are associated with early stage osteogenic differentiation, and OCN and runx2 associated with late stage differentiation, these results suggest that the presence of CA on the surface of a material promotes osteogenic differentiation, and that nanofibrous PLLA/CA scaffolds may be suitable for bone tissue engineering applications.[119]

14.4 Emulsion Templated Porous Polymers (PolyHIPEs)

Emulsion templating offers a route to porous polymers with well-defined morphologies.[129] Emulsions are formed when at least two immiscible liquids are blended to give a heterogeneous suspension of droplets of one liquid inside a continuous phase of the other. If this continuous phase is polymerised, a porous polymer is formed. Emulsions can be described as either oil-in-water (o/w) or water-in-oil (w/o), where the droplet phase is oil or water, respectively.

In order to produce highly porous materials, a certain class of emulsion, known as high internal phase emulsion, or HIPE, is used. HIPEs are defined as having an internal, or droplet, volume phase ratio, ϕ, of 0.74 or greater.[129] A volume fraction of 0.74 represents the maximum volume ratio at which the droplet phase will pack as uniform non-deformable spheres. Values of ϕ up to 0.99 can be observed, indicating that the droplet phase in a HIPE is either non-uniform or that the droplets are deformed into polyhedral ones.[129]

The most commonly used method of forming HIPEs is by slow addition of a porogen (non-continuous phase) to the continuous phase with mixing,[130–132] as demonstrated in Figure 14.6, although other methods can be used.[133] The continuous phase generally consists of a mixture of monomer, co-monomer, and a suitable surfactant; a solvent may also be included in order to reduce the viscosity of the continuous phase. Mixing is generally at a high shear rate and is an important stage in HIPE formation, as it breaks up any larger droplets into smaller ones. Other methods of HIPE formation include the multiple emulsification method and the spontaneous formation method.

HIPEs can be both oil-in-water and water-in-oil emulsions. In w/o emulsions, the continuous phase is the oil phase and the porogen is water; in o/w emulsions, it is the reverse. The type of emulsion formed is dependent on the ratio of each phase and the type of surfactant used.

HIPEs are thermodynamically unstable, but exhibit varying degrees of kinetic stability. The stability of a HIPE is strongly dependent on the internal phase volume ratio, as well as the hydrophilic properties of the monomers, and the type and volume of surfactant used. Increasing the internal phase

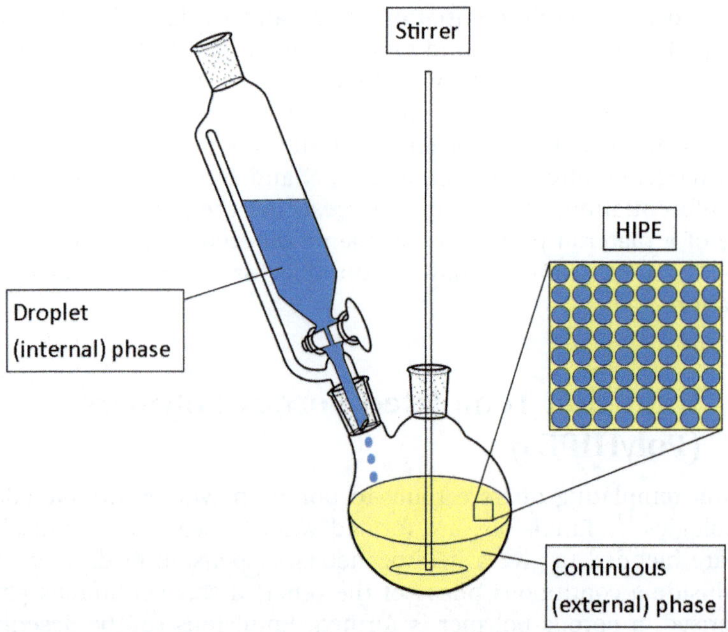

Figure 14.6 Schematic representation of the formation of a high internal phase emulsion.

Figure 14.7 Schematic representation of polyHIPE synthesis.

volume ratio increases the likelihood of droplet coalescence, where droplets merge in order to form larger droplets, and Ostwald ripening,[134] a phenomenon that causes larger droplets to grow at the expense of smaller ones as a result of the high surface energy associated with smaller droplets. The combination of droplet coalescence and Ostwald ripening results in the collapse of the HIPE as the size of the droplets becomes too large for the continuous phase to support them.

Polymerisation, or curing, of the continuous phase of a HIPE gives a porous polymeric material known as a polyHIPE,[135,136] as highlighted in Figure 14.7. The continuous phase of the emulsion must contain a cross-linker in addition to the monomer and surfactant. The cross-linker is needed in order to form the polymer network that makes up the polyHIPE structure.

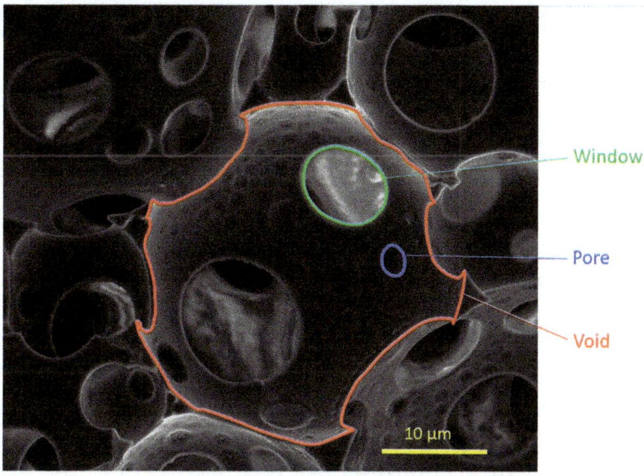

Figure 14.8 SEM image of a typical polyHIPE, highlighting the voids, windows, and pores.

Once cured, the porogen is removed and the porous material is washed by Soxhlet extraction and dried.

The obtained polymer is a highly porous and permeable material, with a complex pore morphology. A SEM image of a typical polyHIPE is shown in Figure 14.8. The spherical cavities shown are referred to as voids, while the smaller interconnecting spheres between voids are known as windows. The much smaller structures within the walls of the polyHIPE are referred to as pores.[129]

The polyHIPE void diameter can be varied from around 1 µm to diameters greater than 100 µm by controlling the diameter of the droplets in the HIPE.[137] The 3D structure of a polyHIPE, including the high degrees of pore interconnectivity that can be obtained, and the ability to tune the structure by varying the properties of the HIPE precursor, makes them particularly attractive as scaffolds for 3D cell culture and tissue engineering applications. The ability to tune the void diameters in a polyHIPE material generally arises by controlling the stability or instability of the HIPE,[137–139] although the rate of shear upon HIPE formation can also have an impact. There are two major factors that affect HIPE stability: Ostwald ripening and droplet coalescence. Ostwald ripening occurs as a direct result of the differences in surface tension and chemical potential between large and small droplets. Smaller droplets experience a higher solubility in the continuous phase as a result of the Kelvin effect,[140] which describes the relationship between the curvature of a liquid's surface and the vapour pressure associated with the liquid. Curved surfaces exhibit a higher vapour pressure than flat surfaces and so smaller droplets have much higher vapour pressures than their larger counterparts. As a result, smaller droplets have a much higher tendency to dissolve and diffuse through the interfacial layer, finally being re-deposited into larger droplets. Droplet coalescence occurs as a result of the thinning and subsequent rupture of the interfacial layer.[141]

Early work focused on the development of non-biodegradable materials for 3D cell culture, fabricated from styrene/divinyl benzene (ST/DVB) polyHIPEs.[142,143] A wide variety of cell lines have been successfully cultured on ST/DVB polyHIPEs, including hepatocytes,[144] osteoblasts,[143] and neural cells.[145] The morphology of hepatocytes cultured on a thin membrane of ST/DVB polyHIPE were found to more closely resemble that of typical liver cells and had significantly more microvilli than the same cells cultured in 2D.[132,146] The cells grown in 3D were also found to synthesize higher concentrations of albumin, a protein found in the liver which plays a critical role in several liver functions, including scavenging free radicals and the binding and transport of drugs. Albumin concentration is commonly used as a marker of hepatocyte metabolic activity,[147] and so the higher concentration of albumin observed in 3D cultures of hepatocytes indicates that growing cells in 3D leads to enhanced cell function.[146] Hepatic cell function can be further explored by monitoring cell sensitivity to drugs metabolised in the liver. Methotrexate is one such drug, which is commonly used in the treatment of solid cancer tumours, such as breast cancer, and leukaemia.[148] Hepatocytes cultured in 2D were found to be sensitive to methotrexate, and the cells' metabolic function was impaired in a dose-dependent manner. As the concentration of methotrexate was increased, a reduction in the number of microvilli was observed and the cells were seen to become flat and eventually disintegrate. When cultured in the 3D environment provided by the ST/DVB scaffold, the cells showed greater viability when exposed to higher concentrations of the drug, and responded to the drug in a manner more similar to that of the liver.[146]

The surface of an ST/DVB polyHIPE is relatively inert and the polymer is highly hydrophobic. The ability to introduce biofunctionality at the polymer surface, either *via* chemical functionalisation or by adsorption of biomolecules onto the surface, can be used to provide the biological cues that cells would normally receive from the extracellular matrix *in vivo*. There are several routes to the preparation of chemically functionalised polyHIPEs, the most commonly used during the preparation of polyHIPEs for 3D cell culture applications is the copolymerisation of a monomer that can be further functionalised post-polymerisation. Acrylic acid can be added to the surface of ST/DVB/2-ethylhexyl acrylate (EHA) through its incorporation into the aqueous phase during HIPE formation.[149] The hydrophilic nature of the acrylic acid monomer results in an increase in the wettability of the scaffold without having a negative impact on hepatocyte adhesion. As a result of these observations, it has been suggested that acrylic acid can be used as a route for further functionalisation of ST/DVB polyHIPEs.[149] A second route to surface functionalisation of ST/DVB polyHIPEs is through the addition of an active ester, name pentafluorophenyl acrylate (PFPA), into the organic phase of the HIPE.[144] The use of active esters, such as PFPA, as a route for the post-polymerisation modification of polyHIPE polymers was first explored in a thiol-ene "click" polyHIPE system.[150] Subsequently, its inclusion in a ST/DVB/EHA polyHIPE was investigated as a route to galactose-functionalised polyHIPEs *via* post-polymerisation functionalisation of the surface bound PFPA.[144] Hepatocyte cells are known to possess a cell-surface asialoglycoprotein receptor[151] (ASGPR), which binds specifically to galactose.[152] This interaction can be targeted in order to improve cell-scaffold binding and cell function. As with previous studies, albumin production was monitored as an indication of the metabolic function of the hepatocyte cells, and was found to be increased in cells cultured on galactose functionalised polyHIPEs. This enhanced albumin production diminished with longer culture times. This decrease is believed to be due to non-specific binding of serum proteins to the polyHIPE surface, and needs to be addressed if galactose-functionalised polyHIPEs are to find widespread application in hepatocyte culture.[144]

ST/DVB scaffolds can be enhanced for 3D cell culture applications through the use of inorganic/organic composite scaffolds.[143] Osteoblast cells cultured on ST/DVB scaffolds exhibited some mineralisation upon culture for 28 and 35 days. In addition, RT-PCR indicated the expression of OCN, OPN, and collagen type I, suggesting that osteogenesis occurs on the polyHIPE scaffolds. When cultured on HA-modified polyHIPEs, there was an increase in the number of cells penetrating into the polyHIPE polymer, and an increase in mineralisation was observed through von Kossa staining, indicating that coating the polyHIPE with HA improves the osteoconductivity of the polyHIPE.[143]

Despite providing an interesting platform from which to study cell behaviour within a 3D environment, ST/DVB polyHIPEs are unsuitable for tissue engineering due to their lack of biodegradability. A number of routes have been explored in order to produce biodegradable polyHIPEs, including the use of vinyl-terminated PLA and PCL, fumarate-based macromers, and by thiol-ene "click" crosslinking chemistry.[131,153,154]

PCL-polyHIPEs were prepared from PCL-diacrylate and either styrene or methyl methacrylate, and found to be partially degradable.[153] Human fibroblasts cultured on these materials for 2.5 days were shown to exhibit the spindle morphology typical of fibroblasts, indicating cell proliferation and tissue growth.[153]

Fully degradable polyHIPEs have been fabricated through the use of thiol-ene "click" crosslinking chemistry between multi-functional thiols and acrylates.[130,155,156] UV initiated crosslinking of a three-armed acrylated PCL and a trithiol yielded polyHIPEs, on which the murine fibroblast cell line L929 was successfully cultured for a period of seven days.[157] The degradation products of the material were confirmed to be biocompatible by MTT assays.

Biodegradable fumarate-based polyHIPEs synthesised from propylene fumarate dimethacrylate have been investigated as injectable scaffolds for the repair of critical sized bone defects.[158] Culture of NIH/3T3 fibroblasts on such polyHIPEs yielded 95% viability after 24 h. Switching from a thermally initiated curing system to redox initiation resulted in a reduction of the polymerisation time.[159] HIPEs with either a reductant or an oxidant could be stored for up to 6 months before mixing the two emulsions to induce polymerisation. hMSCs cultured on the redox cured polyHIPEs were found to have 80% viability after 24 h, as indicated by live/dead staining.[159]

14.5 Breath Figure Method

The breath figure methodology provides a simple and inexpensive route for the preparation of honeycomb patterned porous polymers.[160,161] The term *Breath Figures* refers to the spatial ordering of water droplets as they condense from a highly humid atmosphere onto a cold, uniform surface.[162–164] The formation of breath figures on the surface of a cold solution can be described in three stages: during the initial stage there is little interaction between the water droplets and the droplet diameter is dependent on time; once maximum surface coverage is achieved, the droplets self-assemble into the hexagonal arrays which result in the honeycomb structure observed in the resulting porous polymer. The droplets in these arrays are separated by thin films of liquid, and a narrow dispersity of droplet diameters is observed. In the final stage, droplet coalescence occurs, increasing the observed droplet diameters at a rate that is proportional to time. As the surface of the solution evaporates, the droplets sink in to the solution, imprinting the structure created by the water droplets into the solution, and allowing a new layer of droplets to form on the surface of the solution.[164,165] A schematic representation of the formation of porous polymers *via* the breath figure methodology is shown in Figure 14.9.

If the surface on which the droplets are formed is a solution of a polymer in a highly volatile solvent, porous polymers with highly ordered pore morphologies are produced.[161,167,168] The highly ordered morphologies are a result of two phenomena: precipitation of the polymer at the water/solution interface, minimizing droplet coalescence;[169] and the regular stacking of layers of water droplets in a process driven by Rayleigh–Bernard convection.[163,170,171]

The choice of solvent must be carefully considered as the solvent must have a low boiling point, low solubility in water, a high vapour pressure, and have a higher density than water.[164] The choice of solvent also determines whether a single layer or multiple layers of pores are obtained in two ways: the first of which is the rate at which the solvent evaporates; the second, the interfacial energy between the polymer solution and the water droplets. A less volatile solvent will evaporate slowly, allowing time for the array of water droplets to sink into the polymer solution, resulting in a multi-layered pore structure. The use of fast evaporating solvents will result in a single layer of pores. The interfacial energy between the water droplet and the solvent can be determined from eqn (14.1):

$$z_0 = \frac{\left(\gamma_{\mathrm{w}} - \gamma_{\mathrm{w/s}}\right)}{\gamma_{\mathrm{s}}} \tag{14.1}$$

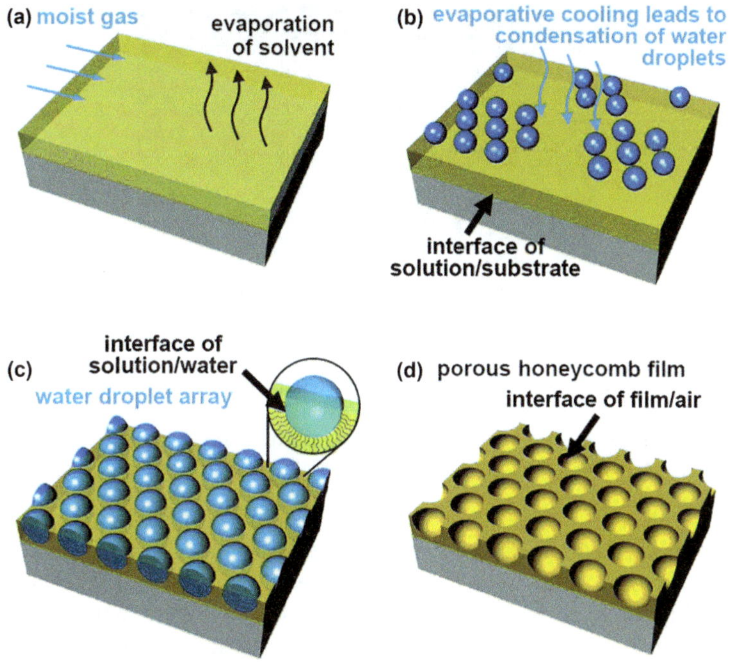

Figure 14.9 Schematic representation of the formation of a porous polymer *via* the breath figure methodology.
Reproduced from ref. 166 with permission from the Royal Society of Chemistry.

where γ_w is the surface tension of the water droplet, γ_s is the surface tension of the polymer solution, and $\gamma_{w/s}$ is the interfacial tension. Multi-layered pore structures are observed in systems where $z_0 > 1$, whereas values of z_0 between −1 and 1 result in single layers or pores.[172]

The diameter of the obtained pores is controlled by two factors: the relative humidity and the concentration/viscosity of the polymer solution.[173–175] A high relative humidity will result in larger pore diameters than those observed at lower humidity, due to the formation of larger water droplets and a decrease in the rate at which the solvent evaporates. This decrease in the rate of solvent evaporation allows for the growth of larger droplets in the initial stages of breath figure formation.[176–178] Highly concentrated polymer solutions will result in pores of smaller diameter due to the high viscosity of the solution, caused by decreased droplet mobility, together with an increased rate of polymer precipitation, preventing droplet coalescence.[178–182] In general, the pore diameter for such honeycomb-structured polymers is on the micron scale, although disordered arrays of 20 nm pores have been reported.[183] A SEM image of a typical honeycomb-structured porous polymer fabricated by breath figure can be seen in Figure 14.10.

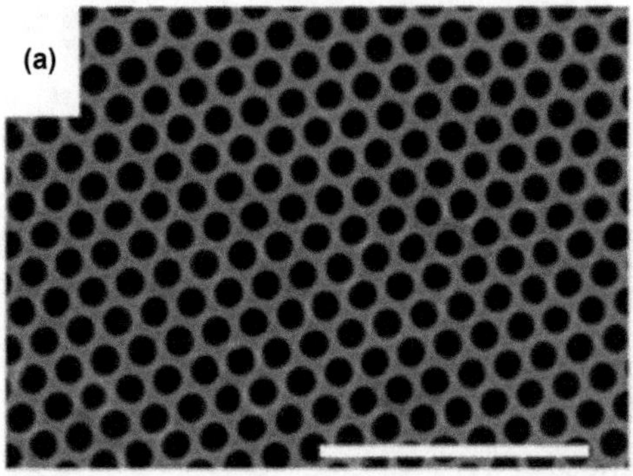

Figure 14.10 SEM image of a honeycomb-structured porous polymer formed by breath figure. Scale bar = 10 μm.
Reproduced from ref. 184 with permission from the Royal Society of Chemistry.

The ease with which the porosity of breath figure honeycomb patterned polymers can be tuned, and the wide variety of polymers from which they can be fabricated, as well as the low cost and ease with which large-scale polymers can be produced, means they have attracted interest as scaffolds for tissue engineering and 3D cell culture.[184–187] There has been particular interest in the use of FDA approved polymers, including PCL,[186] PLA,[174] and PLGA,[188] due to their biocompatibility and biodegradability. The surface properties of such polymers are of great interest as the micron-scale porosity has been shown to have a great influence on the behaviour of cells.

Porous PCL films have received a great deal of interest and have been used to investigate the response of a variety of cell lines to different pore environments.[189–192] By varying the pore diameter of porous films produced from PCL with an amphiphilic copolymer (dodecylacrylamide and ω-carboxyhexylacrylamide) and cast from chloroform, the differentiation of neuronal stem cells (NSCs) can be controlled. NSCs can be maintained in an undifferentiated state when the pore diameter is 3 μm, whereas increasing the pore diameter to greater than 5 μm results in differentiation of the NSCs to immature neurons.[191] Enhanced growth of keratinocytes,[192] fibroblasts,[192] and endothelial cells[190,193,194] has also been observed on honeycomb patterned PCL films.

Casting the polymer films from water miscible tetrahydrofuran (THF), as opposed to the chloroform and benzene used in previous studies, reduces the toxicity of the solvent system to cells. Porous polymers with average pore diameters 10, 6, and 3.5 μm have been produced in this manner and the murine osteoblast cell line MC3T3-E1 was cultured on the materials. When compared to a non-porous PCL control film, the honeycomb patterned polymers were seen to promote the adhesion of cells to the polymer, cell spreading, and proliferation.[186] In addition, gene expression of the bone-specific differentiation markers ALP, collagen type I, OCN, and OPN was promoted and an increase in calcium content was observed. The increase in bone-like activity was more pronounced when the pore size was smallest. Differences in the cytoskeleton of cells cultured on flat films to those cultured on honeycomb PCL films were observed by F-actin staining. MC3T3-E1s cultured on the flat films were observed to have few stress fibres, whereas those cultured on the honeycomb films were found to have a greater number of stress fibres. Stretching the honeycomb polymer resulted in a lengthening of the stress fibres and associated cell elongation as the pores stretched into a groove-like morphology. Cell elongation was particularly pronounced when the pore diameter was 3.5 μm.[186]

In order to produce honeycomb patterned PLA films, additives such as organic surfactants or functional nanoparticles must be added in order to overcome the effects of the high hydrophobicity of PLA. One such example is the use of dendritic amphiphilic block copolymers of poly(L-lysine) and PLLA, synthesised by metal-free ring-opening polymerisation.[195–197] Culture of the human epithelial HeLa cell line showed a distinct increase in cell density after 72 h of culture, indicating that honeycomb patterned polymers produced from such dendrimers are indeed biocompatible.[187]

In addition to changes in cell proliferation, differentiation, and morphology, it is believed that the 3D growth environment provided by porous polymers, such as the honeycomb patterned porous polymers produced by the breath figure methodology, provides a microenvironment that encourages a more physiological response to therapeutics than typical 2D substrates. Combining spin-coating and dip-coating with breath figures allows for control over the pore diameter, with spin-coating resulting in smaller pores due to the increased rate of solvent evaporation. Spin- and dip-coated honeycomb patterned PLGA films have been used as 3D growth substrates for the culture of the breast cancer cell line MCF-7, and the cells' response to the chemotherapeutic doxorubicin was subsequently assessed.[188] Cells cultured on the porous films exhibited more functional growth than those cultured in a 2D environment, with the formation of lobuloalveolar-like structures and branched ductal structures, along with the increased expression of the mammary differentiation genes INTEGB4, GATA3, and EMA. This increase in functional growth was more pronounced on the dip-coated films, and cells cultured on these films exhibit a higher resistance to doxorubicin, comparable to *in vivo* tumour systems.[188]

In addition to pore size, the mechanical strength of polymer substrates has an influence on the behaviour of cells, as cells sense their environment through the focal adhesions that attach the cytoskeleton to the scaffold.[198–202] The mechanical properties of elastomeric poly(1,2-butadiene) (PB) honeycomb patterned films can be controlled by crosslinking the films under UV radiation.[203,204] Crosslinking of the PB films resulted in a 15-fold increase in the elastic modulus but had no effect on the surface properties of the polymer.[205] Murine fibroblasts exhibited stronger adhesions to the crosslinked PB films. When attached to a growth substrate, cells apply a traction force of 30 nN to the surface.[206] Simulation of this force on the surface of both crosslinked and non-crosslinked PB films showed a significant difference in deformation behaviour. Under a 30 nN traction force, the non-crosslinked film elongated and deformed, whereas no deformation was observed when the same force was applied to the crosslinked film.[201] Deformation of the polymer film prevents cell adhesion and spreading, as the cells cannot generate the traction force required to form focal adhesions between the polymer surface and the cell cytoskeleton.[207–210] The crosslinking of honeycomb patterned porous polymers offers a biologically non-invasive method to investigate the behaviour of cells in a variety of microenvironments.[201]

14.6 Conclusions

Over the past few decades there has been significant interest in the field of tissue engineering as the need for replacement tissues and organs increases. As a result, a wide range of scaffold materials has been produced with a variety of fabrication techniques. The morphology, mechanical properties, and degradation profile of a scaffold can be controlled through both the polymer and fabrication technique used. A variety of natural and synthetic polymers have been explored, with synthetic polymers such as PLA, PCL, and PLGA showing excellent potential due to their biocompatibility, suitable mechanical properties, and FDA approval. In addition, various cellular functions can be encouraged through functionalisation of the scaffold at the polymer surface. The incorporation of bioceramics, such as HA or β-TCP, at the polymer surface has been shown to induce osteogenesis,[90,127] and the addition of growth factors, including VEGF and bFGF, has been shown to induce angiogenesis.[79,116] This control over both the physical and chemical properties of a scaffold allows for the design of scaffolds to suit a particular application or tissue type.

While recent advances in scaffold fabrication techniques show great promise, further research is still required in order to both understand and replicate the complex environments found in the native ECM. Closer mimicking of the ECM and integration of more sophisticated biological signalling into synthetic scaffolds will undoubtedly lead to further advances in this field.

Acknowledgements

The authors wish to thank Monash University for funding (PhD scholarship to CRL).

References

1. R. H. Harrison, J.-P. St-Pierre and M. M. Stevens, *Tissue Eng., Part B*, 2014, **20**, 1–16.

2. A. Schober, U. Fernekorn, S. Singh, G. Schlingloff, M. Gebinoga, J. Hampl and A. Williamson, *Eng. Life Sci.*, 2013, **13**, 352–367.
3. S. J. Hollister, *Nat. Mater.*, 2005, **4**, 518–524.
4. R. Langer and J. Vacanti, *Science*, 1993, **260**, 920–926.
5. M. D. Duncan and D. S. Wilkes, *Proc. Am. Thorac. Soc.*, 2005, **2**, 449–455.
6. R. R. Betz, *Orthopedics*, 2002, **25**, S561–S570.
7. Y. Takeuchi and R. A. Weiss, *Curr. Opin. Immunol.*, 2000, **12**, 504–507.
8. G. Chen, Q.-Y. Sun, X.-M. Wang, S.-Q. Shen, H. Guo, H. Wang, Y. Wu, W.-Y. Wang, Y.-L. Xiong and S. Chen, *Xenotransplantation*, 2004, **11**, 123–132.
9. E. Van Rongen, *Ann. N. Y. Acad. Sci.*, 1998, **862**, 177–183.
10. K. Chatterjee, S. Lin-Gibson, W. E. Wallace, S. H. Parekh, Y. J. Lee, M. T. Cicerone, M. F. Young and C. G. Simon Jr, *Biomaterials*, 2010, **31**, 5051–5062.
11. J. Zhu, *Biomaterials*, 2010, **31**, 4639–4656.
12. V. X. Truong, M. P. Ablett, S. M. Richardson, J. A. Hoyland and A. P. Dove, *J. Am. Chem. Soc.*, 2015, **137**, 1618–1622.
13. K. Y. Lee and D. J. Mooney, *Chem. Rev.*, 2001, **101**, 1869–1880.
14. S. Van Vlierberghe, P. Dubruel and E. Schacht, *Biomacromolecules*, 2011, **12**, 1387–1408.
15. A. S. Hoffman, *Adv. Drug Delivery Rev.*, 2012, **64**(Supplement), 18–23.
16. M. S. Silverstein, *Prog. Polym. Sci.*, 2014, **39**, 199–234.
17. H.-Y. Mi, X. Jing and L.-S. Turng, *J. Cell. Plast.*, 2015, **51**, 165–196.
18. Y.-F. Goh, I. Shakir and R. Hussain, *J. Mater. Sci.*, 2013, **48**, 3027–3054.
19. B. Kundu, N. E. Kurland, S. Bano, C. Patra, F. B. Engel, V. K. Yadavalli and S. C. Kundu, *Prog. Polym. Sci.*, 2014, **39**, 251–267.
20. M. Okamoto and B. John, *Prog. Polym. Sci.*, 2013, **38**, 1487–1503.
21. R. J. Wade and J. A. Burdick, *Mater. Today*, 2012, **15**, 454–459.
22. H. K. Noh, S. W. Lee, J. M. Kim, J. E. Oh, K. H. Kim, C. P. Chung, S. C. Choi, W. H. Park and B. M. Min, *Biomaterials*, 2006, **27**, 3934–3944.
23. H. Homayoni, S. A. H. Ravandi and M. Valizadeh, *Carbohydr. Polym.*, 2009, **77**, 656–661.
24. C. Chen, C. Cao, X. Ma, Y. Tang and H. Zhu, *Polymer*, 2006, **47**, 6322–6327.
25. P. Songchotikunpan, J. Tattiyakul and P. Supaphol, *Int. J. Biol. Macromol.*, 2008, **42**, 247–255.
26. K. S. Rho, L. Jeong, G. Lee, B. M. Seo, Y. J. Park, S. D. Hong, S. Roh, J. J. Cho, W. H. Park and B. M. Min, *Biomaterials*, 2006, **27**, 1452–1461.
27. F. Yang, R. Murugan, S. Wang and S. Ramakrishna, *Biomaterials*, 2005, **26**, 2603–2610.
28. S. de Valence, J.-C. Tille, D. Mugnai, W. Mrowczynski, R. Gurny, M. Moeller and B. H. Walpoth, *Biomaterials*, 2012, **33**, 38–47.
29. M. Ghaedi, M. Soleimani, I. Shabani, Y. Duan and A. S. Lotfi, *Cell. Mol. Biol. Lett.*, 2012, **17**, 89–106.
30. D. I. Braghirolli, D. Steffens and P. Pranke, *Drug Discovery Today*, 2014, **19**, 743–753.

31. G. Taylor, *Proc. R. Soc. London, Ser. A*, 1969, **313**, 453–475.
32. J. H. Wendorff, S. Agarwal and A. Greiner, in *Electrospinning*, Wiley-VCH Verlag GmbH & Co. KGaA, 2012, DOI: 10.1002/9783527647705, ch. 2, pp. 29–68.
33. A. Greiner and J. H. Wendorff, *Angew. Chem., Int. Ed.*, 2007, **46**, 5670–5703.
34. V. Sencadas, D. M. Correia, A. Areias, G. Botelho, A. M. Fonseca, I. C. Neves, J. L. Gomez Ribelles and S. L. Mendez, *Carbohydr. Polym.*, 2012, **87**, 1295–1301.
35. Q. P. Pham, U. Sharma and A. G. Mikos, *Tissue Eng.*, 2006, **12**, 1197–1211.
36. S. Agarwal, J. H. Wendorff and A. Greiner, *Adv. Mater.*, 2009, **21**, 3343–3351.
37. S. Agarwal, J. H. Wendorff and A. Greiner, *Polymer*, 2008, **49**, 5603–5621.
38. V. Beachley and X. Wen, *Mater. Sci. Eng., C: Biomimetic Supramol. Syst.*, 2009, **29**, 663–668.
39. S. H. Tan, R. Inai, M. Kotaki and S. Ramakrishna, *Polymer*, 2005, **46**, 6128–6134.
40. J. Doshi and D. H. Reneker, *J. Electrost.*, 1995, **35**, 151–160.
41. N. Charernsriwilaiwat, P. Opanasopit, T. Rojanarata, T. Ngawhirunpat and P. Supaphol, *Carbohydr. Polym.*, 2010, **81**, 675–680.
42. M. Pakravan, M.-C. Heuzey and A. Ajji, *Polymer*, 2011, **52**, 4813–4824.
43. L. Van der Schueren, B. De Schoenmaker, O. I. Kalaoglu and K. De Clerck, *Eur. Polym. J.*, 2011, **47**, 1256–1263.
44. N. Bolgen, Y. Z. Menceloglu, K. Acatay, I. Vargel and E. Piskin, *J. Biomater. Sci., Polym. Ed.*, 2005, **16**, 1537–1555.
45. A. M. Piras, L. Nikkola, F. Chiellini, N. Ashammakhi and E. Chiellini, *J. Nanosci. Nanotechnol.*, 2006, **6**, 3310–3320.
46. D. Srinath, S. Lin, D. K. Knight, A. S. Rizkalla and K. Mequanint, *J. Tissue Eng. Regener. Med.*, 2014, **8**, 578–588.
47. Z. Chen, L. Wang and H. Jiang, *Biofabrication*, 2012, **4**, 035007.
48. N. Kasoju and U. Bora, *Biomed. Mater.*, 2012, **7**, 045004.
49. N. Bhardwaj and S. C. Kundu, *Biotechnol. Adv.*, 2010, **28**, 325–347.
50. Z. M. Huang, Y. Z. Zhang, M. Kotaki and S. Ramakrishna, *Compos. Sci. Technol.*, 2003, **63**, 2223–2253.
51. M. M. Demir, I. Yilgor, E. Yilgor and B. Erman, *Polymer*, 2002, **43**, 3303–3309.
52. W. Bonani, D. Maniglio, A. Motta, W. Tan and C. Migliaresi, *J. Biomed. Mater. Res., Part B*, 2011, **96B**, 276–286.
53. B. L.-P. Lee, H. Jeon, A. Wang, Z. Yan, J. Yu, C. Grigoropoulos and S. Li, *Acta Biomater.*, 2012, **8**, 2648–2658.
54. C. X. Zhang, X. Y. Yuan, L. L. Wu, Y. Han and J. Sheng, *Eur. Polym. J.*, 2005, **41**, 423–432.
55. C. L. Casper, J. S. Stephens, N. G. Tassi, D. B. Chase and J. F. Rabolt, *Macromolecules*, 2004, **37**, 573–578.
56. D. H. Reneker and I. Chun, *Nanotechnology*, 1996, **7**, 216–223.

57. C. Mit-uppatham, M. Nithitanakul and P. Supaphol, *Macromol. Chem. Phys.*, 2004, **205**, 2327–2338.
58. O. Hardick, B. Stevens and D. G. Bracewell, *J. Mater. Sci.*, 2011, **46**, 3890–3898.
59. X. M. Mo, C. Y. Xu, M. Kotaki and S. Ramakrishna, *Biomaterials*, 2004, **25**, 1883–1890.
60. B. Kim, H. Park, S. H. Lee and W. M. Sigmund, *Mater. Lett.*, 2005, **59**, 829–832.
61. A. K. Haghi and M. Akbari, *Phys. Status Solidi A*, 2007, **204**, 1830–1834.
62. A. Sonseca, L. Peponi, O. Sahuquillo, J. M. Kenny and E. Gimenez, *Polym. Degrad. Stab.*, 2012, **97**, 2052–2059.
63. N. E. Zander, J. A. Orlicki, A. M. Rawlett and T. P. Beebe, Jr., *J. Mater. Sci.: Mater. Med.*, 2013, **24**, 179–187.
64. E. Kijenska, M. P. Prabhakaran, W. Swieszkowski, K. J. Kurzydlowski and S. Ramakrishna, *J. Biomed. Mater. Res., Part B*, 2012, **100B**, 1093–1102.
65. J.-P. Chen and C.-H. Su, *Acta Biomater.*, 2011, **7**, 234–243.
66. V. T. Tchemtchoua, G. Atanasova, A. Aqil, P. Filee, N. Garbacki, O. Vanhooteghem, C. Deroanne, A. Noel, C. Jerome, B. Nusgens, Y. Poumay and A. Colige, *Biomacromolecules*, 2011, **12**, 3194–3204.
67. S. D. Sarkar, B. L. Farrugia, T. R. Dargaville and S. Dhara, *J. Biomed. Mater. Res., Part A*, 2013, **101**, 3482–3492.
68. S. He, T. Xia, H. Wang, L. Wei, X. Luo and X. Li, *Acta Biomater.*, 2012, **8**, 2659–2669.
69. G. Pitarresi, C. Fiorica, F. S. Palumbo, S. Rigogliuso, G. Ghersi and G. Giammona, *J. Biomed. Mater. Res., Part A*, 2014, **102**, 1334–1341.
70. W. Yang, F. Yang, Y. Wang, S. K. Both and J. A. Jansen, *Acta Biomater.*, 2013, **9**, 4505–4512.
71. H. Liu, H. Peng, Y. Wu, C. Zhang, Y. Cai, G. Xu, Q. Li, X. Chen, J. Ji, Y. Zhang and H. W. OuYang, *Biomaterials*, 2013, **34**, 4404–4417.
72. W. J. Li, J. A. Cooper, R. L. Mauck and R. S. Tuan, *Acta Biomater.*, 2006, **2**, 377–385.
73. M. B. Chan-Park, J. Y. Shen, Y. Cao, Y. Xiong, Y. Liu, S. Rayatpisheh, G. C.-W. Kang and H. P. Greisler, *J. Biomed. Mater. Res., Part A*, 2009, **88A**, 1104–1121.
74. J. Yu, A. Wang, Z. Tang, J. Henry, B. L.-P. Lee, Y. Zhu, F. Yuan, F. Huang and S. Li, *Biomaterials*, 2012, **33**, 8062–8074.
75. M. B. Browning, D. Dempsey, V. Guiza, S. Becerra, J. Rivera, B. Russell, M. Hoeoek, F. Clubb, M. Miller, T. Fossum, J. F. Dong, A. L. Bergeron, M. Hahn and E. Cosgriff-Hernandez, *Acta Biomater.*, 2012, **8**, 1010–1021.
76. Y. M. Ju, J. S. Choi, A. Atala, J. J. Yoo and S. J. Lee, *Biomaterials*, 2010, **31**, 4313–4321.
77. B. W. Tillman, S. K. Yazdani, S. J. Lee, R. L. Geary, A. Atala and J. J. Yoo, *Biomaterials*, 2009, **30**, 583–588.
78. W. Zheng, Z. Wang, L. Song, Q. Zhao, J. Zhang, D. Li, S. Wang, J. Han, X.-L. Zheng, Z. Yang and D. Kong, *Biomaterials*, 2012, **33**, 2880–2891.

79. L. Ye, X. Wu, Q. Mu, B. Chen, Y. Duan, X. Geng, Y. Gu, A. Zhang, J. Zhang and Z.-G. Feng, *J. Biomater. Sci., Polym. Ed.*, 2011, **22**, 389–406.
80. A. Balguid, A. Mol, M. H. van Marion, R. A. Bank, C. V. C. Bouten and F. P. T. Baaijens, *Tissue Eng., Part A*, 2009, **15**, 437–444.
81. S. J. Eichhorn and W. W. Sampson, *J. R. Soc., Interface*, 2005, **2**, 309–318.
82. Z. Wang, H. Wang, W. Zheng, J. Zhang, Q. Zhao, S. Wang, Z. Yang and D. Kong, *Chem. Commun.*, 2011, **47**, 8901–8903.
83. S. J. Lee, J. Liu, S. H. Oh, S. Soker, A. Atala and J. J. Yoo, *Biomaterials*, 2008, **29**, 2891–2898.
84. N. Thi-Hiep, B. Trinh Quang, I. Park and B.-T. Lee, *J. Biomater. Appl.*, 2013, **28**, 514–528.
85. R. Fang, E. Zhang, L. Xu and S. Wei, *J. Nanosci. Nanotechnol.*, 2010, **10**, 7747–7751.
86. L. Lao, Y. Wang, Y. Zhu, Y. Zhang and C. Gao, *J. Mater. Sci.: Mater. Med.*, 2011, **22**, 1873–1884.
87. J.-J. Lee, H.-S. Yu, S.-J. Hong, I. Jeong, J.-H. Jang and H.-W. Kim, *J. Mater. Sci.: Mater. Med.*, 2009, **20**, 1927–1935.
88. L.-X. Lu, X.-F. Zhang, Y.-Y. Wang, L. Ortiz, X. Mao, Z.-L. Jiang, Z.-D. Xiao and N.-P. Huang, *ACS Appl. Mater. Interfaces*, 2013, **5**, 319–330.
89. E. Baykan, A. Koc, A. E. Elcin and Y. M. Elcin, *Biointerphases*, 2014, **9**, 029011.
90. A. Bianco, E. Di Federico and I. Cacciotti, *Polym. Adv. Technol.*, 2011, **22**, 1832–1841.
91. M. V. Jose, V. Thomas, Y. Xu, S. Bellis, E. Nyairo and D. Dean, *Macromol. Biosci.*, 2010, **10**, 433–444.
92. C. Luo, L. Li, J. Li, G. Yang, S. Ding, W. Zhi, J. Weng and S. Zhou, *J. Mater. Chem.*, 2012, **22**, 15654–15664.
93. T. Chae, H. Yang, V. Leung, F. Ko and T. Troczynski, *J. Mater. Sci.: Mater. Med.*, 2013, **24**, 1885–1894.
94. M. P. Prabhakaran, J. Venugopal and S. Ramakrishna, *Acta Biomater.*, 2009, **5**, 2884–2893.
95. K. Kuroda and M. Okido, *Bioinorg. Chem. Appl.*, 2012, **2012**, 1–7.
96. J. H. Lee, N. G. Rim, H. S. Jung and H. Shin, *Macromol. Biosci.*, 2010, **10**, 173–182.
97. H. Nie, B. W. Soh, Y.-C. Fu and C.-H. Wang, *Biotechnol. Bioeng.*, 2007, **99**, 223–234.
98. H. Nie and C.-H. Wang, *J. Controlled Release*, 2007, **120**, 111–121.
99. X. Yang and Z. Li, *J. Appl. Polym. Sci.*, 2015, **132**, 42249.
100. J. M. Holzwarth and P. X. Ma, *J. Mater. Chem.*, 2011, **21**, 10243–10251.
101. M.-J. Han, *J. Appl. Polym. Sci.*, 2000, **75**, 60–67.
102. P. X. Ma, *Adv. Drug Delivery Rev.*, 2008, **60**, 184–198.
103. P. Vandeweerdt, H. Berghmans and Y. Tervoort, *Macromolecules*, 1991, **24**, 3547–3552.
104. Y. S. Nam and T. G. Park, *J. Biomed. Mater. Res.*, 1999, **47**, 8–17.
105. F. C. Pavia, V. La Carrubba, S. Piccarolo and V. Brucato, *J. Biomed. Mater. Res., Part A*, 2008, **86A**, 459–466.

106. C. A. Martínez-Pérez, I. Olivas-Armendariz, J. S. Castro-Carmona and P. E. García-Casillas, *Advances in Regenerative Medicine*, InTech, 2011, pp. 275–294.
107. J. A. de Lima and M. I. Felisberti, *J. Membr. Sci.*, 2009, **344**, 237–243.
108. B. Luo, Z. Li, J. Zhang and X. Wang, *Desalination*, 2008, **233**, 19–31.
109. B. J. Cha and J. M. Yang, *J. Membr. Sci.*, 2007, **291**, 191–198.
110. B. J. Cha, K. Char, J. J. Kim, S. S. Kim and C. K. Kim, *J. Membr. Sci.*, 1995, **108**, 219–229.
111. W. K. Kim, K. Char and C. K. Kim, *J. Polym. Sci., Part B: Polym. Phys.*, 2000, **38**, 3042–3052.
112. X. Liu and P. X. Ma, *Biomaterials*, 2010, **31**, 259–269.
113. X. Liu and P. X. Ma, *Biomaterials*, 2009, **30**, 4094–4103.
114. J. Zhao, W. Han, H. Chen, M. Tu, R. Zeng, Y. Shi, Z. Cha and C. Zhou, *Carbohydr. Polym.*, 2011, **83**, 1541–1546.
115. H. Shen, Y. Niu, X. Hu, F. Yang, S. Wang and D. Wu, *J. Mater. Chem. B*, 2015, **3**, 4417–4425.
116. J. Guan, J. J. Stankus and W. R. Wagner, *J. Controlled Release*, 2007, **120**, 70–78.
117. G. Jell, R. Verdejo, L. Safinia, M. S. P. Shaffer, M. M. Stevens and A. Bismarck, *J. Mater. Chem.*, 2008, **18**, 1865–1872.
118. J. Jensen, D. C. E. Kraft, H. Lysdahl, C. B. Foldager, M. Chen, A. A. Kristiansen, J. H. D. Roelfing and C. E. Buenger, *Tissue Eng., Part A*, 2015, **21**, 729–739.
119. J. Shao, S. Chen and C. Du, *J. Mater. Chem. B*, 2015, **3**, 5291–5299.
120. P. X. Ma, R. Zhang, G. Xiao and R. Franceschi, *J. Biomed. Mater. Res.*, 2001, **54**, 284–293.
121. J. Guan, K. L. Fujimoto, M. S. Sacks and W. R. Wagner, *Biomaterials*, 2005, **26**, 3961–3971.
122. M. M. L. Deckers, M. Karperien, C. V. D. Bent, T. Yamashita, S. E. Papapoulos and C. W. G. M. Löwik, *Endocrinology*, 2000, **141**, 1667–1674.
123. J. Street, M. Bao, L. deGuzman, S. Bunting, F. V. Peale, N. Ferrara, H. Steinmetz, J. Hoeffel, J. L. Cleland, A. Daugherty, N. van Bruggen, H. P. Redmond, R. A. D. Carano and E. H. Filvaroff, *Proc. Natl. Acad. Sci.*, 2002, **99**, 9656–9661.
124. J. M. Holzwarth and P. X. Ma, *Biomaterials*, 2011, **32**, 9622–9629.
125. P. X. Ma and R. Zhang, *J. Biomed. Mater. Res.*, 1999, **46**, 60–72.
126. K. M. Woo, V. J. Chen and P. X. Ma, *J. Biomed. Mater. Res., Part A*, 2003, **67A**, 531–537.
127. K. M. Woo, J.-H. Jun, V. J. Chen, J. Seo, J.-H. Baek, H.-M. Ryoo, G.-S. Kim, M. J. Somerman and P. X. Ma, *Biomaterials*, 2006, **28**, 335–343.
128. W. Han, J. Zhao, M. Tu, R. Zeng, Z. Zha and C. Zhou, *J. Appl. Polym. Sci.*, 2013, **128**, 1332–1338.
129. N. R. Cameron, *Polymer*, 2005, **46**, 1439–1449.
130. S. Caldwell, D. W. Johnson, M. P. Didsbury, B. A. Murray, J. J. Wu, S. A. Przyborski and N. R. Cameron, *Soft Matter*, 2012, **8**, 10344–10351.

131. W. Busby, N. R. Cameron and A. B. C. Jahoda, *Polym. Int.*, 2002, **51**, 871–881.
132. M. Bokhari, R. J. Carnachan, N. R. Cameron and S. A. Przyborski, *Biochem. Biophys. Res. Commun.*, 2007, **354**, 1095–1100.
133. R. Pons, I. Carrera, P. Erra, H. Kunieda and C. Solans, *Colloids Surf., A*, 1994, **91**, 259–266.
134. K. J. Lissant, *J. Colloid Interface Sci.*, 1966, **22**, 462–468.
135. H. Bartl and W. Vonbonin, *Makromol. Chem.*, 1962, **57**, 74–95.
136. D. Barby and Z. Haq, EP0060138, 1986.
137. J. M. Williams, A. J. Gray and M. H. Wilkerson, *Langmuir*, 1990, **6**, 437–444.
138. J. M. Williams, *Langmuir*, 1991, **7**, 1370–1377.
139. J. M. Williams and D. A. Wrobleski, *Langmuir*, 1988, **4**, 656–662.
140. J. Jiao and D. J. Burgess, *J. Colloid Interface Sci.*, 2003, **264**, 509–516.
141. M. G. Freire, A. M. A. Dias, M. A. Z. Coelho, J. A. P. Coutinho and I. M. Marrucho, *J. Colloid Interface Sci.*, 2005, **286**, 224–232.
142. M. W. Hayman, K. H. Smith, N. R. Cameron and S. A. Przyborski, *Biochem. Biophys. Res. Commun.*, 2004, **314**, 483–488.
143. G. Akay, M. A. Birch and M. A. Bokhari, *Biomaterials*, 2004, **25**, 3991–4000.
144. A. S. Hayward, A. M. Eissa, D. J. Maltman, N. Sano, S. A. Przyborski and N. R. Cameron, *Biomacromolecules*, 2013, **14**, 4271–4277.
145. M. W. Hayman, K. H. Smith, N. R. Cameron and S. A. Przyborski, *J. Biochem. Biophys. Methods*, 2005, **62**, 231–240.
146. M. Bokhari, R. J. Carnachan, N. R. Cameron and S. A. Przyborski, *J. Anat.*, 2007, **211**, 567–576.
147. C. Chatgilialoglu, A. Altieri and H. Fischer, *J. Am. Chem. Soc.*, 2002, **124**, 12816–12823.
148. C. R. UK, Methotrexate (Matrex), http://www.cancerresearchuk.org/about-cancer/cancers-in-general/treatment/cancer-drugs/methotrexate, (accessed 11/5/2015, 2015).
149. A. S. Hayward, N. Sano, S. A. Przyborski and N. R. Cameron, *Macromol. Rapid Commun.*, 2013, **34**, 1844–1849.
150. L. Kircher, P. Theato and N. R. Cameron, *Polymer*, 2013, **54**, 1755–1761.
151. G. Ashwell and A. G. Morell, *Advances in Enzymology and Related Areas of Molecular Biology*, John Wiley & Sons, Inc., 2006, DOI: 10.1002/9780470122860, ch. 3, pp. 99–128.
152. C. S. Cho, S. J. Seo, I. K. Park, S. H. Kim, T. H. Kim, T. Hoshiba, I. Harada and T. Akaike, *Biomaterials*, 2006, **27**, 576–585.
153. W. Busby, N. R. Cameron and C. A. B. Jahoda, *Biomacromolecules*, 2001, **2**, 154–164.
154. E. M. Christenson, W. Soofi, J. L. Holm, N. R. Cameron and A. G. Mikos, *Biomacromolecules*, 2007, **8**, 3806–3814.
155. E. Lovelady, S. D. Kimmins, J. J. Wu and N. R. Cameron, *Polym. Chem.*, 2011, **2**, 559–562.
156. C. R. Langford, D. W. Johnson and N. R. Cameron, *Polym. Chem.*, 2014, **5**, 6200–6206.

157. D. W. Johnson, C. R. Langford, M. P. Didsbury, B. Lipp, S. A. Przyborski and N. R. Cameron, *Polym. Chem.*, 2015, 7256–7263, DOI: 10.1039/C5PY00721F.
158. R. S. Moglia, J. L. Holm, N. A. Sears, C. J. Wilson, D. M. Harrison and E. Cosgriff-Hernandez, *Biomacromolecules*, 2011, **12**, 3621–3628.
159. R. S. Moglia, M. Whitely, P. Dhavalikar, J. Robinson, H. Pearce, M. Brooks, M. Stuebben, N. Cordner and E. Cosgriff-Hernandez, *Biomacromolecules*, 2014, **15**, 2870–2878.
160. U. H. F. Bunz, *Adv. Mater.*, 2006, **18**, 973–989.
161. M. H. Stenzel, C. Barner-Kowollik and T. P. Davis, *J. Polym. Sci., Part A: Polym. Chem.*, 2006, **44**, 2363–2375.
162. B. J. Briscoe and K. P. Galvin, *Colloids Surf.*, 1991, **56**, 263–278.
163. A. V. Limaye, R. D. Narhe, A. M. Dhote and S. B. Ogale, *Phys. Rev. Lett.*, 1996, **76**, 3762–3765.
164. P. Escale, L. Rubatat, L. Billon and M. Save, *Eur. Polym. J.*, 2012, **48**, 1001–1025.
165. M. Hernandez-Guerrero and M. H. Stenzel, *Polym. Chem.*, 2012, **3**, 563–577.
166. L.-S. Wan, L.-W. Zhu, Y. Ou and Z.-K. Xu, *Chem. Commun.*, 2014, **50**, 4024–4039.
167. G. Widawski, M. Rawiso and B. Francois, *Nature*, 1994, **369**, 387–389.
168. B. Francois, O. Pitois and J. Francois, *Adv. Mater.*, 1995, **7**, 1041–1044.
169. O. Pitois and B. Francois, *Eur. Phys. J. B*, 1999, **8**, 225–231.
170. M. Srinivasarao, D. Collings, A. Philips and S. Patel, *Science*, 2001, **292**, 79–82.
171. S. Xu, M. Li, Z. Mitov and E. Kumacheva, *Prog. Org. Coat.*, 2003, **48**, 227–235.
172. A. Bolognesi, C. Mercogliano, S. Yunus, M. Civardi, D. Comoretto and A. Turturro, *Langmuir*, 2005, **21**, 3480–3485.
173. T. Hayakawa and S. Horiuchi, *Angew. Chem., Int. Ed.*, 2003, **42**, 2285–2289.
174. E. Servoli, G. A. Ruffo and C. Migliaresi, *Polymer*, 2010, **51**, 2337–2344.
175. J. Peng, Y. Han, Y. Yang and B. Li, *Polymer*, 2004, **45**, 447–452.
176. J. Li, Q.-L. Zhao, J.-Z. Chen, L. Li, J. Huang, Z. Ma and Y.-W. Zhong, *Polym. Chem.*, 2010, **1**, 164–167.
177. X. J. Wu, M. D. Jones, M. G. Davidson, J. B. Chaudhuri and M. J. Ellis, *Biotechnol. Lett.*, 2011, **33**, 423–430.
178. C. Wang, Y. Mao, D. Wang, Q. Qu, G. Yang and X. Hu, *J. Mater. Chem.*, 2008, **18**, 683–690.
179. W. Dong, Y. Zhou, D. Yan, Y. Mai, L. He and C. Jin, *Langmuir*, 2009, **25**, 173–178.
180. X. Xiong, M. Lin, W. Zou and X. Liu, *React. Funct. Polym.*, 2011, **71**, 964–971.
181. Y. Tian, Q. Jiao, H. Ding, Y. Shi and B. Liu, *Polymer*, 2006, **47**, 3866–3873.
182. H. Ejima, T. Iwata and N. Yoshie, *Macromolecules*, 2008, **41**, 9846–9848.

183. H. Yabu and M. Shimomura, *Chem. Mater.*, 2005, **17**, 5231–5234.
184. T. Kawano, M. Sato, H. Yabu and M. Shimomura, *Biomater. Sci.*, 2014, **2**, 52–56.
185. D. Beattie, K. H. Wong, C. Williams, L. A. Poole-Warren, T. P. Davis, C. Barner-Kowollik and M. H. Stenzel, *Biomacromolecules*, 2006, **7**, 1072–1082.
186. X. Wu and S. Wang, *ACS Appl. Mater. Interfaces*, 2012, **4**, 4966–4975.
187. Y. Zhu, R. Sheng, T. Luo, H. Li, J. Sun, S. Chen, W. Sun and A. Cao, *ACS Appl. Mater. Interfaces*, 2011, **3**, 2487–2495.
188. T. Ponnusamy, G. Chakravarty, D. Mondal and V. T. John, *Adv. Healthcare Mater.*, 2014, **3**, 703–713.
189. M. Tanaka, *Biochim. Biophys. Acta, Gen. Subj.*, 2011, **1810**, 251–258.
190. S. Yamamoto, M. Tanaka, H. Sunami, K. Arai, A. Takayama, S. Yamashita, Y. Morita and M. Shimomura, *Surf. Sci.*, 2006, **600**, 3785–3791.
191. A. Tsuruma, M. Tanaka, S. Yamamoto and M. Shimomura, *Colloids Surf., A*, 2008, **313–314**, 536–540.
192. J. R. McMillan, M. Akiyama, M. Tanaka, S. Yamamoto, M. Goto, R. Abe, D. Sawamura, M. Shimomura and H. Shimizu, *Tissue Eng.*, 2007, **13**, 789–798.
193. M. Tanaka, A. Takayama, E. Ito, H. Sunami, S. Yamamoto and M. Shimomura, *J. Nanosci. Nanotechnol.*, 2007, **7**, 763–772.
194. S. Yamamoto, M. Tanaka, H. Sunami, E. Ito, S. Yamashita, Y. Morita and M. Shimomura, *Langmuir*, 2007, **23**, 8114–8120.
195. Y. Li, Q. Li, F. Li, H. Zhang, L. Jia, J. Yu, Q. Fang and A. Cao, *Biomacromolecules*, 2006, **7**, 224–231.
196. Y. Li, L. Cui, Q. Li, L. Jia, Y. Xu, Q. Fang and A. Cao, *Biomacromolecules*, 2007, **8**, 1409–1416.
197. Y. Li, Y. Zhu, K. Xia, R. Sheng, L. Jia, X. Hou, Y. Xu and A. Cao, *Biomacromolecules*, 2009, **10**, 2284–2293.
198. B. M. Gillette, J. A. Jensen, B. Tang, G. J. Yang, A. Bazargan-Lari, M. Zhong and S. K. Sia, *Nat. Mater.*, 2008, **7**, 636–640.
199. O. Guillame-Gentil, O. Semenov, A. S. Roca, T. Groth, R. Zahn, J. Voros and M. Zenobi-Wong, *Adv. Mater.*, 2010, **22**, 5443–5462.
200. W.-h. Guo, M. T. Frey, N. A. Burnham and Y.-l. Wang, *Biophys. J.*, 2006, **90**, 2213–2220.
201. T. Kawano and S. Kidoaki, *Biomaterials*, 2011, **32**, 2725–2733.
202. A. J. Engler, S. Sen, H. L. Sweeney and D. E. Discher, *Cell*, 2006, **126**, 677–689.
203. H. Yabu, Y. Nakamichi, Y. Hirai and M. Shimomura, *Phys. Chem. Chem. Phys.*, 2011, **13**, 4877–4880.
204. Y. Nakamichi, Y. Hirai, H. Yabu and M. Shimomura, *J. Mater. Chem.*, 2011, **21**, 3884–3889.
205. T. Kawano, Y. Nakamichi, S. Fujinami, K. Nakajima, H. Yabu and M. Shimomura, *Biomacromolecules*, 2013, **14**, 1208–1213.

206. N. Q. Balaban, U. S. Schwarz, D. Riveline, P. Goichberg, G. Tzur, I. Sabanay, D. Mahalu, S. Safran, A. Bershadsky, L. Addadi and B. Geiger, *Nat. Cell Biol.*, 2001, **3**, 466–472.
207. O. F. Zouani, C. Chanseau, B. Brouillaud, R. Bareille, F. Deliane, M.-P. Foulc, A. Mehdi and M.-C. Durrieu, *J. Cell Sci.*, 2012, **125**, 1217–1224.
208. J. Fu, Y.-K. Wang, M. T. Yang, R. A. Desai, X. Yu, Z. Liu and C. S. Chen, *Nat. Methods*, 2010, **7**, 733–736.
209. S. Weng and J. Fu, *Biomaterials*, 2011, **32**, 9584–9593.
210. O. F. Zouani, J. Kalisky, E. Ibarboure and M.-C. Durrieu, *Biomaterials*, 2013, **34**, 2157–2166.

CHAPTER 15

Antimicrobial Polymers and Surfaces – Natural Mimics or Surpassing Nature?

CHRISTIAN KRUMM AND JOERG C. TILLER*

Department of Bio- and Chemical Engineering, TU Dortmund, Emil-Figge-Str. 66, D-44227 Dortmund, Germany
*Email: joerg.tiller@udo.edu

15.1 Introduction

There are several serious challenges with which the modern global society has to cope. These were summarized in the year 1996 under the name "The Millennium Project". In this project, 15 different global challenges the humanity is facing were pointed out. These tasks were summarized in concise questions such as No. 8: "How can the threat of new and reemerging diseases and immune micro-organisms be reduced?" This chapter of the book deals with promising concepts for facing this threat.

Pathogens have been the focus of the development of new, selective, and efficient drugs since Alexander Fleming discovered the first antibiotic. These drugs aim at structural and biosynthetic differences between human cells and pathogens. Biocides and antibiotics were quite effective for a period of time, but the excessive use of antimicrobials and their dilution in the environment, in combination with their constant presence, result in a selection process on the microorganisms.[1] Therefore, certain pathogenic strains have become resistant due to mutations, recombination, and lateral gene transfer.[2]

RSC Polymer Chemistry Series No. 22
Bio-inspired Polymers
Edited by Nico Bruns and Andreas F. M. Kilbinger

Published by the Royal Society of Chemistry, www.rsc.org

Nowadays, germs such as methicillin-resistant *Staphylococcus aureus* (MRSA) dictate the headlines of newspapers again and again.

The targets of antibiotics and the different resistance mechanisms present in bacteria have been summarized by Coates and Henderson.[3,4] Antibiotics mainly focus on the disturbance of the synthesis of DNA, the cell wall, the cell membrane, and proteins.[5–7] Resistant bacteria are equipped with certain defense mechanisms, *e.g.* they decrease the uptake of antibiotics, use highly effective efflux pumps for active transport of the drugs out of the cells, they alter the target size of enzymes or they use enzymes which alter and degrade the antibiotics themselves (Figure 15.1).[8] These different defense systems force the human civilization to continuously establish new, efficient drugs with new modes of action, otherwise humanity will end up in a pre-antibiotic area.[8]

It is obvious that the development of new antibiotics is not expedient, as demonstrated by the following numbers: in 1983–1987, 16 new antimicrobial agents were approved in the United States, whereas in the years 2003–2007 only one per year was approved.[9] This is due to the exploding costs for drug development, clinical tests, and the approval of antimicrobials in general. Therefore, alternatives for antibiotics gain in importance. Polymers are becoming more and more important as one can recognize from the fact that, in 2013, two of the top ten selling drugs in the United States were polymer therapeutics.[10] Antimicrobial polymers are a versatile platform for

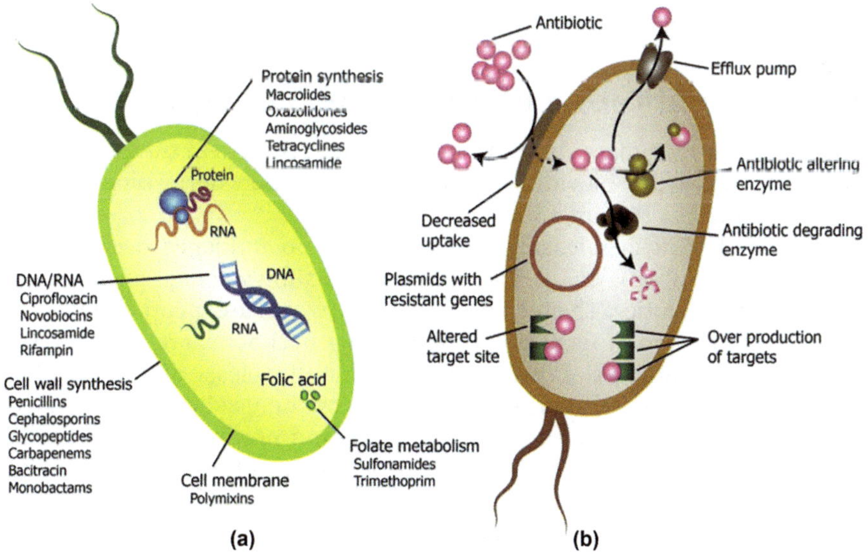

Figure 15.1 (a) Typical antibiotic target sites present in bacterial cells; (b) mechanisms of antibiotic resistance.
Reprinted from M. S. Ganewatta and C. Tang, Controlling macromolecular structures towards effective antimicrobial polymers, *Polymer*, **63**, A1–A29, Copyright 2015 with permission from Elsevier.[8]

the preparation of new and innovative antimicrobial agents. The first examples of biocidal polymers are known since 1965[11] and, until 2014, the number of new publications per year had risen up to nearly 700.[8] There are several very comprehensive reviews, which focus on all kinds of different aspects concerning antimicrobial polymers.[12–25] Recent approaches of developing new antimicrobial polymers have focused on their functional principles and revealed that research has unconsciously mimicked natural motifs such as antimicrobial peptides. Alternatives have also made use of natural concepts, such as the amphoteric behavior of complex molecules like proteins on surfaces. Yet, there is always the question of whether nature can be surpassed by finding working concepts that nature has not "thought" of and might therefore not be able to fight by, *e.g.* building up resistances.

A commonly applied alternative to fighting pathogenic microbes with soluble biocides is to prevent their proliferation, which majorly occurs on surfaces composed of virtually any material in the form of biofilms. Surface modifications that prevent the formation of biofilms are not only hindering the spread of infections, but also protect the materials from deterioration. While nature has applied the concept of non-fouling surfaces, known from the excellent non-fouling activity of albumin, microbe-killing surfaces seem to be a concept beyond nature's "imagination".

Both approaches have a long history, but are still the focus of research. This chapter gives a general overview over the classes, modes of action, and trends in the current field of antimicrobial polymers and surfaces, and brings them into perspective on whether they mimic nature or follow alternative concepts.

15.2 Classification of Antimicrobial Polymers

Antimicrobial polymers have been classified in three different major groups.[20] There are biocide-releasing polymers, polymeric biocides, and biocidal polymers (Figure 15.2).

In general, biocide-releasing polymers have no influence over their intrinsic antimicrobial activity. The polymers are simply acting as carriers for biocides or antibiotics. The biocidal molecules, which are incorporated in the polymer matrix and/or tethered to the polymer backbone, are released. One of the major advantages of these systems is that the release of the embedded antimicrobial active substances is controlled by the used polymeric system. Therefore, the rates of release are adjustable and polymers can release the biocides very close to the cell, which makes them efficient. However, the polymers still release biocides into the environment and will eventually become inactive. Polymeric biocides contain biocidal repeating units. Such macromolecules often show the same mode of action as their repeating units with somewhat lower activity, due to the steric hindrance caused by the polymeric backbone. Biocidal polymers distinguish themselves by the fact that they act as a whole molecule. Further, biocidal polymers have been found to show a lower tendency to build up bacterial resistance.[20]

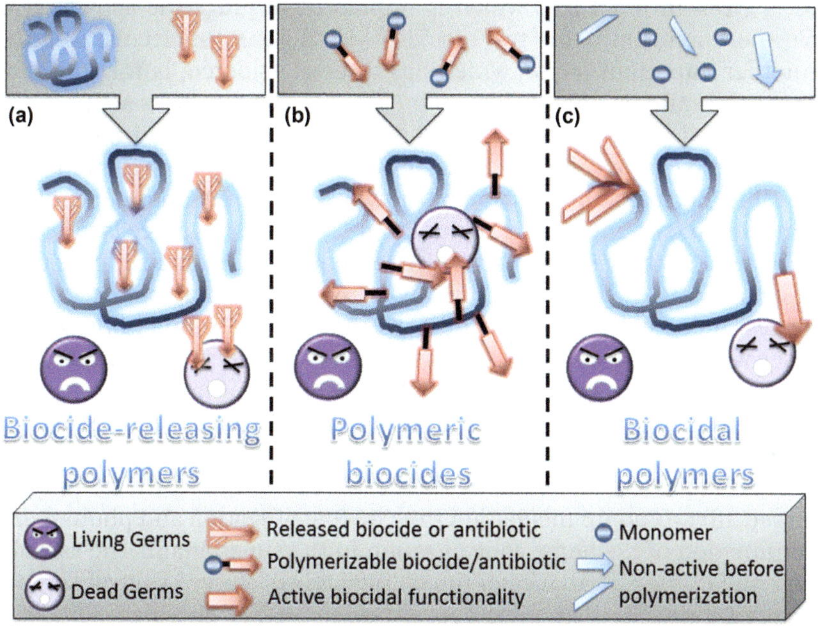

Figure 15.2 General working principles of antimicrobial polymers: (a) biocide-releasing polymers, (b) polymeric biocides, and (c) biocidal polymers. Adapted from ref. 20.

15.2.1 Considerations on the Comparability of Biological Testing of Antimicrobial Polymers

The quantification of the antimicrobial activity of polymers in solution is generally performed by determining the minimal inhibitory concentration (MIC). This concentration is defined as the concentration of biocide at which 99% of all bacteria are inhibited in growth. The MIC-value is determined by preparing a bacteria stock solution in a nutrient broth suitable for the respective strain. Then, a dilution series are prepared starting with a known concentration of the biocidal polymer. After inoculation with a certain number of bacterial cells and incubation, the dilutions/suspensions are photometrically tested for their absorbance. The lowest concentration at which no cell growth is observable is the MIC concentration. Due to the dilution series, the real MIC value is somewhere between this concentration and that of the next overgrown sample. A closer look to the literature reveals that one has to be very careful by comparing different MIC values with each other. The first problem is the non-standardized use of units. The most common concentration is $\mu g\,mL^{-1}$, sometimes given in parts per million (ppm). This concentration refers to the mass of the tested biocides, which is particularly important for formulation purposes in applications. Sometimes, the MIC values are given in $mg\,mL^{-1}$, which leads the reader to overestimate the antimicrobial activity of the investigated polymeric biocide. The molar

MIC is given in μmol L^{-1}, which is important to judge the activity of one molecule when comparing two biocides with different molecular weights.

Another important factor, which has to be considered, is that the inoculum concentrations differ widely. In most cases, the MIC tests are performed with 1×10^4–1×10^8 cells mL^{-1}. Keul and Möller investigated the influence of the used bacteria cell concentration for *S. aureus*.[26] They found that the MIC value increases for the same biocide by up to three times depending on the inoculum concentration, ranging from 6×10^4 to 1×10^8 cells mL^{-1}. This "Inoculum Effect" has been the subject of many studies, mainly investigated for *S. aureus*.[27–31] In general, highly active biocides seem to show the strongest inoculum effect, which means that the determined MIC value increases with the cell concentration by up to two orders of magnitude when increasing the inoculum concentration by 4 orders of magnitude.

The hemocompatibility is determined by using isolated human or mammalian erythrocytes in a sterile buffer. A dilution series of the biocide in the buffer is prepared, inoculated with the blood cells, and the cells lysis is followed after a certain incubation time by centrifugation and photometrical determination of the heme concentration in the supernatant. The sample at which half of the absorbance of the deliberately destroyed control sample is measured gives the hemolytic concentration (HC_{50}) at which 50% of the erythrocytes are lysed. The inoculum effect for red blood cells and the impact on the HC_{50} value has been less often investigated than for MIC investigations. Strassburg *et al.* found that the HC_{50} value is independent in a range of 1×10^7 to 5×10^8 cells mL^{-1} for the antimicrobial polymers they tested.[32]

The selectivity between prokaryotic and eukaryotic cells is expressed by the quotient of HC_{50} and MIC, giving the selectivity index (SI).

15.2.2 Biocide-releasing Polymers

Biocide releasing polymers offer the possibility to adjust the rate of release and therefore the lifetime, efficiency, and bioavailability of the biocide. This is especially important for water soluble agents because they are prone to excessively leaching from the polymer matrices. Of course, the concept of incorporating drugs in macromolecules such as nanoparticles, liposomes, molecular micelles, or polymersomes is very commonly used. The discussion of all these examples is beyond the scope of this chapter, so the reader is referred to recent reviews.[33–37] Moreover, the release of biocides from polymers has been mainly used for the preparation of antimicrobial or nonfouling surfaces. These surfaces will be discussed in detail in Section 15.3. This section focuses on the combination of low molecular weight biocides and polymers as carriers for killing microbial cells in solution. The first example of a biocide releasing polymer was developed by Vogl and Tirrell in 1979.[38] They established the release of salicylic acid by hydrolysis from polyesters. Since then, ester groups have been a very reliable system in the design of polymers for drug release. Another concept was introduced by Worley and coworkers, who used *N*-halamines to bind active chlorine.[39]

The polymers were able to transfer this chlorine atom directly to microbial cells. The chlorine oxidized the phospholipids of the cell membrane, which resulted in the death of the microbial cells. Schoenfisch and coworkers prepared polymers with functional diazeniumdiolate groups, which were able to release NO within several hours.[40,41]

Polymeric mimics of natural polyphenols have found to display antimicrobial activity due to the phenols that are released from the polymers, which kill the microbial cells.[42] Schanze and coworkers prepared poly(phenylene ethynylene) conjugated polymers with quaternary ammonium solubilizing groups for releasing singlet oxygen triggered by illumination.[43] Due to the reactive singlet oxygen species, the polymer even exhibited antiviral properties.[44] Hydrophilic biocides such as sodium benzoate were encapsulated in branched PEI matrices with different molecular weights. Through this strategy, Ritschkoff and coworkers could control the diffusion rate from the polymeric capsules.[45]

Releasing systems based on antibiotics are the focus of modern research. Most commonly, the antibiotics are released from polymers. This is achieved by either using degradable polymers or non-degradable polymers with hydrolysable functional groups. Woo *et al.* prepared a biodegradable polymer system based on 1,6-hexane diisocyanate, polycaprolactone diol and a fluoroquinolone antibiotic, ciprofloxacin (CIP), that released CIP through the full degradation of the polymeric backbone.[46] Vert and coworkers prepared norfloxacin-polymer conjugates based on poly(L-lysine citramide imide), which also released the antibiotic by degradation of the polymer backbone.[47]

Polymers have also been used as non-degradable drug carriers. In these cases, the antibiotics were covalently attached to the polymers with cleavable functional groups. Therefore, antibiotics such as ofloxacin,[48] ciprofloxacin,[49] β-lactam,[50] or lefofloxacin[51] were temporary attached to polymers, such as polysiloxanes or poly(ethylene glycol).[52,53]

All of these systems were prepared with the intention of releasing biocides or antibiotics to the environment. This strategy is very effective for competing against new and reemerging diseases but, unfortunately, this approach is not sufficient for the second part of the challenge: "How can the threat of new and reemerging diseases and immune micro-organisms be reduced?"

15.2.3 Polymeric Biocides

Polymeric biocides are formally composed of repeating units that contain biocidal groups. This is achieved by either polymerizing monomers with biocidal motives or derivatizing polymers with such motives by polymer-analogous reactions. Figure 15.3 depicts a selection of biocidal monomers.

In 1979, Vogl and Tirrell prepared 4-vinylsaliclic acid, 5-vinylsaliclic acid, and additional derivatives of these monomers. The polymers, which were prepared by radical polymerization, showed antimicrobial activity against

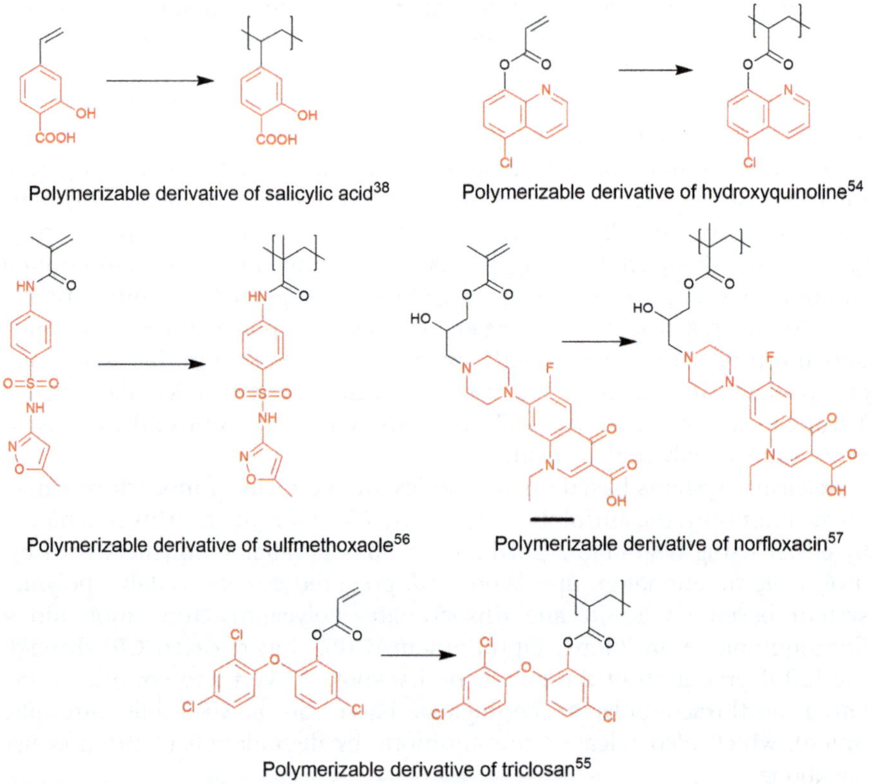

Figure 15.3 Overview of different polymerizable derivatives of biocides and antibiotics.

Escherichia coli and *Staphylococcus aureus*, respectively.[38] Rashkov and coworkers synthesized polymers based on biocidal acrylate-derivatives of 8-hydroxyquinoline. Therefore, this polymer was assigned to the group of polymerized biocides, but the actual antimicrobial activity was found to be caused by the release of 8-hydroxyquinoline from the polymer backbone. With respect to its mode of action, it is rather a biocide releasing polymer.[54] Cho and coworkers functionalized triclosan using acryloyl chloride to introduce a polymerizable functional group. After polymerization, the bacterial activities were excellent against *Pseudomonas aeruginosa* in comparison to non-biocidal control polymers.[55]

Thamizharasi *et al.* started with sulfamethoxazole and introduced methacrylic, acrylic, and maleimide groups. The resulting polymers revealed in some cases even better activities than the low molecular weight monomers.[56]

Similar to biocide-releasing polymers, there is the opportunity to use antibiotics instead of biocides. Moon *et al.* prepared an acryl monomer containing a norfloxacin moiety. This polymer was blended with other common synthetic polymers. These samples reduced the viable cell number of microbial cells in shake flask tests.[57]

Alternatively, polymer conjugates can be obtained by polymer analogous reactions. Nathan *et al.* coupled penicillin V and hydrophilic cephradine to the polymeric backbone of PEG-lysine poly(ether urethane) *via* two different strategies. To this end, reactive groups were introduced post polymerization along the backbone. Then, penicillin V was introduced in a polymer analogous reaction step by a hydrolysable ester group and cephradine was bound by a hydrolytically stable amide bond. Whereas the first example is a non-permanent polymer antibiotic conjugate (PAC), which shows activity after the release of penicillin V, the second permanently bound PAC shows no antimicrobial activity at all, indicating that the polymer itself is not active.[52] Turos *et al.* conjugated the *N*-thiolated β-lactam antibiotic to polyacrylates in order to prepare nanoparticles. The nanoparticles revealed better activity than the antibiotic containing monomers, which is also true for methicillin-resistant *S. aureus* (MRSA).[58] Rimmer and coworkers attached polymyxin B (PMX) covalently to a highly branched poly(*N*-isopropyl acrylamide) (PNIPAM). To this end, the biocidal active functional group of PMX was removed to introduce a potential binding site for PNIPAM. The resulting conjugate was able to precipitate *P. aeruginosa* by binding of the cyclic peptide of PMX to the lipopolysaccharide in the membrane of *P. aeruginosa*. The caused aggregation decreased the lower critical solution temperature of PNIPAM below 37 °C, which led to the precipitation of these aggregates in aqueous solution. This represents a very smart method for the detection of Gram-negative bacteria in aqueous solution.[59]

In conclusion, the activity of polymeric biocides depends strongly on the nature of the respective biocides and the binding strategy. In some cases, the polymers are even more active than their respective monomers.

15.2.4 Biocidal Polymers

The third class of antimicrobial polymers is distinguished by the fact that the repeating units by themselves show no antimicrobial activity. Only the resulting macromolecules display biocidal activity. This class is called biocidal polymers. Over many years, antimicrobial polymers have been designed by random tryouts. However, the comparison of structural motives in the literature has revealed that all these polymers have more or less accidentally mimicking natural motifs. As already pointed out by DeGrado and Tew, amphiphilic polycations may be mimics for membrane-active antimicrobial peptides, such as magainin, which is discussed in more detail below.[13,60] Other hydrophilic motifs may be mimics for hydrophilic antimicrobial polycations, such as chitosan and ε-poly-L-lysine. Other polymers, such as telechelic biocide–polymer conjugates, seem to follow natural folding motifs, such as the amphoteric behavior of surface binding proteins. Altogether, biocidal polymers can be categorized in three major subclasses: amphiphilic polycations, telechelic antimicrobial polymers, and hydrophilic polycations (Figure 15.4).

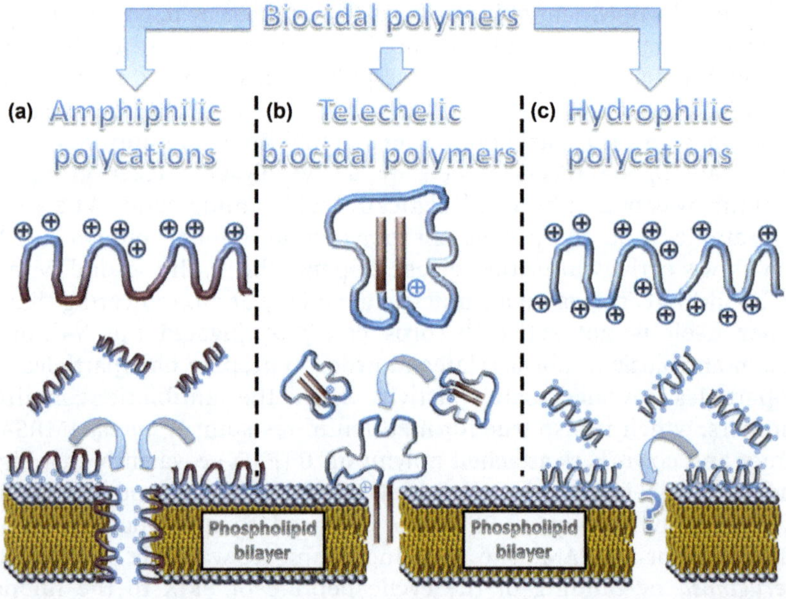

Figure 15.4 Illustration of different subclasses of biocidal polymers: (a) amphiphilic polycations, (b) telechelic biocidal polymers, and (c) hydrophilic polycations.

15.2.4.1 Amphiphilic Polycations

Amphiphilic polycations are the best investigated biocidal polymers by far.[12,15,20,24,61,62] As mentioned above, they are mimics of antimicrobial peptides (AMPs) such as defensin and magainin (Figure 15.5).[60] The known AMPs have the following characteristics: they are folded into a relative ridged structure[20] that separates the surface into a hydrophilic cationic side, formed by locating cationic amino acids such as lysine, arginine, and histidine, and a hydrophobic side that contains amino acids such as valine, phenylalanine, leucine, and isoleucine.[8,63] Based on these characteristics, various modes of action have been proposed, as illustrated in Figure 15.6.[8,64] The barrel stave-, carpet- and worm-hole models are based on the fact that, in the end, the bilayer membrane is disrupted by insertion of the hydrophobic face of the AMPs in the membrane.[64–66]

Amphiphilic polycations have similar properties; they also exhibit multiple positive charges and an amphiphilic moiety as well. Therefore, it was concluded that they act similarly to AMPs.[67,68] It was suggested that amphiphilic polycations first adsorb to the negatively charged cell wall of Gram-positive bacterial cells and then diffuse to the cytoplasmic membrane. In the case of Gram-negative bacteria, it might be sufficient that the polymers interact directly with the outer cell membrane. In both cases, the polymers perforate the cell membrane, causing the lysis and thus death of the bacterial cells.[64–66]

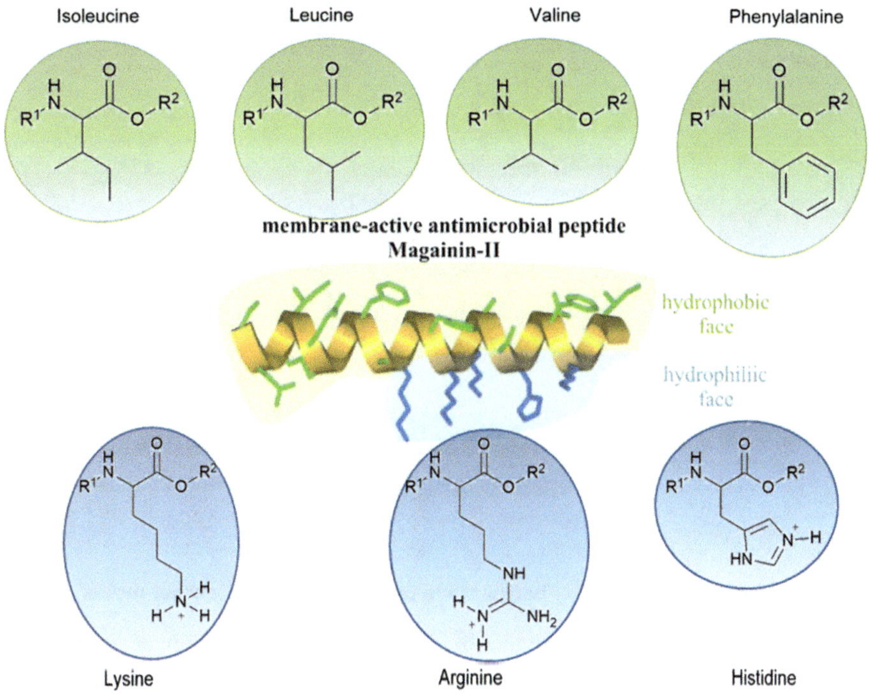

Figure 15.5 Illustration of the antimicrobial peptide magainin. Reprinted and modified from ref. 20.

The destabilization effect is enhanced by the displacement of Ca^{2+} ions, which interact with the negatively charged phospholipid residues of the bacterial membranes.[69,70]

Basically, all amphiphilic polycations are mimics of magainin, with the restriction that they often do not contain a stiff backbone, which results in the fact that the hydrophilic and hydrophobic regions of the macromolecules are not as efficiently separated as they are in magainin.[20] Therefore, synthetic biocidal polymers are not as active as AMPs in most cases.[20]

It has been found that the efficiency of these macromolecules depends mainly on three different properties: the ratio of hydrophilic to hydrophobic components, their molecular weight, and their charge density. The influence of the molecular weight has thoughtfully been explored by different groups. Kuroda and DeGrado investigated amphiphilic copolymers of methacrylate derivatives and found that polymers with lower molecular weight resulted in higher antimicrobial activity and less hemolytic activity. Polymers with the same balance between hydrophobic and hydrophilic blocks revealed MIC-values of 16 $\mu g\,mL^{-1}$ for molecular weights of 1300–1900 g mol^{-1}, whereas polymers with 4500–6000 g mol^{-1} and 7900–10100 g mol^{-1} displayed values of 20 and 40 $\mu g\,mL^{-1}$, respectively.[15,71] With respect to the hemolytic activity, Gellman *et al.* observed that the hemolytic activity of nylon-3 polymers is higher above 2000 g mol^{-1} than below this threshold. In contrast to this

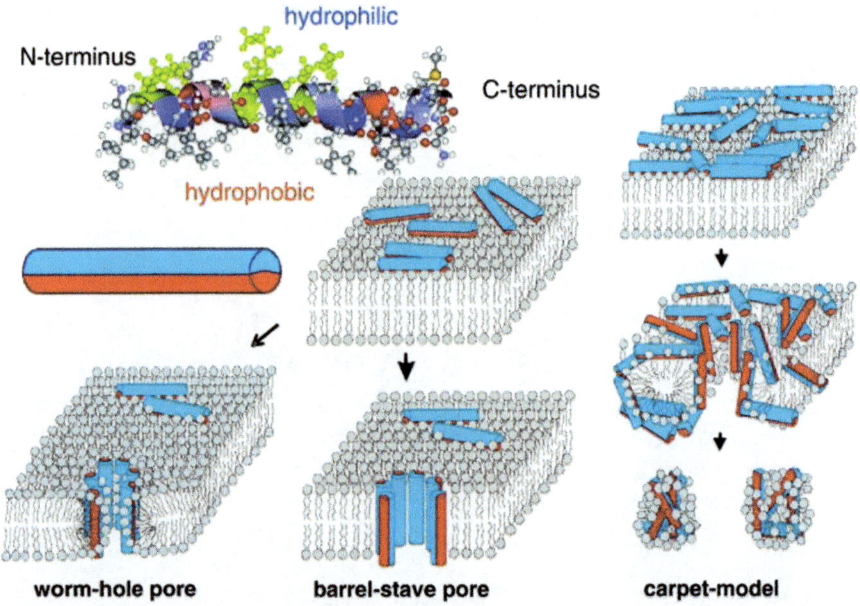

Figure 15.6 Schematically overview of different modes of action of amphiphilic peptides (AMPs).
Reprinted from M. S. Ganewatta and C. Tang, Controlling macromolecular structures towards effective antimicrobial polymers, *Polymer*, **63**, A1–A29, Copyright 2015 with permission from Elsevier.[8]

result, the antimicrobial activity was not strongly influenced by the molecular weight.[72] Another poly(methacrylate) system with quaternary groups was tested by Tang *et al.* and they could also show that low molecular weight polymers revealed higher biocidal activities than the polymers with higher molecular weights.[73] By studying the literature, it becomes clear that this trend is true for many different amphiphilic polycations. So the molecular weight has to be below a certain threshold to gain in antimicrobial activity.[61,74–76]

The balance between hydrophilic and hydrophobic moieties was found to be a very important parameter for the preparation of antimicrobial amphiphilic polycations. On the one hand, the cationic part of the macromolecule is the main reason for electrostatic interactions with the anionic components of the bacterial cell surface.[12] On the other hand, the mode of action of amphiphilic polycations calls for a certain hydrophobicity; otherwise, the amphiphilic polycations would not be able to penetrate the bacterial cell membrane.[12] However, it has been pointed out in various investigations that the higher the hydrophobic part, the higher the hemolysis rate will be.[61] This means that there has to be an optimum value for an amphiphilic polymer system that offers maximum efficacy against bacteria and, simultaneously, minimal hemolytic activity. As pointed out by Palermo and Kuroda, there are mainly three different strategies used to tune the balance between the hydrophilic and hydrophobic components.[12] First of all, the ratio of hydrophobic and cationic side chains is adjusted in random copolymers.[77–82] Another way is to adjust the identity of the hydrophobic parts of the monomers in copolymers or amphiphilic homopolymers.[74,83] Moreover, electrically neutral or hydrophilic polymers are used for conjugation with amphiphilic polycations.[84–86]

The influence of the cationic charge density has been explored by the group of Tew.[87] They polymerized norbonenes with different amounts of primary amino groups (from one to three). The hemolytic activity decreased with the increasing number of amines, whereas the antimicrobial activity stayed on the same level for *S. aureus* and *E. coli*. As mentioned in the literature, it is very difficult to separate the influence of charge and hydrophilicity, because the introduced positive groups automatically enhance the hydrophilicity.[8,61] Additionally, it has been proposed that there is a specific charge threshold that needs to be exceeded to prepare antimicrobial amphiphilic polycations.[74,87–89]

So far, a number of highly active antimicrobial polymers have been developed following the biomimetic concept. They disrupt bacterial cell membranes selectively, because they are negatively charged on the outside. However, amphiphilic antimicrobial polymers are also able to destroy mammalian and human cell membranes due to their amphiphilic character. The tendency of membrane disruption is much lower, because the latter membranes have a neutral surface net charge at the outside and, additionally, contain stabilizing cholesterol. However, amphiphilic polycations are, to some extent, cytotoxic towards mammalian and human cells. So far, selectivity values of hemotoxicity (HC_{50})/antimicrobial activity (MIC) of more than 500 have been achieved.

15.2.4.2 Telechelic Biocidal Polymers

Antimicrobial telechelic polymers have intensively been investigated by our group. In contrast to AMPs, the antimicrobial groups are located at the end of the polymer chains. The polymeric backbone itself exhibits no antimicrobial effect. We have focused on poly(2-oxazoline)s (POx), which are prepared *via* living cationic ring-opening polymerization (CROP) of 2-R-2-oxazolines (R = methyl, ethyl, *etc.*). The resulting non-toxic polymers are adjustable in hydrophilicity through the choice of the R group. Moreover, the living character of the polymerization enables the introduction of specific end groups in the initiating and terminating step of the polymerization.[90–95] End-group conjugation with quaternary ammonium compounds, such as *N*,*N*-dimethyldodecylamine (DDA), is a well-established way of preparing biocidal poly(2-methyloxazoline)s (PMOx).[96]

It was found that the introduction of a second non-biocidal end group strongly influences the antimicrobial activity of the whole macromolecule.[96] This effect is called the satellite group effect. Waschinski and coworkers showed that the activity of polymers could be adjusted over several orders of magnitude by varying this satellite group (SG; ranging from methyl to a hexadecyl groups).[97] The polymeric biocides even surpassed the molar activity of their low molecular weight pendants, in this case *N,N*-dimethyltetradecylammonium chloride (DTAC), against the Gram-positive strain *S. aureus*, as well as the Gram-negative bacterium *E. coli* (Figure 15.7).[97]

The most significant polymers were investigated with regard to the mode of action. After introducing methyl-, decyl- and hexadecyl-satellite groups to PMOx with the biocidal dodecyl ammonium (DDA) group, the interactions with model liposomes were measured. To this end, a fluorescence dye was encapsulated in the liposomes and the release kinetics was investigated in the presence of the different PMOx. It was found that the PMOx with the decyl-satellite group quickly perforated the model liposomes with more efficiency than poly(2-methyloxazoline)s with the methyl SG. A different behavior was found for the poly(2-methyloxazoline)s with the hexadecyl SG. This polymer was bound more strongly to the liposomes but without penetrating the membrane.[98] These results correlate with the observed antimicrobial activities. It was pointed out that the SG supports or weakens the interactions of the biocidal polymer with the membrane and its insertion into the membrane. A mechanistic proposal is shown in Figure 15.8. In solution, the polymers form unimolecular micelles. The SG and the biocidal groups are surrounded by the hydrophilic polymer. After adhesion of the polymer onto the membrane, the polymer refolds and enters the phospholipid membrane.[98] This behavior is typical for some proteins and peptides, such as albumin and antimicrobial peptides. These molecules also refold when in contact with a hydrophobic surface for unspecific binding. Such a behavior is called amphoteric. It is very remarkable that a slight variation of 85 g mol^{-1} in the SG has such a dramatic impact on the macromolecules with molecular weights of several thousand grams per mol.

Fik *et al.* showed that functional SGs can also be used to control the activity of the whole molecule to either increase or even eliminate the biological activity of the whole macromolecule.[99] On basis of these results, Krumm *et al.* prepared a bioswitchable polymer, whose antimicrobial activity was controlled by chemically altering the SG. Therefore, they introduced a hydrolysable ester-containing SG in the initiation step of the polymerization of 2-methyl-2-oxazoline, and terminated this polymer with DDA.[100] It was shown that the hydrophobic ester SG clearly enhanced the antimicrobial activity. Here again, SG with the alkyl residues of C_8 revealed a higher activity against *S. aureus*, compared to ethyl- and tetradecyl-residues (Figure 15.9). After alkaline hydrolysis of all esters, the MIC-values of these polymers rose above 2500 μg mL^{-1}. The most efficient polymer revealed a MIC-value of 39 μg mL^{-1} before hydrolysis. After cleavage, a 30-fold lower antimicrobial activity was obtained.

Additionally, it was shown that the SG could also be cleaved under physiological conditions at pH 7, using the enzyme lipase B from *Candida antarctica*, resulting in a loss of antimicrobial activity as well.[100]

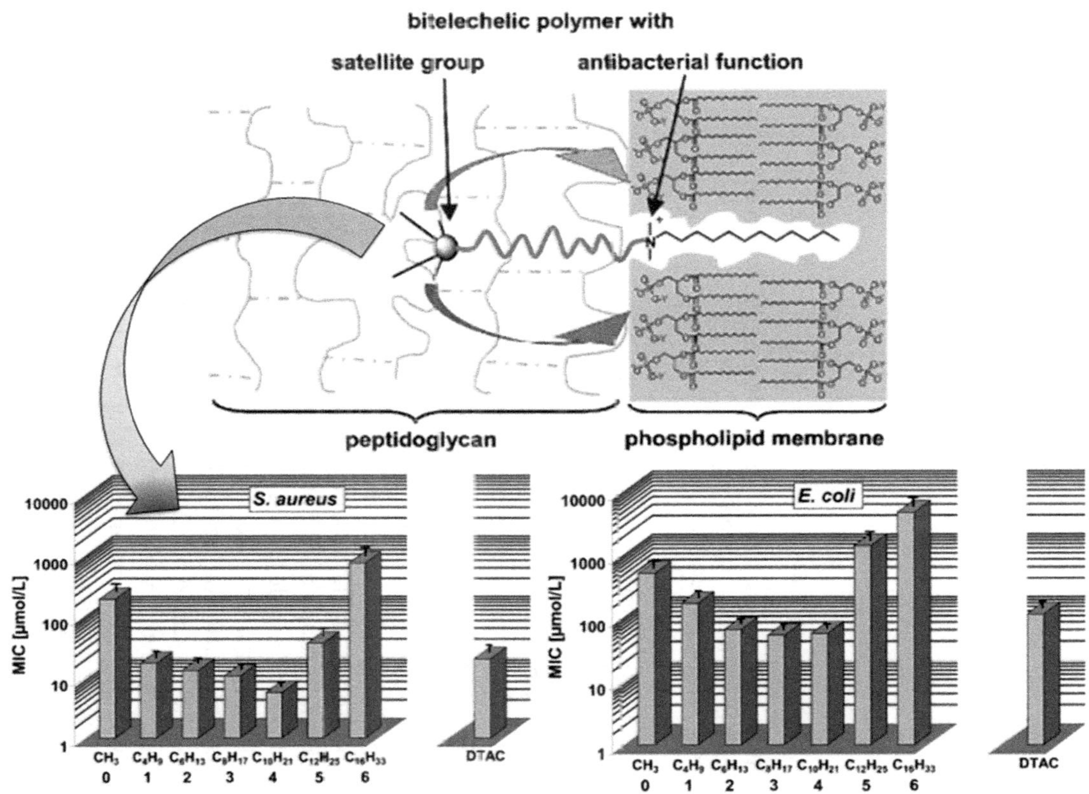

Figure 15.7 Schematic illustration of the proposed actions of telechelic antimicrobial polymers with satellite function (top). MIC values of alkyl-PMOx-DDA derivatives against *S. aureus* (bottom left) and *E. coli* (bottom right).
Reproduced from ref. 97 with permission from John Wiley and Sons. Copyright © 2005 WILEY-VCH Verlag GmbH & Co. KGaA, Weinheim.

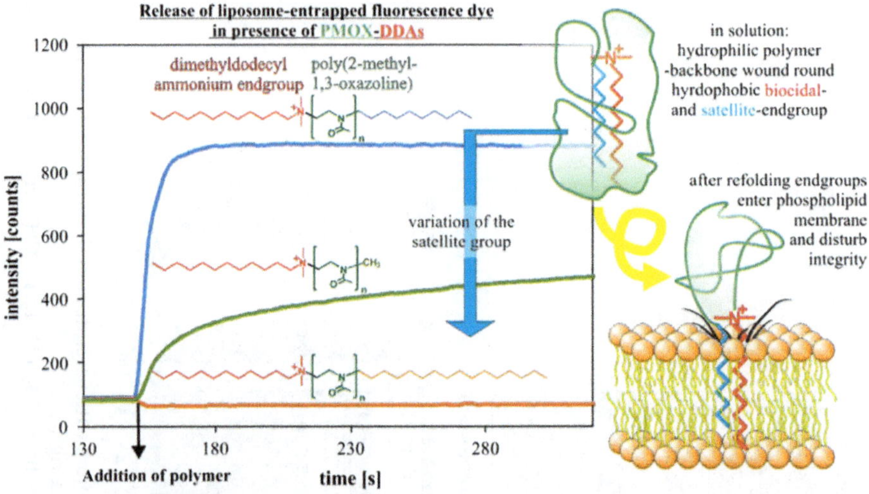

Figure 15.8 Working concepts of telechelic poly(2-oxazoline)s with an antimicrobial end group and a satellite group.
Reprinted from ref. 20.

The SG effect is not restricted to poly(2-oxazoline)s since it was shown by Gellman and coworkers that the N- and C-terminals of nylon-3 random copolymers strongly influenced the bioactivity of these polymers with regard to selectivity and hemolysis.[101] Additional investigations were performed by Griesser and coworkers, who studied the impact of end groups on the hemolysis and antimicrobial activity of antimicrobial polymethacrylates.[102]

As mentioned above, antibiotics are mostly used to prepare biocide releasing polymers systems or polymeric biocides. Telechelic polymers are also used as platforms for the functionalization with different antibiotics of the functional end groups of the biocidal polymers. Du *et al.* coupled tobramycin to PEG *via* amide bonds. Those showed a lower antimicrobial activity in solution against *P. aeruginosa* than free tobramycin, but when these polymer antibiotic conjugates (PACs) were tested against *P. aeruginosa* biofilms, they revealed a 3.2-fold higher activity against the resilient biofilms than tobramycin, which demonstrates the potential of such permanently bound PACs.[103] Lawson *et al.* prepared telechelic non-degradable conjugates of PEG and vancomycin. These conjugates own an additional acrylate end group for later surface modification. The PEG–PACs revealed 6–75 fold decrease in biocidal activity against *Staphylococcus epidermidis* in solution.[104] After surface modification, they found a 7–8 log reduction in bacterial colony forming units. Schmidt *et al.* focused on the preparation of telechelic PACs based on poly(2-oxazoline)s, polyethylene glycol, and CIP.[105] Therefore, they coupled CIP to poly(2-oxazoline)s in two different ways. The direct coupling of poly(2-methyloxazoline) (PMOx) and CIP resulted in low antimicrobial activity. The coupling *via* a

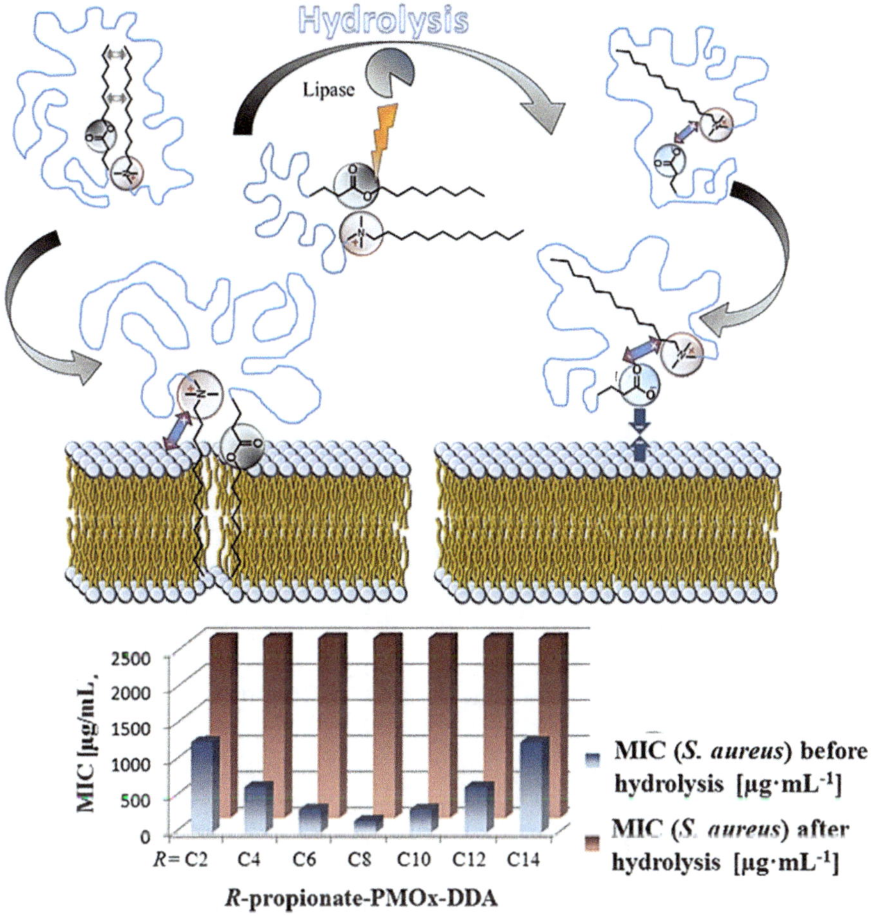

Figure 15.9 Schematic representation of the membrane-active biocidal PMOx controlled by a cleavable satellite group (SG) before and after hydrolysis (top). Minimal inhibitory concentration of antimicrobial telechelic PMOx before and after hydrolysis (bottom).
Adapted from ref. 100 with permission from John Wiley and Sons. Copyright © 2014 WILEY-VCH Verlag GmbH & Co. KGaA, Weinheim.

spacer afforded molecular weight dependent activity, with a molar minimal inhibitory concentration that was even higher than that of pristine CIP (Figure 15.10).

Different Gram-positive and Gram-negative strains were tested and it was found that, in general, the antimicrobial activity of the conjugates increased in the order: poly(2-methyloxazoline) < poly(2-ethyloxazoline) < polyethylene glycol.[105] So far, a satellite group effect has not been found for these systems, most likely due to the fact that CIP does not act at the bacterial cell membrane but as a gyrase inhibitor within the cytoplasm.

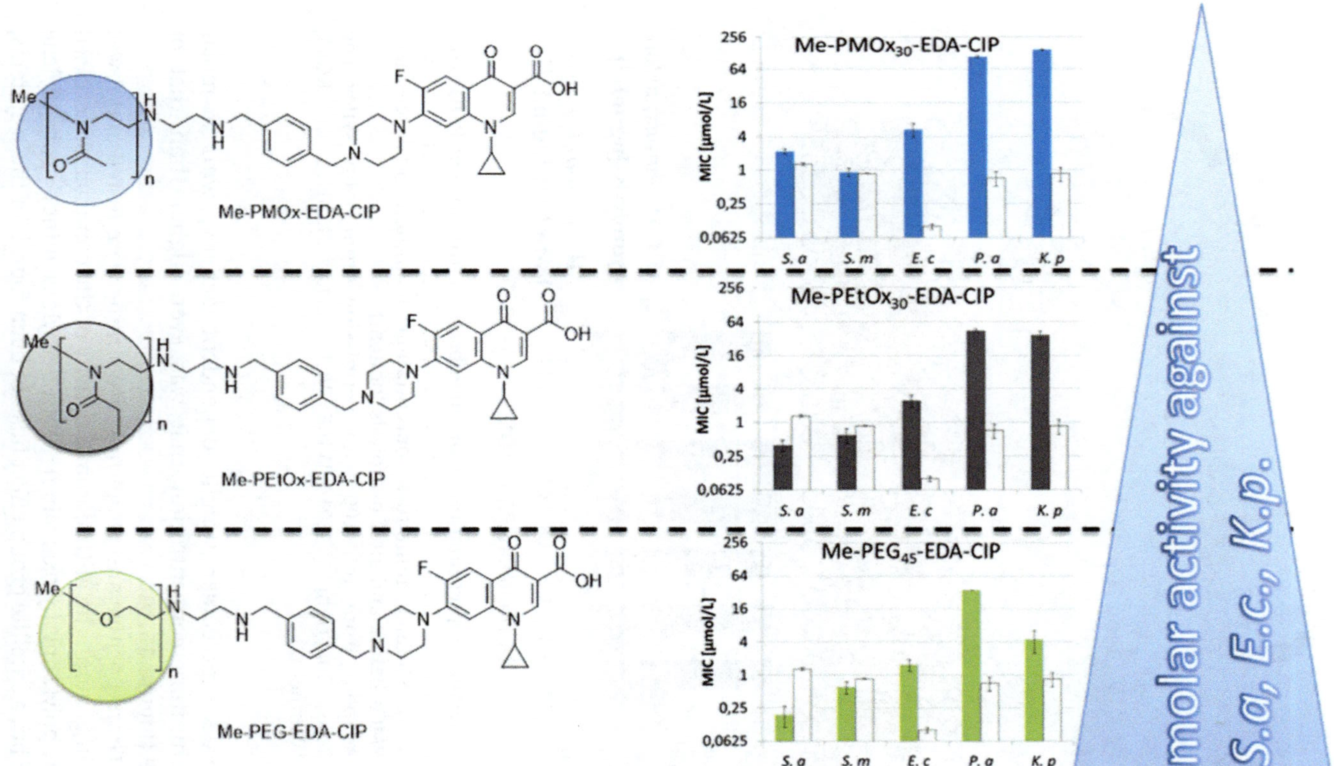

Figure 15.10 Molar MIC values for different polymer antibiotic conjugates (PACs) based on poly(2-methyloxazoline) PMOx, poly(2-ethyloxazoline) (PEtOx), and polyethylene glycol (PEG), and ciprofloxacin (CIP). These were determined against the Gram-positive strains *S. aureus* (*S.a.*) and *S. mutans* (*S.m.*), and the Gram-negative germs *E. coli* (*E.c.*), *P. aeruginosa* (*P.a.*), and *K. pneumoniae* (*K.p.*).
Adapted with permission from M. Schmidt, S. Harmuth, E. R. Barth, E. Wurm, R. Fobbe, A. Sickmann, C. Krumm and J. C. Tiller, *Bioconjugate Chem.*, 2015, **26**, 1950–1962.[105] Copyright (2015) American Chemical Society.

15.2.4.3 Hydrophilic Cationic Polymers

Efforts towards lowering the cytotoxicity of amphiphilic polycations have shown that increasing the hydrophilic character somewhat lowers the antimicrobial activity but also significantly lowers the cytotoxicity (see Section 15.2.4.1). Recent studies have shown that hydrophilic antimicrobial polycations do not destroy the membrane of porcine blood cells even at concentrations of 40 000 $\mu g\,mL^{-1}$. This suggests that their working mechanism might not be based on membrane disruption, which would make them a class of antimicrobial polymers of their own.[32] Hydrophilic polycations were first described in 1933,[106] but their antimicrobial potential was first explored by Ikeda *et al.* in the 1980s. They compared the biocidal efficiency of oligomers and polymers of *N*,*N*-dimethyl hexamethlyeneamine and dimethyliminino-*p*-xylylene.[107] It was found that the activity of these quaternary ammonium salts increases in the order monomer/dimer < trimer < tetramer < polymer. A similar effect was reported for polyhexamethylene biguanides (PHMB) with different chain lengths against *E. coli*.[70]

Different investigations concerning the mode of action of hydrophilic polycations have been performed. All results have in common that the polymers are located in the surroundings of the phosphate ions of the cytoplasmic membrane, lipopolysaccharides, or the peptidoglycan components of the cell wall.[108,109] This results in a displacement of Ca^{2+} ions, which has also been observed for the amphiphilic polycations discussed in Section 2.2.1.4.1.[109] Because of this, it was concluded that the mode of action could be similar to that of amphiphilic polycations.[110] However, because of their low toxicity, the mode of action has to differ from that of amphiphilic polycations in some way, especially due to the lack of hydrophobic domains.

Gilbert and Moore summarized a possible mode of action for PHMB against *E. coli*. When PHMB was concentrated in the surroundings of the cytoplasmic membrane, the hydrophobic domains of PHMB were not able to interact with the hydrophobic core of the cell membrane due to inflexibility.[67] Thus, the polymers were accumulated on the membranes at points with a maximum charge density, *e.g.* integral proteins.[111,112] Therefore, these polycations might disturb the functionality of the integral proteins by inflicting changes in the phospholipid environment and enriching the phospholipid concentration around the proteins.[70,109,113] This leads to a loss of transport functionality, as well as catabolic capability and, in the end, to a damage of the membrane because of the output of micelle-like aggregates consisting of the biocide and parts of the membrane.[67,114] The low cytotoxicity is not explained by this working principle and it does also not explain the rapid killing efficiency of hydrophilic polycations.[32,115] Therefore, the general mode of action for this kind of biocidal polymers has not yet been fully understood.

The group of Agarwal investigated the influence of different spacers between the positively charged groups along the polymeric backbone of different alkyloxyethylammonium ionenes. The polymers were prepared by a

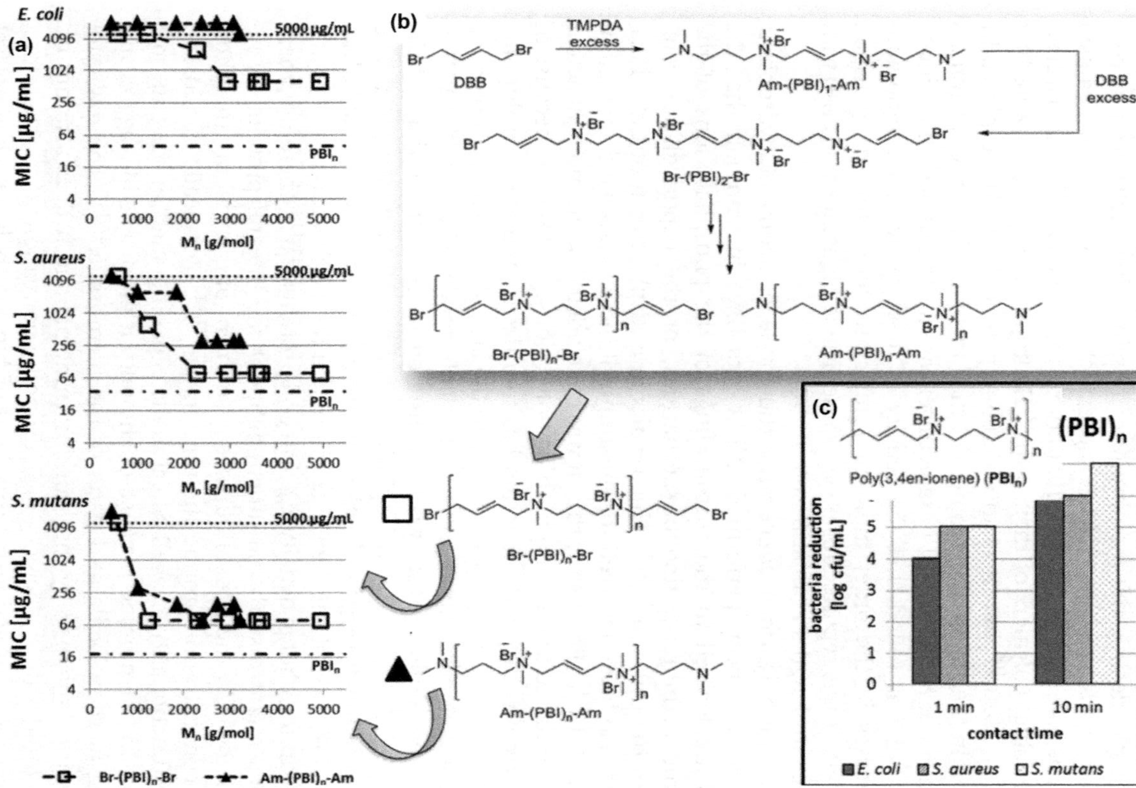

Figure 15.11 (a) Antimicrobial activities (MIC) against bacterial strains *E. coli*, *S. aureus*, and *S. mutans* of Br-(PBI)$_n$-Br and Am-(PBI)$_n$-Am in relation to the molecular weight M_n of the compounds. The MIC of PBI$_n$ prepared by a polyaddition reaction is also shown. (b) Scheme of poly(3,4en-ionene) (PBI) synthesis by sequential monomer addition. (c) Log reduction of different bacterial strains in PBS after 1 and 10 minutes contact time with PBI$_n$ at its MIC.
Adapted from ref. 32 with permission from John Wiley and Sons. Copyright © 2015 Wiley-VCH Verlag GmbH & Co. KGaA, Weinheim.

poly-Menshutkin reaction. The MIC values against *E. coli* were significantly influenced by the spacer length (from 2 to 12 methylene groups). An ethyl spacer between the positive charges revealed a high MIC value, whereas the MIC value changed over two orders of magnitude when the spacer length was increased to dodecylen.[115]

Strassburg *et al.* prepared poly(3,4en-ionene) (PBI) by a sequential synthetic strategy and a poly-Menshutkin reaction. They investigated the influence of the molecular weight and the end group on the antimicrobial impact and hemotoxicity. The MIC values were determined against *E. coli*, *S. aureus*, and *S. mutans*. The PBI prepared through the poly-Menshutkin reaction was also tested for their killing rate and it was found that this polymer is a highly efficient bacteria killer. After one minute of contact at the MIC of PBI, the log reduction of the colony forming units was four against *E. coli* and five against *S. aureus* and *S. mutans* (Figure 15.11). After ten minutes, the log reduction for all tested bacteria was six or higher.[32] The results of sequential syntheses revealed that there is also a relationship between the molecular weight and the antimicrobial activity but, additionally, it was found that the bromine or amine end groups display an additional influence on the antimicrobial impact of these polymers. *E. coli* and *S. aureus* were less susceptible when the sequential PBIs had amine end groups. In the case of *S. mutans*, the end group influence was less pronounced.[32] The best antimicrobial activities of the bromine-terminated sequential PBIs were reached above a molecular weight threshold specific for each bacterial strain (see Figure 15.11a). None of the sequential PBIs was capable of lyzing blood cells at any concentration.

Besides these investigations, it was shown that the selectivity of poly(3,4en-ionene)s is controllable by introducing different alky end groups. Telechelic PBI_6 with end-standing dodecyl groups, for example, was found to be a low toxic, broad-spectrum biocide; whereas a poly(3,4en-ionene) with two butyl end groups was selective for *S. mutans*. In contrast, PBI_6, with two octyl end groups, was selective for both Gram-positive strains *S. mutans* and *S. aureus.*[32] There are natural antimicrobial polymers, such as ε-poly-L-lysine, produced by *Streptomyces albulus via* ε-poly-L-lysine synthase, and the well-known food preservative chitosan, that can also be considered hydrophilic polycations.[116] Thus, even the hydrophilic antimicrobial polymers can be considered a natural mimic.

15.3 Antimicrobial Surfaces

The most critical microbial infection is the biofilm, which is formed by adhesion and proliferation of planktonic microbial cells. Once formed, it is highly resistant, hard to remove, and often a deadly thread to other living forms. To avoid such biofilm formation, nature has developed two major strategies to avoid such infections, microbe killing and repelling surfaces. Most commonly and adapted by nearly all living forms, such as animals, plants, and even bacteria and fungi themselves, is the release of antimicrobial substances, such as antimicrobial peptides and antibiotics. The purpose is to keep surfaces free of biofilms. This approach has been adapted since the very early times of civilization and includes the use of silver ware and coating ships with tar. Presently, these coatings have been developed to show zero order release of silver or antibiotics,[117,118] to release antimicrobial substances on demand,[119,120] and to show multiple antimicrobial actions.[121–123] Recently, numerous examples of antimicrobial coatings that produce antimicrobial reactive oxygen species have been described. They use different energy sources, such as light, chemical, and even biochemical energy. However, the use of such surfaces is limited and lacks the advantages of natural systems that can produce biocides on demand in practically unlimited amounts.[124–127] Additionally, the release of biocides and antibiotics is critical, because they cause massive buildup of resistant microbial strains and pollute the environment in many cases. Alternatively, surfaces that kill microbes on contact without releasing biocides are an intriguing alternative for low contaminated areas. Such surfaces can be created by tethering the above discussed antimicrobial polymers to surfaces. Interestingly, there is no natural motive for such surfaces. Investigations on the mechanism of such surfaces have revealed that they might act differently than the tethered polymers in solution.[20] Besides the idea of carrying the biocidal groups from the surface to the bacterial membrane *via* a polymeric spacer, it has also been proposed that the surface attracts parts of the bacterial cell envelop, such as phospholipids (phospholipid sponge effect)[128] or teichoic acid.[15,129] It has further been proposed that calcium ions might be exchanged, or that the surface adheres the microbial cells so strongly that they are killed due to this intense contact with the cationic charges and rearrangement of the charged and non-charged lipids in their membrane.[69] So far, the true mechanism has not yet been clarified. However, numerous practical solutions have been published recently that permit easy surface modifications, which may allow their industrial use in the near future.[130–132]

One of those approaches includes the formation of antimicrobial fibers from hydrogelators.[133,134] This is supported by surface induced hydrogelation, which allows the concentration of the hydrogelator on affine surfaces, even if they are not forming hydrogels in aqueous solution.[135,136]

The alternative concept of nature is the creation of microbe repelling surfaces. One prominent example is the bio-compatibilization of hydrophobic surfaces by the plasma protein albumin. This protein quickly and strongly adheres to hydrophobic surfaces and renders them microbe repellent.[137] Experimental setups have shown that the self-organized albumin coating lowers the microbial adhesion by 95%. Adapting this concept with Pluronic has shown even better results on silicone.[138] In general, the literature has concluded that microbe-repelling surfaces can either be achieved by ultra-hydrophilic and anionic or by ultra-hydrophobic surface modification. The latter is achieved by either fluorinated surfaces or micro-structuring them by bottom-up, bottom-down, or templating techniques.[139–142] So far, the concept of ultra-hydrophobic surfaces has not proven successful in general, which the exception of lotus-effect surfaces that are frequently washed.

Hydrophilic antimicrobial surfaces are generally achieved by grafting so-called anti-fouling polymers to surfaces. These polymers are discussed in the following section.

15.4 Anti-fouling Polymers

Anti-fouling polymers are designed to prevent non-specific protein adsorption as well as the adhesion of bacterial and mammalian cells or marine organisms on surfaces or particles.[143,144] Therefore, anti-fouling polymers do not kill bacteria, unless they are additionally equipped with an antimicrobial functionality. Polyethylene glycol (PEG) is the most commonly used polymer for preventing protein adsorption to surfaces and is therefore often termed the "gold standard" (Figure 15.12).[145,146]

The modification of surfaces is realized by different grafting techniques in order to enhance the efficiency of PEG coatings.[146–151] The mode of action of PEG is not fundamentally clear.[146,152–154] On the one hand, the repulsion of proteins from the surfaces by PEG is discussed as a working principle.[146,153,154] On the other hand, the hydrophilic character of PEG leads to a hydration layer that should be the reason for the non-accumulation of proteins on the surface.[152]

Poly(glycerol) is a polyether and has a structural similarity to PEG with additional hydroxyl groups.[146] In particular, surface modifications derived from branched poly(glycerol) were found to have protein repelling

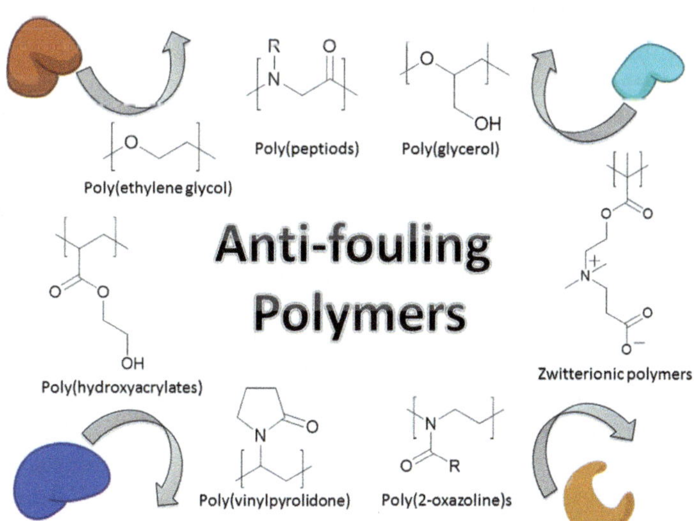

Figure 15.12 Schematic overview of the most commonly used anti-fouling polymers. Reprinted from ref. 146 with permission from the Royal Society of Chemistry.

properties.[155] Surface coatings were prepared by creating self-assembled monolayers (SAM) and grafting methods.[156–162]

Zwitterionic polymers are electrically neutral because they contain homogenously distributed positive and negative charges along their polymeric backbones.[163] They present a class of non-fouling polymers which enbody, in contrast to PEG, an almost inexhaustible source for different polymer designs.[163–165] Mi and Jiang pointed out that there are three main factors for this chemical variability: (a) the choice of the ionic groups, (b) the arrangement of these charged groups in the same polymeric unit, and (c) the different derivatives of commonly used zwitterionic polymers.[145] The charged species appear associated and show a stronger hydration effect than PEG.[163] This is claimed to be the reason for their superior anti-fouling properties.[166,167]

Investigations of Whitesides and coworkers have shown that protein adhesion is reduced even by monolayers of such zwitterionic polymers, *e.g.* self-assembled monolayers (SAM) of non-fouling groups on the surface.[168–171] On the other hand, these surfaces revealed an inadequate efficiency for the repulsion of bacterial cells.[143,145] This effect was explained by very thin layers (1–2 nm) resulting from SAM formation, whereas polymers grafted from surfaces *via* atom transfer radical polymerization (ATRP) had a layer thickness of 20 nm and showed a better performance in bacterial fouling tests. This was attributed to the higher density and stability of non-fouling groups introduced by this method.[172,173] In general, many studies concerning zwitterionic polymers have suggested that they are highly effective protein repelling surface coatings.[146,173–176] This makes them promising candidates for surface modification of implants.[177] They have also been used for the preparation of hydrogels that resisted foreign body reactions for three months in mice due to the prevention of protein adsorption.[178]

Besides these fundamental investigations, anti-fouling polymers have also been designed to respond to external parameters such as the pH.[179,180] Moreover, zwitterionic polymers were combined with biocidal polymers to produce surface modifications with anti-fouling and antimicrobial properties, respectively. Contact killing surfaces suffer from the disadvantage that devitalized bacterial cells accumulate on them. This leads to the development of biofilms on the surface, which deactivates the contact killing biocidal groups. The combination of contact killing and anti-fouling polymers in different layers is a straightforward strategy to circumvent this problem.[121,145,181] However, in this case, the different polymers never demonstrate their properties at the same time. The group of Jiang developed an ester derivative of betaine, which was grafted on a surface.[182] The positively charged surface was able to kill bacteria as long as the ester was present. When the ester was hydrolyzed, the zwitterionic character of betaine led to the repulsion of dead cells. The disadvantage of this system is the fact that the switch between cationic and zwitterionic surfaces is irreversible. Cao *et al.* realized a similar system that could be converted from a cationic to a zwitterionic surface reversibly.[183] To this end, they used a reversible lactonization reaction. The cyclic lactone, used here as a monomer, contains a positively charged quaternary ammonium, resulting in polycations able to kill bacterial cells. After contact with water, the cyclic lactone is converted to its linear zwitterionic form, which is the reason for the repulsion toward dead cells. At low pH, this linear form is dehydrated and the cyclic lactone is formed again. The group of Jiang showed that this cycle could be performed seven times without degradation of the surface coating.

While PEG and zwitterionic polymers account for the majority of used anti-fouling polymers, other polymers have shown their potential as non-fouling polymers as well. Poly(2-oxazoline)s have been shown to be alternatives to PEG as anti-fouling polymers due to the fact that they keep their non-fouling properties for longer times than PEG in oxidative media.[146,184] Poly(2-oxazoline)s were successfully grafted to surfaces *via* different grafting methods.[185–187] They have also been used for the modification of drugs,[188,189] nanocapsules,[190] or enzymes.[95,191] Moreover, they have been used as matrices for bio-calcification.[192]

Electrically neutral and hydrophilic poly-2-hydroxyethylacrylates exhibit anti-fouling properties as well.[146,193–197]

Poly(vinylpyrrolidone) was compared to PEG and poly(2-ethyloxazoline) as surface coating with respect to protein adsorption and was found to be a competitive material for anti-fouling applications.[146,198]

Peptoids represent mimics of natural peptides, whose residues are chemically bonded to the nitrogen instead of the α-carbon atoms.[146] This modification makes them more stable in contrast to peptides.[199] In general, peptoides and peptides have proven themselves as effective non-fouling surface coatings, even for more than five months.[200–204]

Peptides, on the other hand, can be used to trick our immune system. The group of Discher used peptides to equip antigens such as synthetic particles to create stealth properties in the body.[146,205] These modifications resulted in four times better stealth behavior in mice after four hours than the control sample based on PEG.

The major disadvantage of contact-killing as well as antifouling surfaces is that they have to be perfect to be effective. Only a few examples exist that tolerate surface deterioration to some extent.[130]

15.5 Conclusions

Fighting against unfriendly microbial invasion is a problem that all living forms have in common. Many of their strategies have been mimicked by the respective tools developed by mankind. Particularly, the majority of the known antimicrobial polymers – amphiphilic polycations – seem to mimic membrane-active antimicrobial peptides, such as magainin. Hydrophilic polycations are also known to nature and are mimicked in several FDA approved disinfectants, such as polyquaternium-1 and polyhexamethylene biguanidin (PHMB). Since it has become clear that they work differently compared to amphiphilic polycations, the still unclear mechanism will be the focus of future work, as these polymers are not cytotoxic. Even end-group driven antimicrobial polymers seem to follow natural concepts, such as the amphoteric character of surface active peptides, which allows applying the so-called satellite group effect for the enhanced and switchable antimicrobial activity of the polymers. Also, biocide releasing surfaces are mimics of natural systems. The same is true for antifouling surfaces. Only contact-active antimicrobial surfaces with tethered antimicrobial polymers do not mimic natural principles. Whether this makes them superior to natural systems will be clarified in the future. Particularly, the already described tendency of antimicrobial polymeric systems to not build-up microbial resistance as fast as low molecular weight biocides and antibiotics makes them a potent alternative to existing systems.

Acknowledgements

The authors would like to thank the Deutsche Forschungsgemeinschaft (DFG) for financing the project (KR 4700/1-1: Selbst-deaktivierende biozide Polymere).

References

1. M. R. Gillings, *Front. Microbiol.*, 2013, **4**, 4.
2. L. Chow, L. Waldron and M. R. Gillings, *Front. Microbiol.*, 2015, **6**, 10.
3. A. Coates, Y. M. Hu, R. Bax and C. Page, *Nat. Rev. Drug Discovery*, 2002, **1**, 895–910.
4. J. M. Henderson and K. Y. C. Lee, *Curr. Opin. Solid State Mater. Sci.*, 2013, **17**, 175–192.
5. I. Chopra and M. Roberts, *Microbiol. Mol. Biol. Rev.*, 2001, **65**, 232–260.
6. N. J. Johnston, T. A. Mukhtar and G. D. Wright, *Curr. Drug Targets*, 2002, **3**, 335–344.
7. D. J. Tipper, *Pharmacol. Ther.*, 1985, **27**, 1–35.
8. M. S. Ganewatta and C. Tang, *Polymer*, 2015, **63**, A1–A29.
9. H. W. Boucher, G. H. Talbot, J. S. Bradley, J. E. Edwards, D. Gilbert, L. B. Rice, M. Scheld, B. Spellberg and J. Bartlett, *Clin. Infect. Dis.*, 2009, **48**, 1–12.
10. R. Duncan, *J. Controlled Release*, 2014, **190**, 371–380.
11. R. J. Cornell and L. G. Donaruma, *J. Med. Chem.*, 1965, **8**, 388–390.
12. E. F. Palermo and K. Kuroda, *Appl. Microbiol. Biotechnol.*, 2010, **87**, 1605–1615.
13. K. Lienkamp, A. E. Madkour and G. N. Tew, *Adv. Polym. Sci.*, 2013, **251**, 141–172.
14. M. Sobczak, C. Debek, E. Oledzka and R. Kozlowski, *Molecules*, 2013, **18**, 14122–14137.
15. H. Takahashi, E. F. Palermo, K. Yasuhara, G. A. Caputo and K. Kuroda, *Macromol. Biosci.*, 2013, **13**, 1285–1299.

16. A. M. Carmona-Ribeiro and L. D. D. Carrasco, *Int. J. Mol. Sci.*, 2013, **14**, 9906–9946.
17. K. Kuroda and G. A. Caputo, *Wiley Interdiscip. Rev.: Nanomed. Nanobiotechnol.*, 2013, **5**, 49–66.
18. P. Li, X. Li, R. Saravanan, C. M. Li and S. S. J. Leong, *RSC Adv.*, 2012, **2**, 4031–4044.
19. A. Muñoz-Bonilla and M. Fernández-García, *Prog. Polym. Sci.*, 2012, **37**, 281–339.
20. F. Siedenbiedel and J. C. Tiller, *Polymers*, 2012, **4**, 46–71.
21. L. Timofeeva and N. Kleshcheva, *Appl. Microbiol. Biotechnol.*, 2011, **89**, 475–492.
22. G. J. Gabriel, A. Som, A. E. Madkour, T. Eren and G. N. Tew, *Mater. Sci. Eng., R*, 2007, **57**, 28–64.
23. A. Giuliani, G. Pirri, A. Bozzi, A. Di Giulio, M. Aschi and A. C. Rinaldi, *Cell. Mol. Life Sci.*, 2008, **65**, 2450–2460.
24. G. N. Tew, R. W. Scott, M. L. Klein and W. F. Degrado, *Acc. Chem. Res.*, 2010, **43**, 30–39.
25. E.-R. Kenawy, S. D. Worley and R. Broughton, *Biomacromolecules*, 2007, **8**, 1359–1384.
26. Y. C. He, E. Heine, N. Keusgen, H. Keul and M. Möller, *Biomacromolecules*, 2012, **13**, 612–623.
27. F. SORIANO, *J. Antimicrob. Chemother.*, 1992, **30**, 567–569.
28. A. Gehrt, J. Peter, P. A. Pizzo and T. J. Walsh, *J. Clin. Microbiol.*, 1995, **33**, 1302–1307.
29. B. B. Maisuria, M. L. Actis, S. N. Hardrict, J. O. Falkinham Iii, M. F. Cole, R. L. Cihlar, S. M. Peters, R. V. Macri, E. W. Sugandhi, A. A. Williams, M. A. Poppe, A. R. Esker and R. D. Gandour, *Bioorg. Med. Chem.*, 2011, **19**, 2918–2926.
30. K. L. LaPlante and M. J. Rybak, *Antimicrob. Agents Chemother.*, 2004, **48**, 4665–4672.
31. C. J. Ioannou, G. W. Hanlon and S. P. Denyer, *Antimicrob. Agents Chemother.*, 2007, **51**, 296–306.
32. A. Strassburg, F. Kracke, J. Wenners, A. Jemeljanova, J. Kuepper, H. Petersen and J. C. Tiller, *Macromol. Biosci.*, 2015, **15**, 1710–1723.
33. Y. L. Colson and M. W. Grinstaff, *Adv. Mater.*, 2012, **24**, 3878–3886.
34. L. Yang and P. Alexandridis, *Curr. Opin. Colloid Interface Sci.*, 2000, **5**, 132–143.
35. F. H. Meng, Z. Y. Zhong and J. Feijen, *Biomacromolecules*, 2009, **10**, 197–209.
36. R. Cheng, F. Feng, F. H. Meng, C. Deng, J. Feijen and Z. Y. Zhong, *J. Controlled Release*, 2011, **152**, 2–12.
37. R. P. Brinkhuis, F. Rutjes and J. C. M. van Hest, *Polym. Chem.*, 2011, **2**, 1449–1462.
38. O. Vogl and D. Tirrell, *J. Macromol. Sci., Chem.*, 1979, **A13**, 415–439.
39. J. Liang, Y. J. Chen, X. H. Ren, R. Wu, K. Barnes, S. D. Worley, R. M. Broughton, U. Cho, H. Kocer and T. S. Huang, *Ind. Eng. Chem. Res.*, 2007, **46**, 6425–6429.

40. B. V. Worley, K. M. Schilly and M. H. Schoenfisch, *Mol. Pharmaceutics*, 2015, **12**, 1573–1583.
41. N. A. Stasko and M. H. Schoenfisch, *J. Am. Chem. Soc.*, 2006, **128**, 8265–8271.
42. E. Kenawy, A. E. R. El-Shanshoury, N. O. Shaker, B. M. El-Sadek, A. H. B. Khattab and B. I. Badr, *J. Appl. Polym. Sci.*, 2011, **120**, 2734–2742.
43. E. Ji, T. S. Corbitt, A. Parthasarathy, K. S. Schanze and D. G. Whitten, *ACS Appl. Mater. Interfaces*, 2011, **3**, 2820–2829.
44. Y. Wang, T. D. Canady, Z. J. Zhou, Y. L. Tang, D. N. Price, D. G. Bear, E. Y. Chi, K. S. Schanze and D. G. Whitten, *ACS Appl. Mater. Interfaces*, 2011, **3**, 2209–2214.
45. S. Jamsa, R. Mahlberg, U. Holopainen, J. Ropponen, A. Savolainen and A. C. Ritschkoff, *Prog. Org. Coat.*, 2013, **76**, 269–276.
46. G. L. Y. Woo, M. W. Mittelman and J. P. Santerre, *Biomaterials*, 2000, **21**, 1235–1246.
47. S. Gac, J. Coudane, M. Boustta, M. Domurado and M. Vert, *J. Drug Targeting*, 2000, **7**, 393–406.
48. M. Sobczak, G. Nalecz-Jawecki, W. L. Kolodziejski, P. Gos and K. Zoltowska, *Int. J. Pharm.*, 2010, **402**, 37–43.
49. S. P. Parwe, P. N. Chaudhari, K. K. Mohite, B. S. Selukar, S. S. Nande and B. Garnaik, *Int. J. Nanomed.*, 2014, **9**, 1463–1477.
50. H. Namazi and A. Kanani, *Carbohydr. Polym.*, 2009, **76**, 46–50.
51. A. Kugel, B. Chisholm, S. Ebert, M. Jepperson, L. Jarabek and S. Stafslien, *Polym. Chem.*, 2010, **1**, 442–452.
52. A. Nathan, S. Zalipsky, S. I. Ertel, S. N. Agathos, M. L. Yarmush and J. Kohn, *Bioconjugate Chem.*, 1993, **4**, 54–62.
53. R. B. Greenwald, H. Zhao, J. Xia and A. Martinez, *J. Med. Chem.*, 2003, **46**, 5021–5030.
54. M. Bankova, N. Manolova, N. Markova, T. Radoucheva, K. Dilova and I. Rashkov, *J. Bioact. Compat. Polym.*, 1997, **12**, 294–307.
55. S. T. Oh, C. S. Ha and W. J. Cho, *J. Appl. Polym. Sci.*, 1994, **54**, 859–866.
56. S. Thamizharasi, J. Vasantha and B. S. R. Reddy, *Eur. Polym. J.*, 2002, **38**, 551–559.
57. W. S. Moon, J. C. Kim, K. H. Chung, E. S. Park, M. N. Kim and J. S. Yoon, *J. Appl. Polym. Sci.*, 2003, **90**, 1797–1801.
58. E. Turos, J.-Y. Shim, Y. Wang, K. Greenhalgh, G. S. K. Reddy, S. Dickey and D. V. Lim, *Bioorg. Med. Chem. Lett.*, 2007, **17**, 53–56.
59. P. Sarker, J. Shepherd, K. Swindells, I. Douglas, S. MacNeil, L. Swanson and S. Rimmer, *Biomacromolecules*, 2011, **12**, 1–5.
60. G. N. Tew, D. H. Liu, B. Chen, R. J. Doerksen, J. Kaplan, P. J. Carroll, M. L. Klein and W. F. DeGrado, *Proc. Natl. Acad. Sci. U. S. A.*, 2002, **99**, 5110–5114.
61. K. Lienkamp and G. N. Tew, *Chemistry*, 2009, **15**, 11784–11800.
62. F. P. Edmund, V. Satyavani and K. Kenichi, in *Tailored Polymer Architectures for Pharmaceutical and Biomedical Applications*, American Chemical Society, 2013, vol. 1135, ch. 19, pp. 319–330.

63. M. Zasloff, *Proc. Natl. Acad. Sci. U. S. A.*, 1987, **84**, 5449–5453.
64. K. A. Brogden, *Nat. Rev. Microbiol.*, 2005, **3**, 238–250.
65. M. Zasloff, *Nature*, 2002, **415**, 389–395.
66. Y. Shai, *Biochim. Biophys. Acta, Biomembr.*, 1999, **1462**, 55–70.
67. P. Gilbert and L. E. Moore, *J. Appl. Microbiol.*, 2005, **99**, 703–715.
68. E.-R. Kenawy, F. I. Abdel-Hay, A. E.-R. R. El-Shanshoury and M. H. El-Newehy, *J. Polym. Sci., Part A: Polym. Chem.*, 2002, **40**, 2384–2393.
69. A. M. Kelly, V. Kaltenhauser, I. Mühlbacher, K. Rametsteiner, H. Kren, C. Slugovc, F. Stelzer and F. Wiesbrock, *Macromol. Biosci.*, 2013, **13**, 116–125.
70. P. Broxton, P. M. Woodcock and P. Gilbert, *J. Appl. Bacteriol.*, 1983, **54**, 345–353.
71. K. Kuroda and W. F. DeGrado, *J. Am. Chem. Soc.*, 2005, **127**, 4128–4129.
72. B. P. Mowery, A. H. Lindner, B. Weisblum, S. S. Stahl and S. H. Gellman, *J. Am. Chem. Soc.*, 2009, **131**, 9735–9745.
73. Y. Chen, P. A. Wilbon, Y. P. Chen, J. H. Zhou, M. Nagarkatti, C. P. Wang, F. X. Chu, A. W. Decho and C. B. Tang, *RSC Adv.*, 2012, **2**, 10275–10282.
74. K. Lienkamp, A. E. Madkour, A. Musante, C. F. Nelson, K. Nusslein and G. N. Tew, *J. Am. Chem. Soc.*, 2008, **130**, 9836–9843.
75. G. J. Gabriel, J. A. Maegerlein, C. E. Nelson, J. M. Dabkowski, T. Eren, K. Nusslein and G. N. Tew, *Chem. – Eur. J.*, 2009, **15**, 433–439.
76. Y. Qiao, C. Yang, D. J. Coady, Z. Y. Ong, J. L. Hedrick and Y. Y. Yang, *Biomaterials*, 2012, **33**, 1146–1153.
77. K. Kuroda, G. A. Caputo and W. F. DeGrado, *Chem. – Eur. J.*, 2009, **15**, 1123–1133.
78. E. F. Palermo and K. Kuroda, *Biomacromolecules*, 2009, **10**, 1416–1428.
79. B. P. Mowery, S. E. Lee, D. A. Kissounko, R. F. Epand, R. M. Epand, B. Weisblum, S. S. Stahl and S. H. Gellman, *J. Am. Chem. Soc.*, 2007, **129**, 15474–15476.
80. D. Fischer, Y. Li, B. Ahlemeyer, J. Krieglstein and T. Kissel, *Biomaterials*, 2003, **24**, 1121–1131.
81. E. Moreau, M. Domurado, P. Chapon, M. Vert and D. Domurado, *J. Drug Targeting*, 2002, **10**, 161–173.
82. V. Sambhy, B. R. Peterson and A. Sen, *Angew. Chem., Int. Ed.*, 2008, **47**, 1250–1254.
83. M. F. Ilker, K. Nüsslein, G. N. Tew and E. B. Coughlin, *J. Am. Chem. Soc.*, 2004, **126**, 15870–15875.
84. P. H. Sellenet, B. Allison, B. M. Applegate and J. P. Youngblood, *Biomacromolecules*, 2007, **8**, 19–23.
85. T. R. Stratton, J. L. Rickus and J. P. Youngblood, *Biomacromolecules*, 2009, **10**, 2550–2555.
86. S. Colak, C. F. Nelson, K. Nusslein and G. N. Tew, *Biomacromolecules*, 2009, **10**, 353–359.
87. Z. M. Al-Badri, A. Som, S. Lyon, C. F. Nelson, K. Nüsslein and G. N. Tew, *Biomacromolecules*, 2008, **9**, 2805–2810.
88. K. Lienkamp and G. N. Tew, *Chem. – Eur. J.*, 2009, **15**, 11784–11800.

89. M. Pasupuleti, B. Walse, B. Svensson, M. Malmsten and A. Schmidtchen, *Biochemistry*, 2008, **47**, 9057–9070.
90. S. Konieczny, C. Krumm, D. Doert, K. Neufeld and J. C. Tiller, *J. Biotechnol.*, 2014, **181**, 55–63.
91. C. Krumm, C. P. Fik, M. Meuris, G. J. Dropalla, H. Geltenpoth, A. Sickmann and J. C. Tiller, *Macromol. Rapid Commun.*, 2012, **33**, 1677–1682.
92. C. Krumm, S. Konieczny, G. J. Dropalla, M. Milbradt and J. C. Tiller, *Macromolecules*, 2013, **46**, 3234–3245.
93. C. Weber, C. R. Becer, A. Baumgaertel, R. Hoogenboom and U. S. Schubert, *Des. Monomers Polym.*, 2009, **12**, 149–165.
94. L. Tauhardt, M. Frant, D. Pretzel, M. Hartlieb, C. Bucher, G. Hildebrand, B. Schroter, C. Weber, K. Kempe, M. Gottschaldt, K. Liefeith and U. S. Schubert, *J. Mater. Chem. B*, 2014, **2**, 4883–4893.
95. S. Konieczny, M. Leurs and J. C. Tiller, *ChemBioChem*, 2015, **16**, 83–90.
96. C. J. Waschinski and J. C. Tiller, *Biomacromolecules*, 2005, **6**, 235–243.
97. C. J. Waschinski, V. Herdes, F. Schueler and J. C. Tiller, *Macromol. Biosci.*, 2005, **5**, 149–156.
98. C. J. Waschinski, S. Barnert, A. Theobald, R. Schubert, F. Kleinschmidt, A. Hoffmann, K. Saalwächter and J. C. Tiller, *Biomacromolecules*, 2008, **9**, 1764–1771.
99. C. P. Fik, C. Krumm, C. Muennig, T. I. Baur, U. Salz, T. Bock and J. C. Tiller, *Biomacromolecules*, 2012, **13**, 165–172.
100. C. Krumm, S. Harmuth, M. Hijazi, B. Neugebauer, A.-L. Kampmann, H. Geltenpoth, A. Sickmann and J. C. Tiller, *Angew. Chem., Int. Ed.*, 2014, **53**, 3830–3834.
101. J. H. Zhang, M. J. Markiewicz, B. P. Mowery, B. Weisblurn, S. S. Stahl and S. H. Gellman, *Biomacromolecules*, 2012, **13**, 323–331.
102. T. D. Michl, K. E. S. Locock, N. E. Stevens, J. D. Hayball, K. Vasilev, A. Postma, Y. Qu, A. Traven, M. Haeussler, L. Meagher and H. J. Griesser, *Polym. Chem.*, 2014, **5**, 5813–5822.
103. J. Du, H. Bandara, P. Du, H. Huang, K. Hoang, D. Nguyen, S. V. Mogarala and H. D. C. Smyth, *Mol. Pharmaceutics*, 2015, **12**, 1544–1553.
104. M. C. Lawson, R. Shoemaker, K. B. Hoth, C. N. Bowman and K. S. Anseth, *Biomacromolecules*, 2009, **10**, 2221–2234.
105. M. Schmidt, S. Harmuth, E. R. Barth, E. Wurm, R. Fobbe, A. Sickmann, C. Krumm and J. C. Tiller, *Bioconjugate Chem.*, 2015, **26**, 1950–1962.
106. C. F. Gibbs, E. R. Littmann and C. S. Marvel, *J. Am. Chem. Soc.*, 1933, **55**, 753–757.
107. T. Ikeda, H. Yamaguchi and S. Tazuke, *J. Bioact. Compat. Polym.*, 1990, **5**, 31–41.
108. T. Ikeda, B. Lee, H. Yamaguchi and S. Tazuke, *Biochim. Biophys. Acta, Biomembr.*, 1990, **1021**, 56–62.
109. P. Broxton, P. M. Woodcock and P. Gilbert, *Microbios*, 1984, **41**, 15–22.
110. T. Ikeda, H. Yamaguchi and S. Tazuke, *Biochim. Biophys. Acta, Biomembr.*, 1990, **1026**, 105–112.

111. T. Ikeda, A. Ledwith, C. H. Bamford and R. A. Hann, *Biochim. Biophys. Acta, Biomembr.*, 1984, **769**, 57–66.
112. T. Ikeda, S. Tazuke and M. Watanabe, *Biochim. Biophys. Acta*, 1983, **735**, 380–386.
113. P. Broxton, P. M. Woodcock, F. Heatley and P. Gilbert, *J. Appl. Bacteriol.*, 1984, **57**, 115–124.
114. B. Bechinger and K. Lohner, *Biochim. Biophys. Acta, Biomembr.*, 2006, **1758**, 1529–1539.
115. C. Mattheis, M. Zheng and S. Agarwal, *Macromol. Biosci.*, 2012, **12**, 341–349.
116. K. Yamanaka, C. Maruyama, H. Takagi and Y. Hamano, *Nat. Chem. Biol.*, 2008, **4**, 766–772.
117. C. H. Ho, E. K. Odermatt, I. Berndt and J. C. Tiller, *J. Biomater. Sci., Polym. Ed.*, 2013, **24**, 1589–1600.
118. Y. Suzuki, M. Tanihara, Y. Nishimura, K. Suzuki, Y. Kakimaru and Y. Shimizu, *J. Biomed. Mater. Res.*, 1998, **42**, 112–116.
119. X. Ren, L. Kou, H. B. Kocer, C. Zhu, S. D. Worley, R. M. Broughton and T. S. Huang, *Colloids Surf., A*, 2008, **317**, 711–716.
120. M. Tanihara, Y. Suzuki, Y. Nishimura, K. Suzuki and Y. Kakimaru, *Peptides*, 1998, **19**, 421–425.
121. A. M. Bieser, Y. Thomann and J. C. Tiller, *Macromol. Biosci.*, 2011, **11**, 111–121.
122. G. Cheng, H. Xite, Z. Zhang, S. F. Chen and S. Y. Jiang, *Angew. Chem., Int. Ed.*, 2008, **47**, 8831–8834.
123. C. H. Ho, J. Tobis, C. Sprich, R. Thomann and J. C. Tiller, *Adv. Mater.*, 2004, **16**, 957–961.
124. V. Nadtochenko, N. Denisov, O. Sarkisov, D. Gumy, C. Pulgarin and J. Kiwi, *J. Photochem. Photobiol., A*, 2006, **181**, 401–407.
125. S. Noimark, C. W. Dunnill and I. P. Parkin, *Adv. Drug Delivery Rev.*, 2013, **65**, 570–580.
126. P. Tran and T. J. Webster, *Nanotechnology*, 2013, **24**, 155101.
127. B. J. Vartiainen, M. Rättö and S. Paulussen, *Packag. Technol. Sci.*, 2005, **18**, 243–251.
128. A. M. Bieser and J. C. Tiller, *Macromol. Biosci.*, 2011, **11**, 526–534.
129. D. Raafat, K. von Bargen, A. Haas and H. G. Sahl, *Appl. Environ. Microbiol.*, 2008, **74**, 3764–3773.
130. A. D. Fuchs and J. C. Tiller, *Angew. Chem.*, 2006, **45**, 6759–6762.
131. C. J. Waschinski, J. Zimmermann, U. Salz, R. Hutzler, G. Sadowski and J. C. Tiller, *Adv. Mater.*, 2008, **20**, 104–108.
132. F. Siedenbiedel, A. Fuchs, T. Moll, M. Weide, R. Breves and J. C. Tiller, *Macromol. Biosci.*, 2013, **13**, 1447–1455.
133. D. A. Salick, J. K. Kretsinger, D. J. Pochan and J. P. Schneider, *J. Am. Chem. Soc.*, 2007, **129**, 14793–14799.
134. S. Debnath, A. Shome, D. Das and P. K. Das, *J. Phys. Chem. B*, 2010, **114**, 4407–4415.
135. A. M. Bieser and J. C. Tiller, *Chem. Commun.*, 2005, 3942–3944, DOI: 10.1039/b506160a.

136. A. M. Bieser and J. C. Tiller, *J. Phys. Chem. B*, 2007, **111**, 13180–13187.
137. I. H. Prattterpstra, A. H. Weerkamp and H. J. Busscher, *J. Gen. Microbiol.*, 1987, **133**, 3199–3206.
138. M. R. Nejadnik, H. C. van der Mei, W. Norde and H. J. Busscher, *Biomaterials*, 2008, **29**, 4117–4121.
139. C. P. Fik, M. Meuris, U. Salz, T. Bock and J. C. Tiller, *Adv. Mater.*, 2011, **23**, 3565–3569.
140. A. A. Thorpe, V. Peters, J. R. Smith, T. G. Nevell and J. Tsibouklis, *J. Fluorine Chem.*, 2000, **104**, 37–45.
141. N. Yoshino and T. Teranaka, *J. Biomater. Sci., Polym. Ed.*, 1997, **8**, 623–653.
142. W. Barthlott, T. Schimmel, S. Wiersch, K. Koch, M. Brede, M. Barczewski, S. Walheim, A. Weis, A. Kaltenmaier, A. Leder and H. F. Bohn, *Adv. Mater.*, 2010, **22**, 2325–2328.
143. I. Banerjee, R. C. Pangule and R. S. Kane, *Adv. Mater.*, 2011, **23**, 690–718.
144. C. M. Kirschner and A. B. Brennan, in *Annual Review of Materials Research*, ed. D. R. Clarke, Annual Reviews, Palo Alto, 2012, vol. 42, pp. 211–229.
145. L. Mi and S. Y. Jiang, *Angew. Chem., -Int. Ed.*, 2014, **53**, 1746–1754.
146. S. Lowe, N. M. O'Brien-Simpson and L. A. Connal, *Polym. Chem.*, 2015, **6**, 198–212.
147. K. H. A. Lau, C. L. Ren, S. H. Park, I. Szleifer and P. B. Messersmith, *Langmuir*, 2012, **28**, 2288–2298.
148. J. Satulovsky, M. A. Carignano and I. Szleifer, *Proc. Natl. Acad. Sci. U. S. A.*, 2000, **97**, 9037–9041.
149. R. V. Ostaci, D. Damiron, S. Al Akhrass, Y. Grohens and E. Drockenmuller, *Polym. Chem.*, 2011, **2**, 348–354.
150. B. S. Favel, M. Jasieniak, L. Velleman, S. Ciampi, E. Luais, J. R. Peterson, H. J. Griesser, J. G. Shapter and J. J. Gooding, *Langmuir*, 2013, **29**, 8355–8362.
151. S. Jo and K. Park, *Biomaterials*, 2000, **21**, 605–616.
152. S. F. Chen, L. Y. Li, C. Zhao and J. Zheng, *Polymer*, 2010, **51**, 5283–5293.
153. S. I. Jeon, J. H. Lee, J. D. Andrade and P. G. Degennes, *J. Colloid Interface Sci.*, 1991, **142**, 149–158.
154. S. I. Jeon and J. D. Andrade, *J. Colloid Interface Sci.*, 1991, **142**, 159–166.
155. P. Y. J. Yeh, R. K. Kainthan, Y. Q. Zou, M. Chiao and J. N. Kizhakkedathu, *Langmuir*, 2008, **24**, 4907–4916.
156. M. Weinhart, T. Becherer, N. Schnurbusch, K. Schwibbert, H. J. Kunte and R. Haag, *Adv. Eng. Mater.*, 2011, **13**, B501–B510.
157. M. Wyszogrodzka and R. Haag, *Biomacromolecules*, 2009, **10**, 1043–1054.
158. J. I. Paez, V. Brunetti, M. C. Strumia, T. Becherer, T. Solomun, J. Miguel, C. F. Hermanns, M. Calderon and R. Haag, *J. Mater. Chem.*, 2012, **22**, 19488–19497.
159. M. Khan and W. T. S. Huck, *Macromolecules*, 2003, **36**, 5088–5093.
160. T. Weber, M. Bechthold, T. Winkler, J. Dauselt and A. Terfort, *Colloids Surf., B*, 2013, **111**, 360–366.

161. M. Lange, S. Braune, K. Luetzow, K. Richau, N. Scharnagl, M. Weinhart, A. T. Neffe, F. Jung, R. Haag and A. Lendlein, *Macromol. Rapid Commun.*, 2012, **33**, 1487–1492.
162. Q. Wei, T. Becherer, R. C. Mutihac, P. L. M. Noeske, F. Paulus, R. Haag and I. Grunwald, *Biomacromolecules*, 2014, **15**, 3061–3071.
163. A. B. Lowe and C. L. McCormick, *Chem. Rev.*, 2002, **102**, 4177–4189.
164. S. Chen and S. Jiang, *Adv. Mater.*, 2008, **20**, 335–338.
165. S. Jiang and Z. Cao, *Adv. Mater.*, 2010, **22**, 920–932.
166. A. S. Abu Lila, H. Kiwada and T. Ishida, *J. Controlled Release*, 2013, **172**, 38–47.
167. P. Laverman, O. C. Boerman, W. J. G. Oyen, F. H. M. Corstens and G. Storm, *Crit. Rev. Ther. Drug Carrier Syst.*, 2001, **18**, 551–566.
168. M. Mrksich and G. M. Whitesides, *Annu. Rev. Biophys. Biomol. Struct.*, 1996, **25**, 55–78.
169. E. Ostuni, R. G. Chapman, M. N. Liang, G. Meluleni, G. Pier, D. E. Ingber and G. M. Whitesides, *Langmuir*, 2001, **17**, 6336–6343.
170. E. Ostuni, R. G. Chapman, R. E. Holmlin, S. Takayama and G. M. Whitesides, *Langmuir*, 2001, **17**, 5605–5620.
171. R. G. Chapman, E. Ostuni, L. Yan and G. M. Whitesides, *Langmuir*, 2000, **16**, 6927–6936.
172. G. Cheng, Z. Zhang, S. F. Chen, J. D. Bryers and S. Y. Jiang, *Biomaterials*, 2007, **28**, 4192–4199.
173. G. Cheng, G. Z. Li, H. Xue, S. F. Chen, J. D. Bryers and S. Y. Jiang, *Biomaterials*, 2009, **30**, 5234–5240.
174. K. Yoshimoto, T. Hirase, J. Madsen, S. P. Armes and Y. Nagasaki, *Macromol. Rapid Commun.*, 2009, **30**, 2136–2140.
175. R. S. Smith, Z. Zhang, M. Bouchard, J. Li, H. S. Lapp, G. R. Brotske, D. L. Lucchino, D. Weaver, L. A. Roth, A. Coury, J. Biggerstaff, S. Sukavaneshvar, R. Langer and C. Loose, *Sci. Transl. Med.*, 2012, **4**, 153ra132.
176. W. Yang, H. Xue, W. Li, J. L. Zhang and S. Y. Jiang, *Langmuir*, 2009, **25**, 11911–11916.
177. P. S. Liu, E. Domingue, D. C. Ayers and J. Song, *ACS Appl. Mater. Interfaces*, 2014, **6**, 7141–7152.
178. L. Zhang, Z. Q. Cao, T. Bai, L. Carr, J. R. Ella-Menye, C. Irvin, B. D. Ratner and S. Y. Jiang, *Nat. Biotechnol.*, 2013, **31**, 553–556.
179. L. Mi, M. T. Bernards, G. Cheng, Q. M. Yu and S. Y. Jiang, *Biomaterials*, 2010, **31**, 2919–2925.
180. H. S. Sundaram, J. R. Ella-Menye, N. D. Brault, Q. Shao and S. Y. Jiang, *Chem. Sci.*, 2014, **5**, 200–205.
181. P. Zou, W. Hartleb and K. Lienkamp, *J. Mater. Chem.*, 2012, **22**, 19579–19589.
182. G. Cheng, H. Xue, Z. Zhang, S. Chen and S. Jiang, *Angew. Chem.*, 2008, **47**, 8831–8834.
183. Z. Q. Cao, L. Mi, J. Mendiola, J. R. Ella-Menye, L. Zhang, H. Xue and S. Y. Jiang, *Angew. Chem., Int. Ed.*, 2012, **51**, 2602–2605.

184. B. Pidhatika, M. Rodenstein, Y. Chen, E. Rakhmatullina, A. Muhlebach, C. Acikgoz, M. Textor and R. Konradi, *Biointerphases*, 2012, **7**, 15.
185. Y. Chujo, E. Ihara, S. Kure and T. Saegusa, *Macromolecules*, 1993, **26**, 5681–5686.
186. N. Zhang, T. Pompe, I. Amin, R. Luxenhofer, C. Werner and R. Jordan, *Macromol. Biosci.*, 2012, **12**, 926–936.
187. R. Konradi, B. Pidhatika, A. Muhlebach and M. Textort, *Langmuir*, 2008, **24**, 613–616.
188. R. Luxenhofer, A. Schulz, C. Roques, S. Li, T. K. Bronich, E. V. Batrakova, R. Jordan and A. V. Kabanov, *Biomaterials*, 2010, **31**, 4972–4979.
189. C. H. Wang, W. T. Wang and G. H. Hsiue, *Biomaterials*, 2009, **30**, 3352–3358.
190. K. Kempe, S. L. Ng, K. F. Noi, M. Mullner, S. T. Gunawan and F. Caruso, *ACS Macro Lett.*, 2013, **2**, 1069–1072.
191. S. Konieczny, I. Schoenfeld, S. Dech and J. C. Tiller, *Abstr. Pap. Am. Chem. Soc.*, 2014, **248**, 1.
192. N. Rauner, M. Meuris, S. Dech, J. Godde and J. C. Tiller, *Acta Biomater.*, 2014, **10**, 3942–3951.
193. C. Zhao, J. Zhao, X. S. Li, J. Wu, S. F. Chen, Q. Chen, Q. M. Wang, X. Gong, L. Y. Li and J. Zheng, *Biomaterials*, 2013, **34**, 4714–4724.
194. C. Rodriguez-Emmenegger, M. Houska, A. B. Alles and E. Brynda, *Macromol. Biosci.*, 2012, **12**, 1413–1422.
195. C. Zhao, L. Li, Q. Wang, Q. Yu and J. Zheng, *Langmuir*, 2011, **27**, 4906–4913.
196. C. Rodriguez-Emmenegger, O. A. Avramenko, E. Brynda, J. Skvor and A. B. Alles, *Biosens. Bioelectron.*, 2011, **26**, 4545–4551.
197. C. Zhao, L. Y. Li, M. M. Guo and J. Zheng, *Chem. Pap.*, 2012, **66**, 323–339.
198. A. Serrano, O. Sterner, S. Mieszkin, S. Zurcher, S. Tosatti, M. E. Callow, J. A. Callow and N. D. Spencer, *Adv. Funct. Mater.*, 2013, **23**, 5706–5718.
199. S. M. Miller, R. J. Simon, S. Ng, R. N. Zuckermann, J. M. Kerr and W. H. Moos, *Drug Dev. Res.*, 1995, **35**, 20–32.
200. R. Chelmowski, S. D. Koster, A. Kerstan, A. Prekelt, C. Grunwald, T. Winkler, N. Metzler-Nolte, A. Terfort and C. Woll, *J. Am. Chem. Soc.*, 2008, **130**, 14952–14953.
201. M. Schneider, C. Fetsch, I. Amin, R. Jordan and R. Luxenhofer, *Langmuir*, 2013, **29**, 6983–6988.
202. A. R. Statz, R. J. Meagher, A. E. Barron and P. B. Messersmith, *J. Am. Chem. Soc.*, 2005, **127**, 7972–7973.
203. K. H. A. Lau, C. L. Ren, T. S. Sileika, S. H. Park, I. Szleifer and P. B. Messersmith, *Langmuir*, 2012, **28**, 16099–16107.
204. R. Luxenhofer, C. Fetsch and A. Grossmann, *J. Polym. Sci., Part A: Polym. Chem.*, 2013, **51**, 2731–2752.
205. P. L. Rodriguez, T. Harada, D. A. Christian, D. A. Pantano, R. K. Tsai and D. E. Discher, *Science*, 2013, **339**, 971–975.

CHAPTER 16

Superwettability of Polymer Surfaces

XI YAO, JIE JU AND LEI JIANG*

Beijing National Laboratory for Molecular Science (BNLMS) Key Laboratory of Organic Solids, Institute of Chemistry, Chinese Academy of Sciences, Zhongguancun North 1st street, Beijing 100190, P. R. China
*Email: jianglei@iccas.ac.cn

16.1 Introduction

The development of superwettability arises from pure inspiration from nature. For instance, water beads on lotus leaves roll off freely because the surface of lotus leaves is superhydrophobic. The same occurs with the underwater superoleophobicity of fish scales. Through the development of both theoretical insights[1–3] and vast experimental observations,[4,5] we know now that there are basically two factors that governs the superwettability on solid surfaces: the chemical composition and the surface structure.[6] As shown in Figure 16.1, on a typical solid surface with ideal smoothness, the wettability is solely determined by the chemical composition of that surface. For surfaces of organic solids, when the water contact angle (CA_w) is greater than 65°, the surface is defined as hydrophobic, whereas it is defined as hydrophilic when CA_w is less than 65°. However, there is no ideal solid surface in real situations. That is, in the real world, solid surfaces are always rough. According to theory, a rough structure amplifies the intrinsic wettability.[7,8] Therefore, with an appropriate rough structure, the superhydrophilicity or superhydrophobicity on solid surfaces can be achieved. For a long time, researchers have strived to discover all kinds of interfacial

RSC Polymer Chemistry Series No. 22
Bio-inspired Polymers
Edited by Nico Bruns and Andreas F. M. Kilbinger

Published by the Royal Society of Chemistry, www.rsc.org

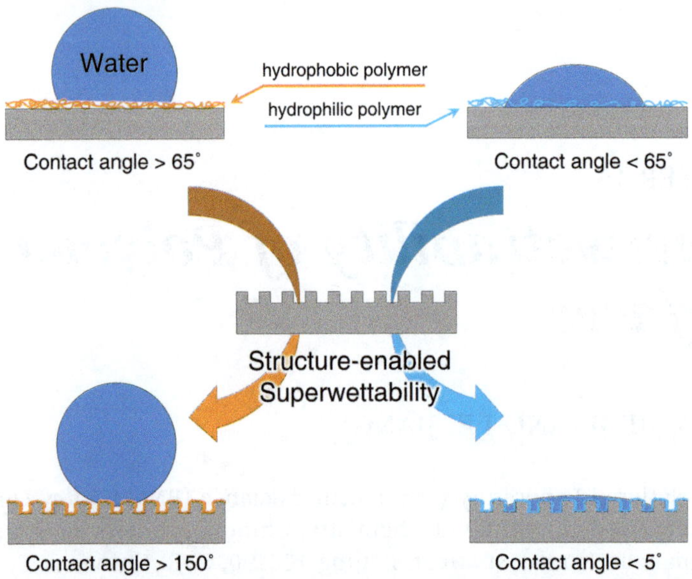

Figure 16.1 The water contact angle (CA_w) at 65° divides a smooth polymer surface as hydrophobic ($CA_w > 65°$) or hydrophilic ($CA_w < 65°$). The superwettability is achieved *via* introduction of proper rough structures on the surface.

materials and develop the relationship between chemical property and structure towards superwettability, among which polymers are without a doubt the most significant materials with broad applications in different areas.

Polymers are a class of organic materials, chain-like molecules composed of repeating monomer units. A surface made of polymeric material exhibits certain interfacial properties that resemble those on the surface of living creatures. In fact, many biological surfaces that show superwettability are covered by polymers. For instance, the biological hydrogels covering fish scales are comprised of collagen, which is a biological polymer found in nature. The epithelial wax of the plant cuticle covering lotus leaves and other plants' superhydrophobic surfaces is another example of oligomers, which are composed by relatively few repeating monomer units (Figure 16.2). In fact, the vast development of polymer synthesizing technology has promoted the emerging of polymer materials with different molecular structures and properties, many of which have mimicked and sometimes even surpassed the functions of biological polymers.

In this chapter, recent developments on different polymer surfaces with superwettability will be introduced. As mentioned above, the superwettability on a solid surface is not solely determined by the chemical composition of that surface, but is also a result of some roughness arrangement. Theoretically speaking, all polymers can be either made superhydrophilic or superhydrophobic with the help of proper surface structures. As a result, it is

Figure 16.2 Different superwettability on bio-surfaces. (A) Water on lotus leaves exhibits superhydrophobicity and can easily slide off by barely tilting the surface. (W. Barthlott, Purity of the sacred lotus, or escape from contamination in biological surfaces, *Planta*, 303, 1997, with permission of Springer[5]). (B) and (C) are the rose petals and peanuts leaves, respectively, which exhibit high adhesion to water drops despite the very large CA_w values. ((B) Adapted with permission from L. Feng, Y. Zhang, J. Xi, Y. Zhu, N. Wang, F. Xia and L. Jiang, *Langmuir*, 2008, **24**, 4114–4119.[10] Copyright (2008) American Chemical Society, and (C) Adapted from ref. 11 with permission from John Wiley and Sons. © 2013 WILEY-VCH Verlag GmbH & Co. KGaA, Weinheim.)

difficult to categorize superwetting polymers according to different chemical compositions. However, the area of application of the superwetting phenomena from polymer materials can be easily categorized. As a result, it is a more reasonable fashion to introduce superwetting polymers according to different fields of application.

16.2 Self-cleaning Polymer Surfaces

16.2.1 Polymer Surfaces in Air

The lotus leaf, as a symbol of sacredness in the world's classical literacy, has been described countless times regarding its ability to stay clean despite mostly rooting from muddy water. Researchers found that the water drops that are deposited on the leaf surface maintain a nearly spherical shape. These water "beads" can easily roll on the surface with merely a few degrees of inclination of the leaf, like mercury drops rolling on the ground. The rolling water beads can remain non-wetting to the surface while picking up all the dust and contaminants on the surface of the leaves.[5] The self-cleaning mechanism of lotus leaves is clear. It is also easy to understand that the cuticle wax covering the leaf surface is responsible for the low attraction to water. However, this structure–function relationship remained obscure until Feng *et al.* reported in the year 2002 that the superhydrophobicity of the lotus leaf, together with other biological plant surfaces, is the result of a synergetic effect by the combination of micron- and nano-scale rough structures.[6] Such combination benefits the amplification of surface roughness and the resulting superhydrophobicity. Such rationale has had a profound influence on the design of artificial superhydrophobic surfaces. Many superhydrophobic polymer surfaces showing excellent anti-fouling properties have been developed based on this principle.[9]

Unlike the lotus leaf, rose petals[10] and peanut leaves[11] show high adhesion and large contact angle to water drops simultaneously (Figure 16.2). Using in-depth observation, it was found that the micron scale structures are similar to those on the lotus leaves, but the nanometer scale structures create a situation where the water is sucked into the "valleys" by a capillary effect. Consequently, although the water drops are spherical in shape on these bio-surfaces, the drops cannot slide away even when the surface is placed vertically. These hanging water beads are believed to be helpful in attracting insects to the roses. As for the peanut leaves, a similar drop accumulating mechanism might facilitate the self-irrigation to help survival in drought environments.

In conventional notions, heterogeneous structures and a low surface free energy must be present simultaneously on a solid surface so as to obtain in-air self-cleaning properties. However, a recent report by Wong *et al.* introduced a new wetting state,[12] the slippery state, which shows excellent self-cleaning and self-healing properties and is inspired from the lubricating mechanism of the pitcher plant surface.[13,14] According to their report, the slippery liquid-infused porous surface (SLIPS) was prepared by impregnating perfluoropolyether (PFPE) into the porous structures of certain solid substrates with substantially low surface free energy (Figure 16.3A). The resulting SLIPS possesses a liquid surface stabilized by the solid rough structure. As concluded by Lafuma *et al.*,[15] the condition required for a given liquid to float on SLIPS is interpreted as follows:

$$\Delta\gamma = \gamma_1 \cos\theta_1 - \gamma_2 \cos\theta_2 - \gamma_{12} > 0 \tag{16.1}$$

Here, γ_1 and γ_2 are the surface tension of the lubricant layer and the floating drops. θ_1 and θ_2 are the measurable contact angles of the respective liquid 1 and 2 on the smooth surface of the materials used as porous media. The most prominent feature of the slippery state is that the lubricant layer on top

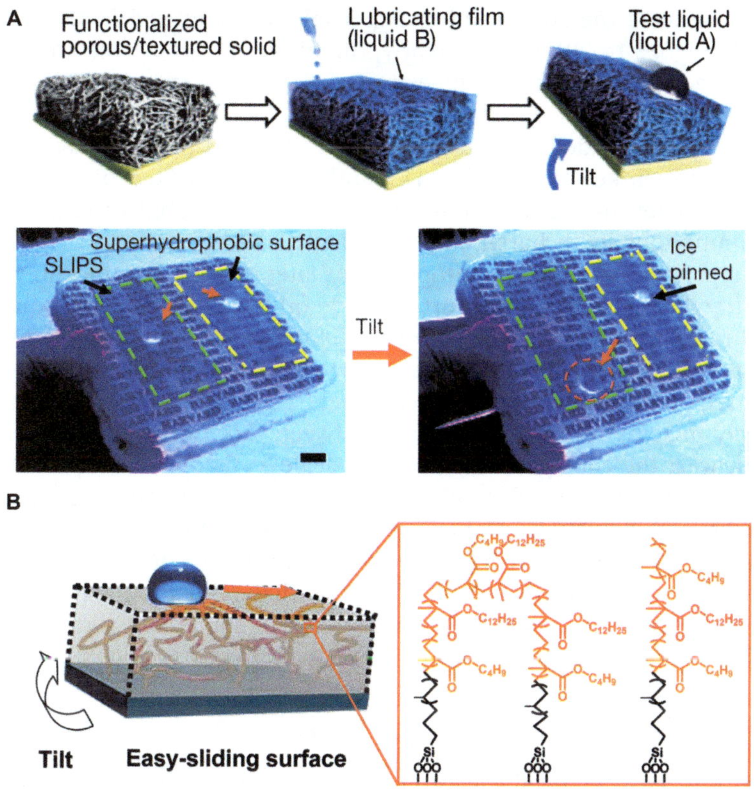

Figure 16.3 SLIPS and organogel materials with self-cleaning properties. (A) The originally reported SLIPS material that not only exhibits low adhesion to water and oil, but also shows low adhesion to ice formed on its surface. (Reprinted by permission from Macmillan Publishers Ltd: *Nature*,[12] Copyright (2011)). (B) A silicone-oil suffused organogel that can be used to self-clean with water. (Adapted from ref. 16 with permission from John Wiley and Sons. Copyright © 2013 WILEY-VCH Verlag GmbH & Co. KGaA, Weinheim.)

of the solid surface plays a role in preventing the solid surface from directly contact the object above (water, oil, solids, *etc.*). Wong's work provides a composite material that exhibits multiple functions including enhanced transparency, pressure stability, excellent repellency to both water and most oils, and the surface omniphobicity is also self-healable. From the preparation technique it can be clearly seen that SLIPS is a composite material. The oil floats on top and inside a porous medium, which poses a threat when such material is used in liquid flows or harsh environments such as in the rain. The lubricant would be depleted by constant shearing. Therefore, it is an urgent need to develop SLIPS comprised of a homogeneous phase. Liu *et al.* reported that the supporting solid porous media can be replaced by a crosslinking network that holds the lubricant liquid in place.[16] In the polymer network, the silicon oil acted as the lubricant which found better stability because the silicone oil was partially immobilized in the polymer network (Figure 16.3B). Consequently, the high water repellency of the material surface remains after being exposed to air for a long time. In addition, such technique allows the application of SLIPS on various inorganic solid surfaces. Yao extended this idea to the PFPE-based SLIPS system by preparing a crosslinked fluorogel.[17] The author demonstrated bio-antifouling on the fluorogel surface, which may have the potential ability to exhibit long term omniphobicity in harsh environments.

The organogel self-cleaning property was later extended to the anti-waxing capability during tunnel transport of crude oil in the petroleum industry.[18] The ultralow adhesion of the organogel material to the solidified paraffin wax was different from that in previous works. Ultralow adhesion was realized based on the prevention of heterogeneous nucleation of wax crystals at a relatively miscible diffusion interface between the lubricant layer and the paraffin wax; the thus generated micron-scale structure at the wax surface diminished the direct solid–solid contact with the underlying substrate. The lubricant depletion problem was addressed by a self-replenishing mechanism of the organogel material, in which the lubricant could be continuously replenished from the crude oil flow. This innovation could be potentially extended to deal with a wide range of solid-deposition problems in industry and the daily life.

16.2.2 Polymer Surfaces under Water

Besides the plants living above water, creatures living underwater also utilize biological polymers to generate self-cleaning surfaces to deal with underwater contamination, which are generally oil-based contaminants. Liu *et al.* reported that a wide variety of fish exhibits underwater oil-repellency on their scales due to the excellent superhydrophilicity of the scale surface.[19] The authors successfully proved this concept by developing an oil-repellent surface using superhydrophilic materials (Figure 16.4A). The substance on the fish scales that shows superoleophobicity was later identified as collagen, a type of biological hydrogel comprised of proteins.[20] In recent years, hydrogel materials with underwater superoleophobicity have been reported. Lin *et al.* reported that clay-enhanced hydrogels showed stable oil-repellency under pressure.[21] As shown in Figure 16.4B, the clay-mediated hydrogel network showed enhanced mechanical strength, which made the hydrogel materials more durable to pressure. Durability is a crucial factor for application in deep sea conditions since, due to its soft nature, the hydrogel network tends to collapse under high pressure, which would sabotage the anti-fouling function under such conditions.

Besides self-cleaning, the application of hydrogels can also be extended to other areas such as capture and regulation of oil drops underwater. The key question is to realize hydrogels with stimulus-responsive adhesion properties. Chen *et al.* reported that a poly(*N*-isopropylacrylamide) (PNIPAAm) hydrogel exhibited a wettability and underwater oil adhesion switch

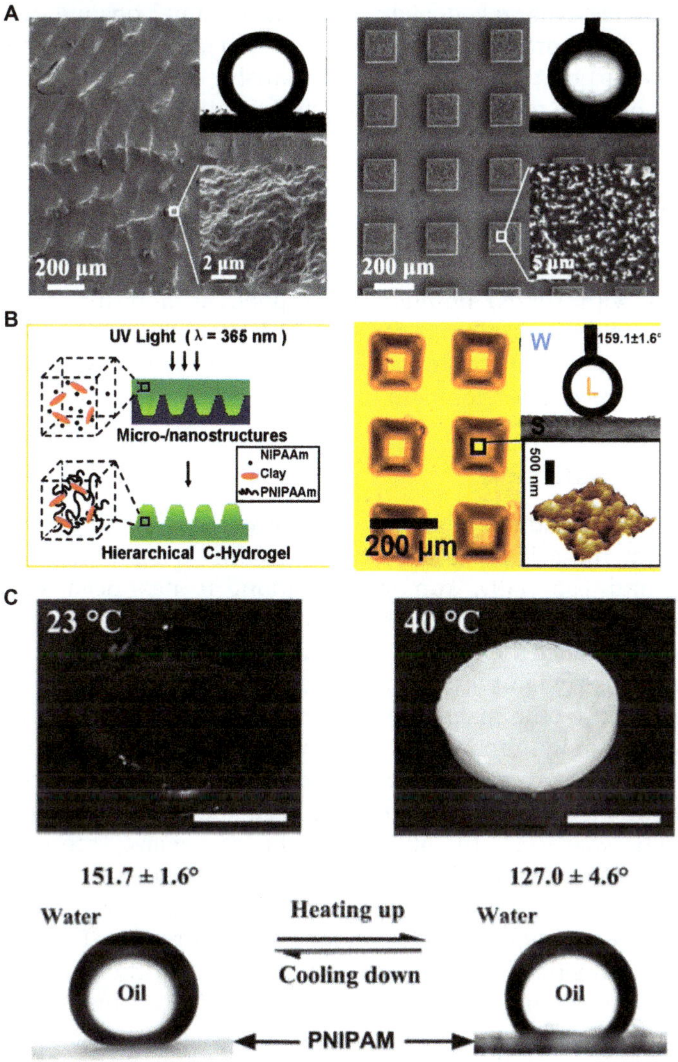

Figure 16.4 (A) The underwater superoleophobicity of fish scales and a superhydrophilic surface makes them ideal anti-fouling surfaces. (Adapted from ref. 19 with permission from John Wiley and Sons. Copyright © 2009 WILEY-VCH Verlag GmbH & Co. KGaA, Weinheim.) (B) Exfoliated clay enhances the stable superoleophobicity of hydrogels under pressure. (Adapted from ref. 21 with permission from John Wiley and Sons. Copyright © 2010 WILEY-VCH Verlag GmbH & Co. KGaA, Weinheim.) (C) Poly(*n*-isopropylacrylamide) (NIPAAm) hydrogel exhibits distinctive wettability differences at different temperatures. (Adapted from ref. 22 with permission from the Royal Society of Chemistry.)

according to the temperature.[22] At temperatures below the lowest critical solution temperature (LCST) of PNIPAAm, the hydrogel was hydrated, fully transparent, and showed superoleophobicity; whereas at temperatures above the LCST, the hydrogel dehydrated and became opaque and white (Figure 16.4C). Importantly, the hydrogel in such conditions showed certain adhesion toward oil drops, which is a promising result for applications in underwater oil drop manipulation, as well as controlled capture and release of oil-based objects.[23]

16.3 Special Adhesion

Due to their soft matter nature, polymer materials present various types of interactions with solids, liquids, and vapors. As a result, superwetting polymers have been exploited for cell and protein capture, solvent absorption, as well as underwater gas collection. In the following section, we will introduce recent development of polymers in these application fields.[24]

16.3.1 Cell-adhesion

Cells live in a soft and wet environment. The soft surrounding is in fact very important for the cell processes including adhesion, proliferation, differentiation, and especially, tumor generation. It has been increasingly acknowledged that mimicking the tumor growth environment is a key prerequisite to realize the high throughput capture of circulating tumor cells (CTCs), which is very important for the early stage-diagnosis and therapy of cancer. Inspired by the macrophage's ability to identify cancer cells in the circulatory system in human bodies, Huang and coworkers developed a CTC capture-and-release system by inactivated and magnetized macrophages.[25] As shown in Figure 16.5A, the macrophage was first impregnated with paramagnetic particles to obtain magnet-responsive properties. The macrophage was then silicified to keep only the surface morphology of it. This step also brought convenience to the following surface functionalization with anti-epithelial cell adhesion molecule antibodies. The as-prepared macrophage-mimetic silica particles were released to the simulated circulatory system to capture CTCs. After a certain period of time, the capturers were retrieved simply using a magnet. The CTCs could be detached from the biomimetic silica particle by breaking the disulfide bond in the linker between the silica and the antibody. Consequently, the CTCs were released for further study. This research provides a complete method to capture and release CTCs from blood, which is of great importance for current clinical application in cancer diagnostics and therapy.

Liu *et al.* developed a simple method to fabricate polystyrene (PS) nanotube templates to mimic a soft environment for capturing breast cancer cells from whole blood.[26] First, PS was dissolved in a solvent, forming a solution. Then, anionic aluminum oxide (AAO) was made to have intimate contact with such solution. Through capillary forces, the solution was absorbed into the nanopores of AAO. After complete evaporation of the solvent, all components solidified. Then, AAO was etched off with NaOH, leaving a template covered by PS nanotubes. As shown in Figure 16.5B, such soft structure was

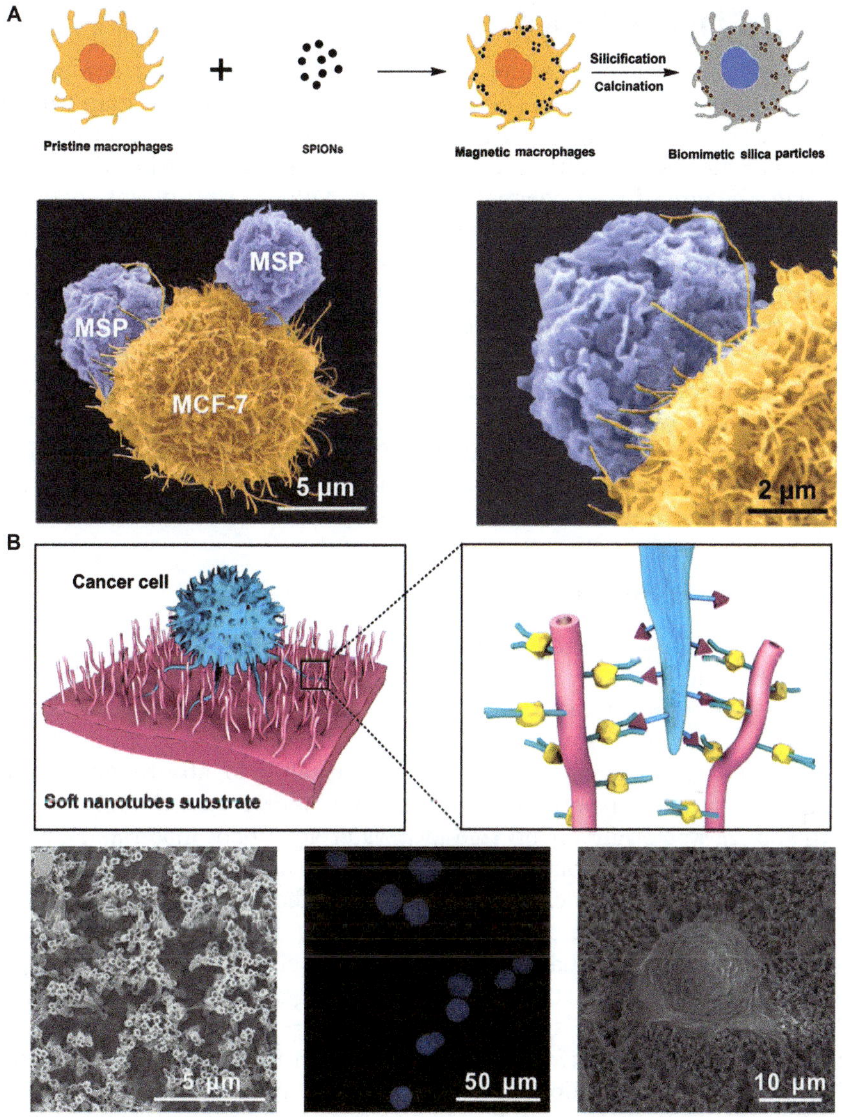

Figure 16.5 (A) The magnetization of macrophages followed by silicification makes them bio-mimicking cancer cell capturers. (Adapted from ref. 25 with permission from John Wiley and Sons. Copyright © 2014 WILEY-VCH Verlag GmbH & Co. KGaA, Weinheim.) (B) The soft polystyrene (PS) nanotube surface is functionalized and used to capture circulating tumor cells. (Reprinted by permission from Macmillan Publishers Ltd: *NPG Asia Mater.*,[26] Copyright 2013).

able to extract cancer cells from a whole blood sample. Essentially, the PS nanotubes formed a soft forest that provided a 3D morphology to favor the capture and immobilization of cell filopodia.

16.3.2 Liquid-adhesion

Some polymer materials show good affinity towards certain solvents. As defined in polymer science, solvents that make the polymer chain fully extended are good solvents for the polymer. When the polymer is crosslinked into a polymer network, however, the polymer can no longer be "dissolved". Instead, the crosslinked polymer is swollen. Namely, the solvent enters the polymer network and forms a polymer gel. Vast research efforts have been devoted to optimize the uptake capacity of crosslinked polymer networks toward certain solvents. The materials that can absorb solvents up to tens of times their weight are called super absorbents.[27–30]

Super absorbents are applied to collect solvent spills in labs, restore water resources, or deal with crude oil spills on the sea surface (Figure 16.6A). To maximize the solvent capacity, a low crosslinking density is required. However, when the crosslinking density is too low, the mechanical strength of the swollen gel becomes very weak, which is inconvenient when people try to collect them from the sea.

16.3.3 Air-adhesion

Methane, as a powerful greenhouse gas which is 25 times more potent than carbon dioxide, generally exists as a solid gas-hydrate on the ocean floor.[31,32] Therefore, an effective solution to confine this potential threat and even to make good use of it as an energy resource is necessary. Our group has developed a superhydrophobic sponge that is able to absorb air bubbles underwater and transport the air to facilities above water.[33] Polyurethane (PU) sponge was originally utilized as a water absorbent material. However, after surface hydrophobization treatment, the macro porous sponge was turned superhydrophobic. Due to the low surface free energy, the superhydrophobic sponge exhibits super-aerophilicity underwater (Figure 16.6B). It is worth mentioning that the authors found a simple method to transport collected methane inside the sponge by attaching a rigid tube at any point on the upper surface of the sponge. Since the differential pressure propels the methane gas to escape *via* any open passage on the same horizontal level, the physical principle allows the idea of setting up a transport channel regardless of the position and abundance of methane on the sea floor.

16.4 Oil/water Separation

The separation of oil/water mixtures is a common problem faced in many industrial applications, such as crude oil exploitation, crude oil transport, and decontamination of oil spills on the sea surface. A material that allows the separation of water from oil or *vice versa* should exhibit high affinity toward either water or oil. Water and oil are generally immiscible. When either water or oil is wetting the surface of a material, it becomes difficult for the other liquid to penetrate the pores or slits of the media made by such

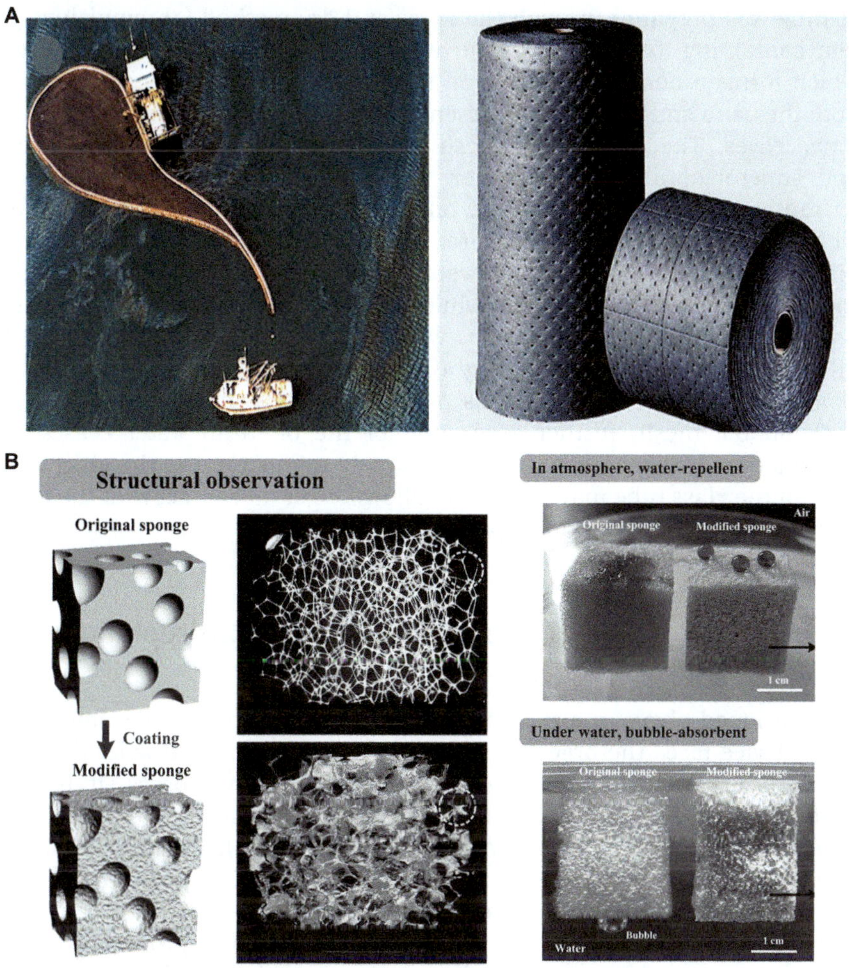

Figure 16.6 (A) Super-absorbent made of liquid-affinitive gels with a lightly cross-linked polymer network has broad applications in collecting or retrieving natural resources. (B) A hydrophobized PU sponge able to collect methane largely distributed beneath the sea floor, which is a threat to accelerate global warming.
(Adapted from ref. 33 with permission from John Wiley and Sons. Copyright © 2012 WILEY-VCH Verlag GmbH & Co. KGaA, Weinheim.)

material. This is the principle of separating water/oil mixtures based on superwetting materials.

16.4.1 Oil/water Separation based on Superhydrophobic Materials

In 2004, Feng *et al.* reported a novel method to separate oil from water with the help of a poly(tetrafluoroethylene) (PTFE)-coated steel mesh.[34] The mesh coating was prepared through the spray-and-dry method (an emulsion system containing Teflon, surfactant, water-based glue, and distilled water), which forms a dense PTFE layer on steel with heterogeneous structures at both the nano and micro-scales, resembling the heterogeneous structures on lotus leaves. The as-prepared mesh exhibited stable superhydrophobicity and superoleophilicity underwater. The separation device was prepared by sandwiching the coated mesh on one side of a cylindrical glass tube, which became the fundamental prototype of a superwetting-material-based oil/water separator. When an oil/water mixture was poured onto the mesh, the oil was allowed to penetrate while water was blocked on top. There is a practical issue about this design, which is the difference in density. As oils have typically a lower density than water, they float on top of water. Consequently, once the water has made complete contact with the superhydrophobic mesh, further separation of the oil from water ceases. To address this drawback, the mesh is obliquely sandwiched in the glass tube, or even the glass tube may be tilted to allow longer oil contact with the mesh surface for high-throughput separation.

16.4.2 Oil/water Separation based on Superhydrophilic Materials

As mentioned above, the superhydrophobic-material-based oil/water separation device faces the practical issue that water is usually heavier than oil, which limits the productivity during the separation process. It is thus rational to develop a separation device on the basis of superhydrophilic materials. Among all the superhydrophilic materials, hydrogels exhibit the most robust underwater superoleophobicity and therefore, they are the most ideal candidates for such separation systems.

Inspired by the superhydrophilic and superoleophobic properties of fish scales, Xue *et al.* fabricated a hydrogel-coated mesh capable of separating water from oil/water mixtures.[35] As shown in Figure 16.7, an oligomer of polyacrylamide (PAM) was coated on the copper mesh surface, which had been treated with oxygen plasma. Then, PAM formed a hydrogel coating on the mesh surface, which showed superhydrophilicity. Once the hydrophilic "sieve" was placed in the separating device, the oil/water mixture could be easily separated by pouring it in. Moreover, successive oil/water separations could be realized on the device because whenever the oil/water mixture was poured into the glass tube, water would eventually reach the hydrogel-coated mesh at the bottom and penetrate through the mesh. The hydrogel-based separation system paved the avenue for the next generation of oil/water separators. Since its publication, many works related to superhydrophobic material-mediated oil/water separation have been reported,[36–38] including stimulus-responsive switching liquid adhesion for controlled oil/water separation[39] and robust materials with potential application in harsh environments.[40]

Figure 16.7 A hydrogel-coated copper mesh was demonstrated to be used as an oil/water separator with superior separation ability as the hydrogel superhydrophilicity improved the throughput of the separation system. (Adapted from ref. 35 with permission from John Wiley and Sons. Copyright © 2011 WILEY-VCH Verlag GmbH & Co. KGaA, Weinheim.)

16.5 Liquid Collection and Transport

16.5.1 Spider-silk-inspired Fog Collection System

Spider silk is a composite fibril material spun by spiders that is mainly comprised of fibrin protein. Inspired by morning dews hanging on spider silk (Figure 16.8A), researchers have developed artificial spider silk to collect water from fog.[41] Through careful observation, the authors found that there are nanoscale fibrins that assemble into a bundle of silk fiber, where the puffs are evenly distributed. In humid air, the puffs on the silk turn into spindle-knots in which the nano-fibrins are randomly distributed, whereas the rest of the section turns into joint sites between the spindle-knots, and the nano-fibrins are oriented in the direction of the silk axis. As the water droplets from fog make contact with the spider silk, they start to accumulate at all spots. As they grow larger, however, adjacent drops collide, which generates an excess of energy that drives those drops to move and stick to the spindle-knots. This process gives rise to the beautiful morning dews hanging on spider nets. Inspired by this efficient fog collection mechanism, Zheng, Bai, *et al.* proposed a method to fabricate artificial spider silk with the same spindle-knot structure as that of its natural counterpart. More importantly,

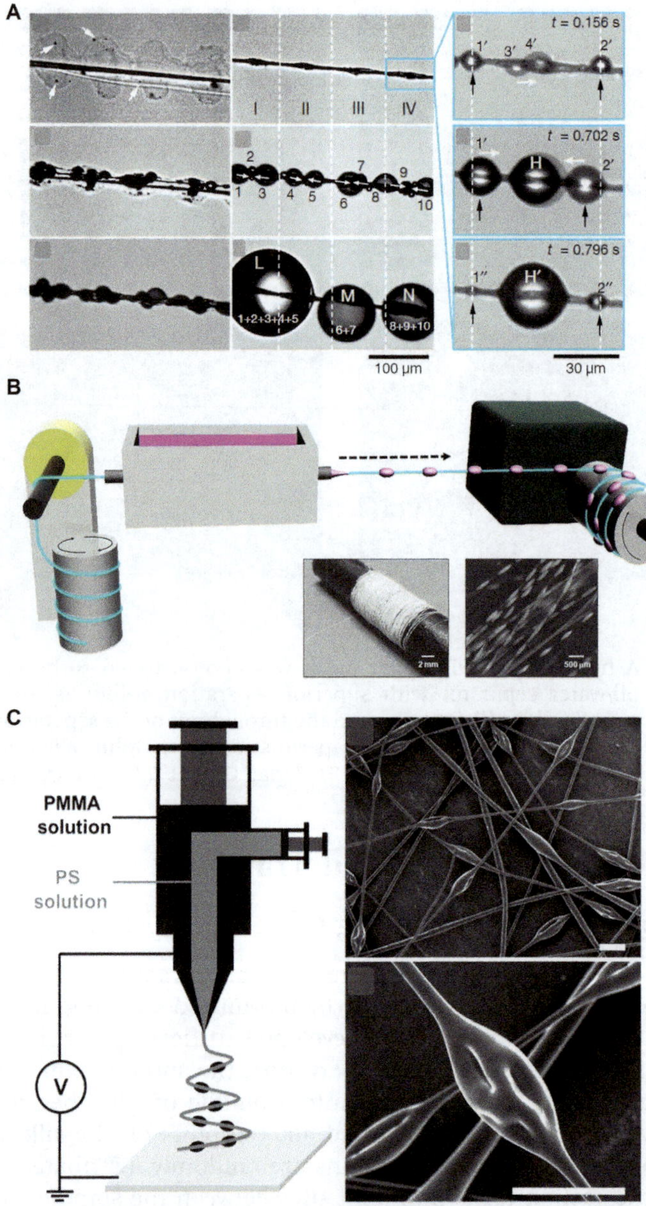

Figure 16.8 (A) Spider silk possesses the ability to collect, transport, and store water drops at the spindle knots. (Reprinted by permission from Macmillan Publishers Ltd: *Nature*,[41] Copyright (2010).) (B) A continuous production line was devised for the up-scale manufacture of artificial spider silk. (Adapted from ref. 43 with permission from John Wiley and Sons. Copyright © 2011 WILEY-VCH Verlag GmbH & Co. KGaA, Weinheim.) (C) The electrospinning technique is also a good way for the large scale production of artificial spider silk for various applications. (Adapted from ref. 49 with permission from John Wiley and Sons. Copyright © 2012 WILEY-VCH Verlag GmbH & Co. KGaA, Weinheim.)

the spindle-knot structure on the artificial silk fiber exhibited the same ability to collect fog water, and therefore provided a new way to collect and preserve water resources from humid air. The spindle-knots were prepared utilizing the Rayleigh instability of a thin liquid (an organic solution of poly(methyl methacrylate) (PMMA)) covering the fiber surface.[42]

As shown in Figure 16.8B, the authors later perfected this strategy into a whole system for large-scale manufacture of artificial spider silk.[43] Although the water transport on natural and artificial spider silk is uniformly directed to the spindle-knots, there is also a mechanism to reverse this tendency, *i.e.*, the water drops can be transported away from the spindle-knots. The key is to control the surface wettability of the spindle-knots. On the conventional silk composed by fibrin protein or PMMA, the spindle-knots are hydrophilic. Bai *et al.* showed that when the spindle-knots are made of polystyrene (PS), which is a typical hydrophobic material, the water drops captured from fog stay away from the spindle-knots.[44] According to the theory, the direction of the Laplace pressure experienced by a water drop on a conical surface depends on the wetting state,[45] and the PS spindle-knots give a Laplace pressure that drives the colliding water away from the knots. This result has enriched our knowledge about the directional water transport on spider-silk-inspired fibril materials.

It is worth mentioning that the spindle-knot structure on fibril materials has also inspired applications other than water collection. Bai *et al.* reported that the special curvature arrangement between adjacent spindle-knots forms a claw-like morphology that favors the capturing of large water drops.[46] To release the drops, one only needs to relax the formerly stringent fiber, which results in a morphology change that no longer favors the attachment of the large water drop. According to a review by the same authors,[47] the spindle-knot structures are also able to generate long-range directional water transport by applying a shape gradient or a chemical gradient to the spindle-knots. As reported by Chen *et al.*, long-range water transport was realized by improving the original method to prepare artificial spider silk.[48] The authors obliquely pulled the polymer strings out of the bath of a polymer organic solution. Due to the inclination of the string, a gradient in thickness of the solution layer was formed on the string. As a result, a gradient in size of the spindle-knots was generated along the fiber. Experimental results showed that the water droplets collected were transported across multiple spindle-knots due to their size gradient, providing long-range water transport.

Besides the pulling out method, electrospinning technology has also been used to fabricate artificial spider silk. Dong *et al.* prepared large scale artificial spider silk by coaxial electrospinning PMMA silk decorated by PS spindle-knots.[49] A major advantage of this technology is that the silk was then able to be assembled or knitted into a silk mat in one step, which makes it ready to be used to collect fog water or as a platform for lab-on-fiber catalytic studies.

16.5.2 Cactus-inspired Fog Collection System

Water crisis is an urgent problem faced by the modern world, which has drawn increasing attention of researchers. A special type of cacti living in the Chihuahua desert in Mexico, *Opuntia Microdasys*, has caught our eyes because of its exceptional ability to collect water from air.[50] Chihuahua desert has the typical climate where there is very little water fall in the whole year, however provides abundant water resources in the form of fog every morning. Therefore, creatures that are able to utilize this type of water are able to survive in the drought desert, and *Opuntia Microdasys* is one of these survivors.

To unravel the water collection mechanism, researchers took a close look at the micro and nano scale surface morphology of the plant. Surprisingly, heterogeneous structures on the spines of the cactus were spotted. Through fog simulations, it was found that different levels of structures have different roles during fog water collection (Figure 16.9A). Specifically, the micro barbs reversely distributed at the tip zone of the spines are the primary water droplet deposition sites because they penetrate the flow boundary which favors the collision of those tiny water droplets on to the cactus. Besides, the conical shapes of the barbs also act to transport water drops to the main body of the spines. On the spines, the barbs are oriented towards the opposite direction to that of the spine, which benefits the tip-to-base directional water transport. When the water drops are transported on the spines, micro-grooves aligning the axis of the spine can minimize the kinetic energy dissipation caused by transverse or rotational motion of the water drops. As the water drops are transported to the base of a spine cluster by Laplace pressure,[51] the superhydrophilic trichomes absorb the water into the succulent stem of the cactus. More importantly, after water is collected, the spine is ready to transport new water drops again, forming a continuous fog water collection system. This work has attracted wide attention from the world. Since its publication, the cactus fog collection system has become a research hot spot for the scientific community.

Ju *et al.* later investigated the influence of surface wettability on a single spine model.[52] They found that, on a hydrophilic spine, the volume of the water drop collected grows slowly but they move fast to the base of the spine whereas, on a hydrophobic spine surface, large water drops can be accumulated quickly on the tips, however it is difficult to transport them to the base. To combine the advantages of the two situations, the authors created a wettability gradient with the tip being hydrophobic and the rest being hydrophilic. The gradient was realized by immersing-and-withdrawing the conical copper spine repeatedly from a grafting solution. The resulting spine inherited both the fast transport feature from the hydrophilic spine and the large drop accumulation of the hydrophobic spine. Therefore, the spine with the wettability gradient performed significantly better than the previous two materials. Besides this study on a single spine, Ju *et al.* also tried to mimic the small spine arrays on the cactus stem.[53] The artificial spine arrays were

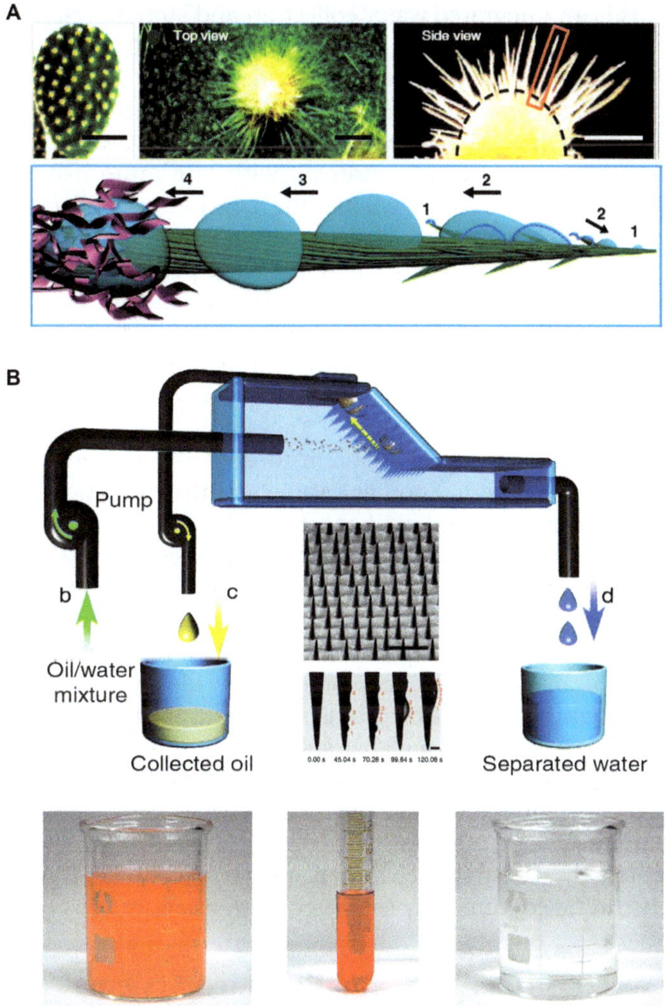

Figure 16.9 (A) The heterogeneous structures on spines of the cactus *Opuntia microdasys* were discovered to have an integrated fog water collection system. (Reprinted by permission from Macmillan Publishers Ltd: *Nat. Commun.*,[50] Copyright (2012).) (B) Inspired by the cactus spines' ability to collect water from fog, an artificial spine array was used to successfully collect micron-sized oil droplets from the flow of oil/water mixtures. (Reprinted by permission from Macmillan Publishers Ltd: *Nat. Commun.*,[54] Copyright (2013).)

made of poly(dimethylsiloxane) (PDMS) elastomers and prepared on a plane surface using the puncturing mold replication method developed by Li *et al.*[54] The spine arrays were in rectangular and hexagonal arrangements. The results showed that the hexagonal spine arrays exhibited a higher water collection efficiency, which was attributed to the maximized turbulence when the fog flow passed by the arrays.

The cactus is an integrated water collection and storage system. To mimic its whole functionalities, several details must be taken into account. First, spines with similar heterogeneous structures to the natural counterpart must be created. Second, the arrangement of spines into clusters should be presented so as to make the collection efficient. Third, a water reservoir that absorbs the collected water and stores it with minimum losses should be presented to store water and ensure the continuous collection process. Considering these aspects, Cao *et al.* assembled a cactus-inspired fog water collector.[55] The authors dispersed paramagnetic particles in a PDMS oligomer. In response to the magnetic field, the mixture formed a cluster of spines. After solidification of the PDMS spines, they were strapped on a clay sphere, whose interior was stuffed with superhydrophilic fibril cotton. The results showed that the cactus-inspired fog collector's efficiency reached 18 mL h^{-1}.

Water collection from fog is a system to extract a discrete phase from a continuous phase. There is another case where the extraction of a discrete phase from a continuous phase has practical application value, which is oil/water separation. The separation device mentioned above was founded on a mesh surface, and the lowest-sized oil droplets that could be collected strongly depended on the size of the mesh in use. Li *et al.* prepared a separation device that could separate oil droplets much smaller in size than those separated by the mesh separator.[54] The idea is similar to the fog collection mechanism, *i.e.*, the oil droplets are captured and transported by the spine arrays. PDMS arrays on a planar substrate were prepared *via* puncturing mold replication. As shown in Figure 16.9B, since the oil droplets are carried by the water flow and subjected to the tilted PDMS cone arrays, they collide on the cones, accumulate, and are transported to the cone base *via* the Laplace pressure difference. Using the separation apparatus shown in Figure 16.9B, the oil droplets can be continuously separated until clean water is obtained.

16.5.3 Water Strider Legs and Extended Systems

The ability to stand, run, and jump on free water surfaces makes the water strider a star in the scientific community. An early report on the ability of the water strider pointed out that the heterogeneous structures on its legs enable it to stay above water.[56] As shown in Figure 16.10A, despite the secreted wax coating on the water striders' legs, the finely distributed microsetaes together with the nanoscale grooves on those microsetaes on the legs are responsible for the excellent water-repellency[57] (Figure 16.10A).

In the past decade, many studies have been carried out to understand the details of the interaction between water and water strider legs. In Feng's work, the role of the nanogrooves on the microsetaes was emphasized.[58] It was theoretically simulated that the nanostructures imparted the legs with robust superhydrophobicity, which makes the pressure needed to drown the water strider to be as high as 750 dyn (over 60 times of the insect's body weight). With sufficient knowledge on the superhydrophobic legs, people

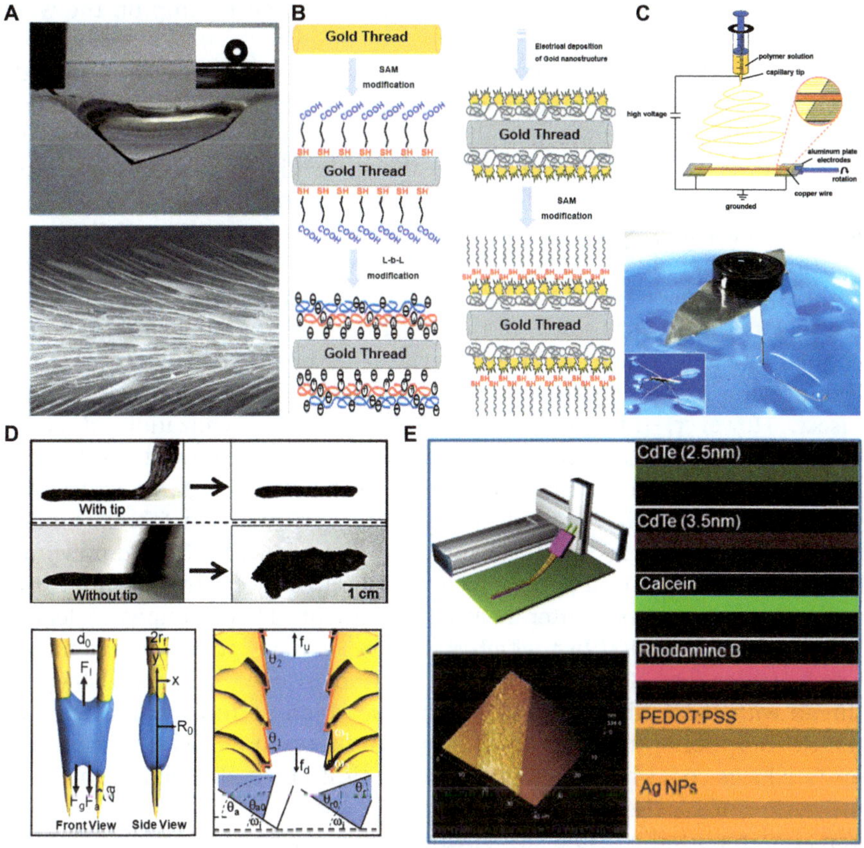

Figure 16.10 (A) The water-repellency of the water strider legs comes from the heterogeneous structures on the legs. (Reprinted by permission from Macmillan Publishers Ltd: *Nature*,[56] Copyright (2004).) (B) Through the layer-by-layer deposition technique, artificial water strider legs can be achieved. (Adapted from ref. 59 with permission from John Wiley and Sons. Copyright © 2005 WILEY-VCH Verlag GmbH & Co. KGaA, Weinheim.) (C) Electrospinning of polymer nanofilaments makes the surface morphology of artificial strider legs closer to their natural counterpart and allows artificial insects to stand on water. (Adapted with permission from F. Bai, J. Wu, G. Gong and L. Guo, *ACS Appl. Mater. Interfaces*, 2014, **6**, 16237–16242.[60] Copyright (2014) American Chemical Society.) (D) Research reveals the hydrostatic mechanism of Chinese brushes. (Adapted from ref. 64 with permission from John Wiley and Sons. Copyright © 2014 WILEY-VCH Verlag GmbH & Co. KGaA, Weinheim.) (E) Inspired by the Chinese brush, researchers invented a simple prototype of capillary pen to directly write nano lines. (Adapted with permission from Q. Wang, Q. Meng, P. Wang, H. Liu and L. Jiang, *ACS Nano*, 2015, **9**, 4362–4370.[65] Copyright (2015) American Chemical Society.)

started to design insect robots that can walk and even jump on the water surface.

To obtain a superhydrophobic artificial water strider leg, Shi *et al.* developed a combined technique which consisted first on covering the gold thread surface with poly(diallyldimethylammonium chloride) (PDDA) *via* layer-by-layer (L-b-L) deposition.[59] As shown in Figure 16.10B, polyelectrolytes were deposited on a gold thread surface by L-b-L deposition. Then, gold nanoparticles were sputtered onto the polyelectrolyte surface, followed by grafting alkyl chains onto the gold nanoparticle surface. According to the authors, the reason why L-b-L was utilized here is that this technique does not pick the geometric status of the substrate and so, the technique can be applied to a wide range of materials. The L-b-L-based artificial water strider legs provide a sophisticated method to mimic the behavior of the water insect. However, this method only aimed at the realization of superhydrophobicity on the metal thread material surface, without regard to true mimicking of the geometric features of the water strider legs. To achieve this goal, they later utilized the electrospinning technique to fabricate nanofibrin covered artificial legs.[60] As shown in Figure 16.10C, PS and polyimide (PI) were coaxially mixed to form a composite fiber by electrospinning. Since PI is thermally stable, after heating treatment, PS decomposes, leaving nanostreaks on the PI fiber, which resemble the nanogrooves on the natural water strider legs. Electrospinning was used to cover the metal (copper) thread with these fibers with nanogrooves, which perfectly mimic the geometric morphology of water strider legs. An insect model with four of these artificial legs could bear a load up to seven times its body weight. Due to the utilization of PI, the water-repellent ability showed high thermal stability. Yao *et al.* reported a stable superhydrophobic surface comprised of vertical ribbed nanoneedle arrays, which was inspired by the microsetaes on the water strider legs.[61] Due to the nanogrooves on those needles, the superhydrophobicity was so stable that a water drop squeezed on the surface would return to the original ultralow adhesion state after the pressure was unloaded. Notably, the authors set up a new standard to measure the superhydrophobicity of a solid surface, which is to test the water adhesion according to a load/unload normal pressure. A stable superhydrophobicity should display minimum unloading adhesion hysteresis compared to the force curve during the loading process. It is worth noting that, similar to the principle to achieve water-repellent properties by superhydrophobic materials, the idea for stay-on-oil function underwater based on superoleophobic materials is also valid. Liu *et al.* reported an artificial water strider leg prepared by a wet etching method, which endowed copper wires with nano-scale structures.[62] The insect model was able to "stand" on the oil surface in underwater conditions. Recently, Koh *et al.* designed a simple robot insect to mimic the jumping behavior of water striders on the water surface.[63] Through observation and analysis of the jumping process of water striders, it was found that, in order to achieve the maximum momentum transfer, the driving force functions at the arthropod tips should be smaller than a certain value regarding the water surface tension. As such, the arthropods never break or penetrate the water and perform the jumping motion since, when the water surface is penetrated, the upward force will be drastically reduced. Based on this principle, the authors successfully built an insect robot that was able to perform the jumping behavior on the water surface with optimized efficiency. Such a robotic design has a profound value for the design of semi-aquatic robots and facilities.

The extraordinary water repellency on water strider legs is not necessaryly limited to mimic insect motion. There are other application opportunities regarding what we can find during water condensation on their legs. The microsetaes are arranged on the leg in a manner by which the water drops accumulated there can only be transported away from the leg. That is, the microsetaes have the ability to prevent the leg from being wetted by water vapor or fog. On the contrary, the Chinese brush utilizes the equilibrium between gravity and the capillary force acting on the ink between the tip hairs so as to retain ink while performing smooth writing, as reported by Wang, *et al.*[64] Considering the capillary effect produced by the hair, the conical shape of the hair tips plays a key role. As shown in Figure 16.10D, the brushes are vertically held, the hydrophilic conical hairs generate an upward Laplace pressure which drives the liquid to remain at a high position. The gaps between adjacent hairs generate a secondary capillary force which also propels the liquid to stay at a high position. As a comparison, a Chinese brush without conical tips is not able to hold ink any more. This discovery inspired a novel water reservoir to be used to directly write micro and nanoscale patterns[65] (Figure 16.10E).

16.5.4 Water Transport on SLIPS and Organogels

A major contribution of the SLIPS is that it brings a new wetting state, the slippery state,[66] which both enables target liquids with good mobility and provides a relatively large contact area between the target and the SLIPS.

To turn the SLIPS into a platform for water transport and manipulation, researchers put efforts into extending SLIPS in different forms. Yao *et al.* proposed that, by integrating SLIPS with an elastomer substrate, the water adhesion on SLIPS can be tuned by simply stretching or bending the composite substrate.[67] As shown in Figure 16.11A, when the elastomer is relaxed, the PFPE liquid is evenly distributed on the porous surface, forming a smooth liquid surface, and water or oil droplets can slide on it under this condition. However, when the substrate is stretched, the pore size is enlarged, which gives rise to insufficient PFPE liquid on the SLIPS. Due to the capillary effect, the PFPE liquid retracts into the pores, exposing the porous surface of the solid substrate. The target liquid under this condition makes contact with the porous solid surface, thus experiences high adhesion due to Wenzel state contact. More importantly, it is the reversibility of the slipper/Wenzel state transition that makes the current SLIPS system a good platform for the manipulation of liquid drops.

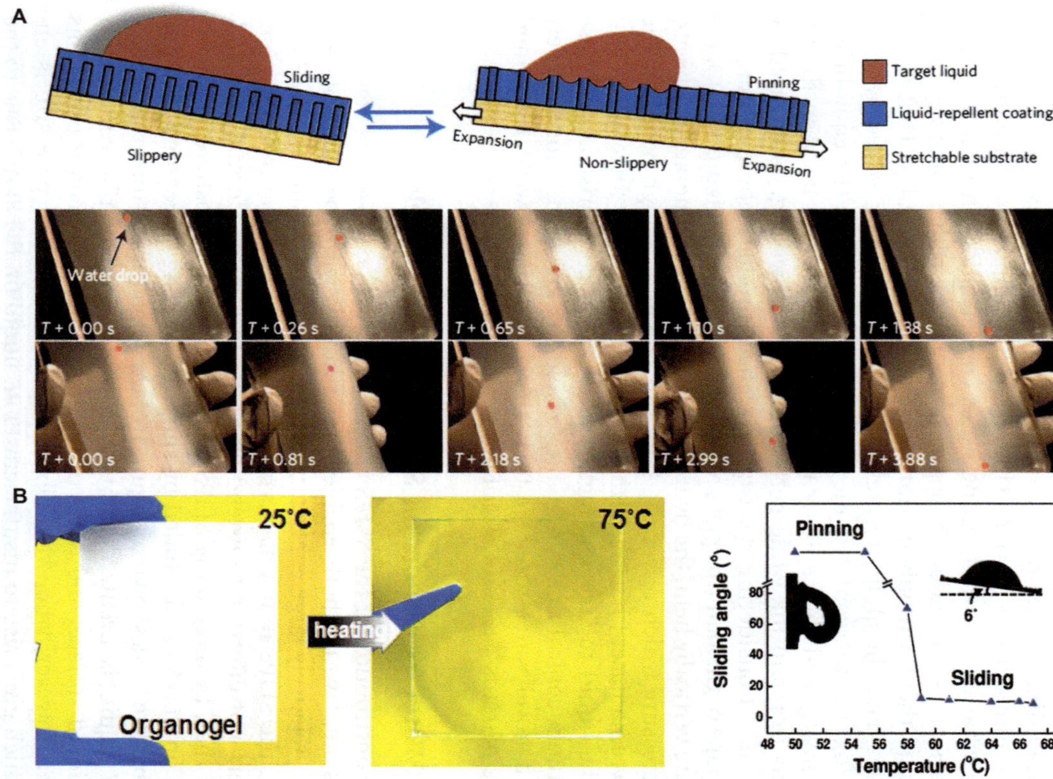

Figure 16.11 (A) SLIPS based on an elastic substrate was utilized to manipulate the drop motion. (Reprinted by permission from Macmillan Publishers Ltd: *Nature Mater.*,[67] Copyright (2013).) (B) A phase-change organogel is able to reversible switch water adhesion in response to the temperature. (Adapted from ref. 68 with permission from John Wiley and Sons. Copyright © 2013 WILEY-VCH Verlag GmbH & Co. KGaA, Weinheim.)

Similar to the original SLIPS system, organogel materials are also capable of generating ultralow adhesion of liquids by virtue of slippery lubrication. What distinguishes this from the SLIPS is that, on an organogel surface, no pre-designed micro or nanostructures are needed. Because organogels are a type of oil swollen crosslinked polymer network, the solid adaptability is superior to the conventional porous solid in the following aspects. First, organogels contain larger amounts of oil, which is beneficial for long term applications. Second, the elasticity makes the organogels shrink as the lubricant is depleted, squeezing more oil to the surface while returning to a smaller pore size. Third, organogels are initially elastic and ready to fit into complex shapes.

Yao *et al.* developed an organogel material by immersing PDMS rubber in a hot bath of melted paraffin wax.[68] Since PDMS shows a certain affinity towards paraffin wax, it becomes swollen, forming an organogel. After taking the organogel out to room temperature, the paraffin wax inside the organogel network solidifies. The gel in this condition is opaque and with the same color as paraffin wax whereas, at a temperature higher than the melting point of the wax, the organogel becomes fully transparent (Figure 16.11B). Moreover, the water adhesion on the organogel surface varies with the temperature. As seen in Figure 16.11B, at low temperature, water is retained firmly on the surface even when the surface is held vertically. However, water demonstrated high mobility, like moving on a SLIPS, at temperatures above the melting point. The working mechanism of such adhesion switching is related to the Wenzel/slippery wetting transition in response to the temperature. Since paraffin wax exhibits a reversible phase transition according to ambient heat, water manipulation on the organogel can be controlled in multiple cycles. In addition, the transparency change mentioned above has potential applications in smart thermal sensitive windows and lenses.

To investigate the influence of surface structures on the organogel surface wetting properties, Zhang prepared organogels with micro-grooved structures through mold-replication.[69] Different from the aforementioned organogel, the current one was prepared from a poly(butyl methacrylate-*co*-lauryl methacrylate) (P(BMA-*co*-LMA)) crosslinked network swollen by silicone oil. As mentioned above, no structures are needed to achieve a slippery state on the organogel surface. In fact, it was shown in this study that the mobility of liquids on the organogels is compromised if there are structures present on its surface. First, water drops slide freely on a non-structured organogel surface. Second, if there are pillar arrays on the organogel, water drops cannot slide when the surface is slightly tilted. Third, if there are microgrooves and they are orthogonal to the sliding direction of the water drop, the drops will not slide. And finally, only if the microgrooves align with the sliding direction, the drops will slide. Therefore, similar to the Wenzel state contact between a liquid and a solid surface, the mobility of water depends on the fluctuation of the triple phase contact lines: the less fluctuation, the higher the mobility.[70] To make use of these conclusions, the authors demonstrated controlled drop sliding motion by stretching the upper or lower part of the organogel with microgrooves aligned with the sliding direction. The sliding motion of the water drops could be stopped by stretching the lower part of the organogel and resumed by stretching the upper part. This could also be interpreted as the lubricant layer having broken up into discrete drops by stretching, stopping the liquid drop sliding motion by destroying the slippery state. Further investigations on this issue should be carried out.

16.6 Superwetting Polymers in Functional Nanochannels and Nanopores

The manipulation of substance transport across biological interfaces is crucially important for the proper functionality of living cells. During this process, the main access to substance transport is through the biological channels in the proteins that are distributed across the cell membrane.[71] The proteins utilize different strategies to control the entrance and exit of substances to the cell entity, inspired by which, people have developed many different nanochannels and nanopores to regulate substance diffusion and transport. In this section, we focus on the applications of superwetting polymers in nanochannels.

16.6.1 Superwetting Polymers in Nanochannels to Control Ion Transport

To prepare membranes with artificial nanochannels, one can start with several polymer materials such as polyimide (PI),[72] polycarbonate (PC),[73] and polyethylene terephthalate (PET).[74] A prerequisite to apply these kinds of polymers for nanochannel fabrication is that they are easy to process and susceptible to different etching techniques to produce nano-sized channels of different morphologies. Current methods for this purpose include ion-track-etching,[75] electrochemical etching,[76] and laser techniques.[77] Once the nanochannels have been produced, one needs to functionalize the channel inner wall to obtain different functions. Following are some examples of polymer decorated nanochannels.

As reported by Hou *et al.* G-Quadruplex (G4) DNA molecules were immobilized on the inner wall of a conical nanochannel prepared through an ion-track-etching technique.[78] The G4-DNA exhibited conformational switching with the presence of potassium ions. Consequently, this G4-DNA-incorporated nanochannel responded to the presence of potassium ions in a certain concentration range. The most significant advance of this work was to closely imitate the biological process in a confined space, which provides a new platform for studying ion transport behavior in living organisms.[79] According to the authors, this strategy may also boost the development of biosensors, molecular sieves, and nanofluidic devices.

An early attempt on stimulus-responsive polymer-incorporated nanochannels was conducted by Yameen *et al.* The authors started with PI microfilms at the thickness of ~12 μm.[72] The conical nanochannels were fabricated through the ion-etching technique as mentioned above. In the nanochannels, the PNIPAAm polymer brushes were produced *via* surface-initiated atom transfer radical polymerization. As PNIPAAm responds to thermal stimuli in the form of extension and retraction, the functionalized nanochannels can be opened/closed according to the environment temperature.

Hou *et al.* tried dual-responsive surface decoration on either side of an hour-glass-shaped single nanochannel.[80] As seen in Figure 16.12A, superior to a single responsive nanochannel, the current nanochannel has an hour-glass shape which is one step closer to the biological ion channels with sophisticated morphologies.[71] The asymmetric decoration of the channel inner walls also enriched the responding patterns. Zhang *et al.* prepared a cigar-shaped nanochannel that was decorated with different stimulus-responsive polymers, poly(acrylic acid) (PAA) and poly(vinylpyridine) (PVP) on either end of the channel.[81] PAA and PVP perform opposite conformational changes in response to pH. As a result, the PAA-cigar-PVP channel exhibits three different types of response according to symmetric and asymmetric external stimuli (Figure 16.12B). Double gating is the most prominent feature of the current nanochannel system, which endows a unique property never realized before, that is, the fail-safe ion gate mechanism, which closely mimicks the fail-safe ion pumping feature in living organisms. Specifically, when one gate is opened, certain ions are allowed to enter the channel. Then, the gate closes, creating a sealed chamber. Finally, the other gate is opened on demand, releasing the formerly trapped ions to the medium on the other side. PAA and PVP are certain types of stimulus-responsive materials that display opposite charges according to external pH, which essentially promote unidirectional transport of ions with opposite charges; the fail-safe ion pumping provides an extra mechanism to ensure the pumping direction according to specific needs.

Besides the decoration inside the nanochannels, the crosslinked polymer layer outside the nanochannels is able to enhance, weaken, or neutralize the intrinsic ion selectivity. As reported by Wang *et al.* the agar hydrogel was incorporated to either or both sides of the conical nanochannel on a PET thin film.[82] Upon application of a voltage bias, the negatively charged nanochannel initially has the ability to generate directed ion transfer. By incorporating the hydrogel, which is also negatively charged, a different situation arises. When the hydrogel is attached to the thinner side of the channel, it enhances the transfer of positively charged ions to the broader side of the channel, because it provides an accumulated ion resource to transfer from the thinner side. However, when the hydrogel is attached to the broad side of the channel, it reduces its transfer ability. This is probably because, when the positively charged ions are transferred out of the nanochannel, they are absorbed through electrostatic interactions into the

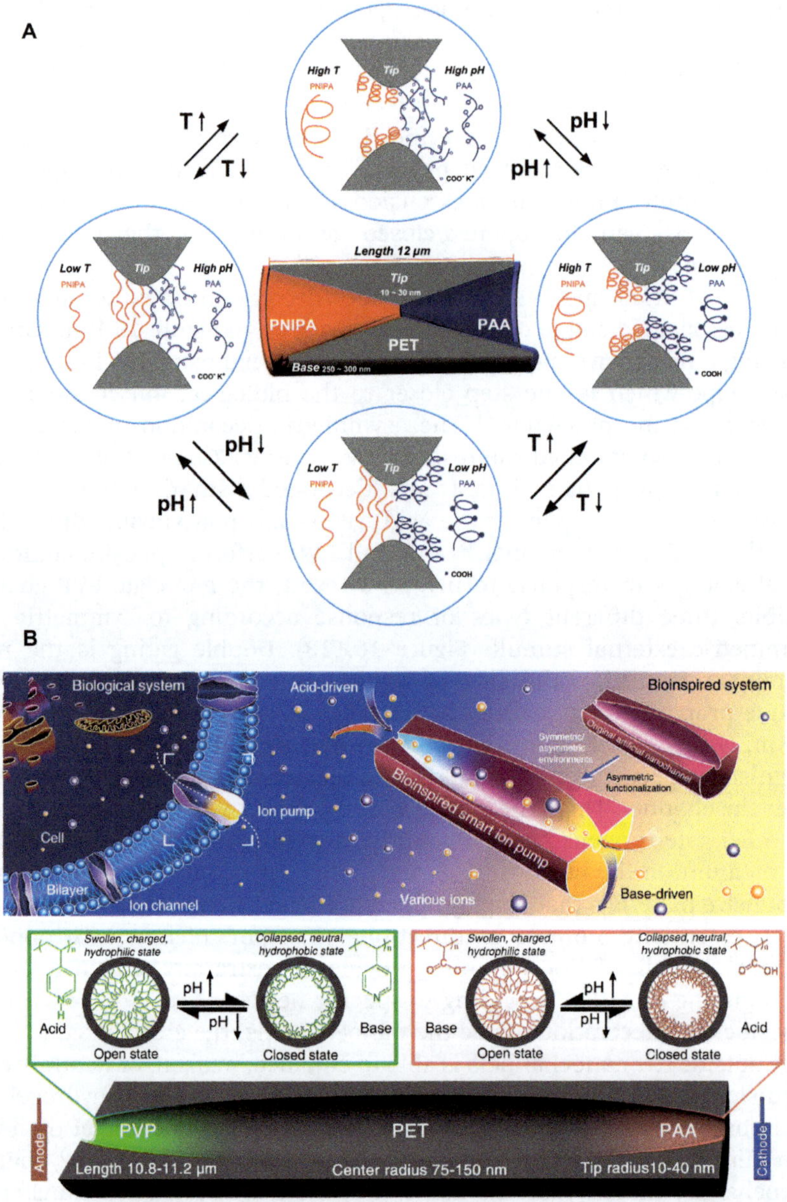

Figure 16.12 (A) Hour-glass shaped nanochannels that incorporate dual stimulus-responsive polymers exhibit efficient ion gating properties. (Adapted from ref. 79 with permission from John Wiley and Sons. Copyright © 2012 WILEY-VCH Verlag GmbH & Co. KGaA, Weinheim.) (B) Opposite termini of a cigar-shaped nanochannel decorated with a pair of stimulus-responsive polymers, poly(acrylic acid) PAA and poly(vinyl-pyridine) (PVP), which respond oppositely to pH changes. This biomimetic pore gives rise to diverse ion gating patterns. (Adapted with permission from H. Zhang, X. Hou, L. Zeng, F. Yang, L. Li, D. Yan, Y. Tian and L. Jiang, *J. Am. Chem. Soc.*, 2013, **135**, 16102–16110.[81] Copyright (2013) American Chemical Society.)

negatively charged hydrogel network, which weakens the detection signal of the electrode at the outer surface of the hydrogel layer. When the hydrogels are attached on both sides of the nanochannel, the direction transfer of ions is neutralized, because the ion transporting rate through the nanochannel is no longer the decisive factor that affects the output data from the detector but, rather, the ion transfer through the hydrogel layer on both sides will ultimately report to the electrodes. Since the negatively charged hydrogels are homogeneous everywhere in the network, they possess no transporting preference through their bulk entities.

16.6.2 Superwetting Polymers in Nanochannels for Energy Conversion

Generating electricity through nanochannels is possible because, when a liquid is driven to pass through such channels, a net electrical current and electric potential is generated and accumulated at the end of the channel.[83] As estimated by van der Heyden and coworkers, a maximum electric power can be converted from hydrostatic energy at low salt concentrations. With all the existent theoretical conclusions and deductions, the authors hypothesized a device which could be used to generate electricity through electrokinetic propulsion. For a 10×10 cm^2 effective area, the energy conversion efficiency could be up to 50% and the device could achieve 2.6 W of energy output from an input pressure of 5 bar.

There is an example of photoelectric energy conversion from nature, the retina in our eyes and its mechanism to transfer light into bioelectric signals to our brains. The key substance during the energy conversion process is rhodopsin, a light-sensitive receptor protein which decomposes upon light irradiation, opening a cation channel. It has inspired people to grasp the core functional entity of an energy conversion system, the photosensitive molecules that reversibly release and capture ions in response to light irradiation. Wen *et al.* used the malachite green carbinol base (MGCB), which releases hydroxyl ions and becomes positively charged when being shone upon in the first chamber with an electrode.[84] Here, the conical shaped nanochannels in the ultrathin polymer film play a key role in separating a second chamber from the first one, and to translocate the as-produced hydroxyl ions to the second chamber that contains the other electrode. When the nanochannels manage to separate the oppositely charged ions into the two chambers, the two electrodes are able to generate an electric current in an outside circuit. According to the authors, the energy conversion efficiency of the apparatus was up to 36% based on the calculation of the Gibbs free energy diffusion, which warrants its application for future energy demands. Notably, the authors also tried 8-hydroxylpyrene-1,3,6-trisulfonate (HA), a photo-sensitive molecule, as the source of counterions.[85]

Inspired by electric eels, Guo *et al.* developed a graphene-hydrogel membrane (GHM), which was negatively charged and allowed only cations to

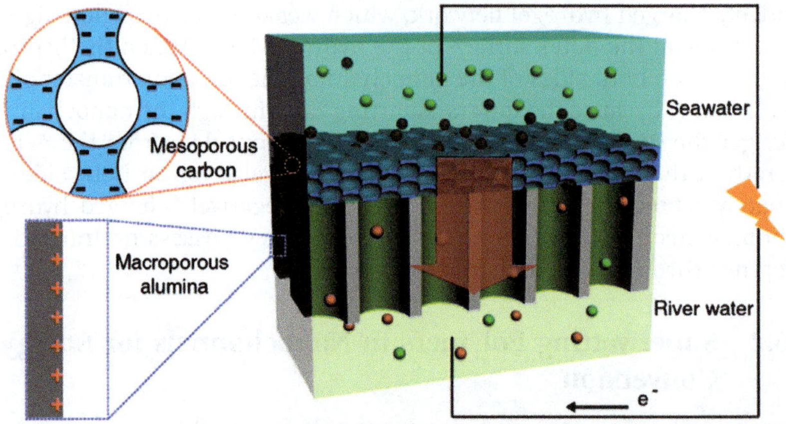

Figure 16.13 Electric power generation was realized with an ionic diode membrane that comprised a heterogeneous junction between microporous carbon and macroporous aluminum. The membrane harvested energy from aqueous media with a salinity gradient with high efficiency. (Adapted with permission from J. Gao, W. Guo, D. Feng, H. Wang, D. Zhao and L. Jiang, *J. Am. Chem. Soc.*, 2014, **136**, 12265–12272.[87] Copyright (2014) American Chemical Society.)

pass through the membrane upon pressure.[86] This pressure driven process induced a charge distribution between the sides of the GHM, which was able to generate electric power for *in situ* application in biosensing and micromachinery. This device protocol is particularly promising to harness energy resources in constant flows or alternate flows, such as the blood flow in our bodies.

Recently, Gao *et al.* reported an ion diode membrane (IDM) system to generate electric power from a salinity concentration gradient in aqueous media.[87] As shown in Figure 16.13, the idea was to create a heterojunction between mesoporous carbon (MesoC) (~7 nm of pore size and negatively charged) and macroporous aluminum (MacroA) (~80 nm in pore size and positively charged). Upon subjecting salinized water to the MesoC side, due to electrostatic interactions, the cations were induced to diffuse through the layer. The opposite situation occurred to the opposite side of the IDM, that is, the anions were attracted and allowed to penetrate through the MacroA membrane. However, due to the size difference between MacroA and MesoC, as well as the existence of a saline concentration gradient stronger at the MesoC side and weaker at the other side, the anion translocation on the MacroA side was much weaker than at the opposite side. As a result, a net translocation of ions through the IDM was generated, which was converted to electric power. In fact, the optimized IDM system obtained an energy output of 3.46 W m^{-2}, which outperforms some state-of-the-art commercially available ion exchange membranes. It is worth mentioning that the utilization of MacroA also helped to prevent concentration polarization, which also contributed to the high-efficiency energy output.

16.7 Concluding Remarks

In this chapter, we introduced different types of superwetting polymer surfaces according to their different application fields, from scientific research to industry. Thanks to the vast development of polymer syntheses and processings, more and more advanced functions are introduced into superwetting polymer surfaces. The ever increasing knowledge about the interfacial properties of polymer materials allows people to integrate multiple functions in a single polymer material system and to make major breakthroughs in the scientific realm, which was once difficult to achieve. We hope that, through the introduction from this chapter, the reader will have a good comprehension of the subject of superwettability on polymer surfaces and may be able to extend their application to more and more fields in science and the daily life.

References

1. A. B. D. Cassie and S. Baxter, *Trans. Faraday Soc.*, 1944, **40**, 546–551.
2. A. B. D. Cassie and S. Baxter, *Nature*, 1945, **155**, 21–22.
3. D. Quéré, *Annu. Rev. Mater. Res.*, 2008, **38**, 71–99.
4. Y.-T. Cheng and D. E. Rodak, *Appl. Phys. Lett.*, 2005, **86**, 144101.
5. W. Barthlott and C. Neinhuis, *Planta*, 1997, **202**, 1–8.
6. L. Feng, S. Li, Y. Li, H. Li, L. Zhang, J. Zhai, Y. Song, B. Liu, L. Jiang and D. Zhu, *Adv. Mater.*, 2002, **14**, 1857–1860.
7. G. H. Findenegg and S. Herminghaus, *Curr. Opin. Colloid Interface Sci.*, 1997, **2**, 301–307.
8. P.-G. d. Gennes, F. Brochard-Wyart and D. Quéré, *Capillarity and Wetting Phenomena: Drops, Bubbles, Pearls, Waves*, Springer, New York, 2004.
9. M. Nosonovsky and B. Bhushan, *Curr. Opin. Colloid Interface Sci.*, 2009, **14**, 270–280.
10. L. Feng, Y. Zhang, J. Xi, Y. Zhu, N. Wang, F. Xia and L. Jiang, *Langmuir*, 2008, **24**, 4114–4119.
11. S. Yang, J. Ju, Y. Qiu, Y. He, X. Wang, S. Dou, K. Liu and L. Jiang, *Small*, 2014, **10**, 294–299.
12. T. Wong, S. Kang, S. K. Y. Tang, E. J. Smythe, B. D. Hatton, A. Grinthal and J. Aizenberg, *Nature*, 2011, **477**, 443–447.
13. H. F. Bohn and W. Federle, *Proc. Natl. Acad. Sci. U. S. A.*, 2004, **101**, 14138–14143.
14. U. Bauer and W. Federle, *Plant Signaling Behav.*, 2009, **4**, 1019–1023.
15. A. Lafuma and D. Quéré, *Europhys. Lett.*, 2011, **96**, 56001.
16. H. Liu, P. Zhang, M. Liu, S. Wang and L. Jiang, *Adv. Mater.*, 2013, **25**, 4477–4481.
17. X. Yao, S. S. Dunn, P. Kim, M. Duffy, J. Alvarenga and J. Aizenberg, *Angew. Chem., Int. Ed.*, 2014, **53**, 4418–4422.
18. X. Yao, S. Wu, L. Chen, J. Ju, Z. Gu, M. Liu, J. Wang and L. Jiang, *Angew. Chem., Int. Ed.*, 2015, **54**, 8975–8979.

19. M. Liu, S. Wang, Z. Wei, Y. Song and L. Jiang, *Adv. Mater.*, 2009, **21**, 665–669.
20. T. Ikoma, H. Kobayashi, J. Tanaka, D. Walsh and S. Mann, *Int. J. Biol. Macromol.*, 2003, **32**, 199–204.
21. L. Lin, M. Liu, L. Chen, P. Chen, J. Ma, D. Han and L. Jiang, *Adv. Mater.*, 2010, **22**, 4826–4830.
22. L. Chen, M. Liu, L. Lin, T. Zhang, J. Ma, Y. Song and L. Jiang, *Soft Matter*, 2010, **6**, 2708–2712.
23. T. Sun, G. Wang, L. Feng, B. Liu, Y. Ma, L. Jiang and D. Zhu, *Angew. Chem., Int. Ed.*, 2004, **116**, 361–364.
24. J. B. Williams, C. Clarkson, C. Mant, A. Drinkwater and E. May, *Water Res.*, 2012, **46**, 6319–6328.
25. C. Huang, G. Yang, Q. Ha, J. Meng and S. Wang, *Adv. Mater.*, 2015, **27**, 310–313.
26. X. Liu, L. Chen, H. Liu, G. Yang, P. Zhang, D. Han, S. Wang and L. Jiang, *NPG Asia Mater.*, 2013, **5**, e63.
27. X. Hong-Quan, Z. Wei and X. Dong, *J. Appl. Polym. Sci.*, 2008, **107**, 2681–2688.
28. T. Wan, J. Yao and X. Ma, *J. Appl. Polym. Sci.*, 2008, **110**, 3859–3864.
29. Q. Tang, J. Wu, J. Lin, S. Fan and D. Hu, *J. Mater. Res.*, 2009, **24**, 1653–1661.
30. T. Ono, M. Ohta, K. Iseda and K. Sada, *Soft Matter*, 2012, **8**, 3700–3704.
31. E. D. Sloan Jr., *Nature*, 2003, **426**, 353–363.
32. I. A. Pecher, *Nature*, 2002, **420**, 622–623.
33. X. Chen, Y. Wu, B. Su, J. Wang, Y. Song and L. Jiang, *Adv. Mater.*, 2012, **24**, 5884–5889.
34. L. Feng, Z. Zhang, Z. Mai, Y. Ma, B. Liu, L. Jiang and D. Zhu, *Angew. Chem., Int. Ed.*, 2004, **43**, 2012–2014.
35. Z. Xue, S. Wang, L. Lin, L. Chen, M. Liu, L. Feng and L. Jiang, *Adv. Mater.*, 2011, **23**, 4270–4273.
36. F. Lu, Y. Chen, N. Liu, Y. Cao, L. Xu, Y. Wei and L. Feng, *RSC Adv.*, 2014, **4**, 32544–32548.
37. W. Zhang, Y. Cao, N. Liu, Y. Chen and L. Feng, *RSC Adv.*, 2014, **4**, 51404–51410.
38. Y. Cao, Y. Chen, N. Liu, X. Lin, L. Feng and Y. Wei, *J. Mater. Chem. A*, 2014, **2**, 20439–20443.
39. Z. Xue, Z. Sun, Y. Cao, Y. Chen, L. Tao, K. Li, L. Feng, Q. Fu and Y. Wei, *RSC Adv.*, 2013, **3**, 23432.
40. Z. Xue, Y. Cao, N. Liu, L. Feng and L. Jiang, *J. Mater. Chem. A*, 2014, **2**, 2445–2460.
41. Y. Zheng, H. Bai, Z. Huang, X. Tian, F. Q. Nie, Y. Zhao, J. Zhai and L. Jiang, *Nature*, 2010, **463**, 640–643.
42. H. Bai, J. Ju, R. Sun, Y. Chen, Y. Zheng and L. Jiang, *Adv. Mater.*, 2011, **23**, 3708–3711.
43. H. Bai, R. Sun, J. Ju, X. Yao, Y. Zheng and L. Jiang, *Small*, 2011, **7**, 3429–3433.

44. H. Bai, X. Tian, Y. Zheng, J. Ju, Y. Zhao and L. Jiang, *Adv. Mater.*, 2010, **22**, 5521–5525.
45. S. Michielsen, J. Zhang, J. Du and H. J. Lee, *Langmuir*, 2011, **27**, 11867–11872.
46. X. Tian, Y. Chen, Y. Zheng, H. Bai and L. Jiang, *Adv. Mater.*, 2011, **23**, 5486–5491.
47. H. Bai, J. Ju, Y. Zheng and L. Jiang, *Adv. Mater.*, 2012, **24**, 2786–2791.
48. Y. Chen, L. Wang, Y. Xue, L. Jiang and Y. Zheng, *Sci. Rep.*, 2013, **3**, 2927.
49. H. Dong, N. Wang, L. Wang, H. Bai, J. Wu, Y. Zheng, Y. Zhao and L. Jiang, *Chemphyschem*, 2012, **13**, 1153–1156.
50. J. Ju, H. Bai, Y. Zheng, T. Zhao, R. Fang and L. Jiang, *Nat. Commun.*, 2012, **3**, 1247.
51. D. Quéré, J.-M. D. Meglio and F. Brochard-Wyart, *Science*, 1990, **249**, 1256–1260.
52. J. Ju, K. Xiao, X. Yao, H. Bai and L. Jiang, *Adv. Mater.*, 2013, **25**, 5937–5942.
53. J. Ju, X. Yao, S. Yang, L. Wang, R. Sun, Y. He and L. Jiang, *Adv. Funct. Mater.*, 2014, **24**, 6933–6938.
54. K. Li, J. Ju, Z. Xue, J. Ma, L. Feng, S. Gao and L. Jiang, *Nat. Commun.*, 2013, **4**, 2276.
55. M. Cao, J. Ju, K. Li, S. Dou, K. Liu and L. Jiang, *Adv. Funct. Mater.*, 2014, **24**, 3235–3240.
56. X. F. Gao and L. Jiang, *Nature*, 2004, **432**, 36.
57. F. Shi, J. Niu, J. Liu, F. Liu, Z. Wang, X. Q. Feng and X. Zhang, *Adv. Mater.*, 2007, **19**, 2257–2261.
58. X. Q. Feng, X. Gao, Z. Wu, L. Jiang and Q. S. Zheng, *Langmuir*, 2007, **23**, 4892–4896.
59. F. Shi, Z. Wang and X. Zhang, *Adv. Mater.*, 2005, **8**, 1005–1009.
60. F. Bai, J. Wu, G. Gong and L. Guo, *ACS Appl. Mater. Interfaces*, 2014, **6**, 16237–16242.
61. X. Yao, Q. Chen, L. Xu, Q. Li, Y. Song, X. Gao, D. Quere and L. Jiang, *Adv. Funct. Mater.*, 2010, **20**, 656–662.
62. X. Liu, J. Gao, Z. Xue, L. Chen, L. Lin, L. Jiang and S. Wang, *ACS Nano*, 2012, **6**, 5614–5620.
63. J.-S. Koh, E. Yang, G.-P. Jung, S.-P. Jung, J. H. Son, S.-I. Lee, P. G. Jablonski, R. J. Wood, H.-Y. Kim and K.-J. Cho, *Science*, 2015, **349**, 517–521.
64. Q. Wang, B. Su, H. Liu and L. Jiang, *Adv. Mater.*, 2014, **26**, 4889–4894.
65. Q. Wang, Q. Meng, P. Wang, H. Liu and L. Jiang, *ACS Nano*, 2015, **9**, 4362–4370.
66. T. Wong, T. Sun, L. Feng and J. Aizenberg, *MRS Bull.*, 2013, **38**, 366–371.
67. X. Yao, Y. Hu, A. Grinthal, T. Wong, L. Mahadevan and J. Aizenberg, *Nat. Mater.*, 2013, **12**, 529–534.
68. X. Yao, J. Ju, S. Yang, J. Wang and L. Jiang, *Adv. Mater.*, 2014, **26**, 1895–1900.

69. P. Zhang, H. Liu, J. Meng, G. Yang, X. Liu, S. Wang and L. Jiang, *Adv. Mater.*, 2014, **26**, 3131–3135.
70. L. Gao, A. Y. Fadeev and T. J. McCarthy, *MRS Bull.*, 2008, **33**, 747–751.
71. E. Gouaux and R. Mackinnon, *Science*, 2005, **310**, 1461–1465.
72. B. Yameen, M. Ali, R. Neumann, W. Ensinger, W. Knoll and O. Azzaroni, *Small*, 2009, **5**, 1287–1291.
73. C. C. Harrell, P. Kohli, Z. Siwy and C. R. Martin, *J. Am. Chem. Soc.*, 2004, **126**, 15646–15647.
74. W. Guo, H. W. Xia, F. Xia, X. Hou, L. X. Cao, L. Wang, J. M. Xue, G. Z. Zhang, Y. L. Song, D. B. Zhu, Y. G. Wang and L. Jiang, *Chemphyschem*, 2010, **11**, 859–864.
75. P. Y. Apel, Y. E. Korchev, Z. Siwy, R. Spohr and M. Yoshida, *Nucl. Instrum. Methods Phys. Res., Sect. B*, 2001, **184**, 337–346.
76. S. E. Létant, T. W. v. Buuren and L. J. Terminello, *Nano Lett.*, 2004, **4**, 1705–1707.
77. S. I. Kudryashov, G. Mourou, A. Joglekar, J. F. Herbstman and A. J. Hunt, *Appl. phys. lett.*, 2007, **91**, 141111.
78. X. Hou, W. Guo, F. Xia, F. Q. Nie, H. Dong, Y. Tian, L. Wen, L. Wang, L. Cao, Y. Yang, J. Xue, Y. Song, Y. Wang, D. Liu and L. Jiang, *J. Am. Chem. Soc.*, 2009, **131**, 7800–7805.
79. X. Hou, H. Zhang and L. Jiang, *Angew. Chem., Int. Ed.*, 2012, **51**, 5296–5307.
80. X. Hou, F. Yang, L. Li, Y. Song, L. Jiang and D. Zhu, *J. Am. Chem. Soc.*, 2010, **132**, 11736–11742.
81. H. Zhang, X. Hou, L. Zeng, F. Yang, L. Li, D. Yan, Y. Tian and L. Jiang, *J. Am. Chem. Soc.*, 2013, **135**, 16102–16110.
82. L. Wang, H. Zhang, Z. Yang, J. Zhou, L. Wen, L. Li and L. Jiang, *Phys. Chem. Chem. Phys.*, 2015, **17**, 6367–6373.
83. F. H. J. v. d. Heyden, D. J. Bonthuis, D. Stein, C. Meyer and C. Dekker, *Nano Lett.*, 2006, **6**, 2232–2237.
84. L. Wen, X. Hou, Y. Tian, J. Zhai and L. Jiang, *Adv. Funct. Mater.*, 2010, **20**, 2636–2642.
85. L. Wen, Y. Tian, Y. Guo, J. Ma, W. Liu and L. Jiang, *Adv. Funct. Mater.*, 2013, **23**, 2887–2893.
86. W. Guo, C. Cheng, Y. Wu, Y. Jiang, J. Gao, D. Li and L. Jiang, *Adv. Mater.*, 2013, **25**, 6064–6068.
87. J. Gao, W. Guo, D. Feng, H. Wang, D. Zhao and L. Jiang, *J. Am. Chem. Soc.*, 2014, **136**, 12265–12272.

CHAPTER 17

Bio-mimetic Structural Colour using Biopolymers

ROX MIDDLETON,[a] ULLRICH STEINER[b] AND SILVIA VIGNOLINI*[a]

[a] Department of Chemistry, University of Cambridge, CB2 1EW, UK
[b] Adolphe Merkle Institute, CH-1700 Fribourg, Switzerland
*Email: sv319@cam.ac.uk

17.1 Introduction

The most striking and brilliant colours in nature are obtained by structuring materials on the scale of the wavelength of light,[1] especially when these are combined with absorbing pigments. Such colourations, called structural, are the result of the constructive interference of reflected light with periodically organised interfaces within nanostructured materials. This means colour can be created from any material, transforming otherwise transparent or slightly absorbing polymers and composites into evolutionary assets of shimmering colours and hues.

Natural polymers are assembled in an incredible variety of complex architectures in the plant and animal kingdom.[2] The diversity of the mechanisms used in nature for the production of structural colour has been intensively covered in several books and reviews, which have appeared during the recent maturation of this field.[1,3,4] Although the formation/development of these structures in nature is far from being understood,[5] they are an excellent source of inspiration for emerging technologies, even though the artificial production of structural colour is still in its infancy in comparison to the advanced structures found in nature. Most bio-mimetic

RSC Polymer Chemistry Series No. 22
Bio-inspired Polymers
Edited by Nico Bruns and Andreas F. M. Kilbinger

Published by the Royal Society of Chemistry, www.rsc.org

photonic structures that have been produced so far were obtained by using synthetic polymers, photoresist, or inorganic materials.[6–9] The slow progress in developing processes capable of rivalling natural effects is largely due to the complexity of the natural structures, where a large variety of complex three-dimensional architectures are observed. A striking example of such complexity is the single gyroid structure found in the wing of various *papilionid* and *lycaenid* butterflies.[10] Natural gyroids have been replicated on different scales both with bottom-up and top-down fabrication. However, both approaches require either the use of synthetic polymers[11,12] or inorganic materials,[13] and the fabrication of such architectures using natural building blocks remains still an open challenge.

In contrast, simpler architectures, such as one-dimensional photonic structures, *e.g.* the multilayers found in insects exocuticles[14] or plants epidermis,[15] have been replicated using a plethora of methodologies and materials,[16–19] including also natural polymers.[20]

In this chapter, we will therefore focus on this latter group of structures, in particular helicoidal stacks, where the optical properties are well understood, and whose artificial production using biopolymers has been demonstrated.

After a brief introduction of helicoidal architecture and their optical properties, we will describe examples observed in nature. Then, we will discuss liquid crystals and biomimetic photonics. Finally, we will discuss the future of the field, its extension into templating, sensors, and the use of cellulose derivatives.

In order to progress beyond our current limits, it seems necessary to better study natural processes and materials in order to incorporate them into the production of more bio-compatible and sustainable biopolymer-based photonic structures with novel functionalities.[21]

17.2 Helicoidal Structures and their Optical Response

17.2.1 An Intuitive Explanation of the Optical Effect

Lyotropic mesophases have been observed in different biopolymers including polysaccharides, DNA, and proteoglycans.[20] Among the different phases observed, the chiral nematic (or cholesteric) phase is quite common on a wide range of length scales.[22] In such architectures, also commonly referred to as helicoidal, rod-like fibrils are organised in quasi-planes where the fibrils are all aligned in one direction (Figure 17.1). Such planes are then piled on top of each other so that the directionality identified by the fibrils rotates from plane to plane, by a few degrees, thereby forming a helicoidal stack.[23] Helicoids can be described by the distance between two planes with the same orientation of the fibrils, defined as pitch. When such distance is comparable to the wavelength of light, it can be reflected as in the case of standard Bragg multilayers.

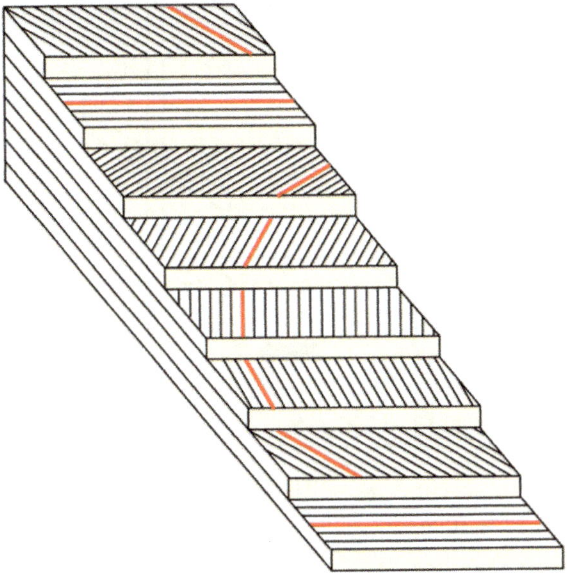

Figure 17.1 A diagram showing a helicoidal stack. The red lines are Bouligand arcs resulting from an oblique section.
Reproduced from B. D. Wilts *et al.*, Natural Helicoical Structures: Morphology, Self-assembly and Optical Properties, *Mater. Today*, **1**, 177–185, Copyright 2014 with permission from Elsevier.[22]

17.2.1.1 Bragg Reflection

A standard Bragg multilayer consists of repeating layers with alternating refractive indexes. Through constructive interference from refractive index boundaries, a single wavelength band is reflected.[24]

For a sufficient number of layers, the total reflection of light of wavelength λ occurs when

$$n\,\lambda' = 2\,d\,\sin\theta$$

where d is the pitch. For normally incident light, $n=1$, $\theta=0$. λ' is the wavelength of the light in the material. This yields the reflection condition

$$\lambda' = 2\,d.$$

17.2.1.2 Helicoidal Stacks

Rod-like particles are birefringent. In a helicoidal stack, their rotating arrangement therefore gives rise to refractive index changes, resulting in a periodically varying refractive index. This implies that the Bragg condition applies.

The Bragg condition for light reflection for the wavelength that matches the material pitch, however, only applies for circularly polarised light that turns in the opposite direction to the twist of the material, as shown in

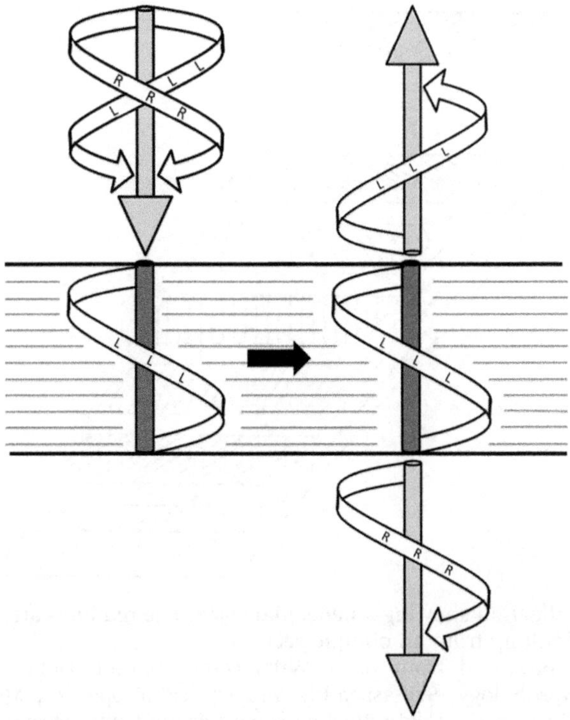

Figure 17.2 Schematic showing the reflection of left circularly polarised light by a left-handed chiral material. The image illustrates the interplay of structural and optical chirality.

Figure 17.2. Light of the same polarisation direction passes through. In contrast to standard Bragg reflectors, chiral multilayers therefore only reflect one-half of the intensity of unpolarised light.

The handedness of light is defined by the receiver (*i.e.* light coming towards the observer), whilst the material handedness is defined by the chirality moving away from the observer. For this reason, left handed materials have opposite chirality to, and reflect, left-handed light.

17.2.2 Mathematical Description

The intuitive description above shows that a helicoidal stack may be considered in terms of a standard Bragg reflector in the rotating frame of circularly polarised light. In more detail, its optical response may be demonstrated analytically using Maxwell's equations to show the existence of a photonic bandgap. The key results of the analytic treatment are summarised below.

Since chiral nematic liquid crystals phases have a very similar structural morphology (*i.e.* rod-like molecules with a rotating director), the method developed to describe chiral nematic liquid crystals can be applied to the case of a solid helicoidal film. This method examines the electromagnetic response of materials with screw-like rotating electrical axes.[25] The use of transfer matrices, which can be constructed to express the effect on a wave vector as it passes through a material, were developed for general materials[26–28] and later for stratified anisotropic materials.[28,29]

The full derivation of this approach can be found in ref. 29. In summary, from Maxwell's equations we first consider the general case of a 4×4 transfer matrix for electromagnetic materials, followed by the specific case of a screw-like electrical axis.

Maxwell's equations may be written in rectangular co-ordinates

$$\begin{bmatrix} 0 & 0 & 0 & 0 & -\frac{\partial}{\partial z} & \frac{\partial}{\partial y} \\ 0 & 0 & 0 & \frac{\partial}{\partial z} & 0 & -\frac{\partial}{\partial x} \\ 0 & 0 & 0 & -\frac{\partial}{\partial y} & \frac{\partial}{\partial x} & 0 \\ 0 & \frac{\partial}{\partial z} & \frac{\partial}{\partial x} & 0 & 0 & 0 \\ -\frac{\partial}{\partial z} & 0 & \frac{\partial}{\partial y} & 0 & 0 & 0 \\ \frac{\partial}{\partial y} & \frac{\partial}{\partial x} & 0 & 0 & 0 & 0 \end{bmatrix} \begin{bmatrix} \boldsymbol{E}_x \\ \boldsymbol{E}_y \\ \boldsymbol{E}_z \\ \boldsymbol{H}_x \\ \boldsymbol{H}_y \\ \boldsymbol{H}_z \end{bmatrix} = \frac{1}{c}\frac{\partial}{\partial t} \begin{bmatrix} \boldsymbol{D}_x \\ \boldsymbol{D}_y \\ \boldsymbol{D}_z \\ \boldsymbol{B}_x \\ \boldsymbol{B}_y \\ \boldsymbol{B}_z \end{bmatrix}, \tag{17.1}$$

where ***E***, ***H***, ***D***, and ***B*** are the electromagnetic field vectors. The 6×6 matrix is symmetrical with elements only in the first (upper right) and third (lower left) quadrant. The non-zero elements are the **curl** operators.

This can be rewritten as

$$\boldsymbol{RG} = \frac{1}{c}\frac{\partial}{\partial t}\boldsymbol{C}, \tag{17.2}$$

where $\boldsymbol{R}$ is the 6×6 differential matrix. By defining the time dependence of a light wave by the exponential factor $e^{-i\omega t}$ (ω is the wave frequency) and, in a first approximation, ignoring nonlinear effects (*i.e.* assuming a linear relationship between $\boldsymbol{G}$ and $\boldsymbol{C}$), this is rewritten as

$$\boldsymbol{R\Gamma} = \frac{-i\omega}{c}\boldsymbol{M\Gamma}, \tag{17.3}$$

where $\boldsymbol{\Gamma}$ is the spatial part of $\boldsymbol{G}$. In this construction

$$\boldsymbol{M} = \begin{pmatrix} \boldsymbol{\varepsilon} & \boldsymbol{\rho} \\ \boldsymbol{\rho}' & \boldsymbol{\mu} \end{pmatrix}, \tag{17.4}$$

where each of the four quadrants represent a 3×3 tensor (*i.e.* 3 spatial dimensions), $\boldsymbol{\varepsilon}$ is the dielectric tensor, $\boldsymbol{\mu}$ the permeability tensor, and $\boldsymbol{\rho}$ and $\boldsymbol{\rho}'$ the optical rotation tensors.

In the case of a plane wave incident obliquely in the x, z plane on a region, $\boldsymbol{M}$ is a function only of z. By expanding eqn (17.3) in terms of the 6×6 matrix

$\boldsymbol{R}$, two linear algebraic equations in terms of the six components of $\boldsymbol{\Gamma}$ are found. Two components may thus be expressed in terms of the other four, resulting in four first-order linear differential equations for the other four variables, and two linear algebraic equations. These differential equations may be written in 4×4 matrix form as

$$-\frac{ic}{\omega}\frac{\partial}{\partial z}\begin{pmatrix} 0 & 0 & 0 & 1 \\ 0 & 0 & -1 & 0 \\ 0 & -1 & 0 & 0 \\ 1 & 0 & 0 & 0 \end{pmatrix}\begin{pmatrix} E_x \\ E_y \\ H_x \\ H_y \end{pmatrix} = \boldsymbol{S}\begin{pmatrix} E_x \\ E_y \\ H_x \\ H_y \end{pmatrix}, \tag{17.5}$$

which, by multiplying by the inverse 4×4 matrix and rearranging, is equivalent to

$$\frac{\partial}{\partial z}\begin{pmatrix} E_x \\ H_y \\ E_y \\ -H_x \end{pmatrix} = \frac{i\omega}{c}\begin{pmatrix} S_{41} & S_{44} & S_{42} & -S_{43} \\ S_{11} & S_{14} & S_{12} & -S_{13} \\ -S_{31} & -S_{34} & -S_{32} & S_{33} \\ S_{21} & S_{24} & S_{22} & -S_{23} \end{pmatrix}\begin{pmatrix} E_x \\ E_y \\ H_x \\ H_y \end{pmatrix}. \tag{17.6}$$

In abbreviated form,

$$\frac{\partial}{\partial z}\psi = \frac{i\omega}{c}\boldsymbol{\Delta}\psi. \tag{17.7}$$

If $\boldsymbol{\Delta}$ is approximately independent of z over a small distance δz, the four periodic solutions to eqn (17.7) are

$$\psi_j(\delta z) = e^{-iq_j\delta z}\psi_j(0). \tag{17.8}$$

The four eigenvectors and corresponding eigenvalues are found analytically by solving the characteristic equation

$$\mathrm{Det}\left(\frac{\omega}{c}\boldsymbol{\Delta} - q\boldsymbol{I}\right) = 0, \tag{17.9}$$

where $\boldsymbol{I}$ is the 4×4 Identity matrix. If $\boldsymbol{\Delta}$ is independent of z over a distance h, integrating eqn (17.7) gives

$$\psi(z+h) = \boldsymbol{P}(h)\psi(z). \tag{17.10}$$

$\boldsymbol{P}(h)$ can be found by taking the matrix of eigenvectors $\boldsymbol{\Psi}$ and using it in eqn (17.7):

$$\boldsymbol{\Psi K}(h) = \boldsymbol{P}(h)\boldsymbol{\Psi}. \tag{17.11}$$

Here, $\boldsymbol{K}(h)$ is a diagonal matrix with elements

$$K_{jj} = e^{iq_jh}. \tag{17.12}$$

With this,

$$\boldsymbol{P}(h) = \boldsymbol{\Psi K}(h)\boldsymbol{\Psi}^{-1}. \tag{17.13}$$

Exact analytic solutions can be found for cases where $\boldsymbol{M}$ is periodic, as in a single domain of a chiral nematic structure illuminated along its axis of helicity. This method then yields eigenmodes which are permitted in a material.

In this material the optical activity in each plane is assumed to be negligible in comparison to the optical rotation of the structure. Oseen's model[30] has unit permeability, zero optical rotation and

$$\boldsymbol{\varepsilon}=\begin{pmatrix} \epsilon+\delta\cos 2\beta z & \delta\sin 2\beta z & 0 \\ \delta\sin 2\beta z & \epsilon-\delta\cos 2\beta z & 0 \\ 0 & 0 & \epsilon_3 \end{pmatrix}. \tag{17.14}$$

This represents a helical twist in the dielectric values with an x value of $\epsilon+\delta$ at $z=0$. The pitch of the helical spiral is

$$\lambda=2\pi/\beta. \tag{17.15}$$

Using eqn (17.14), the components of $\boldsymbol{S}$ can be derived, so that

$$\boldsymbol{\Delta}(z)=\begin{pmatrix} 0 & 1-X^2/\epsilon_3 & 0 & 0 \\ \epsilon+\delta\cos 2\beta z & 0 & \delta\sin 2\beta z & 0 \\ 0 & 0 & 0 & 1 \\ \delta\sin 2\beta z & 0 & \epsilon-X^2\quad\delta\cos 2\beta z & 0 \end{pmatrix} \tag{17.16}$$

$$X=\frac{c\xi}{\omega}, \tag{17.17}$$

where ξ is the component of the propagation vector in the x direction.

The secular equation can be solved numerically, providing results that fit well the observed behaviour in chiral nematic solids.[31] This method was used to produce the model of the flat chiral nematic multilayer shown in Figure 17.1.

When X is zero, the characteristic equation can be solved analytically for the eigenvectors and eigenvalues. In this case E_x, H_y are the same, as are E_y, $-H_z$, except for a phase difference at $z=\lambda/4$. The eigenvectors are

$$\psi=e^{i(q+\beta)z}\begin{pmatrix} A \\ A' \\ -iA \\ -iA' \end{pmatrix}+e^{i(q-\beta)z}\begin{pmatrix} B \\ B' \\ -iB \\ -iB' \end{pmatrix}, \tag{17.18}$$

where the A and B values have to be determined.

Matrix multiplication when X is zero gives

$$\boldsymbol{\Delta}\psi=e^{i(q+\beta)z}\begin{pmatrix} A \\ \epsilon A'+\delta B \\ -iA \\ -i(\epsilon A'+\delta B) \end{pmatrix}+e^{i(q-\beta)z}\begin{pmatrix} B \\ \epsilon B'+\delta A \\ -iB \\ -i(\epsilon B'+\delta A) \end{pmatrix} \tag{17.19}$$

and by differentiation,

$$-i\frac{c}{\omega}\frac{\partial\psi}{\partial z} = e^{i(q+\beta)z}\begin{pmatrix} A \\ A' \\ -iA \\ -iA' \end{pmatrix}\frac{c}{\omega}(q+\beta) + e^{i(q-\beta)z}\begin{pmatrix} B \\ B' \\ -iB \\ -iB' \end{pmatrix}\frac{c}{\omega}(q-\beta). \quad (17.20)$$

Eqn (17.7) implies that the left hand sides of the last two equations are equal, so by equating the terms, four equations are obtained, two of which provide non-trivial solutions for the case when q has one of four eigenvalues,

$$q = \pm\left(\beta^2 \pm 2\beta\frac{\omega}{c}\sqrt{\epsilon + \frac{\frac{\delta^2}{\epsilon}\left(\frac{\omega}{c}\right)^2}{4\beta^2}} + \left(\frac{\omega}{c}\right)^2\epsilon\right)^{\frac{1}{2}}, \quad (17.21)$$

which provides a full solution with either A or B specified.

The important aspect about the solution for the eigenvectors is that, for small values of ω, the eigenvalues of q are approximately circularly polarised, affected by the chiral material as if it was optically active. In addition, when

$$\omega \approx \frac{c\beta}{\sqrt{\epsilon}} \quad (17.22)$$

i.e. when the wavelength in the medium is near half the pitch length, the eigenvectors are no longer approximately circularly polarised. Furthermore, one pair of eigenvalues is imaginary over a frequency band and is attenuated. In real materials, the wavelength of light exciting this mode is totally reflected for samples consisting of many perfectly ordered layers. The non-attenuated modes are transmitted through the material.

This single reflection band for light breaks up at oblique incidence. The solutions when X is non-zero give further weak bands near higher multiples of the fundamental band frequency.[32]

17.2.3 Examples

The plots in Figure 17.3 were created using the Berreman 4×4 matrix method, simulating the optical response of a multilayer. Variations in the anisotropy of the refractive index, the pitch of the layers, the number of layers, and the average refractive index are shown. The changes shown by the Berreman simulation match the intuitive understanding of ordinary Bragg reflectors.

17.3 Helicoidal Structures in Nature

Helicoidal architectures of polysaccharides, like chitin and cellulose are found both in the animal and plant kingdoms with a great variety range.[33] Since the pioneering work on these structures, mainly preformed in the

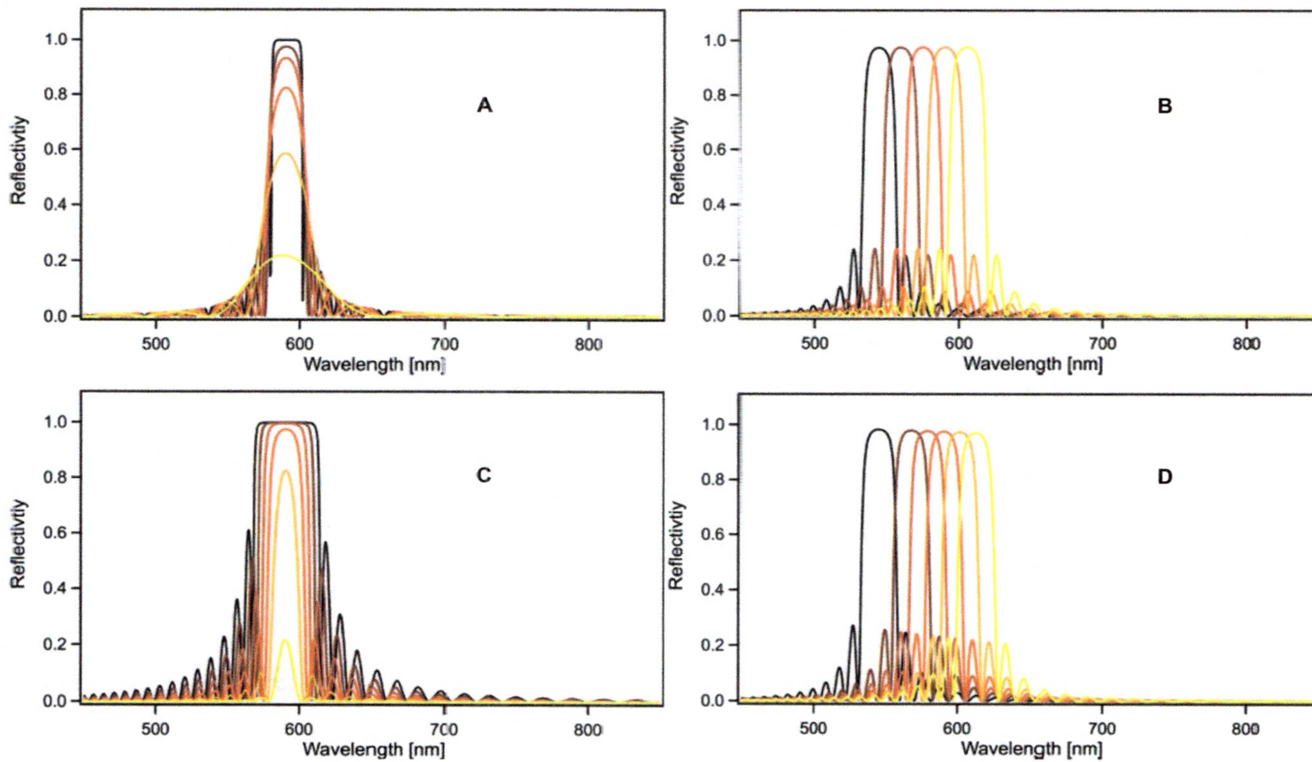

Figure 17.3 Variation of parameters in a chiral nematic multilayer numerically solving the Berreman equation. (A) Variation in the number of layers: 5 (yellow), 10, 15, 20, 25, 50 (black). (B) Variation in the chiral pitch: 390 nm (yellow), 380 nm, 370 nm, 360 nm, 350 nm (black). (C) Variation in the refractive index contrast: 0.01 (yellow), 0.02, 0.03, 0.04, 0.05, 0.06 (black). (D) Variation in the average refractive index: 1.62 (yellow), 1.59, 1.56, 1.53, 1.50, 1.47 (black).

1970s by Neville,[34–38] they have been shown to be extremely widespread, probably because of their mechanical performance.[39–42] Cellulose and chitin are, in fact, called structural polysaccharides. They can be found in every plant cell wall and in the cuticle of different genera.[39,43] For this reason, they are the most abundant renewable organic materials produced in the biosphere.[44,45] The mechanical properties of cellulose and chitin are a consequence of their super-molecular architecture.[46] In nature, both cellulose and chitin organise into fibrillar structures mainly *via* inter-molecular hydrogen bonding.[47] The fact that the elementary fibrils contain interconnected amorphous and crystalline regions result in good stiffness and elastic properties. The size of these elementary fibrils has been extensively researched. It strongly depends on the polysaccharide source.[39,48] The dimensions of the crystalline parts also vary between species.[44,49] A schematic illustration of the organisation of these polysaccharides in nature is shown in Figure 17.4.

These hierarchical fibrils are extruded from the external plasma membrane of the cell into the cell wall by groups of synthase molecules arranged in 'rosettes'.[51,52] The parallel synthesis of cellulose molecules results in the growth of the polymorph *I* instead of the otherwise energetically more favourable polymorph *II*. This second polymorph is the form of cellulose synthesised *in vitro*.[53]

The extruded fibrils are continuously wrapped around the cell[54] to build the growing cell wall.[51,55,56] The arrangement of fibrils is adapted to the cell type, and it may be a random, axial, transverse, crossed, helical, or helicoidal arrangement.[56]

To form helicoids, each individual nano-fibril is arranged parallel to another one in stacked planes. From plane to plane, the orientation of the fibrils changes by a few degrees to obtain the helicoidal structure. While in many instances the dimensions of such helicoidal structures are not on the right length-scale to produce structural colour, there are many examples where such hierarchical structures provide a vivid and strong iridescent colouration.

17.3.1 Cellulose-based Helicoidal Structures for Structural Colour

Cellulose helicoidal structures can be found in algae and land plants.[22] Cellulose is one of the polysaccharides composing the cell wall and it is synthesised by large membrane complexes, which extrude the fibril from the cell. Cellulose is composed by (1,4)-linked β-D-glucan chains in which every unit is corkscrewed 180° with respect to its neighbours, see Figure 17.4. Each cellulose chain possesses a directional chemical asymmetry with respect to the termini of its molecular axis: one end has a chemically reducing functionality (*i.e.*, a hemiacetal unit) and the other has a pendant hydroxyl group, the nominal non-reducing end. During biosynthesis, cellulose chains aggregate into microfibrils with cross-sectional dimensions ranging from 2 to

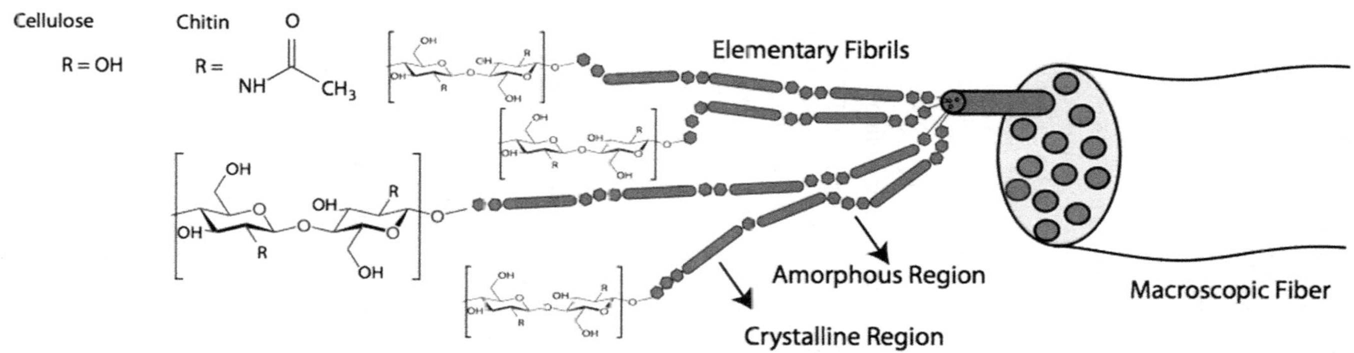

Figure 17.4 Hierarchical structure of cellulose and chitin from molecules to macroscopic fibres. Adapted from N. Lavouire *et al.*, Microfibrillated cellulose – Its barrier properties and applications in cellulosic materials: A review, *Carbohydr. Polym.*, **90**, 735–764, Copyright 2012 with permission from Elsevier.[50]

Figure 17.5 Structurally coloured helicoids in plants. (a) *Pollia condensata* and (b) *Elaeocarpus* fruits. (c) *Danaea nodosa* ferns.
Image (a) courtesy of Silvia Vignolini, University of Cambridge, and images (b,c) courtesy of David Lee, Florida International University. Reproduced from B. D. Wilts *et al.*, Natural Helicoical Structures: Morphology, Self-assembly and Optical Properties, *Mater. Today*, **1**, 177–185, Copyright 2014 with permission from Elsevier.[22]

20 nm, depending on the source of the cellulose.[44] Cellulose can be found in the primary and secondary cell walls, but helicoidal structures that display structural colour are observed mainly in the primary cell wall.

Helicoidal structures that produce selective reflection of the blue-ultraviolet wavelength band have been observed in leaves of many different rain-forest understorey plants (Figure 17.5). Examples are the leaves of *Danaea nodosa* and *Diplazium tomentosum.*[4,7,15,21,57,58] In the case of *Danaea nodosa*, the blue colour is observed only in young leaves, as a result of ordered layers with a pitch of about 133 nm in the adaxial cell walls of the adaxial epidermis, but the helicoidal structure is maintained also in the mature tissue. Similar cellulose-based helicoids are observed also in flowering plants, like *Phyllagathis rotundifolia*[4,58] and in the tropical understorey herb *Mapania caudate.*[59] In the latter, in addition to the blue reflection of left-handed circularly polarised light from a cellulose-helicoid structure, silica granules accumulate at uniform positions within the helicoids, contributing to the optical appearance of the tissue.

Helicoidal architectures have also been found in the epicarp of different fruits (Figure 17.5). The first observation of this phenomenon was reported by D. Lee in 1991 in *Elaeocarpus.*[60] Recently, similar structures have been observed in the fruit of *Margaritaria nobilis* and *Pollia condensate.*[15,61] While all the other published examples of helicoidal architectures (both in cellulose and chitin) are left-handed, the fruit of *Pollia condensata* has also cell walls with right-handed orientation of the cellulose structures.

17.3.2 Chitin-based Helicoidal Structures for Structural Colour

Chitin is a fairly completely acetylated polysaccharide akin to cellulose. The monosaccharide units are joined by 1,4 links that make the molecule very straight and ribbon-like. Within the body of the cuticle, chitin is assembled into nanofibres with diameters and lengths that vary in different organisms.[49] Helicoidal plywoods are particularly widespread in the exocuticle of different species of beetles.[14,39] The exocuticle is one of the layers composing the exoskeleton of beetles (their hardened wings). In this layer, chitin microfibrils of very small diameter (2.8 nm)[62] are arranged in a helicoidal architecture. Such chitin microfibrils are embedded in a matrix of proteins, which are stiffened by small, reactive aromatic molecules which cross-link the protein chains. The first observation of this architecture to produce structural colour dates back to 1911 when Albert Abraham Michelson discovered that the scarab beetle *Plusiotis resplendens* had a left-handed circularly polarised component.[63]

Since then, many examples in the jewelled beetles (family Scarabaeidae) and its subfamilies Rutelinae, Scarabaeinae, and Cetoniinae have been reported[14,39] (Figure 17.6). About 20% of the species in both the Rutelidae and Cetoniidae families have an entirely or partially metallic-shiny exocuticle, but not all metallic body parts reflect circularly polarised light.[64]

In the beetle *Chrysina gloriosa*, left-handed green colour reflection is observed from the exoskeleton, but without colour variation, and patterns are structurally and optically analogous to the focal conic domains formed spontaneously on the free surface of a cholesteric liquid crystal.[65] More sophisticated helicoidal structures can be found in *Plusiotis resplendens*. The latter species is capable of reflecting both left and right circularly polarised light.[66] In contrast to *Pollia condensate*, the right-handed circular

Figure 17.6 Three representative beetles featuring strong reflectance of circularly polarised light. (a) *Anomala dimidata*, (b) *Plusiotis (Chrysina) gloriosa*, and (c) *Plusiotis (Chrysina) optima*.
Reproduced from B. D. Wilts *et al.*, Natural Helicoical Structures: Morphology, Self-assembly and Optical Properties, *Mater. Today*, **1**, 177–185, Copyright 2014 with permission from Elsevier.[22]

polarisation in this case is obtained by using a half-wave retardation plate that is sandwiched between two left handed helicoids.

While there are not conclusive studies on the biological significance of structural colouration produced by helicoids, it is commonly accepted that such a reflection mechanism might play a role in mate signalling and/or camouflage.[67,68]

While extremely widespread in beetles, other chitin architectures are less prominent in other species, and only a few examples are reported in the literature, probably because helicoidal components in some Crustacea (crayfish, crabs) are not individual chitin crystallites, but bundles with a diameter of 25–50 nm.[62] Such filaments form non-optically active helicoid with pitches on the order of microns. However, in the case of *Panulirus argus,* chitin calcified architectures produce green colouration.[69]

17.4 Liquid Crystals as Helicoidal Structures

All of the examples of helicoidal architectures given in the previous section are created in living organisms by self-replicating and active processes. This presents an engineering challenge for biomimesis to develop an artificial helicoidal architecture without self-replicating cells. The similarity of biological helicoidal stacks and chiral nematic liquid crystals, however, suggests that biomimeic approaches are possible.

The extent to which different biologically grown tissues make use of this liquid crystalline self-assembly, (a passive, entropy-driven process, as described below) has been the subject of much debate for decades.[33,70,71] Although liquid crystalline morphologies in many ways strongly resemble helicoidal biological tissue, there are significant differences in the materials and conditions present in passive liquid crystals and living cells.[72–78] The evidence is now in favour of a 'directed self-assembly' mechanism in organisms, in which the chiral nematic is formed by active directed deposition *in situ* with a passive, liquid-crystalline relaxation.[79]

Liquid crystals were first identified in 1888[80] in a molecule extracted from carrots. The first observed phase was the cholesteric or chiral nematic phase, whose shape is the subject of this chapter. Liquid crystals are made of molecules of particular shapes which, under certain conditions, align to form the phase shown in Figure 17.7. The first synthetic liquid crystals were manufactured in 1889.[80] Liquid crystalline phases have been observed in many biopolymers.[20]

The constituent parts that form liquid crystals can exist over a wide range of sizes, from molecules to colloids. At each size scale different forces control the interactions of the liquid crystal constituents, but the principles of liquid crystalline formation remain the same. In order to understand the use of liquid crystals in biomimetic structural colour for helicoidal architectures, a brief introduction into the huge and varied field of liquid crystals is given here.

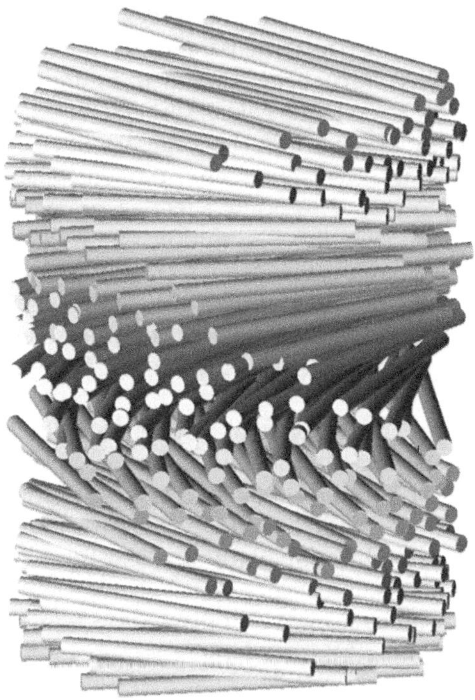

Figure 17.7 Schematic showing the liquid crystal chiral nematic phase. Image courtesy of Gen Kamita, University of Cambridge.

Liquid crystal (LC) phases can occur when particles in a liquid have orientational order through anisotropy, either in effective shape or charge, which tends to favour a certain directionality when packing the particles. The simplest of these particles, with one dimension of anisotropy, are rod-like particles.[81]

For these rod-like particles, there is a critical threshold, which according to the system may be either in the concentration of the particles in the system (lyotropic), or in the temperature (thermotropic).[79] At this concentration or temperature, the interaction of the particles in the liquid is such that entropic evaluation dominates the behaviour.

The entropic description of liquid crystalline phases was first made by Onsager.[82] In Onsager's description, bringing the particles into a closely packed system, or a liquid crystal, decreases the excluded volume between the particles and creates a two phase system. This reduction of excluded volume is sufficient to overcome the orientational entropy favouring orientational disorder in both rotation and translation.

This description also takes into account the electrostatic interaction, which tends to repel particles from each other, and to increase the excluded volume. This balance between entropic crystallisation and electrostatic repulsion is the key to a stable liquid crystal, which neither disperses nor flocculates; it may be affected in real systems by the addition of ions on the particles and in solution. Onsager used the idea of an effective diameter to take into account the change in ionic strength in the system.

Onsager's theory has since been extended to include chiral nematic phases, notably in the Stroobants–Lekkerkerker–Odijk (SLO) description.[83] By quantitatively taking into account the tendency of electrostatic repulsion to favour perpendicular orientation, and the entropic consideration to favour parallel orientation, a twisting effect is predicted, caused by the charges present. This allows the description of the chiral nematic phase through the use of two parameters; the effective diameter, and the twisting action. This account of chiral nematic phases has been shown to provide a good prediction of experimental results, although discrepancies have been demonstrated and explained as due to flexibility[84,85] and inhomogeneity[86] in non-ideal systems.

In a liquid crystal phase, the system remains a colloidal liquid, but the arrangement of the particles becomes ordered and periodic, and exhibits many of the attributes of a traditional solid crystal, despite having an order parameter far lower than that in a solid crystal. This difference can be seen in the latent heat of crystallisation, with a typical value for solid crystals of 250 J g^{-1}, while liquid crystals have 5 J g^{-1}.[87] Understanding the control of these systems allows us to exploit chiral nematic liquid crystals to create artificial structural colour with biopolymers.

17.5 Helicoidal Architectures in Biomimetic Photonics

Using self-assembly of a chiral nematic phase in a biopolymer liquid crystal of cellulose nanocrystals, a well-controlled technique has been developed to create solid helicoidal architectures for structural colour and for further functionalisation. This section describes the self-assembly process and the control parameters of tuneable helicoidal cellulose films, and the prospects for future development.

Currently, this method relies on the biological origin of cellulose nanoparticles, since cellulose synthesised *in vitro* forms a different polymorph, as described in Section 17.3.1. This is, however, not a limiting factor, since cellulose has been processed from its natural sources for centuries. Cellulose nanocrystals are extracted and stabilised in suspension, which is then dried to form a helicoidal film, as shown in Figure 17.8.

Through controlled treatment of this suspension before and during drying, the parameters of the resulting structurally coloured film can be tuned. These steps are discussed in the next sections along with the prospect for subsequent functionalisation and transfer to other materials.

The extraction of cellulose nanocrystals or CNCs (also called (nano-/micro-)crystallites, whiskers, nanorods, and nanowires[88]) was first demonstrated in 1952[89] using acid hydrolysis of cellulose fibres. However, interest in the technique remained low until 1992, when their chiral

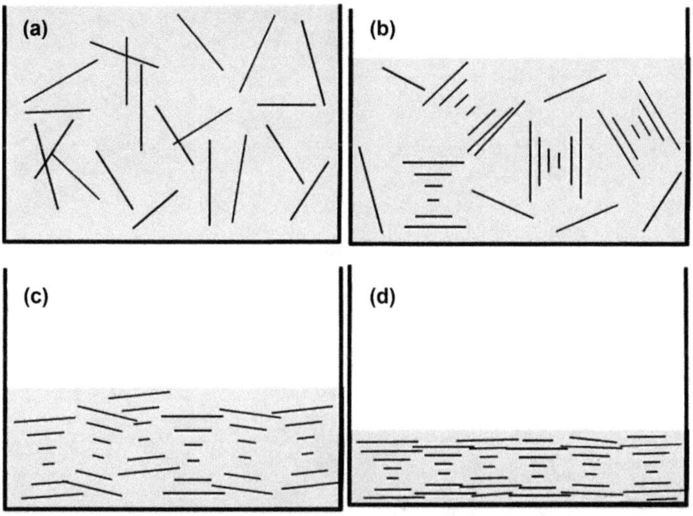

Figure 17.8 Schematic of film drying. (a) Isotropic suspension. (b) Emergence of self-assembled tactoids at a critical concentration. (c) Alignment of tactoids in a highly concentrated solution. (d) Dried chiral nematic film.

nematic properties were first reported.[52,90–93] A patent on the production of structurally coloured cellulose films was filed in 1995.[92] Since then, interest in the characteristics of CNCs has grown,[52] especially in the context of mechanical reinforcement.[94]

17.5.1 Artificial Helicoidal Cellulose Films

17.5.1.1 Cellulose Nanocrystal Extraction by Hydrolysis

The first part of the process consists on the extraction of rod-like particles of cellulose, which can form a liquid crystal phase. In microfibrils, these crystalline regions are connected by amorphous regions, as shown in Section 17.3. The amorphous parts are less well-packed, and therefore less dense than the crystalline regions. This low density makes them more vulnerable to acid hydrolysis. When the fibrils are exposed to sulphuric acid, the amorphous parts are stripped away first, leaving only the crystalline regions.[90] The suspension is sonicated to break apart the crystallites, and filtered to remove large aggregations. The hydrolysis can be controlled by the concentration of sulphuric acid and the length of time of acid exposure.[44] After hydrolysis, the cellulose suspension is cooled and dialysed for several days until the acid is removed from the suspension.

This process produces rod-like or whisker-like nanocrystals of 3–30 nm[95] in width and between 50 nm to several microns in length.[96] The size and aspect ratio is influenced by the hydrolysis and sonication parameters which break down the fibrils to different extents.[89,97,98] The more intense the

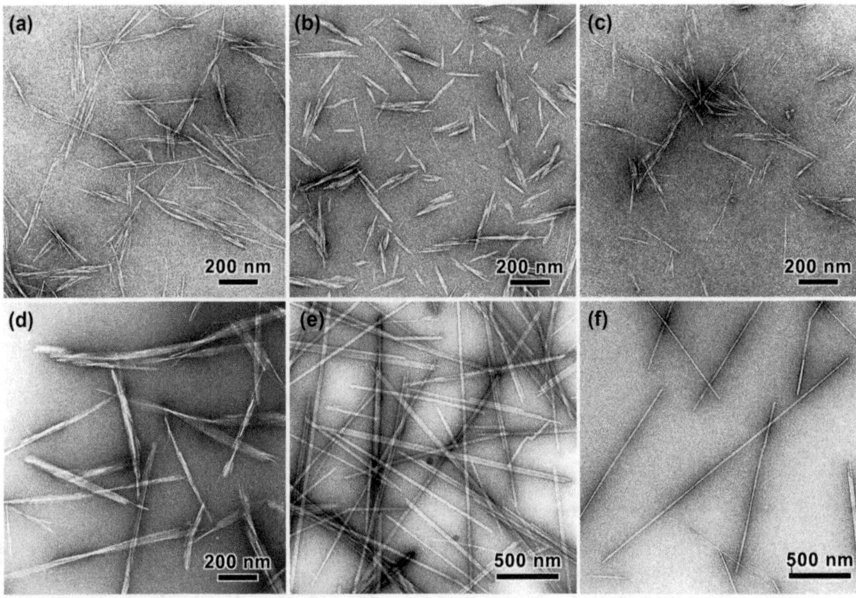

Figure 17.9 TEM micrographs of CNCs taken from (a) wood, (b) cotton, (c) bamboo, (d) bacteria, (e) algae, and (f) tunicates.
Reproduced with permission from Kaushik *et al.*[99] Copyright 2015, Intech.

hydrolysis, the narrower and shorter the crystallites become.[89] However, the form of the crystallites is more strongly influenced by the biological source of the cellulose fibrils,[95] as can be seen in the wide variety of cellulose forms in the transmission electron microscopy (TEM) images in Figure 17.9.

17.5.1.2 Suspension Parameters for Chiral Films

After extraction, further changes to the CNC suspension can be made by heating, sonication, and addition or removal of salts. These changes can influence the liquid crystalline phase according to the SLO theory,[83] as explained in the discussion of liquid crystals. Furthermore, these changes can be used to control the parameters of the final structure.

The assembly of the chiral nematic structure happens in two steps. As the solvent evaporates from the suspension, the critical concentration of the lyotropic liquid crystal phase is reached. The CNCs form discrete localised clusters of chiral nematic phase, called tactoids, as predicted by the phase separation of liquid crystals. These tactoids can be observed between cross-polarisers under a microscope. The first reported image of tactoids in a cellulose suspension[90] is reproduced in Figure 17.10. The tactoids are visible from the birefringent stripes, which represent axes of crystallite arrangement parallel to the stripes.[90]

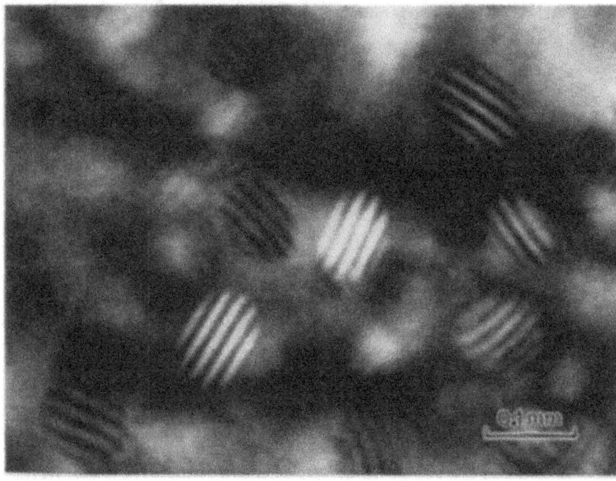

Figure 17.10 Reproduction of the first recorded tactoids of cellulose nanocrystals in suspension. Birefringent bands are visible between crossed polars.
Reproduced from the Revol *et al.*, Helicoidal self-ordering of cellulose microfibrils in aqueous suspension, *International J. Biol. Mol.*, **14**, 170–172, Copyright 1992 with permission from Elsevier.[90]

The second step is the evaporation of the solvent, compressing, and aligning the tactoids to form a dry film. The rate of evaporation from the suspension at this stage is crucial, as the self-assembly can only occur provided the crystals are not forced to freeze in the isotropic arrangement as happens when the evaporation rate is too high. With a sufficiently slow evaporation rate, the interaction between the CNCs is mediated by the charges on the surfaces of the crystals and those in solution. The tactoids in suspension grow to form the characteristic 'fingerprint texture' shown in Figure 17.10.[90]

Several parameters in the processing of the CNCs influence the degree of chirality and tactoid pitch. A decrease in the width of the individual CNCs has been shown to cause a decrease in the final pitch of the film. Less intuitively, a decrease in the length of the individual CNCs also causes a decrease in pitch.[100]

The effect of sonication has also been extensively investigated and has been shown to increase the pitch of the final film.[101] Sonication breaks apart aggregations of crystals, as well as breaking the weak links remaining between the crystallites, and releases ions into the solution. The release of ions from the crystallites is the mechanism proposed to explain the increase in pitch.[101]

The ion concentration may also be changed by direct addition of electrolytes into the suspension. By increasing the surface charges, it has been shown[86,102] that the chirality of the suspension can be increased, *i.e.* the pitch decreased. It has again been suggested that the ions increase the effective diameter and the twist on the CNCs and make the chiral stacking interactions stronger. This effect can be demonstrated to be even more pronounced by the inclusion of surfactants, which effectively screen the charge of the nanocrystals and allow a much closer packing, decreasing the pitch to a tenth of its previous value.[103]

17.5.1.3 Drying Parameters for Chiral Films

The final part of the formation of a helicoidal film is the transition from the hydrated chiral nematic phase suspension to the solid dry film.

Further evaporation of the chiral nematic suspension after the chiral nematic phase has formed, compresses the tactoids and brings the alignment of their planes closer to the substrate plane.[104,105a] The pitch changes during this process according to the forces perpendicular and parallel to the plane in which the crystals lie, and the duration of the drying process.[106] By changing the parameters of the evaporation environment, the film's characteristics may be tuned.

These forces are mostly mediated by surface tension acting on the diminishing volume of the suspension. This evaporation increases the concentration, which tends to compress the hydrated cholesteric structure. A further effect, the so-called coffee-ring effect, is caused by the geometry of an evaporating liquid, which has increased evaporation at the edges, where the suspension is pinned to the substrate (*e.g.* the contact-line of a drop). This increased evaporation causes fluid flow towards the edges, causing increased particle deposits at the edges.[107–109]

The rate with which the evaporation and dehydration of the film takes place plays a vital role in the film pitch. Slower evaporation rates have been shown to produce blue-shifted, smaller-pitch films.[101,110] This is caused by the increased time allowing the movement of particles into a more tightly packed and twisted arrangement.[111] An even faster evaporation rate can however prevent the formation of tactoids altogether by jamming nanocrystals in their low-interaction state to produce gelation or a glassy configuration. This generally also occurs at the edges of films, due to the coffee-ring effect.[106] Increasing the CNC concentration in the initial suspension, a parameter that also influences the evaporation rate, increases the pitch (redshifts the film) through an increase in the proportion of the anisotropic phase in the suspension.[112] A higher temperature, in contrast, has been shown to decrease the pitch. This is explained by an increase in the flexibility of the rods which weakens the interaction between rods.[112] The electrostatic interaction of the film with its substrate has also been demonstrated as a control variable.[110]

In addition to these parameters, extensive control has been shown using fields.[111] Helicoids may be unwound, aligning the nanocrystals in an electric field.[113] Furthermore, the director of the helicoid itself may be oriented by a magnetic field.[91,114] A similar alignment of the helicoidal director has been reported using hydrodynamic shear forces.[115]

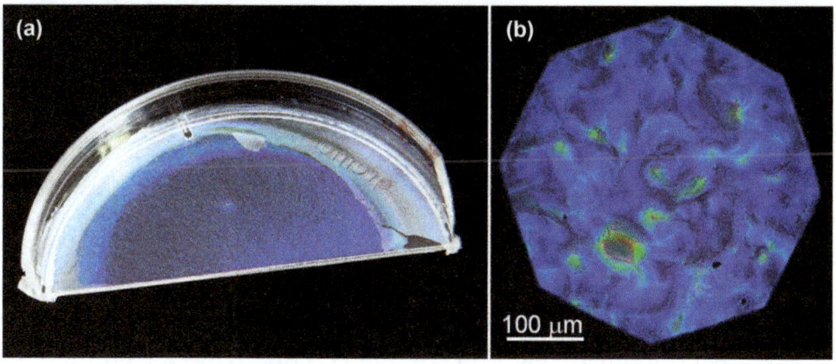

Figure 17.11 Photograph of a chiral nematic cellulose film (a) and its optical microscopy image in cross polarisation configuration (b).
Images courtesy of Silvia Vignolini, University of Cambridge. Reproduced under terms of creative commons license from Dumanli *et al.*[105]

17.5.1.4 Cellulose Film Characterisation

CNC films (Figure 17.11) have been produced and characterised using electron microscopy and spectrometry, and the chiral Bragg reflector effect was confirmed, with the observed reflection of light corresponding to the helicoidal structure, as predicted by the theory.[116] Evidence of digitalisation in the observed colours within one film (*i.e.* discrete jumps in the spectral response) has also been reported,[105a] which correlated with different discrete numbers of layers found within a uniformly thick film. The films have been shown to strongly resemble optically responsive helical plywoods found in nature using electron microscopy.[61]

17.5.2 Cellulose as a Template – Extending Functionality

So far, the creation of structural colour by helicoidal architecture has been described. This unusual architecture has however much potential in extended applications beyond its optical characteristics, and huge advantages remain to be exploited. Pathways to these functionalities lie in the inclusion of other materials in the helicoidal structure, and the transfer of the structure into entirely new materials through the creation of porosity, and hard and soft templating of the original structure.

17.5.2.1 Cellulose Film Composites

The inclusion of other materials within structurally coloured CNC films is of huge importance for the production of structurally coloured films. Of particular interest is the inclusion of plasmonic particles,[117–122] whose directed inclusion has been especially studied, as well as latex nanoparticles.[123] There has also been significant interest in the inclusion of polymers in the films to improve their mechanical flexibility.[124,125] The inclusion of these molecules also causes a redshift by acting as a spacer between the cellulose nanocrystals, similar to the inclusion of a smaller molecule, D-(+)-glucose.[106]

17.5.2.2 Templating

Templating is an emerging field that promises to transfer the attractive architectural characteristics of self-assembly to other materials, as illustrated in Figure 17.12.

A key development in the use of helicoidal films as template materials is the use of silica templates. In this method, silica is grown into the matrix of a chiral nematic crystal using sol–gel chemistry, employing the chiral nematic structure as a soft template (Figure 17.13). Removal of the cellulose then leaves an inverse helicoidally patterned template in the silica. Kresge's[127] demonstration of this technique has since been applied in soft-templating techniques using liquid crystals derived from cellulose.[128] This template has also been used to demonstrate hard templating of helicoidal patterns into titanium dioxide and other metal oxides,[111,126,129–131] by refilling the silica replica, followed by etching away the silica.

The use of other materials has been proposed in order to create robust and high-refractive index structural colour, and it has also been proposed as a way of constructing metamaterials.

Soft-templating has been recently reported by the same group,[132] where CNCs were embedded into an amino resin polymer. The obtained composite films were highly flexible and their colouration could be permanently changed by simply applying pressure.

17.5.3 Sensors

An attractive and demonstrably viable application for these template structures is as sensors. The strong sensitivity of structural colours to very slight changes in scale or shape and the visible and obvious effect on their optical response make them excellent candidates for sensors. This has been demonstrated in chemical,[133] temperature,[126] and pressure[134] sensing. In each of these cases, the infiltration of another molecule, or the expansion of the existing structure causes the Bragg layer to expand, reflecting light of a longer wavelength. This simple technique can be used to create cheap and simple sensors which are non-toxic and biocompatible.

17.5.4 Hydrogels and Cellulose Derivatives

Chiral nematic phases can also be achieved in polymeric systems, employing polymers that incorporate motives that form a liquid crystalline phase. Liquid crystallinity occurs in polymers either when they are dissolved in a solvent (lyotropic liquid crystal) or for temperatures above their glass or melt transition points (thermotropic liquid crystal polymers). Two examples of liquid crystal polymers are hydroxypropylcellulose (HPC) and ethyl cellulose.

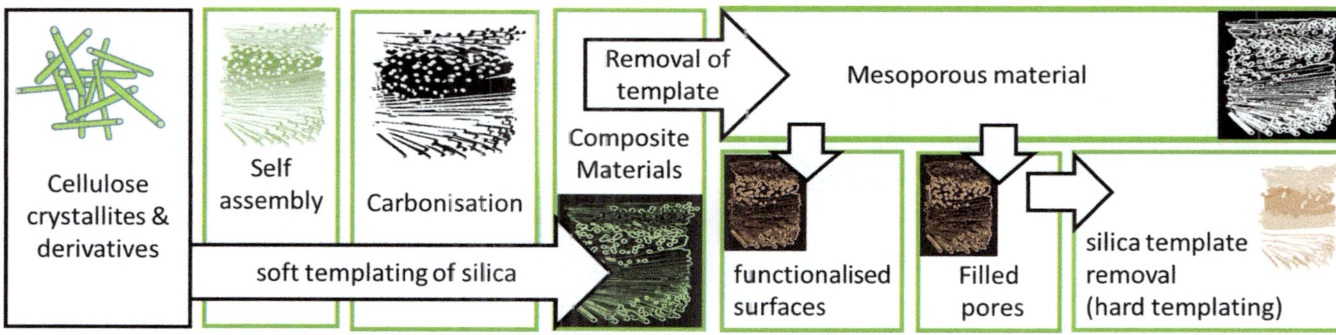

Figure 17.12 Different routes to functional nanostructured materials using CNC and HPC films as templating materials. Information adapted from Giese *et al.*[126]

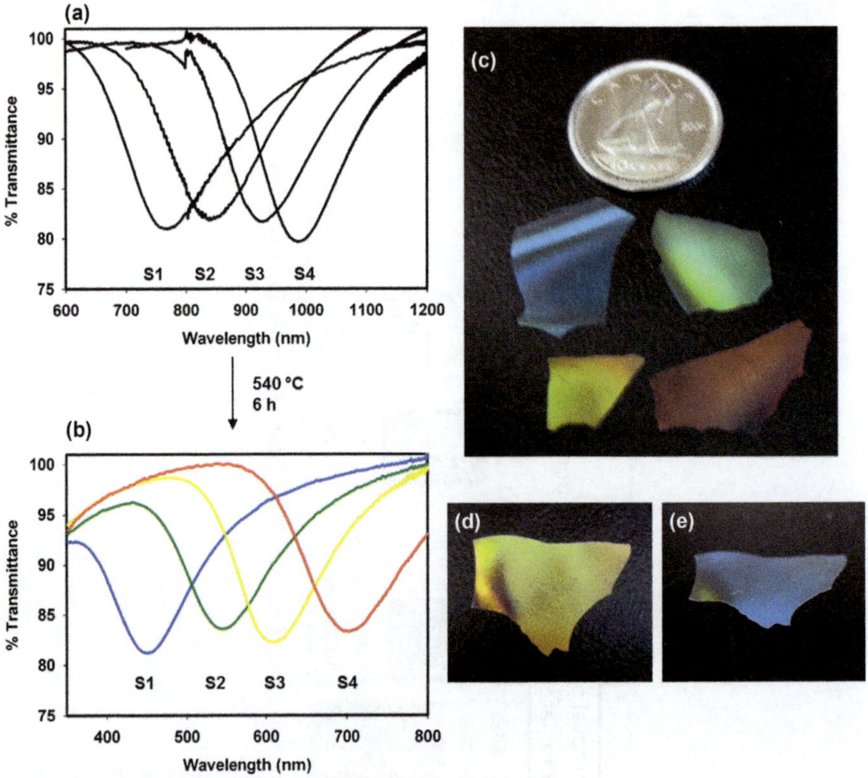

Figure 17.13 Silica free-standing films created by replicating the CNC chiral phase using sol–gel chemistry. (a) and (b) Transmission spectra before and after the removal of the CNCs. (c) Samples with a variety of colours produced by this technique. (d) and (e) The same flake of patterned silica seen at different angles.
Reprinted by permission from Macmillan Publishers Ltd: *Nature*,[128] Copyright 2010.

These polymers are synthetic cellulose derivatives prepared by alkaline treatment to yield alkali cellulose, followed by reaction with alkyl chlorides.[111,126,135–137] The first observation of colour production from HPC suspensions was reported by Werbowyj *et al.* in 1976.[138,139] Since then, a plethora of works have studied the possibility to tune the colouration in suspension[140–142] and its preservation in dry films.[143,144]

17.5.5 Chitin

From the beginning, the production of artificial helicoidal structural colour has also attempted to use the same technique for the helicoidal assembly of chitin.[145–147] The use of chitin to assemble into a chiral nematic phase in suspension was demonstrated soon after it had been demonstrated in cellulose,[148] with no obvious differences. The parameters of chiral nematic assembly have been demonstrated[149] in suspension and in a gel.[150] At the drying stage, chitin however produced a white or colourless film in every study. Even though the helicoidal structure is maintained during the drying of the film, its pitch was larger than the visible wavelengths of light. A chitin helicoidal structure has been demonstrated by co-assembly with $CaCO_3$, showing prominent helicoidal layers.[151] The reason why visibly reflecting chitin films have not yet been obtained is not clear. As discussed in Section 17.3, visibly reflecting helicoidal chitin structures are abundant in nature. Furthermore, cellulose films can be made from cellulose extracted from non-iridescent fibres. An intact crab cuticle has even been stripped of material to produce a visibly reflecting helicoidal material using biologically assembled materials.[152] Nevertheless, the mechanical and structural properties of chitin have been extensively explored[43,45,153–155] and it promises to establish a complementary system to helicoidal cellulose films in the future.

17.6 Conclusions

The development in the last century of a theory of liquid crystalline behaviour and of transfer matrices to understand optical equations, along with the use of computer simulations, has allowed us to understand the theory of these chiral materials. Furthermore, technological advances including electron microscopes have allowed us to unlock the structures of the natural world that create our most vivid materials. We now have the opportunity to capitalise on these in the creation of biocompatible and biomimetic materials using the natural phenomenon of self-assembly observed for abundant and easily processable biopolymers.

Cellulose nanocrystals are a fascinating avenue into the construction of novel chiral and structurally coloured materials. Research in the field has matured to the point where the technology can be used to create sensors and functionalised structures using new materials with properties extending beyond those found in the natural world.

There is still much progress to be made for the full control of the process and its extension to other materials with applications that have not yet been designed. Furthermore, the self-assembly of other biopolymers such as chitin to produce colour, although prevalent in the natural world, has not yet been demonstrated, and the field of cellulose derivatives remains to be fully explored. Biopolymers will certainly be at the forefront of the future use of structural colour in technological materials.

Acknowledgements

R. M. thanks NanoDTC Cambridge (EP/G03/7221/1) for financial support. S. V. thanks ERC Starting Grant (639088), BBSRC (BB/K014617/1) for financial support. U.S. thanks the National Center of Competence in Research (NCCR) Bio-Inspired Materials for financial support. All authors acknowledge Giulia Guidetti, Gen Kamita, and Bodo Wilts for contributing images and discussion.

References

1. S. Kinoshita, *Structural Colors in the Realm of Nature*, World Scientific Publishing Co., Singapore, 2008.
2. C. Sanchez, H. Arribart and M. M. G. Guille, *Nat. Mater.*, 2005, **4**, 277–288.
3. P. Vukusic and J. R. Sambles, *Nature*, 2003, **424**, 852–855.
4. D. Lee, *Nature's Palette*, Chicago, 2007.
5. V. Saranathan, A. E. Seago, A. Sandy, S. Narayanan, S. G. J. Mochrie, E. R. Dufresne, H. Cao, C. O. Osuji and R. O. Prum, *Nano Lett.*, 2015, **18**, 3735–3742.
6. L. P. Lee and R. Szema, *Science*, 2005, **310**, 1148–1150.
7. O. Karthaus, *Biomimetics in Photonics*, CRC Press, 2012.
8. K. Yu, T. Fan, S. Lou and D. Zhang, *Prog. Mater. Sci.*, 2013, **58**, 825–873.
9. A. R. Parker and H. E. Townley, *Nat. Nanotechnol.*, 2007, **2**, 347–353.
10. G. E. Schröder-Turk, S. Wickham, H. Averdunk, F. Brink, J. D. Fitz Gerald, L. Poladian, M. C. J. Large and S. T. Hyde, *J. Struct. Biol.*, 2011, **174**, 290–295.
11. T. H. Epps, E. W. Cochran, T. S. Bailey, R. S. Waletzko, C. M. Hardy and F. S. Bates, *Macromolecules*, 2004, **37**, 8325–8341.
12. M. D. Turner, M. Saba, Q. Zhang, B. P. Cumming, G. E. Schröder-Turk and M. Gu, *Nat. Photonics*, 2013, **7**, 801–805.
13. C. Mille, E. C. Tyrode and R. W. Corkery, *Chem. Commun.*, 2011, **47**, 9873–9875.
14. A. E. Seago, P. Brady, J.-P. Vigneron and T. D. Schultz, *J. R. Soc. Interface*, 2009, **6**(Suppl. 2), S165–S184.
15. S. Vignolini, E. Moyroud, B. J. Glover and U. Steiner, *J. R. Soc., Interface*, 2013, **10**, 20130394.
16. M. Kolle, A. Lethbridge, M. Kreysing, J. J. Baumberg, J. Aizenberg and P. Vukusic, *Adv. Mater.*, 2013, **25**, 2239–2245.
17. M. Mitov and N. Dessaud, *Liq. Cryst.*, 2007, **34**, 183–193.
18. A. Matranga, S. Baig, J. Boland, C. Newton, T. Taphouse, G. Wells and S. Kitson, *Adv. Mater.*, 2013, **25**, 520–523.
19. L. P. Biró, K. Kertész, E. Horváth, G. I. Márk, G. Molnár, Z. Vértesy, J.-F. Tsai, a Kun, Z. Bálint and J. P. Vigneron, *J. R. Soc., Interface*, 2010, **7**, 887–894.
20. I. W. Hamley, *Soft Matter*, 2010, **6**, 1863–1871.
21. S. Z. M. Diah, S. B. Karman and I. C. Gebeshuber, *J. Nanomater.*, 2014, **2014**, 1–15.
22. B. D. Wilts, H. M. Whitney, B. J. Glover, U. Steiner and S. Vignolini, *Mater. Today Proc.*, 2014, **1**, 177–185.
23. M. M. Giraud-Guille, E. Belamie and G. Mosser, *Comptes Rendus – Palevol*, 2004, **3**, 503–513.
24. M. Born and E. Wolf, *Principles of Optics*, Cambridge University Press, 1999.
25. H. de Vries, *Acta Crystallogr.*, 1951, **4**, 219–226.
26. H. Hurwitz and R. C. Jones, *J. Opt. Soc. Am.*, 1941, **31**, 493–499.

27. R. C. Jones, *J. Opt. Soc. Am.*, 1941, **31**, 488–493.
28. S. Teitler and B. W. Henvis, *J. Opt. Soc. Am.*, 1970, **60**, 830–834.
29. D. W. Berreman, *J. Opt. Soc. Am.*, 1972, **62**, 502–510.
30. Oseen, *Trans. Faraday Soc.*, 1932, **29**, 833.
31. H. G. Yoon and H. F. Gleeson, *J. Phys. D. Appl. Phys.*, 2007, **40**, 3579–3586.
32. D. Berreman, *J. Opt. Soc. Am.*, 1972, **62**, 502–510.
33. Y. Bouligand, *Tissue Cell*, 1972, **4**, 189–217.
34. A. C. Neville, *BioEssays*, 1985, **3**, 4–8.
35. A. C. Neville, *J. Theor. Biol.*, 1988, **131**, 243–254.
36. A. C. Neville and S. Caveney, *Biol. Rev. Cambridge Philos. Soc.*, 1969, **44**, 531–562.
37. A. C. Neville and S. Levy, *Planta*, 1984, **162**, 370–384.
38. A. C. Neville and B. M. Luke, *J. Cell Sci.*, 1971, **8**, 93–109.
39. A. C. Neville, *Biology of Fibrous Composites*, Cambridge University Press, 1993.
40. J. C. Weaver, G. W. Milliron, A. Miserez, K. Evans-Lutterodt, S. Herrera, I. Gallana, W. J. Mershon, B. Swanson, P. Zavattieri, E. DiMasi and D. Kisailus, *Science*, 2012, **336**, 1275–1280.
41. T. Zhang, Y. Ma, K. Chen, M. Kunz, N. Tamura, M. Qiang, J. Xu and L. Qi, *Angew. Chem., Int. Ed.*, 2011, **50**, 10361–10365.
42. S. Weiner and H. D. Wagner, *Annu. Rev. Mater. Sci.*, 1998, **28**, 271–298.
43. R. A. A. Muzzarelli, *Nanochitins and Nanochitosans, Paving the Way to Eco-Friendly and Energy-Saving Exploitation of Marine Resources*, Elsevier B.V., 2012, vol. 10.
44. Y. Habibi, L. A. Lucia and O. J. Rojas, *Chem. Rev.*, 2010, **110**, 3479–3500.
45. R. A. A. Muzzarelli, J. Boudrant, D. Meyer, N. Manno, M. Demarchis and M. G. Paoletti, *Carbohydr. Polym.*, 2012, **87**, 995–1012.
46. J. F. V. Vincent, *Mater. Today*, 2002, **5**, 28–41.
47. M. Jarvis, *Nature*, 2003, **426**, 611–612.
48. H. Chanzy, K. Imada, A. Mollard, R. Vuong and F. Barnoud, *Protoplasma*, 1979, **100**, 303–316.
49. J. F. V. Vincent and U. G. K. Wegst, *Arthropod Struct. Dev.*, 2004, **33**, 187–199.
50. N. Lavoine, I. Desloges, A. Dufresne and J. Bras, *Carbohydr. Polym.*, 2012, **90**, 735–764.
51. T. H. Giddings and L. A. Staehelin, *Planta*, 1988, **173**, 22–30.
52. R. J. Moon, A. Martini, J. Nairn, J. Simonsen and J. Youngblood, *Cellulose Nanomaterials Review: Structure, Properties and Nanocomposites*, 2011, vol. 40.
53. A. O'sullivan, *Cellulose*, 1997, **4**, 173–207.
54. A. M. C. Emons and H. Kieft, *Protoplasma*, 1994, **180**, 59–69.
55. D. P. Delmer and Y. Amor, *Plant Cell*, 1995, **7**, 987–1000.
56. M. C. a, Emons and B. M. Mulder, *Trends Plant Sci.*, 2000, **5**, 35–40.
57. B. J. Glover and H. M. Whitney, *Ann. Bot.*, 2010, **105**, 505–511.
58. R. M. Graham, D. W. Lee and K. Norstog, *Am. J. Bot.*, 2015, **80**, 198–203.
59. G. Strout, S. D. Russell, D. P. Pulsifer, S. Erten, A. Lakhtakia and D. W. Lee, *Ann. Bot. (Oxford, U. K.)*, 2013, **112**, 1141–1148.

60. D. Lee, *Nature*, 1991, **349**, 260–262.
61. S. Vignolini, P. J. Rudall, A. V. Rowland, A. Reed, E. Moyroud, R. B. Faden, J. J. Baumberg, B. J. Glover and U. Steiner, *Proc. Natl. Acad. Sci. U. S. A.*, 2012, **109**, 15712–15715.
62. M. N. Ravi Kumar, *React. Funct. Polym.*, 2000, **46**, 1–27.
63. A. A. Michelson, *Philos. Mag. Ser. 6*, 1911, **21**, 554–567.
64. R. Hegedüs, G. Szél and G. Horváth, *Vision Res.*, 2006, **46**, 2786–2797.
65. V. Sharma, M. Crne, J. O. Park and M. Srinivasarao, *Mater. Today Proc.*, 2014, **1**, 161–171.
66. S. Caveney, *Proc. R. Soc. London Ser. B*, 1971, **178**, 205–225.
67. P. Brady and M. Cummings, *Am. Soc. Nat.*, 2010, **175**, 614–620.
68. M. Srinivasarao, *Chem. Rev.*, 1999, **99**, 1953–1961.
69. A. C. Neville and B. M. Luke, *J. Insect Physiol.*, 1971, **17**, 519–526.
70. G. W. Gray, *Molecular Structure and Properties of Liquid Crystals*, Academic Press, London, 1962.
71. G. D. Mazur, J. C. Regier and F. C. Kafatos, *Insect Ultrastructure*, Springer-US, 1982, pp. 150–185.
72. D. Reis, B. Vian, H. Chanzy and J. C. Roland, *Biol. Cell*, 1991, **73**, 173–178.
73. B. Vian, D. Reis, D. Darzens and J. C. Roland, *Protoplasma*, 1994, **180**, 70–81.
74. S. Mizuta, A. Watanabe, S. Kimura and K. Yoshida, *Protoplasma*, 1994, **180**, 82–91.
75. J. C. Roland, D. Reis, B. Vian, B. Satiat-Jeunemaitre and M. Mosiniak, *Protoplasma*, 1987, **140**, 75–91.
76. B. Satiat-Jeunemaitre, *Tissue Cell*, 1992, **24**, 315–334.
77. J. Chan, *J. Microsc.*, 2012, **247**, 23–32.
78. M.-M. Giraud-Guille, *Curr. Opin. Solid State Mater. Sci.*, 1998, **3**, 221–227.
79. A. D. Rey, *Soft Matter*, 2010, **6**, 3402.
80. M. Mitov, *ChemPhysChem*, 2014, **15**, 1245–1250.
81. D. Demus, *Mol. Cryst. Liq. Cryst. Inc. Nonlinear Opt.*, 1988, **165**, 45–84.
82. L. Onsager, *Ann. N. Y. Acad. Sci.*, 1949, 627–659.
83. A. Stroobants, H. N. W. Lekkerkerker and T. Odijk, *Macromolecules*, 1986, **19**, 2232–2238.
84. S. T. Fraden, G. Maret, D. L. D. Caspar and R. B. Meyer, *Phys. Rev.*, 1989, **63**, 2068–2071.
85. T. E. Strzelecka and R. L. Rill, *Macromolecules*, 1991, **24**, 5124–5133.
86. X. M. Dong, T. Kimura and D. G. Gray, *Langmuir*, 1996, **12**, 2076–2082.
87. P. J. Collings and M. Hird, *Introduction to Liquid Crystals*, Taylor & Francis, 1997.
88. H. Charreau, M. L. Foresti and A. Vázquez, *Recent Pat. Nanotechnol.*, 2013, **7**, 56–80.
89. B. G. Ranby and E. A. Immergut, *Engineering*, 1951, 1183–1189.
90. J. F. Revol, H. Bradford, J. Giasson, R. H. Marchessault and D. G. Gray, *Int. J. Biol. Macromol.*, 1992, **14**, 170–172.

91. J. Revol, L. Godbout, X. Dong, D. G. Gray, H. Chanzy and G. Maret, *Liq. Cryst.*, 1994, **16**, 127–134.
92. D. Godbout and D. Gray, *US Pat.* 5629055, 1997, **28**, 9.
93. D. G. Gray, *Carbohydr. Polym.*, 1994, **25**, 277–284.
94. M. A. S. A. Samir, A. Dufresne and F. Alloin, *Biomacromolecules*, 2005, **6**, 612–626.
95. S. J. Eichhorn, *Soft Matter*, 2011, **7**, 303.
96. S. Elazzouzi-Hafraoui, Y. Nishiyama, J. L. Putaux, L. Heux, F. Dubreuil and C. Rochas, *Biomacromolecules*, 2008, **9**, 57–65.
97. S. M. Mukherjee and H. J. Woods, *Biochim. Biophys. Acta*, 1953, **10**, 499–511.
98. X. M. Dong, J.-F. Revol and D. G. Gray, *Cellulose*, 1998, **5**, 19–32.
99. M. Kaushik, C. Fraschini, G. Chauve, J. Putaux and A. Moores, *The Transmission Electron Microscope – Theory and Applications*, 2015, pp. 129–163. Available at http://dx.doi.org/10.5772/60985.
100. S. Beck-Candanedo, M. Roman and D. G. Gray, *Biomacromolecules*, 2005, **6**, 1048–1054.
101. S. Beck, J. Bouchard and R. Berry, *Biomacromolecules*, 2011, **12**, 167–172.
102. J. Araki and S. Kuga, *Langmuir*, 2001, **17**, 4493–4496.
103. L. Heux, G. Chauve and C. Bonini, *Langmuir*, 2000, **16**, 8210–8212.
104. J.-F. Revol, L. Godbout and D. G. Gray, *J. Pulp Pap. Sci.*, 1998, **24**, 146 149.
105. (a) A. G. Dumanli, H. M. van der Kooij, G. Kamita, E. Reisner, J. J. Baumberg, U. Steiner and S. Vignolini, *ACS Appl. Mater. Interfaces*, 2014, **6**, 646–650; (b) A. G. Dumanli, G. Kamita, J. Landman, H. M. van der Kooij, B. J. Glover, J. J. Baumberg, U. Steiner and S. Vignolini, *Adv. Opt. Mater.*, 2014, **2**, 646–650.
106. X. Mu and D. G. Gray, *Langmuir*, 2014, **30**, 9256–9260.
107. R. D. Deegan, O. Bakajin, T. F. Dupont, G. Huber, S. R. Nagel and T. A. Witten, *Phys. Rev. E.*, 2000, **62**, 756–765.
108. X. Mu and D. G. Gray, *Cellulose*, 2015, **22**, 1103–1107.
109. S. Choi, S. Stassi, A. P. Pisano and T. I. Zohdi, *Langmuir*, 2010, **26**, 11690–11698.
110. T.-D. Nguyen, W. Y. Hamad and M. J. MacLachlan, *Chem. Commun.*, 2013, **49**, 11296–11298.
111. J. P. F. Lagerwall, C. Schütz, M. Salajkova, J. Noh, J. Hyun Park, G. Scalia and L. Bergström, *NPG Asia Mater.*, 2014, **6**, e80.
112. J. Pan, W. Hamad and S. K. Straus, *Macromolecules*, 2010, **43**, 3851–3858.
113. Y. Habibi, T. Heim and R. Douillard, *J. Polym. Sci.*, 2008, **46**, 1430–1436.
114. F. Kimura, T. Kimura, M. Tamura, A. Hirai, M. Ikuno and F. Horii, *Langmuir*, 2005, **21**, 2034–2037.
115. J. H. Park, J. Noh, C. Schütz, G. Salazar-alvarez, G. Scalia, L. Bergström and J. P. F. Lagerwall, *ChemPhysChem*, 2014, **15**, 1477–1484.
116. J. Majoinen, E. Kontturi, O. Ikkala and D. G. Gray, *Cellulose*, 2012, **19**, 1599–1605.

117. A. Querejeta-Fernandez, B. Kopera, K. S. Prado, A. Klinkova, M. Methot, G. Chauve, J. Bouchard, A. S. Helmy and E. Kumacheva, *ACS Nano*, 2015, **9**(10), 10377–10385.
118. G. Chu, X. Wang, T. Chen, J. Gao, F. Gai, Y. Wang and Y. Xu, *ACS Appl. Mater. Interfaces*, 2015, **7**, 11863–11870.
119. Q. Liu, M. G. Campbell, J. S. Evans and I. I. Smalyukh, *Adv. Mater.*, 2014, **26**, 7178–7184.
120. M. Schlesinger, M. Giese, L. K. Blusch and Y. Hamad, *Chem. Commun.*, 2014, **51**, 530–533.
121. H. Qi, K. E. Shopsowitz, W. Y. Hamad and M. J. Maclachlan, *J. Am. Chem. Soc.*, 2011, **133**, 3728–3731.
122. A. Lukach, H. Therien-Aubin, A. Querejeta-ferna, N. Pitch, G. Chauve, M. Methot, J. Bouchard and E. Kumacheva, *Langmuir*, 2015, **31**, 5033–5041.
123. H. Thérien-aubin, A. Lukach, N. Pitch and E. Kumacheva, *Nanoscale*, 2015, **7**, 6612–6618.
124. C. C. Y. Cheung, M. Giese, J. A. Kelly, W. Y. Hamad and M. J. Maclachlan, *ACS Macro Lett.*, 2013, **2**, 1016–1020.
125. R. Bardet, N. Belgacem and J. Bras, *Appl. Mater. Interfaces*, 2015, **7**, 4010–4018.
126. M. Giese, L. K. Blusch, M. K. Khan and M. J. MacLachlan, *Angew. Chem., Int. Ed.*, 2015, **54**, 2888–2910.
127. C. T. Kresge, M. E. Leonowicz, W. J. Roth, J. C. Vartuli and J. S. Beck, *Nature*, 1992, **359**, 710–712.
128. K. E. Shopsowitz, H. Qi, W. Y. Hamad and M. J. Maclachlan, *Nature*, 2010, **468**, 422–425.
129. K. E. Shopsowitz, A. Stahl, W. Y. Hamad and M. J. MacLachlan, *Angew. Chem. Int. Ed.*, 2012, **51**, 6886–6890.
130. K. E. Shopsowitz, J. a. Kelly, W. Y. Hamad and M. J. MacLachlan, *Adv. Funct. Mater.*, 2014, **24**, 327–338.
131. K. E. Shopsowitz, W. Y. Hamad and M. J. MacLachlan, *J. Am. Chem. Soc.*, 2012, **134**, 867–870.
132. M. Giese, M. K. Khan, W. Y. Hamad and M. J. Maclachlan, *ACS Macro Lett.*, 2013, **2**, 818–821.
133. Y. P. Zhang, V. P. Chodavarapu, A. G. Kirk and M. P. Andrews, *Sens. Actuators, B*, 2013, **176**, 692–697.
134. H. Coles, *Handbook of Liquid Crystals*, Wiley-VCH Verlag GmbH, 1998, pp. 335–409.
135. Q. Liao, Q. Shao, H. Wang, G. Qiu and X. Lu, *Carbohydr. Polym.*, 2012, **87**, 2648–2654.
136. G. S. Rekhi and S. S. Jambhekar, *Drug Dev. Ind. Pharm.*, 1995, **21**, 61–77.
137. D. G. Gray, in *J. Appl. Polym. Sci.: Appl Polym. Symp.*, J. Wiley, New York, 1982.
138. R. S. Werbowyj and D. G. Gray, *Mol. Cryst. Liq. Cryst.*, 1976, **34**, 97–103.
139. R. S. Werbowyj and D. G. Gray, *Macromolecules*, 1980, **13**, 69–73.
140. Y. Onogi and Y. Nishijima, *Kobunshi Ronbunshu*, 1986, **43**, 223–229.

141. Y. Nishio, R. Chiba, Y. Miyashita and K. Oshima, *Polym. J.*, 2002, **34**, 149–157.
142. R. S. Werbowyj and D. G. Gray, *Macromolecules*, 1984, **17**, 1512–1520.
143. G. Charlet and D. G. Gray, *Macromolecules*, 1987, **20**, 33–38.
144. S. N. Bhadani and D. G. Gray, *Mol. Cryst. Liq. Cryst.*, 1984, **102**, 255–260.
145. R. H. Marchessault, F. F. Morehead and N. M. Walter, *Nature*, 1959, **184**, 632–633.
146. K. Kurita, *Prog. Polym. Sci.*, 2001, **26**, 1921–1971.
147. S. Ifuku and H. Saimoto, *Nanoscale*, 2012, **4**, 3308–3318.
148. J. F. Revol and R. H. Marchessault, *Int. J. Biol. Macromol.*, 1993, **15**, 329–335.
149. E. Belamie, P. Davidson and M. M. Giraud-Guille, *J. Phys. Chem. B*, 2004, **108**, 14991–15000.
150. M. V. Tzoumaki, T. Moschakis and C. G. Biliaderis, *Biomacromolecules*, 2010, **11**, 175–181.
151. S. Matsumura, S. Kajiyama, T. Nishimura and T. Kato, *Small*, 2015, **11**(38), 5127–5133.
152. T. Nguyen and M. J. Maclachlan, *Adv. Opt. Mater.*, 2014, **2**, 1031–1037.
153. R. A. A. Muzzarelli, *Chitin*, Pergamon Press, 2011, pp. 1–35.
154. K. Gopalan Nair and A. Dufresne, *Biomacromolecules*, 2003, **4**, 657–665.
155. J. D. Goodrich and W. T. Winter, *Biomacromolecules*, 2007, **8**, 252–257.

Subject Index